AF556761

# Photochemistry and Photobiology of Nucleic Acids

## Volume I

## Chemistry

## Contributors

Michel Charlier
Malcolm Daniels
Dov Elad
Catherine Fenselau
G. J. Fisher
William W. Hauswirth
Claude Hélène
D. P. Hollis
H. E. Johns
Isabella L. Karle
Thérèse Montenay-Garestier
G. Scholes
Shih Yi Wang

# Photochemistry and Photobiology of Nucleic Acids

## Volume I

## Chemistry

EDITED BY

*Shih Yi Wang*

Department of Biochemistry
School of Hygiene and Public Health
The Johns Hopkins University
Baltimore, Maryland

ACADEMIC PRESS New York San Francisco London 1976

*A Subsidiary of Harcourt Brace Jovanovich, Publishers*

ACADEMIC PRESS, INC.
111 Fifth Avenue, New York, New York 10003

*United Kingdom Edition published by*
ACADEMIC PRESS, INC. (LONDON) LTD.
24/28 Oval Road, London NW1

**Library of Congress Cataloging in Publication Data**

Main entry under title:

Photochemistry and photobiology of nucleic acids.

Includes bibliographies and index.
CONTENTS: v. 1. Chemistry.–v. 2. Photobiology.
1. Nucleic acids. 2. Photochemistry. 3. Photobiology. I. Wang, Shih Yi, (date)
QD433.P48 547′.596 75-26528
ISBN 0–12–734601–5 (v. 1)

PRINTED IN THE UNITED STATES OF AMERICA

# Contents

## Chapter 10 Nuclear Magnetic Resonance of Photoproducts

*D. P. Hollis*



## Chapter 11 Crystal and Molecular Structure of Photoproducts from Nucleic Acids

*Isabella L. Karle*



## Chapter 12 The Radiation Chemistry of Pyrimidines, Purines, and Related Substances

*G. Scholes*

# List of Contributors

Numbers in parentheses indicate the pages on which the authors' contributions begin.

Michel Charlier (381), Centre de Biophysique Moleculaire, Orleans, Cedex, France

Malcolm Daniels (23, 109), Chemistry Department, Oregon State University, Corvallis, Oregon

Dov Elad (357), Department of Chemistry, The Weizmann Institute of Science, Rehovot, Israel

Catherine Fenselau (419), Department of Pharmacology and Experimental Therapeutics, The Johns Hopkins University, School of Medicine, Baltimore, Maryland

G. J. Fisher (169, 225),* Department of Medical Biophysics, University of Toronto, and The Ontario Cancer Institute, Toronto, Ontario, Canada

William W. Hauswirth (109), Department of Biochemical and Biophysical Sciences, The Johns Hopkins University, School of Hygiene and Public Health, Baltimore, Maryland

Claude Helene (381), Centre de Biophysique Moleculaire, Orleans, Cedex, France

D. P. Hollis (447), Department of Physiological Chemistry, The Johns Hopkins School of Medicine, Baltimore, Maryland

H. E. Johns (169, 225), Department of Medical Biophysics, University of Toronto, and The Ontario Cancer Institute, Toronto, Ontario, Canada

Isabella L. Karle (483), Laboratory for the Structure of Matter, Naval Research Laboratory, Washington, D. C.

* Present address: Department of Nuclear Medicine and Radiobiology, Centre Hopitalier Universitaire, Universite de Sherbrooke, Sherbrooke, Quebec, Canada.

Thérèse Montenay-Garestier (381), Laboratoire de Biophysique, Muséum National d'Histoire Naturelle, Paris, France

G. Scholes (521), Laboratory of Radiation Chemistry, University of Newcastle upon Tyne, England

Shih Yi Wang (1, 295), Department of Biochemistry, School of Hygiene and Public Health, The Johns Hopkins University, Baltimore, Maryland

# Preface

The prospect of understanding living processes through physical and chemical principles has attracted many physical scientists to the study of molecular biology. Since the mid 1930's, their enthusiasm plus that of biologists has brought about unparalleled advances in many areas of biology. One of the areas which requires the actual interaction of physicists, chemists, and biologists is photobiology. It is this interaction which is particularly fascinating and has drawn many to the study of photobiology. Such a trend proved to be fruitful in the study of photochemistry of nucleic acid components, and early work led to the isolation and characterization of ultraviolet photoproducts of pyrimidines. This discovery, in turn, touched off a high level of activity in the study of photobiology of nucleic acids at the molecular level. Current activity in this area is unprecedented in the history of photobiological studies. This intensive and concerted effort has resulted in the recognition not only of processes of repair of and of protection against irradiation damage in biological systems but also of the possible relevancy of pyrimidine photoproducts to mutagenesis and carcinogenesis.

This two-volume treatise provides a judicious review of these exciting developments with up-to-date information as well as the necessary background knowledge. The chapters, each a self-contained entity, are concise and authoritive reviews by a group of researchers who are active in their respective areas. Since this is a vigorous field of study, the range of statements necessarily extends from those with a high probability of continuing certainty to those that are speculative but potentially of heuristic value. This should convey to our readers the explorative attitude of the investigators and the excitement inherent in this research. While the orientation of each chapter naturally reflects the interests and viewpoint of the author, there has been a genuine effort to present a critical treatment of the existing data.

Volume I is concerned with the UV-induced physical and chemical alterations in nucleic acid components including pyrimidines, purines, their nucleosides and nucleotides, and related compounds. In addition, chapters on mass and nuclear magnetic resonance spectrometry and crystal and molecular structural determinations by x-ray diffraction are included. Together with the pertinent examples, a brief discussion of the theory and techniques is also presented in each chapter. This should be of considerable help to those who, although not personally involved in the interpretation of data, may wish to understand and possibly evaluate reported findings. Also, the close relationship between the chemical effects of UV light and x- or $\gamma$-radiation prompted the inclusion of a chapter on radiation chemistry for the purpose of comparison.

Volume II is concerned with the biological effects due to stable UV-induced alterations in critical cellular macromolecules including cell death, growth delay, mutagenesis, and carcinogenesis. The fact that the problem has been most successfully pursued by assuming DNA to be the macromolecule most relevant to cell pathology is reflected in most of the chapters. It is also necessary to consider the photochemical and photobiological properties of RNA's which are also essential in cellular functions. Although knowledge about protein and amino acid photochemistry is less advanced than that of nucleic acids, a chapter dealing with the UV-induced cross-linkings of proteins with nucleic acids is appropriate. This knowledge may be required for a full understanding of the mode of action of UV on cells, since it is improbable that biological effects can be explained solely in terms of damage to nucleic acids. Later chapters delve into the mechanisms that provide some protection against and are capable of repairing damage caused by UV photons and by ionizing radiation (also chemical mutagens) in organisms ranging from viruses to mammalian cells. These repair processes which were initially of concern only to photobiologists have gained the interest of investigators in other areas of molecular biology. Apparently, repair processes play a role in monitoring and preserving the structural integrity of DNA during physiological processes such as replication and transcription. Because of this widespread interest, research of these repair processes has mushroomed in recent years. Although the study of photoreactivation may have been effectively covered in a single chapter, such a treatment of the vast existing literature on "dark repair" would have been insufficient. Additionally, knowledge concerning these complex processes is in a state of flux. For these reasons such a review has not been included in these volumes.

This treatise should serve as an authoritative and important reference work for researchers active in the study of photochemistry and photobiology in nucleic acids as well as for advanced undergraduate and graduate students interested in this field. Since this is an interdisciplinary area, an attempt has been made to direct our writings to an audience which includes physicists, chemists, biologists, and physicians.

An attempt was made to further a balanced viewpoint by requesting two or more scholars, active in the respective areas, to review each chapter. I wish to thank all of these reviewers, some of whom are contributors, for their most helpful criticisms, comments, and suggestions. These reviewers include the late Ruth F. Hill (York University); A. A. Lamola (Bell Laboratory); E. Fahr (Universität Würzburg); J. E. Cleaver (Imperial Cancer Research Fund Laboratory); M. W. Logue (University of Maryland, Baltimore County); J. R. Williams (Temple University); B. A. Bridges (Medical Research Council); J. F. Ward (University of California, Los Angeles); W. A. Summers (University of Oklahoma); H. Werbin, W. Harm, C. S. Rupert (The University of Texas at Dallas); L. Brand, T. Merz, J. L. Alderfer, R. M. Herriott, P. C. Huang, J. Scocca (The Johns Hopkins University); and, in particular, William Hauswirth who read and commented on major portions of the text.

Finally, I wish to thank my colleagues in this endeavor. Particular gratitude goes to Drs. John Jagger and Michael H. Patrick for discussions on the organization of Volume II for which Dr. Patrick has served as coordinator; to Elizabeth Hopkins Roth, Patricia Whiting Linton, and Jane Entwisle Shipley for their editorial assistance; to Sally Vasek who dealt effectively with the typing and illustrations of all of the edited manuscripts; and to the staff of Academic Press for their cooperation and efficient processing of this publication.

*Shih Yi Wang*

# Contents of Volume II

# 1 Introductory Concepts for Photochemistry of Nucleic Acids

*Shih Yi Wang*



## A. Introduction

This chapter serves as an introduction to these volumes. It is a brief discussion of the basic aspects of photochemistry and, hopefully, will help to introduce basic concepts to readers who are new to the field. However, since this book is aimed at a diversified audience, many readers may find this review superfluous. For a more detailed introduction to photochemistry, the reader may take advantage of some of the excellent monographs published recently, many of which have been beneficial to this author. Some of these are listed as general references at the end of the chapter. This list is a partial one since no attempt was made to compile an exhaustive bibliography.

Since our primary concern is nucleic acids and their ultraviolet (UV) radiation behavior, an introduction to the fundamentals of these two aspects should be included. This is treated in Chapter 1, Volume II and Chapter 4, Volume II, respectively, where it is germane, rather than at the outset. However, two points must be kept in mind from the beginning. First, the properties of individual bases or nucleotides may be significantly altered when incorporated in large polymers (Chapter 1, Volume II); and second, *in vitro* studies may or may not yield clues to *in vivo* processes. Such an outlook will be helpful in analyzing the findings discussed throughout these volumes.

While these three chapters (this chapter and Chapters 1 and 4, Volume II) attempt to present some basic theoretical knowledge in this area, information concerning UV radiation techniques is also essential, particularly in the planning and interpretation of experiments. An instruction manual for such a purpose is available (Jagger, 1967). It provides beginners with the most basic practical information but also has sufficient references for more detailed sources. However, brief discussions of experimental approaches will be found occasionally in various chapters.

## B. Basic Aspects of Photochemistry

Photobiological phenomena, such as bioluminescence, circadian rhythms, photodamage, photomorphogenesis, photoperiodism, photosynthesis, phototaxism, phototropism, and vision, are the manifestations of photochemical effects on biological materials and systems. It is easily perceived that only light which is absorbed can cause photochemical effects—this is the first law of photochemistry (the Grotthus-Draper Principle). In order to understand these photochemical effects, one should ask three questions: What is the nature of light? What range of light is involved? What happens when a molecule absorbs light?

### 1. Nature of Light

The present concept of the nature of light is that it has dual properties; i.e., it is necessary to speak of light both as an electromagnetic wave and as a stream of particles. Each of these two viewpoints partially explains the characteristics of light and does not exclude the other. Indeed, the Planck quantum theory suggests that light energy can be absorbed or emitted only in discrete packets (as particles) whose energy is proportional to the frequency of the radiation (as waves). This relationship may be written as

$$E = h\nu \tag{1}$$

in which $E$ is the energy of 1 quantum of frequency $\nu$ and $h$ is a basic numerical universal constant, known as Planck's constant, equal to $6.63 \times 10^{-27}$ erg sec.

*a. Light as Waves*

It has been shown experimentally that the velocity of visible light in a vacuum ($c = 3 \times 10^{10}$ cm/sec) is exactly the same for other wavelengths of radiation which are electromagnetic in nature. This finding leads to the conclusion that light is a form of electromagnetic radiation. This radiation is a moving wavelike force field similar to that generated by an oscillating electric dipole with positively and negatively charged poles or regions (Fig. 1) which also acts on the charges it encounters as it propagates. At any point, the radiated electric and magnetic fields are perpendicular to each other as well as to the direction of propagation. The arrows within the sine curves in Fig. 1 represent electric or magnetic vectors indicating both the direction and the magnitude of the respective force fields, and the two fields are in phase. The wave as a whole propagates in various directions from the oscillating dipole with the speed of light. Its amplitude ($A$) decreases with increasing distance along any given direction, but the total energy integrated over all directions is constant. It can also be seen from Fig. 1 that the frequency $\nu$ (cycles/cm) can be ascertained by the simple relationship

$$\nu = c/\lambda \tag{2}$$

in which $\lambda$ is the wavelength, i.e., the distance between two successive antinodes. However, in media other than a vacuum, the speed of light becomes $c/n$, $n$ being the refractive index which is unity in a vac-

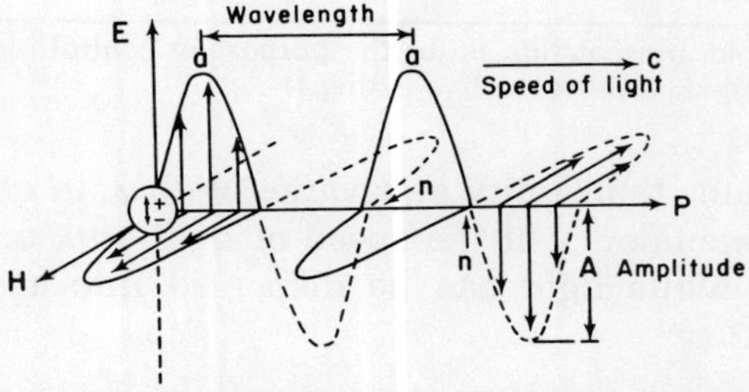


**Fig. 1.** *The instantaneous electric* (E) *and magnetic* (H) *field strength vectors of a light wave radiated by an oscillating electric dipole as a function of position along the axis of propagation* (P). *The positions of maximal amplitude* (A) *are designated as antinodes* (*a*) *and of minimal or zero amplitude are designated as nodes* (*n*). [*Adapted from Calvert and Pitts* (1966) *and from Clayton* (1970).]

**Table 1** Significant Developments Concerning Light and Photochemistry

| Significant developments[a] | Investigator | Year |
|---|---|---|
| Discovery of visible light spectrum by prismatic dispersion | I. Newton | 1666 |
| Discovery of infrared light by thermometer measurement | W. Herschel | 1800 |
| Discovery of UV light by blackening of silver chloride | J. W. Ritter | 1801 |
| Postulation of first law of photochemistry | C. J. D. Grothus, J. W. Draper | 1843, 1817 |
| Suggestions of laws of absorption of light | P. Bouquer, A. Beer, J. H. Lambert | 1852, 1729 |
| Postulation of wave theory of electromagnetism | J. C. Maxwell | 1873 |
| Determination of the speed of light by interferometer | A. A. Michelson | 1879 |
| Discovery of x-ray by penetration of sheet metal | W. K. Roentgen | 1895 |
| Discovery of $\gamma$-ray in radioactive decay products | M. Curie | 1896 |
| Explanation of black-body spectrum by quantum hypothesis | M. Planck | 1901 |
| Explanation of photoelectric effects | A. Einstein | 1905 |
| Postulation of stimulated emission | A. Einstein | 1917 |
| Postulation of second law of photochemistry | J. Stark, A. Einstein | 1921, 1919 |
| Analysis of electron–photon scattering | A. H. Compton | 1923 |
| Postulation of "vertical" electronic transition | J. Franck, E. U. Condon | 1926 |
| Development of flash photolysis technique | R. G. W. Norrish, G. Porter | 1950's |
| Discovery of laser by coherent stimulated emission of ruby crystals | C. H. Townes, N. G. Basov, A. M. Prokhorov | 1950's |

[a] This compendious presentation is for the purpose of controlling the length of the text but is not intended as an exhaustive listing.

uum but is greater than unity and varies with $\lambda$ in other media. By virtue of this dependency of the speed of light on $\lambda$ or $\nu$, the various components of white light can be dispersed into a spectrum by a prism (Table 1).

*b. Light as Particles*

The effect of light is usually not perceived until light interacts with matter by absorption. Under such conditions, experimental results often cannot be explained in terms of the wave theory of light. For instance, illumination of a metal plate with light of sufficiently short wavelength (with frequency above a certain threshold frequency) causes the liberation of electrons. This is recognized as the photoelectric effect. It was found that the number of electrons emitted is proportional to the light intensity. However, the velocity of the emitted electrons is independent of intensity but does depend on the wavelength of the light, provided that the frequency is above the threshold frequency. Einstein postulated that photons, small packets of light energy, may transfer their energy, which is proportional to the frequency and is independent of the light intensity, to the electron. This transfer is an "all-or-none" process—the electron either gets all the photon's energy or none. Once absorbed, the photon then simply ceases to exist. The energy acquired by the electron may enable it to escape from the metal surface. Since some energy, $w_0$, called the work function, is necessary to remove an electron from the metal, the kinetic energy of the photoelectrons ejected by light is

$$E = mv^2/2 = h\nu - w_0 \tag{3}$$

The magnitude of $w_0$ depends on the physical environment from which the electrons escape. Interestingly, Einstein did not get his Nobel prize in 1921 for his 1919 proposal of the now famous relativity theory, but for his 1905 explanation of this photoelectric effect which was experimentally confirmed by Millikan in 1916. Einstein's photoelectric equation [Eq. (3)] is an outgrowth of Planck's quantum expression [Eq. (1)]. The understanding of the photoelectric effect provides the experimental verification of the existence of light quanta and, in turn, the quantum theory. Thus, radiation consists of discrete energy packets or particles which are so tiny that the light appears to be continuous except under unusual conditions. (In a very dark room, the "splotchy" appearance of a uniform wall is caused by the quantum nature of the eye's receptor mechanism.)

### 2. Interaction of Light with Chemical Systems

#### *a. Action of Photons and Quantum Yields*

In the study of photochemistry, light must be considered to be composed of individual photons ($h\nu$) because the essential features of photochemical reactions cannot be explained by the wave theory. A photochemical reaction begins with the direct or indirect absorption of a photon by a molecule (M) resulting in the formation of an excited-state molecule ($M^*$):

$$M + h\nu \rightleftharpoons M^* \longrightarrow \text{product}$$

M* is distinct from M and has acquired new physical and chemical properties. Thus, a light-initiated reaction must differ from a dark reaction by the fact that the light reaction proceeds through M* rather than M. Light should by no means be considered as a catalyst in a reaction, even though on occasion the same product is obtained from both a light and a dark reaction. Once photons are absorbed they become nonexistent, unlike a true catalyst which should remain unaltered at the completion of a reaction.

Also, it is important to note the distinction between dark and light reactions in the consideration of product yields. In ordinary chemical reactions the yield of products can be defined as the percentage of M, the reactants, transformed into products. In light reactions, the logical analog is the percentage of M*, the excited-state molecules, transformed into products. Here we have to invoke the Stark-Einstein photochemical equivalence law, the second law of photochemistry, in order to define the so-called quantum yields ($\phi$) for photochemical reactions. This law states that each photon or quantum of radiation absorbed by a molecule activates one molecule in the primary step of a photochemical process. Thus, the number of M* must be equal to the number of quanta absorbed by M, and quantum yields may be defined as follows:

$$\phi = \frac{\text{no. of molecules reacted}}{\text{no. of photons absorbed}}$$

It is often possible to increase total yields by prolonging the irradiation period, especially when the quantum yields are very low.

This second law remains valid as long as relatively modest levels of light intensity are used ($< 10^{15}$ quanta absorbed/ml/sec), such as those found in normal photochemical and photobiological studies. In these cases there is a very low concentration of M*, and it is highly improbable for M*, which has a very short lifetime, to absorb a second quantum of light. However, biphotonic excitation has been observed in several systems. In flash photolysis, a high-intensity flash of light ($\sim 10^{18}$–$10^{23}$ quanta/ml/msec or $\mu$sec) generates a relatively high concentration of M* or other transients. When a second short flash of light follows immediately, some M* may absorb a second photon. Similar biphotonic processes have been claimed in laser experiments in which an exceedingly high-intensity monochromatic light is attainable. This is to be expected especially when M* has a relatively long lifetime, as in the case of triplet-state molecules (see Chapter 2).

### b. *Energies, Wavelengths, and Frequencies*

Since the characteristics of M* are dependent on the nature or the energy of electromagnetic radiation, we should consider the energy of photons in relation to wavelengths and frequencies. By substituting $c/\lambda$ for $\nu$ into Eq. (1), the following equation is obtained:

$$
\begin{aligned}
E &= hc/\lambda \\
E\ (\text{eV}) &= 1240/\lambda \\
E\ (\text{kcal/einstein}) &= 28{,}600/\lambda \quad (1 \text{ einstein} = 1 \text{ mole of quanta})
\end{aligned}
\tag{4}
$$

With $\lambda$ expressed in nanometers (1 nm $= 10^{-9}$ m $= 1$ m$\mu$ $= 10$ Å), the energy of some typical electromagnetic radiations may be calculated and are given in Table 2. While there is no well-defined borderline between the various radiations described in Table 2 or Fig. 2, three distinct radiation effects, namely molecular vibrations, electronic excitation, and ionization, may be produced. At wavelengths longer than 1000 nm ($<29$ kcal/einstein), the energy imparted is sufficient only for molecular vibrations with no apparent chemical changes. At wavelengths shorter than 100 nm ($>290$ kcal/einstein), all three radiation effects may result. The energy delivered by light of wavelengths between 1000 and 100 nm is insufficient to cause ionization in most organic compounds but can produce, in addition to molecular vibration, electronic excitation which affects photochemical changes. The 100 nm line was chosen arbitrarily to separate photobiology from radiobiology and, thus, photochemistry of nonionizing radiation from radiation chemistry of ionizing radiation. Based on the foregoing, we may define photochemistry as the study of the effects of molecular electronic excitation initiated by light in approximately the 1000 to 100 nm region, or in the region of UV and visible light (Fig. 2).

In the study of photochemistry and photobiology of nucleic acids, one is concerned primarily with UV effects on biological systems. The region of 100–190 nm is called the vacuum- (or Schumann) UV region because experiments in this region usually must be conducted under vacuum conditions since both air and water absorb UV radiation of wavelengths shorter than 190 nm. The remaining UV region is separated into far-UV (190–300 nm) and near-UV (300–380 nm) regions. Incidentally, atmospheric ozone serves as a filter to cut off solar UV radiation of wavelengths shorter than 300 nm. This cut-off wavelength coincides with the point of demarcation for these two UV regions. Also, the screening action of ozone provides the means to protect life on earth from damage by far-UV radiation and is of environmental importance commanding considerable current interest.

**Table 2** The Wavelength, Frequency, and Energy of Typical Electromagnetic Radiation[a]

| Approximate description | Typical wavelength (nm) | Frequency (cycles/sec) | Wave number ($cm^{-1}$) | Energy (eV) | Energy (kcal/einstein) |
|---|---|---|---|---|---|
| *Radiowave* | | | | | |
| Long | $1.0 \times 10^{12}$ (1000 m) | $3.00 \times 10^{5}$ (300 kc) | $1.00 \times 10^{-5}$ | $1.24 \times 10^{-9}$ | $2.86 \times 10^{-8}$ |
| Short | $1.0 \times 10^{10}$ (10 m) | $3.00 \times 10^{7}$ (30 Mc) | $1.00 \times 10^{-3}$ | $1.24 \times 10^{-7}$ | $2.86 \times 10^{-6}$ |
| *Microwave* | $1.0 \times 10^{7}$ (1 cm) | $3.00 \times 10^{10}$ | 1.00 | $1.24 \times 10^{-4}$ | $2.86 \times 10^{-3}$ |
| *Infrared* | | | | | |
| Far | $1.0 \times 10^{4}$ (10 $\mu$m) | $3.00 \times 10^{13}$ | $1.00 \times 10^{3}$ | $1.24 \times 10^{-1}$ | 2.86 |
| Near | $1.0 \times 10^{3}$ (1 $\mu$m) | $3.00 \times 10^{14}$ | $1.00 \times 10^{4}$ | 1.24 | 28.6 |
| *Visible Light* | | | | | |
| Red | $7.0 \times 10^{2}$ (700 nm) | $4.28 \times 10^{14}$ | $1.43 \times 10^{4}$ | 1.77 | 40.8 |
| Orange | $6.2 \times 10^{2}$ | $4.84 \times 10^{14}$ | $1.61 \times 10^{4}$ | 2.00 | 46.1 |
| Yellow | $5.8 \times 10^{2}$ | $5.17 \times 10^{14}$ | $1.72 \times 10^{4}$ | 2.14 | 49.3 |
| Green | $5.3 \times 10^{2}$ | $5.66 \times 10^{14}$ | $1.89 \times 10^{4}$ | 2.34 | 53.9 |
| Blue | $4.7 \times 10^{2}$ | $6.38 \times 10^{14}$ | $2.13 \times 10^{4}$ | 2.64 | 60.8 |
| Violet | $4.2 \times 10^{2}$ | $7.14 \times 10^{14}$ | $2.38 \times 10^{4}$ | 2.95 | 68.1 |
| *Ultraviolet Light* | | | | | |
| Near | $3.0 \times 10^{2}$ | $1.00 \times 10^{15}$ | $3.33 \times 10^{4}$ | 4.13 | 95.3 |
| Far | $2.0 \times 10^{2}$ | $1.50 \times 10^{15}$ | $5.00 \times 10^{4}$ | 6.20 | 142.9 |
| Vacuum | $1.5 \times 10^{2}$ | $2.00 \times 10^{15}$ | $6.67 \times 10^{4}$ | 8.27 | 190.6 |
| *X-ray* | | | | | |
| Long | $3.0 \times 10$ | $1.00 \times 10^{16}$ | $3.33 \times 10^{5}$ | 41.3 | 953.0 |
| Short | $1.0 \times 10^{-1}$ | $3.00 \times 10^{18}$ | $1.00 \times 10^{8}$ | $12.4 \times 10^{2}$ | $286 \times 10^{3}$ |
| *γ-ray* | $1.0 \times 10^{-3}$ | $3.00 \times 10^{20}$ | $1.00 \times 10^{10}$ | $12.4 \times 10^{4}$ | $286 \times 10^{5}$ |

[a] Taken largely from Calvert and Pitts (1966) by courtesy of John Wiley & Sons, Inc.

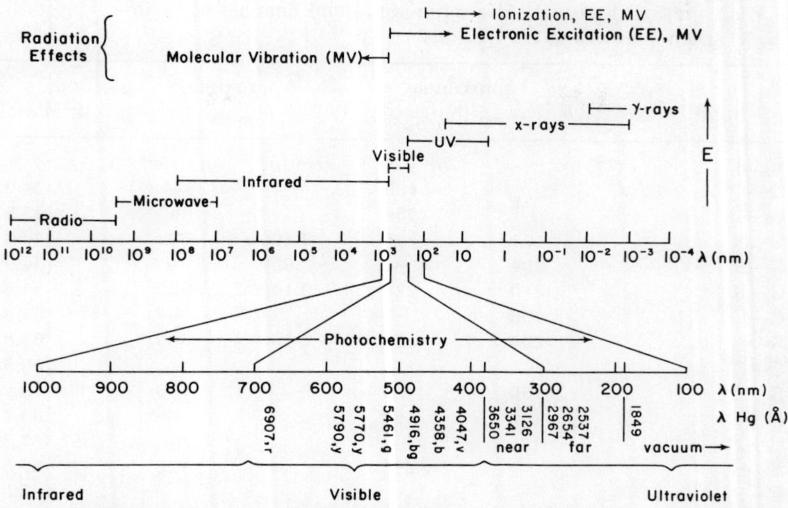


**Fig. 2.** *Wavelengths of electromagnetic radiation on a logarithmic scale and of photochemical radiation on a linear scale. Above these scales indicate various regions and the possible radiation effects. Below are indicated the wavelengths of the emission spectrum of the low-pressure mercury lamps. With medium- and high-pressure mercury lamps, self-absorption of resonance lines at 2537 and 1849 Å results in the absence of emission at both wavelengths.* [*Adapted from Jagger (1967) and from Suppan (1973).*]

## 3. Absorption and Kinetics

### *a. Chromophores*

Since a photochemical reaction is brought about by the molecular absorption of light quanta, one must ask how a molecule selects a particular wavelength of light. The chromophore in a molecule is responsible for such a selection. In 1876, Witt coined the term chromophore for unsaturated groups, such as C═C, C═O, and C═N, thought to be essential for color of the molecules. Table 3 lists the absorption maxima for some basic chromophoric groups, several amino acids, and purine (Pur) and pyrimidine (Pyr) bases, together with the molar extinction coefficients, $\epsilon$ (see Table 3). It can be seen that the selective high absorption exhibited by Pur and Pyr bases, components of nucleic acids, is in the 260 nm region. Since low-pressure mercury resonance lamps which emit mainly at 253.7 nm (Fig. 2) are readily available for photobiological studies, major biological effects resulting from UV radiation have been found to occur with nucleic acids.

### *b. Absorption*

In general a beam of parallel monochromatic (single frequency) light is absorbed exponentially. This means, for example, that if the first

**Table 3** Characteristics of UV Absorption and Bond Energies of Various Chromophores and Compounds

| Bond | Approximate $\lambda_{max}$ (nm) | | Approximate $\epsilon_{max}$ | | Bond energy (kcal/mole)[a] |
|---|---|---|---|---|---|
| $\sigma$ (single) | $\sigma \rightarrow \sigma^*$ | $n \rightarrow \sigma^*$ | $\sigma \rightarrow \sigma^*$ | $n \rightarrow \sigma^*$ | |
| O—O | | $<400$ | | — | 34.9 |
| C—I | — | 258 | — | 365 | 45.5 |
| C—N | 174 | 215 | 2,200 | 600 | 48.6 |
| C—Br | 204 | — | 2,000 | — | 54.0 |
| C—S | 210 | 229 | 2,140 | 620 | 54.5 |
| C—C | 135 | — | 11 | — | 58.6 |
| S—S | 205 | 250 | 2,100 | 500 | 63.8 |
| C—Cl | — | 173 | — | — | 66.5 |
| C—O | 150 | 184 | — | 150 | 70.0 |
| N—H | 152 | 194 | — | — | 83.7 |
| C—H | 122 | — | — | — | 87.3 |
| S—H | — | 225 | — | 126 | 87.5 |
| C—F | | | | | 107.0 |
| O—H | 150 | 183 | 1,900 | 200 | 110.2 |
| $\pi$ (multiple) | $\pi \rightarrow \pi^*$ | $n \rightarrow \pi^*$ | $\pi \rightarrow \pi^*$ | $n \rightarrow \pi^*$ | |
| C=N | 190 | | 5,000 | | 94 |
| C=C | 165 | 183 | 10,000 | 250 | 100 |
| C=S | | 330 | | 5 | 103 |
| C=C | 173 | — | 6,000 | — | 123 |
| C=O | 190 | 280 | 900 | 30 | 147 |
| CONH | 190 | 220 | 7,000 | 63 | — |
| C=C—C=N | 219 | — | 25,000 | — | — |
| C=C—C=C | 209 | — | 25,000 | — | — |
| C=C—C=O | 215 | 330 | 20,000 | 30 | — |
| *Compound*[b] | $\pi \rightarrow \pi^*$ | | $\pi \rightarrow \pi^*$ | | |
| Phenylalanine | 257.4 | | 197 | | |
| Tyrosine | 274.6 | | 1,420 | | |
| Tryptophane | 279.8 | | 5,600 | | |
| Cystine | 260 ($n \rightarrow \sigma^*$) | | 280 ($n \rightarrow \sigma^*$) | | |
| Acetylcysteine | 280 ($n \rightarrow \pi^*$) | | 5 ($n \rightarrow \pi^*$) | | |
| Adenine (Ade) | 260.5 | | 13,400 | | |
| Cytosine (Cyt) | 267 | | 6,100 | | |
| Guanine (Gua) | 246, 276 | | 10,700, 8,150 | | |
| 4-Thiouracil (Sra) | 328 | | 16,600 | | |
| Thymine (Thy) | 264.5 | | 7,900 | | |
| Uracil (Ura) | 259.5 | | 8,200 | | |

[a] Largely taken from L. Pauling, "The Nature of the Chemical Bond." Cornell Univ. Press, Ithaca, New York, 1948.

[b] Taken from the "Handbook of Biochemistry" (R. C. Weast, ed.), Chem. Rubber Publ. Co., Cleveland, Ohio, 1968.

millimeter of a material absorbs 50% of the radiation, then each succeeding millimeter will absorb 50% of the remaining radiation. At the end of 10 mm, ~0.1% of the initial radiation will emerge. In theory, no matter how thick the absorber is, the radiation would never be totally absorbed. In the case of solutions, this situation may be expressed as follows:

$$I/I_0 = 10^{-\mathrm{OD}} \tag{5}$$

in which $I_0$ is the incident intensity per unit area per unit time, $I$ is the emergent intensity per unit area per unit time, and OD is the optical density of the solution. Taking the logarithm of Eq. (5) gives

$$\log_{10}(I_0/I) = \mathrm{OD} \tag{5a}$$

an expression that has the familiar form of the Beer-Lambert law, i.e.,

$$\log_{10}(I_0/I) = \epsilon cl \tag{6}$$

in which $c$ is the concentration of the solute in moles/liter; $l$ is the thickness of the cuvette in cm, and $\epsilon$ is the molar extinction coefficient in liters/mole cm. For a given absorber, $\epsilon$ is a constant at a given wavelength and is a measure of the probability of the absorption of the photon in a photon–molecule interaction.

### c. *Photochemical Kinetics*

At this juncture, it seems proper to consider some aspects of the chemical kinetics of photoreactions, since they are closely related to light absorption by the reaction solution.

Let us consider a simple case in which the reactant (M) is the only light-absorbing substance, and both the solvent and the product (P) are nonabsorbing. For a monomolecular reaction,

$$\mathrm{M} \xrightarrow{h\nu} \mathrm{P}$$

the number of moles reacted per unit time is

$$-d\mathrm{M}/dt = E\phi \tag{7}$$

in which $E$ is the energy absorbed per unit time, and $\phi$ again is the quantum yield. Since

$$E = AI_{\mathrm{abs}}$$

in which $A$ is the front surface area of the reaction vessel, and $I_{abs}$ is the incident intensity absorbed per unit area per unit time,

$$\begin{aligned} E &= A(I_0 - I) \\ &= AI_0(1 - 10^{-\text{OD}}) \end{aligned}$$

Thus, Eq. (7) may take on the form

$$-d\text{M}/dt = AI_0(1 - 10^{-\text{OD}})\phi \tag{8}$$

In Eq. (8), only OD, the absorption of the solution at the wavelength of the incident radiation, is time dependent. When OD is small, such that OD << 1, $1 - 10^{-\text{OD}} \cong \text{OD} \ln 10$ and Eq. (8) becomes

$$-d\text{M}/dt = k[\text{OD}] = k_1[\text{M}] \tag{9}$$

which is linear in OD. This implies that at relatively low concentrations (OD << 1/ln 10) the absorption of incident light cannot be complete and must depend on the OD, i.e., the concentration of the solution, [M]. This absorption decreases steadily in proportion to the decreasing [M] during the course of the reaction. Thus, the photoreaction should be strictly first order.

When OD is large, i.e., OD > 2, $1 - 10^{-\text{OD}} \cong 1$ and is independent of OD. Equation (8) is reduced to a pseudo zero-order kinetics, i.e.,

$$-d\text{M}/dt = k_2 \tag{10}$$

Indeed, at relatively high concentrations of the reaction solutions, the "complete" absorption of the incident light is expected. This should result in a constant decrease of the reactant and, thus, a constant production of M* and the product during most of the course of the reaction. Under these conditions, such photolysis must yield strictly zero-order kinetics.

If photolysis is carried out with the OD of the solution ranging between 0.4 and 2, the photoreaction should exhibit rate constants between zero- and first-order kinetics. Thus, it is evident that the absorbed light intensity is most likely to be the limiting factor which controls the reaction rate and that a change in the kinetic order of the reaction resulting from a change in concentration is characteristic of a light-dependent reaction.

It shoud be noted that the foregoing discussion is of interest in reference to another kinetic treatment (for Pyr photohydrates) in Chapter 4. While there are characteristic differences between these two approaches, a similar expression may be reached from both. This viewpoint is supported by the fact that there is an excellent agreement between the quantum yields (of photolysis of $Me_2Ura$) determined by S. Y. Wang in the above manner and by H. Johns and associates using the other procedure (see Chapter 4).

*d. Inner Filters*

Sometimes one may be confronted with a reaction solution containing more than one species or compound absorbing the particular radiation wavelength. If $\epsilon_1$ and $\epsilon_2$ are the molar extinction coefficients and $c_1$ and $c_2$ are the concentrations for compounds $M_1$ and $M_2$, respectively, the fraction of the OD for each will be

$$OD_{M_1} = \frac{\epsilon_1 c_1}{\epsilon_1 c_1 + \epsilon_2 c_2} \qquad OD_{M_2} = \frac{\epsilon_2 c_2}{\epsilon_1 c_1 + \epsilon_2 c_2}$$

and the Beer-Lambert law assumes the following form:

$$I/I_0 = 10^{-(\epsilon_1 c_1 + \epsilon_2 c_2 + \cdots)l}$$

More often than not, only one of the absorbing compounds is under study; therefore, only the fraction of light that is absorbed by this particular compound should be taken into consideration. In the event that only one of these compounds is photoactive, the other inert compounds are commonly called the "inner filter"; they absorb certain fractions of incident light but produce no apparent photochemical changes.

## 4. Nature of Electronic Excitation

*a. Atomic Excitation*

Upon absorbing a photon, the ground-state molecule M is promoted to M* as the result of a redistribution of electric charges in the molecule. A qualitative semiclassical description of such a change should be helpful. Let us consider first the simplest atom, which has one electron traveling in an orbit around the nucleus, i.e., the hydrogen atom. The electron represented by ↑, which also indicates its direction of spin, is held fixed in its orbit (*s* orbital) by the balancing of the centrifugal force of the revolving electron against the force of attraction between the negatively charged electron and the positive nucleus (see Fig. 3). (The dashed circle indicates that the probability of finding the electron is zero.)

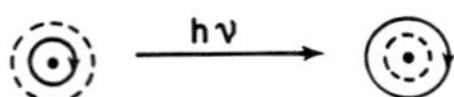


**Fig. 3.** *Electronic configurational change of an atomic orbital upon excitation.*

If this atom is exposed to light with $\nu$ proportional to the energy required to excite the electron from one level to the next, i.e., $\Delta E = E_2 - E_1 = h\nu$, the electron may be promoted accordingly. This resulting atom may be said to be in the electronically excited state.

### b. *Molecular Excitation:* $\sigma \rightarrow \sigma^*$, $\pi \rightarrow \pi^*$, *and* $n \rightarrow \pi^*$ *Transitions*

Now, let us consider the simplest molecule, hydrogen. As two hydrogen atoms approach each other, the nucleus of each will begin to attract the electron originally associated solely with the other. When the distance separating the nuclei is at or near the bonding distance, the two electrons in this system are associated with both nuclei; instead of the original "atomic orbital" on each atom, we now have a "molecular orbital" resulting from the combination of two atomic orbitals. The process of "adding" (+) two atomic orbitals may be represented graphically in several ways. One simple way is to consider the boundary surface of the molecular orbital as having been created from the overlap of the two boundary surfaces of the individual atomic orbitals. The probability of finding the electrons is high within the boundary surface. This orbital is referred to as a "bonding orbital" (Fig. 4).

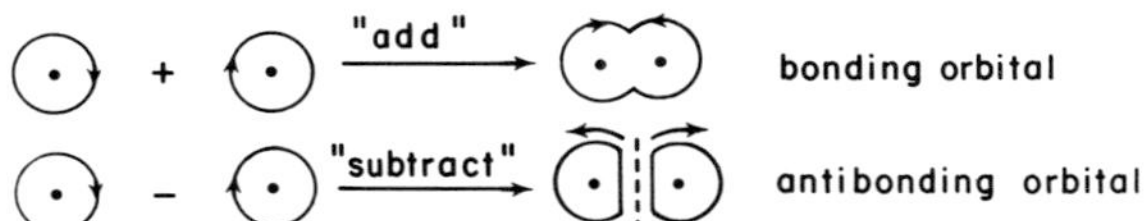


**Fig. 4.** *The addition and subtraction of two 1s atomic orbitals.*

A second linear combination of hydrogen atomic orbitals can be obtained by "subtracting" (−) one atomic orbital from the other. This process can also be represented by the boundary surface diagram (Fig. 4). In this molecular orbital, the probability of finding the electrons at exactly half the distance between the nuclei is zero*; there is a "nodal

* In the detailed quantum theory, the molecular orbital amplitudes are obtained by adding or subtracting the atomic orbital amplitudes. The probability of finding an electron is given by the absolute square of the amplitude. Thus, negative probability is never encountered.

plane," represented by the dashed line in Fig. 4. On the average, both electrons are farther from either of the nuclei than they would be in the bonding orbital. This orbital is referred to as an "antibonding orbital." Since the attractive forces between the nuclei and the electrons are increased in the bonding orbital and decreased in the antibonding orbital with little change in repulsive forces, the bonding orbital is a lower energy state than the antibonding orbital. Consequently, in a molecular bond both electrons would populate the bonding orbital resulting in an "empty" higher-energy antibonding orbital. In fact, the presence of an electron in such a molecular orbital would cause the separation of the two atoms rather than the formation of a bond. The process of the combination of two atomic orbitals to form two molecular orbitals can be represented by an energy diagram (Fig. 5). The two lateral lines on the left represent two equal-energy and isolated atomic orbitals in close proximity, each with an electron which have spins of opposite directions. The two horizontal lines in the middle are equally spaced (for the hydrogen molecule) from the atomic orbitals; the upper is the "antibonding orbital" with no electrons, and the lower is the "bonding orbital" with two "paired electrons" having opposite spins. This process results in the formation of a stable *single* bond ($\sigma$ bond) with a lower energy level. If a photon with a proper $\nu$ or $\lambda$ is absorbed by this molecule, one of the "paired electrons" may be promoted to the antibonding orbital ($\sigma^*$). This excitation mode is designated as a $\sigma \rightarrow \sigma^*$ transition. Since $\sigma$ bonds absorb energy with wavelengths mostly in the vacuum-UV region (Table 3), this $\sigma \rightarrow \sigma^*$ excitation is less important in our consideration of photochemistry and photobiology of nucleic acids but assumes greater significance in radiation chemistry and related studies (Chapter 12; Chapter 9, Volume II).

In contrast, the $\pi \rightarrow \pi^*$ and $n \rightarrow \pi^*$ transitions are more often encountered in the study of nucleic acid chemistry and biology because Pyr bases having C=C, C=N, and C=O double bonds ($\pi$ bonds) absorb strongly in the near-UV regions. In the case of the $\pi \rightarrow \pi^*$ transitions, we can also outline a linear combination of a pair of $p$ or $sp^2$

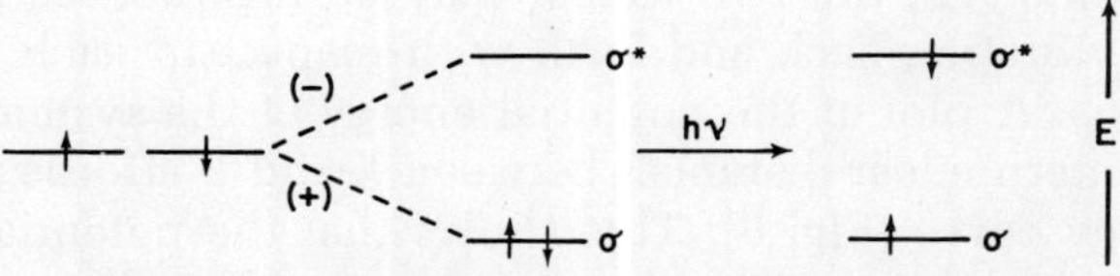


**Fig. 5.** *The molecular orbital energy diagram for $H_2$ and its electronic configurational change upon excitation.*

atomic orbitals by addition and subtraction to form these double bonds. However, such a qualitative description must be lengthy due to the complexity of these orbitals, and it results in an analogous energy diagram. That is, the formation of a $\pi$ bond from a pair of $p$ and $sp^2$ orbitals creates two molecular orbitals. Similarly, by radiation, one of the electrons in the bonding orbital is promoted to the "empty" antibonding orbital and thus produces a $\pi \rightarrow \pi^*$ transition.

As for the $n \rightarrow \pi^*$ transition, we are concerned in this case with a third type of molecular orbital, i.e., the nonbonding orbital ($n$ orbital). Electrons occupying such an orbital make no contribution to the bonding process or to the binding energy of the molecule. These are the lone-pair electrons of heteroatoms, such as O, N, and S atoms. Although a pair of nonbonding electrons will occupy a molecular orbital, its characteristics will be those of a pure atomic orbital or a hybrid atomic orbital of a single atom only, having produced *no* $n^*$ molecular orbital. Therefore, the electronic transition in promoting an electron from a $n$ orbital by absorption of light energy must involve a "borrowed" antibonding orbital of the molecule. In the case of Pyr bases, an electron from the $n$ level on an oxygen or a nitrogen atom is more likely to be promoted to the antibonding orbitals of $\pi^*$ than $\sigma^*$. This $n \rightarrow \pi^*$ transition requires the least energy. For example, the $n \rightarrow \pi^*$ transition of the C=O system in Ura and Thy is expected to give an absorption band at ~280 nm with a relatively low molar extinction coefficient ($\epsilon \cong 100$). This absorption band appears at the long wavelength end of the absorption spectra of Pyr with the high $\epsilon$ absorption maxima of the $\pi \rightarrow \pi^*$ transition band as the characteristic $\lambda_{max}$ (Table 3). This and other properties of these $\sigma \rightarrow \sigma^*$, $\pi \rightarrow \pi^*$, and $n \rightarrow \pi^*$ transitions will be considered in Chapter 2.

## 5. Photodissociation and Franck-Condon Principle

Since electronic distributions of excited-state molecules differ from those of the ground state, one may ask how this difference affects the nuclear configurations of the molecules. The vibrational motions of a molecule persist whether it is in the ground or excited state. For a diatomic molecule XY, the two nuclei may be regarded as joined by a spring and vibrating back and forth with respect to each other along the bond axis. A plot of the potential energy of the system as a function of the internuclear distances between X and Y affords a Morse potential energy curve (Fig. 6). This shows that the potential energy of the system is at a minimum when the internuclear distance ($r_{XY} = r_e$) or the configuration of the nuclei is at equilibrium. If the distance is

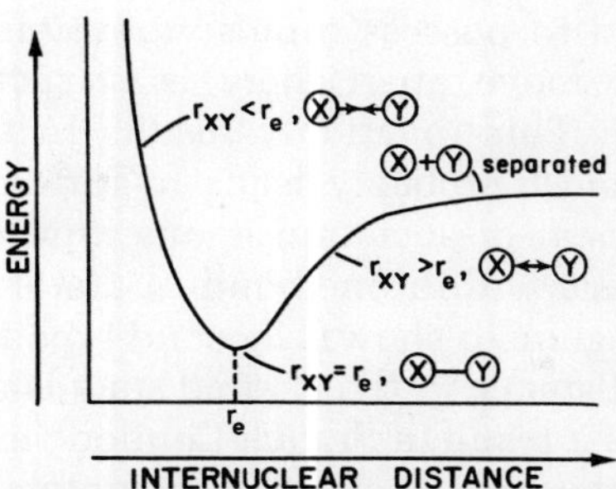


**Fig. 6.** *Potential energy curve of a diatomic molecule XY. [Adapted from Orchin and Jaffe (1967) and from Turro (1965).]*

decreased, $r_{XY} < r_e$, the potential energy of the system increases rapidly due to electronic and internuclear repulsions. If the internuclear separation is increased, $r_{XY} > r_e$, the potential energy also increases, though perhaps less rapidly, as a result of stretching of the X—Y bond. It can also be seen that a critical internuclear distance, the X—Y bond would break, resulting in the separation of X and Y.

Electronic transitions occur so rapidly ($\sim 10^{-15}$ sec) upon absorption of a photon that there is no time for nuclear motion ($\sim 10^{-12}$ sec). Thus, the potential energy of the molecule rises abruptly, but the nuclear configurations remain unchanged. According to the Franck-Condon principle, transitions tend to occur between vibrational levels of two electronic states for which the nuclear configurations are the same. This is best represented by a vertical jump in the energy of the system from a ground- to an excited-state Morse curve; there are three typical examples as shown in Fig. 7. The type illustrated in Fig. 7a shows an excited-stage curve lacking a potential energy well. Many

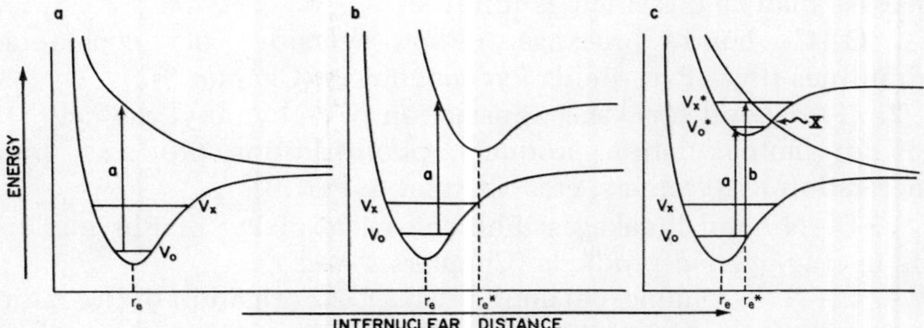


**Fig. 7.** *Potential energy curves showing various photodissociations: (a) optical dissociation, (b) photodissociation, and (c) predissociation. (See text for details.) [Adapted from Orchin and Jaffe (1967) and from Wayne (1970).]*

compounds are said to possess repulsive excited states such that as soon as the atoms move apart there is no potential minimum to reverse the motion. This invariably results in fragmentation of the molecules, which most probably leads to further chemical changes. This process of optical dissociation is one kind of photodissociation which may take place within one bond in diatomic as well as polyatomic molecules. Figure 7b shows a second type in which the equilibrium internuclear distance in the excited state is much greater than in the ground state. As a result, a Franck-Condon vertical transition from the ground state intersects above the asymptote of the excited-state potential curve. Momentarily, the molecule experiences vigorous vibration due to nuclear repulsions, but in such a high energetic state with no restoring force that it may simply collapse. This may be considered another type of photodissociation. Predissociation is a third kind of photodissociation which sometimes occurs in complex molecules. Upon excitation these molecules usually have a number of attractive and repulsive states; Fig. 7c shows the simplest version of such a situation. The excited-state potential curves intersect at X. Upon excitation to a vibrational level lower than X (arrow a), the molecule behaves in a nondissociative manner (see Sect. 6). However, upon excitation to a level higher than X (arrow b), the vibrational motions, at times, cause the molecule to cross over at X. This crossing from attractive to repulsive electronic states would lead to the dissociation of the molecule. Predissociation commonly takes place in heavy diatomic and polyatomic molecules.

Having examined the nature of photodissociation processes in the simplest diatomic molecules, we may reasonably assume that one or several of these processes may occur simultaneously in polyatomic Pyr and Pur derivatives. Indeed, photodissociation occupies an exceptional place in photodamage and photorepair of nucleic acids. These processes may be classified as follows:

1. C—C bond breakage: Photoreversion of cyclobutadipyrimidines (Pyr◇Pyr) yields Pyr monomers (Chapter 5).

2. C—H bond breakages: Formation of $\alpha$-thyminyl radical ($Thy^{\alpha}\cdot$) leads to photooxidation products, photoaddition products, photodimerization to tetramers, etc. (Chapter 6, Part B).

3. C—N bond breakages: Photocleavage of Pyr or Pur nuclei results in ring-opened products (Chapters 4 and 7).

4. C—X (X = halogens) bond breakages: Formation of Pyr radicals (Pyr-5·) gives coupled products (Chapter 6, Part A).

5. N—H bond breakage: The apparent photooxidation of an amino moiety forms oxidation products (Chapter 7).

## 6. Photophysical Deactivation Processes

If the excitation energy of M* is dissipated before its atoms have a chance to move far enough apart for photodissociation, nondissociative deactivation processes of a photochemical and photophysical nature may take place. Radiative and nonradiative transitions constitute photophysical processes. Two radiative transitions have been discerned. Fluorescence, a short-lived (usually $10^{-9}$ to $10^{-6}$ sec) emission, is usually associated with the excited singlet (two electrons with opposite spins) to ground-state singlet transitions. Phosphorescence, a somewhat longer-lived (usually $10^{-3}$ to several sec) emission, is related to excited triplet (two electrons with parallel spins) to ground-state singlet transitions. Thus, fluorescence is the emission between states of the same multiplicity, and phosphorescence occurs between states of different multiplicity. Similarly, there are two nonradiative deactivation processes. One is designated as internal conversion. It concerns the loss of the excess vibrational energy as heat through collisional transfer resulting in "vibrational cascade" from a higher to the lowest vibrational level of an excited state (usually $\sim 10^{-12}$ sec) or from a higher to a lower excited state (usually $10^{-13}$ to $10^{-14}$ sec) or ground state (usually $10^{-9}$ to $10^{-12}$ sec). The other, designated as intersystem crossing, concerns the interconversion of states of different multiplicities or between singlet and triplet states [usually $10^{-5}$ to $10^{-6}$ sec except for heavy-atom-substituted-molecules ($\sim 10^{-9}$ sec)].

In many instances, the species that absorbs a photon ($D_s$, a molecule of the singlet ground state) is not the reacting molecule ($M_s$) but rather serves as an energy donor. $D_s$ upon absorption of a photon becomes electronically excited and may transfer its excitation energy to $M_s$. Although either excited singlets ($D_s$*) or excited triplets ($D_T$*) may undergo such a process, the short lifetime of $D_s$* as compared to that of $D_T$* results in a much less efficient energy transfer process for the singlets.

$$D_s \xrightarrow{h\nu} D_s^* \xrightarrow[\text{crossing}]{\text{intersystem}} D_T^*$$

$$D_T^* + M_s \longrightarrow D_s + M_T^*$$

Therefore, the most efficient and important process is the triplet–triplet energy transfer, as shown above. This transfer requires intermolecular collisions or near collisions; therefore, it is a diffusion-controlled process. The donor molecule $D_s$ is commonly called a photosensitizer and, therefore, this process is designated as photosensitization. One particular case, which is an extension of photosensitization, having biological importance is photodynamic action. This process involves the oxidation of biomolecules by oxygen in the presence of a visible-light-absorbing dye as the sensitizer. Photosensitization reactions in the presence or absence of oxygen in *in vitro* and *in vivo* systems will often be encountered throughout these volumes.

These processes may be conveniently summarized in a Jablonski diagram which is presented in Chapter 2 together with a lucid discussion of these and related phenomena. On account of special features of the monomeric moieties in DNA polymers, a modified Jablonski diagram together with a succinct analysis has been included in Chapter 2, Volume II.

### 7. Remarks

A brief consideration of the photochemistry of nonradiative deactivation processes including photodissociations of Pyr and Pur derivatives as monomers or in polymers perhaps is in order here. However, the succeeding chapters will contain judicious discussions of characterizations and reaction mechanistic studies of various photoproducts and their photobiological consequences.

Undoubtedly, the study of photochemistry and photobiology of nucleic acids has been fascinating and has met with a great deal of success. The field may even have reached its climax in popularity in recent years. However, in closing this introductory chapter, it seems fitting and desirable to emphasize that the accomplishments of the past are recorded in this monograph for the motivation of new conceptual models and novel laboratory methods in order to provide fresh avenues of approach which will eventually lead study in this field to its acme.

## References

Bowen, E. J. (1946). "Chemical Aspects of Light." Oxford Univ. Press, London and New York.

Calvert, J. G., and Pitts, J. N., Jr. (1966). "Photochemistry." Wiley, New York.

Chapman, O. L. (1967). "Organic Photochemistry." Arnold, London.

Clayton, R. K. (1970). "Light and Living Matters," Vol. 1. McGraw-Hill, New York.

Jagger, J. (1967). "Introduction to Research in Ultraviolet Photobiology." Prentice-Hall, Englewood Cliffs, New Jersey.

Kan, R. O. (1966). "Organic Photochemistry." McGraw-Hill, New York.

LeGrand, Y. (1967). "An Introduction to Photobiology." Amer. Elsevier, New York.
McLaren, A. D., and Shugar, D. (1964). "Photochemistry of Proteins and Nucleic Acids." Pergamon, Oxford.
Neckers, D. C. (1967). "Mechanistic Organic Photochemistry." Van Nostrand-Reinhold, Princeton, New Jersey.
Orchin, M, and Jaffe, H. H. (1967). "The Importance of Antibonding Orbitals." Houghton, Boston, Massachusetts.
Rollefson, G. K., and Burton, M. (1939). "Photochemistry and the Mechanism of Chemical Reactions." Prentice-Hall, Englewood Cliffs, New Jersey.
Seliger, H. H., and McElroy, W. D. (1965). "Light: Physical and Biological Action." Academic Press, New York.
Smith, K. C., and Hanawalt, P. C. (1969). "Molecular Photobiology: Inactivation and Recovery." Academic Press, New York.
Suppan, P. (1973). "Principles of Photochemistry." Chemical Society, London.
Turro, N. J. (1965). "Molecular Photochemistry." Benjamin, New York.
Wayne, R. P. (1970). "Photochemistry." Amer. Elsevier, New York.

# 2 Excited States of the Nucleic Acids: Bases, Mononucleosides, and Mononucleotides

*Malcolm Daniels*



## A. Introduction

### 1. Aims

In order to understand the ultimate photochemistry of DNA's and RNA's it is necessary to have quantitative knowledge about the excited states produced by the absorption of a photon, for along with photochemistry, the many possible electronic deactivation processes of excited states include internal conversions ($S_2 \rightarrow S_1$), intersystem crossing ($S_1 \rightarrow T_1$), fluorescence and phosphorescence emission, and

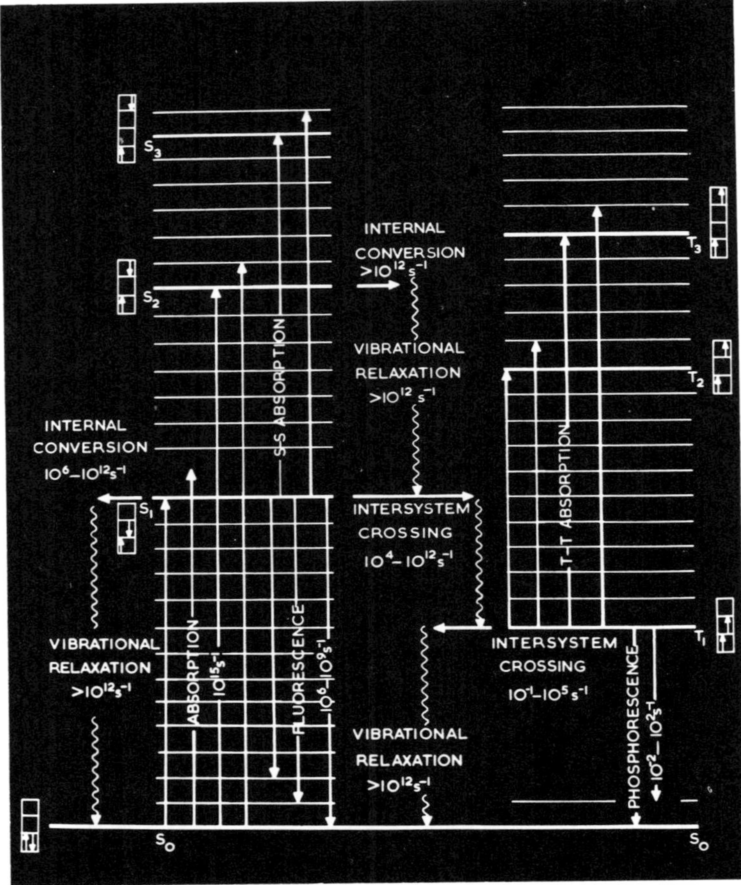


**Fig. 1.** *A generalized schematic of the lower energy levels of an organic molecule with an even number of electrons and no orbital degeneracy.*

energy transfer. The vocabulary of this conceptual framework is explained in Fig. 1. We must first characterize the transitions in absorption from the ground state to the lower excited singlet levels. In the case of complex molecules, such as the Pur and Pyr bases, the few absorption bands commonly observed in the UV are probably merely envelopes of several electronic transitions which frequently overlap. Excitation not only of bonding electrons ($\pi \rightarrow \pi^*$) but also of nonbonding electrons ($n \rightarrow \pi^*$) can occur. Specifically we should aim to describe the probability of an electronic transition (characterized by its oscillator strength, $f_{osc}$), the energy at which the transition occurs,

$^1E(0\text{–}0')$, the molecular direction of the transition given by the orientation of the transition dipole moment between the connecting states, and the band vibronic structure (or an analytical description of the vibronic band envelope).

We then need to know the properties of the excited states so produced. Some of the properties can be considered pseudo-stationary within the lifetime of the excited state, e.g., dipole moment, polarizability, and vibrational spectrum, but the kinetic behavior of excited states is of predominant importance for eventual chemical change. Consequently, in the case of the excited singlet state, we try to define rate constants for internal conversion, $k_{ic}$, fluorescence emission, $k_f$, intersystem crossing, $k_{isc}$, and unimolecular chemical processes $k_{chem}$. Other properties, such as rate constants for excimer formation, can be defined as necessary.

Similarly, we need to know the probability of populating the triplet state, $\phi_{isc}$, the energy of the state $^3E(0\text{–}0')$, and its nature ($\pi\pi^*$ or $n\pi^*$, dipole moment, ZFS parameters, and vibronic characteristics). The rates of the various first-order processes which depopulate the triplet level, e.g., radiative intersystem crossing to the ground state (phosphorescence), $k_p(T_1 \rightarrow S_0)$, and nonradiative intersystem crossing, $k_{isc}(T_1 \rightarrow S_0)$, and also rates of the second-order processes, such as quenching $k(T_1 + Q)$ and chemical reaction $k(T_1 + X)$, should be determined. We may also have to consider reexcitation processes such as thermal repopulation of the singlet level, $T_1 + kT \rightarrow S_1$, excitation to higher triplet levels (triplet–triplet absorption: $T_1 + h\nu \rightarrow T_x$), and second-order triplet–triplet interaction ($T_1 + T_1 \rightarrow X^*$).

Although many of the parameters are intrinsic to the molecule, most are determined in a certain medium, not in the gas phase or matrix-isolated state, and the study of medium effects is consequently of considerable importance. This is perhaps as it should be because the photochemistry we are concerned with occurs at approximately room temperature and at neutral pH in a medium which is predominantly aqueous and is (in the cell) viscous or gel-like in character.

The fulfillment of aims such as these is a difficult and time-consuming task, but considerable progress has been made during the twenty years since the formulation of the Crick-Watson model for the secondary structure of DNA. The earlier work was naturally concerned with and limited by techniques available at the time. Thus, absorption spectroscopy was mainly concerned with solutions, and fluorescence and triplet state studies were carried out at low temperatures in glasses ranging in composition from ethylene glycol/water to hydrocarbons. However, since 1968 significant advances have been made in all three areas. In particular we will draw attention in Section B to the increased activity in studies of higher-resolution polarized reflectance and polarized absorption spectroscopy of single crystals since in 1969. The properties of excited singlet states are obtained (with the exception of $S_1 \rightarrow S_x$ absorption spectra) from fluorescence measurements, and the most recent development in this area is the ability to observe fluorescence emissions at room temperature (Section C). Section D is concerned with triplet states, and techniques in this regard have also been extended to aqueous media at room temperature, so that triplet quantum yields ($\phi_{isc}$), T–T absorption spectra, triplet lifetimes, and rate constants have been determined. Research in these areas is still active, but the findings are sufficiently advanced to allow review.

### 2. Historical Development

Detailed study of the excited electronic states of nucleic acid bases can perhaps be considered to have started with the work of Mason (1954) on the solution UV absorption spectra of the purines. Previous reports (e.g., Shugar and Fox, 1952) were mainly concerned with the use of absorption spectra to determine pKs and tautomeric structures. Mason's work, which is still the starting point for current spectroscopic discussions, was essentially a study of substituent and medium effects on the transitions in some 34 purines. A later series of papers (Mason, 1959) was concerned with substituent and solvent effects on the $^1(n\pi^*)$ and the $^1(\pi\pi^*)$ states of the azines, the vibrational structure of the $(n\pi^*)$ band of *s*-tetrazine, and the rotational structure of the vibrationless band. This produced the first experimental evidence that an $n \rightarrow \pi^*$ transition is polarized out-of-plane. Accounts of this work have been presented by Mason (1962, 1963).

The study of luminescence emissions was spurred by the discovery of phosphorescence from Ade, its nucleoside and nucleotide in glycerol solutions at 77°K, and from DNA and RNA (Steele and Szent-Györgyi, 1957). Previous reports of fluorescences (Euler *et al.*, 1935; Stimson and Reuter, 1941) are suspect since the solutions were transparent at the reported wavelength of excitation, 366 nm. Bersohn and Isenberg (1964) investigated the phosphorescence spectra of Pur nucleosides and nucleotides in detail and found considerable vibrational structure. Fluorescence was observed in purines and pyrimidines, but no phosphorescence could be detected from the pyrimidines, and the phosphorescence of DNA resembled a mixture of Ade- and Gua-like emissions. Strong quenching by paramagnetic ions was found.

The problem of determining the orientations of the transition moments of the bases was undertaken by Stewart and Davidson (1963), who measured the polarized absorption spectra of thin sections of 1-MeUra and 9-MeAde crystals and their H-bonded complex. Polarization of the fluorescence from Gua, 9-EtGua, and Ade at −195°K was investigated as a function of exciting frequency by Callis *et al.* (1964) with the aim of resolving overlapping transitions. Polarization of the phosphorescence from Pur, Ade, 9-MeAde, and 9-nBuAde at 77°K was studied by Cohen and Goodman (1965). A spate of papers on low-temperature fluorescence and phosphorescence appeared in the years 1966 to 1968. Longworth *et al.* (1966) studied the effects of ionization (protonic) and tautomerization on the fluorescence and phosphorescence yields of the Pur and Pyr bases, and Rahn and Shulman (1966) studied the triplet states by ESR. Parameters of the singlet and triplet states at 77°K were determined for the nucleotides by Guéron *et al.* (1967a,b), and these, together with other data on the di-, oligo-, and polynucleotides and DNA, all at 77°K, have been reviewed extensively by Eisinger (1968) and Eisinger and Shulman (1968). At about the same time, photoionization of Ado, Guo, and Pur by a biphotonic process was demonstrated by phosphorescence and esr measurements (Hélène *et al.*, 1966) and by delayed luminescence (Guermanprez *et al.*, 1967). The triplet level was implicated as the intermediate state which absorbed the second photon although the T–T absorption spectra were not known at the time. Brynda (1971) extended this work by studying the biphotonic thermoluminescence. Phosphorescence of the pyrimidines, not detected by Bersohn and Isenberg (1964) or by Guéron *et al.* (1967a), was reported by Imakubo *et al.* (1967), and there are interesting differences among the phosphorescence spectra of DNA from these three groups. Extensive studies on the effects of solvent and temperature on the absorption and emission spectra of substituted purines and pyrimidines, including polarization measurements, were carried out by Drobnik and Augenstein (1966a,b) and by Kleinwächter *et al.* (1966, 1967a,b).

Since 1968 the major developments have been in single crystal studies by polarized reflectance and absorption (Chen and Clark, 1969; Eaton and Lewis, 1970; Callis and Simpson, 1970; Lewis and Eaton, 1971) and in studies of triplet and singlet states at room temperature, mostly in aqueous solution. From continuous beam photokinetics, a measure of $\phi_{isc}$ was obtained for Ura (Brown and Johns, 1968) and Thy (Fisher and Johns, 1970). A more direct estimation of $\phi_{isc}$ for all the Pur and Pyr nucleotides of the nucleic acids has been provided by Lamola and Eisinger (1971) who counted the triplets by energy transfer to $Eu^{3+}$. Triplet–triplet absorption spectra of Ura and Thy (Whillans and Johns, 1971) have been determined in aqueous solution by flash photolysis. In nonaqueous solution T–T spectra have been obtained by Hayon (1969) using pulse radiolysis and by Szabo *et al.* (1970) by flash photolysis. At the singlet level, the room temperature fluorescence of the Pur and Pyr bases in neutral aqueous solution has been detected by a signal accumulation technique and the corrected emission and excitation spectra and quantum yields have been given (Daniels and Hauswirth, 1971; Hauswirth and Daniels, 1971a). Confirmation of these fluorescences was made by a photocounting technique (Vigny, 1971a) and was extended to the nucleosides and nucleotides (Vigny, 1971b). Hauswirth and Daniels (1971b) showed how these results lead directly to estimates of $k_{ic}$, $k_f$, $k_{chem}$, and $k_{isc}$ for Thy and Ura.

### 3. Factors Involved in the Absorption and Emission of Radiation

The primary factor governing the absorption and emission of radiation is that the energy level of the photon must correspond to a difference between energy levels of the molecular system.

$$E = h\nu = hc/\lambda = hc\bar{\nu}$$

For transitions between electronic energy levels in the UV and visible region, $\bar{\nu}$ is commonly $\gtrsim$ 10,000 $cm^{-1}$, vibrational transitions usually have $\bar{\nu} \sim 1000\ cm^{-1}$, and rotational transitions have $\bar{\nu} \sim 50\ cm^{-1}$. Rotational transitions are usually only observed in the gas phase and vibrational structure is often lost due to solute–solvent interaction so that only a composite electronic-vibrational (vibronic) transition is seen.

The energy levels of electrons in molecules may be classified in increasing order of energy, and their contribution to the total energy is shown in Fig. 2.

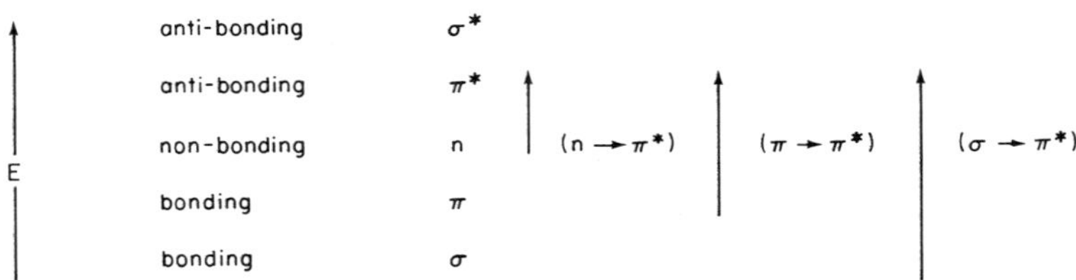


**Fig. 2.** *Approximate sequence of energy levels.*

Within the framework of simple molecular orbital (MO) concepts, $\sigma$ MO functions are symmetric and $\pi$ functions are antisymmetric with respect to reflection through the molecule plane. A distribution of electrons over MOs possessing a well-defined total spin constitutes a configuration, and although the wave functions of states can only be approximated by a single configuration [because of the neglect of electron–electron interaction (configuration interaction) which is particularly important for excited states] transitions and excited states are still conveniently categorized in terms of the most important MO involved. Transitions between states are then discussed as if they were entirely $n \rightarrow \pi^*$ or $\pi \rightarrow \pi^*$. For simple molecules it is apparent from Fig. 2. that the $n \rightarrow \pi^*$ transitions require less energy and hence lie at longer wavelengths than the $\pi \rightarrow \pi^*$ transitions. Comparable in energy with $\pi \rightarrow \pi^*$ and $\sigma \rightarrow \sigma^*$, $n \rightarrow \sigma^*$ are found only in the far UV. In more complex molecules the highest occupied $\pi$ level is less strongly bound with increasing conjugation and moves to higher energies. Ultimately this level overlaps the nonbonding levels which are unaffected by conjugation and consequently the $n \rightarrow \pi^*$ absorption bands become hidden under the intrinsically stronger (more intense) $\pi \rightarrow \pi^*$ band. In this way we encounter one of the main problems of Pur and Pyr spectroscopy.

The probability of light being absorbed is extracted from intensity measurements (at a particular wavelength) on the incident and transmitted beams, using the Beer-Bouguer-Lambert Law in the form log $(I_0/I_t)_\lambda = \epsilon_\lambda cl$ in which $c$ is the (molar) concentration of the absorbing species (presumed known) and $l$ is the optical path length. The constant $\epsilon$, known variously as the (molar) extinction coefficient, absorbance, or absorptivity, depending upon the recommendations of various committees, has then the units $M^{-1}$ $cm^{-1}$.* Values range from $10^5$ to $10^{-2}$, the former indicating a strongly allowed transition, and the latter a forbidden transition. Intermediate values of 100 to 1000 cannot be used per se to assign degrees of forbiddenness and are usually discussed on an *a priori* basis.

Whereas for an atomic system satisfaction of the resonance condition leads to a line absorption corresponding to a pure electronic transition, for a polyatomic system the electronic transition can be accompanied by vibrational (and rotational) transitions, which may or may

* These units are also $cm^2$/mole, and on this basis the probability is sometimes referred to as a cross-section for absorption, given another symbol $\sigma_a$ and often expressed in units of $cm^2$/molecule. The distinction between $\epsilon$ and $\sigma$ is trivial, and, as almost all spectroscopic data for the UV and visible regions are reported in terms of $\epsilon$, the use of $\sigma_a$ is unnecessary and possibly confusing.

not be resolved, leading to the appearance of band spectra. In this case a better measure of the intensity of the complete vibronic transition is given by the band area

$$\int_{\bar{\nu}_1}^{\bar{\nu}_2} \epsilon(\bar{\nu}) d\bar{\nu}$$

For historical reasons, band intensities are described in terms of the oscillator strength, $f_{osc}$, defined from classical theory as

$$f_{osc} = \frac{2303\ m_e c^2}{\pi e^2 N n} \int_{\bar{\nu}_1}^{\bar{\nu}_2} \epsilon(\bar{\nu})\ d\bar{\nu} = \frac{4.35 \times 10^{-9}}{n} \int_{\bar{\nu}_1}^{\bar{\nu}_2} \epsilon(\bar{\nu})\ d\bar{\nu}$$

in which $n$ is the refractive index of the medium, usually a slowly varying function over the width of the band. The oscillator strength, a dimensionless quantity, is of the order 1 for strongly allowed transition having $\epsilon_{max} \sim 10^5$. Forbidden transitions, for example, a singlet → triplet absorption, would have $f_{osc} \sim 10^{-6}$. A practical point in the determination of $f_{osc}$ concerns the location of the limits of integration $\bar{\nu}_1$ and $\bar{\nu}_2$ in the case of overlapping bands. Indeed, in heavily overlapping bands the resolution of individual transitions may not be apparent and may have to be based on properties other than direct absorption, such as fluorescence or polarized excitation spectra. Examples of this will be presented in Section B.

The intrinsic probability that a molecule in a particular state $l$ (lower) will absorb a photon of frequency $\nu$, reaching an u (upper) state, is derived in terms of the Einstein coefficients for (stimulated) absorption ($B_{l\to u}$), stimulated emission ($B_{u\to l}$), and spontaneous emission ($A_{u\to l}$), for the situation in which the molecule is imagined to be immersed in a radiation bath of density $\rho(\nu)$, in which $\nu$ is given by $(E_u - E_l)/h$ (Fig. 3).

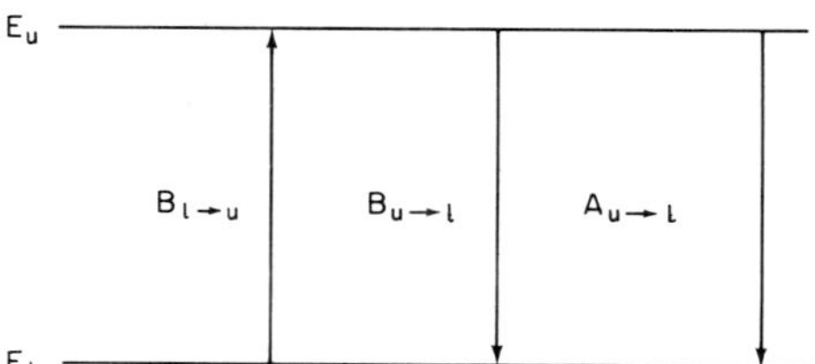


**Fig. 3.** *Definition of Einstein coefficients.*

It is then shown quite simply (see, for example, Barrow, 1962) that

$$B_{l\to u} = B_{u\to l}$$

and

$$A_{u\to l} = \frac{8\pi h^3}{c^3} \nu^3 n^3 B_{u\to l}$$

Connection with experiment comes from the relation of $B_{l\to u}$ to the oscillator strength:

$$B_{l\to u} = \frac{2303C}{Nhn} \int \frac{\epsilon(\bar{\nu})}{\bar{\nu}} \, d\bar{\nu}$$

$$f_{osc} = \frac{mhc}{\pi e^2} \bar{\nu} B_{l\to u}$$

$B_{l\to u}$ is in turn related to the microscopic description of the molecular mechanism of radiation absorption through time-dependent perturbation theory.

The molecular wave functions determine the value of any physically observable property of a molecule and with each observable property there is associated an operator $O$ such that the mean value of the observable property in the state $\psi_l$ is given by the integral

$$\int \psi_l^* |O| \psi_l \, d\tau \equiv \langle \psi_l | O | \psi_l \rangle$$

The form of the operator is usually obtained from the corresponding classical expression for the process to be described. For example, to describe the dipole moment of the state the operator $O$ is $\Sigma \mathbf{r}_i z_i$ in which $\mathbf{r}_i$ is the position vector of the $i$th particle and $z_i$ is the charge number of the particle, so that the dipole moment of the ground state would be given by

$$\mu_l = e \, \langle \psi_l | \sum_i z_i \mathbf{r}_i | \psi_l \rangle$$

in which $e$ is the electronic charge.

When the physically observable process is light absorption, the most common mechanism involves the interaction between the electrons of the molecule and the electric vector $\mathbf{E}$ of the radiation, usually evaluated in the electric dipole approximation. The process is characterized by a transition dipole moment between states, $M_{l-u}$, commonly called the transition moment,

$$M_{l \ \mathrm{u}} = e \ \langle \psi_l | \sum_i z_i \mathbf{r}_i | \psi_\mathrm{u} \rangle$$

and the time-dependent perturbation theory (see Eyring *et al.*, 1944) shows that

$$B_{l \to \mathrm{u}} = \frac{8\pi^2}{3h^2} \ |M_{l \to \mathrm{u}}|^2$$

Two other mechanisms of light absorption are to be considered. Magnetic dipole radiation, due to interaction of the magnetic vector **H** of the radiation with the electrons, is characterized by the angular momentum operator $|\mathbf{rxp}|$, and electric quadrupole radiation is characterized by the operator $|\mathbf{rr}|$. The relative probabilities of these processes are estimated as electric dipole: magnetic dipole: electric quadrupole, $1:10^{-5}:10^{-7}$. Consequently, magnetic dipole or electric quadrupole transitions will be significant only for those cases in which the electric dipole matrix element becomes zero through the particular symmetry properties of the states $l$ and u. The best known cases of magnetic dipolar radiation are found in emissions from the transition metal ions, but it is mentioned here because an $n \to \pi^*$ transition which is for symmetry reasons forbidden as an electric dipole transition is allowed as a magnetic dipole (and electric quadrupole) transition.

The transition dipole moment $M$ thus determines the probability of a transition occurring in a molecule. In turn $M$ is determined to a considerable extent by the form of the molecular wave functions and of the associated operator. The transition dipole moment itself is a real physical property of the molecule, and, as such, must be invariant with respect to any rotation, reflection, or inversion of the molecular symmetry axes. The vectors **r** have the symmetry property of a translational motion along one or more of the molecular axes, and, therefore, the matrix elements are nonzero only if the direct product of the upper and lower state functions contains a term with the symmetry properties of a translational motion (Sponer and Teller, 1941). The direction of the motion gives the orientation of the electronic dipole transition moment. Since a molecule will absorb only that part of the incident radiation which has an electric vector parallel to the transition moment, the direction of the transition moment can be determined by measuring the spectrum of an oriented molecule with polarized light. This is the basis of some of the methods discussed in Section B for determining transition moment directions.

Approximate selection rules, which predict from the symmetry properties of the wave functions and from the operator whether or not the transition moment can have a nonzero value, are based on the assumption that a molecule has a fixed shape which does not change on excitation. However, there are always some vibrations which periodically change the shape of a molecule, such as out-of-plane vibrations. If this happens, the direct product of the symmetries of the ground and excited state vibronic functions can contain a term with the symmetry properties of a translation, while the product of the pure electronic functions cannot. A nominally forbidden transition then becomes allowed to the extent that the perturbing vibration changes the electronic wave function. This process is known as vibronic intensification and is treated by first-order perturbation theory. A pertinent example is the $n \rightarrow \pi^*$ transition of the carbonyl group, for which, for the 0–0′ transition, the products of the irreducible representations of the excited state $(A_2)^*$, the electric dipole moment operator

$$\begin{pmatrix} B_1 \\ B_2 \\ A_1 \end{pmatrix}$$

and ground state $(A_1)$ do not contain the totally symmetric representation; that is, for $\langle \phi_u \, |e\mathbf{r}| \, \phi \rangle$

$$\Gamma(\phi_u)\,\Gamma\begin{pmatrix} x \\ y \\ z \end{pmatrix}\Gamma(\phi_l) = A_2 \times \begin{pmatrix} B_1 \\ B_2 \\ A_1 \end{pmatrix} \times A_1 = \begin{pmatrix} B_2 \\ B_1 \\ A_2 \end{pmatrix}$$

Hence the transition $n \rightarrow \pi^*$ is not allowed. However coupling with vibration of appropriate symmetry, for example $B_2$, can make it allowed $(B_2 \times B_2 = A_1)$.

In symbolic form a first approximation for vibrational coupling can be expressed as

$$\psi_n(Q) = \psi_n\,(0_n) + \sum_{n \neq m} C_{nm}\,(Q)\,\psi_m\,(0_m)$$

which describes the mixing of an unperturbed electronic function $\psi_m(0_m)$ ($0_m$ represents the zero displacement of the $m$th nuclear equilibrium configuration) with an equilibrium configuration $\psi_n(0_n)$ and in which Q is a vibrational displacement of the normal coordinates from equilibrium. The mixing coefficient $C_{nm}$ is determined by the energy

* The symmetry of the excited state can be seen to be $A_2$ from

$$\Gamma(\phi_u) = \Gamma(n) \times \Gamma\,(\pi^*) = \Gamma(y) \times \Gamma\,(x) = B_2 \times B_1 = A_2$$

splitting and the Hamiltonian operator for the electronic–vibrational coupling

$$C_{nm} = \frac{1}{E_m - E_n} \langle \psi_n(0_n) | \mathscr{H}_{\mathrm{ev}}(Q) | \psi_m(0_n) \rangle$$

The integral has to conform to symmetry selection rules, i.e., the integrand has to be totally symmetric under the symmetry operation of the molecule, and $\mathscr{H}_{\mathrm{ev}}$ (Q) transforms just as do the coupling vibrations $\chi(Q)$. As a result of contributions such as these, in an allowed electronic band the totally symmetric vibrations (which do not change the shape of the molecule) can often be seen as progressions starting from the zero vibrational level. For forbidden transitions made allowed by a nontotally symmetric vibration, the absorption is characterized by the absence of a band origin for the pure electronic transition, followed by a vibrational progression of out-of-plane bending, or ring-angle bending modes. For more details, Hochstrasser (1966) is recommended.

The equilibrium shapes of molecules may also change in the excited state and again a forbidden transition may become allowed through the change in symmetry. However, the operation of the Franck-Condon principle militates against this effect in absorption process and in any case a significant change in the shape of molecules the size of DNA bases is not anticipated.*

Just as an appropriate vibrational coupling can remove restrictions on a symmetry-forbidden transition, the change-of-multiplicity restriction forbidding singlet–triplet transitions can be weakened by the magnetic interaction between spin and orbital angular momenta known as spin–orbit coupling. This mechanism, which offers the explanation of the intersystem crossing process ($S_1 \rightarrow T_1$), of phosphorescence ($T_1 \rightarrow S_0 + h\nu$), and $S_0 \rightarrow T_1$ absorption, has the effect of mixing some triplet character into a singlet state and vice versa so that the multiplicity of either state is not strictly defined and the transition may be regarded as occurring between the triplet component of the (mainly) singlet state and the (mainly) triplet state.

In formal terms the process is described by the relations

$$\phi_{T_1} \sim {}^3\phi_{1m} + C_{1mn}{}^1\phi_n = \psi_1{}^3\Omega_m + C_{1mn}\, \psi_n{}^1\Omega_n$$

in which $\phi$'s are zero-order electron wave functions including spin,

* For the azines, Innes *et al.* (1967) conclude that there is no strong evidence that they are nonplanar in any state.

$\psi$'s are pure electronic functions, and $\Omega$'s are spin functions. The mixing coefficient again is given by

$$C_{1mn} = \frac{1}{E(^1\phi_n) - E(^3\phi_{1m})} \langle ^3\phi_{1m} | \mathscr{I}_{L-S} | ^1\phi_n \rangle$$

in which $\mathscr{H}_{L-S}$ is the spin–orbit coupling Hamiltonian. The form and magnitude of the Hamiltonian are given (see McGlynn *et al.*, 1969) in the one-electron approximation by $\mathscr{H}_{L-S} \sim \text{const} \ (\mathbf{r} \times \mathbf{ps}) = \kappa \mathbf{ls}$, in which $l = \mathbf{rxp}$ is the one-electron orbital angular momentum operator and $\kappa$ is the spin–orbit coupling factor, dependent on the atomic number $Z$. Neglecting vibrational coupling, the spin–orbit operator has the symmetry properties of rotations $R_x$, $R_y$, and $R_z$ (see Hochstrasser, 1966) and consequently, for conjugated molecules which remain planar when excited, it can mix $(\pi\pi^*)$ configurations with $(\sigma\pi^*)$ or $(\pi\sigma^*)$. Triplet spin functions also transform the same way so that for the 0–0′ transition of phosphorescence emission $T_1 \rightarrow S_0$, selection rules for the mixing of $T_1$ with $S_2$ or $T_2$ with $S_0$ can be expressed by the following relation between the direct products of the irreducible representations

$$\Gamma(R_i) \times \Gamma(X_j) = \Gamma(\psi_1)$$

in which $\psi_1$ is the orbital function of $^3\phi_1$ and $X_j$ is the polarization direction. This same condition is fulfilled for several directions because three components of $\mathscr{H}_{L-S}$ and three components of $^3\Omega$ are to be considered.* A second order treatment includes the effects of vibrations on spin–orbit coupling. Selection rules and polarizations for these conditions can be found, together with an extensive discussion, in Hochstrasser (1966). We can again only indicate some of the consequences of these symmetry considerations for some simple azines related to the DNA bases.

The allowed $^1(n \rightarrow \pi^*)$ transitions of the azines are known, from rotational analysis of their vapor spectra (Mason, 1959), to be out-of-plane polarized, and the symmetry of these allowed singlets corresponds to the $B_2$ representation in a $C_{2v}$ point group. Now the singlet spin function belongs to the totally symmetric representation $A_1$ and, since the symmetry of a complete wave function is a product of the symmetries of the spin and orbital functions, a singlet state of orbital

* Resolution of the three triplet components can only be observed at very low temperatures (e.g., that of liquid He) because of the magnitude of the splittings, which is $\sim 1$ $\text{cm}^{-1}$ at 15,000 G. For an account of recent developments, see El-Sayed (1971).

symmetry $\Gamma(S)$ will have the same symmetry when spin is included: $\Gamma(S)A_1 = \Gamma(S)$. But the three components of the triplet state will have total symmetries $\Gamma(S)R_x$, $\Gamma(S)R_y$, and $\Gamma(S)R_z$. Since $R_x$ corresponds to $B_2$, $R_y$ to $B_1$, and $R_z$ to $A_2$, the three components of the corresponding $^3(n\pi^*)$ should belong to $A_1$, $A_2$, and $B_1$ representations respectively. Of these, $A_1 \rightleftarrows A_2$ is symmetry forbidden, and both of the other transitions, $A_1 \rightleftarrows A_1$ and $B_1 \rightleftarrows A_1$ are predicted to be in-plane polarized, in contrast to the $^1(n\pi^*)$ case. From a slightly different point of view, mixing of states can only occur between states of the same symmetry. Consequently, we have

$$^1(n\pi^*) \not\longleftrightarrow {}^3(n\pi^*) \quad \text{but } ^1(\pi\pi^*) \longleftrightarrow {}^3(n\pi^*)$$
$$\text{and } ^3(\pi\pi^*) \longleftrightarrow {}^1(n\pi^*)$$

including ground-state mixing. Then the symmetry and polarization of $T_1 \rightleftarrows S_0$ will be determined by $^1(\pi\pi^*)$ transitions of $A_1 \leftrightarrow A_1$ and $A_1 \leftrightarrow B_1$ types; since $^1(\pi\pi^*)$ transitions are in-plane polarized, then $^3(n\pi^*)$ should be also. The validity of this conclusion for symmetries lower than $C_{2v}$ may be accepted if an appropriate local symmetry is defined. The direct consequences of these conclusions, for aid in identifying transitions, are that if the excitation is into an $^1(n\pi^*)$ transition and if emission is observed from an $^3(n\pi^*)$, it will be polarized perpendicularly to the absorption and be characterized by a negative value of the degree of polarization $P = (I_{\parallel} - I_{\perp})/(I_{\parallel} + I_{\perp})$. If excitation is into a $^1(\pi\pi^*)$ transition, $P$ can be negative or positive depending on the angle between the two in-plane transitions (Krishna and Goodman, 1962). Conversely, when the emitting transition is of the $^3(\pi\pi^*)$ type, then the polarization direction will be out-of-plane (El-Sayed and Brewer, 1963), and P will be positive if the absorbing transition is $^1(n\pi^*)$ and negative if it is $^1(\pi\pi^*)$. Clearly, the use of polarization measurements in characterizing emitting states relies on a valid assignment of the transition responsible for absorption, and it becomes interesting to consider other criteria by which the $^3(\pi\pi^*)$ and $^3(n\pi^*)$ states may be distinguished.

Primarily because of the magnitude of spin–orbit coupling, together with intensification from $^1(\pi\pi^*)$ states, the transition $^3(n\pi^*) \rightarrow S_0$ is approximately $10^2$ to $10^3$-fold faster than the corresponding $^3(\pi\pi^*) \rightarrow S_0$. Consequently, while the phosphorescence decay times in aromatic hydrocarbons of $\sim 1$ sec are characteristic of $^3(\pi\pi^*)$ transitions, $\tau_p \leqslant 10^{-2}$ sec are indicative of $^3(n\pi^*)$ emitting states. Related to this is the fact that ESR methods could only detect $^3(\pi\pi^*)$ states so far because of the short lifetime of the $^3(n\pi^*)$ states and the currently poor time resolution of most esr arrangements. The characteristic† ZFS parameter **D*** for $^3(\pi\pi^*)$ is usually 0.2 $cm^{-1}$. Recent developments in the time resolution of ESR (Atkins *et al.*, 1970) should allow detection of $^3(n\pi^*)$ states and their **D*** parameter.

The energy gap between singlet and triplet $(n\pi^*)$ states, i.e., $E^1(n\pi^*)$ to $E^3(n\pi^*)$, is usually quite small and in the range 2000 to 6000 $cm^{-1}$, whereas $E^1(\pi\pi^*)$ to $E^3(\pi\pi^*)$ is usually much larger, i.e., 20,000 $cm^{-1}$. However this depends both on the extent of the mixing of states and on the successful identification of the $^1(n\pi^*)$ transition.

Criteria for the differentiation of the corresponding singlet states were developed much earlier (see Kasha, 1961) and are rather more qualitative. The absorption band considered to belong to an $^1(n\pi^*)$ transition should not, of course, be found in the corresponding hydrocarbon. In general, the $(n\pi^*)$ bands show a blue shift in going from nonpolar solvents; coupled with the red shift often found for $^1(\pi\pi^*)$ states, this means that the $^1(n\pi^*)$ band often "disappears" in polar solvents. In nonpolar solvents and in the vapor phase the $^1(n\pi^*)$ are found at lower energies than the $^1(\pi\pi^*)$ bands and they are generally weak, being of various degrees of forbiddenness, $\epsilon \sim 10$–100 and $f_{osc} \sim 10^{-2}$–$10^{-4}$. Perhaps the most distinctive criterion, that $^1(n\pi^*)$ transitions are expected to be polarized out-of-plane, has not received much experimental attention since it is more difficult to determine than other parameters. An earlier criterion that $^1(n\pi^*)$ states do not fluoresce has fallen with the discovery of fluorescence from diazaphenanthrene and the diazines, Pyr and pyridazine, among others, but it is being replaced by the confirmation from solvent effects that the dipole moment of $^1(n\pi^*)$ states is less than for the ground state (Baba *et al.*, 1966, Mugiya and Baba, 1967), whereas it is common for $^1(\pi\pi^*)$ states that $\mu_u \geqslant \mu_l$.

In view of the position of Pyr as a parent compound for the nucleic acid bases and the detailed study which it has received, the excited

† ZFS stands for zero field splitting and is a measure of the interaction, in the absence of an external field, of magnetic dipoles associated with the electronic spins of triplet states. Such interactions remove the triplet degeneracy and lead to anisotropic ESR spectra. **D*** is then the ZFS dipolar coupling tensor. In terms of the principal axes which diagonalize the zero field tensor, the zero field splitting can be rewritten in terms of two independent constants $D$ and $E$, so that the spin Hamiltonian takes the form

$$\mathscr{H}_d = D(S_z^2 - \tfrac{1}{3}S^2) + E(S_x^2 - S_y^2)$$

$D$ and $E$ thus indicate the anisotropy of the triplet wave functions (Carrington and McLachlan, 1967).

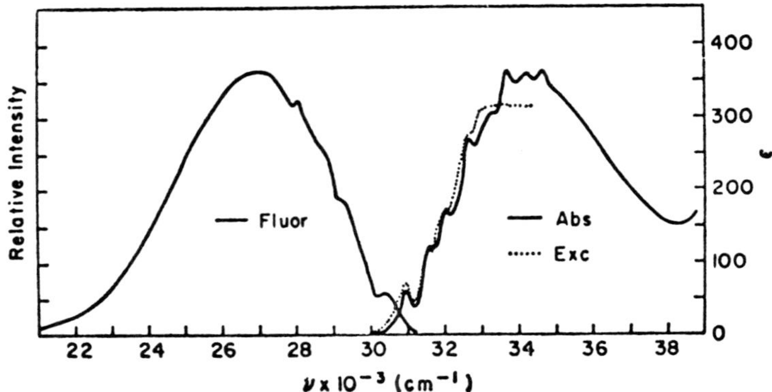


**Fig. 4.** *Pyrimidine* (n$\pi^*$) *fluorescence, absorption, and excitation spectra in isooctane at 298°K* (*Cohen* et al., *1965*).

states of this molecule together with pyrazine and pyridazine are worth considering.

The fluorescence of Pyr, discovered by Börresen (1963), was shown to be an ($n\pi^*$) emission by comparison of the corrected emission spectrum with the ($n\pi^*$) absorption band, agreement of the excitation and absorption spectra (Fig. 4), and positive polarization with respect to absorption into the ($n\pi^*$) band (Cohen *et al.*, 1965). However, Pyr with two azine nitrogens should have two ($n\pi^*$) transitions, and the interaction between these transitions should give symmetry-allowed and symmetry-forbidden states. Evidence for the location of these states has been obtained by El-Sayed and Robinson (1961) who observed weak absorption bands in matrix isolation experiments at 4°K. Consequently, the energy level diagrams (Fig. 5) for the diazines are

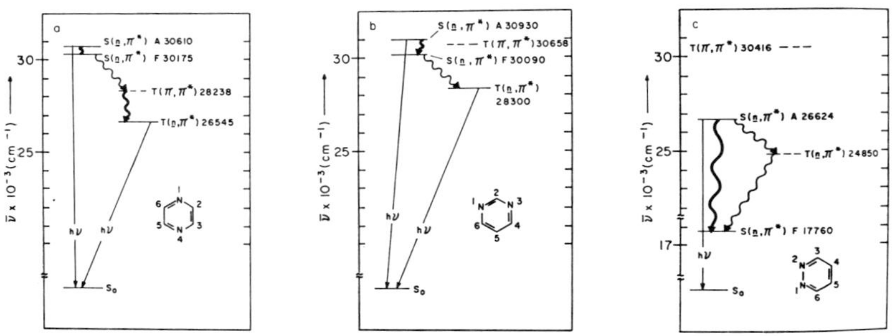


**Fig. 5.** (*a*) *Energies of the low-lying states of pyrazine.* (b) *Energies of the low-lying states of pyrimidine.* (c) *Energies of the low-lying states of pyridazine. State energies which have been experimentally determined are indicated by solid lines. Dashed lines denote calculated values. Fast radiative* (→) *and radiationless* (⇝) *processes are denoted by dense lines* (*Cohen and Goodman, 1967*).

discussed in terms of two singlet $(n\pi^*)$ states with splittings which decrease in the sequence

$$\begin{array}{ccccc} \text{pyridazine} & > & \text{pyrimidine} & > & \text{pyrazine} \\ 8864 \text{ cm}^{-1} & & 840 \text{ cm}^{-1} & & 435 \text{ cm}^{-1} \end{array}$$

as the N atoms move further apart. Two triplet levels are considered, $^3(n\pi^*)$ (allowed) and $^3(\pi\pi^*)$, but the latter has not been detected. The assignment of the fluorescing state to the allowed $^1(n\pi^*)$ level is made on the basis of the observed energy gaps between the 0–0′ band in absorption (presumed to be the allowed state) and the first vibronic band of the emission, compared with the gaps to be expected if emission were from the forbidden $(n\pi^*)$ state. The observed gaps are essentially constant at $\sim 500 \text{ cm}^{-1}$ and bear no relation to the gaps between the allowed and forbidden levels. This assignment (Fig. 5), which is essentially that fluorescence occurs as an $S_2 \rightarrow S_0$ process, relies heavily on the results of El-Sayed and Robinson, which would seem to warrant confirmation.* [Another unusual feature in the case of pyridazine is that $E(T_1)(n\pi^*) > E(S_1)(n\pi^*)$ by $\sim 7000 \text{ cm}^{-1}$.]

From measurements of $\phi_f$, $\phi_p$ (at 77°K), and $\phi_{isc}$ (at 298°K) and neglecting photochemistry, the rate constants for the various radiative and radiationless processes have been estimated for the following sequence (for Pyr)

$$S_0 \xrightarrow{h\nu} \underset{\text{(allowed)}}{S_2(n\pi^*)}$$

$$S_2 \xrightarrow{k_f} S_0 + \text{fluorescence}$$

$$S_2 \xrightarrow{k_{ic}} \underset{\text{(forbidden)}}{S_1(n\pi^*)}$$

$$\underset{\text{(forbidden)}}{S_1(n\pi^*)} \xrightarrow{k_{isc}} \underset{\text{(allowed)}}{T_1(n\pi^*)}$$

$$\underset{\text{(allowed)}}{T_1(n\pi^*)} \xrightarrow{k_p} S_0 + \text{phosphorescence}$$

$$\underset{\text{(allowed)}}{T_1(n\pi^*)} \xrightarrow{k_{nr}} S_0 + \text{heat}$$

* Absorption by the lowest $^1(n\pi^*)$ forbidden state of Pyr has been observed in 3-MP at 77°K (Li and Lim, 1971), but no fluorescence is detected from this state.

**Table 1** Pyrimidine

| | | Excited state data at | | |
|---|---|---|---|---|
| | | 77°K | 298°K | |
| | $\phi_f$ | $5.8 \times 10^{-3}$ [a] | $2.9 \times 10^{-3}$ [b] | |
| | $\phi_p$ | 0.14[a] | — | |
| | $\phi_{isc}$ | — | 0.12[c] | |
| | | Excited state parameters[d] | | |
| $k_f$ | $k_{ic}$ | $k_{isc}$ | $k_p$ | $k(T_1 \rightarrow S_0)$ |
| $5 \times 10^6$ | $9.1 \times 10^8$ | $1.2 \times 10^8$ | 50 | 0 |

[a] In *n*-pentane/methylcyclohexane (1:2).
[b] In isooctane.
[c] In *n*-hexane.
[d] All in units of $sec^{-1}$.

The relevant data are shown in Table 1 for the purpose of comparison with the results for the nucleic acid bases discussed in Section C. It is surprising that singlet and triplet state yields seem hardly to vary with temperature, and it may be noted that $\phi_{ic} + \phi_{chem} \cong 0.88$. There is also uncertainty about whether the pathway for internal conversion of $S_2$ proceeds directly to $S_0$ or via $S_1$. The importance of securely establishing the process among the lowest levels is shown by the excitation spectrum of Börresen (1963) from which it is apparent that efficient internal conversion occurs from the $^1(\pi\pi^*)$ level into the $^1(n\pi^*)$ states.

The phosphorescence of Pyr and 2-ClPyr has been studied by Nishi *et al.* (1970) from 77°K to 1.5°K, and their results modify somewhat the picture given by Cohen and Goodman. A detailed vibrational analysis suggests that the emission arises from two components of the emitting triplet. Consistent with this Nishi *et al.* found that the decay from the 0–0′ level can be resolved into two exponential components, while decay from the phosphorescence maximum has a third component which becomes very weak at very low temperatures. From a comparison of the observed and calculated lifetimes and the relative intensities, they concluded that the lowest triplet is an $(n\pi^*)$ state, $^3B_1$, the 0–0′ emission coming from the $B_2$ and $A_1$ components. In contrast to Fig. 5b, the next triplet level, a $(\pi\pi^*)$ state, $^3A_1$, lies below the singlet $(n\pi^*)$ states $^1B_1$ and $^1A_2$, and the major intersystem crossing pathways are from both of the singlet $(n\pi^*)$ states to the triplet $(\pi\pi^*)$ state. However, the lowest $^1(n\pi^*)$ state (forbidden) still remained undetected in these experiments.

At the other end of the temperature spectrum, the work on the gas phase photochemistry of Pyr and pyrazine by Magat *et al.* (1969) may be noted.

To bring these considerations for the diazines into perspective, it is instructive to consider the structures of the major bases of the nucleic acids. The commonly accepted tautomeric forms given below correspond to the predominant H-bonded forms of DNA (see Tinoco and Holcombe, 1964).

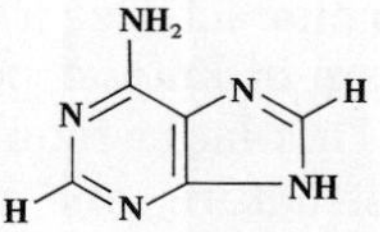

Adenine 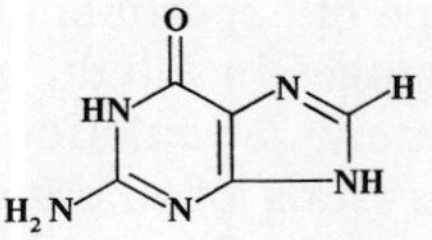 Guanine

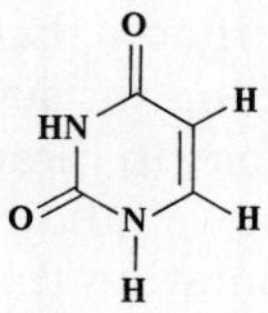

Uracil 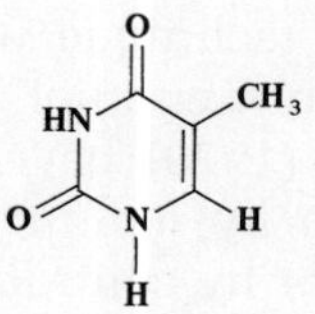 Thymine 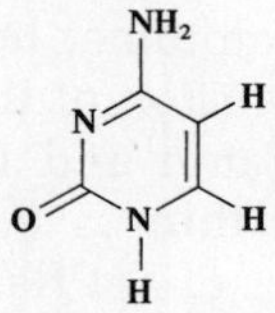 Cytosine

In a simple way Ade affords the possibility of three azine $(n\pi^*)$ transitions, while Gua shows two azine and one carbonyl transition. Ura and Thy show only the carbonyl type $(n\pi^*)$, but Cyt has an azine as well as a carbonyl type. Tautomeric forms increase the number of possibilities, for example, the conversion of lactam (—N—C=O) to lactim (—N=C—OH) would cause Gua to lose its carbonyl-type $(n\pi^*)$ state and would lead to the appearance of azine $(n\pi^*)$ transitions in Ura and Thy. It is apparent that the identification and location of $(\pi\pi^*)$ and $(n\pi^*)$ states in these bases, particularly in hydroxylic media, is of major importance in understanding their behavior under the action of radiation. If, in the following sections, the behavior of the nucleic acid bases seems much simpler than that of Pyr and pyrazine, it would be wise to conclude that the complexities are hidden rather than nonexistent.

## B. Electronic Transitions from Ground State

### 1. Polarized Reflectance and Absorption Spectroscopy

Each individual electronic transition is expected to have its own characteristic energy location (although degeneracies are important), oscillator strength, and orientation within the molecule. Accordingly, the most direct type of experiment to characterize a transition consists of determining changes in a light beam of known polarization caused by molecules of known orientation. This latter requirement is best attained by using crystals of known structure, and while the obvious property to measure is direct absorption, this has some problems and limitations. First, the extinction coefficients of the bases are so high ($10^3$–$10^4$) that extremely thin crystals ($\sim 0.1\ \mu m$) must be used. In their pioneering work Stewart and Davidson (1963) were able to use an ultramicrotome, but the technique was so difficult that confirmation or extension of their work was not forthcoming for several years. Recently Eaton and Lewis (1970) have been able to prepare single crystals as thin as $0.05\ \mu m$ by melting and recrystallizing between quartz discs of $\lambda/4$ flatness or by controlled evaporation from aqueous solution. The lower wavelength limit to this direct technique is $\sim 230$–$240$ nm and hence it usually gives information only for the first absorption bands of the bases.

A potentially less limited technique, polarized reflectance spectroscopy measures the absorption characteristics indirectly. It has long been recognized that the three processes of reflection, dispersion, and absorption of light are intimately related (Ditchburn, 1963). The reflection coefficient $R$ (measured at normal incidence) is related to the absorption process through the complex amplitude $[R(\omega)]^{1/2}e^{i\theta(\omega)}$ ($\theta$ is the phase) as $(n^* - 1)/(n^* + 1) = [R(\omega)]^{1/2}e^{i\theta(\omega)}$, in which $n^*$ is the complex index of refraction. However, the amplitude $[R(\omega)]^{1/2}$ and the phase $\theta(\omega)$ are not independent but are related by a Kramers-Kronig dispersion relationship (Kronig, 1926; Kramers, 1927) which may be written

$$\theta(\omega_0) = -\frac{\omega_0}{\pi}\int_0^\infty \frac{\ln[R(\omega)/R(\omega_0)]}{\omega^2 - \omega_0^{\,2}}\, d\omega$$

Consequently, from a measured reflectance spectrum, $\theta(\omega_0)$ can be calculated for each frequency of interest. Then, since $n^* = n + ik$, the absorption index $k$ can be obtained from $\theta$ and $R$. The drawback to this is that the range of the integration is from 0 to $\infty$ (that is, the absorption at a particular frequency depends on the reflectivity at all frequencies), while measurements are carried out in a finite range. The extension from the lower frequency limit (usually $\sim$14,000 $cm^{-1}$) to zero presents little trouble but the treatment of the high-frequency range has to take account of poorly characterized contributions. This is done empirically using various trial functions subject to the constraint that the computed absorbance in the region of known transparency be zero. Chen and Clark (1969) indicated that for Pur, where the upper limit of measurement is $54 \times 10^3$ $cm^{-1}$, the oscillator strengths of the lower energy transitions are only slightly dependent ($\pm$5%) on the trial functions, while the higher energy bands at $> 50 \times 10^3$ $cm^{-1}$ show about $\pm$15% variation. The situation is, however, inherently capable of improvement as the range of experimental measurement can be extended farther into the vacuum UV. Reflectance data for Pur are shown in Fig. 6A and the corresponding absorption spectra in Fig. 6B. Subsequent treatment of the data involves resolution of the curves of Fig. 6B into their overlapping components. Such procedures are quite empirical and the separations shown in Fig. 6C are only claimed to be reasonable. It is clear, however, that these separations have a considerable degree of internal consistency, and no unusual separations are proposed.

Last, there remains the problem, common to all pure crystal work, of evaluating the effects of crystal structure, i.e., the intermolecular interactions which are often manifested as both intensity differences and frequency shifts between the isolated molecule spectrum (vapor phase or diluted solution) and the observed crystal spectrum. The general treatment of these effects is involved and in practice the approach used is to proceed by successive approximations. For a planar isolated molecule all transition moments are considered to be either in the plane or are perpendicular. When there is more than one molecule per unit cell there are generally two in-plane directions consistent with a given dichroic ratio* measured on one crystal face. An additional complication is that there are in-plane directions which give the dichroic ratio expected for an out-of-plane transition. Consequently, data from two crystal faces are preferred but this situation is often difficult to achieve experimentally. Given experimental data

* The ratio of the extinction coefficients along appropriate axes for a given frequency, e.g., $(E_x/Ey)_\nu$.

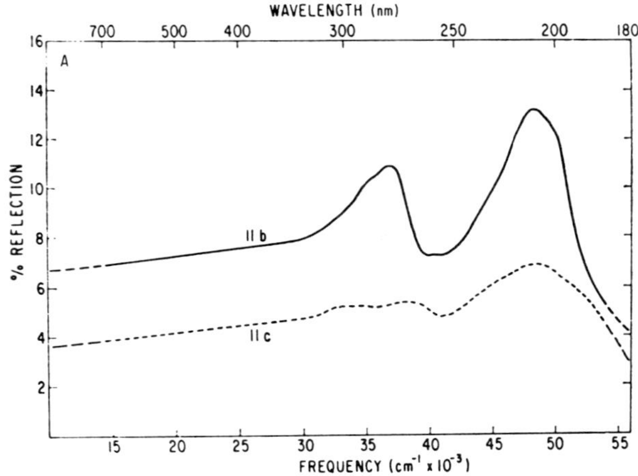


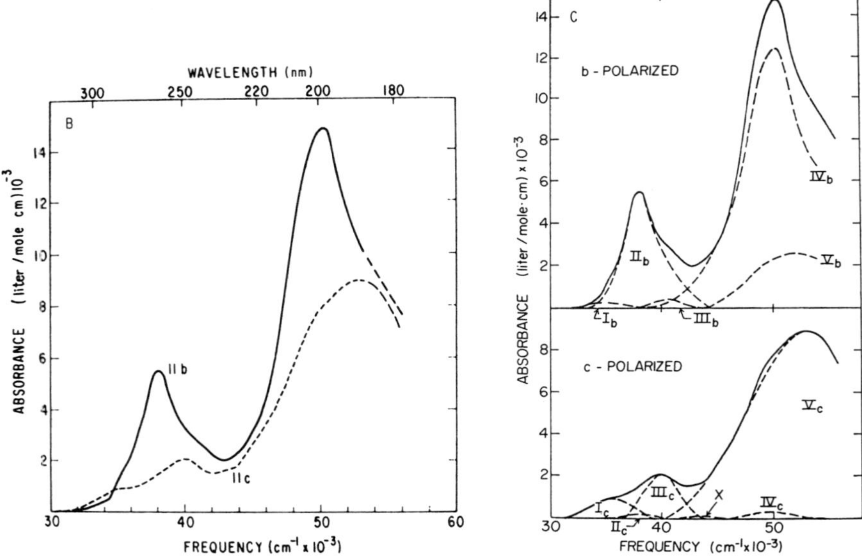


**Fig. 6.** (*A*) *Reflection spectra of the* (100) *face of purine. The solid line is for incident radiation polarized along the* b *axis, and the dashed line is for radiation polarized along the* c *axis. The broken line extensions on either side are part of the extrapolation used in the Kramers-Kronig transformations* (*Chen and Clark, 1969*). (*B*) *Computed absorption spectra of the* (100) *face of purine.* (*C*) *Resolution into components I-V of the b- and c-polarized absorption spectra.*

for various axes, the common procedure is to calculate the total oscillator strengths for the various transitions. First, the molecules are assumed to be noninteracting (the so-called oriented gas model) and compared with the isolated molecule spectrum. Then changes due to intermolecular interactions are estimated on the basis of various approximations and the resultant assignments again are compared with the experiment. For 1-MeUra Eaton and Lewis (1970) found a small excitation splitting, $\pm 100\ cm^{-1}$, for the 0–0′ band of the lowest ($\pi \rightarrow \pi^*$) transition, which is close to that expected from calculations using the point–dipole approximation to the intermolecular potential in the weak coupling situation. The isotropic oscillator strength, calculated directly from the oscillator strengths for the crystal axes, is, within experimental error, the same as the solution value, and DeVoe (1971) showed that the crystal shift of $\sim 2$ kK is probably due to the background polarizability. In the case of Pur, Chen and Clark (1969) considered a one-dimensional dipole–dipole interaction in the stacking direction of the molecular planes (which are separated by only 3.66°) using an iterative procedure to calculate changes in the oscillator strengths until the calculated oscillator strengths reproduced the observed crystal spectrum when mixed. In this way substantial changes were found in the component intensities. New dichroic ratios were thus obtained and used in turn to deduce free molecule polarization directions.

The problems of crystal interactions between the same molecular species can be avoided by dispersing the bases in a solid matrix, though it should not be forgotten that matrix–solute interactions can be significant. Two types of studies have been carried out on such dilute solid solutions. Transparent films of polyvinyl alcohol have been prepared containing some common bases and their nucleosides (Fucaloro and Forster, 1971). By stretching such films it is considered (though apparently not proven) that the solute molecules tend to align themselves with an orienting axis along the stretch direction. In this condition the absorption becomes anisotropic and dichroic spectra can be obtained. The dichroic ratio is clearly a function of the degree of orientation and expressions have been derived whereby the angle between the orienting axis and the transition moment can be related to the dichroic ratio and the stretch ratio. Since the orienting axis itself cannot be determined relative to the molecular coordinates, the method is restricted to evaluating the angle between two oscillators. The spectra obtained are very similar to those in aqueous solution but the observed dichroic ratios are quite small, reaching only 2.0 in cases of Gua and 9-MeAde.

**Table 2** Electronic Transitions of the Purine Bases

| Transition no[a] | $f_{osc}$ | $\bar{\nu}^{b}_{max}$ | Polarization[c] | Orientation | Assignment | Notes and references[d] |
|---|---|---|---|---|---|---|
| Purine ($N_7$—H) | | | | | | Chen and Clark (1969) |
| | | | | | | Polarized reflectance on *bc* face, 700–185 nm |
| I | $3.5 \times 10^{-3}$ | 34 (294) | o.p. | | $n \rightarrow \pi^*$ | Observed DR (b/c) = 1:5 DR recalculated for one-dimensional interaction = 1.75; theoretical = 1:8.3 |
| II | 0.13 | 38 (263) | i.p. | 48° | $\pi \rightarrow \pi^*$ | Observed DR = 22.4:1 |
| III | 0.03 | 40 (250) | o.p. | | $n \rightarrow \pi^*$ | |
| IV | 0.35 | 50 (200) | i.p. | 51° | $\pi \rightarrow \pi^*$ | Observed DR = 44:1 |
| V | 0.19 | 53 (189) | o.p. | | ? | Recalculated DR = 1:7 |
| 9-Methyladenine | | | | | | Stewart and Davidson (1963); Stewart and Jensen (1964) |
| | | | | | | Polarized absorption |
| I | 0.28 | 36.4 (275) | i.p. | $-3 \pm 3°$ | $\pi \rightarrow \pi^*$ | DR = 5.1 (±0.5):1 |
| | | | | | | Crystal absorption max. are given |
| II | $8 \times 10^{-3}$ | 39.2 (255) | i.p. | Approx. ⊥ to *I* | | |
| III | 0.62 | 48.5 (206) | i.p. | Approx. ⊥ to *I* | | |

| | | | | | | |
|---|---|---|---|---|---|---|
| Adenine hydrochloride ($N_1$—$H^+$) | | | | | | Chen and Clark (1973) |
| | | | | | | Polarized reflectance |
| I | 0.08 | 36.6 (274) | i.p. | −28° | $\pi \rightarrow \pi^*$ | First absorption band |
| II | 0.20 | 38.9 (257) | i.p. | 100° | $\pi \rightarrow \pi^*$ | |
| III | 0.16 | 48.5 (206) | i.p. | 15° | $\pi \rightarrow \pi^*$ | |
| IV | 0.51 | 51.0 (196) | i.p. | 120° | $\pi \rightarrow \pi^*$ | |
| Guanine hydrochloride | | | | | | L. B. Clark (unpublished results, 1972) |
| I | 0.15 | 34 (294) | i.p. | 2° | $\pi \rightarrow \pi^*$ | |
| II | 0.20 | 41.5 (241) | i.p. | −82° | $\pi \rightarrow \pi^*$ | |
| III | 0.48 | 51.0 (196) | i.p. | −75° | $\pi \rightarrow \pi^*$ | |
| 9-Ethylguanine | | | | | | |
| I | 0.16 | 36.0 (278) | i.p. | −4° | $\pi \rightarrow \pi^*$ | L. B. Clark (unpublished results, 1972) |
| II | 0.25 | 39.3 (253) | i.p. | −75° | $\pi \rightarrow \pi^*$ | |
| III | 0.41 | 49.0 (204) | i.p. | −74° | $\pi \rightarrow \pi^*$ | |
| IV | 0.50 | 53.0 (189) | i.p. | −6° | | |
| I | — | — | i.p. | −14° or 44° | | Callis *et al.* (1971) |
| | | | | | | Polarized reflectance |
| II | — | — | i.p. | −65° or 95° | | Fulcaloro and Forster (1971) |
| III | — | — | i.p. | ‖ll to *II* | | Polarized absorption in PVA |

[a] Transitions are numbered in order of increasing energy of maxima.
[b] Transition locations are given in kK ($cm^{-1} \times 10^{-3}$); for convenience, equivalent values in nm are given in parenthesis.
[c] o.p. = out-of-plane; i.p. = in-plane.
[d] DR = dichroic ratio.

The other procedure for obtaining dispersed systems has been to cool solutions in an appropriate solvent until a glass is obtained. Common glasses are ethylene glycol/water and isopropanol/isopentane. Although the molecules do not have any preferential orientation in the glass, they can be photoselected by polarized exciting radiation (Albrecht, 1970), and the polarization $P$ of the subsequent fluorescence can be determined as a function of exciting wavelength. The background to the interpretation of these measurements is as follows. When the oscillator for absorption is parallel to the oscillator for emission, then excitation by polarized radiation will lead to emission polarized in the same sense. This is the situation for the lowest excited state, and it can be shown that the limit value for $P$ (the degree of polarization, defined earlier) is then +0.5. In practice, various depolarizing effects which are difficult to eliminate result in limiting values ~0.4–0.5. If, however, the absorption oscillator is at an angle to the emission oscillator, as may be the case for higher excited states, then the emission will be depolarized relative to the exciting radiation. Limiting values for this case approach −0.33 for mutually perpendicular oscillators. Results from this technique have been obtained for Gua, 9-MeGua, Ade (Callis *et al.*, 1964), and 5-MeGua (Callis and Simpson, 1970). For Gua and 9-MeGua at 195°K, the polarization of the emission was about +0.4 and was appreciably constant across the first absorption band but decreased to about −0.2 in and across the second band. This is fairly clear evidence that the two absorption bands correspond to separate transitions which are orientated approximately perpendicularly to one another. Results in the case of Ade are not so clear. The polarization is ~0.4 across most of the band but on the high energy side it decreases to ~0.2. This indicates the presence of a second transition within the band envelope but not much more can be deduced. If the second transition is perpendicular to the first, it must be weak; but this alone cannot be distinguished from the possibility that the polarization corresponds to a strong transition somewhat less than perpendicular to the first. Other experimental information needed to resolve this point is perhaps provided by the work of Stewart and Davidson (1963) (see Table 2). This very same point is exemplified by the results for 5-MeCyt. In this case $P$ has the value +0.4 in the first band and decreases to +0.2 in the second. Since the bands are not overlapping, there is no ambiguity in this case and it is concluded that the second transition is at an angle of 25° to the first (Table 2). The weakness of the fluorescence from Thy, Ura, and Cyt has so far prevented this technique from being applied in these cases.

The results obtained by these techniques are summarized in Table 2 for the purines and in Table 3 for the pyrimidines. By a commonly ac-

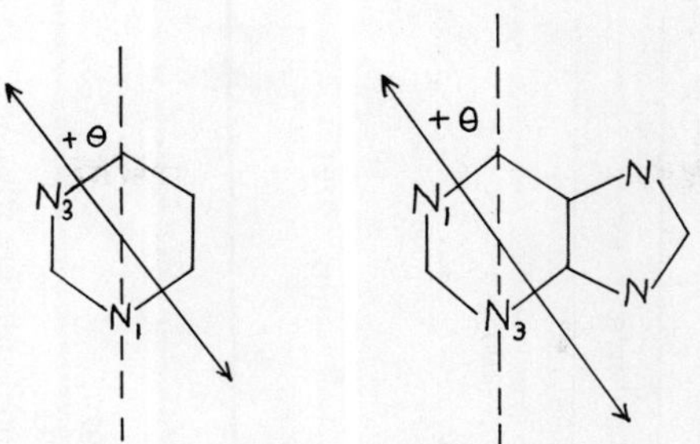


**Fig. 7.** *Sign conventions for transition moment directions of purines and pyrimidines.*

cepted convention (Fig. 7), orientations of transition moments are counted positive with respect to counterclockwise rotation from the C(4)–C(5) axis (short axis) for the purines and from the N(1)–C(4) axis for the pyrimidines (DeVoe and Tinoco, 1962). Some emphasis is given to Pur itself as a parent compound. The existence of three transitions perpendicular to the molecular plane should be noted, these alternating with strong in-plane transitions. Two of the perpendicular bands can easily be identified as $n \rightarrow \pi^*$ and the in-plane transitions are $\pi \rightarrow \pi^*$. The lowest energy transition is thus a weak $n \rightarrow \pi^*$ and the first major absorption band contains only one $\pi \rightarrow \pi^*$ transition. These relations are shown in Figs. 6 and 8. For 9-MeAde the interesting result is the evidence for two in-plane transitions in the first absorption band envelope, polarized perpendicularly to one another. The longer wavelength transitions are much more intense and are essentially short-axis polarized. As has been mentioned, the polarization excitation spectra of Callis *et al.* (1964) are consistent with this

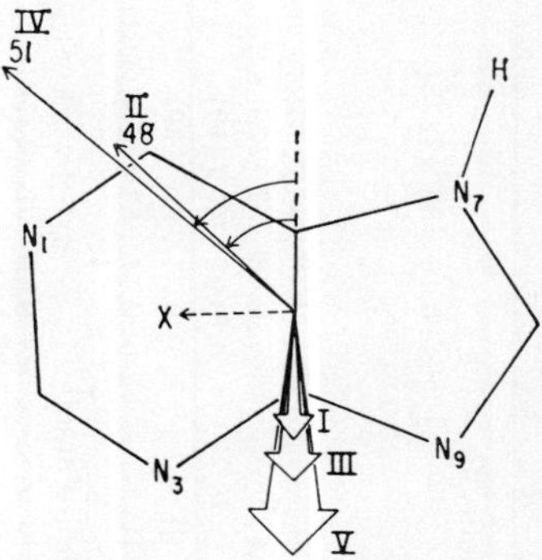


**Fig. 8.** *Final transition moment direction assignments of the five bands of purine (Chen and Clark, 1969).*

**Table 3** Electronic Transitions of the Pyrimidine Bases

| Transition No.[a] | $f_{osc}$ | $\bar{\nu}^{b}_{max}$ | Polarization[c] | Orientation | Assignment | Notes and references[d] |
|---|---|---|---|---|---|---|
| 1-Methylthymine | | | | | | |
| I | 0.19 | 36.3 (275) | i.p. | −19° | $\pi \rightarrow \pi^*$ | Stewart and Davidson (1963)<br>Polarized absorption<br>DR = 11 · 2:1 |
| II | 0.28 | 48.3 (207) | i.p. | Approx. ⊥ to *I* | | |
| I | 0.18 | 37 (270) | i.p. | −20° | $\pi \rightarrow \pi^*$ | L. B. Clark (unpublished results, 1972)<br>Polarized reflectance |
| II | 0.29 | 47 (213) | i.p. | −53° | $\pi \rightarrow \pi^*$ | |
| III | 0.31 | 56 (179) | i.p. | −36° | $\pi \rightarrow \pi^*$ | |
| IV | 0.17 | 61 (164) | i.p. | −42° | $\pi \rightarrow \pi^*$ | |
| 1-Methyluracil | | | | | | |
| I | 0.19 | 36.3 (276) | i.p. | 0° or 7° | | Eaton and Lewis (1970)<br>Polarized absorption<br>DR = 250:1; vibrational structure with $\Delta\nu$ ~ 750 cm |
| II | — | — | i.p. | Approx. ⊥ to *I* | | |
| I | 0.19 | 37 (270) | i.p. | −9° | $\pi \rightarrow \pi^*$ | L. B. Clark (unpublished results, 1972)<br>Polarized reflectance |
| II | 0.06 | 45 (222) | o.p. | | $n \rightarrow \pi^*$ | |
| III | 0.27 | 47 (213) | i.p. | −53° | $\pi \rightarrow \pi^*$ | |
| IV | 0.11 | 55 (184) | o.p. | | $n \rightarrow \pi^*$ | |
| V | 0.30 | 56 (179) | i.p. | −33° | $\pi \rightarrow \pi^*$ | |
| VI | 0.17 | 61 (164) | i.p. | −40° | $\pi \rightarrow \pi^*$ | |

| | | | | | | |
|---|---|---|---|---|---|---|
| Cytosine[d,e] | | | | | | |
| I | 0.13 | 37.0 (270) (0 0′ at 35.5) | i.p. | 14° | $\pi \rightarrow \pi^*$ | Lewis and Eaton (1971) Polarized absorption by cytosine hydrate Vibrational structure with $\Delta\bar{\nu} \sim 750$ cm$^{-1}$ DR = 48:1 |
| II | | 44 (227) | i.p. | −5 ± 3° | $\pi \rightarrow \pi^*$ | DR = 8.5:1 |
| III | | 51 (196) | | | | No evidence of $n \rightarrow \pi^*$ below 44 kK |
| I | | | i.p. | 12 ± 3° | | Callis and Simpson (1970) |
| II | | | i.p. | (−11° ↔ 9°) | | Polarized reflectance of cytosine hydrate and 1-methylcytosine; polarized fluorescence of 5-methylcytosine gives P(I) ≃ +0.4 P(II) ≃ +0.2 |

[a] Transitions are numbered in order of increasing energy.
[b] Transitions are located in kK (cm$^{-1}$ × 10$^{-3}$); for convenience equivalent values are given in nm in parentheses.
[c] i.p. = in-plane; o.p. = out-of-plane.
[d] DR = dichroic ratio; P = degree of polarization of emission.
[e] A weak peak or shoulder is observed in aqueous solutions of cytosine and 5-methylcytosine at ~ 42 kK and is most probably due to a minor tautomer; transition II is considered to be intrinsic to the major tautomer of cytosine (see text).

but conflict with the room temperature polarized absorption work on polyvinyl alcohol (PVA) films (Fucaloro and Forster, 1971). While the absorption spectra for Ade appear the same in ethylene glycol/water glass (7:3) at 195°K as in PVA at room temperature, the PVA spectra show a constant dichroism from 37,000 to 41,000 $cm^{-1}$, which is precisely the region in which fluorescence polarization decreases from +0.4 to +0.2 (Fig. 9a and b). Furthermore, some significance is attached to the ~3% increase in dichroism in the *C* region (Fig. 9b) although no change is seen in *P* in this region. The complexity of the Ade first absorption band is shown by the different spectra obtained in PVA for 9-MeAde and Ado. For the protonated Ade ($H^+$ probably at N(1)) L. B. Clark (unpublished results, 1972) found that the intensities of the first two transitions are reversed when compared with 9-MeAde, while the polarization and directions of the transition moments remain largely unchanged, that is, the lowest energy transition is still essentially short-axis polarized. While there is no doubt that there are at least two transitions intrinsic to the Ade first absorption band (which would correspond to the aromatic states $^1L_a$ and $^1L_b$), the precise description of the band remains in doubt because of the sensitivity of the parameters of one or both of these states to substitutional (9-Me or 9-Rib) or protonation effects (probably at N(1)). For Gua and its derivatives the situation is rather clearer primarily because of the separation of bands I, II, and III. Each absorption band is due primarily to one transition, and there is only moderate overlap. By all techniques, transition I is found to be in plane but perpendicular to transitions II and III. The polarized reflectance work of L. B. Clark (unpublished results, 1972) shows that on protonation at N(7) the first transition retains its intensity and short-axis polarization but is red-shifted ~ 2000 $cm^{-1}$ from 9-EtGua. The first transitions are thus similarly polarized in both Ade and Gua but behave differently on protonation.

The pyrimidines present a simpler situation than the purines. Thy, Ura, and Cyt each are considered to have only one transition (which is $\pi \rightarrow \pi^*$) in the first absorption band and all are polarized fairly close to the N(1)–C(4) axis (Fig. 10). This is in contrast to the spectral correlation work of Clark and Tinoco (1965) which led to the suggestion that the first absorption band of Thy and Ura might contain two merged transitions (derived from the benzene $^1L_a$ and $^1L_b$ states). It seems that the $^1L_a$ transition might be found in the second absorption band ~210 nm. Two general comments on the Thy and Ura results seem pertinent. First, methods based on polarization can only distinguish transitions of different molecular orientation and the possibility

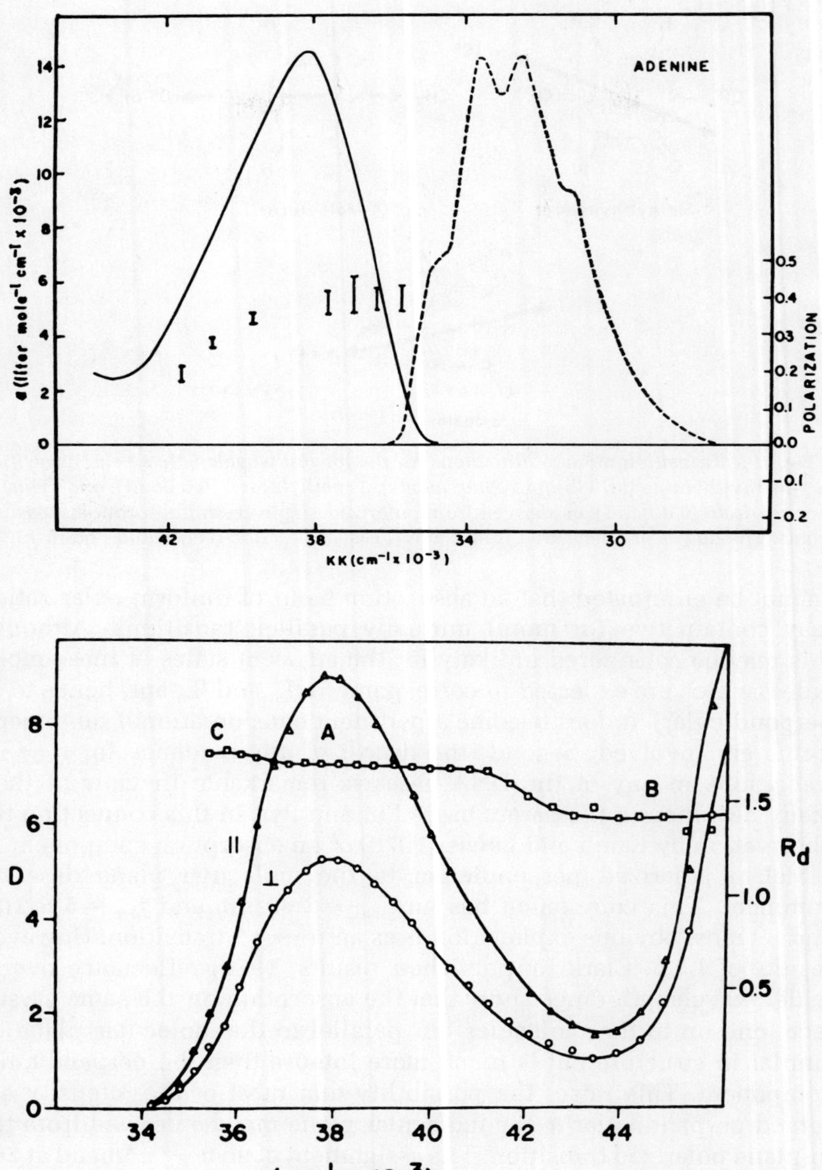


**Fig. 9.** *(a) Absorption (solid line), fluorescence (dashed line), and polarization P of adenine (0.1 mM) in EG/water (7:3) glass at 195°K (Callis* et al.*, 1964). (b) Polarized absorption spectrum of adenine in stretched* PVA: *$R_d$ = dichroic ratio (Fucaloro and Forster, 1971).*

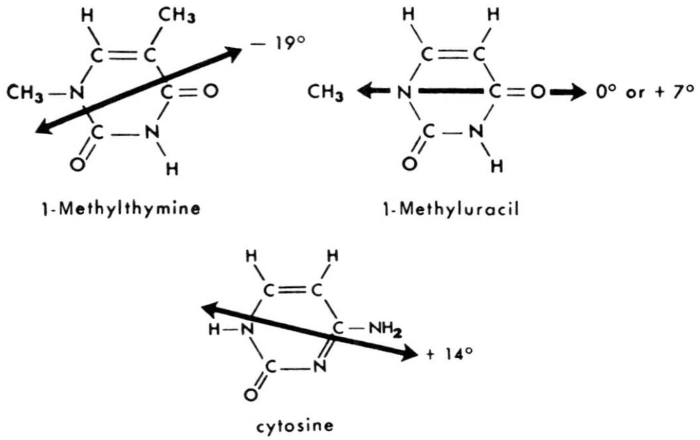

**Fig. 10.** *Transition moment directions for the longest wavelength $\pi \rightarrow \pi^*$ transition in 1-methylthymine (all vibronic components), 1-methyluracil (0,0 band), and cytosine monohydrate (0,0 band), as derived from polarized single-crystal absorption measurements. The angles are measured from the N(1)–C(4) direction (Lewis and Eaton, 1971).*

cannot be eliminated that an absorption band of uniform polarization may contain two (or more) mutually parallel transitions. Although this may be considered unlikely for the adjacent states of these molecules (which are *expected* to correspond to $^1L_a$ and $^1L_b$ and hence to be perpendicular), it does become a pertinent consideration if tautomeric forms are involved. Second, the dearth of any evidence for $n \rightarrow \pi^*$ transitions in any of the DNA bases is remarkable in view of their ready detection in the parent bases Pur and Pyr. In this connection the observation by Eaton and Lewis (1970) of an absorption component in 1-MeUra polarized perpendicular to the molecular plane deserves comment. This component has an $\epsilon_{max} \cong 260$ nm and $f_{osc} = 5 \times 10^{-3}$ and its most obvious explanation is as an $n \rightarrow \pi^*$ transition. However, results of L. B. Clark (unpublished results, 1972) reflectance over a wider wavelength range show that the absorption on the same crystal face (end-on to the molecules but parallel to the molecular plane) is similar in structure but is much more intense than the perpendicular component. This raises the possibility that most of the intensity observed perpendicular to the molecular plane may be derived from the in-plane polarized transition. The assignment of an $n \rightarrow \pi^*$ band at 264 nm is hence dubious and, in fact, Clark only finds evidence for a distinct $n\ \pi^*$ band beginning at 42,000 cm$^{-1}$.

The case of Cyt is interesting in that transition II is oriented very close to transition I and if it were an overlapping band, resolution by polarization would be somewhat ambiguous. However, the only

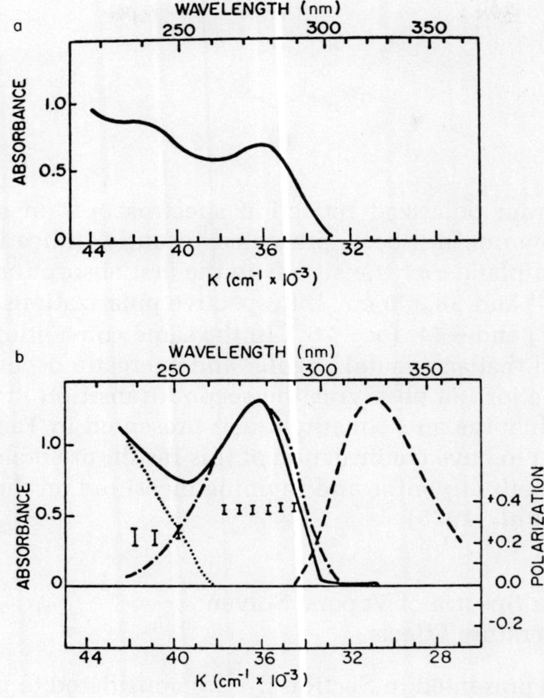


**Fig. 11.** *(a) Room-temperature absorption spectrum of 5-methylcytosine. (b) Absorption (solid line), fluorescence (dash), and fluorescence polarization (vertical bars) of 5-methylcytosine in isopropyl alcohol/isopentane at liquid air temperature. The dot-dashed line represents the mirror image of the spectrum and the dotted line represents the difference between the absorption spectrum and the mirror image of the fluorescence (Callis and Simpson, 1970).*

problem in this connection has been that the precise location of transition II has caused some confusion because of the presence under certain conditions of a weak band or shoulder observable at $\sim$42,000 $cm^{-1}$ in the absorption spectrum of Cyt. This absorption band is solvent and temperature dependent (compare, for example, the spectrum of 5-MeCyt at room temperature in aqueous solution with the absorption in isopropanol/isopentane at liquid air temperature, Fig. 11). The investigations by Morita and Nagakura (1968) of solvent effects at room temperature, temperature effects in aqueous solution, and comparison with spectra of known structures led to the assignment of the absorption to the 2-keto-6-imino tautomer. This tautomer is not found in crystals and the transition II in Table 3 is intrinsic to the amino form of Cyt.

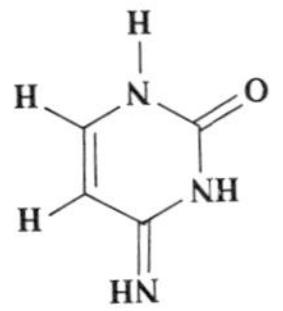

Evidence from polarized reflection spectroscopy on single-crystal anhydrous thymine has been presented recently indicating the presence of two in-plane $\pi\pi^*$ transitions in the first absorption band, lying at 36,500 $cm^{-1}$ and 38,300 $cm^{-1}$. Respective polarizations are given as $+14°$ (or $+46°$) and $-44°$ (or $-76°$), in the same convention as Table 3. It is suggested that accidental angular and energetic degeneracies may be responsible for not observing the second transition in earlier work on 1-methylthymine and 1-methyluracil presented in Table 3. Clearly it is important to have confirmation of this result, preferably by having spectra of 1-methylthymine and thymine measured on the same apparatus (Anex *et al.*, 1975).

## 2. Absorption Spectra of Vapors; Solvent and Temperature Effects

The results presented in Section B,1 are considered to pertain to the isolated molecule, although most of them are derived from measurements on molecular crystals. It is clearly desirable to attempt to obtain as much data as possible from isolated studies in order to avoid the approximations and uncertainties inherent in the treatment of the crystal data.

Furthermore, since the end result we need is an understanding of the bases in an aqueous environment, the effect of various solvents becomes interesting so that the different factors involved (polarity, hydrogen bonding, etc.) may be sorted out.

Isolated molecule studies are most obviously carried out in the gas phase in which the possibility exists that significant spectral resolution might be obtained (as was observed for the diazines) particularly in view of the suggestion of resolution in non-hydrogen-bonding solvents. In this respect the results have been disappointing. Clark *et al.* (1965) were able to measure the vapor spectra of Ade, Gua, Ura, Cyt, some methyl derivatives, and 9-MeHyp at temperatures ranging from 71°C for $Me_2Ura$ to 237°C for 9-MeHyp. Only in the cases of Gua and 9-MeHyp were some structures observed in the first absorption band. However, some other interesting effects were reported. For the adenines, Hyp, and the uracils, the aqueous spectrum is shifted about

10 nm to the red of the vapor spectrum; this is the not unexpected general dispersion effect (Basu, 1964) as is indicated by the fact that the shift is almost the same in trimethyl phosphate. However, for Cyt and Gua, $\lambda_{max}$ seems to be at 290 nm for the vapor and in aqueous solution shifts are to the blue by approximately 20 nm. This might indicate considerable ground-state–solvent interaction for these molecules and indeed it may be noted that Johnson *et al.* (1971), working with substituted cytidines allowing no possibility of tautomerism, conclude that the observed spectral shifts are indeed due to solvent-solute hydrogen-bonded complex, ruling out the possibility of solvent-shifted tautomerization. However, the situation may be complicated by temperature-dependent tautomerism. Thus, the $\lambda_{max}$ for Cyt and 3-MeCyt are the same in the vapor phase but are quite different in aqueous solution in which Cyt has $\lambda_{max} = 267$ nm and exists largely as the amino tautomer while 3-MeCyt has $\lambda_{max}$ at 296 nm (in its uncharged form). A similar alternative is not so apparent for Gua which may, thus, interact quite strongly (~ 5.5 kcal/mole) with solvent water, and it is interesting to note that Bunville and Schwalbe (1966) have observed a deuterium effect on the UV absorption spectrum of aqueous Gua. Another interaction, not well understood, is indicated by the increase in oscillator strength for 9-MeAde from 0.16 in the gas phase to 0.28 in aqueous solution. Significant solvent difference at room temperature is found for dGuo in dioxane, isopropanol, and glycerol (Drobnik and Augenstein, 1966b) but band resolution will be necessary before these shifts can be interpreted in terms of intensity changes and frequency shifts.

In attempts to obtain resolution, spectra have been measured at various low temperatures. Ade in ispropanol and ethylene glycol/water and Ade in isopropanol are reported at 150°K (Drobnik and Augenstein, 1966b), with various shoulders and shifts but with nothing like the well-defined structure seen in emission (Section C). The absorption spectra of the five common nucleotides in 1:1 ethylene glycol/water at 77°K were presented by Gueron *et al.* (1967a) with remarkably little change from room temperature in water, despite the possibilities for aggregation at the concentrations used (~ 3 m*M*). In these circumstances the only remaining step seems to be to attempt matrix isolation spectroscopy (cf. El-Sayed and Robinson, 1961). Although undoubtedly difficult, this would have a twofold reward; it would give badly needed vibronic spectra in an inert medium and it would allow a study of the medium effects with relatively simple solute molecules such as He, $H_2$, $N_2$, and $CH_4$. Another attraction is the possibility that evidence may be found for forbidden, low-lying ($n \rightarrow \pi^*$) transitions, as in diazines.

## C. Properties of Excited Singlet States

Following the ideas expressed in the Franck-Condon principle, absorption of a photon leads to vertical excitation, i.e., the initially formed excited state has the same internuclear configuration as the ground-state molecule. In general such a state will be vibrationally excited relative to its own potential minimum and will undergo vibrational relaxation at a rate usually considered to be $\geqslant 10^{12}$ $sec^{-1}$ to give the equilibrium excited state. Insofar as fluorescence lifetimes are commonly measured in the range 100–1 nsec, it is clear that excited-state properties determined through fluorescence measurements usually refer to this equilibrium state (Fig. 12). In solution this simple

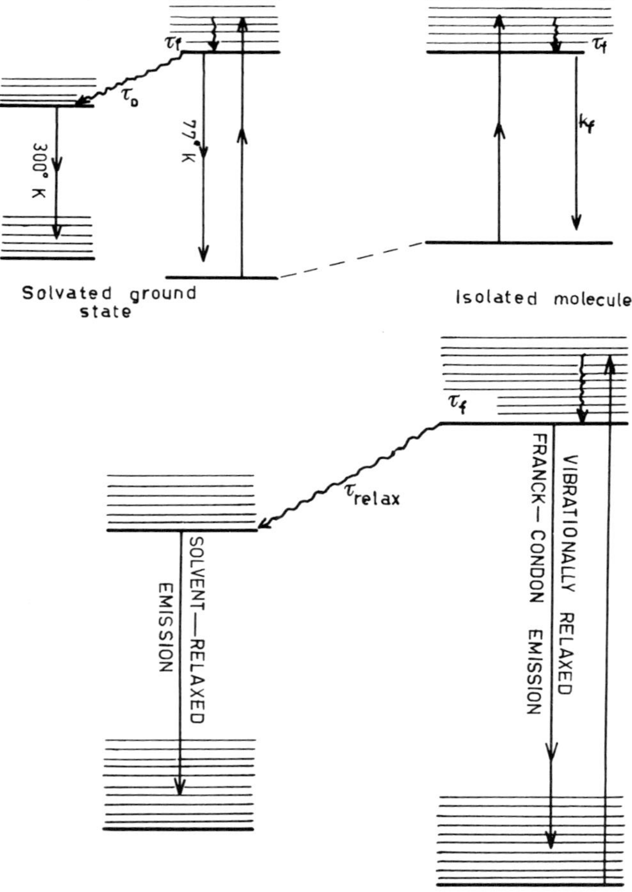


**Fig. 12.** *Solvent effects on the ground state and the time dependency of such effects in the excited state.* $\tau_D$ *is the relaxation time of the solvent around the excited molecule.*

conceptual framework must be modified. First, the absorption spectrum will reflect primarily solvent interaction with the ground state. Second, the equilibrium excited state usually has properties quite different from the ground state and its interaction with the solvent will be different. When this is manifested by dipolar orientation effects, the emission spectrum can show significant solvent shifts depending on the magnitude of the dielectric relaxation time ($\tau_D$) and the lifetime of the excited state ($\tau_f$). If $\tau_f >> \tau_D$, then complete solvent relaxation will occur before emission and the emission will refer to a completely solvated excited state. If $\tau_D >> \tau_f$, the emission will occur from a state in which the dipolar solvation interaction will be primarily that of the ground state. This latter situation will apply in general to solutions at 77°K. Figure 12 illustrates this situation for these limiting cases. The properties of the excited singlet states determined via fluorescence are conveniently considered in three categories, first the low-temperature (~ 77°K) results in a rigid glass, then the studies of the temperature dependence, and finally studies in fluid medium at room temperature.

### 1. Fluorescence Behavior at ~ 77°K

#### *a. Emission Spectra*

Despite numerous investigations, reports of emission spectra corrected for monochromator spectral transmission and photomultiplier spectral response are rare. Recognizing this deficiency, Guéron *et al.* (1967a) determined the corrected spectra for all five common nucleotides in 1:1 ethylene glycol/water glass at ~ 77°K. These data are presented in Figs. 13 and 14, and several features may be noted. The spectra of the pyrimidines are all broad, with half-widths. $\Delta\bar{\nu}_{1/2} \sim 4000\ \text{cm}^{-1}$ and $\bar{\nu}_{max} \sim 31{,}000\ \text{cm}^{-1}$, and show no sign of structure. The purines, AMP and GMP, are somewhat different; AMP shows signs of structure, and both have long low-energy tails, that of GMP extending to 22,000 $\text{cm}^{-1}$. All of the bases exhibit an extensive red shift of the fluorescence with respect to the absorption curve. As shown in Fig. 12, this Stokes shift may reflect the change in geometry between the ground state and the equilibrium excited state and the extent of vibrational coupling with the solvent ground state, depending on how it is defined. Resonance emission occurs at the 0–0′ frequency, i.e., $^1E(0\text{–}0')$, and the Stokes shift may then be defined (Ermolaev, 1964) as the difference between this and the mean frequency $M(\nu_f)$ of the fluorescence curve,

$$\Delta\bar{\nu}\ (\text{Stokes shift}) = \bar{\nu}_0 - M(\bar{\nu}_f)$$

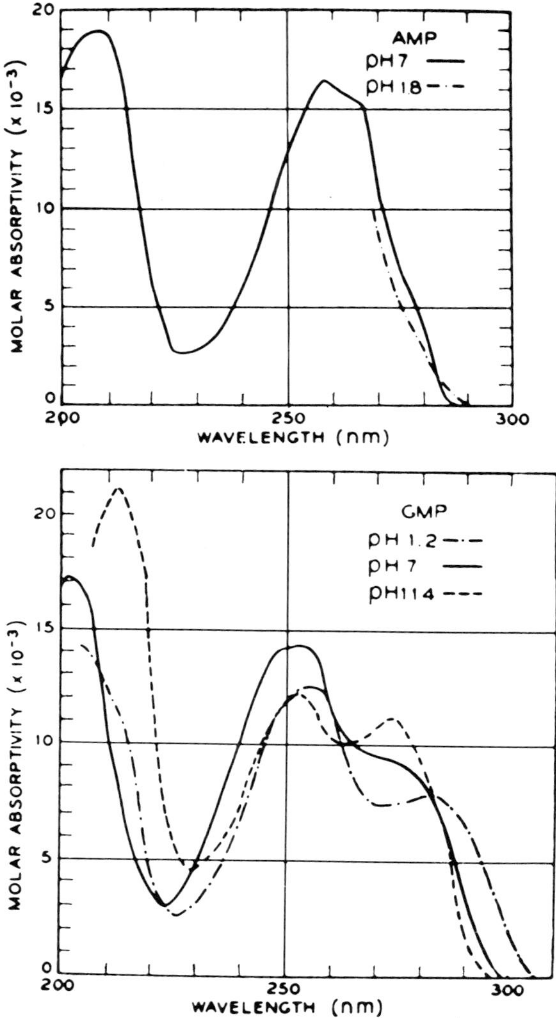


**Fig. 13.** *Absorption spectra of the major nucleotides at* ~77°K *in buffered EG/$H_2O$* (1 : 1) *glass* (Guéron et al., 1967*a*).

[in which $\bar{\nu}_0$ is estimated from the intersection of the absorption and emission curves, and $M(\bar{\nu}_f) = \int \bar{\nu} f(\bar{\nu})\, d\bar{\nu} / f(\bar{\nu})\, d\bar{\nu}$ in which $f(\bar{\nu})$ is the fluorescence flux (in quanta) per unit wave number; $M(\bar{\nu}_f)$ is thus the first moment of the fluorescence curve]. Defined in this way, the Stokes shift represents only the energy loss to solute–solvent vibrational levels by whatever mechanism. Because of experimental un-

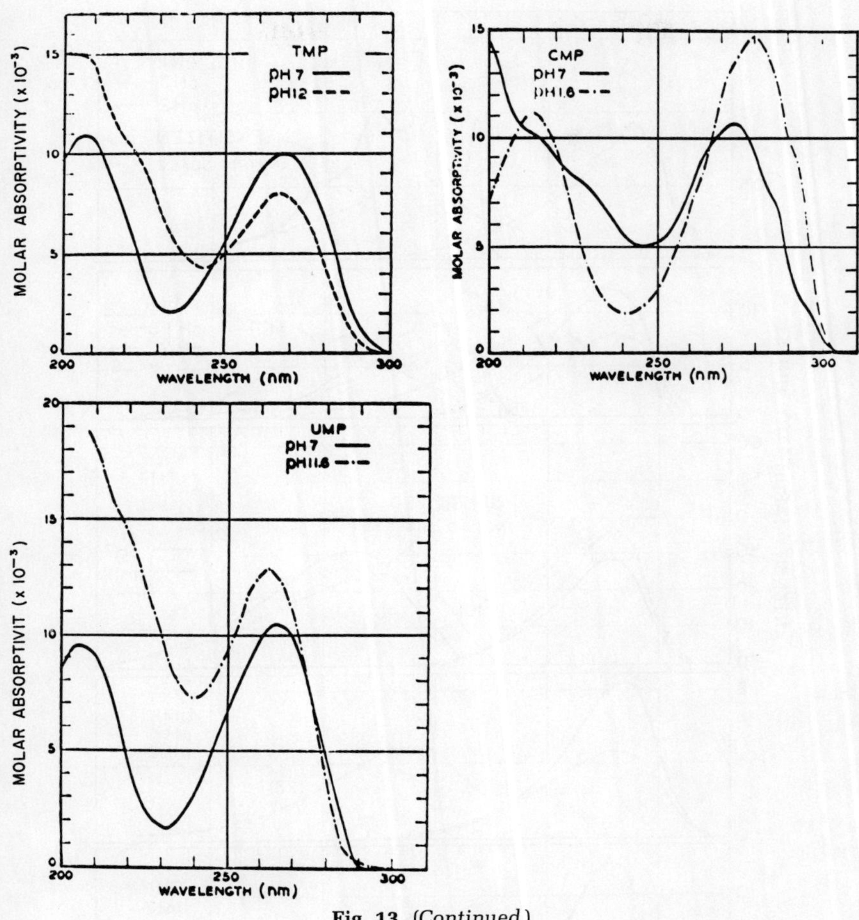


Fig. 13. (*Continued*)

certainties in locating $\bar{\nu}_0$, a Stokes loss can also be defined as the difference between the mean absorption frequency and the mean fluorescence frequency,

$$\Delta\bar{\nu} \text{ (Stokes shift)} = M(\bar{\nu}_a) - M(\bar{\nu}_f)$$

or absorption and emission maxima can be used. This sort of definition may then involve factors reflecting the change in geometry between the ground and excited state. Guéron *et al.* (1967a) chose to use the maxima and obtained Stokes shifts of 5 to 6 kK. It is usually considered that such large values are indicative of considerable sol-

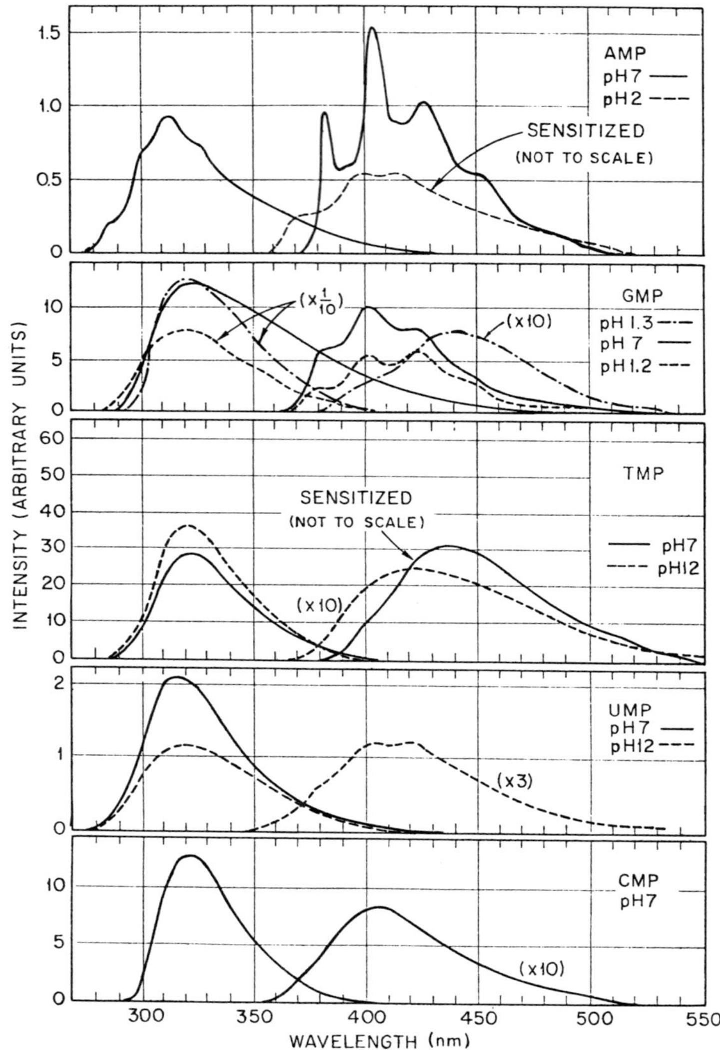


**Fig. 14.** *Corrected fluorescence and phosphorescence spectra of the major nucleotides under the same conditions as Fig. 13.* $\lambda_{ex}$ *was 265 nm (Guéron* et al., *1967a).*

vent interaction such as extensive hydrogen bonding. Because of this, solvent studies should be very interesting. Kleinwächter *et al.*, (1967a) reported such studies for Ade and its derivatives and Drobnik and Augenstein (1966b) for Gua but gave only one spectrum, Ade in isopropanol/isopentane (1:1) (Fig. 15) which shows little difference

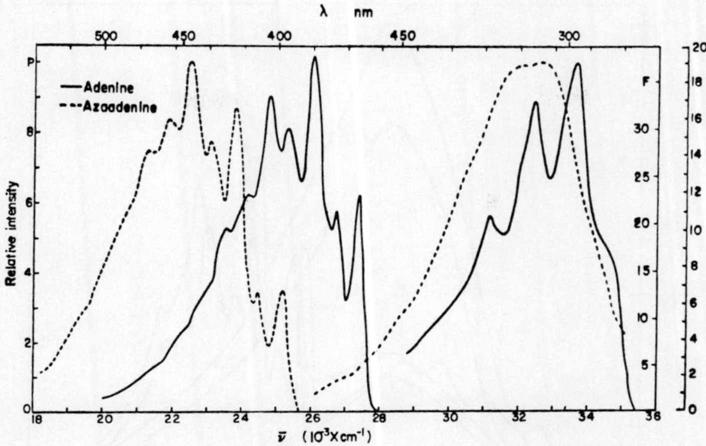


**Fig. 15.** *The emission spectra of adenine and azaadenine in isopropanol/isopentane (1:1). The phosphorescence has been normalized to 10 at the maximum; the intensity of fluorescence relative to this is given on the right (adenine, 0–35; azaadenine 0–20) (Kleinwächter, Drobnik, and Augenstein, 1967).*

from the ethylene glycol/water (EG/water) glass. Eastman and Rosa (1968) presented the fluorescence spectrum of Ade at 170°K in 7:3 EG/water and in isobutanol, showing very little change with $\bar{\nu}_{f(max)}$ at 33,400 and 32,300 $cm^{-1}$, but in hydrocarbon media at 77°K, 9-BuAde emits with $\bar{\nu}_{f(max)}$ at 28,600 $cm^{-1}$ (Cohen and Goodman, 1965), blue shifting by $\sim 1000^{-1}$ in a butanol/isopentane (3:7) solvent (Fig. 16).

The vibration structure seen most strongly in Ade seems to be related to the solvent, being observed only in the hydroxylic glasses (Longworth *et al.*, 1966). However, one feature which must be remarked upon, and which does not seem to have been given a satisfactory explanation, is the loss of vibrational structure in passing from Ade to Ado and AMP (all in the same hydroxylic solvent). Before further use can be made of absorption and emission spectra, it is of considerable importance to determine if they are symmetric. Guéron *et al.* (1967a) showed that the absorption and emission spectra should be presented in the forms $\epsilon(\bar{\nu})/\bar{\nu}$ and $f(\bar{\nu})/(\bar{\nu})^3$, respectively, in which $\epsilon(\bar{\nu})$ is the extinction coefficient at $\bar{\nu}$, and $f(\bar{\nu})$ as above is the relative photon emission per unit wave number interval (Fig. 17a). They found the rather interesting result that whereas for CMP (and, it is stated, for UMP and TMP) the fluorescence emission is a good mirror image of the first absorption band, this is not the case for AMP. However, the situation is rather different for Ade in isobutanol, in which case Eastman (1969) showed more similarity between the absorption and

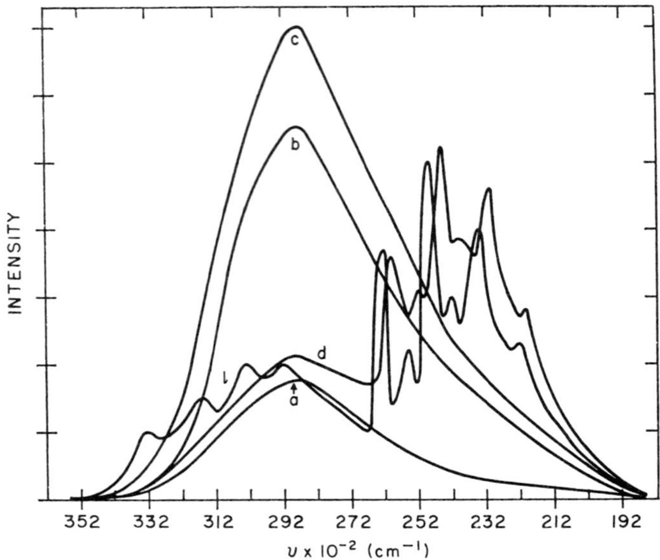


**Fig. 16.** *Total emission of 9-n-butyladenine at 77°K in (a) isopentane/methylcyclohexane (5:1), (b) hydrocarbon/10% diethyl ether, (c) hydrocarbon/1% triethylamine, (d) hydrocarbon/10% triethylamine, and (e) isopentane/butanol (7:3) (Cohen and Goodman, 1967).*

emission spectra (Fig. 17b). This could be construed as evidence that substitution at N(9) in Ade changes the component transition intensities and position in this complex band.

It should be pointed out that corrected emission spectra are still not available for Gua and Guo, Cyt and Cyd, or Ura and Urd.

### b. *Excitation Spectra*

When the frequency of excitation is varied over an absorption band, then the intensity of the emission will usually vary. In a simple way, the intensity of emission can be written as

$$I_f = \phi_f \, I_{abs}{}^{\lambda} = \phi_f \, I_0{}^{\lambda} \, (1 - 10^{-\epsilon_\lambda cl})$$

in which $I_{abs}{}^{\lambda}$ is the rate of absorption of exciting light, $I_0{}^{\lambda}$ is the incident intensity, $\epsilon_\lambda$ the molar extinction coefficient, $c$ the molar concentration of solute, and $l$ the pathlength. When the concentration is so chosen that $\epsilon_\lambda cl$ is sufficiently small, this expression will be approximated by

$$I_f = \phi_f \, I_0{}^{\lambda} \, 2.303 \, \epsilon_\lambda cl$$

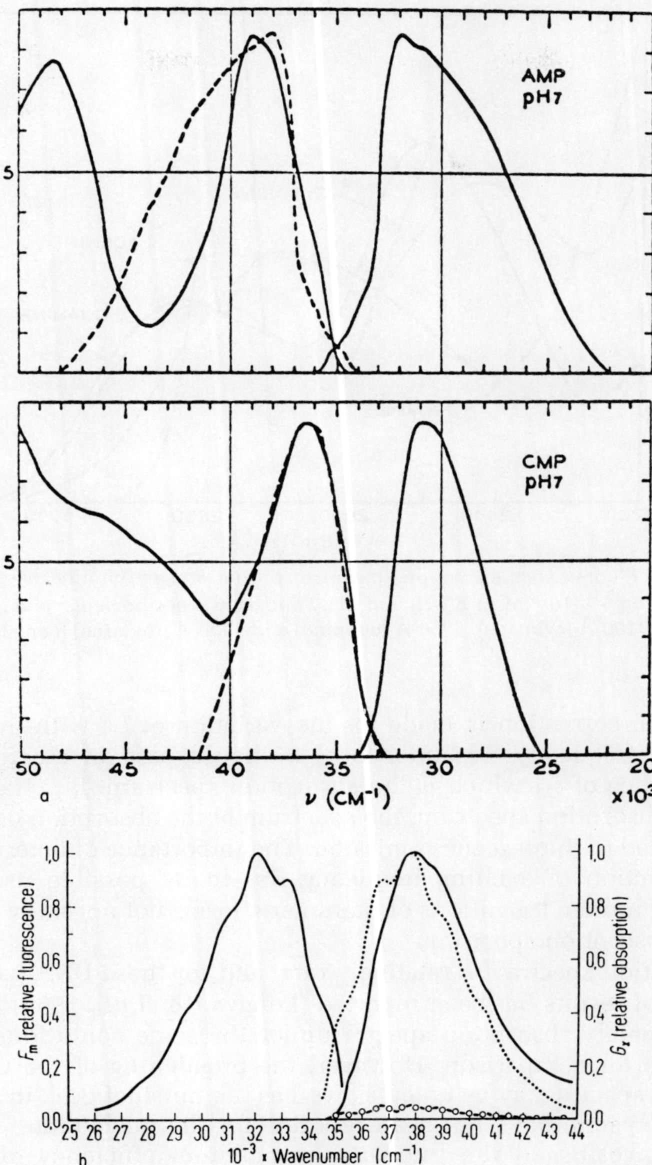


**Fig. 17.** (*a*) *Symmetry between absorption and fluorescence for AMP and CMP. The solid lines are plots of* $\epsilon(\bar{\nu})/\bar{\nu}$ *and* $f(\bar{\nu})/\bar{\nu}^3$ *with peak values equalized. The dotted curve is the mirror image of* $f(\bar{\nu})/\bar{\nu}^3$ *about the presumed 0–0′ frequency* (Guéron et al., 1967*a*). (*b*) *Absorption (solid line) and fluorescence (solid line) spectra of adenine in isobutanol at 170°K. The dotted line represents is the reflection of the fluorescence about the 0–0′ energy* (Eastman, 1969).

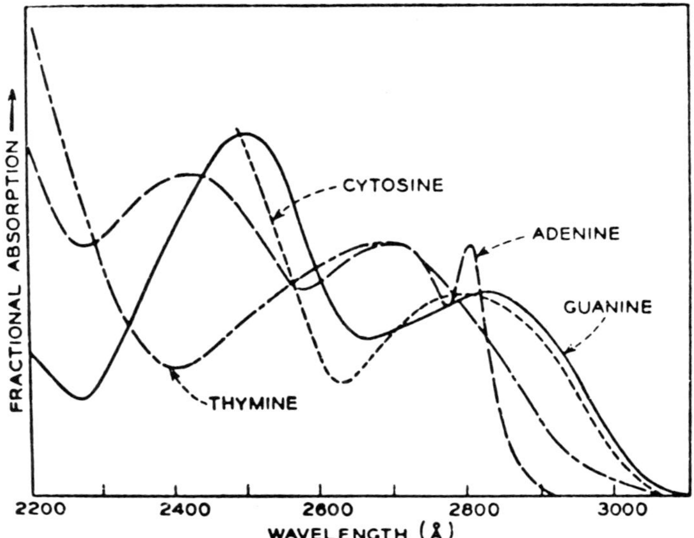


**Fig. 18.** *Fluorescence excitation spectra of purine and pyrimidine bases at 77°K, concentration* $5 \times 10^{-5}$ M *in EG/$H_2O$ at pH7. Emission measurements made at 3200 Å (guanine), 3100 Å (cytosine), 3250 Å (adenine) and 3300 Å (thymine) (Longworth* et al., *1966).*

and when correction is made for the variation of $I_0^{\lambda}$ with frequency, then the frequency dependence of $I_f$ should parallel the frequency dependence of $\epsilon_{\lambda}$, which is the absorption spectrum. $I_f$ is then an effective absorption spectrum, the spectrum of the absorption oscillators which lead to fluorescence emission. The importance of determining $I_f$ as a function of exciting frequency lies in its possible use in uncovering hidden transitions or tautomeric forms not normally revealed in the absorption spectrum.

Excitation spectra are relatively rare, and for these DNA bases only one set of results has been reported (Longworth *et al.*, 1966) (Fig. 18). Unfortunately absorption spectra under the same conditions are not available for comparison. However, the broadening of the Cyt, Gua, and Thy spectra may be noted as well as the unusual peak in Ade. Excitation spectra have not been reported for the nucleotides. Eastman (1969) investigated the fluorescence quantum efficiency of Ade in isobutanol and EG/water at 170°K as a function exciting frequency and found a threefold variation of $\phi_f$ from onset at 35,000 to 39,000 $cm^{-1}$ (the absorption maximum). This implies, of course, that the excitation spectrum does not coincide with the absorption spectrum (Fig. 17b). By analogy with the differences in quantum efficiency of 7-MeAde and 9-MeAde, Eastman suggested that the fluorescence of Ade is due to the 7-H tautomer, which is present only to the extent of ~6%. Temperature-jump studies on the ground-state tautomerism of Ade around room temperature are reported by Dreyfus *et. al.* (1975). Though plausible, the evidence is not complete in that the excitation spectrum of 7-MeAde was not determined for comparison with its absorption spectrum and with the excitation spectrum of Ade. Other tautomers are possible and the possibility that the differences are due to two states in Ade (as indicated by absorption studies; see Section B and Section C,1,f) was not considered.

#### c. Polarization Studies

The considerations involved in polarized excitation are outlined in Section B,1. Results for Gua, 9-EtGua, and Ade (Callis *et al.*, 1964) and for 5-MeCyt (Callis and Simpson, 1970) are shown in Figs. 9a and 11. In all cases the fluorescence, monitored presumably at the emission maximum, was strongly positive with respect to excitation from onset to near the band maximum, suggesting clearly that the emission was from a ($\pi\pi^*$) state. Decreasing values of polarization at higher frequencies showed the second absorption band of Gua to be approximately perpendicular to the first band and the single band of Ade to contain two transitions. For 5-MeCyt the first band contains only one transition and the second band is rather less than perpendicular to it. However, these results cannot determine whether the second transition is perpendicular in-plane or out-of-plane.

Variation of polarization of emission across the emission band has been reported by Kleinwächter *et al.* (1967a) for dAdo in an isopropanol/isopentane glass (1:7), but the excitation wavelength was not given.

#### d. Fluorescence Quantum Yields

Quantum yields are notoriously difficult to establish under ordinary circumstances, and their determination at 77°K in EG/water glass seems even more uncertain. Available data are presented in Tables 4 and 5. Moderate agreement, within a factor of 2, is noted for AMP; the range of the values for GMP varies by a factor of 4 and for TMP by a factor of 3. These differences, not normally significant, tend to hide considerable internal differences in the results of any one worker which cannot be accounted for simply in terms of different quantum standards. For example, Longworth *et al.* (1966) reported a 20-fold decrease in $\phi_f$ from Ade to Ado, whereas Hønnas and Steen (1970) found only a 2.5-fold decrease. Similarly Longworth *et al.* reported $\phi_f$(Ado) ~ $\phi_f$(AMP), whereas Hønnas and Steen found $\phi_f$

**Table 4** Fluorescence Quantum Yields for the Nucleotides[a] at 77°K

| Compound | Longworth *et al.*[b] (1966) | Guéron *et al.*[c] (1967a) | Eisinger and Shulman (1968) | Hønnas and Steen (1970)[d] |
|---|---|---|---|---|
| AMP | $4 \times 10^{-3}$ | $4.3 \times 10^{-3}$ | $1 \times 10^{-2}$ | $5 \times 10^{-3}$ |
| GMP | $3 \times 10^{-2}$ | $6.6 \times 10^{-2}$ | 0.13 | 0.17[e] |
| TMP | — | $9.7 \times 10^{-2}$ | 0.16 | 0.30 |
| CMP | — | $4.2 \times 10^{-2}$ | $5 \times 10^{-2}$ | $5 \times 10^{-2}$ [e] |
| UMP | — | $7 \times 10^{-3}$ | $\sim 1 \times 10^{-2}$ | — |

[a] At pH 7 in 1:1 EG/$H_2O$ glass.
[b] Relative to a tryptophan quantum yield at 298°K of 0.19; nucleotide concentration 1 m*M*.
[c] Relative to a *p*-terphenyl quantum yield at 300°K of 0.93; nucleotide concentration 3 m*M*, $\lambda_{ex}$ 265 nm.
[d] Relative to 1 m*M* tryptophan in EG/water glass at 77°K, $\phi_f + \phi_p = 1.0$; nucleotide concentration 4 m*M*, $\lambda_{ex} = 270$ nm.
[e] Relative to $\phi_f + \phi_p = 0.85$ for tryptophan (Steen, personal communication).

(Ado) ~ $10\phi_f$ (AMP). Although the quantum yields may vary with frequency of excitation (see Section C,1,b), the difference caused by a change from 265 to 270 nm in these cases will probably be slight. There is also evidence for oxygen quenching at 170°K (Eastman and Rosa, 1968) and this may account, in part, for the higher results of Hønnas and Steen, who degassed their samples. The effect of light intensity has not been investigated. A more intensive systematic study of the quantum yields seems to be necessary before they can be put to quantitative use.

In the meantime, the following qualitative features may be noted. The fluorescence of the Ade group seems to be quenched, more or less severely, on substitution at N(9). This is not the case for Gua; indeed, the Eisinger and Shulman (1968) value for GMP shows an increase. The most strongly fluorescent molecule at 77°K is Thy, which also

**Table 5** Fluorescence Quantum Yields of the Bases and Nucleosides at 77°K

| Compound | Longworth *et al.* (1966) | Hønnas and Steen (1970) |
|---|---|---|
| Ade | $6 \times 10^{-2}$ | 0.1 |
| Ado | $3 \times 10^{-3}$ | $4 \times 10^{-2}$ |
| Gua | $6 \times 10^{-2}$ | — |
| Guo | $2 \times 10^{-2}$ | — |
| Thy | — | 0.21 |
| Thd | — | 0.22 |
| Cyt | $6 \times 10^{-2}$ | — |
| Cyd | $3 \times 10^{-2}$ | — |
| Ura | $8 \times 10^{-4}$ | — |
| Urd | $1 \times 10^{-3}$ | — |

shows indication of $\phi_f$ increasing when N(1) is substituted. The general sequence of quantum yields seems to be

$$\phi(\text{Thy}) \sim 0.3 > \phi(\text{Gua}) \sim 0.1 > \phi\ (\text{Cyt}) \sim 5 \times 10^{-2} > \phi(\text{Ura}) \sim 1 \times 10^{-2} > \phi(\text{Ade}) \sim 5 \times 10^{-3}$$

*e. Fluorescence Lifetimes*

The quantum yields of fluorescence emission can be expressed in terms of rate constants for various radiative and nonradiative processes (Fig. 1) as

$$\phi_f = \frac{k_f}{k_f + k_{ic} + k_{isc} + k_{chem}} = \frac{\tau_f}{\tau_f^0}$$

in which $\tau_f$ is the experimentally measured fluorescence lifetime and $\tau_f^0$ is the intrinsic radiative lifetime. Direct measurement of $\tau_f$ thus provides an independent parameter whereby the quantum yields and intrinsic lifetimes may be related ($\tau_f^0$ may also be calculated from the shapes of the absorption and corrected emission spectra). Some fluorescence lifetimes, determined by the single-photon counting technique, have been briefly reported for EG/water glass at 77°K (Eisinger and Shulman, 1968; Blumberg *et al.*, 1968) (see Table 6) and are discussed in the next section.

*f. Derived Quantities*

First we note that when the absorption and fluorescence spectra have clearly defined leading edges, the position of the 0–0′ transition

**Table 6**[a] Excited State Parameters at 77°K

| Compound | $f_{osc}$ | $^1E(0–0')$ (kK) | $\tau_f^0$ (nsec)[d] | $\tau_f$ (nsec) | $\tau_f^0$ (nsec)[f] |
|---|---|---|---|---|---|
| Ade | 0.3[b] | 34.95[c] | 2.9 | — | — |
| Ado | — | — | — | 4.3[e] | — |
| AMP | 0.3 | 35.2 | 3 | 2.8 | 280–560 |
| GMP | 0.08 | 34.0 | 12 | ~5 | 38–167 |
| TMP | 0.22 | 34.1 | 4.5 | 3.2 | 10–30 |
| CMP | 0.18 | 33.7 | 5.5 | — | — |
| UMP | 0.2 | 34.9 | 4.5 | — | — |

[a] All data from Eisinger and Shulman (1968) except where noted.
[b] Eastman and Rosa (1968).
[c] Eastman (1969).
[d] Intrinsic radiative lifetimes calculated from absorption and emission spectra by the Strickler-Berg equation.
[e] Blumberg *et al.* (1968).
[f] Calculated from the quantum yields and fluorescence decay times.

can be evaluated from the overlap of the spectra. Experimental uncertainty is chiefly introduced by the contribution of Raman scattering to the emission spectrum, and this may be diminished by exciting at a higher frequency (provided that the emission spectrum does not change). The values in Table 6 are bunched within a range of ~1500 $cm^{-1}$, and in this range the sequence is AMP ~ UMP > TMP ~ GMP ~ CMP.

From the oscillator strength and the corrected fluorescence spectrum the intrinsic radiative lifetime of the excited state may be calculated with the Strickler-Berg equation. The use of this equation implies that single electronic states are involved and that the observed fluorescence and absorption are indeed directly related. Values for $\tau_f^0$ calculated in this way are given in column 4 of Table 6 and are not unreasonable. However, as indicated in Section C,1,e $\tau_f^0$ may also be calculated from the experimental quantum yields and the experimental lifetimes. Despite the uncertainties in $\phi_f$ noted earlier, the values obtained for $\tau_f^0$ are much longer than those calculated from the absorption and emission spectra, by an order of magnitude for TMP and GMP and more for AMP. Eisinger suggested that the reason for this may be the existence of hidden low-lying levels of long lifetime from which the emission actually occurs. Although attractive in relation to the energy level schemes for diazines (see Section B), there is at present no other evidence to support this hypothesis and alternative explanations can be proposed. Thus, noting the evidence from polarized absorption of 9-MeAde (Stewart and Davidson, 1963), supported by the fluorescence-polarized excitation spectrum (Callis *et al.*, 1964) for a weak transition on the high energy side of the Ade absorption band, it is clear that if this is the state responsible for fluorescence, then the experimental results may be reconciled. A value for the effective $f_{osc} < 0.3$ has the effect of increasing the intrinsic lifetime calculated from the Strickler-Berg relation. At the same time this implies that the true fluorescence quantum yield is greater than that calculated from the total light absorption; a larger $\phi_f$ then reduces the value of $\tau_f^0$ calculated from the measured fluorescence decay time and it is easy to calculate that both experimental approaches can be brought into agreement if $f_{osc} \cong 2 \times 10^{-2}$. This should be compared with the value for 9-MeAde crystal at room temperature of $8 \times 10^{-3}$. The corresponding values for intrinsic lifetime and the true quantum yield than becomes $\tau_f^0 \sim 40$ nsec and $\phi_f \sim 7 \times 10^{-2}$.

An explanation of this type may also be possible for GMP, in which bands I and II overlap, but this is not the case for TMP, which, in the crystal, contains only one transition in the first absorption band.

However, evidence has been presented (Daniels, 1972) that at room temperature Thy and Ura exist in tautomeric forms in aqueous solution, only one of which is responsible for the fluorescence. If this situation is true for TMP in the hydroxylic glass at 77°K, then it affords the possibility of an explanation in terms of a decreased effective $f_{osc}$.

These lifetime studies represent one of the major problems in the experimental characterization of the excited states of the DNA bases, and they place a new emphasis on the interpretation of fluorescence data if, indeed, it should turn out that the Ade fluorescence characterizes a state of only minor importance in absorption and the Thy fluorescence characterizes a tautomer.

The resolution of these problems is necessary before we can achieve an understanding of the energy degradation pathways from the singlet levels.

### 2. Studies of Temperature Dependence

The situation concerning the fluorescence behavior of the DNA bases in neutral aqueous solution at room temperature is simply stated: none of the bases fluoresce at sufficient intensity to be detectable without using signal accumulation techniques (Daniels and Hauswirth, 1971). The quantum yields at room temperature are now known to be $\sim 10^{-4}$, and it becomes of interest to investigate the mechanisms whereby the yields shown in Table 5 are quenched by a factor of $\sim 10^{-2}$.

Lamola and Eisinger (1971) presented some results for TMP in neutral EG/water glass from 250° to $\sim$80°K (Fig. 19). Although there seem to be two temperature-dependent quenching processes operative, they analyzed their results in terms of one temperature-dependent process and a temperature-independent yield considered to be reached at 80°K, and then evaluated the rate constant and activation energy for the quenching process (from 250° to 150°K) from the equation

$$k_q(T) = C\left[\frac{1}{I_f(T)} - \frac{1}{I_f(0)}\right] = ce^{-E/RT}$$

obtaining $E \sim 3$ kcal/mole. The isotope effect ($D_2O/H_2O$) is small but it seems to be compatible with the observed at 300°K (see Section C,3,c). Between 150° and 200°K, the range of maximum quenching, the wavelength of the emission maximum shifts from 321 to 326 nm, and this is thought to be indicative of solvent relaxation as the matrix soft-

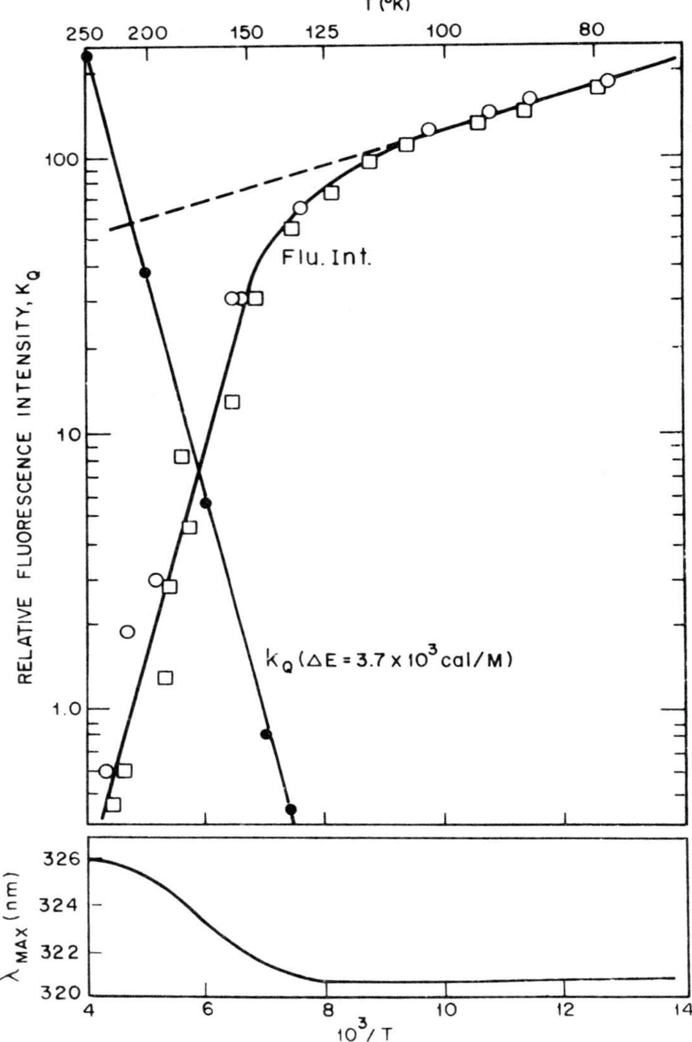


**Fig. 19.** *Temperature dependence of the fluorescence intensity and the nonradiative quenching rate* ($k_Q$) *of TMP in EG/$H_2O$ (○) and in deuterated EG/$H_2O$ (□). The lower figure shows the wavelength of the fluorescence maximum as a function of 1/T (Eisinger, 1968).*

ens. Similar but more extensive shifts ($\sim 6000\ cm^{-1}$) were reported for Ade in acid solution (Eisinger, 1968) and for Guo and Gua (Börresen, 1965b). In both systems the quenching and spectral shift phenomena are quite well separated but there is a striking difference between them. For protonated Ade the first effect of lowering the temperature is that the emission maximum shifts from ~370 to 312 nm, with only minor changes in quantum yield. This process is completed at ~120°K. Thereafter as the temperature is lowered to 80°K the emission spectrum remains constant and the quantum yield increases 5-fold. For protonated Gua the sequence of events is quite the opposite. The first effect of lowering the temperature is to increase the quantum yield without any attendant spectral shift. Only after the quantum yield has reached its maximum at ~160°K does the emission spectrum blue shift; this process is completed at ~120°K. The spectral shift thus occurs in roughly the same temperature range for each system, which is reasonable if it is due to a solvent reorientation process, the two solvents being EG/water (1:1) at pH 2 and methanol/water (9:1) at pH 1.33. However, the quenching process seems to be independent of solvent relaxation and different for Ade and Gua. Little is known about the actual quenching mechanisms.

Eastman and Rosa (1968) investigated the temperature and solvent dependence of the fluorescence of neutral Ade but did not report whether the emission shifts with temperature in these cases. Viscosities were measured for their solvents but no simple dependence on $T/\eta$ was observed. No study of Cyt derivatives has been reported.

### 3. Fluorescence Studies at Room Temperature

Until quite recently the fluorescence of the bases at room temperature could only be measured at extremes of pH and was attributed to the protonated cationic forms in the case of Ade and Gua and their derivatives and to anionic forms in alkaline solution (Gua and Thy). The fluorescences of the Pur bases under these conditions have been studied in detail by Börresen (1963) and the fluorescence of Thy and its methylated derivatives has been studied by Gill (1968) and by Berens and Wierzchowski (1969). Although such pH studies are not of direct concern to this review, it is worthwhile to point out another striking difference between the behavior of Ade and Gua. At room temperature the fluorescence of both Gua and Ade increases with acidity, but at 77°K the fluorescence of Gua increases and that of Ade decreases with increased acidity. This difference is at present unaccounted for.

In neutral solution at room temperature, the fluorescence of the bases is observed if a signal accumulation technique is used. Daniels and Hauswirth (1971) applied such a technique in conjunction with a Turner Model 210 spectrofluorimeter and were able to determine corrected emission spectra, corrected excitation spectra, and overall quantum yields for all the major DNA bases. They have also initiated an extended study of tautomeric, solvent, and isotope effects. When working with such low quantum yields ($\phi_f \sim 10^{-4}$), the question of genuineness of the emission must always be considered. The following considerations can be used to provide evidence on this point, although none are absolute criteria. First the emission spectra should be similar to known emissions of related compounds. In the present instance spectra were compared when possible with the (corrected) spectra of the same bases at 77°K, the identity of which is presumably unquestioned. Only in the case of Ura was no spectrum available, but for Cyt and Gua only uncorrected spectra could be found. However, bearing in mind the differences of temperature and solvent, satisfactory agreement was obtained. The second consideration is that the fluorescence intensity should increase linearly with solute concentration (at low enough concentrations). This has been observed, but it should be noted that this only rules out the possibility of the fluorescence arising from solvent impurity or cell contamination. Third, the excitation spectrum of the emitter should parallel the absorption spectrum for a simple molecular species. The observed excitation spectra are significantly different from the absorption spectra even when measured on the same instrument. However, this does not necessarily prove the emission to be false, for the results can be considered, *inter alia,* as evidence for tautomerism (see below). Finally, the spectra should smoothly change into those known for acid and basic forms as the pH changes, provided that the complications of pK* are taken into account. In the present work this approach could not be used, as the broader and much more intense emissions of the acid and alkaline forms completely overwhelm the emission of the neutral species at intermediate pHs. In brief, it seemed that there was no strong reason to doubt the genuineness of the spectra, and it is gratifying to note that essential confirmation by a photon-counting technique has been obtained independently (Vigny, 1971a,b).

*a. Emission Spectra*

The corrected spectra of the Pur and Pyr bases shown in Fig. 20 are compared with those at 77°K. It can be seen that the spectra are sufficiently similar to serve to establish identity, but some interesting differences are noted. The Ade spectrum has, not surprisingly, lost its vibrational structure and is significantly broadened. A long tail is found from 350 to 450 nm; this feature is also found in the spectra of Vigny. The feature of spectral broadening, which is particularly pro-

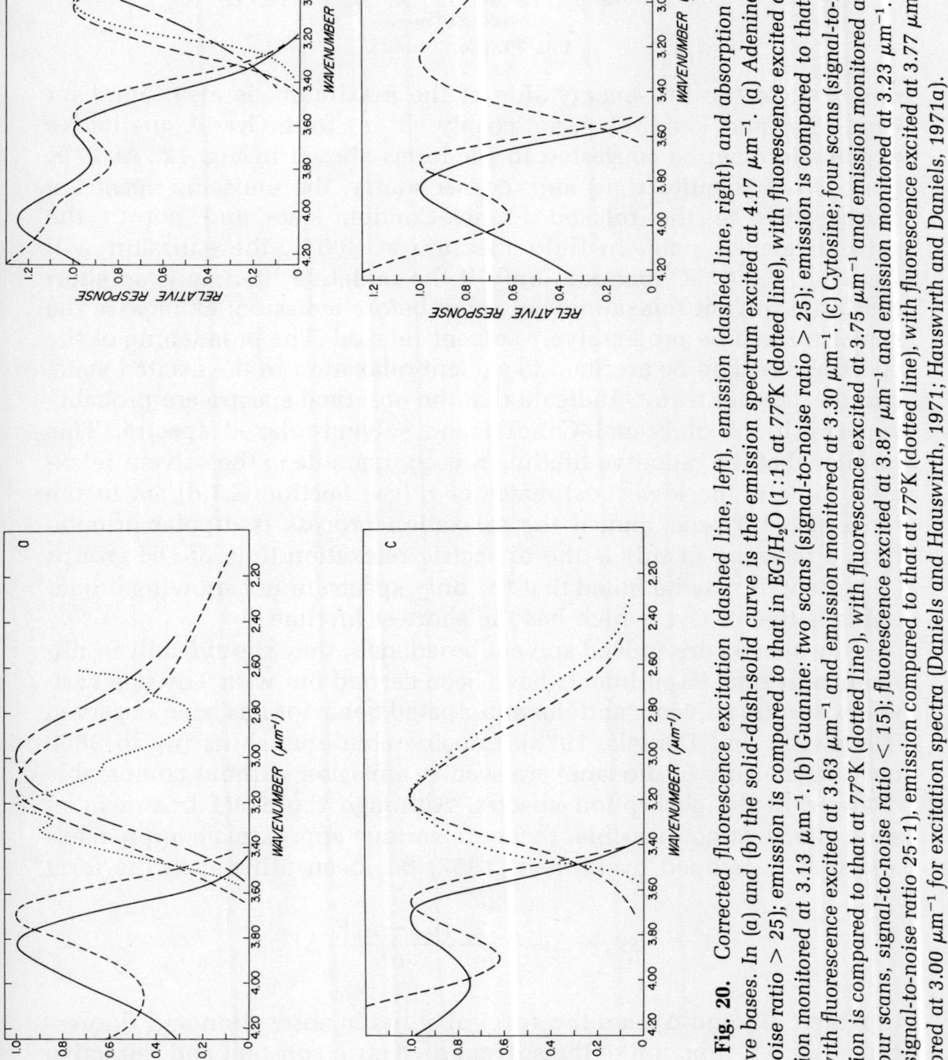


**Fig. 20.** Corrected fluorescence excitation (dashed line, left), emission (dashed line, right), and absorption (solid line) spectra for five bases. In (a) and (b) the solid-dash-solid curve is the emission spectrum excited at 4.17 $\mu m^{-1}$. (a) Adenine: two scans (signal-to-noise ratio > 25); emission is compared to that in EG/$H_2O$ (1:1) at 77°K (dotted line), with fluorescence excited at 3.83 $\mu m^{-1}$ and emission monitored at 3.13 $\mu m^{-1}$. (b) Guanine: two scans (signal-to-noise ratio > 25); emission is compared to that at 195°K (dotted line), with fluorescence excited at 3.63 $\mu m^{-1}$ and emission monitored at 3.30 $\mu m^{-1}$. (c) Cytosine; four scans (signal-to-noise ratio > 15); emission is compared to that at 77°K (dotted line), with fluorescence excited at 3.75 $\mu m^{-1}$ and emission monitored at 3.13 $\mu m^{-1}$. (d) Uracil: four scans, signal-to-noise ratio < 15); fluorescence excited at 3.87 $\mu m^{-1}$ and emission monitored at 3.23 $\mu m^{-1}$. (e) Thymine: six scans (signal-to-noise ratio 25:1), emission compared to that at 77°K (dotted line), with fluorescence excited at 3.77 $\mu m^{-1}$, and emission monitored at 3.00 $\mu m^{-1}$ for excitation spectra (Daniels and Hauswirth, 1971; Hauswirth and Daniels, 1971a).

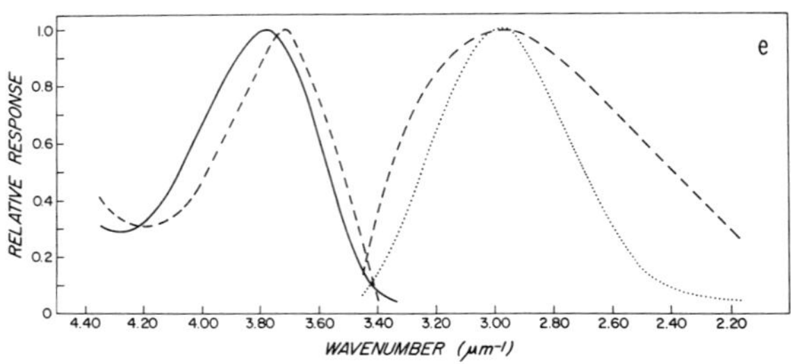


**Fig. 20.** (*Continued*)

nounced on the low-energy side of the maximum, is also found for Gua, Thy, and Ura, but is noticeably absent from Cyt. A qualitative explanation can be suggested in the terms shown in Fig. 12. At 77°K the solvent is quite rigid and, consequently, the emission spectrum corresponds to the relaxed Franck-Condon state and not to the solvent-relaxed state. In fluid medium at 300°K, the emission will resemble the 77°K spectrum only if the radiative lifetime is so short that little solvent relaxation can occur before emission. Otherwise the spectrum will be progressively solvent relaxed. The broadening of the spectra may then be ascribed to solvent relaxation in the excited state, and the present results indicate that the observed spectra are probably superpositions of Franck-Condon and solvent-relaxed spectra. This implies that the radiative lifetime $\tau_f$ is comparable to the solvent relaxation time $\tau$; the lowest estimates of $\tau_f$ (see Section C,3,d) are in the range 0.2–1.5 psec, and, if the relaxation process is dipolar orientation, a measure of this is the dielectric relaxation time of OH groups (~ 3 psec). It may be noted that the only spectrum not showing broadening is that of Cyt, which has the shortest lifetime.

If the spectra are indeed solvent broadened, they should shift in different solvents. Experiments have been carried out with Thy in a variety of aprotic solvents and the anticipated behavior has been observed (Hauswirth and Daniels, 1973). Considerable blue shifts (up to 3600 $cm^{-1}$ in the case of dioxane) are seen in emission without comparable changes in the absorption spectra. Although the exact treatment of such effects is not possible, there are various approximate approaches and that developed by Lippert (1957) has been utilized in the form

$$\Delta\bar{\nu}_a - \Delta\bar{\nu}_f = \frac{2}{hc}\,\frac{(\mu_e - \mu_g)^2}{a^3}\,\Delta f$$

in which $\Delta\bar{\nu}_a$ and $\Delta\bar{\nu}_f$ are the spectral shifts in absorption and fluorescence, $\Delta f$ is a function of the solvent dielectric constant and refractive index,

$$\Delta f = \frac{\epsilon - 1}{2\epsilon + 1} - \frac{n^2 - 1}{2n^2 + 1}$$

$a$ is the cavity radius (approximately equal to the molecular radius), and $\mu_e$ and $\mu_g$ are the dipole moments of the solute in the excited and ground states respectively.

The range of suitable solvents was limited by the region of excitation (261 nm), by the requirement for lack of solvent fluorescence at this wavelength, and by the solubility of Thy. Despite this, data have been obtained which show a reasonable fit to the Lippert equation, and using the cavity radius of Pyr (3.9 Å, Baba *et al.*, 1966), the dipole moment of Thy in the excited state is found to be given by $\mu_e = \mu_g \pm 4.1$ D. The direction of the spectral shifts in emission together with the lack of effect on the absorption spectrum is consistent with the process in which there is excitation of a $\pi$ state and emission from a $\pi^*$ state and in which the sign of $\mu$ is positive. The ground-state dipole moment is reported to be 3.95 D (Mauret and Fayet, 1967) so that in the excited state the dipole moment approximately doubles.

By measuring absorption spectrum changes due to an applied electric field (electrochromism) Seibold and Labhart (1971) have been able to deduce excited-state dipole moments for uracil and thymine in dioxane solution, with the following results:

$$\text{Ura: } \mu_e - \mu_g = 3.3 \pm 0.4 \text{ D}$$
$$\text{Thy: } \mu_e - \mu_g = 1.65 \pm 0.3 \text{ D}$$

The Ura spectra were measured over the range 250–275 nm and were observed to be irregular below 255 nm. The Thy spectra were measured from 255 to 290 nm and were irregular below 275 nm. The authors suggest that the irregularities may be indications of overlapping transitions, but again the possible role of tautomerism cannot be disregarded. It is entirely feasible that the species observed in fluorescence are not the major contributions to the electrochromism.

Although these sorts of experiments have so far only been carried out for Thy and Ura, extension to other bases should not present major difficulties, and thus we may expect to have excited singlet state dipole moments for Ade, Gua, Cyt, and Ura.

Reverting to aqueous solutions, we may note that a determination of the emission spectrum of Thy in EG/water (1:1), carried out for the purpose of comparison with the spectrum at 77°K, showed a blue shift of the maximum by 1400 $cm^{-1}$ (together with a quantum yield change, see Section C,2). Vigny (1971b) presented emission spectra for all nucleosides and nucleotides of the common bases on excitation at 250 nm. The spectra are in general agreement with those of the bases; the unusual spectrum which he previously found for Cyt (Vigny, 1971a) was probably caused by some impurity. The broad spectrum of Guo, GMP, and Gua, with a $\lambda_{max} \sim 350$ nm, seems to be quite different from that reported by Daniels and Hauswirth (1971), which is narrow with $\lambda_{max} \sim 318$ nm, for, although Vigny's spectral data are uncorrected, the corrections are stated to be small. Further investigation seems warranted.

*b. Excitation Spectra*

In all cases investigated the excitation spectrum corrected for the finite absorbance of the solution did not coincide with the absorption spectrum (Fig. 20). The differences are not simply of peak:valley ratios, as are usually found in spectra of different resolution, but consist of distinct spectral shifts. Though completely unexpected, these results correlate well with previous reports under other conditions. Thus, for Ade at room temperature the relative quantum yields at 40, 37.3, and 35.7 kK are calculated from the excitation spectra as 0.54:1.0:2.7. Eastman (1969) in EG/water (7:3) at 170°K found the same trend and with the ratios 0.6:1.0:1.9. Excitation spectra differing from absorption spectra were also found for Gua at acid and alkaline pHs (Börresen, 1965a), for Ade at acid pH (Börresen, 1965b), and for Thy at alkaline pH (Gill, 1968; Berens and Wierzchowski, 1969). At least for the DNA bases, coincidence of excitation and absorption spectra seems to be the exception rather than the rule. In explanation of such behavior all workers considered the role of tautomeric structures, but such hypotheses are notably difficult to substantiate and other processes must be considered. Hauswirth and Daniels (1971b) discussed the excitation spectrum of Thy in terms of increasing probability of deactivation from the higher vibrational levels of $S_1$ competing with fluorescence emission. Such a model corresponds with that of Johns *et al.* (see Section D,2,a) for the mechanism of populating the triplet state if the deactivation process is equated with intersystem crossing. Unfortunately, a quantitative correlation of the fluorescence and triplet yield data cannot be found by this model and it must be discarded. Daniels and Hauswirth (1971) also suggested the possibility that emission comes from a state hidden in the red edge of the absorption band. The only base for which independent evidence for such a state exists is Ade (Section B), so that although this mechanism cannot be general, it may be valid for Ade. A requirement of this model is that the other transitions in the Ade band envelope must not be molecularly connected with the emitting transition, i.e., denoting the two levels $S_1$ and $S_2$, then the internal conversion rate $S_2 \rightarrow S_1$ must be quite inefficient, otherwise the excitation spectrum would coincide with the absorption spectrum. On the basis of the present evidence it is not possible to decide between the tautomeric model (Börresen, Eastman) and the two-state model (Daniels and Hauswirth). On the other hand, in the case of Ura and Thy, only one state is known in the first absorption band, and it has been proved possible to treat the fluorescence excitation spectra and triplet cross-section yields quantitatively (see Section D,2,a) and to reconcile them with the absorption spectra, if it is assumed that each originates in different tautomers (Daniels, 1972). This successful correlation of independent data must be regarded as strong evidence for the existence of Ura and Thy in tautomeric forms in a neutral aqueous solution. The most likely structure for the fluorescing tautomer is the N(3)–C(4) lactim form,

while the observed triplets originate from the normal diketo or lactam form.

Further work is needed to substantiate the explanation for Ura and Thy and to allow a distinction to be made in the case of Ade.

*c. Quantum Yields*

Quantum yield determinations have been carried out using excitation at $\lambda_{max}$ in absorption for all bases, and at other selected wavelengths for Ade and Thy, with reference to 1,5-diphenyloxazole (PPO) in nitrogen-flushed cyclohexane assuming $\phi_f(\text{PPO}) = 1.0$ (Berlman, 1965). There has been some discussion (Birks, 1970) about the correctness of this value, which is based on $\phi_f$(9,10-diphenylanthracene, DPA) $= 1.0$ in cyclohexane, the principal point being that the value $\phi_{isc} = 0.13$ has been obtained for DPA in liquid paraffin. However, this question cannot be settled until $\phi_{isc}$ and $\phi_f$ are measured in the identical system, and in the meantime $\phi_f(\text{PPO}) = 1.0$ is retained as a reference point. In view of the large differences in quantum yields of standard and the DNA bases, an intermediate reference was used, namely Gua at pH 10.9, for which $\phi_f$ was found to be

**Table 7** Experimental Parameters of the Fluorescences of the Bases,[a] Nucleosides,[b] and Nucleotides[b] in Neutral Aqueous Solution at 300°K

| Compound | Excitation energy (kK) | $\phi_f \times 10^4$ | $^1E(0\text{–}0')$ (kK) |
|---|---|---|---|
| Ade | 38.3 | 2.6 | 35.6 |
| Ado | 40 | 0.08 | 35.2 |
| AMP | 40 | 0.03 | 35.1 |
| Gua | 36.3 | 3.0 | 33.4 |
| Guo | 40 | 0.06 | 33.75 |
| GMP | 40 | 0.02 | 33.3 |
| Thy | 37.7 | 1.02 | 34.5 |
| Thd | 40 | 0.4 | 34.0 |
| TMP | 40 | 0.4 | 34.0 |
| Cyt | 37.5 | 0.82 | 34.9 |
| Cyd | 40 | 0.4 | 34.0 |
| CMP | 40 | 0.3 | 34.0 |
| Ura | 38.7 | 0.45 | 35.7 |
| Urd | 40 | 0.1 | 35.1 |
| UMP | 40 | 0.1 | 34.9 |

[a] All data for the bases from Daniels and Hauswirth (1971).
[b] All data for the nucleosides and nucleotides from Vigny (1971b).

$1.65 \times 10^{-2}$ (relative to PPO). This reference enabled the same bandwidths to be used for the standard and the bases.

The values thus obtained for quantum yields are shown in Table 7 and range from $5 \times 10^{-5}$ for Ura to $3 \times 10^{-4}$ for Gua. In view of the uncertainties arising from the excitation spectra, it is not yet possible to give any molecular interpretation to these results; for example, if the Ade emission should prove to originate from a minor amount of tautomer, then the molecular quantum yield from the tautomer could easily be an order of magnitude greater than the experimentally observed value. Until this problem is resolved the quantum yields must be regarded as solely experimental parameters. Overall quantum yields for emission from the nucleosides and nucleotides in aqueous solution have been reported by Vigny (1971b) relative to the values for the bases reported by Daniels and Hauswirth (1971). It is interesting to note that the yields from the Pur nucleosides and nucleotides are 100-fold lower than the yields from the bases. This is similar to the behavior at 77°K but is more pronounced. On the other hand, the yields from the Pyr nucleosides and nucleotides are only decreased by a factor of 2 to 4.

More extensive quantum yield determinations have been carried out on Thy in a variety of hydroxylic solvents and aprotic solvents (Table 8) (Hauswirth and Daniels, 1973). There is a significant sol-

**Table 8** Thymine Fluorescence in Different Solvents at 300°K

| Solvent | $\phi_f \times 10^4$ | ${}^1E(0\text{–}0')$ (kK) |
|---|---|---|
| $H_2O$ | 1.04 | 34.5 |
| $D_2O$ | 1.48 | 34.3 |
| Methanol | 1.41 | 34.6 |
| EG/$H_2O$ | 2.53 | 34.3 |
| Acetonitrile | 1.12 | 34.8 |
| $(CH_3)_2SO$ | 2.47 | 34.3 |
| Dioxane | 1.57 | 35.2 |

vent deuterium effect with the quantum yield increasing by a factor very close to $\sqrt{2}$, which is the ratio of the normal stretching vibrations of OH and OD bonds. This is particularly interesting in view of the solvent deuterium effects on the photochemistry of the Pyr bases—hydration decreases with deuteration (Wierzchowski and Shugar, 1957) while dimer formation increases in $D_2O$ (Nnadi and Wang, 1969). Further work is planned in which the effect of $D_2O$ on the nonradiative rates may be elucidated. This emphasizes the point that solvent effects on quantum yields cannot be analyzed in isolation, although, for example, it might be tempting to examine the data for hydroxylic solvents in terms of H-bonding or viscosity/relaxation time effects. The data must be broken down into individual rate constants such as $k_f$ and $k_{isc}$ before analysis is possible. However, two results are worth noting. First, EG/$H_2O$ (1:1) causes a significant increase in $\phi_f$; we do not yet know what effect it has on $\phi_{isc}$. Second, the lack of change of $\phi_f$ in acetonitrile contrasts strongly with the large increase (to 0.18) found for $\phi_{isc}$ in this solvent (Lamola and Mittal, 1966).

*d. Derived Quantities and Other Results*

The 0–0′ level of the singlet state transition may be evaluated from the overlap of the absorption and corrected emission spectra. These results are shown in Table 7, and for the bases they fall in the sequence* Ade ~ Ura > Cyt > Thy > Gua. Essentially the same result is obtained for the nucleosides and nucleotides, and the significant difference from the sequence determined at 77°K is the position of Gua, which at 77°K does not have the lowest-lying level. Such a change is unexpected (though not unknown), and, in view of the

* The 0–0′ values obtained from the overlap of excitation spectra and emission spectra, which may reflect more correctly the true molecular situation, show the same sequence.

narrow spread of energies (2200 $cm^{-1}$), consideration has to be given to uncertainties in estimation of these values at 77°K and 300°K. The region of the overlap lies in that part of the spectrum in which the Raman scattering is the strongest, and hence the means of correcting it become important. In both studies the corrections are essentially digital at 300°K, whereas corrections at 77°K were probably analog. However, this in itself should not change the position in a sequence of one base at 77°K. For the present we conclude that the effect, in which the 0–0′ level of Gua red-shifts by 700 $cm^{-1}$ from 77° to 300°K, is probably real, and we may speculate that the origin of the effect lies in an increase in Gua solvent interaction at 300°K during the lifetime of the excited state.

At present, excited-state lifetimes at 300°K must be obtained indirectly, and two sets of estimates are available. From the corrected fluorescence and absorption spectra the mean intrinsic radiative lifetime, $\tau_f^0$, may be calculated using an expression such as the Strickler-Berg equation, and then from the quantum yield of fluorescence the actual radiative lifetime is given by $\tau_f = \phi_f \tau_f^0$. Such values are given for the bases in column 3 of Table 9. Uncertainties stem from the problem caused by the excitation spectra; thus $\tau_f^0$ may be in error on this account, and similarly the true quantum yields may be larger than those shown in Table 9. Consequently the values listed in column 3 of Table 9 should be considered a lower limit.

Singlet lifetimes have also been obtained by a different technique. Lamola and Eisinger (1971) investigated the transfer of excitation

**Table 9** Singlet Lifetimes at 300°K

| Compound | pH (pD) | $^1\tau$ (psec) | |
|---|---|---|---|
| | | *a* | *b* |
| Ade | 7.3 | 1.0 | |
| AMP | 5.0 | | 1.3 |
| Gua | 6.3 | 1.4 | |
| GMP | 5.5 | | 4.4 |
| Thy | 6.7 | 0.9 | |
| TMP | 5.5 | | 18.0 |
| Cyt | 6.5 | 0.2 | |
| CMP | 5.0 | | 3.6 |
| Ura | 6.8 | 0.7 | |
| UMP | 5.0 | | 23.0 |

[a] Calculated from fluorescence data, neglecting tautomeric effects.
[b] Determined from the slope of the plot from the $Eu^{3+}$ transfer method, $^1k_t{}^1\tau$, assuming $^1k_t = 5 \times 10^9\ M^{-1}sec^{-1}$ in all cases; all $Eu^{3+}$ results obtained in $D_2O$.

energy from singlet and triplet states of the nucleotides to $Eu^{3+}$ in aqueous solution, the process being followed by the emission from the excited $Eu^{3+}$. At lower concentrations of $Eu^{3+}$ the triplet states are scavenged, while at higher concentrations the singlet states are scavenged. Under favorable conditions these two processes may be observed separately. These investigations of the transfer and quenching mechanisms lead to the following equation for the scavenging of singlet states:

$$\frac{\beta}{\phi_f(\mathrm{Eu})} = 1 + \frac{{}^1k_q}{{}^1k_t} + \frac{1}{{}^1k_t\,{}^1\tau\,[\mathrm{Eu}^{3+}]}$$

in which $\beta$ is the probability that an excited $Eu^{3+}$ ion will emit (presumed independent of $[Eu^{3+}]$), $\phi_f(Eu)$ is the quantum yield of emission from $Eu^{3+}$ per photon absorbed by the donor, ${}^1k_q$ is the rate constant for quenching of the donor by $Eu^{3+}$, ${}^1k_t$ is the rate constant for transfer of singlet energy from the donor to $Eu^{3+}$, and ${}^1\tau$ is the singlet lifetime of the donor. Thus the slope of the linear relation between $\beta/\phi_f(Eu)$ and $1/[Eu^{3+}]$ allows the evaluation of ${}^1k_t\,{}^1\tau$, and ${}^1\tau$ is obtained by assuming ${}^1k_t = 5 \times 10^9\ M^{-1}\ \mathrm{sec}^{-1}$. The authors point out two limitations in their results. The first is that steeper slopes may be obtained at higher concentrations of $Eu^{3+}$ (though the slope should be independent of concentration), and the second is that ${}^1k_t$ could well be less than $5 \times 10^9$ which is the diffusion-controlled limit (no experimental values for ${}^1k_t$ are known). As these two considerations oppose each other, it is not possible to say whether the results (Table 9, column 4) represent upper or lower limits. Perhaps the most interesting result is the magnitude of the difference in Pyr lifetimes determined by the fluorescence and europium transfer methods.

A kinetic account of the singlet state following excitation describes its behavior in terms of all of the processes resulting in depopulation of the state. Within the conceptual framework of Fig. 1 the deactivation pathways can be represented as in Fig. 21. It must be emphasized that this approach considers only the minimum adequate representation. Problems stemming from tautomerization, from the frequency dependence of $\phi_{isc}$ (see Section D), and from the effect of relaxation on internal conversion and intersystem crossing processes are all ignored for the purpose of demonstrating that a complete, if approximate, account of the fundamental processes of Fig. 21 can now be given (Hauswirth and Daniels, 1971b). Future improvements in data will probably change the constants by less than an order of magnitude. According to the scheme in Fig. 21, three experimental quantities are

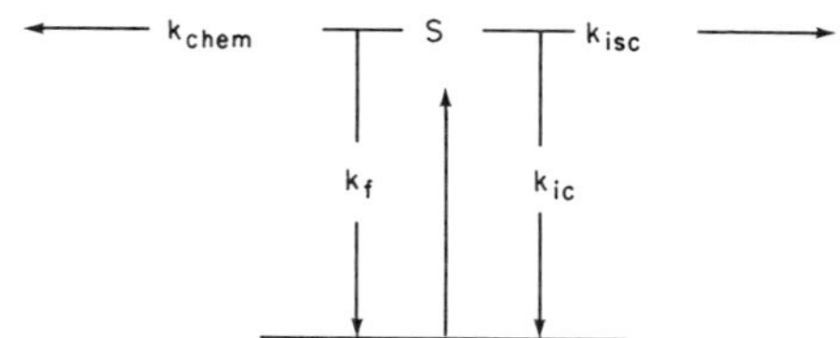


**Fig. 21.** *Deactivation pathways for an excited state S.*

defined: the photochemical quantum yield for reaction from the singlet state level,

$$\phi_{\text{chem}} = k_{\text{chem}}/\Sigma k$$

in which $\Sigma k = k_f + k_{ic} + k_{\text{chem}} + k_{isc}$; the fluorescence quantum yield,

$$\phi_f = k_f/\Sigma k$$

and the intersystem crossing yield (triplet formation),

$$\phi_{isc} = k_{isc}/\Sigma k$$

However, four rate constant parameters are involved. Solution is possible if $k_f$ is calculated from the absorption and emission spectra (Strickler-Berg equation), which give $\tau_f^0 = 1/k_f$. Values for $k_{isc}$, $k_{ic}$, and $k_{\text{chem}}$ are then derived from the quantum yield ratios. Reasonably complete data are available only for Ura and Thy as shown in Table 10.

As expected, the internal conversion process is dominant, and the interesting question is whether this process is intramolecular or solvent coupled. There are several indications that it may be the latter, such as the change in fluorescence quantum yields with solvent and isotope, and more particularly the large value of $\phi_{isc}$ (0.2–0.4) reported by Lamola and Mittal (1966) in acetonitrile as solvent. The rather high values of $k_{isc}$ suggest little prohibition on the intersystem crossing

**Table 10** Rate Constants for Processes Occurring from the 0′ Level of Ura and Thy at 300°K[a]

| Compound | $k_f$ | $k_{isc}$ | $k_{\text{chem}}$ | $k_{ic}$ |
|---|---|---|---|---|
| Ura | $1.2 \times 10^8$ | $4.8 \times 10^9$ | $2.1 \times 10^9$ | $10.3 \times 10^{11}$ |
| Thy | $1.0 \times 10^8$ | $7.1 \times 10^8$ | $1.8 \times 10^8$ | $5.7 \times 10^{11}$ |

[a] All data are in $\sec^{-1}$.

process. These results should be compared with those for Pyr in Table 1b, noting, however, that $k_{isc}$ seems to be temperature dependent for Ura and Thy, whereas yields for Pyr are not.

## D. Properties of the Triplet States

The discussion of the triplet states of these molecules will be arranged in the sequence of formation characteristics followed by properties of the states. However, as most of the available experimental information has been obtained at either 77°K or at room temperature, with no studies of temperature dependence, it will be convenient to separate the discussion in the same way.

### 1. Low-Temperature Studies

*a. Formation of the Triplet State*

Under most experimental conditions the triplet state is populated from the singlet state by the process of intersystem crossing. Specifically, we disregard the processes of formation by direct singlet → triplet absorption and those involved in the use of ionizing radiation, interesting and significant though they are. With reference to the simple scheme of Fig. 1, the probability of populating the triplet state following excitation into the singlet can be expressed as

$$k_{isc}/\Sigma' k = \phi_{isc} = [^3X]/I_{abs}$$

in which $\Sigma' k$ represents all of the pathways responsible for depopulating the singlet state and $[^3X]$ represents the stationary triplet concentration under continuous excitation. Determination of $\phi_{isc}$ then involves measurement of $[^3X]$ under conditions in which the rate of photon absorption is known and is constant. Since triplet states are paramagnetic, they have been studied by ESR absorption since the work of Hutchinson and Magnum (1958) and Van der Waals and de Groot (1959). Such studies can elucidate the magnitude and character of the splitting of the triplet level due to magnetic dipole–dipole interaction of the unpaired spins or to the spin–orbit interaction (zero field splitting constants). In the present context the triplet population can be determined from the signal line intensity by comparison with a sample containing a known concentration of unpaired electrons. The method was first proposed by Aleksandrov and Pukhov (1964) who obtained a working relationship

$$\frac{[\mathrm{R}]}{[\mathrm{T}]} = \frac{8}{15} \frac{(D^2 + 3E^2)}{(\hbar\,\omega_0)^2} \frac{I_\mathrm{R}}{I_T}$$

where R is the reference unpaired electron concentration (a stable free radical R), T is the triplet concentration, $D$ and $E$ are the ZFS constants, and $I_\mathrm{R}/I_\mathrm{T}$ is the ratio of the integrated intensities of both signals measured at the same frequency. The relation can also be put in the form

$$\frac{[\mathrm{R}]}{[\mathrm{T}]} = \frac{56}{12} \frac{\mu_1}{2\omega_0} \frac{I_\mathrm{R}}{I_\mathrm{T}}$$

which then requires only the measurement of the position of the line centroid relative to the frequency $2\omega_0 = 2g\beta H_0/\hbar$ ($\mu_1$ is the first moment of the absorption line). Dilute solutions should be used to obtain a uniform distribution of triplet states.

When applied to the DNA bases, the method was modified by using substances of known $\phi_{\mathrm{isc}}$, triphenylene and naphthalene, as standards. Because the $\Delta m = 2$ lines are broad, only the absorption edge was observable and the modified method using optically dense samples has been described (Guéron *et al.*, 1967b, 1968). Available results are presented in Table 11. It should be noted that these data have been obtained using broad-band excitation (260–280 nm), and in view of the unusual excitation spectra for Ade discussed earlier, the wavelength dependence of $\phi_{\mathrm{isc}}$ for the pyrimidines at room temperature (discussed in Section D,2), and the possibility of biphotonic effects the interpretation of the values of $\phi_{\mathrm{isc}}$ in Table 11 as molecular parameters should be viewed with caution. In conjunction with singlet lifetimes at 77°K, the magnitudes of $\phi_{\mathrm{isc}}$ (77°K) indicate values for $k_{\mathrm{isc}}$ from $10^6$ to $10^8$ $\mathrm{sec}^{-1}$. With the above limitations, these values will be discussed in conjunction with $\phi_{\mathrm{isc}}$ (300°K).

The intersystem crossing yield $\phi_{\mathrm{isc}}$ gives the probability of the process $S_1 \rightarrow T_1$. It is of particular interest to know the origin of the triplet state, i.e., which particular state $S_1$ is involved in the process. Very little direct information is available. Cohen and Goodman (1965) have carried out measurements of the polarized phosphorescence excitation spectra of Pur, Ade, and 9-BuAde at 77°K in Ether/isopentane/alcohol (Fig. 22). While the behavior of Pur shows nicely the expected change in the leading edge of the absorption spectrum which corresponds to the $(n\pi^*)$ transition, Ade and 9-BuAde show a constant degree of polarization from 285 to 245 nm. This contrasts strongly with the changes observed in fluorescence polarization exci-

**Table 11** Triplet State Data at 77°K

| | $\phi_p$[a] | | | | | $\phi_{isc}$ |
|---|---|---|---|---|---|---|
| Compound | Longworth *et al.*[b] (1966) | Guéron *et al.* (1967a) | Eisinger and Shulman[c] (1968) | Imakubo (1968) | Hønnas and Steen[d] (1970) | Guéron *et al.*[e] (1967b) |
| AMP | $4 \times 10^{-3}$ | $6.1 \times 10^{-3}$ | $1.5 \times 10^{-2}$ | $4 \times 10^{-3}$ | $1.0 \times 10^{-2}$ | $2 \times 10^{-2}$ |
| Ado | $8 \times 10^{-3}$ | — | — | — | $3 \times 10^{-3}$ | — |
| Ade | $3.5 \times 10^{-2}$ | — | — | — | $6 \times 10^{-2}$ | — |
| GMP | $4 \times 10^{-2}$ | $4.2 \times 10^{-2}$ | $7 \times 10^{-2}$ | $2 \times 10^{-2}$ | 0.13 | 0.15 |
| Guo | $3 \times 10^{-2}$ | — | — | — | — | — |
| Gua | $7 \times 10^{-2}$ | — | — | — | — | — |
| CMP | — | $5 \times 10^{-3}$ | $1 \times 10^{-2}$ | — | $2 \times 10^{-2}$[f] | $\sim 3 \times 10^{-2}$[g] |
| Cyd | — | — | — | — | — | — |
| Cyt | — | — | — | — | — | — |
| TMP | — | — | — | $4 \times 10^{-4}$ | $8 \times 10^{-3}$ | $< 3 \times 10^{-3}$[g] |
| Thd | — | — | — | — | $6 \times 10^{-3}$ | — |
| Thy | — | — | — | — | $6 \times 10^{-3}$ | — |

[a] All measurements were made in neutral EG/$H_2O$ (1:1) solution.
[b] Relative to $\phi_f$(tryptophan) = 0.19 at 300°K.
[c] Relative to $\phi_f$(p-terphenyl) at 300°K = 0.93.
[d] Relative to $\phi_p + \phi_f = 1.0$ for tryptophan in EG/$H_2O$ at 77°K.
[e] Relative to $\phi_{isc}$(triphenylene in EPA) = 0.92 and $\phi_{isc}$(naphthalene in EPA) = 0.45.
[f] H. B. Steen, personal communication (1972) based on $\phi_f + \phi_p = 0.85$ for tryptophan at 77°K.
[g] Lamola and Eisinger (1971).

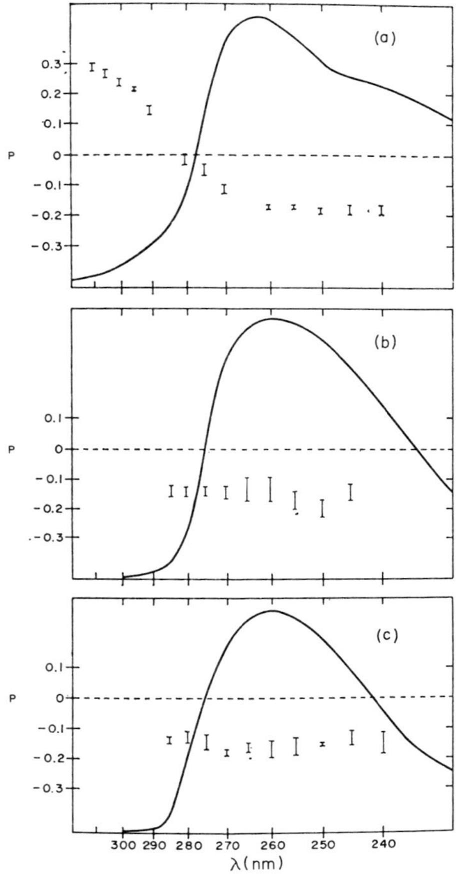


**Fig. 22.** *Absorption spectra in EPA at room temperature and polarized phosphorescence excitation spectra in EPA at 77°K corrected for instrumental and scatter effects of (a) purine, (b) adenine, and (c) 9-n-butyladenine. Degree of polarization (P) is defined as* $P = (I_{\parallel} - I_{\perp})/(I_{\parallel} + I_{\perp})$ *where* $I_{\parallel}$ *and* $I_{\perp}$ *refer to the emission intensity with perpendicular and parallel electric vectors relative to the exciting light electric vectors. The values represent the scatter of three measurements* (Cohen and Goodman, 1965).

tation (Callis *et al.*, 1964) but is not unexpected, for although there are two mutually perpendicular transitions in Ade, they are both in-plane and polarized perpendicular to the phosphorescence emission (which is out-of-plane, see Section D,1,b). This experiment thus does not allow any discrimination between these two transitions.

An experiment which would be of great interest would be the determination of $\phi_{isc}$ for each of the bases as a function of exciting wavelength (triplet excitation spectrum) but there are considerable technical difficulties involved, not the least of which is the low sensitivity of the ESR method coupled with the low quantum yield.

There are no studies of the effect of solvents on $\phi_{isc}$, although the effect is well known in other systems (Horrocks *et al.*, 1967; Kearwell and Wilkinson, 1969).

*b. Properties and Characteristics*

Perhaps the most prominent feature of triplet states at low temperatures is phosphorescence emission, a multiplicity-forbidden radiative transition from the lowest triplet level to the singlet ground state. In the earliest work (Steele and Szent-Györgi, 1957), phosphorescence emission was readily observed from the Pur bases and their derivatives and was recognized by being red-shifted from the fluorescence and by having a decay time of the order of seconds. Emission from the Pyr bases was more difficult to detect, but Guéron *et al.* (1967a) were successful with CMP, and Imakubo (1968) and Hønnas and Steen (1970) gave spectra and quantum yields for the Thy derivatives.

Available data for the quantum yields of phosphorescence, $\phi_p$, are summarized in Table 11 and are interesting in that comparison of $\phi_p$ and $\phi_{isc}$ allows an evaluation of the extent of nonradiative intersystem crossing, $T_1 \rightarrow S_0$, under conditions in which collisional quenching is minimal:

$$\phi_p = \phi_{isc} \frac{k_p}{k_p + \Sigma k_d}$$

in which $k_p$ is the rate constant for phosphorescence emission, and $\Sigma k_d$ represents all other triplet deactivating processes, so that $\phi_p/\phi_{isc}$ represents the probability that the triplet emits phosphorescence. Like the fluorescence yields, there is a fair amount of scatter in $\phi_p$, and taken in conjunction with the insensitivity of the $\phi_{isc}$ determinations, clear conclusions are difficult to make. In general, $\phi_p < \phi_{isc}$, but only Hønnas and Steen determined $\phi_p$ in the absence of oxygen; their values are the highest and in the case of Thy, they find $\phi_p > \phi_{isc}$! This may indicate significant oxygen quenching, but this point has not been directly investigated.

The shapes of the phosphorescence emission curves are characteristic. The purines, Ade and Gua, show considerable vibrational structure in polar solvents whereas Cyt and Thy emissions are smooth. A vibrational classification of the spectrum of Pur itself has been given by Drobnik and Augenstein (1966a) who showed the existence of 2

**Table 12** Triplet Energy Levels and Singlet–Triplet Splitting at 77°K

| Compound | $^3E(0\text{–}0')$ (kK) Guéron *et al.*[a] (1967a) | $\Delta E_{S-T} = {}^1E(0\text{–}0') - {}^3E(0\text{–}0')$ (kK) | |
|---|---|---|---|
| | | Guéron *et al.* (1967a) | Cohen and Goodman[b] (1965) |
| AMP | 26.7 | 8.5 | — |
| Ade | — | — | 13.9 |
| 9-MeAde | — | — | 13.6 |
| 9-*n*-BuAde | — | — | 13.7 |
| GMP | 27.2 | 6.8 | — |
| CMP | 27.9 | 5.8 | — |
| TMP | 26.3 | 7.8 | — |

[a] In EG/$H_2O$ (1:1).
[b] In EPA.

progressions of 1300 $cm^{-1}$, with a displacement of 500 $cm^{-1}$ between the series. A similar behavior was noted for Ade and its derivatives (Kleinwächter *et al.*, 1967a) except that the 0–0′ band was not detected. This was also the conclusion of Cohen and Goodman (1965) and is important both for the interpretation of the spectra and for the use of the spectra to determine the triplet energy levels. With reference to Fig. 1 it can be seen that the triplet state energy, $^3E(0\text{–}0')$, relative to the ground state should be given by the shortest wavelength emission, i.e., it can be estimated from the blue edge of the emission. This method has been used by Guéron *et al.* (1967a) to obtain the values given in Fig. 12. However, if the 0–0′ band is not observed this can lead to an underestimate of $^3E(0\text{–}0')$ and hence to an overestimate of the singlet–triplet splitting, as was recognized by these authors. Cohen and Goodman estimated $^1E(0\text{–}0') - {}^3E(0\text{–}0')$ as $\bar{\nu}_{abs}(\max) - \bar{\nu}_p(\max)$ which is even more of an overestimate. Clearly neither of these methods is satisfactory, and it is not known how much reliance can be placed on these present values. Furthermore since the possible uncertainty in the triplet levels is of the order 1000–2000 $cm^{-1}$ and is comparable to the spread of values in Table 12, even the order of triplet energies presently suggested may be incorrect.

It would be desirable, though undoubtedly difficult, to obtain singlet–triplet absorption spectra by a perturbation technique or a phosphorescence excitation method. The small magnitudes of the S–T splittings raise some problems if the intersystem crossing process is a $^1\pi^* \rightarrow {}^3\pi^*$.

The $(\pi\pi^*)$ nature of the phosphorescent state has been shown by two different techniques. The polarization of the Ade emission has been found to be negative (Fig. 23), although the magnitude of the

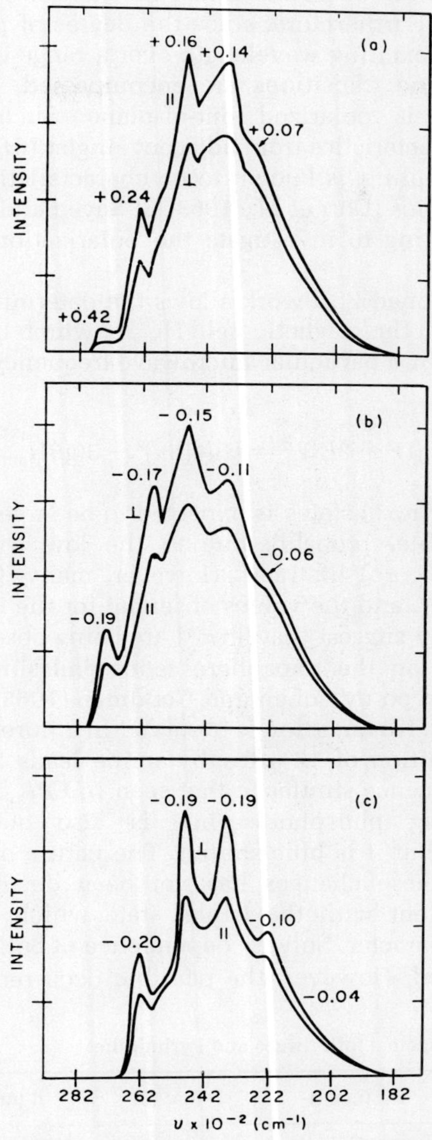


**Fig. 23.** *Polarized phosphorescence emission spectra in EPA at 77°K of (a) purine with respect to excitation at 300 nm, (b) adenine with respect to excitation at 280 nm, and (c) n-butyladenine with respect to excitation at 280 nm, corrected for instrumental and scatter effects. The degree of polarization is shown for the principal peaks (Cohen and Goodman, 1965).*

degree of polarization decreases across the emission band (Cohen and Goodman, 1965). Furthermore, since the degree of polarization is independent of the exciting wavelength over a range in which two perpendicular in-plane transitions are encompassed, this implies that phosphorescence is polarized out-of-plane. Such a polarization, acquiring its characteristics from adjacent singlet $(n\pi^*)$ or $(\sigma\pi^*)$ states by spin–orbit coupling, is known to be characteristic of $^3(\pi\pi^*)$ states of N-heteroaromatics (Dörr *et al.*, 1963; El-Sayed and Brewer, 1963). It would be interesting to investigate the polarization behavior of the other bases.

The aforementioned ESR work allows the determination of the ZFS parameter $\mathbf{D}^*$ from the magnetic field $H_{\text{min}}$ at which the $\Delta m = 2$ transition is observed for a particular microwave frequency, using the equation

$$\mathbf{D}^* = (D^2 + 3E^2)^{1/2} = [3/4(h\nu)^2 - 3(g\beta H_{\text{min}})^2]^{1/2}$$

Although $\mathbf{D}^*$ for $^3(n\pi^*)$ states is expected to be large, not much information is available, probably due to the low triplet populations caused by short $^3(n\pi^*)$ lifetimes. However, many $^3(\pi\pi^*)$ states have small values of $\mathbf{D}^*$, and the values observed for the Pur and Pyr bases (Table 13) strongly suggest that $^3(\pi\pi^*)$ are being observed.

Solvent effects on the phosphorescence emission spectrum have been clearly observed by Cohen and Goodman (1965) in the case of 9-*n*-BuAde (Fig. 16). No emission is observed in a pure hydrocarbon solvent, but the addition of 10% triethylamine leads to a strong structured phosphorescence similar to that seen in EPA but red-shifted by 400 $cm^{-1}$. Strong phosphorescence is also obtained in *n*-butanol/isopentane but it is blue-shifted. The nature of the interactions responsible for these changes has not been determined, but they would be consistent with the triplet state which has considerable charge-transfer character. Solvent dependence of both $^3E(0\text{–}0')$ and $k_{\text{isc}}$ may be implicated. However, the possible occurrence of biphotonic

**Table 13** ZFS[a] Parameters for Purines and Pyrimidines

| Compound | $\mathbf{D}^*$ ($cm^{-1}$) | $D$ ($cm^{-1}$) | $E$ ($cm^{-1}$) | pH |
|---|---|---|---|---|
| AMP | 0.126 | 0.121 | 0.027 | 7 |
| GMP | 0.145 | 0.141 | 0.017 | 7 |
| Thy | 0.198 | — | — | 7 |
| Thd | 0.198 | — | — | 7 |
| 5-MeCyt | 0.192 | — | — | <12 |

[a] Zero field splitting.

effects (Section D,1,c) in these experiments must be investigated so that genuine solvent interactions can be clearly delineated. It would be interesting to know if similar solvent effects occur on the triplet states of the other Pur and Pyr bases.

*c. Biphotonic Effects*

The behavior of the singlet and triplet states presented so far has been analyzed on the assumption that all excitation processes are first order in light intensity and that the simple sequence of Fig. 1 applies. Delayed fluorescence following bimolecular triplet recombination (annihilation) has not been observed in these systems. However, there is clear evidence that biphotonic processes are important at 77°K. Hélène *et al.* (1966) first found that for broad-band excitation ($\lambda > 250$ nm) of Ado and Gua in ethanol the intensity of phosphorescence decreases as a function of time of excitation, and the rate of decrease depends on the intensity of radiation to the power 1.6. Fluorescence is not affected. Studies by ESR over the same time range showed that a singlet was produced which could be bleached by visible light. With increasing concentration of ethanol, the spectrum of the $\alpha$-ethanol radical was superimposed on the singlet, and in pure ethanol only the ethanol radical (which does not photobleach) was seen. At low intensities the rate of radical production was proportional to $I^2$, and some typical results (for Pur) are shown at higher intensities in Fig. 24. It

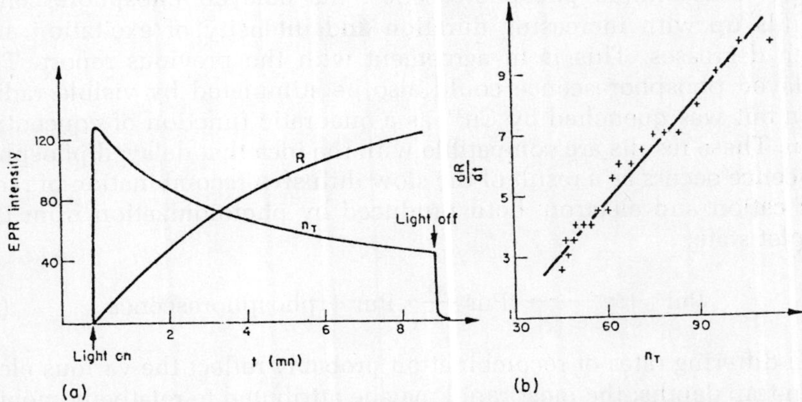


**Fig. 24.** *(a) Variation, with time of irradiation, of the concentrations in purine triplet-state molecules ($n_T$) and ethanol radicals (R) studied by EPR at a high incident radiation level. (b) Rate of ethanol radical formation dR/dt vs. triplet-state molecule concentration ($n_T$) (deduced from the curves in a). Solvent: ethanol. Purine concentration: 1 mM. Similar curves were obtained with guanosine in ethanolic solutions. The triplet-state population was too low in the case of adenosine to allow such quantitative measurements (Hélène et al., 1966).*

seems clear that the processes leading to lowering of the triplet yield and to the formation of ethanol radicals are competitive. Much weaker effects were observed with the pyrimidines.

The authors interpreted their results by supposing the second photon to be absorbed by the triplet state of the Pur, causing photoionization:

$$^{3}\text{Pur} \xrightarrow{h\nu_2} \text{Pur}^{+} + e^{-} \tag{a}$$

Direct evidence for this step was not obtained; that is, the excitation spectrum for the second photon was not determined (it should coincide with the T–T absorption spectrum), and although the authors suggested that the ESR singlet (observed in the absence of ethanol) can be attributed to both $\text{Pur}^{+}$ and $e^{-}$, this has not been definitely established. No absorption spectra attributable to $\text{Pur}^{+}$, $e^{-}$, or the electron adduct $\text{Pur}^{-}$ were observable. However, later work extends the observations of biphotonic phenomena and strengthens the interpretation in terms of photoionization.

Guermanprez *et al.* (1967) reported the observation of delayed emissions from Gua at 77°K, the major part occurring during a period of ~40 sec after termination of excitation and a slower one from 40 to ~$10^{3}$ sec. No spectral analysis was given, but the emission was considered to be phosphorescence on the grounds of instrumental sensitivity. Like simple phosphorescence, the delayed phosphorescence builds up with increasing duration and intensity of excitation and then decreases. This is in agreement with the previous report. The delayed phosphorescence could also be stimulated by visible radiation but was quenched by $Cu^{2+}$ as a quadratic function of concentration. These results are compatible with the idea that delayed phosphorescence occurs as a result of the slow diffusive recombination of radical cation and electron, both produced by photoionization from the triplet state:

$$\text{Pur}^{+} + e^{-} \longrightarrow {}^{3}\text{Pur} \longrightarrow \text{Pur} + \text{phosphorescence} \tag{b}$$

The differing rates of recombination probably reflect the various electron trap depths; the most rapid may be attributed to relatively mobile electrons or to those located not too far from the cation, whereas the very slow luminescence may reflect those located out of range of the coulombic field of the cation but capable of thermal activation at 77°K. Optical stimulation then activates electrons from deeper traps.

The thermoluminescence reported by Guermanprez *et al.* has been

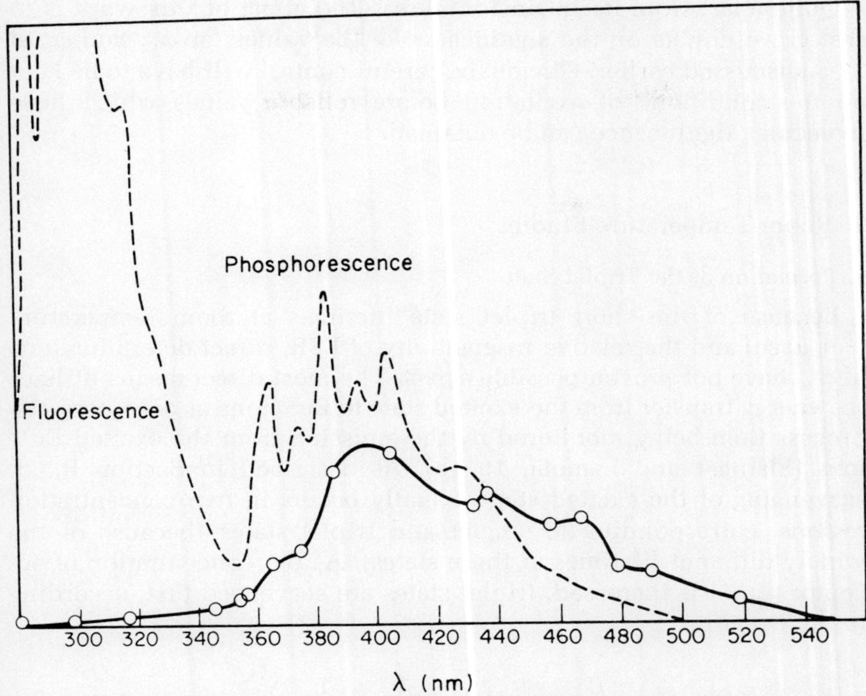


**Fig. 25.** *Emission spectrum and UV thermoluminescence from adenine 1* mM *in EG/$H_2O$ (2:1) (solid line, thermoluminescence, dashed line, fluorescence and phosphorescence) (Brynda, 1971.)*

investigated in detail by Brynda (1971). Using monochromatic excitation (254 nm) and solutions of Ado and Guo in EG/$H_2O$ (2:1), Brynda found two emission peaks at 95°K and 135°K which are indicative of two characteristic trap depths. At low concentrations (50 $\mu M$), low intensities, and short excitation times (so that the number of vacant traps always exceeded the number of trapped electrons) a good quadratic dependence of the total emission on the light intensity was observed, and from the build-up of total emission with time, Brynda showed that the lifetime of the state absorbing the second photon is $2.3 \pm 0.45$ sec for Ade. Comparison with the phosphorescence lifetime of 2.2 sec establishes the intermediate state as the triplet. Furthermore, he showed that the thermoluminescence emission spectrum is quite similar to the phosphorescence spectrum, indicating that the emitting state is the triplet (Fig. 25). Consequently, although direct evidence for $Pur^+$ and $e_{tr}^-$ is still lacking, all experimental evidence is consistent with the sequence (a) and (b).

Quite apart from its intrinsic interest, the effect of this work is to cast grave doubts on the significance of the values for $\phi_{isc}$ and $\phi_p$ at 77°K discussed earlier. Obviously, careful control will have to be kept on the conditions of excitation before reliable values which have molecular significance can be obtained.

## 2. Room-Temperature Studies

### *a. Formation of the Triplet State*

Because of the short triplet state lifetimes at room temperature ($\sim 1\ \mu$sec) and the relative insensitivity of ESR, direct determinations of $\phi_{isc}$ have not proven possible as yet. The most direct means utilizes the energy transfer from the excited state to $Eu^{3+}$ ions as acceptors, the process then being monitored by the emission from the excited $Eu^{3+}$ ions (Eisinger and Lamola, 1971b). As described in Section B,3,d, scavenging of the excited states usually occurs in two concentration regions, corresponding to singlet and triplet states (because of the widely different lifetimes of these states). As the concentration of acceptor ($Eu^{3+}$) is increased, triplet states are scavenged first, according to the equation

$$\frac{\beta}{\phi} = \frac{1}{\phi_{isc}} \left(1 + \frac{1}{{}^3k_t{}^3\tau} \frac{1}{[Eu^{3+}]}\right)$$

where the symbols are as stated earlier, ${}^3k_t$ being the rate constant for triplet energy transfer from the excited base to $Eu^{3+}$ and ${}^3\tau$ the lifetime of the base triplet under the particular experimental conditions. Graphing $\beta/\phi$ as a function of $1/[Eu^{3+}]$ then allows the determination of $\phi_{isc}$ by extrapolation. Calibration is against acetophenone for which $\phi_{isc} = 1.0$. Results obtained for the nucleotides by this method are presented in Table 14.

Intersystem crossing yields have also been estimated for some pyrimidines by chemical methods; indeed, this was the first method by which $\phi_{isc}$ was estimated in aqueous solution (Brown and Johns, 1968). The method was based on the evidence, reviewed in detail in Chapter 5, that in dilute solution the well-known dimerization of pyrimidines occurs as a triplet state reaction,

$$\text{Pyr} \xrightarrow{h\nu} {}^3\text{Pyr}\ (\phi_{isc})$$

$${}^3\text{Pyr} + \text{Pyr} \xrightarrow{k_1} \text{Pyr}\diamond\text{Pyr}$$

**Table 14** Intersystem Crossing Yields, $\phi_{isc}$, at 300°K[a]

| Compound | Photochemical kinetics[b] | Energy transfer to $Eu^{3+}$ [c] | *Cis-trans* isomerization (in acetonitrile)[d] |
|---|---|---|---|
| Thy | $8 \times 10^{-4}$ (254); $4.7 \times 10^{-4}$ (265) | — | 0.18 (254) |
| Thd | $5.6 \times 10^{-4}$ (265) | — | 0.12 (254) |
| TMP | — | $8 \times 10^{-3}$ (265) | — |
| Ura | $1.6 \times 10^{-2}$ (254)*; $8.3 \times 10^{-3}$ (254) | — | — |
| Urd | $5.0 \times 10^{-3}$ (265) | — | 0.40 (254) |
| UMP | — | $7.3 \times 10^{-3}$ (265) | 0.30 (254) |
| Cyt | — | — | — |
| Cyd | — | $1.5 \times 10^{-3}$ (265) | — |
| CMP | — | — | — |
| Ade | — | — | — |
| Ado | — | — | — |
| AMP | — | $3.7 \times 10^{-4}$ (265) | — |
| Gua | — | — | — |
| Guo | — | — | — |
| GMP | — | $4.6 \times 10^{-4}$ (265) | — |

[a] $\lambda_{excit}$ values are given in parentheses.
[b] All values represent $\rho\phi_{isc}$ (where $\rho \leqslant 1.0$, see text), except*. Data sources: Thy and Thd, Fisher and Johns (1970); Ura, Brown and Johns (1968);* Jellineck and Johns (1970).
[c] From Lamola and Eisinger (1971).
[d] From Lamola and Mittal (1966); relative to $\phi_{isc}$(triphenylene) = 0.95; $\phi_{isc}$(1,3-$Me_2$Thy) = 0.02; $\phi_{isc}$(1,3-$Me_2$Ura) = 0.08.

which can be inhibited by oxygen quenching of the triplet state:

$$^3\text{Pyr} + O_2 \xrightarrow{k_2} \text{Pyr}$$

First-order decay of the triplet

$$^3\text{Pyr} \xrightarrow{k_3} \text{Pyr}$$

is included to complete the scheme. Brown and Johns demonstrated that the quantum yields of dimer, determined as functions of oxygen and Pyr concentrations, lead to values of $\phi_{isc}$ by extrapolation to zero oxygen and infinite Pyr concentrations. However, it was demonstrated shortly thereafter that in acetonitrile solution there is an additional pathway for triplet deactivation which is not measurable by the dimer formation method. Wagner and Bucheck (1968) measured the dimer formation and determined triplet yields independently by isomerization of *cis*-1,3-pentadiene, thus extending the original work of Lamola and Mittal (1966). In this way, considerable discrepancies were observed between $\phi_{isc}$ and $\phi_{dimer}$. Since dimers were the only product, this could only be accounted for if the ground-state Pyr could quench the triplet without forming dimer. This process can be written (although little is known of the mechanism) as

$$^{3}\text{Pyr} + \text{Pyr} \xrightarrow{k_1'} 2\ \text{Pyr}$$

so that the limiting yields determined by the method of Johns really represent $\phi_{isc}[k_1/(k_1 + k_1')] = \rho\phi_{isc}$ in which $\rho$ is the probability that quenching of the triplet by ground-state Pyr leads to stable dimer. Such values are shown in Table 14. It has been possible to evaluate $\rho$ directly for aqueous solution in only one case. Jellinek and Johns (1970) investigated the photoinduced reaction between Ura and cysteine and showed that the formation of Cys(S-5)hUra and of hUra is competitive with dimer formation. The following reaction scheme

$$\text{Pyr} \longrightarrow {}^{3}\text{Pyr} \quad (\phi_{isc})$$

$$^{3}\text{Pyr} \xrightarrow{k_1} \text{Pyr}$$

$$^{3}\text{Pyr} + \text{Pyr} \xrightarrow{k_2} \text{Pyr}_2{}^{*}$$

$$\text{Pyr}_2{}^{*} \xrightarrow{k_3} \text{Pyr}_2 \qquad \text{Pyr}_2{}^{*} \xrightarrow{k_4} 2\text{Pyr}$$

$$^{3}\text{Pyr} + \text{RSH} \xrightarrow{k_5} \text{Pyr—H} + \text{RS}\cdot$$

$$\text{Pyr—H} + \text{RSH} \xrightarrow{k_6} \text{hPyr—H}_2 + \text{RS}\cdot$$

leads to the equation

$$\frac{1}{\phi(\text{hPyr})} = \frac{1}{\phi_{isc}} + \frac{k_1}{\phi_{isc}k_5(\text{RSH})} + \frac{k_2(\text{Pyr})}{\phi_{isc}k_5(\text{RSH})}$$

In the limit of high [RSH], dimer formation ($k_2$) is eliminated and $\phi_{isc}$ is obtained from the yield of hUra. By simultaneously measuring dimer formation at other concentrations of RSH, $\rho$ can be determined from

$$\frac{\phi(\text{hPyr}_2)}{\phi(\text{Pyr}_2)} = \frac{k_5(\text{RSH})}{k_2(\text{Pyr})} \cdot \frac{1}{\rho}$$

In this way $\rho$ has been determined to be 0.56, and although many other estimates have been suggested in the literature, this seems to be the only experimentally determined value of $\rho$. It does, however, depend on the completeness of the reaction mechanism.

The many deficiencies in Table 14 are apparent. The only complete set of data is for the nucleotides and these values must be approached with caution. First, it should be noted that all values are determined in $D_2O$, and, as has been mentioned, there are significant deuterium solvent effects in the photochemistry of the singlet state (photohydration), of the triplet state (dimerization in dilute solution), and in the fluorescence quantum yields at room temperature. Accordingly, it would not be unreasonable to expect a deuterium effect on $\phi_{isc}$. Second, the experiments were carried out at pD ~ 5, and in the case of three of the solutes (AMP, GMP, and CMP) this is uncomfortably close to the ground state pK. Clearly investigation of p(H,D) effects is needed which may yield values of $\phi_{isc}$ for the protonated and the neutral forms of the bases, as well as elucidate the H/D effect on $\phi_{isc}$ of the neutral and acid forms, and also lead to values of pK($^3$Base).

A major feature of $\phi_{isc}$ for the pyrimidines was the discovery by Brown and Johns that $\rho\phi_{isc}$ for Ura was strongly dependent on excitation wavelength in the first absorption band, increasing by a factor of ~ 13 from 280 to 240 nm. Similar behavior has been found for Oro (both in neutral and ionized forms) and for Thy, and the effect has been confirmed for Oro and TMP by Lamola and Eisinger (1971) using the $Eu^{3+}$ transfer technique. This is the corollary to the wavelength dependence of the fluorescence quantum yields noted in Section C,3,b, and any satisfactory explanation of this behavior must be applicable to both the fluorescence and the triplet yield results.

The currently favored explanation is based on the proposition that intersystem crossing can occur from the higher vibrational levels of the first excited singlet state as well as from the lowest level (Fig. 26).

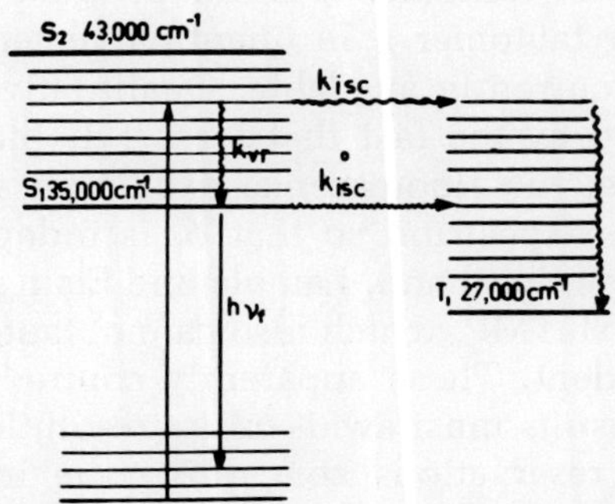


**Fig. 26.** *Energy levels and rate constants for a model of triplet level population in which intersystem crossing is competitive with vibrational relaxation.*

Denoting the triplet yield as $\phi_{isc}$ when excitation energy exceeds the 0–0′ level and the yield from the 0–0′ level as $\phi_{isc}(0)$, it is easy to see that

$$\begin{aligned}\phi_{isc} &= \left(\frac{k'_{isc}}{k'_{isc}+k_{vr}}\right) + \left(\frac{k_{vr}}{k'_{isc}+k_{vr}}\right)\phi_{isc}(0)\\ &= f'_{isc} + (1 - f'_{isc})\phi_{isc}(0)\end{aligned}$$

in which $f'_{isc}$ is the probability of intersystem crossing at the higher energy. Similarly, for the same molecule, the fluorescence yield may be written

$$\phi'_f = \left(\frac{k_{vr}}{k'_{isc}+k_{vr}}\right)\phi_f(0) = (1 - f'_{isc})\,\phi_f(0)$$

and the intersystem crossing yields and the fluorescence yields should thus be related through the common factor $f'_{isc}$. Results sufficiently detailed to allow the evaluation of $f'_{isc}$ are only available for Ura; from $\phi_{isc}$ data for Ura, $f'_{isc} \cong 0.6 \times 10^{-2}$ whereas from $\phi_f$ data, $f'_{isc} \cong 0.7$. This discrepancy casts considerable uncertainty on the validity of the mechanism. An alternative explanation is based on the suggestion that the fluorescence and triplet yields do not pertain to the same molecular species, i.e., that the bases exist in aqueous solution in two tautomeric forms, one of which has a higher $k_{isc}$ than the other. Accordingly, its $\phi_{isc}$ will be higher and its $\phi_f$ will be lower than the other tautomer. Detailed numerical considerations presented elsewhere (Daniels, 1972) show that on this basis, the triplet formation cross section $\sigma_t$ reported by Brown and Johns, the fluorescence excitation spectrum of Daniels and Hauswirth, and the experimental UV absorption spectrum are mutually consistent. As a consequence, it can be shown that $\phi_{isc}$ for tautomer II is independent of excitation wavelength, as is $\phi_f$ for tautomer I. In quantitative terms, the tautomeric model is the best currently available. Qualititatively, the tautomeric model is supported by the fact that for 2,4-diethoxythymine, which cannot tautomerize, the fluorescence excitation spectrum coincides with the absorption spectrum so that $\phi_f$ is independent of exciting wavelength. On the other hand, Lamola and Eisinger reported that for 1,3-$Me_2$Thy and 3-MeTMP, which also cannot tautomerize, $\phi_{isc}$ is still wavelength dependent. These apparently contradictory fluorescence and triplet yield results must await future resolution.

In view of the reservations concerning the intersystem crossing yields, only a preliminary discussion can be given about two impor-

**Table 15** Intersystem Crossing Parameters for the Nucleotides

| Compound | $\phi_{isc}/\phi_f$ (77°K)[a,b] | $\phi_{isc}/\phi_f$ (300°K)[a,b] | $k_{isc}(300)/k_{isc}$ (77°K)[c,d] | $E^{\ddagger}_{isc}$ (kcal/mole) |
|---|---|---|---|---|
| AMP | 4 | $1.2 \times 10^2$ | $0.3 \times 10^2$ (<30) | 0.5 |
| GMP | 1.5 | $2.3 \times 10^2$ | $1.5 \times 10^2$ (1.3) | 0.8 |
| CMP | 0.6 | $0.5 \times 10^2$ | $1 \times 10^2$ (~5) | 0.7 |
| UMP | <0.3 | $7 \times 10^2$ | $>2 \times 10^3$ (>30) | 1.2 |
| TMP | $<1 \times 10^{-2}$ | $2 \times 10^2$ | $>2 \times 10^4$ (>530) | 1.6 |

[a] $\phi_{isc}$ at 77°K and 300°K from Lamola and Eisinger (1971).
[b] $\phi_f$ at 77°K from Table 4 and at 300°K from Table 7.
[c] Expressing $k_{isc}$ as $k_{isc} = Ae^{-E^{\ddagger}/RT}$ gives log $A \cong 10.3$–11.8.
[d] Values in parentheses are from Lamola and Eisinger (1971) (see text).

tant features, namely, the effect of temperature on $k_{isc}$ (the intersystem crossing rate constant) and the effect of solvent on $k_{isc}$ and $k_{ic}$ at room temperature. Table 15 shows the ratios $\phi_{isc}/\phi_f$ for all the nucleotides at two temperatures. From these two ratios may be obtained $k_{isc}(300)/k_{isc}(77)$, column 4; $k_f$ is independent of temperature. For comparison, the values calculated by Lamola and Eisinger (1971) are included in parentheses. These values were obtained using the relation $k_{isc} = \phi_{isc}/{}^1\tau$, the singlet lifetimes being measured directly at 77°K (except for UMP and CMP) and the room temperature lifetimes being estimated by the $Eu^{3+}$ method as described in Section C. The differences between the two sets of calculations is a measure of the unsatisfactory state of quantitative determinations in this field. However, there does seem to be agreement on an order of magnitude difference between the temperature coefficients of TMP and UMP and the other nucleotides. Representing the rate constants by exponential activation functions leads to the activation energies in column 5. If $\phi_{isc}$ were to undergo most of its change between 150°K and 300°K, as does $\phi_f$ for TMP, then the activation energies would range from 2 to 6 kcal/mole, to be compared with the 3 kcal/mole estimated by Lamola and Eisinger from fluorescence quenching. The limited data do not allow us to decide if increased intersystem crossing is the process responsible for fluorescence quenching.

Finally, attention is drawn to the considerably higher triplet yields for Thy and Ura found in acetonitrile solution (Table 14), which contrast with the much smaller effect of acetonitrile on the fluorescence of Thy. Since Lamola and Mittal found that no photochemistry occurs from the singlet level of Thy in acetonitrile, $\phi_f$, $\phi_{isc}$, and ${}^1\tau_0$ constitute a complete set of parameters. Therefore, following the procedure outlined in Section C, these values are obtained: $k_{isc} = 1 \times 10^{10}$ sec$^{-1}$ and $k_{ic} = 4 \times 10^{10}$ sec$^{-1}$. When compared with the aqueous solution data in Table 10, $k_{isc}$ seems to be increased by an order of magnitude and $k_{ic}$ to be decreased correspondingly. Only highly speculative suggestions can be made about the mechanisms of such effects and the data are subject to the uncertainty that $\phi_{isc}$ and $\phi_f$ may both be functions of exciting wavelength in acetonitrile, as in aqueous solution.

*b. Triplet State Properties at 300°K*

The major uses of phosphorescence emission are to allow estimation of triplet energy levels, triplet lifetimes, and vibrational structure. Phosphorescence emission is not observable at room temperature, nor can ESR measurements presently be carried out. What little is known of triplet state properties has come through transient absorption spectroscopy (flash photolysis).

The earliest searches for transient species after flash excitation using discharges of 1300 J with an effective resolution time of ~75 μsec gave negative results for Thy (Daniels and Stott, 1968), with the long-lived transients from Ura not behaving as triplets. Successful identification of the Ura, and later Thy, triplets in aqueous solution came with the development of an apparatus discharging ~300 J in a flash of half-width ~1 μsec (Whillans and Johns, 1971). The earlier technical problems were due to a combination of low quantum yield, short lifetime decreased by diffusion-controlled self-quenching, and probably to a significant amount of biphotonic photolysis at the higher flash energies. Since this work, which gives important information on the reactivities of the Pyr triplet states, is presented in detail in Chapter 5, only a brief account of certain aspects is given here to complete the survey of excited states of the bases.

The identification of the observed transients as triplet states is indirect but convincing. The transient species undergo first-order decay and are self-quenched and oxygen quenched with rate constants which agree with those determined from steady-state photokinetics for dimer formation in dilute solution, which in turn is considered to proceed through the triplet state. In addition, it has been possible to sensitize the transient formation from Thy by first exciting acetone, presumably to its triplet state. The Thy transient kinetics were then identical to those due to direct excitation and the process in the presence of acetone is ascribed to triplet–triplet transfer. Acetone sensitization has also been demonstrated for Thy in acetonitrile solution (Szabo *et al.*, 1970) and effective quenching by 2,4-hexadien-1-ol, a triplet state quencher, has been found.

An interesting method of populating the triplet state at room temperature is triplet–triplet transfer following pulsed 2.3 MeV electron excitation (Hayon, 1969). Electron excitation of organic liquids causes formation of excited states of the solvent which can then transfer their energy to appropriate solutes; e.g., naphthalene triplets can be observed in acetonitrile in this way. Hayon reported transient absorption spectra for Ura in acetonitrile, DMF, and isopropanol (Fig. 27a), which are essentially identical with $\lambda_{max}$ at 289 nm. The absorbances are proportional to the yield of naphthalene triplets in the same solvent; for Thy $\lambda_{max}$ is $\sim 300$ nm. These spectra seem to be significantly different from those reported for aqueous solution (Fig. 27b) particularly in the absence of the peaks at 340 nm (Thy) and 370 nm (Ura).

The absorbance scales of Fig. 27 are relative. The only reported determination of an extinction coefficient has come from the pulse electron excitation. Assuming that transfers from the solvent are equal, comparing the absorption by triplet Ura with that of naphthalene under similar conditions gives the extinction coefficient of $^3$Ura at 289 nm as $\geq 91{,}000$ relative to a value of 22,600 $M^{-1}$cm$^{-1}$ for the naphthalene triplet. Such a high value, if confirmed and extended to other bases, emphasizes the latent possibilities of biphotonic effects through triplet state absorption, especially in flash photolysis studies but possibly also under certain circumstances during continuous irradiation experiments. Further developments in this area may be expected.

## E. Future Work

Despite the extensive and intensive work of the past decade on the excited states of the nucleic acid bases, it will be readily apparent that our knowledge is far from satisfactory and that until it is improved our understanding of the behavior of the dinucleotides, polynucleotides, and nucleic acids themselves will be resting on insecure foundations. The following is a minimal list of topics to be investigated and of questions which must be answered concerning the behavior of the excited states of the bases. Other questions will no doubt occur to the reader as, it is hoped, will new ideas on how to tackle these problems.

1. Singlet States at 77°K. Precise determinations of $\phi_f$ are needed as a function of exciting wavelength. More extensive lifetime determinations are needed, and quantum yields and lifetimes must be correlated with each other and with the absorption spectra.

2. Triplet States at 77°K. Determination of $\phi_{isc}$ by ESR as a function of exciting wavelength is needed. Concurrent processes of bipho-

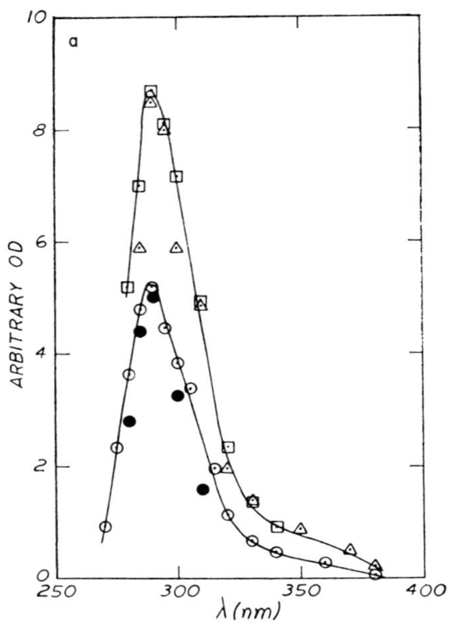


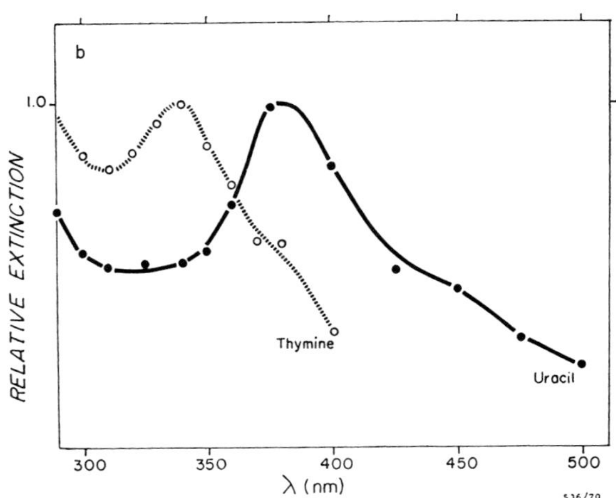


**Fig. 27.** *(a) Transient optical spectra attributed to T–T absorption obtained on pulse radiolysis of oxygen-free solutions of uracil in acetonitrile ,○, 0.2 mM), in dimethylformamide (□, 0.7 mM), and in isopropyl alcohol (△, 0.2 mM). The OD was measured at ±0.2 μsec after a 30-nsec pulse of electrons (dose per pulse ~28 krads). Solid circles (●) were obtained from flash photolysis of oxygen-free uracil in acetonitrile 0.02 mM), OD normalized to pulse data (Hayon, 1969). (b) T–T absorption spectra obtained by flash photolysis of aqueous solutions (Whillans and Johns, 1971).*

tonic photoionization and recombination must be monitored by T–T absorption and photoconductivity measurements.

3. Singlet States at Room Temperature. More extensive investigation of the wavelength dependence of $\phi_f$ is called for, particularly in view of the recent interpretation in terms of tautomerism. Solvent studies will be important for evaluating excited-state dipole moments and for clarifying the role of solvent on the radiative and nonradiative processes.

4. Triplet States at Room Temperature. The wavelength dependence of $\phi_{isc}$ has only been observed for the pyrimidines. If this is a general effect, is should also be found for the purines. The determination of $\phi_{isc}$ is needed in various solvents, and the wavelength dependence in these solvents should be investigated also.

## References

Albrecht, A. C. (1970). *Prog. React. Kinet.* **5,** 301.

Aleksandrov, I. V., and Pukhov, K. K. (1964). *Opt. Spectros. (USSR)* **17,** 513.

Anex, B. G., Fucaloro, A. F., and Dutta-Ahmed, A. (1975). *J. Phys. Chem.* **79,** 2636.

Atkins, P. W., McLauchlan, K. A., and Simpson, A. F. (1970). *J. Phys. E.* **3,** 547.

Baba, H., Goodman, L., and Valenti, P. C. (1966). *J. Amer. Chem. Soc.* **88,** 5410.

Barrow, G. M. (1962). "Introduction to Molecular Spectroscopy," Chapter IV. McGraw-Hill, New York.

Basu, S. (1964). *Advan. Quantum Chem.* **1,** 145.

Berens, K., and Wierzchowski, K. L. (1969). *Photochem. & Photobiol.* **9,** 433.

Berlman, I. B. (1965). "Handbook of Fluorescence Spectra of Aromatic Molecules." Academic Press, New York.

Bersohn, R., and Isenberg, I. (1963). *Biochem. Biophys. Res. Commun.* **13,** 205.

Bersohn, R., and Isenberg, I. (1964). *J. Chem. Phys.* **40,** 3175.

Birks, J. B. (1970). "Photophysics of Aromatic Molecules," Wiley (Interscience), New York.

Börresen, H. C. (1963). *Acta Chem. Scand.* **17,** 921.

Börresen, H. C. (1965a). *Acta Chem. Scand.* **19,** 2089.

Börresen, H. C. (1965b). *Acta Chem. Scand.* **19,** 2100.

Blumberg, W. E., Eisinger, J., and Navon, G. (1968). *Biophys. J.* **8,** A-106.

Brown, I. H., and Johns, H. E. (1968). *Photochem. & Photobiol.* **8,** 273.

Brynda, E. (1971). *Photochem. & Photobiol.* **14,** 713.

Bunville, L. G. and Schwalbe, S. J. (1966) *Biochem.* **5,** 3521.

Callis, P. R., and Simpson, W. T. (1970). *J. Amer. Chem. Soc.* **92,** 3593.

Callis, P. R., Rosa, E. J., and Simpson, W. T. (1964). *J. Amer. Chem. Soc.* **86,** 2292.

Callis, P. R., Franconi, B., and Simpson, W. T. (1971). *J. Amer. Chem. Soc.* **93,** 6697.

Carrington, A., and McLachlan, A. D. (1967). "Introduction to Magnetic Resonance." Harper, New York.

Chen, H. H., and Clark, L. B. (1969). *J. Chem. Phys.* **51,** 1862.

Chen, H. H., and Clark, L. B. (1973). *J. Chem. Phys.* **58,** 2593.

Clark, L. B., and Tinoco, I., Jr. (1965). *J. Amer. Chem. Soc.* **87,** 11.

Clark, L. B., Peschel, G. G., and Tinoco, I., Jr. (1965). *J. Phys. Chem.* **69,** 3615.
Cohen, B. J., and Goodman, L. (1965). *J. Amer. Chem. Soc.* **87,** 5487.
Cohen, B. J., and Goodman, L. (1967). *J. Chem. Phys.* **46,** 713.
Cohen, B. J., Baba, H., and Goodman, L. (1965). *J. Chem. Phys.* **43,** 2902.
Daniels, M. (1972). *Proc. Nat. Acad. Sci. U.S.* **69,** 2488.
Daniels, M. (1973). *In* "Physicochemical Properties of Nucleic Acids" (J. Duchesne, ed.), Vol. 1, Chapter 4. Academic Press, New York.
Daniels, M., and Hauswirth, W. (1971). *Science* **171,** 675.
Daniels, M., and Stott, D. A. (1968). *Abstr., Int. Photobiol. Congr., 5th, 1968,* p. 18.
DeVoe, H. (1971). *J. Phys. Chem.* **75,** 1509.
DeVoe, H., and Tinoco, I., Jr. (1962). *J. Mol. Biol.* **4,** 500.
Ditchburn, R. W. (1963). "Light," Vol. II. Wiley (Interscience), New York.
Dörr, F., Gropper, H., and Mika, N. (1963). *Ber. Bunsenges. Phys. Chem.* **67,** 202.
Drobnik, J., and Augenstein, L. (1966a). *Photochem. & Photobiol.* **5,** 13.
Drobnik, J., and Augenstein, L. (1966b). *Photochem. & Photobiol.* **5,** 83.
Dreyfus, M., Dodin, G., Bensaude, O., and DuBois, J. E. (1975). *J. Amer. Chem. Soc.* **97,** 2369.
Eastman, J. W. (1969). *Ber. Bunsenges. Phys. Chem.* **73,** 407.
Eastman, J. W., and Rosa, E. J. (1968). *Photochem. & Photobiol.* **7,** 189.
Eaton, W. A. and Lewis, T. P. (1970). *J. Chem. Phys.* **53,** 2164.
Eisinger, J. (1968). *Photochem. & Photobiol.* **7,** 597.
Eisinger, J., and Lamola, A. A. (1971a). *In* "Excited States of Proteins and Nucleic Acids" (R. F. Steiner and I. Weinryb, eds.), p. 107. Plenum, New York.
Eisinger, J., and Lamola, A. A. (1971b). *Biochim. Biophys. Acta.* **240,** 299.
Eisinger, J., and Shulman, R. G. (1968). *Science* **161,** 1311.
El-Sayed, M. A. (1971). *Accounts Chem. Res.* **4,** 23.
El-Sayed, M. A., and Brewer, R. G. (1963). *J. Chem. Phys.* **34,** 1622.
El-Sayed, M. A., and Robinson, G. W. (1961). *J. Chem. Phys.* **35,** 1897.
Ermolaev, V. L. (1964). *Optika i Spectroskopiya* **16,** 704.
Euler, H., Brandt, K. M., and Neumuller, G. (1935). *Biochem. Z.* **281,** 206.
Eyring, H., Walter, J., and Kimball, G. E. (1964). "Quantum Chemistry," Chapter VIII. Wiley, New York.
Fisher, G. J., and Johns, H. E. (1970). *Photochem. & Photobiol.* **11,** 429.
Fucaloro, A. F., and Forster, L. S. (1971). *J. Amer. Chem. Soc.* **93,** 6443.
Gill, J. E. (1968). *J. Mol. Spectrosc.* **27,** 539.
Guermanprez, R., Hélène, C., and Ptak, M. (1967). *J. Chim. Phys.* **64,** 1376.
Guéron, M., Eisinger, J., and Shulman, R. G. (1967a). *J. Chem. Phys.* **47,** 4077.
Guéron, M., Eisinger, J., and Shulman, R. G. (1967b). *Bull. Amer. Phys. Soc.* [2] **12,** 401.
Guéron, M., Eisinger, J., and Shulman, R. G. (1968). *Mol. Phys.* **14,** 111.
Hauswirth, W., and Daniels, M. (1971a). *Photochem. & Photobiol.* **13,** 157.
Hauswirth, W., and Daniels, M. (1971b). *Chem. Phys. Lett.* **10,** 140.
Hauswirth, W., and Daniels, M. (1973). In preparation.
Hayon, E. (1969). *J. Amer. Chem. Soc.* **91,** 5397.
Hélène, C., Santus, R., and Douzou, P. (1966). *Photochem. & Photobiol.* **5,** 127.
Hochstrasser, R. M. (1966). "Molecular Aspects of Symmetry." Benjamin, New York.
Hønnas, P. I., and Steen, H. B. (1970). *Photochem. & Photobiol.* **11,** 67.
Horrocks, A. R., Medinger, T., and Wilkinson, F. (1967). *Photochem. & Photobiol.* **6,** 21.
Hutchinson, C. A., Jr., and Mangum, B. W. (1958). *J. Chem. Phys.* **29,** 952.

Imakubo, K. (1968). *J. Phys. Soc. Jap.* **24,** 143.
Imakubo, K., Higashimura, T., and Sidei, T. (1967). *J. Phys. Soc. Jap.* **22,** 339.
Innes, K. K., Byrne, J. P., and Ross, I. G. (1967). *J. Mol. Spectrosc.* **22,** 125.
Jellinek, T., and Johns, R. B. (1970). *Photochem. & Photobiol.* **11,** 349.
Kasha, M. (1961). *In* "Light and Life" (W. D. McElroy, and B. Glass, eds,), p. 31. Johns Hopkins Univ. Press, Baltimore, Maryland.
Kearwell, A., and Wilkinson, F. (1969). *In* "Transitions Non-radiatives dans les Molecules," 20th Reunion Soc. Chim. Phys., p. 125.
Kleinwächter, V., Drobnik, J., and Augenstein, L. (1966). *Photochem. & Photobiol.* **5,** 579.
Kleinwächter, V., Drobnik, J., and Augenstein, L. (1967a). *Photochem. & Photobiol.* **6,** 133.
Kleinwächter, V., Drobnik, J., and Augenstein, L. (1967b) *Photochem. & Photobiol.* **6,** 147.
Kramers, H. A. (1927). *Atti Congr. Int. Fisi., Como* **2,** 545.
Krishna, V. G., and Goodman, L. (1962). *J. Chem. Phys.* **36,** 2217.
Kronig, R. de L. (1926). *J. Opt. Soc. Amer.* **12,** 547.
Lamola, A. A., and Eisinger, J. (1971). *Biochim. Biophys. Acta* **240,** 313.
Lamola, A. A., and Mittal, J. P. (1966). *Science* **154,** 1560.
Lewis, T. P., and Eaton, W. A. (1971). *J. Amer. Chem. Soc.* **93,** 2054.
Li, Y. H., and Lim, E. C. (1971). *Chem. Phys. Lett.* **9,** 574.
Lippert, E. (1957). *Z. Elektrochem.* **61,** 692.
Longworth, J., Rahn, R. O., and Shulman, R. G. (1966). *J. Chem. Phys.* **45,** 2930.
McGlynn, S. P., Azumi, T., and Kinoshita, M. (1969). "Molecular Spectroscopy of the Triplet State," Chapter V. Prentice-Hall, Englewood Cliffs, New Jersey.
Magat, M., Ivanoff, N., Lahamani, F., and Pileni, M.-P. (1969). "Transitions Non-radiatives dans les Molecules," 20th Reunion Soc. Chim. Phys., p. 212.
Mason, S. F. (1954). *J. Chem. Soc. London* p. 2071.
Mason, S. F. (1959). *J. Chem. Soc., London* p. 1240.
Mason, S. F. (1962). *In* "The Pyrimidines" (D. J. Brown, ed.), p. 477. Wiley (Interscience), New York.
Mason, S. F. (1963). *In* "Physical Methods in Heterocyclic Chemistry" (A. R. Katritzky, ed.), Vol. 2, p. 1. Academic Press, New York.
Mauret, P., and Fayet, J-P (1967). *C. R. Acad. Sci., Ser. C* **264,** 2081.
Montenay-Garestier, T., and Hélène, C. (1970). *Biochemistry* **9,** 2865.
Morita, H., and Nagakura, S. (1968). *Theor. Chim. Acta* **11,** 279.
Mugiya, C., and Baba, H. (1967). *Bull. Chem. Soc. Jap.* **40,** 2201.
Nishi, N., Shimada, R., and Kanada, Y. (1970). *Bull. Chem. Soc. Jap.* **43,** 41.
Nnadi, J., and Wang, S. Y. (1969). *Tetrahedron Lett.* **27,** 2211.
Seibold, K., and Labhart, H. (1971). *Biopolymers* **10,** 2063.
Shugar, D., and Fox, J. J. (1952). *Biochim. Biophys. Acta* **9,** 199.
Steele, R. H., and Szent-Györgyi, A. (1957). *Proc. Nat. Acad. Sci. U.S.* **43,** 477.
Stewart, R. F., and Davidson, N. (1963). *J. Chem. Phys.* **39,** 255.
Stewart, R. F., and Jensen, L. H. (1964). *J. Chem. Phys.* **40,** 2071.
Szabo, A. G., Riddel, W. D., and Yip, R. W. (1970). *Can. J. Chem.* **48,** 694.
Tinoco, I., Jr., and Holcombe, D. N. (1964). *Annu. Rev. Phys. Chem.* **15,** 371.
Van der Waals, J. H., and de Groot, M. S. (1959). *Mod. Phys.* **2,** 333.
Vigny, P. (1971a). *C. R. Acad. Sci. Ser. D* **272,** 2249.

Vigny, P. (1971b). *C. R. Acad. Sci., Ser. D* **272,** 3206.
Wagner, P. J., and Bucheck, D. J. (1968). *J. Amer. Chem. Soc.* **90,** 6530.
Warshaw, M. M., Bush, C. A., and Tinoco, I., Jr. (1965). *Biochem. Biophys. Res. Commun.* **18,** 633.
Whillans, D. W., and Johns, H. E. (1971). *J. Amer. Chem. Soc.* **93,** 1358.
Wierzchowski, K. L., and Shugar, D. (1957). *Biochim. Biophys. Acta* **25,** 355.

# 3 Excited States of the Nucleic Acids: Polymeric Forms

*William W. Hauswirth and Malcolm Daniels*



For a photochemist or photobiologist, the behavior of the excited states of nucleic acid bases serves primarily to aid in understanding photophysical events occurring in whole DNA. From this viewpoint, the most critical structural feature of DNA, as distinguished from its monomeric units, is its closely stacked and paired base configuration which allows interbase electronic perturbations not possible in dilute monomeric solutions. Since such effects may significantly alter the chemical fate of radiant energy absorbed in DNA, their existence and relative magnitudes are a key link in extrapolating from free base excited-state properties to those of DNA.

Base–base interactions and their influence on electronic processes initiated in the ground state have received a great deal of theoretical and experimental attention in the past eighteen years. Optical methods used for investigating these interactions include hypo- and hyperchromicity, circular and linear dichroism, and optical rotary dispersion. Additionally, nuclear magnetic resonance, x-ray crystallography, and osmotic pressure techniques have been used to estimate the strength and orientation of these interactions, particularly on model oligonucleotides. Although it is beyond the scope of this chapter to review this work extensively, an experimental foundation is provided from which current theories of stacking interactions have been derived and which to some degree have been extended to nucleic acid excited states. Therefore, pertinent results and theories derived from these methods will be selectively discussed in subsequent sections as an introduction to our survey of excited-state properties.

## A. Dinucleoside Phosphates

### 1. Ground-State Optical Properties and Base Stacking Effects

Relative to nucleic acid structures the simplest units in which electronic base–base interactions are possible are the dinucleoside phosphates. In a sufficiently dilute solution other effects commonly observed in polynucleotides, such as hydrogen-bonded base pairing, can be neglected and solvent exclusion is minimal. Stacking of the bases occurs independently of exciting or monitoring radiation and is considered to have its origin in hydrophobic effects. The extent of association correlates with the polarizabilities of the bases (Ts'o, *et al.*, 1963), and temperature studies of the association constants give free energies of interaction of about $-1$ kcal/mole and $\Delta H$ about $-3$ to $-5$ kcal/mole. Theoretical considerations (Sinanŏglu and Abdulnar, 1965) indicate that the free energy of association should be much less in nonaqueous media. When thus brought into a suitable geometry by stacking or association, optical interaction can occur between the bases.

Qualitatively the spectral shape of the low-energy absorption band of dinucleoside phosphates closely resembles an equimolar mixture of its constituent nucleosides, but reduction in the band intensity (hypochromism) has been observed in all sixteen common ribodinucleoside phosphates (Warshaw and Tinoco, 1965, 1966) and in all sixteen common deoxyribodinucleoside phosphates except TpdC (Cantor *et al.*, 1970). Compared to the magnitude of hypochromicity observed in most helical polynucleotides ($\%H \sim 30\%$), these models exhibit considerable hypochromism ($\%H = 1$–$11\%$). The fact that hypochromism is primarily a nearest-neighbor effect has been shown by the experimental and theoretical sequence dependence of DNA hypochromism (Pysh and Richards, 1972). For our purposes, the significance of hypochromism in di- and oligonucleotides arises from the origin of the effect (for a comprehensive review of current theories, see Weissbluth 1971). Qualitatively, if one considers the two bases as possessing independent transition dipoles (i.e., matrix elements of the dipole moment operator between the ground and excited states) the major interaction between dipoles can be characterized by a coulombic dipole–dipole term* (Tinoco, 1960; Rhodes, 1961). Superposition of fields due to polarization of the chromophore by the incident light electric field and dipoles induced in the adjacent chromophore results first in an exciton-type splitting of the absorption band of the monomer into allowed and forbidden components. Wave functions have the form

$$\Psi_{\pm} = \frac{1}{\sqrt{2}}\,(\psi_1{}^{a}\psi_1{}^{0} \pm \psi_1{}^{0}\psi_2{}^{a})$$

giving energies

$$E_{+} = E_{a} - (|\mu_{e}| \cdot |\mu_{g}| + \mu_{t}{}^{2})/r^{3}$$

and

$$E_{-} = E_{a} - (|\mu_{e}| \cdot |\mu_{g}| - \mu_{t}{}^{2})/r^{3}$$

in which $\mu_g$ is the ground-state dipole moment, $\mu_e$ the excited-state dipole moment, and $\mu_t$ the transition dipole moment. The strength of

* This approach ignores specific electronic interaction such as charge-transfer complexes (ground state or excited state) or exciplex formation (excited state) which necessarily involve electronic overlap and hence modify the assumption of independent transition dipoles.

the interaction (the splitting) depends strongly on the interplanar distance $E_+ - E_- = 2\mu_t^2/r^3$. If only one transition is considered in each monomer, then intensity (oscillator strength) is conserved and no hypochromism is observed. But if two (or more) transitions are involved in each monomer, then hypo- and hyperchromism can occur between major absorption bands. For the interplanar distance of the DNA helix (3.4 Å) the magnitude of the exciton splitting is estimated as $\sim 500$ to 1000 $cm^{-1}$. The weak coupling approximation holds, and a consequence of the weakness of the forbidden component and the small splitting is that no spectral resolution is observed, there being only a minor shift of the first absorption band of DNA to the blue. However, intensity changes are strong; the first two bands of DNA at 260 and 210 nm are hypochromic and it is inferred that the "lost" intensity will be found in higher energy transitions in the vacuum UV.

The resonant interactions of transition dipoles which lead to exciton splitting of degenerate energy levels in a dinucleoside phosphate are also manifested phenomenologically in alterations of optical rotatory dispersion and circular dichroism spectra compared to mononucleotides.* For recent comprehensive reviews of theory and experiment, the reader is referred to Ts'o (1974) and Bush (1974) and references therein. Optical rotatory dispersion (ORD) (Warshaw and Tinoco, 1965, 1966) and circular dichroism (CD) (Brahms *et al.*, 1967; Warshaw and Cantor, 1970) of all the ribo- and deoxyribodinucleoside phosphates (Cantor *et al.*, 1970) have been measured. Using exciton theory and $\pi \rightarrow \pi^*$ in-plane-polarized transitions only, Bush and Tinoco (1967) calculated the equal but opposite rotational strengths of the two exciton levels in the ribodinucleoside phosphates and hence theoretical ORD spectra.† Agreement of theoretical spectra with experimental findings is reasonably good in some cases, but generally suffers from lack of sufficient information on mononucleotide transition dipoles‡ (see Chapter 2, Section B) and their relative orientation

* Although the free base is not optically active and the sugar moiety does not absorb above 56 kK (180 nm), ORD and CD curves appear in nucleosides because of interaction between the base transition dipole and the polarizable asymmetric sugar environment.

† In actual analysis (Tinoco, 1968) two contributions to the CD spectrum appear; those due to resonance (excitonic) interaction which give equal and opposite bands (conservative CD) and nonresonance (dispersive) interaction which gives a CD band roughly the shape of the absorption band (nonconservative CD). CD bands are then transformable to ORD using the Kramers-Kronig relationship (Moscowitz, 1957).

‡ In part to avoid this problem Bush and Tinoco (1967) used a scaled monopole–monopole interaction arguing that in a Watson-Crick geometry strict dipole–dipole interaction may not be valid. The monopoles themselves are calculated from first-order wave functions.

in the dinucloside models. More recent molecular orbital approaches apparently suffer from similar problems (see, for example, Bailey, 1972).

In attempting to anticipate possible excited-state interactions, CD and hypochromicity data give useful indications concerning the magnitude of stacking interactions prior to excitation. Using the magnitudes of both hypochromicity and ORD relative to mononucleotides, Warshaw and Tinoco (1965, 1966) ordered ribodinucleoside phosphates into those which were either stacked or unstacked. Although some degree of stacking is apparent in all cases, UpU, UpA, UpC, GpU, and UpU are generally less stacked than the others at pH 7. The tendency to stack generally seems to decrease in the order purine–purine > purine–pyrimidine > pyrimidine–pyrimidine, which is also the order observed for base–base association in aqueous solution (Ts'o *et al.*, 1963). Offsetting this simple tendency is a wide variation in stacking as a function of temperature (Brahms *et al.*, 1967; Cantor *et al.*, 1970). This introduces uncertainties into any interpolation between two temperatures of properties dependent on base stacking.

The strong geometric influence on the electronic interaction is demonstrated by the much larger exciton splitting ($\sim 3600\ cm^{-1}$) observed for $Me_2$Thy molecular pairs when they are produced in a constrained situation by photolysis of the dimer ($Me_2$Thy$\diamond Me_2$Thy) in a rigid glass at 77°K (Eisinger and Lamola, 1969; Lamola and Eisinger, 1968). Assuming the point dipole approximation to be applicable, it is calculated that the interplanar separation is $\sim 2.8$ Å. This point emphasizes the different behavior of excited molecular pairs at different separation distances. Cyclobutyl dimerization is known to take place efficiently from the singlet excited state in concentrated solutions of Thy in which stacking occurs (Lisewski and Wierzchowski, 1969; Fisher and Johns, 1970). Consequently, starting with an internuclear separation of 3.4 Å, the photochemical end result of excitation is an internuclear distance of 1.5 Å. Evaluation of the present status of our knowledge concerning the nature of the intermediate stages and their time dependence is the recurrent theme of this part of the chapter.

### 2. Excited-State Properties: Low-Temperature Studies

#### *a. Fluorescence*

As with the monomers, room-temperature emission from the common dinucleoside phosphates is very weak and in fact has not been detected, although emissions have been observed from concentrated solutions of Ado, Urd, and Cyd in which stacking by association is known to occur (Daniels, 1973). Therefore, except for some fluorescent base analogs, most direct excited-state data have been gathered in glassy solvents at liquid nitrogen temperatures (77°–80°K), where emission quantum yields are several orders of magnitude larger. Figure 1 shows the corrected emission spectra of the sixteen ribodinucleoside phosphates, along with spectra of some of the corresponding alternating copolymers. When compared to the fluorescence

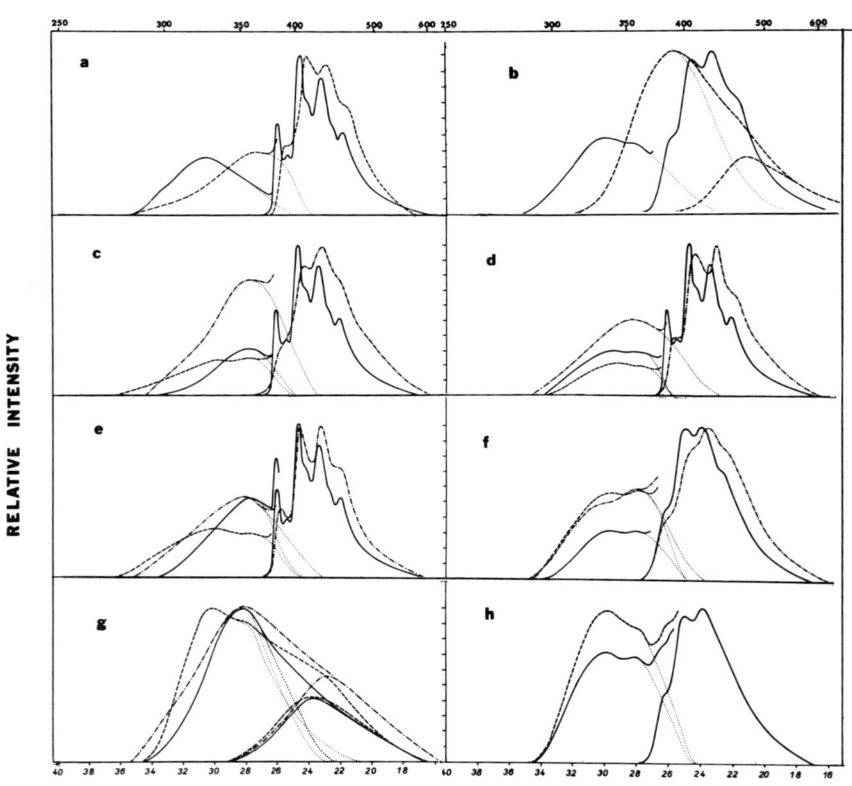


Fig. 1. *77°K luminescence spectra resolved into fluorescence (higher energy) and phosphorescence (lower energy) components. All spectra are in a 0.1* M *phosphate (pH 7) propylene glycol glass (v/v). (From Kleinwächter and Koudelka, 1972.) (a) ApA—; ApA in 0.3M NaCl, ---. (b) GpG—; CpC ---. (c) ApC—; CpA ---; poly(rA-rC)-·-·-. (d) ApG—; GpA ---; poly(rA-rG)-·-·-. (e) ApU—; UpA ---; poly(rA-U)-·-·-. (f) GpC—; CpG ---; poly(rG-rC)-·-·-. (g) CpU—; UpC ---; poly(rC-U)-·-·-. (h) GpU—; UpG ---.*

maximum of the appropriate equimolar mixture of mononucleotides, in all cases (except UpU, which shows no detectable emission) the dinucleoside phosphates show a variable proportion of red-shifted unstructured fluorescence (Table 1; see also Table 7, Chapter 2 for comparison with mononucleotides). This effect has been generally attributed to fluorescence from a complex formed between the excited state of one base and the ground state of the other leading to a lower energy "dimeric" excited state, usually termed an "exciplex."* (Eisinger *et al.*, 1966; Imakubo, 1968; Kleinwächter and Koudelka, 1972).

The theory of exciplex states has been developed primarily for aromatic hydrocarbons (reviewed by Birks, 1970); however, the salient features are applicable to the present case. Basically the excimer interaction potential [$V(\mathrm{r})$] is determined from its wave function formed by configuration interaction between (1) a "charge-resonance state" [$\alpha\psi(A^+)\psi(B^-) + \beta\psi(A^-)\psi(B^+)$] involving ionization potentials and electron affinities of monomers A and B, and coulombic interaction between the complementary molecular ions, and (2) and "exciton-resonance state" [$\gamma\psi(A^*)\psi(B) + \delta\psi(A)\psi(B^*)$] involving transition dipole–dipole interactions as previously discussed. Hence the relative stacking of bases and their abilities to act as charge donors and acceptors are expected to play dominant roles in determining the extent of exciplex formation in a given dinucleoside phosphate. As the distance between chromophores decreases, a repulsive term due to van der Waals forces [$R(\mathrm{r})$] must also be considered in both the ground and excited states. Because the ground state is dissociative at close approach of bases, structured emission due to transitions between vibronic states is lost in the exciplex. The resultant energy [$D(\mathrm{r})$] is then

$$D(\mathrm{r}) = V(\mathrm{r}) + R(\mathrm{r})$$

A schematic potential energy diagram of excimer formation is shown in Fig. 2.

The existence of exciplex emission is normally indicated experimentally if a structureless, red-shifted fluorescence is observed *without* alteration of the absorption spectrum. For exciplex formation between free bases, a concentration dependence is also observable. Most dinucleoside phosphates at 77°K exhibit this fluorescent criterion (Fig. 1 and Table 1). It is unfortunate, however, that low-temperature absorption and fluorescence excitation spectra are not available

* The term "exciplex" refers to excited-state interaction between any two chromophores; the term "excimer" will be reserved for the case of exciplex interaction between identical chromophores.

**Table 1** Low-Temperature Singlet Properties of Some Dinucleoside Phosphates[a]

| Compound | Reference[b] | $\phi_f(\times 10^2)$ | $\epsilon_{max}$ (kK) Monomer band | Excimer band | % Excimer[c] |
|---|---|---|---|---|---|
| ApA | a | 7 | 30.40 | (28.09)[d] | 90[d] |
| | b,c | 4 | 31.5 | — | 0 |
| AMP + GMP | a | 10 | 29.76 | — | — |
| | d | 4 | 30.6 | — | — |
| ApG | a | 4 | 29.41 | 28.09 | 40, 80[d] |
| | d | 8 | 29.9 | — | — |
| GpA | a | 4 | 29.41 | 27.93 | 30, 80[d] |
| | d | 3 | — | 27.8 | 100 |
| AMP + CMP | a | 5 | 31.45 | — | — |
| | d | 2 | 31.3 | — | — |
| ApC | a | 7 | — | 27.90 | 80, 90[d] |
| | d | 6 | — | 27.9 | 100 |
| CpA | a | 3 | 29.85 | 27.80 | 60, 70[d] |
| | d | 4 | 31.3 | 26.3 | 10 |
| dAMP + TMP | d | 6 | 31.1 | — | — |
| dApT | d | 6 | — | 28.2 | 100 |
| TpdA | d | 5 | 30.3 | — | — |
| AMP + UMP | a | 2 | 33.3 | — | — |
| | d | 1 | 32.1 | — | — |
| ApU | a | 3 | — | 27.86 | 80, 90[d] |
| | d | 3 | — | 27.6 | 100 |
| UpA | a | 3 | 29.94 | 27.90 | 30, 50[d] |
| | d | 1 | — | 27.8 | 100 |
| GpG | a | 7 (10)[d] | 30.30 | 28.57 (29.74)[d] | 50, 80[d] |
| GMP + CMP | a | 13 | 30.12 | — | — |
| | e,f | — | 30.3 | — | — |
| GpC | a | 5 | 29.67 | 27.93 | 50, 50[d] |
| | e,f | — | 29.9 | 27.8 | ~50 |
| CpG | a | 9 | 29.59 | 27.93 | 50, 50[d] |
| | b,e,f | — | 30.3 | 27.8 | ~50[d] |
| GMP + UMP | a | 10 | 29.76 | — | — |
| | d | 5 | 30.77 | — | — |
| GpU | a | 7 | 29.76 | 28.33 | 40, 50[d] |
| | d | 1 | 30.10 | 24.69 | 30 |
| UpG | a | 3 | 29.76 | 28.33 | 40, 50[d] |
| CpC | a | 12 (3)[d] | — | 28.09 (29.50)[d] | 100, 100[d] |
| | d | 9 | — | 28.09 | 100 |
| CMP + UMP | a | 3 | 31.06 | — | — |
| CpU | a | 4 | — | 28.40 | 90, 40[d] |
| UpC | a | 2 | 30.30 | 28.01 | 50, 100[d] |
| TpT | d | 14 | 30.1 | — | 0 |

[a] Data in parentheses are from aggregates of the mononucleotides at 77°K (Kleinwächter, 1972a).

[b] References:

a. Kleinwächter and Koudelka (1972); solvent: propylene glycol/0.1 *M* phosphate, pH7 (1:1); spectra corrected.

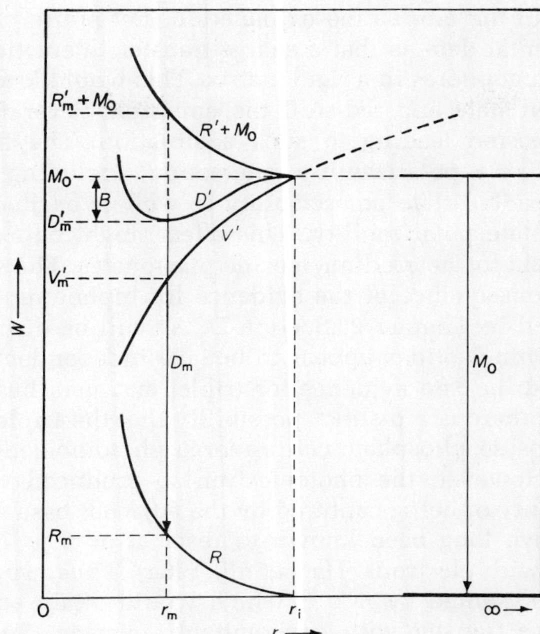


**Fig. 2.** *Schematic potential energy* (W) *diagram of pair of parallel molecules,* $^1M^*$ *and* $^1M$, *as a function of intermolecular separation* (r). R, R′, *repulsive potentials in ground* ($^1M$) *and excited* ($^1M^*$) *states;* V′, *excimer interaction potential;* D′ (= V′ + R′), *resultant excimer energy;* $M_0$, *molecular 0–0′ transition;* $D_m$, *peak excimer transition;* $R_m$, $R_m'$, $V_m'$, $D_m'$, *potentials at equilibrium excimer separation,* $r_m$. (*From Birks, 1970.*)

to test the second criterion. However, the assignment of this red-shifted structureless emission is not entirely unambiguous. Recall that V(r) is a function of two types of interaction, charge transfer and exciton. Usually the later mechanism is proposed to explain exciplex

---

b. Koudelka and Augenstein (1968); solvent: ethylene glycol/$H_2O$; spectra uncorrected.
c. Brown *et al.* (1968); solvent: ethylene glycol/$H_2O$ (1 : 1); spectra corrected.
d. Eisinger and Shulman (1968); solvent: ethylene glycol/$H_2O$ (1 : 1); spectra corrected.
e. Imakubo (1968); solvent: ethylene glycol/$H_2O$ (1 : 1); spectra uncorrected.
f. Hélène and Michelson (1967); solvent: propylene glycol/$H_2O$ (1 : 1); spectra uncorrected.

[c] Excimer emission was estimated as the ratio (×100) of the observed $\phi_f$ for the red-shifted fluorescence to $\phi_f$ of the monomeric system. This is, of course, only strictly valid if all excimer excited-state processes are comparable to those in the monomeric system.

[d] The solvent was 0.1 *M* phosphate (pH 7)/0.05 *M* sodium acetate/0.25% glucose rather than that indicated in Ref. a.

formation, but the alternative explanation, not entirely ruled out by the experimental data, is that a charge-transfer interaction may occur between chromophores in a rigid matrix. This would lower the energy of the excited state and red-shift the emission. Several mechanisms may be suggested leading to such a situation. The first could be through a dipole–polarizability interaction, including ground-state dipole and excited-state polarizability as well as excited-state dipole and ground-state polarizability. This effect might be expected to be more important for heterodinucleoside phosphates. The second mechanism is a consequence of the evidence for biphotonic photoionization discussed in Chapter 2, Section D. As will be discussed, triplet states in polynucleotides appear to behave independently once they are populated; i.e., no evidence for triplet excimers has been found. Accordingly, there is a distinct possibility that the triplet of one base in a dinucleoside phosphate can undergo photoionization via triplet absorption. However, the photoelectron so produced would have a high probability of being captured by the adjacent base since Pur and Pyr bases have long been known to react at or near diffusion-controlled rates with electrons (Hart *et al.*, 1964). Thus, we would arrive at a state represented by $B^+p^-B$ which would attain equilibrium by reverse charge transfer with concomitant emission. Such a process seems entirely feasible and could be subjected to several experimental tests (see Sec. B).

Ground-state charge-transfer effects seem to occur near the pK of Cyd (Montenay-Garestier and Hélène, 1970). At pHs ~4, CMP in aggregates as well as CpC and poly(rC) in glasses at 77°K exhibit a sharp increase in total luminescence (fluorescence and phosphorescence) with reduced emission at either more basic or acidic pHs. This has been interpreted as evidence for formation of a ground-state charge-transfer complex between one protonated and one neutral Cyd residue in pairs. Similarly, at pH 3.5 ApA shows a marked red-shift in fluorescence which could have resulted from either ground- or excited-state charge-transfer interaction leading to ApA excimer formation (Kleinwächter and Koudelka, 1972).

The third test of exciplex formation, concentration dependence, does not, of course, apply directly to dinucleoside phosphates. However, excimer-like emission has been observed for the mononucleotides in nonglassy aqueous solvents (Hélène and Michelson, 1967; Kleinwächter *et al.*, 1968). These conditions may promote formation of microcrystalline aggregates (Wang, 1965) and therefore effectively increase local solute concentration. For comparison, some of these data are listed in Table 1. Whereas no consistent trend in fluorescent quantum yields (aggregates *vs.* homodinucleoside phosphates) is apparent,* some interesting effects appear concerning the fluorescence band. No emission is detectable for UpU while aggregates of UMP exhibit a slightly red-shifted but easily detectable emission ($\phi_f = 0.02$) (Kleinwächter, 1972a). The fluorescence of ApA is also slightly red-shifted ($\sim 1000$ cm$^{-1}$). ApA is the most strongly stacked at room temperature (Warshaw and Tinoco, 1965, 1966), and it was suggested that fluorescence may be from an exciton level, since the shift is equal to the calculated exciton splitting (Guéron *et al.*, 1967a; Warshaw *et al.*, 1965). Aggregates of AMP, however, have a much more red-shifted band ($> 4000$ cm$^{-1}$).

The origin of such spectral differences in dinucleosides compared to mononucleotide aggregates probably arises from orientational differences between chromophores. Dinucleosides in a polyalcohol matrix show predominantly vertical base stacking, while in an aqueous matrix, microcrystals can have both vertical and horizontal interactions. The existence of the appropriate crystal orientations has been demonstrated recently by crystal x-ray diffraction studies on dinucleoside phosphates (Rosenberg *et al.*, 1973; Day *et al.*, 1973). Some data for aggregates of dinucleosides are also included in Table 1; however, interpretation is clouded by similar considerations, the net spectra being the result of a mixture of inter- and intramolecular base interactions.

Steric requirements for shifted fluorescence seem amenable to only limited generalization. The 3′-ribose–phosphate linkage appears important for excimer interaction since 2′,5′-ApC and GpA show fluorescence spectra intermediate between a mixture of mononucleotides and the corresponding 3′,5′-dinucleosides which are totally excimeric (Eisinger and Shulman, 1968). Studies on synthetic models containing two nucleosides linked by —$(CH_2)_2$—, —$(CH_2)_3$—, or —$(CH_2)_6$— bridges suggested that the two-methylene bridge is too short to allow base interaction, 6 methylenes allow too much free rotation, while 3 methylenes are optimal for red-shifted fluorescence (Brown *et al.*, 1968). Other conclusions based on primary structure are not possible, most likely because a major factor determining the fluorescence character of dinucleoside phosphate emission seems to be the solvent rather than base content (see later discussion).

While the red-shifted structureless fluorescence has generally been assumed to be an exciton–exciplex emission, there are several experi-

* Quantum yields in nonglassy solvents at 77°K are notoriously difficult to reproduce.

mental and theoretical considerations which cast doubt on this assignment. First we note that while at the internucleotide separation of 3.4 Å no exciton splitting is detectable in absorption, the "normal" fluorescence of ApA is red-shifted by an amount which is calculated to be of the magnitude expected at this internucleotide separation for an exciton shift ($\sim 500\ cm^{-1}$; Eisinger and Lamola, 1971). Second, at an internucleotide separation of $\sim 2.8$ Å, the exciton splitting for $Me_2Thy$ is $\sim 1900\ cm^{-1}$ which should then give a similar fluorescence red-shift (assuming that the other factors affecting the Stokes shift remain unchanged). The red-shift observed for most dinucleoside phosphates is considerably in excess of this and would require, on the excimer model, an even shorter internucleotide distance. However, at this 2.8 Å separation the $Me_2Thy$ dimerizes with a quantum yield $\sim 1$ and fluorescence is not observed, although $\phi_f \sim 10^{-2}$ would be detectable.

On the basis of the admittedly sparse information presently available, it appears that the internucleotide separations required by coulombic approximation to the exciton–exciplex model are unrealistic. A further conceptual difficulty arises concerning the physical feasibility of applying this model to the present conditions. At 77°K the matrix is considered to be a rigid glass. The excimer model requires that the internucleotide separation decrease by somewhere up to $\sim 1.9$ Å (the limit of contraction is assumed to be covalent dimer formation with a C—C bond distance $\sim 1.5$ Å; this is a reduction of 1.9 Å from 3.4 Å). For other systems polarization studies have shown that molecular motion is severely limited in such matrices. Not only is it difficult to conceive how such movement might happen, but it must also occur within the lifetime of the excited singlet state, which at 77°K is $\sim 5 \times 10^{-9}$ sec. The internal contradiction of the model is simply stated: the necessary molecular condition for the emission (decreased interbase separation) is not feasible for the experimental conditions under which it is observed. Furthermore, should the red-shifted emissions observed from concentrated solutions of the nucleosides at room temperature be shown to be intrinsic, this would tend to suggest that emission occurs from a state of unchanged interbase separation, since the singlet lifetime is so short (psec) that insignificant movement could occur in the formation of the emitting state. The interactions involved would thus seem to be of longer range than the exciton–exciplex model ($1/r^3$) and may be indicative of charge-transfer effects ($1/r^2$).

In conclusion, it is most prudent to consider that the origin of the red-shifted structureless fluorescence of dinucleoside phosphates (and stacked structures in general) should remain an open question until more critical experiments have been carried out. These would include low-temperature absorption and fluorescence excitation spectra, studies of polarization, especially at the red edge of the absorption band, and fluorescence lifetime studies, with a search in particular for rise-time effects.

*b. Phosphorescence*

Figure 1 shows corrected phosphorescence spectra which were resolved from low-temperature fluorescence. Other physical properties of the triplet states are listed in Table 2. From the blue edge of nucleotide phosphorescence Guéron *et al.* (1967a) established a relative order of triplet energies, Urd (anion) > Cyd > Gua > Ade > Thd. Triplet emissions from the dinucleoside phosphates are seen to be predominately those characteristic of the nucleotide possessing the lower-lying triplet level. The only exceptions, at neutral pH, are those containing Ade and Gua. At 80°K the triplet decay curves for ApG and GpA can be resolved into two exponential components, the lifetimes of which correspond to those of the constituent nucleotides. This gives an estimate of the relative contribution of each base to the net triplet emission (Kleinwächter and Koudelka, 1972; Table 2). Due to their shorter lifetimes and smaller phosphorescence yields, it is more difficult to resolve possible contributions of Pyr triplets; and this may introduce uncertainty as to the "purity" of an apparent single component decay.

These results raise an immediate question concerning the mechanism of preferential population of the lower nucleotide triplet. First, the fact that dinucleoside phosphates and mononucleotide $^3\tau$'s are essentially identical suggests that whatever the mechanism, it must occur much faster than triplet decay ($<10^{-1}$ sec.). Basically two mutually complementary proposals have been forthcoming. (1) A suggestion of preferential intersystem crossing from an exciplex singlet to a mononucleotide triplet was based on a fluorescence and phosphorescence excitation study of ApC (Guéron *et al.*, 1966). It was found that excitation of low-temperature red-shifted fluorescence could be best described as a function of *both* Ade and Cyd absorption, thus suggesting a bichromophoric excited singlet. Also, the ratio of fluorescence to phosphorescence intensity as a function of excitation energy was constant, suggesting the precursor to the Ade triplet is this same singlet. In view of the uncertainty in the nature of the excited singlet, it cannot be concluded whether an initial "exciplex" or normal mononucleotide triplet is formed as the initial result of intersystem crossing. (2) It was later proposed that both monomeric triplet states

**Table 2** Low-Temperature Triplet Properties of Some Dinucleoside Phosphates

| Compound | Reference[a] | $\phi_p(\times 10^2)$ | $^3\tau$ (sec) | Type of emission |
|---|---|---|---|---|
| ApA | a | 13 | 2.6 | A |
| | b,c | 5.5 | — | A |
| AMP + GMP | a | 12 | 2.7 + 1.2 | A (35%) + G (65%) |
| | d | 3 | — | A + G |
| ApG | a | 8 | 2.7 + 1.1 | A (90%) + G (10%) |
| | d | 11 | — | A |
| GpA | a | 11 | 2.6 + 1.1 | A (90%) + G (10%) |
| | d | 11 | — | A |
| AMP + CMP | a | 3 | 2.7 + 0.6 | A (80%) + C (20%) |
| | d | 1 | — | A + C |
| | e | — | 2.7 | A |
| ApC | a,d,e | 17,7,— | 2.6,f,3.0 | A |
| CpA | a,d,e | 7,1,— | 2.6,f,2.7 | A |
| AMP + UMP | a,d,e | 1.5,0.6,— | 2.6,f,2.7 | A |
| ApU | a,d,e | 4.0,4.0,— | 2.6,f,2.65 | A |
| UpA | a,d,e | 4.0,3.0,— | 2.5,f,2.5 | A |
| GpG | a | 11 | 1.4 | G |
| GMP + CMP | a,e | 13,— | 1.3,1.25 | G |
| GpC | a,e | 13,— | 1.3,1.25 | G |
| CpG | a,e | 12,— | 1.3,1.2 | G |
| GMP + UMP | a,d,e | 9,3,— | 1.3,—,1.25 | G |
| GpU | a,d,e | 8,0.9,— | 1,2,—,1.2 | G |
| UpG | a,d,e | 2.5,—,— | 1.2,—,1.15 | G |
| CpC | a,d | 3.1 | 0.7,— | C |
| CMP + UMP | a | 1.5 | 0.6 | C |
| CpU | a | 2.0 | 0.6 | C |
| UpC | a | 1.0 | 0.7 | C |
| TpT | d | 1 | — | T |
| TMP + dAMP | d | 1.6 | — | A |
| TpdA | d | 1 | — | T |
| dApT | d | 2 | — | T |

[a] References:
a, b, c, and d as in Table 1.
e. Hélène and Michelson (1967).
f. 2.4–3.0 sec by ESR signal decay (Guéron *et al.*, 1966).

are populated and subsequent triplet–triplet energy transfer populates predominately the base with the lower triplet level (Eisinger and Lamola, 1971). Contrasted to (Förster) singlet–singlet transfer, triplet–triplet interaction proceeds primarily through an electron exchange interaction and therefore requires electronic overlap and chromophores in close proximity (Sommer and Jortner, 1968).

The rate expression for triplet-triplet transfer takes the form

$$N_{t \to t} = \frac{2\pi}{h} \mathbf{M}^2 \int \text{overlap}$$

in which the overlap integral is between the phosphorescence spectrum of the donor ($T_1 \to S_0$) and the appropriate absorption spectrum of the acceptor ($S_0 \to T_1$). **M** is the exchange matrix element between donor–acceptor pairs.

Using an estimate of $\mathbf{M} \cong 10\ \text{cm}^{-1}$, Eisinger and Lamola (1971) calculated theoretical triplet–triplet transfer rates between the bases at 77°K. One obvious difficulty is that nucleotide $S_0 \to T_1$ absorption spectra are not known, and must be estimated. "Downhill" transfer (higher triplet → lower triplet) and transfer in homodinucleoside phosphates are estimated to have reasonable overlap integrals and the predicted transfer rates are all $>10^7$ sec.$^{-1}$ With $^3\tau > 10^{-1}$ sec, triplet transfer, in theory, should be very efficient in these cases. The frequencies of back transfer (i.e., energetically "uphill" transfer) are temperature dependent, but are at least a factor of $10^5$ slower at 77°K and therefore preferential population of the lower triplet is predicted.

Again, the geometry and degree of base stacking are critical factors in determining triplet properties. For example, when stacking is disrupted by ionic repulsion, as in ApT at pH 11.5 where Thd is singly ionized, both fluorescence and phosphorescence become characteristic of an equimolar mixture of the nucleotides: no red-shifted fluorescence or triplet transfer is apparent (Eisinger, 1969). At the other extreme, the model dinucleoside, A—$(CH_2)_3$—*iso*-Pen A, gives both a shifted unstructured phosphorescence and fluorescence, implying such strong stacking that even the triplet emission is from an interacting state (Leonard *et al.*, 1969). This would seem to be the only example of an "exciplex" triplet. Formation of monomeric aggregates also favors triplet transfer since phosphorescence from Ado aggregates can be entirely quenched by $Co^{2+}$ at a concentration implying that each ion has affected 110 adenosines (Hélène and Montenay-Garestier, 1968). Also in Thd–Ade and Gua–AcCyt aggregates only Thd and Gua phosphorescences are observed. In this case additional triplet transfer between hydrogen-bonded base pairs must be considered. The predicted rates are $10^3$ sec$^{-1}$ which are 10 to 100 times faster than back transfer. Thus base-pair transfer may also compete for the donor triplet (Eisinger and Lamola, 1971).

*c. Summary*

Taken together, low-temperature singlet and triplet excited-state processes in dinucleoside phosphates can be summarized as follows:

(1) A rigid solvent matrix locks in a certain distribution of stacked conformations, some of which may be favorable to excitation resonance or charge-transfer interactions. (2) The excited singlet is characterized by two fluorescing populations of varying proportion, one characteristic of rather unperturbed mononucleotide singlets and the other, a lower-energy state, characteristic of either a charge-transfer or an exciplex state involving base interaction. (3) Intersystem crossing gives ultimately mononucleotide triplets and also possibly, in strongly stacked cases, a lower-energy triplet again due to base interaction. (4) In times less than mononucleotide triplet lifetimes ($>10^{-1}$ sec), the base possessing the lower-energy triplet is populated, by direct intersystem crossing from the singlet and/or by triplet energy transfer between bases via exchange interaction. (5) Phosphorescence occurs solely from terminally populated mononucleotide triplet(s).

At the moment these qualitative statements are the only general description available. There are no reports as yet of exciplex absorption spectra, T–T absorption spectra, or photoionization effects. It would, of course, be highly informative to be able to estimate dinucleoside phosphate $\phi_{isc}$'s for $S_1 \rightarrow T_1$ since this process is critical in triplet population and hence important in determining the ultimate photochemical pathways. However, in extrapolating data to room temperature, changes in the physical environment may well alter or eliminate such processes observed at 77°K. Thus, in the future, it will probably be more fruitful to focus on room-temperature properties of dinucleoside phosphates, even though special techniques for fluorescent detection are necessary and triplet parameters cannot be directly observed.

## 3. Excited-State Properties: Room-Temperature Studies

### *a. Temperature and Solvent Effects: Extrapolation from 77°K to 300°K*

General processes responsible for quenching radiative transitions and spectral shifts accompanying the transition from a rigid matrix at 77°K to liquid solution at 300°K have been discussed in Chapter 2 for the bases and their monomeric derivatives. Although comparable studies on the dinucleoside phosphates have not been done, base stacking studies have received considerable attention. Theories on ground-state stacking equilibria as a function of temperature have been developed primarily through effects on CD and ORD, and Fig. 3 is a typical example. Two models, representing extremes in the number of observable states, have been proposed. In the noncooperative two-state model (Van Holde *et al.*, 1965) an equilibrium is es-

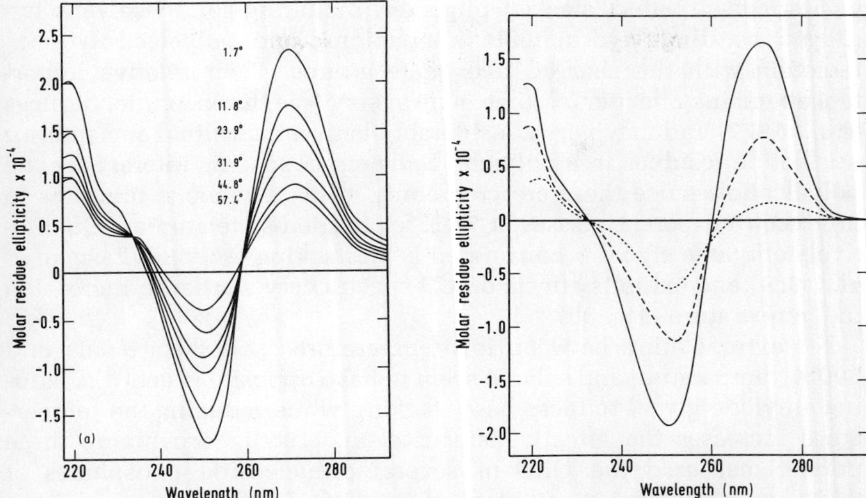


**Fig. 3.** *(a) Circular dichroism of ApA as a function of temperature (°C) in 1 M NaCl. (b) Circular dichroism of ApA as a function of added ethylene glycol [—, 0%, ---, 40%, . . . ., 80% (v/v)] in 1 M NaCl at 10°C. (From Powell et al., 1972).*

tablished between stacked and unstacked conformations according to a van't Hoff relationship,

$$K(T) = \exp(-\Delta H^0/RT) \exp(\Delta S^0/R),$$

in which $K(T)$ is the ratio of fraction stacked to fraction unstacked. In the other model (Glaubiger *et al.*, 1968), temperature-dependent torsional oscillations occur around an equilibrium stacked position (which is independent of temperature), creating a continuum of states. A decrease in the optical rotation or CD $[\phi(T)]$ at higher temperatures is considered to be due to a higher oscillatory frequency according to

$$\phi(T) = \phi_0 \exp(-2kT/\kappa)$$

in which $\kappa$ is a net torsional force constant. Recently, Powell *et al.* (1972) critically examined both models for several dinucleoside phosphates and concluded that although the oscillator model cannot be ruled out in reality, in an operational sense, optical measurements are completely amenable to a two-state equilibrium. However, it would seem that critical experiments could be designed to test the oscillator model.

The effects of solvents on stacking interactions are at present not completely understood. A number of factors contribute to the stability of a single-stranded, stacked oligomer, including solute–solvent hydrogen bonding, hydrophobic interactions, and polyelectrolyte interaction with the charged phosphate groups. Their relative importance remains a matter of some controversy (see, for example, Scruggs *et al.*, 1972) and may vary considerably between terminal and interior nucleotide residues in a polymer. Solvents of specific interest are the polyalcohols since these are commonly used in aqueous mixtures to produce transparent glasses at 77°K for excited-state studies. Qualitatively ethylene glycol is considered a "destacking" solvent (Fasman *et al.*, 1964) and exhibits effects on CD spectra very similar to increasing the temperature (Fig. 3b).

For extrapolation between low-temperature excited-state data and 300°K, temperature and solvent seem to have opposite effects (i.e., adding ethylene glycol reduces base stacking while reducing the temperature increases the effect). Eisinger *et al.* (1966), recognizing these points, measured the ORD of several dinucleoside phosphates at −30°C or −25°C in 50% ethylene glycol/$H_2O$. In every case, considerable stacking was indicated whereas at room temperature very little stacking was expected in this mixed solvent. Since the solvent is not entirely rigid at −30°C, some additional interaction might be expected upon further cooling to 77°K. The critical comparison, however, is between low-temperature glasses and an aqueous solution at 300°K. In this case no quantitative comparison of stacking is available, but one might expect the degree of stacking in aqueous solution (from CD data at 300°K) to be indicative of possible excited-state interaction analogous to similar correlations at 77°K. However, other considerations contradict this hypothesis. Lack of major absorption spectra alterations compared to the nucleotides eliminates the possibility of extensive ground-state charge-transfer or exciton interactions at 300°K. Hence, only exciplex formation remains a possibility and the likelihood of its formation is worth considering. By analogy with nucleotide fluorescence, nucleotide oligomers should have $\phi_f$'s of $\sim 10^{-4}$, implying $^1\tau$'s of $\sim 10^{-12}$ sec. Exciplex formation requires interbase contraction, but estimates as to the rate of contraction are unavailable. This rate is probably no faster than vibrational relaxation ($10^{12}$ sec$^{-1}$) and is likely about the same as solvent relaxation ($10^{10}$–$10^{12}$ sec$^{-1}$). Therefore, mononucleotide fluorescence might well occur before exciplex formation. Corroboration of such predictions will have to await experimental data (see also Section 2a).

*b. Fluorescent Analogs of Dinucleoside Phosphates*

In view of the obvious importance of base interactions for excited-state properties at 77°K, the difficulty in confidently extrapolating these data to room temperature, and the experimental inaccessibility of fluorescence in the normal dinucleosides at 300°K, it is rather surprising that only limited effort has been directed toward synthesizing oligomers containing fluorescent base analogs (for a compilation of possibly useful modified bases, see Eisinger and Lamola, 1971). Leng *et al.* (1968) studied the fluorescence at 20°C of $Me^7Gua$ in several ribodinucleoside phosphates. $Me^7Gua$ itself has an uncorrected emission maximum at 26.0 kK (excited at 34.2 kK) which is essentially unchanged in energy in 3′,5′- or 5′,3′-dinucleoside phosphates containing Ade, Cyt, or Ura. Although hypochromicity and ORD indicated various degrees of stacking, and marked temperature and "destacking" solvent effects were observed on the fluorescent *intensity,* no alteration in the *energy* was reported. No evidence for excimer formation at 20°C could be found. Although direct comparison of these results to emission at 77°K was not reported, this study suggests that fundamental differences exist between excited-state processes in rigid matrices and liquid solution, even though considerable stacking is present under both conditions.

The often quoted argument (Guéron and Shulman, 1968; Kleinwächter and Koudelka, 1972) that the similarity between mononucleotide emissions at room and low temperature justifies general extrapolation of excited-state processes between temperatures must therefore be viewed cautiously. The apparent fact that stacking effects alone cannot predict excited state interaction in dinucleoside phosphates coupled with the increased efficiency of radiationless deactivation modes (Chapter 2, Section C) confirms the need for restraint in making such conclusions.

*c. Photochemical Studies*

Photochemical processes (see Fig. 1, Chapter 2) compete with emission processes for deactivation of an excited state, and hence may provide additional information as to its nature. The photochemistry of Pyr dinucleoside phosphates is reviewed in Chapters 4 and 5. Since these studies deal with only relatively stable end products of a light-initiated chemical process, they are at best indirect indications of excited-state properties. Nevertheless, photochemical yields as well as sensitization and quenching studies are often the only source of excited-state data available at room temperature, particularly for triplet states.

From mononucleotide studies, it is generally concluded that Pyr photohydration proceeds via an excited singlet (Chapter 4) while Pyr photodimerization occurs from the triplet, although in concentrated solutions favoring stacked nucleotides, singlet dimerization occurs.

Data concerning the nature of the excited-state precursors to these products in dinucleoside phosphates are sparse. The lack of significant excitation wavelength effects on the yield of photohydration in UpU (Brown *et al.*, 1966) and CpC (Hariharan and Johns, 1968) suggests a singlet precursor, while dimerization may occur from either the singlet or triplet, depending on the extent of stacking (Chapter 5). The geometric effects of stacking in favoring dimerization are well appreciated, whereas the role of stacking interactions in altering excited singlet and triplet states and their subsequent reactions is often overlooked and requires further examination.

A comparison of photohydration quantum yields is shown in Table 3 for Ura- and Cyt-containing dinucleotides and dinucleoside phosphates along with the corresponding monomeric derivatives. No large differences exist in the yield of this singlet-derived product, implying that either exciplex formation does not significantly compete with direct mononucleotide–solvent processes in the excited singlet or that the probability of hydrate formation from an excimeric state is not significantly different than from the mononucleotide. In Pur–Pyr dinucleoside phosphates, singlet energy absorbed by an unreactive Pur may also contribute to Pyr photohydration via energy transfer or excimer hydration, thus increasing an apparent quantum yield. Such processes may also introduce a wavelength dependence into the yield of dinucleoside phosphate photohydration and should be further investigated.

**Table 3** Room-Temperature Photohydration Quantum Yields for Some Dinucleoside Phosphates and Their Constituents[a]

| Compound | Irradiation ($\lambda$, nm) | $\phi \times 10^2$ (hydration) | Reference |
|---|---|---|---|
| U | 230–280 | 0.7 | Brown and Johns (1968) |
| pU | 254 | 2.2 | Sinsheimer (1954) |
| UpU | 280 | 0.9 | Brown *et al.* (1966) |
| dUpU | 280 | 1.8 | Helleiner *et al.* (1963) |
| pC | 265 | 1.6 | Johns (1971) |
| Cp | 265 | 0.86 | Becker *et al.* (1967) |
| CpC | 254 | 0.6 | Hariharan and Johns (1968) |
| CpCp | 254 | 0.8 | Wierzchowski and Shugar (1962) |
| ApCp | 254 | 0.8 | Wierzchowski and Shugar (1962) |
| GpCp | 254 | 0.6 | Wierzchowski and Shugar (1962) |
| TpC | 240–280 | 0.6 | Haug (1964) |

[a] Data are selected from the review of Fisher and Johns (Chapter 4) and are at neutral pH.

Triplet sensitization with acetone and acetophenone at 300°K reveals the relative order of triplet energies. Assuming that the low-temperature order (acetone > CPM > UMP > GMP > AMP > acetophenone > TMP) also holds at room temperature, acetone sensitization should and does give Pyr nucleotide cyclobutyl dimers in concentrated Pyr solutions (1 m*M*) (Chapter 5), while acetophenone is expected to sensitize dimers only in TMP. The same trend is seen for TpT, TpdC, dCpT, and CpC (W. Hauswirth, unpublished results, 1973) where dimerization could be sensitized in all compounds by acetone but not in CpC if acetophenone is used. Therefore, energy levels for population of triplets in dinucleoside phosphates appear to be little changed in relative ordering and magnitude compared to mononucleotides. However, photochemical sensitization results do not exclude the involvement of processes occurring subsequent to triplet population which may alter a photochemical yield. For example, sensitization by acetone should populate the triplets of CpC, CpU, UpC and UpU about equally efficiently, but the net sensitized dimerization is almost 20 times greater for UpU compared to CpC (Kucan *et al.*, 1972).

In summary, the lack of direct data at room temperature (e.g., fluorescence, polarization studies) on dinucleoside phosphates precludes extended discussion, but, based on available information, it can be stated that excited singlet interaction is probably not as extensive at 300°K as at 77°K and that relative triplet energy levels are largely unaltered. Finally, it should be noted that all excited-state data, except for the Thy-containing models, have been gathered on *ribodinucleoside* phosphates. The conformational effects of the *deoxyribose* (Cantor *et al.*, 1970) on excited-state properties is unknown.

## B. Oligonucleotides

Since base stacking appears to be primarily a noncooperative process, the interaction of each pair of adjacent bases in a polymer is expected to be controlled by independent equilibria. It should then be possible to calculate absorptive and dispersive properties of oligomeric nucleotides from known dimeric values as long as short-range effects predominate. Such a nearest-neighbor approach has been shown to be in excellent agreement with the experimental ORD spectra of ribotrinucleotides (Cantor and Tinoco, 1967). With some modification, to allow for the difference between terminal and internal bases, this analysis also holds for hypochromicity in deoxy- and ribooligomers at least up to $n \sim 5$ (summarized by Tazawa *et al.*, 1972) (Fig. 4). Thus, non-base-paired oligonucleotides generally

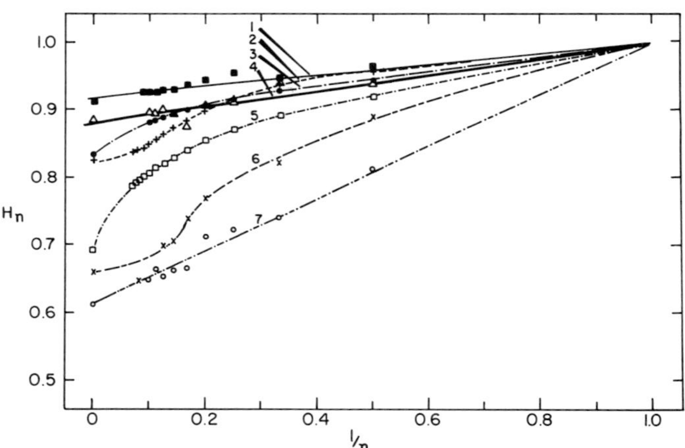


**Fig. 4.** *Monomer normalized extinction coefficient* ($H_n = \epsilon_n/\epsilon_{monomer}$) *as a function of inverse polynucleotide chain length* [*curve 1, poly(U); curve 2, poly(rI); curve 3, poly(dC); curve 4, poly(dT); curve 5, poly(rC); curve 6, poly(rA); curve 7, poly(dA)*]. (From *Tazawa* et al. *1972.*)

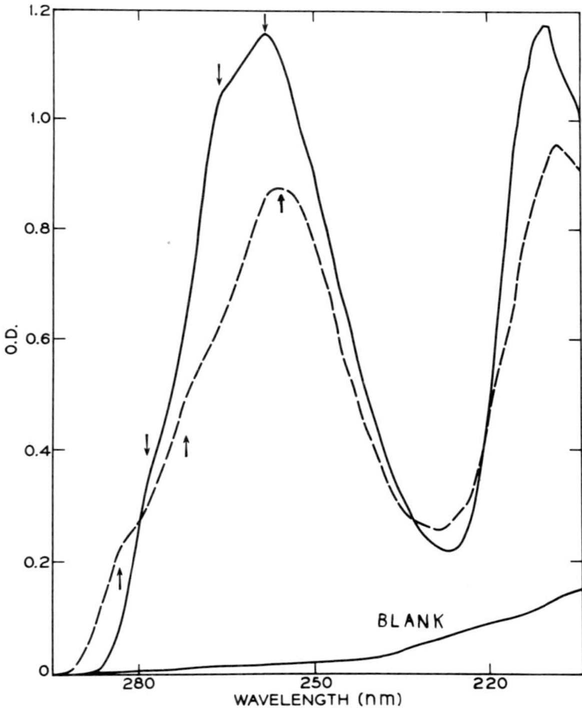


**Fig. 5.** *Comparison of 77°K absorption of AMP* (—) *and poly* (rA) (---). *Arrows indicate the approximate position of individual components within the bonds.* (From Guéron et al., 1973.)

exhibit conformationally controlled ground-state electronic effects similar to the dinucleoside phosphates.

Oligonucleotide emission properties are also expected to be generally similar to a sum of their constituent dinucleotides; however, some limited differences have been recorded. Montenay-Garestier *et al.* (1969) observed small, progressive red-shifts in low-temperature fluorescence and phosphorescence spectra as a function of chain lengths for both 3′,5′-($n = 2,5,12$,poly) and 2′,5′-($n = 2,3,4,8$) oligo(rA). Although Kleinwächter and Koudelka (1972) found no difference in emission spectra between ApA and ApApA, this is not necessarily contradictory to the earlier observations since hypochromicity at room temperature (Fig. 4) suggests no alteration in the degree of stacking per dinucleotide until $n > 5$, at which point certain polynucleotides, including poly(rA), deviate toward some degree of cooperative stacking. In a similar series, the absorption *peak* blue-shifted (Guéron and Shulman, 1968), but a red-shift occurred in the lower energy shoulders (Fig. 5). One explanation is that the progressively less polar environment, attendant with longer polymers, red-shifts weak $n \rightarrow \pi^*$ transitions and blue-shifts the dominant $\pi \rightarrow \pi^*$ bands which make up the composite Ade absorption. The emission would red-shift if it originated from an $n \rightarrow \pi^*$ band, but this would seem to contradict the evidence that the mononucleotide emission is ($\pi\pi^*$). An alternative approach is that with increasing chain length exciton effects become more prominent.

Emission of several trinucleoside diphosphate sequence isomers containing Ade and Ura has also been studied (Kleinwächter and Koudelka, 1972). Compounds containing the sequence —ApU— show an exciplex red-shift characteristic of ApU, the third nucleotide being of little consequence. Unexplainably $\phi_f$'s of all trimeric models are less than half those expected from constituent dimeric $\phi_f$'s. Phosphorescence spectra are essentially identical, with triplet lifetimes generally similar to ApA or AMP. Oligomers of Ade and Cyt, $A_n$pC ($n = 5$–$12$), exhibit luminescence characteristic only of oligo(rA) (Montenay-Garestier *et al.*, 1969).

## C. Single-Stranded Polynucleotides

### 1. Homopolymers

#### *a. Poly(rA)*

The excited states of poly(rA), of all the synthetic polynucleotides, have received by far the most attention because of poly(rA)'s relatively strong emission at 77°K and its structural properties. Hypochromicity and CD both indicate an extensively stacked single-stranded helix at neutral pH (Holcomb and Tinoco, 1965). At low temperature (77°K), this conformation is retained (Rahn *et al.*, 1966a) and the absorption spectrum exhibits a 600 $cm^{-1}$ blue-shifted absorption peak with a corresponding red-shift in two higher-energy shoulders compared to AMP under identical conditions (Fig. 5). Extensive ground-state–exciton interaction is ruled out since nonpolar organic solvents have similar effects on the AMP absorption spectrum (Chapter 2, Section B) and this is consistent with a model of the stacked bases in poly(rA) in a relatively nonpolar environment.

A pronounced peak in the 77°K fluorescence excitation spectrum of Ade at 35.6 kK (Chapter 2, Section B and Fig. 18) and solvent-induced shifts suggest that fluorescence originates from a partially hidden transition. Excitation spectra are not available for poly(rA) but a similar situation may apply as well. The fluorescence maximum is only slightly red-shifted from ApA which itself is only ~1000 $cm^{-1}$ lower than AMP (Fig. 6). This is significantly smaller than a normal "excimer" shift, and therefore, consistent with absorption analysis, may simply reflect emission predominately from the low-energy exciton band (Guéron *et al.*, 1973). The fluorescence lifetime has been reported as 2.8 nsec (Blumberg *et al.*, 1968), but it is not known if it varies with excitation wavelength. Polarization studies would also be of considerable interest.

Interest in the triplet of poly(rA) began with the phosphorescence quenching studies of Bersohn and Isenberg (1964). The fact that much

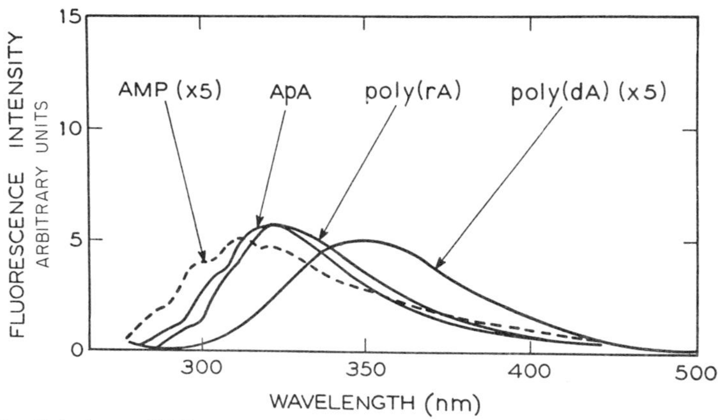


**Fig. 6.** *Relative 77°K fluorescence spectra of AMP, ApA, poly(rA), and poly (dA). (From Guéron* et al.*, 1973.)*

less than stoichiometric amounts of bound cation ($Mn^{2+}$ at ~1 mole%) were needed to quench emission indicated that the triplet could migrate freely along the poly(rA) strand. Other paramagnetic divalent ions ($Co^{2+}$, $Ni^{2+}$, $Cu^{2+}$, bound to the phosphates) were also found to be effective quenchers and did not appreciably affect the fluorescence yield (Eisinger and Shulman, 1966). Using a diffusion-controlled quenching model, the best fit to experimental data gave a lifetime between triplet transfer events of $4 \times 10^{-4}$ sec, implying that ~80 bases could be traversed by each triplet. This was consistent with a low-concentration limit for extent of quenching of ~100 bases, which implies triplet migration of 50 bases.

It was also observed, however, that although $\phi_p$ was greatly reduced, ${}^3\tau$ (2.5 sec measured by both phosphorescence and ESR decay) was unaffected by quenching (Eisinger and Shulman, 1968). Quenching and hence triplet migration must be so fast that only a class of unquenchable triplet is observable. The actual triplet jump time then must be shorter than $4 \times 10^{-4}$ sec. Rahn *et al.* (1966a) proposed a lower limit of $> 2 \times 10^{-10}$ sec based on the lack of spin–spin motional ESR averaging and subsequent theoretical triplet–exciton calculations estimated an actual jump time of $\sim 10^{-9}$ sec (Sommer and Jortner, 1968). Since the triplet diffusion model had to be rejected, it was proposed that structural barriers to transfer existed (probably local unstacked regions or "kinks") with triplet trapping occurring at the barriers.

More recent work has shown these processes to be at least this complicated. Longworth and Battista (1970) reported delayed fluorescence (DF) spectrally identical to prompt fluorescence (Fig. 7). Long range triplet transfer and delayed fluorescence appear to be general properties of stacked adenine residues since aggregates of adenosine in frozen solution exhibit the same properties (Montenay-Garestier, 1973; Montenay-Garestier and Hélène, 1973).

In this and a later work (Hélène and Longworth, 1972) several other properties of delayed fluorescence are reported: (1) The DF intensity is proportional to the square of the phosphorescence intensity, except at high incident light intensity. (2) DF decay is nonexponential. (3) The phosphorescence rise has an exponential time constant progressively smaller than phosphorescence decay as a function of incident intensity, while ${}^3\tau$ itself is unchanged. Since the absorption sites are not depleted at these incident intensities, this must mean that triplet emission sites can be saturated. (4) DF rise time is also exponential, having the same time constant as phosphorescence rise. A model is proposed in which free triplets (excitons) and triplets trapped at structural defects coexist, the later populated either by direct absorption or by rapid transfer from free triplets. Only the traps emit, and since they are in relatively small concentration, a saturation of phosphorescence intensity will appear at high incident intensities. Delayed fluorescence arises from fusion of a trapped triplet and a free triplet.

Recently, however, Bazin *et al.* (1973) noted excitation wavelength and cation salt effects on delayed fluorescence which could not be explained by triplet–triplet fusion. They proposed, instead, that DF results from triplet–triplet absorption (steps 1, 2, 3) resulting in photoionization of the adenylate triplet (step 4) which may lead to a radical cation $A^{\dot{+}}$ and solvated electron $e_{aq}^-$ or a radical cation – anion pair if the ejected electron is captured by a neighboring unexcited. Ade. Evidence for this scheme rests primarily on the similarity of the excitation wavelength dependent quantum efficiencies for delayed fluorescence and stimulated phosphorescence produced by visible light excitation (546 nm) of solvent-trapped electrons and recombination with $A^{\dot{-}}$ (steps 7 to 10).

$$S_0 \xrightarrow{h\nu_1} S_1 \tag{1}$$

$$S_1 \xrightarrow{\text{isc}} T_1 \tag{2}$$

$$T_1 \text{ (trapped)} \xrightarrow{h\nu_2} T_n \tag{3}$$

$$T_n \xrightarrow{\text{ionization}} A^{\dot{+}} + e^-_{\text{free}} \text{ (or } A^{\dot{+}} + A^{\dot{-}}) \tag{4}$$

$$A^{\dot{+}} + A^{\dot{-}} \xrightarrow{\text{recombination}} S_0 + S_n \quad \text{(or } T_n) \tag{5}$$

$$S_n \longrightarrow S_0 + h\nu_3 \text{ (delayed fluorescence)} \tag{6}$$

$$e^-_{\text{free}} \longrightarrow e^-_{\text{trapped}} \tag{7}$$

$$e^-_{\text{trapped}} \xrightarrow[\text{light}]{\text{visible}} e^-_{\text{free}} \tag{8}$$

$$e^-_{\text{free}} + A^{\dot{+}} \xrightarrow{\text{recombination}} S_m \text{ or } T_m \tag{9}$$

$$T_m \longrightarrow S_0 + h\nu_4 \text{ (stimulated phosphorescence)} \tag{10}$$

Since this wavelength dependence similarity breaks down at excitation energies $<280$ nm, DF is postulated to come only from recombination (steps 5 and 6).

The fact that phosphorescence can be stimulated by visible light strongly suggests, by analogy with other systems known to exhibit T–T absorption and ionization, that steps 3 through 10 do indeed occur in poly(rA). There is no direct evidence however that the $A^{\dot{+}}A^{\dot{-}}$ pair is formed from an initial $e^-A^{\dot{+}}$ pair. Conductivity measurements

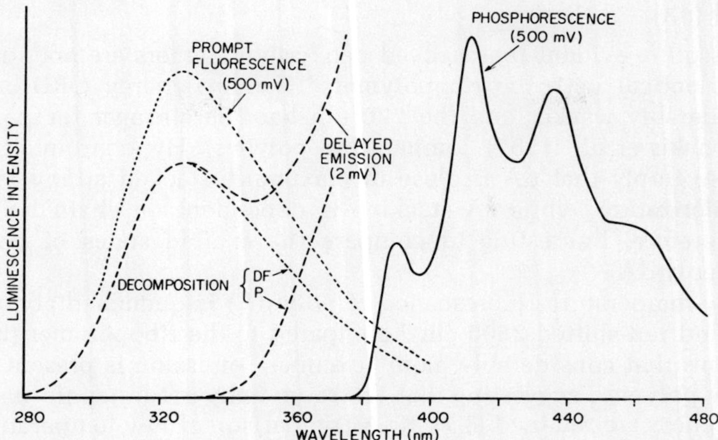


**Fig. 7.** *Relative luminescence spectra of poly(rA) resolved into prompt fluorescence, delayed fluorescence, and phosphorescence components. (From Hélène and Longworth, 1972.)*

and determination of triplet–triplet absorption spectra may give more direct information on this point. More perplexing, the quantum yield dependence on wavelength is very similar to the low-temperature absorption spectrum (Fig. 5). This poses an additional problem since the *quantum yields* of such processes are not normally expected to vary with wavelength. If variations are observed, there is no theoretical reason to expect the resulting wavelength dependence to mirror so closely $S_0 \rightarrow S_1$ absorption. In the absence of more information we cannot resolve the point except to conclude that more data are required.

In summary, the exact situation concerning the poly(rA) triplet is not entirely clear, but the basic processes are as follows:

(1) There is relatively little excited singlet–base interaction.

(2) Rapid triplet migration ($>100$ bases) follows intersystem crossing.

(3) Phosphorescence occurs from a triplet trap of as yet unknown nature, which is populated quickly ($<10^{-5}$ sec; Guéron *et al.*, 1973) from a migrating triplet exciton.

(4) Delayed fluorescence takes place via exciton–triplet fusion and/or triplet–triplet absorption.

(5) All emission disappears above the glass–liquid transition of the solvent (Hélène and Longworth, 1972), and the extent of these processes is unknown at room temperature.

*b. Poly(dA)*

It is quite evident that stacked deoxyribopolymers are not structurally identical with the ribopolymers. The low-energy ORD band is considerably weaker and the 220 nm band is stronger for poly(dA) (Vournakis *et al.*, 1967) compared to poly(rA). Hypochromicity data (Fig. 4) imply that dA is close to maximally stacked at any level of polymerization, while rA stacking is dependent on chain length. It is, therefore, interesting to compare the excited states of poly(rA) with poly(dA).

Low-temperature fluorescence of poly(dA) is quenched about fivefold and red-shifted 2500 $cm^{-1}$ compared to the ribopolymer (Fig. 6). It seems that considerably more "excimer" emission is present in the deoxy polymer, suggesting that the room temperature conformational differences are retained at 77°K. A comparison of low-temperature absorption and fluorescence excitation spectra should be informative in this respect, as would a comparison between dApdA and ApA excited states.

Delayed fluorescence is also observed for poly(dA) (Hélène and Longworth, 1972) and is identical to prompt fluorescence. The triplet lifetime is 1.95 sec (although the first 30% is nonexponential!) and delayed fluorescence decay is again nonexponential. Similar triplet processes are thus observed for both adenylates despite their very different excited singlets. Apparently intersystem crossing from either polymer gives essentially the same triplet-state exciton. DF studies comparable to those for poly(rA) have not yet been done. Poly(dA) is, unfortunately, the only single-stranded deoxyribopolymer for which some comparison of the excited state with that of its ribo counterpart has been reported.

*c. Other Homopolymers*

Table 4 summarizes the low-temperature emission data for the remaining homopolynucleotides poly(rC), poly(rG), poly(rU), and poly(T). No spectra or yields have been reported for the latter two polymers and they are generally considered "nonluminescent." Kleinwächter *et al.* (1968) reported spectrally identical fluorescence at 77°K from poly(rC) (aqueous solution + 0.25% glucose) at pH 7 and pH 3.5, while Eisinger *et al.* (1966) reported a similar spectrum, ~600 $cm^{-1}$ red-shifted, in ethylene glycol/$H_2O$ at pH 5.0. The critical structural feature is that between pH ~3.5 and 5.5, poly(rC) is a semiprotonated double helix, poly($rC \cdot rCH^+$) (Hartman and Rich, 1965), while it is single-stranded at neutral pH. Montenay-Garestier and Hélène (1970) studied the pH dependence of luminescence intensity in more detail and reported a sharp increase in both fluorescence and phosphorescence efficiency between pH 3 and 4, although neither quantum yields nor full spectra are presented. From the observation that the maximal emission intensity occurs at the same pH as the ground-state pK (determined from an absorption titration), they proposed a charge-transfer interaction between stacked Cyd pairs, one protonated the other unprotonated, to account for the apparently increased yields. There is some uncertainty in this because the room-temperature absorption is red-shifted $\sim 1400\ cm^{-1}$ upon formation of poly(rC-rCH$^+$) while no shift in the low-temperature fluorescence or phosphorescence maxima is apparent (Kleinwächter *et al.*, 1968). Exciplex formation via charge transfer should reflect at least some of this shift in the emission spectra. Additionally, emission from strongly interacting bases is generally decreased (Tables 1 and 2) while in this case a two- to threefold increase is observed. The precise reasons for this excited-state perturbation in poly(rC) therefore remain unclear, but it might just as likely arise from the unique hydrogen bonding in poly(rC·rCH$^+$) as from a stacked-base interaction. Independent evidence for a unique structural relationship between adjacent Cyd residues at pH = pK comes from Rhoades and Wang (1971a,b) who find a sharp peak in the photoadduct yield of poly(rC) *vs.* pH at pH = 4.0 and from room temperature fluorescence measurements on poly(rC·CH$^+$) (Vigny and Favre, 1974 and see Section E).

Rahn *et al.* (1964) first reported luminescence from poly(rG) in a glassy matrix; its most interesting property was a nonexponential triplet decay which could be resolved into two components (Table 4). The longer-lived one corresponded to GMP, and the short-lived one was assigned to aggregated (interacting) GMP (recently confirmed by Kleinwächter, 1972b). Relative to GMP, the red-shifted emission maximum of poly(rG) is not reflected in a corresponding absorption shift, and hence excited-state interaction is indicated. The major interest in poly(rG) rests in a comparison of its excited-state properties with the hydrogen-bonded duplex, poly(rG·rC), and will be fully discussed in the next section.

### 2. Random Copolymers: rA-rC, rA-rG, rA-U, rC-U, rG-rC, dG-dC, and rA-rC-rG-U

Structurally the nucleotide copolymers can be separated into three general groups depending on the extent of hydrogen-bonded interaction between polymers or within a polymer strand: (1) those containing non-hydrogen-bonding bases, poly(rA-rC) and poly(rC-U), will

**Table 4** Low-Temperature Singlet and Triplet Properties of Polynucleotides[a]

| | | Fluorescence | | | Phosphorescence[b] | | | |
|---|---|---|---|---|---|---|---|---|
| Compound | Reference[c] | Solvent[d] | $\epsilon_{max}$ (kK) | $\phi_f$ ($\times 10^2$) | $\epsilon$ (kK) | $\phi_p$ ($\times 10^2$) | ${}^3\tau$ (sec)$_e$ | Other references[f] |
| poly(rA) | k | E/W | 30.0 | — | (24.3) | — | 2.6 | d,e,f,m,n,p,q |
| poly(dA) | k | E/W | 27.4 | — | — | — | 1.95 | — |
| poly(rC) | f | G/W | 28.17 | — | 23.81 | — | 0.67 (i) | c,h,i |
| poly(rG) | b | E/W | 28.90 | 14 | (23.42) | 8 | 0.5 (45%), 1.5 | d,i |
| poly(rU) | — | — | — | weak | — | — | — | (a,d,f,i,p) |
| poly(T) | — | — | — | weak | — | — | — | (m) |
| poly(rA-rC) | a | P/W | 27.75 | 6 | (23.15) | 6 | 0.6 (25%), 2.5 | a,e,q |
| poly(rA-rG) | a | P/W | 28.33 | 10 | (23.04) | 12 | 1.3 (30%), 2.7 | a |
| poly(rA-U) | a | P/W | 28.17 | 4 | (23.04) | 4 | — | a,h,n,p,q |
| poly(rC-U) | a | P/W | 28.10 | 4 | 22.95 | 2 | 0.5 | a |
| poly(rG-rC) | b | E/W | 28.09 | 2 | (23.53) | 3 | 0.5 (20%), 1.5 | a,b,h |
| poly(dG-dC) | — | — | — | weak | — | — | — | (h) |
| poly(dA-T · dA-T) | a | P/W | 28.00 | 6 | 21.79 | 2 | 0.5 | a,g,h,i,j,l,o |
| poly(rA-rC-rG-U) | a | P/W | 27.80 | 5 | (23.36) | 4 | 0.6 (25%), 2.4 | a |
| poly(rA · U) | f | G/W | 30.30 | — | 23.40 | — | 0.5 (30%), 2.3 | d,q |
| poly(rA · rI) | f | G/W | 30.30 | — | 23.10 | — | 0.6 (40%), 2.5 | d,q |
| poly(rC · rI) | — | — | — | weak | — | — | — | (f) |
| poly(rG · rC) | b | E/W | 26.95 | 10 | (27.83) | 4 | 0.8[g] | b,h,i |

[a] Only the most recent or representative values are given; other references are listed under the column so labelled.

[b] Phosphorescence spectra which show some degree of structure (e.g., Fig. 1) have the energy of the highest intensity peak in parentheses.

[c] References:

a. Kleinwächter and Koudelka (1972).
b. Kleinwächter (1972b).
c. Montenay-Garestier and Hélène (1970).
d. Rahn *et al.* (1964).
e. Montenay-Garestier *et al.* (1969).
f. Kleinwächter *et al.* (1968).
g. Eisinger and Shulman (1967).
h. Eisinger *et al.* (1966).
i. Rahn *et al.* (1966b).
j. Lamola *et al.* (1967).
k. Hélène and Longworth (1972); Longworth and Battista (1970).
l. Szerenyi and Dearman (1972).
m. Rahn *et al.* (1970).
n. Eisinger and Shulman (1966).
o. Isenberg *et al.* (1967).
p. Douzou *et al.* (1961).
q. Rahn *et al.* (1966a).

[d] Solvents were E/W, ethylene glycol/$H_2O$ (1:1); G/W, 0.25% glucose in $H_2O$; P/W, propylene glycol/$H_2O$ (1:1).

[e] If two exponential components were observed, both are reported with their proportion in parentheses. Only data for 1:1 copolymers is listed.

[f] A reference quoted twice for the same compound indicates data in nonglassy solvent conditions are also available. References for the copolymers and double-stranded homopolymers are often not for molecules containing a one-to-one ratio of bases; refer to text and the original references for further details.

[g] Nonexponential triplet decay could not be resolved into two components.

exist in neutral solution and polyalcohol matrices as single strands, (2) those containing hydrogen-bonding bases in random distribution or in an unequal molar distribution will have small double-stranded regions and possibly single-stranded loops separated by noninteracting single-stranded regions [poly(rA-U), poly(rG-rC), poly(dG-dC), poly(rA-rC-rG-U)], and (3) an alternating, hydrogen-bonding copolymer forming a double-stranded helix, poly(dA-T · dA-T), which will be discussed in the next section.

77°K emission data for the ribocopolymers of classes (1) and (2) are presented in Fig. 1 and Table 4. Fluorescence spectral analysis in terms of dinucleoside phosphates applies fairly well to the copolymers except for an additional small red-shift in most cases. Phosphorescence spectra are dominated by emission from the base having the lower-lying triplet. Triplet decay is generally composed of two components; the shorter (0.5 to 0.6 sec) is probably characteristic of mutually interacting stacks of chromophores while the longer is that of the noninteracting dominant phosphorescent base (Kleinwächter and Koudelka, 1972). An alternative explanation, not entirely ruled out, is that at least part of the short phosphorescence component is due to intrinsic Cyd emission since its $^3\tau$ is of this same magnitude (note that the short component appears mainly in Cyd-containing copolymers). The phosphorescence lifetimes for poly(rA-rG) are suggestive of independent Ade and Gua emission, similar to ApG and GpA, and do not have the short-lived component. Random copolymers of Ade and Ura have the emission properties of Ade [component triplet lifetimes of 0.8 (25%) and 2.7 sec in an ice matrix], while in the double-stranded homopolymer, poly(rA·$U_2$), phosphorescence is completely lost (Rahn *et al.*, 1966a). This suggests quenching of Ade via its hydrogen-bonded complex with Ura, and possible mechanisms will be discussed in the next section. These considerations, along with the lack of triplet quenching data, prevent general assessment of the existence and range of triplet migration in copolymers at this time.

## D. Double-Stranded Polynucleotides

### 1. Homopolynucleotides (77°K)

The effect of associating one luminescent polynucleotide [poly(rA) or poly(rC)] with its nonluminescent hydrogen-bonded complement [poly(U) or poly(rI)] was studied first by Douzou *et al.* (1961) and later by Rahn *et al.* (1964, 1966a) and Kleinwächter *et al.* (1968). There is general agreement that both the fluorescence and phosphorescence are progressively quenched as a function of nonluminescent polynucleotide (or mononucleotide) complexed. Neither the spectral shape nor triplet lifetimes are substantially altered. The exact quenching mechanism is unclear. Rahn *et al.* (1966a) proposed quenching at the singlet level via an excited-state proton transfer from Ade to Ura, but excited-state pK calculations do not favor this process (Lamola *et al.*, 1967). Even in highly quenched polynucleotides, quenching at the singlet level remains most probable, in view of the constant $^3\tau$'s (Kleinwächter *et al.*, 1968). Triplet sensitization experiments comparing single- and double-stranded polymers might clarify this point.

The situation for poly(rG·rC) is even more uncertain because the originally reported tenfold luminescence quenching in this double-stranded polymer compared to the free nucleotides (Eisinger *et al.*, 1966; Rahn *et al.*, 1966b) is not confirmed by the recent report of Kleinwächter (1972b), who found only a small reduction (0–30%) in the fluorescence or phosphorescence yields. Kleinwächter was careful to eliminate structural artifacts due to the solvent matrix by finding little difference for spectra taken in pure water, 0.25% glucose/water, or ethylene glycol/water (1:1), the latter solvent being identical to that used in the original reports. Also, the spectra are sufficiently different (red-shifted 900 to 1400 $cm^{-1}$) from poly(rG) or poly(rC) alone to eliminate the possibility of strand dissociation and emission from single strands. The triplet lifetimes are indicative of a 1:1 mixture of Cyd and Gua and also suggest minimal quenching. The data of Kleinwächter are self-consistent, but the discrepancy between this and early work may be critical in evaluating excited-state processes in DNA and deserves resolution by other investigators. Recently, the efficiency of singlet photochemistry in poly(dC), assayed by pyrimidine adduct fluorescence (see Chapter 6) or by light-induced exchange at C(5) of Cyd (Hauswirth and Wang, 1975a), was observed to be largely unaltered when in double-stranded complexes with poly(dI) or poly(dG) (Hauswirth and Wang, 1975b) and this appears to confirm Kleinwächter's observations.

### 2. Poly(dA-T · dA-T)

Singlet emission from double-stranded poly(dA-T·dA-T) resembles that of dApT, with absorption by either Ade or Thy giving identical spectra (Eisinger and Shulman, 1967). Fluorescence from a common singlet state is indicated. The triplet of poly(dA-T·dA-T) has been characterized by ESR (see Sec. F,2) and phosphorescence studies which show that only Thy emits (Lamola *et al.*, 1967). This is consistent with the ordering of mononucleotide triplet energies and with recent excited-state molecular orbital calculations on the Ade–Thy base pair (Danilov *et al.*, 1971).

Extensive triplet energy transfer similar to poly(rA) was not observed by Isenberg *et al.* (1967), who found that triplet quenching by $Mn^{2+}$ was ineffective beyond six base residues. Even this short range could be explained in terms of Förster transfer rather than intrastrand triplet migration. The interposition of Ade between Thy moieties also severely inhibits Thy–Thy triplet migration. This is contrasted to the efficient triplet transfer implied from the enhancement of $\phi_f$ by $Hg^{2+}$ bound to poly(T) (Rahn *et al.*, 1970).

## E. Polynucleotide Studies at Room Temperature

None of the common polynucleotides emit at room temperature. However, at a pH = pK (Cyd), fluorescence has been observed from poly(rC·rCH$^+$) (Favre, 1972); poly(rI·rCH$^+$) (Thiele *et al.*, 1972) and neutral CppC (Vigny and Favre, 1974). For the former, $\phi_f$ is $4 \times 10^{-3}$, and this, in contrast to CpC which is nonfluorescent at any pH, implies that a specific arrangement of bases along or between semiprotonated poly(rCH$^+$) chains is responsible for the higher $\phi_f$. The large Stokes shift (~11 kK) for poly(rC·rCH$^+$) also implies considerable excited-state interaction and is suggestive of the charge-transfer complex proposed by Montenay-Garestier and Helene (1970) for poly(rC) under similar pH conditions at low temperature. In the present case possible ground-state interaction is also indicated since the corrected excitation spectrum is red-shifted 500 $cm^{-1}$ compared to absorption.

Fluorescence has also been reported for poly(rG·rC) at pH 2.6 (Michelson and Pochon, 1969). In this case, protonation of Gua at N(7) produces emission similar to protonated GMP, hence the fluorescent polymer must be poly(rGH$^+$·CH$^+$). As these polynucleotides fluoresce only at nonphysiological pHs, their relevance to nucleic acid excited states is questionable.

We now turn to model synthetic polynucleotides which emit at neutral pH. Me$^7$Ino is fluorescent at 300°K and Leng *et al.* (1968) found that although the emission yield is reduced more than twofold progressing through the homologous series (Me$^7$Ino)$_n$ ($n$ = 1,2,3,4,5, poly), the spectral shape is unchanged. A red-shifted fluorescence, characteristic of excimeric interaction, is absent at room temperature. The fluorescence is also unaltered when $Me^7$Ino is incorporated into poly($Me^7$Ino·rC) (Pochon *et al.*, 1968). Similar results were reported for ribocopolymers of the fluorescent nucleotide analogs, formycin ($Am^6$Pur) and $Am^2$Pur, alternating with Urd, rThd, 5-FlUrd, and 5-BrUrd, except that a hundredfold fluorescence quenching occurred (Ward *et al.*, 1969). In both studies it would have been interesting to look at the 77°K luminescence spectra for a red-shifted fluorescence so that by analogy a more direct conclusion could be drawn concerning the presence or lack of excited-state interaction in natural polynucleotides at 300°K compared to 77°K. As it stands, the results suggest little interaction, but there is a possibility of a fundamentally different fluorescent transition in these models because of their low-energy fluorescence excitation spectrum compared to the natural nucleotides.

More recently Gill (1970, 1971, 1973) synthesized a series of alternating copolymers of the fluorescent $Me^5$dCyd ($\phi_f = 3 \times 10^{-4}$) with Ino and Guo as well as the corresponding double-stranded homopolymers. Compared to the mononucleotide, the absorption and fluorescence of $Me^5$dCyd is qualitatively unaltered in poly(5MdC), poly(5MdC·dI), or poly(dI-5MdC·dI-5MdC), again suggesting no excited-state interaction. On the other hand, both absorption and fluorescence of $Me^5$dCyd in poly(dG-5MdC·dG-5MdC) and poly(T-dG·dA-5MdC) are 4000 $cm^{-1}$ *blue*-shifted. This higher-energy shift cannot be due to enhanced base interaction and therefore is probably associated with a specific hydrogen-bonding effect on $Me^5$Cyd transitions.

The recent synthesis of a fluorescent derivative of poly(rA) opens new possible approaches (Steiner *et al.*, 1973). Treatment of poly(rA) with chloroacetaldehyde gives 3β-D-ribofuranosylimidazol (2,1-i) purine, commonly termed ε-Ade (Barrio *et al.*, 1972), randomly distributed in the polymer. Room-temperature and 77°K fluorescence are characteristic of ε-Ade, while 77°K phosphorescence is characteristic of both Ade (excitation at 38.5 kK) and ε-Ade (excitation at 29.5 kK) in poly($rA_{0.82} - \epsilon - rA_{0.18}$). It might be possible, therefore, to use ε-Ade as a "reporter group" for the room-temperature processes of Ade by comparison to 77°K data in triplet quenching experiments. Additionally, the synthesis of poly(2-aza-ε-rA) has been reported by Tsou and Yip (1974).

In summary, direct room-temperature fluorescence at low pH or from polynucleotide analogs gives no evidence for singlet exciplex formation and as yet nothing can be concluded about the triplet state.

## F. DNA

Consideration of the electronic states of DNA and transitions between these states introduces another level of complexity, namely that one must consider independent transitions of the four major bases and all the ensuing interactions, i.e., ground–ground, ground–excited, excited–excited state interactions of each spin multiplicity for both stacked and hydrogen-bonded conformations. Although the model studies discussed above fall considerably short of providing a clear picture of what to expect in a biological nucleic acid, a number of excited-state phenomena are sufficiently similar in DNA to allow construction of a preliminary picture of the fate of absorbed energy in DNA.

### 1. Ground-State Properties

It is not possible here to review the great number of works pertaining to nucleic acid structure. The reader is referred to the reviews of Cantor and Katz (1971) and Arnot (1970) as well as Chapter 1, Volume II. The atomic coordinates of native nucleic acids have been continually refined in the past decade, primarily by fiber x-ray studies, and recent stereoscopic drawings of the A and B forms of DNA are shown (Chapter 1, Volume II). Since the main structural determinant for excited-state properties is the spatial arrangement of transition dipoles, it seems possible from these data and from individual transition moment orientation to define a priori many electronic properties of nucleic acids. Indeed, this has been accomplished to a degree of success for the CD and ORD of dinucleoside phosphates, as mentioned previously. Additionally, assuming the dominance of nearest-pair interactions (Gray and Tinoco, 1970), reconstruction of CD (Allen *et al.*, 1972; Gray *et al.*, 1973) and hypochromism (Pysh and Richards, 1972; Brown and Pysh, 1972) of natural and synthetic nucleic acids is also possible. These results lead to the expectation that excited-state processes in DNA and RNA should be not much different from those observed in small oligomeric models. However, such calculated properties must ultimately base their validity on measured *solution* properties, hence solution methods complementary to crystal and fiber x-ray methods are of utmost importance in a complete understanding of macromolecular electronic states.

Toward this end, recent work on the linear dichroism and birefringence of oriented nucleic acids in solution holds promise (see reviews by Charney, 1971; Wada, 1972). The basic concept is that an *oriented* polymer containing a regular arrangement of chromophores will exhibit anisotropic absorption or refraction of incident light depending on whether the electric vector is aligned parallel or perpendicular to the main orientational axis, which in DNA is the axis of helical symmetry. In a manner analogous to CD, a linear dichroic spectrum is defined within an absorption band as $\Delta\epsilon(\bar{\gamma})/\epsilon(\bar{\nu})$ (the reduced dichroism) in which $\Delta\epsilon = \epsilon_{\parallel} - \epsilon_{\perp}$, the difference in extinction coefficients for the two polarized light beams. Linear birefringence is then the counterpart of ORD and measures the difference in refraction between mutually perpendicular linearly polarized light beams passing through the oriented polymer solution. Linear birefringence varies monotonically outside absorption bands and shows inflections (Cotton effects) for each transition within an absorption band. Since linear dichroism and birefringence are related through Kramers-Kronig transforms, as are CD and ORD, they give the same structural information.

For our purposes the most important information to be gained is an estimate of the orientation of transition moments relative to the helical axis, the degree of regularity of this orientation, and evidence for weak $n \rightarrow \pi^*$ transitions, possibly intensified relative to the monomer. Other data of less direct interest are macromolecular hydrodynamic properties including relaxation processes and the binding characteristics of smaller chromophores. The utility of dichroic spectra in determining the polarization of a transition moment in films and crystals of oriented bases and nucleotides has been outlined in Chapter 1, Section B. To a limited extent these solid-phase techniques have been extended to polynucleotides. In an oriented film of poly(rC) Rich and Kasha (1960) observed a small band to the red of the main birefringence band which they suggested may be an $n \rightarrow \pi^*$ transition because it was polarized perpendicular to the plane of the bases. However, no such band was found in films of poly(dA-T) (Gellert, 1961) or in DNA (Wada, 1964; Gray and Rubenstein, 1968). Additional dichroic studies pertaining to this still unresolved question are discussed below.

A significant difference in dichroic studies between nucleotides and polynucleotides is that physical methods of orienting relatively rigid polymers in solution, i.e., hydrodynamic flow or electric fields, provide information in biologically more relevant physical conditions. In some instances solution techniques have been extended to orientation of encapsulated nucleic acids (Cram and Deering, 1970; Gabler and Bendet, 1972). A major difficulty in solution, however, lies in accurate knowledge of the distribution of polymer orientation. In the case of flow dichroism, critical factors seem to be the details of the hydrodynamic molecular model, especially for somewhat flexible molecules (Wada, 1972). Polymer alignment by electric field circumvents many but not all of these problems, particularly if a pulsed field is used (Ding *et al.*, 1972; Yamaoka and Charney, 1973).

A theoretical plot of the parallel ($\Delta\epsilon_{\parallel}/\epsilon$), perpendicular ($\Delta\epsilon\ /\epsilon$), and net reduced dichroism as a function of the angle $\theta$ between the transition moment and orientational axis (assumed coincident to the molecular symmetry axis) is shown in Fig. 8a (Yamaoka and Charney, 1973). Since all known major transitions in the first nucleotide absorption bands are $\pi \rightarrow \pi^*$ polarized in the base plane (Chapter 2, Section B) and the normal conformations of nucleic acids do not have bases tilted very far from 90° to the symmetry axis it is expected and found that $\epsilon_{\parallel} > \epsilon_{\perp}$; thus, reduced dichroism is always negative. For the extreme case in which all polymers are fully oriented and all transition moments are perpendicular to the axis of orientation, $\Delta\epsilon/\epsilon = -1.5$. In fact, values less than $-0.8$ for DNA are rarely reported due to base tilting, incomplete molecular alignment, or lack of complete helicity.

Experimental values for the reduced dichroism have varied from $\sim -0.10$ to $-0.80$ depending, in part, on the structural differences of the DNAs used. A recent example of a reduced dichroism spectrum for native DNA, oriented by an electric field, is shown in Fig. 8b. The variation in dichroism from 300 to 220 nm indicates at least three and possibly four transitions. It is not possible at the moment to assign exact base orientations from such data because the true limiting value of $\Delta\epsilon/\epsilon$ at saturating electric fields is uncertain and the positions, bandwidths, and extinction coefficients of individual base transitions are in a state of continual refinement. Thus, from an extrapolated limiting value of $-1.35$, Ding *et al.* (1972) estimated $\theta \geq 80°$, consistent with either the A or B form of DNA. Also, possible out-of-plane ($n \rightarrow \pi^*$) transitions cannot yet be assigned, but such transitions, if they are weak and dominated by a coincidental in-plane $\pi \rightarrow \pi^*$ transition, might be inferred from the positive-trending dichroism at 230–240 nm and $>280$ nm.

## 2. Excited Singlet Properties—Low Temperature

The characteristic features of singlet emission from native double-stranded DNA are its red-shifted spectra relative to mixtures of monomers and approximately tenfold intensity loss (Eisinger *et al.*, 1966; Imakubo, 1968). The spectrum is similar to the double-stranded

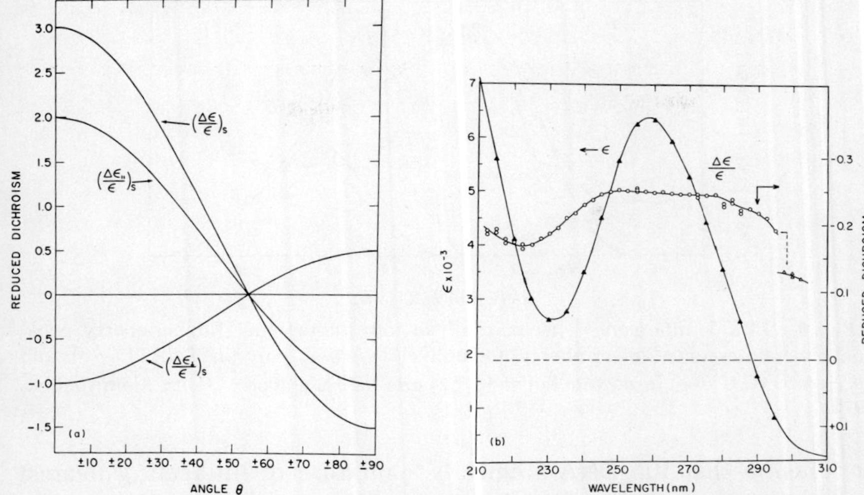


**Fig. 8.** *(a) Theoretical curves for the reduced ($\Delta\epsilon/\epsilon$), parallel ($\Delta\epsilon_{\parallel}/\epsilon$), and perpendicular ($\Delta\epsilon_{\perp}/\epsilon$) dichroism at a saturating electric field strength. $\theta$ is the transition dipole–molecular symmetry axis angle. (b) Reduced electric dichroism spectrum of native calf thymus DNA (0.24 mM in 1 mM NaCl; open circles). The open triangles are for a 3.6 mM DNA sample; filled triangles are an isotropic absorption spectrum. (From Yamaoka and Charney, 1973.)*

model poly(dA-T·dA-T) and to the stacked model, poly(rA-G-C-U) (Fig. 9), again with a lower $\phi_f$ (Table 5). These facts, together with previous considerations on oligomeric emission, lead to the general

**Table 5** Representative Low-Temperature Excited-State Properties of DNA

| Compound | Reference | Source | Solvent[b] | Fluorescence $\epsilon_{max}$ (kK) | Fluorescence $\phi_f$ | Phosphorescence $\epsilon_{max}$ (kK) | Phosphorescence $\phi_p$ | Phosphorescence $^3\tau$ (sec) |
|---|---|---|---|---|---|---|---|---|
| Native DNA | a | Calf thymus | G/W | 28.17 | 0.01 | 21.98 | 0.003 | 0.5 |
| | b | Calf thymus | EG/W | 28.4 | 0.005 | 22.3 | 0.002 | 0.3 |
| Denatured DNA | a | Calf thymus | G/W | 28.17 | 0.02 | 21.60 | 0.006 | 0.5 |
| | b | Calf thymus | EG/W | 29.8 | 0.02 | 23.8 | 0.02 | 0.3, 1.9 |

[a] References:
a. Kleinwächter *et al.* (1968).
b. Eisinger *et al.* (1966); Rahn *et al.* (1966b).

[b] G/W, 0.25% glucose/water; EG/W, ethylene glycol/water (1:1).

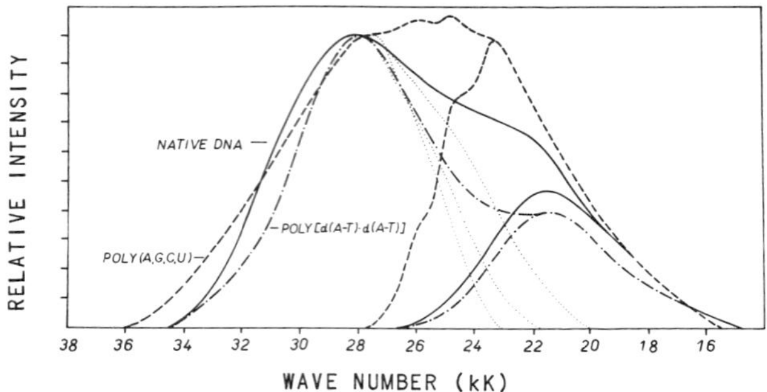


**Fig. 9.** *77°K luminescence spectra resolved into fluorescence (higher-energy peak) and phosphorescence (lower energy) for native DNA (—), poly (dAT·dAT) (-··-·-), and poly(rA,rG,rC,U) (---) in acetate buffer (pH 7) and 0.25% glucose. (From Kleinwächter, 1972c.)*

conclusion that the DNA singlet is a mixture of interacting nearest neighbors, each undergoing its own radiative and nonradiative processes. Unfortunately, an analysis of DNA fluorescence in terms of dinucleoside phosphate fluorescence is not possible from spectra alone because all spectra are broad and lie within ~500 $cm^{-1}$ of that for DNA itself. However, data possibly more characteristic of individual pairs, such as emission and excitation polarization, when available, may facilitate such analysis. At the moment other, more indirect, data must serve to define further the nature of the fluorescent singlet.

The emission spectrum of poly(rG·rC) is ~1000 $cm^{-1}$ lower than most other polymers and a comparable signal is generally absent in DNA. The reasons for this are still a point of some disagreement, arising from the question as to whether fluorescence is or is not quenched in poly(rG·rC) (Section D). Guéron *et al.* (1973), who maintained that emission is quenched in the model, proposed that all Gua or Cyt absorption events are quenched at the singlet level due to hydrogen bonding in DNA. Kleinwächter (1972c), who claimed to have observed relatively unquenched emission from the model, ascribed a low-energy shoulder in DNA fluorescence (Fig. 9) to weak contributions of Gua and Cyt. Furthermore, partial quenching of Gua–Cyt is ascribed not to hydrogen bonding but to unspecified base sequence effects. A preliminary report of an increase in fluorescence yield with the Ade–Thy content of DNA (Lamola *et al.*, 1967) supports either idea, but more quantitative data are not available.

The difference in fluorescence energy between Ade–Thy and Gua–Cyt base pairs might allow emission polarization studies to resolve this problem by locating the position of each component in a DNA spectrum. In a complementary manner, comparison of excitation and absorption spectra may also be helpful. It should be noted that such experiments involving radiative properties can only deduce the source of emission and not whether less efficiently emitting centers may have contributed to fluorescence by energy transfer to centers of higher intrinsic $\phi_f$. In theory this process may be detected from singlet quenching data, but the short lifetimes of the bases and nucleotides (Chapter 2, Section C) makes it difficult in a practical sense. Nevertheless, some attempts have been made at room temperature using intercalated molecules as quenchers; these will be discussed later.

At the moment, the only means we have to estimate the possibility of singlet energy transfer in DNA comes from the theoretical calculations of Guéron *et al.* (1967a,b). Within the framework of Förster's "very weak" coupling case (Förster, 1965), two singlet transfer mechanisms are distinguished depending on whether the transfer rate is faster or slower than relaxation from the vibrationally excited singlet, $S_1$ (termed respectively "before" and "after relaxation" transfer).

In both cases transfer efficiency is proportional to the overlap integral of the donor emission and acceptor absorption. "Before relaxation emission" $[S_1(\nu \neq 0) \rightarrow S_0]$ is assumed to have a spectral distribution similar to absorption with its blue edge at the excitation energy (see Fig. 1, Chapter 2). Hence this sort of transfer will be a function of excitation energy with maximum efficiency at $\bar{\nu}_{\text{excitation}} = \bar{\nu}_{\text{max}}$ (absorption). The derived rate expression is,

$$N_b = \frac{2\pi}{h} U^2 \int \text{overlap}$$

Assuming that the interaction potential $U$* is determined by dipole–dipole interaction within the confines of a Watson-Crick geometry, transfer rates are $10^{12}$ to $10^{13}$ $\text{sec}^{-1}$ for most pairs of nucleotides. Since vibrational relaxation rates are of the same order (Chapter 2, Section A), "before relaxation transfer" is considered possible at both room and low temperature in DNA but could not extend for more than a few bases.

"After relaxation" transfer, developed within the same framework, corresponds to the well-known Förster singlet transfer mechanism. In terms of a distance $R_0$ at which transfer is 50% efficient:

* U's of several hundred $\text{cm}^{-1}$ were estimated and are comparable to monopole calculations of Bush and Tinoco (1967).

$$R_0^6 = (\text{constant})\mathbf{K}^2 \int \phi_f F_0(\bar{\nu})\epsilon_A(\bar{\nu}) \frac{d\bar{\nu}}{\bar{\nu}^4}$$

in which **K** is a transition dipole orientation factor between donor and acceptor, $\phi_f$ is the donor fluorescence quantum yield in the absence of acceptor, $F_0(\bar{\nu})$ is the donor fluorescence spectrum (normalized to unity), and $\epsilon_A$ is the acceptor absorption spectrum in terms of its molar extinction coefficient. At 77°K, calculated $R$'s for $\sim 70\%$ of the interbase transfer combinations give values greater than the Watson-Crick separation of 3.4 Å and are, therefore, theoretically possible. However, at room temperature, with $\phi_f$ a factor of $\sim 10^3$ smaller, Förster transfer is not expected. Also, the possibility of singlet transfer via short-range exchange interactions (see section A) cannot be eliminated since the interaction potential ($M \cong 10\ \text{cm}^{-1}$) is only an order of magnitude less than at 77°K.

In view of the uncertainty in the exact spectral origin of the fluorescent transitions in the nucleotides (see Chapter 2) contrasted to the assumption implicit in the above calculations that fluorescence is derived from the entire low-energy nucleotide absorption band, it seems best to consider these calculations a first estimate requiring further refinement and experimental verification, particularly at room temperature.

### 3. Triplet Properties—Low Temperature

Historically, DNA phosphorescence was the first nucleic acid excited-state property to be reported (Steele and Szent-Györgyi, 1957) and has held the attention of many laboratories since. The primary problems have been the source of triplet emission and the processes leading to or quenching the radiative state or states. The answer to the first problem has evolved, through several explanations, to the presumably correct answer at present, while data bearing on the second problem come mainly from the model studies discussed previously and DNA quenching experiments, and many details remain uncertain.

Agroskin *et al.* (1960, 1961) first proposed that phosphorescence occurs only from Pur residues in DNA since apurinic acid shows no luminescence, apyrimidinic acid shows enhanced luminescence, and, at pH 1, DNA luminescence is solely characteristic of Gua. Subsequently, Longworth (1962) observed measurable phosphorescence from the free Pur bases at neutral pH and Bersohn and Isenberg (1963, 1964) attributed structured DNA phosphorescence to a summation of Pur spectra. However, the same groups also observed that $\phi_p$ was approximately an order of magnitude lower for DNA than for a mixture of purines and that triplet decay was nonexponential, but with a major component of ~0.3 sec, much shorter than Pur lifetimes (see Table 11, Chapter 2; and Tables 2 and 5, this chapter).

Douzou *et al.* (1961), to account for the short DNA $^3\tau$, proposed that Pyr moieties serve as triplet traps via Pur → Pyr triplet transfer. Subsequently, the similarity in ESR signals and phosphorescence between singly ionized TMP (pH 12) and native DNA (Table 6) (Rahn *et al.*, 1965) led to the proposal that DNA phosphorescence was primarily from ionized Thy produced by excited-state proton exchange with Ade, simultaneously quenching the Ade triplet.

Indications that this explanation was also unsatisfactory came from the observation that phosphorescence could be observed from 1,3-$Me_2$Thy, which cannot ionize (Hélène, 1966), and also from Thy itself (Kleinwächter *et al.*, 1966). By triplet sensitization with acetone or acetophenone, the triplet properties of neutral TMP were finally accessible (Table 6) (Lamola *et al.*, 1967). Based on the very closely similar ESR and luminescent properties between neutral TMP and DNA, an unfavorable energy barrier to Ade–Thy proton transfer, the fact that TMP has the lowest-energy triplet of all nucleotides, and a linear correlation of DNA Ade + Thy content with the Thy ESR signal intensity (% Ade + % Thy < 65%), these authors proposed neutral

**Table 6** 77°K ESR Properties of DNA and Some Constituent Mono- and Polynucleotides

| Molecule | Reference[a] | pH | $H_{min}$ (G) | $D^*$ ($cm^{-1}$) | $\tau$ (sec) |
|---|---|---|---|---|---|
| DNA[b] | a | 7 | 1054 | 0.199 | 0.30 |
| DNA[b]($D_2O$) | a | 7 | — | — | 0.45 |
| Poly(dA-T·dA-T) | a | 7 | 1040 | 0.201 | 0.30 |
| Poly(rA) | c | 7 | — | 0.126 | 2.6 |
| Poly(rC) | b | 7 | 1121 | — | 0.6 |
| Poly(rG) | c | 7 | — | 0.145 | 1.26 |
| TMP[b] | a | 7 | 1050 | 0.200 | 0.33 |
| TMP[b]($D_2O$) | a | 7 | — | — | 0.52 |
| TMP | a | 12 | 1065 | 0.198 | 0.50 |
| TMP($D_2O$) | a | 12 | — | — | 0.55 |

[a] References:
a. Lamola *et al.* (1967).
b. Rahn *et al.* (1966b).
c. Rahn *et al.* (1964).
[b] Acetone-sensitized triplet.

rather than anionic Thy as the source of DNA phosphorescence [similar conclusions were also later reached by Kleinwächter *et al.* (1968) and Imakubo (1968)]. The only evidence contrary to this most recent conclusion is the structured DNA phosphorescence observed by Bersohn and Isenberg compared to the less structured spectra generally reported (e.g., Fig. 9). Kleinwächter (1972c) suggested that, in the solvent used by Bersohn and Isenberg (95% glycerol), disruption of base interactions could inhibit triplet transfer to Thy, thus allowing more individual Pur emission and hence more structured phosphorescence. A similar increase in structure and $\phi_p$ was observed in denatured DNA compared to native DNA (Rahn *et al.*, 1966b; Isenberg *et al.*, 1967).

Assuming that the DNA triplet is indeed localized on Thy, the important question as to the mechanism of triplet energy localization remains. Paramagnetic ion quenching (Isenberg *et al.*, 1965, 1967; Rahn *et al.*, 1966b) of DNA phosphorescence is effective up to a range of 10 to 20 bases depending on the ion used. As in quenching of poly(rA), neither 77°K fluorescence nor the triplet lifetime is affected. However the range of quenching is much smaller than for poly(rA), probably due to the interposition of Ade or other bases between Thy triplets as discussed previously. The unchanged $^3\tau$ again suggests that transfer occurs much faster than phosphorescence.

Sensitized phosphorescence from 9-aminoacridine intercalated into DNA provides similar results (Galley, 1968). Direct excitation of the bound dye (at energies where DNA does not absorb) gives a characteristic p/f (phosphorescence/fluorescence) value which is increased eightfold when excitation is absorbed first by DNA and then transferred to the dye. Only DNA triplet → dye triplet transfer will increase p/f; thus p/f measures the extent of DNA triplet quenching by the dye. A limiting value of p/f is reached at approximately one intercalated dye for every 30 bases, implying a quenching range of 15 bases, similar to metal ion quenching results.

Since the effective range of a quencher is expected to vary according to the nature of the quencher and the type of binding to DNA, the significance of a quenching range relative to the extent of actual triplet migration in DNA is uncertain. For example, a metal ion bound to the phosphates of DNA may be in contact with one or two phosphates on the same strand but remote from the bases, while an intercalated dye will have electronic interaction with the bases on one strand in addition to causing some structural alteration. Therefore, the coincidence of quenching range from these two methods may be fortuitous, and it is not possible to evaluate an accurate triplet migration range from quenching data without more extensive knowledge of the structural changes involved.

Recently a more theoretical approach to the extent of DNA triplet migration has been elaborated (Eisinger and Lamola, 1971; Guéron *et al.*, 1973). Using calculated triplet–triplet transfer rates between bases, as discussed in Chapter 2, Section A, the probability of triplet migration in a given sequence can be determined. The migration range varies depending on the sequence and hence is a function of the Ade–Thy content. Calculated triplet delocalization is quite limited (2 to 5 bases) regardless of the type of DNA, but may be larger if interstrand transfer is included. We emphasize that there are many untested assumptions in this calculation, but the general approach should be a useful guide for future analysis.

The temperature dependence of DNA phosphorescence decay between 4.2° and 160°K was recently investigated (Szerenyi and Dearman, 1972). At any temperature, the triplet decay can be described by two first-order components; one, ~0.3 sec, corresponds to the commonly observed decay (Table 5) and the other is considerably shorter, ~0.06 to 0.09 sec. Both become progressively shorter at higher temperatures and $\phi_p$ undergoes a 20-fold decrease between 4.2° and 160°K. While no analytical description is possible, these trends are not inconsistent with a proposed thermal activation of the DNA triplet (presumably involving only Thy) inducing back migration to the higher triplets of neighboring bases. Subsequent triplet quenching could explain the shortened lifetime and decreased $\phi_p$. The sensitivity of $^3\tau$ and $\phi_p$ to temperature further emphasizes the problem of reproducibility of these parameters.

## 4. Room-Temperature Studies

The quantum yields for radiative processes in DNA suffer large reductions, similar to all other models, upon going from 77°K to room temperature. Native double-stranded DNA, having $\phi_f$ and $\phi_p$ an order of magnitude lower than other polynucleotides at 77°K (Table 5), might also be expected to be the least fluorescent at room temperature. Based on recently determined $\phi_f$'s for the free bases and nucleotides ($\sim 10^{-4}$) and assuming an order of magnitude reduction similar to that at 77°K for the double-strandedness, as well as a further reduction commonly associated with fluorescence from interacting bases, native DNA should have $\phi_f \leq 10^{-6}$.

Despite this consideration, one group claims to have observed DNA fluorescence at room temperature (Pisarevskii *et al.*, 1968). Using photon counting techniques, denatured calf thymus DNA was reported to have $\phi_f \sim 10^{-5}$ while native DNA has $\phi_f < 10^{-6}$. However, the reported spectral characteristics are inconsistent with more recent room-temperature data on the constituent bases (Daniels and Hauswirth, 1971; Hauswirth and Daniels, 1971; Vigny, 1971a,b) and suggest a possible artifactual origin for the observed signals. Specifically, denatured DNA showed fluorescence bands* at 33.33 and 25.97 kK with excitation maxima at 36.76 and ~31.0 kK respectively. Monomeric constituents have fluorescent maxima between ~30.0 and 32.5 kK (Chapter 2, Section C) and should red-shift under the stacked-base conformation of DNA. Hence, DNA emission at ~26 kK is not illogical except that the excitation maximum for this band is at 31.0 kK, entirely outside DNA absorption, even for an $n \rightarrow \pi^*$ transition (~34.0 kK). The higher-energy emission band, peaking at 33.3 kK, exhibits a Stokes shift of only 3.43 kK compared to the normal ~6 kK shift for monomers. This shift is almost coincident with a Raman scattering band shift of 3.39 kK observed for aqueous solutions (Parker, 1968), and the possibility that the reported emission band is uncompensated Raman scattering must be considered. More recently, using the techniques developed for the study of the bases at room temperature, the fluorescence emission spectrum of salmon sperm DNA has been determined at room temperature (Daniels, 1973). Compared with the spectra of the bases, the DNA emission was found to be broader and somewhat red-shifted, suggesting the occurrence of both monomer and excimer charge-transfer components. The quantum yield ($\phi_f \simeq 2 \times 10^{-5}$) was an order of magnitude greater than expected. These are no more than preliminary results, and further developments must await refinements in technique which will allow determination of excitation spectra at higher resolution and the determination of polarization characteristics.

Analysis of the distribution of direct photochemically produced pyrimidine dimers and UV-produced defects in double-stranded DNA may have relevance to the question of absorbed energy migration in DNA. Early evidence indicating a clustering of dimer repair sites on DNA (Brunk and Hanawalt, 1969) and a preference for dimer formation in triplet pyrimidine sequences (Setlow *et al.*, 1964) was confirmed recently by Brunk (1973) who found that the efficiency of dimerization increased with the length of a pyrimidine run. This suggested that non-random photodimerization occurred in DNA

* The spectra are assumed to be uncorrected since no statement to the contrary is apparent.

which might be explained by energy migration into pyrimidine tracts. This conclusion was independently reached by Shafranovskaya *et al.* (1973) using the kinetic formaldehyde method for determining the number of UV induced defects (see Vol. II, Chapter 3 and Frank-Kamenetskii and Lazurkin (1974) for details of this technique). However the critical observation that the number of dimers per defect increases with UV exposure has been recently disputed (Jonker and Blok, 1975) and resolution of this question awaits a better understanding of the relationship between a pyrimidine dimer and a DNA defect as detected by formaldehyde denaturation. In any event the estimated energy migration distance of $10^3$-$10^4$ bases (Shafranovskaya *et. al.*, 1973) seems considerably too high. At present the best explanation of dimer clustering is a combination of short range energy migration into pyrimidine tracts and preferential dimer formation within single-stranded regions of DNA.

## 5. Photoconductivity of DNA

The early suggestions of Riehl (1940) and Szent-Györgyi (1941) that the concept of the electronic band structure of periodic solids might be pertinent to biological phenomena received little attention before the elaboration of the helical structure of DNA. But as the relation between the genetic role of DNA and its structure developed, it became apparent that this periodic structure could possibly provide a molecular framework for electronic conductivity, which in turn might be highly relevant to considerations of energy transfer and storage. DC semiconductivity (i.e., thermally excited conductivity) was found to be a property of DNA by Eley (1959) and by Duchesne *et al.* (1960), and work has continued at a slow but steady pace since then. We shall summarize these results and discuss some of the experimental and conceptual problems encountered in this work, with particular reference to understanding the processes of photoconductivity.

Phenomenologically, semiconductivity characterizes a class of materials whose electrical behavior lies between insulators ($R > 10^{17}$ ohm cm) and conductors ($R < 10^{-4}$ ohm cm), described by a specific conductivity $\sigma$ [ohm cm)$^{-1}$] which has a positive temperature coefficient. A major concern is to determine if the conductivity is extrinsic to the material (i.e., due to an impurity such as metal ions or an adsorbant such as oxygen) or intrinsic. All of the studies reported so far (with the exception of those of O'Konski) assume at the outset that the semiconductivity of DNA is intrinsic and, applying the concepts of energy band theory as developed for materials such as silicon and germanium, interpret the thermal activation energy as the separation between intrinsic ground and conductive bands according to the relation

$$\sigma = \sigma_0 \exp\left(-\frac{\Delta E_g}{2kT}\right)$$

There seems to be reasonable agreement for dry DNA in the absence of $O_2$ that $\Delta E_g \simeq 2.4$ eV (Burnel *et al.*, 1969). Subertova *et al.* (1969) reported 2.04–2.25 eV using oriented DNA (orientation not specified) and Snart (1969) reported a strong orientation effect, finding the resistance across the DNA fibers a hundredfold greater than along the fiber axis (the dependence of $\Delta E_g$ on fiber orientation was not reported). It should be noted that the actual experimental dark activation energy, $\Delta E_d$, not presupposing any model, is thus ~1eV.

Conductivity is further characterized by the mechanism of charge transport, and it is necessary to attempt to establish whether conduction occurs by an electronic or ionic mechanism, or indeed by both. The role of ionic conduction should be indicated by the occurrence of electrolysis. No experiments have been reported for dry DNA and only a very brief mention exists (O'Konski *et al.*, 1964) of electrolysis of a sample held at 93% relative humidity. In general, the conductivity is determined by three factors: (1) the charge of the carriers, (2) the density of carriers, and (3) the mobility of carriers, expressed in the relation

$$\sigma = [\Sigma Z_i n_i \mu_i + n_e \mu_e]$$

in which $i$ refers to the ion of charge $Z_i$, the subscript $e$ refers to the electron or positive hole, and $\mu$ is the carrier mobility. If ionic conductivity can be disregarded, then the problem of determining the nature and sign of the charge carriers is usually carried out by measurements of the Hall effect or Seebeck effect. Only one determination of the Hall effect has been carried out, at microwave frequencies, (Trukhan, 1966), with the result that for dry T4 phage DNA, a $p$-type carrier is indicated with a mobility $\mu$ of 0.5 cm$^2$/volt sec, a value similar to those found for anthracene and phthalocyanine (single crystal) and considerably different from the free proton ($23.6 \times 10^{-3}$ at 25°C) and $Na^+$ ($5.2 \times 10^{-4}$). Positive hole conduction is thus indicated under these conditions, although "ice-type" protonic conduction is not excluded. There has been no measurement of the temperature dependence of this mobility.

There are several lines of evidence supporting the idea that semiconductivity can occur by an electronic mechanism in dry DNA. First, a DC conductivity is observed which is steady in time and which exhibits ohmic behavior (Eley and Spivey, 1962; Snart, 1963; O'Konski *et al.*, 1964; Subertova *et al.*, 1969). This DC conductivity has an overall activation energy of 1.0–1.2 eV. Investigation by AC methods (O'Konski *et al.*, 1964) has shown an extensive dispersion of the conductivity at frequencies up to $10^9$. The high-frequency conductivity is independent of water content at low degrees of hydration and is also largely independent of the nature of the counterion ($Na^+$, $Li^+$, $Mg^{2+}$, $Ca^{2+}$). Thus, under conditions in which electrode polarization and electrolysis are negligible, the conductivity is probably electronic in nature. However, the high-frequency activation energy is much lower than the DC activation energy at $\Delta E_d \sim 0.2$ eV (O'Konski, 1963), and this suggests that the mechanism of conductivity is mixed. A working model can be described as follows. Positive hole migration occurs over a limited spatial extent in a band structure characterized by a mobility $\mu \sim 1$ cm$^2$/volt sec$^{-1}$ and an activation energy $\sim 0.2$ eV. Such a migration encounters barriers—for example, at the ends of macromolecules, at kinks in the molecule, at microcrystal surfaces—such that the overall DC activation energy is much higher at $\sim 1.2$ eV. It is unfortunate that the DC mobility is unknown.

In the dry state, DNA is photoconductive: i.e., at constant temperature the current passed increases considerably when light impinges on the sample (Liang and Scalco, 1963, 1964). Snart (1968) quoted a 500-fold decrease in resistance on illumination. The photocurrent shows an exponential temperature dependence,

$$i = i_0 \exp\left(\frac{-\Delta E}{kT}\right)$$

with an activation energy $\Delta E \sim 0.9$–$1.2$ eV (Liang and Scalco, 1964; Snart, 1969; Subertova *et al.*, 1969). Only a small change in magnitude is found for the high-frequency mobility of the major charge carrier under illumination ($\mu \sim 0.85$ cm$^2$/volt sec; Trukhan, 1966) but no change in sign. The temperature dependence of a photoconductor under constant irradiation could be due to several causes, among them (1) a temperature dependence of carrier mobilities, (2) a shifting of the Fermi level in a logarithmic distribution of traps, or (3) it could correspond to the difference between the energy of the hole–electron pair and the photon which would be the dissociation energy of the initial excited state, whatever its nature. No distinction can be made among these alternatives in the present state of experimental information; however, it would be very interesting to determine the activation energy for high-frequency photoconductivity and the temperature dependence of DC mobility; together these may provide a distinction between (1) and (2). The precise relation between the energy of the photon and the activation energy raises the question of the excitation spectrum for photoconductivity and here some unusual results are reported. Liang and Scalco (1963) induced photoconductivity using a tungsten source, a blackbody emitter with most of its output at ~1000 nm and very little, if any, in the DNA absorption region. Subertova *et al.* (1969) essentially confirmed this, finding photoconductivity excited by radiation passing a filter transmitting from 450 to 600 nm. This seems to constitute clear evidence that the generation of photoconductivity can be extrinsic, by impurity excitation. Generation can, however, also be intrinsic: Snart (1968) determined an (uncorrected) excitation spectrum which corresponds to the UV absorption of DNA and Subertova *et al.* (1969) found that a filter transmitting from 300 to 400 nm (i.e., in the tail of the DNA absorption region) is also effective. The interesting point is that the activation energy for the photocurrent seems to be independent of photon energy, and this would seem to rule out (3) above.

Further complexity in the mechanism of photoconductor carrier generation is shown in the dependence of photocurrent on the intensity of exciting light (Snart, 1973). Using the unfiltered beam from a mercury medium-pressure arc, the photocurrent shows two dependencies on $I_0$. At high $I_0$ the relation is $i \propto I_0^{0.2}$ whereas at low intensities $i \propto I_0^{1.5}$. The latter behavior recalls the biphotonic photoionization discussed in Chapter 2, Section D, whereas the former may be related to a concentration-dependent mobility (cf. phthalocyanine).

It must be reemphasized that all the previous discussions concern the semiconductivity and photoconductivity of *dry* DNA in the absence of oxygen. The association of water with DNA, which is so important for the maintenance of the helix geometry, causes profound changes in the conductivity of DNA, and we consider next the behavior of DNA in the presence of water. At the outset it must be noted that the presence of water in an ionized polyelectrolyte immediately raises the possibility of conductance by impurity ions, protons, or the alkali counterion. In addition, there are experimental problems of electrode polarization, non-ohmic relations, and net electrolysis. DC methods are inherently unsuitable for this sort of situation, but the only work in which semiconductivity and photoconductivity have been compared on the same samples has used the DC method.

There are two noticeable features of the effect of water on the dark conductivity. First, whereas the resistance in the dry state is established within a few seconds of applying the voltage, for a wet specimen the time required to reach steady resistance is much greater (Liang and Scalco, 1964, found 200 sec at 15% relative humidity and 700 sec at 44% relative humidity—these values are obviously specific for a particular experimental geometry). While the slow change of resistance may indicate ionic conduction or electrode polarization, these workers suggested that the initial value largely represents electronic conduction. Second, it is then found that the initial resistance decreases by many orders of magnitude with increasing water content of the sample. Temperature dependence studies of this decrease in resistance then show that it is essentially due to a lowering of the activation energy and not a change in the preexponential factor. It is rather remarkable that the same behavior is observed for the photoconductivity, the ratio $\Delta E_d/\Delta\epsilon$ being almost independent of water content. The lowest values given by Liang and Scalco (1964) after only one day of dehydration are $\Delta E_d \sim 0.8$ eV and $\Delta\epsilon \sim 0.7$ eV. No values seem to be available for nondehydrated material. Although these results are rather limited, they can be seen to be compatible with the working model for dry DNA sketched above. Let it be supposed that water does not affect the process of generation of charge carriers but has its main effect on the mechanism of conduction. The effect of water is then probably to lower the barrier to DC charge transport, and the magnitude of the change is such that it may be a dielectric effect on the polarization energy of the barrier. If the barrier is ionic in nature (or just represents a local charge), then, following the Born charging process, the change in its polarization energy on hydration can be written

$$\Delta P = \frac{e^2}{2r}\left(\frac{1}{K} - \frac{1}{K'}\right)$$

in which r is the effective cavity radius of the charge, and $K$ and $K'$ are the low-frequency ("static") dielectric constants for the dry and wet states, respectively. Experimentally the initial resistance of DNA decreased by a factor of $10^{12}$ on going from the dry state to a state at 90% relative humidity (Liang and Scalco, 1964), and as in this limit $1/K \gg 1/K'$, then

$$10^{12} = \exp\left(\frac{e^2}{2kTrK}\right)$$

For $K = 3$ (O'Konski *et al.*, 1964), r may be estimated as 3.3 Å, and from these numbers, the decrease in activation energy is $\sim 0.7$ eV; this is not in disagreement with experiment as it is apparent from the findings of Liang and Scalco that the lower limit of $\Delta E_d$ has not been reached. The residual activation energy ($\sim 1.0$ eV $- \sim 0.7$ eV $=$ 0.3 eV) is then not too different, perhaps not significantly different from the high-frequency activation energy.

The above discussion has been kept rather brief in view of its speculative nature. Clearly more extensive experimentation is called for and some of the required experiments are listed.

1. The DC mobility of the majority charge carrier needs to be determined for dry DNA. Flash photoconductivity would seem to be an appropriate method.
2. The temperature dependence of the DC mobility would distinguish between temperature-dependent generation and temperature-dependent transport.
3. A corrected action spectrum is needed for excitation from 250 to 1000 nm, together with a good determination of the quantum yield at 500 nm.
4. The identity of the impurity which seems to be responsible for thermal and photon charge generation must be determined.
5. The molecular identity of the barrier to conduction needs elucidation.
6. Data are needed on helical DNA at 94% relative humidity clearly delineating the roles of electronic, proton, and ionic conduction.

The chemical and biological consequences of photoconductivity of DNA are not entirely clear at this time. However, it does seem significant that there is an inverse correlation between luminescence and photoconductivity. Whereas direct excited-state processes are so heavily quenched at room temperature as to be very difficult to detect, photoconductivity, having a positive temperature coefficient, may be the predominant physical process in DNA at room temperature, particularly in wet or hydrated DNA. It is also noteworthy that the efficiency of photochemical dimerization of thymine in DNA increases by an order of magnitude between −196°C and room temperature (Rahn and Hosszu, 1968). Only future work can show if there is a relation between energy transfer by photoconduction and photochemical reactivity.

## G. RNA

The intrinsic excited-state properties of RNA (Steiner *et al.*, 1967) have received little attention compared to those of DNA, primarily as a result of interest in DNA generated by its dominant role in cellular UV inactivation. It is important to note, however, that photochemical effects in viral RNA and certain transfer RNAs can also be linked to loss in biological function (see Chapter 7, Volume II for a full discussion); hence, RNA excited states are in the same sense relevant to cellular UV effects. The majority of model polynucleotide emission studies bear more directly on RNA excited states since, most commonly, the ribopolymers were used. The model RNA, poly(rA-rC-rG-U), has a usual red-shifted singlet emission at 77°K and a structured phosphorescence suggesting principally triplet emission from Ade, which for RNA bases has the lowest triplet level (Fig. 9; Kleinwächter and Koudelka, 1972). Triplet lifetimes confirm this, and $\phi_f$ and $\phi_p$ are the magnitudes expected from those of the model two-base copolymers (Table 4). Contrasted to this single-stranded model, a natural RNA molecule may have many hydrogen-bonded, double-stranded regions which may additionally quench emission processes of the common bases (see Section D). Quenching in single-stranded RNA regions may also be possible by energy transfer from these sequences to adjacent double-stranded regions, although such processes have not been specifically investigated.

The fact that several tRNAs contain "odd" bases which dominate the room- and low-temperature emission spectra has prompted their use as structural probes. Specifically, room-temperature fluorescence is observed from 7-MeGua (Leng *et al.*, 1968), the so-called Y base (Yoshikami *et al.*, 1968; Eisinger *et al.*, 1970) which has been identified as a 7-substituted 4,9-dihydro-4,6-dimethyl-9-oxo-1H-imidazo[1,2-*a*]purine (Nakanishi, *et al.*, 1970; Kasai *et al.*, 1971; Blobstein *et al.*, 1973) located at the 3′-end of the anticodon (Rajbhandary *et al.*, 1967; Dudock *et al.*, 1969), and Sra* (Favre *et al.*, 1969) located in the eighth nucleotide from the 5′-end (Holley *et al.*, 1965). These bases absorb and emit to the red of the four common bases (Table 7) and this fact has been used to selectively excite the latter two for use as "reporter" groups in intact tRNA.

Beardsley and Cantor (1970) observed room-temperature singlet energy transfer from the Y base in $tRNA^{Phe}$ to acridine dyes covalently attached to the 3′-terminus. From the Förster equation a donor–acceptor distance of $>40$ Å was calculated. Assuming that this corresponds to separation between the anticodon region and the amino acid accepting end, only certain types of tRNA tertiary structures were concluded to be allowed.

* The question of the origin of room temperature emission from Sra, whether fluorescence or phosphorescence, is discussed at the end of this section.

**Table 7** Room-Temperature Emission Properties of Some Odd Bases of tRNA.

| | Fluorescence $\epsilon_{max}$ (kK) | | $\phi_f$ | |
|---|---|---|---|---|
| Compound | Monomer | In tRNA | Monomer | In tRNA |
| 7Me-Gua[a] | 26.2 | — | .035 | — |
| Y-base[b] | 22.0 | 23.1 | .035 | .07 |
| Srd[c] | ~18.2[d,e] | 18.9[d] | $<10^{-4}$ [e] | .0015[d] |

[a] Guéron *et al.* (1973).
[b] Eisinger *et al.* (1970); $tRNA^{Phe}$ in 0.01 *M* $Mg^{2+}$.
[c] In Shalitin and Feitelson (1973) evidence is presented suggesting that Srd emission is room-temperature phosphorescence (see text).
[d] Pochon *et al.* (1971); $tRNA^{Val}$ in 0.01 *M* $Mg^{2+}$.
[e] Shalitin and Feitelson (1973).

The Y base, being adjacent to the anticodon, also shows small fluorescence blue-shifts when the RNA binds a natural or synthetic codon, hence binding constants could be evaluated (Eisinger *et al.*, 1970). The fluorescence is additionally sensitive to anticodon conformation since divalent cations, known to bind to tRNA and induce structural changes, increase $\phi_f$ as much as fourfold (Beardsley *et al.*, 1970; Eisinger *et al.*, 1970; Zimmerman *et al.*, 1970).

Sra photochemistry has received considerable attention both in tRNA and in free solution (see Chapter 6 for a detailed discussion). Irradiation of tRNA at ~30 kK excites only Sra producing a photoadduct between itself and a Cyd residue of the thirteenth nucleotide in $tRNA^{Val}$ (Favre *et al.*, 1969) as well as in other tRNAs containing these bases (Yaniv *et al.*, 1969; Chaffin *et al.*, 1971), or the analogous product in frozen aqueous mixtures of the two bases (Bergström and Leonard, 1972). Interestingly, the structure of this product is identical to the photoadduct obtained from poly(rC) (Rhoades and Wang, 1971a,b); the 4-thio group is lost during formation of the bipyrimidine linkage in tRNA.

Recently Shalitin and Feitelson (1973) examined more closely the emission properties of Srd. Based on its unusually large Stokes shift (~11.6 kK), a diffusion-controlled oxygen quenching, emission sensitization with 4-methoxyacetophenone (via triplet–triplet energy transfer), and low-temperature lifetime of 300 $\mu$sec, they concluded that the weak Srd emission at ~18kK is a rare example of room-temperature phosphorescence rather than fluorescence as was concluded earlier. Additionally, at low temperature, a second, stronger phosphorescence is observed at ~21 kK, with a lifetime of 960 $\mu$sec. The exact nature of these two triplets remains open, but it is suggested that the

~21 kK state is a "Franck-Condon" triplet of Srd while the ~18 kK state is either the solvent-relaxed triplet or the triplet of a transient formed by Srd photochemistry. The phosphorescence of yeast $tRNA^{Phe}$ has been recently studied in low temperature glasses at 77°K and at 1.2°K (Hoover *et al.*, 1974). At the higher temperature only the Y-base is phosphorescent while at 1.2°K only Ade phosphorescence is observable. Adenine emission is quenched by codon binding or removal of $Mg^{2+}$, both of which alter the anticodon conformation, suggesting that emission arises from within the anticodon ($G_m$-A-A). The quenching of the adenine triplet upon shifting to 77°K is attributed to both energy transfer to the Y base and thermal quenching, but the nature of the transfer process, whether singlet-singlet or triplet-triplet, is not determined. The relationship between the two adenine triplet lifetimes (1.33s and 9.25s) resolved at 1.2°K and the one (2.6s.) commonly observed at 77°K is not apparent and calls for a study of the temperature dependence of $\tau_3$.

## References


Agroskin, L. S., Korolev, N. V., Kulayev, I. S., Meisel, M. N., and Pomashchinkova, N. A. (1960). *Dokl. Akad. Nauk SSSR* **131,** 1440.

Agroskin, L. S., Korolev, N. V., Kulayev, I. S., and Pomashchinkova, N. A. (1961). *Dokl. Akad. Nauk SSSR* **136,** 226.

Allen, F. S., Gray, D. M., Roberts, G. P., and Tinoco, I., Jr. (1972). *Biopolymers* **11,** 853.

Arnot, S. (1970). *Progr. Biophys. Mol. Biol.* **21,** 267.

Bailey, M. (1972). *Biopolymers* **11,** 1091.

Barrio, J. R., Secrist, J. A., and Leonard, N. J. (1972). *Biochem. Biophys. Res. Commun.* **46,** 597.

Bazin, M., Santus, R., and Hélène, C. (1973). *Chem. Phys.* **2,** p. 119.

Beardsley, K., and Cantor, C. R. (1970). *Proc. Nat. Acad. Sci. U.S.* **65,** 39.

Beardsley, K., Tao, T., and Cantor, C. R. (1970). *Biochemistry* **9,** 3524.

Becker, H., LeBlanc, J. C., and Johns, H. E. (1967). *Photochem. & Photobiol.* **6,** 733.

Bergstrom, D. E., and Leonard, N. J. (1972). *Biochemistry* **11,** 1.

Bersohn, R., and Isenberg, I. (1963). *Biochem. Biophys. Res. Commun.* **13,** 205.

Bersohn, R., and Isenberg, I. (1964). *J. Chem. Phys.* **40,** 3175.

Birks, J. B. (1970). "Photophysics of Aromatic Molecules," Wiley, New York.

Blobstein, S. H., Grunberger, D., Weinstein, I. B., and Nakanishi, K. (1973). *Biochemistry* **12,** 188.

Blumberg, W. E., Eisinger, J., and Navon, G. (1968). *Biophys. J.* **8,** A-106.

Brahms, J., Maurizot, J. C., and Michelson, A. M. (1967). *J. Mol. Biol.* **25,** 481.

Brown, D. T., Eisinger, J., and Leonard, N. J. (1968). *J. Amer. Chem. Soc.* **90,** 7302.

Brown, E., and Pysh, E. S. (1972). *J. Chem. Phys.* **56,** 21.

Brown, I. H., Freeman, K. B., and Johns, H. E. (1966). *J. Mol. Biol.* **15,** 640.

Brown, I. H., and Johns, H. E. (1968). *Photochem. & Photobiol.* **8,** 273.

Bruice, T. C., and Butler, A. R. (1965). *Fed. Proc., Fed. Amer. Soc. Exp. Biol.* **24,** S-45.

Brunk, C. F. (1973). *Nature New Biology,* 241, 74.

Brunk, C. F., and Hanawalt, P. C. (1969). *Rad. Res.* **30,** 285.

Burnel, M. E., Eley, D. D., and Subramanyan, V. (1969). *Ann. N.Y. Acad. Sci.* **158,** 191.
Bush, C. A. (1974). *In* "Basic Principles of Nucleic Acid Chemistry" (P. O. P. Ts'o, ed.). Academic Press, New York **2,** Ch. 2.
Bush, C. A., and Tinoco, I., Jr. (1967). *J. Mol. Biol.* **23,** 601.
Cantor, C. R., and Katz, L. (1971). *Ann. Rev. Phys. Chem.* **1,** 1.
Cantor, C. R., and Tinoco, I., Jr. (1967). *Biopolymers* **5,** 821.
Cantor, C. R., Warshaw, M. M., and Shapiro, H. (1970). *Biopolymers* **9,** 1059.
Chaffin, L., Omilianowski, D. R., and Bock, R. M. (1971). *Science* **172,** 854.
Charney, E. (1971). *Progr. Nucl. Acid Res.* **2,** 176.
Cram, L. S., and Deering, R. A. (1970). *Biophys. J.* **10,** 413.
Daniels, M. (1972). *Proc. Nat. Acad. Sci U.S.* **69,** 2488.
Daniels, M. (1973). *In* "Physicochemical Properties of Nucleic Acids" (J. Duchesne, ed.), Vol. 1, Chapter 4. Academic Press, New York.
Daniels, M., and Hauswirth, W. (1971). *Science* **171,** 675.
Danilov, V. I., Zheltovsky, N. V., Ogloblin, V. V., and Pechenaya, V. I. (1971). *J. Theor. Biol.* **30,** 559.
Day, R. O., Seeman, N. C., Rosenberg, J. M., and Rich, A. (1973). *Proc. Nat. Acad. Sci. U.S.* **70,** 849.
Ding, D., Rill, R., and Van Holde, K. E. (1972). *Biopolymers* **11,** 2109.
Douzou, P., Franq, J., Hanss, M., and Ptak, M. (1961). *J. Chim. Phys.* **58,** 926.
Duchesne, J., Depireux, J., Bertinchamps, A., Cornet, N., and Van der Kaa, J. (1960). *Nature (London)* **188,** 405.
Dudock, B. S., Katz, G., Taylor, E. K., and Holley, R. W. (1969). *Proc. Nat. Acad. Sci. U.S.* **62,** 941.
Eisinger, J. (1969). *Photochem. Photobiol.* **9,** 247.
Eisinger, J., and Lamola, A. A. (1969). *Mol. Photochem.* **1,** 209.
Eisinger, J., and Lamola, A. A. (1971). *In* "Excited States of Proteins and Nucleic Acids" (R. F. Steiner and I. Weinryb, eds.), Ch. 3. Plenum, New York, 1971.
Eisinger, J., and Shulman, R. G. (1966). *Proc. Nat. Acad. Sci. U.S.* **55,** 1387.
Eisinger, J., and Shulman, R. G. (1967). *J. Mol. Biol.* **28,** 445.
Eisinger, J., and Shulman, R. G. (1968). *Science* **161,** 1311.
Eisinger, J., Guéron, M., Shulman, R. G., and Yamane, T. (1966). *Proc. Nat. Acad. Sci. U.S.* **55,** 1015.
Eisinger, J., Feuer, B., and Yamane, T. (1970). *Proc. Nat. Acad. Sci. U.S.* **65,** 638.
Eley, D. D. (1959). *Research* **12,** 293.
Eley, D. D., and Spivey, D. I. (1962). *Trans. Faraday Soc.* **58,** 411.
Fasman, G. D., Lindblow, C., and Grossman, L. (1964). *Biochemistry* **3,** 1015.
Favre, A. (1972). *FEBS (Fed. Eur. Biochem. Soc.) Lett.* **22,** 280.
Favre, A. (1974). *Photochem. & Photobiol.* **19,** 15.
Favre, A., Yaniv, M., and Michelson, A. M. (1969). *Biochem. Biophys. Res. Commun.* **37,** 226.
Fisher, G. J., and Johns, H. E. (1970). *Photochem. & Photobiol.* **11,** 429.
Förster, T. (1965). *In* "Modern Quantum Chemistry" (O. Sinanŏglu, ed.), Part III, p. 93. Academic Press, New York.
Frank-Kamenetskii, M. D. and Lazurskin, Y. S. (1974). *Ann. Rev. Biophys. Bioeng.* **3,** 127.
Gabler, R., and Bendet, I. (1972). *Biopolymers* **11,** 2393.
Galley, W. C. (1968). *Biopolymers* **6,** 1279.
Gellert, M. (1961). *J. Amer. Chem. Soc.* **83,** 4664.
Gill, J. E. (1970). *Photochem. & Photobiol.* **11,** 259.

Gill, J. E. (1971). *Biochem. Biophys. Res. Commun.* **44,** 779.
Gill, J. E. (1973). *Biophys. Soc. Abstr. p. 25a.*
Glaubiger, D., Lloyd, D. A., and Tinoco, I., Jr. (1968). *Biopolymers* **6,** 409.
Gray, D. M., and Rubenstein, I. (1968). *Biopolymers* **6,** 1605.
Gray, D. M., and Tinoco, I. (1970). *Biopolymers* **9,** 223.
Gray, D. M., Ratliff, R. L., and Williams, D. L. (1973). *Biopolymers* **12,** 1233.
Guéron, M., Shulman, R. G., Eisinger, J. (1966). *Proc. Nat. Acad. Sci. U.S.* **56,** 814.
Guéron, M., Eisinger, J., and Lamola, A. A. (1973). *In* "Basic Principles of Nucleic Acid Chemistry" (P.O.P. Ts'o, ed.). Academic Press, New York, Vol. 1, Ch. 4.
Guéron, M., Eisinger, J., and Shulman, R. G. (1967a). *J. Chem. Phys.* **47,** 4077.
Guéron, M., Eisinger, J., and Shulman, R. G. (1967b). *Bull. Amer. Phys. Soc.* [2] **12,** 401.
Guéron, M., and Shulman, R. G. (1968). *Ann. Rev. Biochem.* **37,** 571.
Hariharan, P. V., and Johns, H. E. (1968). *Can. J. Biochem.* **46,** 911.
Hart, E. J., Gordon, S., and Thomas, J. K. (1964). *J. Phys. Chem.* **68,** 127.
Hartman, K. A., and Rich, A. (1965). *J. Amer. Chem. Soc.* **87,** 2033.
Haug, A. (1964). *Photochem. & Photobiol.* **3,** 207.
Hauswirth, W., (1973) unpublished results.
Hauswirth, W., and Wang, S. Y. (1975a). *Biophys. J.* **15,** D-10.
Hauswirth, W., and Wang, S. Y. (1975b). *5th Int. Biophys. Cong.* Copenhagen, Abst. P-108.
Hauswirth, W., and Daniels, M. (1971). *Photochem. & Photobiol.* **13,** 157.
Hélène, C. (1966). *Biochem. Biophys. Res. Commun.* **22,** 237.
Hélène, C., and Longworth, J. W. (1972). *J. Chem. Phys.* **57,** 399.
Hélène, C., and Michelson, A. M. (1967). *Biochim. Biophys. Acta* **142,** 12.
Hélène, C., and Montenay-Garestier, T. (1968) *Chem. Phys. Lett.* **2,** 25.
Helleiner, C. W., Pearson, M. L., and Johns, H. E., (1963). *Proc. Nat. Acad. Sci. U.S.* **50,** 761.
Holcomb, D. N., and Tinoco, I., Jr. (1965). *Biopolymers* **3,** 121.
Holley, R. W., Apgar, J., Everett, G. A., Madison, J. T., Marquisee, M., Merrill, S. H., Penswick, J. R., and Zamir, A. (1965). *Science* **147,** 1462.
Hoover, R. J., Luk, K. F. S., and Maki, A. H. (1974). *J. Mol. Biol.* **89,** 363.
Imakubo, K. (1968). *J. Phys. Soc. Jap.* **24,** 143.
Isenberg, I., Baird, S. L., Jr., and Rosenbluth, R. (1965). *Science* **150,** 1179.
Isenberg, I., Rosenbluth, R., and Baird, S. L., Jr. (1967). *Biophys. J.* **7,** 365.
Johns, H. E. (1971). *In* "Creation and Detection of the Excited State," A. A., Lamola, Ed., Marcel Dekker, New York, p. 123.
Jonker, L. S., and Blok, J. (1975). *Nature* **255,** 245.
Kasai, H., Goto, M., Takemura, S., Goto, T., and Matsaura, S. (1971). *Tet. Lett.* **29,** 2725.
Kaufman, M., and Weill, G. (1971). *Biopolymers* **10,** 1983.
Kleinwächter, V. (1972a). *Collect. Czech. Chem. Commun.* **37,** 1622.
Kleinwächter, V. (1972b). *Collect. Czech. Chem. Commun.* **37,** 2333.
Kleinwächter, V. (1972c). *Stud. Biophys.* **33,** 1.
Kleinwächter, V., and Koudelka, J. (1972). *Collect. Czech. Chem. Commun.* **37,** 3433.
Kleinwächter, V., Drobník, J., and Augenstein, L. (1966). *Photochem. & Photobiol.* **5,** 579.
Kleinwächter, V., Drobník, J., and Augenstein, L. (1968). *Photochem. & Photobiol.* **7,** 485.
Koudelka, J., and Augenstein, L. (1968). *Photochem. & Photobiol.* **7,** 613.
Kucan, Z., Waits, H. P., and Chambers, R. W. (1972). *Biochem.* **11,** 3290.
Lamola, A. A., and Eisinger, J. (1968). *Proc. Nat. Acad. Sci. U.S.* **59,** 46.
Lamola, A. A., Guéron, M., Yamane, T., Eisinger, J., and Shulman, R. G. (1967). *J. Chem. Phys.* **47,** 2210.

Leng, M., Pochon, F., and Michelson, A. M. (1968). *Biochim. Biophys. Acta* **169,** 338.
Leonard, N. J., Iwamura, H., and Eisinger, J. (1969). *Proc. Nat. Acad. Sci. U.S.* **64,** 35.
LePecq, J., and Paoletti, C. (1967). *J. Mol. Biol.* **27,** 87.
Lerman, L. S. (1963). *Proc. Nat. Acad. Sci. U.S.* **49,** 94.
Liang, C. Y., and Scalco, E. G. (1963). *Nature* **198,** 86.
Liang, C. Y., and Scalco, E. G. (1964). *J. Chem. Phys.* **40,** 919.
Lisewski, R., and Wierzchowski, K. L. (1969). *Chem. Commun.* 348.
Longworth, J. W. (1962). *Biochem. J.* **84,** 104P.
Longworth, J. W., and Battista, M. D. C. (1970). *Photochem & Photobiol.* **11,** 207.
Michelson, A. M., and Pochon, F. (1969). *Biochim. Biophys. Acta* **174,** 604.
Montenay-Garestier, T. (1973). *J. Chem. Phys.* **10,** 1379.
Montenay-Garestier, T., and Hélène, C. (1973). *J. Chim. Phys.* **10,** 1391.
Montenay-Garestier, T., and Hélène, C. (1970) *Biochemistry* **9,** 2865.
Montenay-Garestier, T., Hélène, C. and Michelson, A. M. (1969). *Biochim. Biophys. Acta* **182,** 342.
Moscowitz, A. (1957). Ph.D. Thesis, Harvard University, Cambridge, Massachusetts.
Nakanishi, K., Furutachi, N., Funamizu, M., Grunberger, D., and Weinstein, I. B. (1970). *J. Amer. Chem. Soc.* **92,** 7617.
O'Konski, C. T. (1963). *Rev. Mod. Phys.* **35,** 722.
O'Konski, C. T., Moser, P., and Shirai, M. (1964). *Biopolym. & Lymp.* **1,** 479.
Parker, C. A. (1968). "Photoluminescence of Solutions" Elsevier, Amsterdam.
Pisarevskii, A. N., Spitkovskii, D. M., Cherenkevich, S. N., Adrianov, V. T., and Soshina, N. V. (1968). *Biofizika* **13,** 413.
Pochon, F., Leng, M., and Michelson, A. M. (1968). *Biochim. Biophys. Acta* **169,** 350.
Powell, J. T., Richards, E. G., and Gratzner, W. B. (1972). *Biopolymers* **11,** 235.
Pysh, E. S., and Richards, J. L. (1972). *J. Chem. Phys.* **57,** 3680.
Rahn, R. O., and Hosszu, J. L. (1968). *Photochem. & Photobiol.* **8,** 53.
Rahn, R. O., Longworth, J. W., Eisinger, J., and Shulman, R. G. (1964). *Proc. Nat. Acad. Sci. U.S.* **51,** 1299.
Rahn, R. O., Shulman, R. G., and Longworth, J. W. (1965). *Proc. Nat. Acad. Sci. U.S.* **53,** 893.
Rahn, R. O., Yamane, T., Eisinger, J., Longworth, J. W., and Shulman, R. G., (1966a). *J. Chem. Phys.* **45,** 2947.
Rahn, R. O., Shulman, R. G., and Longworth, J. W. (1966b) *J. Chem. Phys.* **45,** 2955.
Rahn, R. O., Battista, M. D. C., and Landry, L. C. (1970). *Proc. Nat. Acad. Sci. U.S.* **67,** 1390.
Rajbhandary, U. L., Chang, S. H., Stuart, A., Faulkner, R. D., Hoskinson, R. M., and Khorana, H. G. (1967). *Proc. Nat. Acad. Sci. U.S.* **57,** 751.
Rhodes, W. (1961). *J. Amer. Chem. Soc.* **83,** 3609.
Rhoades, D. W., and Wang, S. Y. (1971a). *Biochemistry* **9,** 4416.
Rhoades, D. W., and Wang, S. Y. (1971b). *J. Amer. Chem. Soc.* **93,** 3779.
Rich, A., and Kasha, M. (1960). *J. Amer. Chem. Soc.* **82,** 6197.
Riehl, N. (1940). *Naturwissenschaften* **28,** 601.
Rosenberg, J. M., Seeman, N. C., Kim, J. J. P., Suddath, F. L., Nicholas, H. B., and Rich, A. (1973). *Nature (London)* **243,** 150.
Scruggs, R. L., Achter, E. K., and Ross, P. D. (1972). *Biopolymers* **11,** 1961.
Shalitin, N., and Feitelson, J. (1973). *J. Chem. Phys.* **59,** 1045.
Setlow, R. B., Carrier, W. L., and Bollum, F. J. (1964). *Biochim. Biophys. Acta.* **91,** 446.
Shafranovskaya, N. N., Trifonov, E. N., Luzurkin, Yu. S., and Frank-Kamenetskii, M. D. (1973). *Nature New Biology* **241,** 58.

Sinanõglu, O., and Abdulnar, S. (1965). *Fed. Proc., Fed. Amer. Soc. Exp. Biol.* **24,** 12.
Sinsheimer, R. L. (1954). *Rad. Res.* **1,** 505.
Snart, R. S. (1963). *Trans. Faraday Soc.* **59,** 754.
Snart, R. S. (1968). *Biopolymers* **6,** 293.
Snart, R. S. (1973). *Biopolymers* **12,** 1493.
Sommer, B., and Jortner, J. (1968). *J. Chem. Phys.* **49,** 3919.
Steele, R. H., and Szent-Györgyi, A. (1957). *Proc. Nat. Acad. Sci. U.S.* **43,** 477.
Steiner, R. F., Millar, D. B., and Hoerman, K. C. (1967). *Arch. Biochem. Biophys.* **120,** 464.
Steiner, R. F., Kinnier, W., Lunasin, A., and Delac, J. (1973). *Biochim. Biophys. Acta* **294,** 24.
Subertova, E., Prosser, V., and Drobník, J. (1969). *Biopolymers* **8,** 421.
Sutherland, B. M., and Sutherland, J. C. (1969a). Biophys. J. *8,* 490.
Sutherland, B. M., and Sutherland, J. C. (1969b). *Biophys. J.* **9,** 1045.
Sutherland, B. M., and Sutherland, J. C. (1970). *Biopolymers* **9,** 639.
Szerenyi, P., and Dearman, H. H. (1972). *Chem. Phys. Lett.* **15,** 81.
Szent-Györgyi, A. (1941). *Nature (London)* **148,** 157.
Tazawa, S., Tazawa, I., Alderfer, J. C., and Ts'o, P. O. P. (1972). *Biochemistry* **11,** 3544.
Thiele, D., Guschlbauer, W., and Favre, A. (1972). *Biochim. Biophys. Acta* **272,** 22.
Tinoco, I., Jr. (1960). *J. Chem. Phys.* **33,** 1332.
Tinoco, I., Jr. (1968). *J. Chim. Phys.* **65,** 91.
Trukhan, E. M. (1966). *Biophysics (USSR)* **11,** 468.
Ts'o, P. O. P., Melvin, J. S., and Olson, A. C. (1963). *J. Amer. Chem. Soc.* **85,** 1289.
Ts'o, P. O. P. (1974). *In,* "Basic Principles of Nucleic Acid Chemistry," P. O. P. Ts'o, Ed. Academic Press, New York, Vol. 2, Chapter 5.
Tsou, K. C., and Yip, K. F. (1974). *Biopolymers* **13,** 987.
Van Holde, K. E., Brahms, J., and Michelson, A. M. (1965). *J. Mol. Biol.* **12,** 726.
Vigny, P. (1971a). *C. R. Acad. Sci., Ser. D* **272,** 2249.
Vigny, P. (1971b). *C. R. Acad. Sci., Ser. D* **272,** 3206.
Vigny, P., and Favre, A. (1974). *Photochem. & Photobiol.* **20,** 345.
Vournakis, J. N., Poland, D., and Scheraga, H. A. (1967). *Biopolymers* **5,** 403.
Wada, A. (1964). *Biopolymers* **2,** 361.
Wada, A. (1972). *Appl. Spectrosc. Rev.* **6,** 1.
Wang, S. Y. (1965). *Fed. Proc., Fed. Amer. Soc. Exp. Biol.* **24,** S-71.
Ward, D. C., Reich, E., and Stryer, L. (1969). *J. Biol. Chem.* **244,** 1228.
Warshaw, M. M., and Cantor, C. R. (1970). *Biopolymers* **9,** 1079.
Warshaw, M. M., and Tinoco, I., Jr. (1965). *J. Mol. Biol.* **13,** 54.
Warshaw, M. M., and Tinoco, I., Jr. (1966). *J. Mol. Biol.* **20,** 29.
Warshaw, M. M., Bush, C. A., and Tinoco, I., Jr. (1965). *Biochem. Biophys. Res. Commun.* **18,** 633.
Weill, G., and Calvin, M. (1963). *Biopolymers* **1,** 401.
Weissbluth, M. (1971). *Quart. Rev. Biophys.* **4,** 1.
Wierzchowski, K. L., and Shugar, D. (1962). *Photochem. & Photobiol.* **1,** 21.
Yamaoka, K., and Charney, E. (1972). *J. Amer. Chem. Soc.* **94,** 8963.
Yamaoka, K., and Charney, E. (1973). *Macromolecules* **6,** 66.
Yaniv, M., Favre, A., and Varrell, B. G. (1969). *Nature (London)* **223,** 1331.
Yoshikami, D., Katz, G., Keller, E. B., and Dudock, B. S. (1968). *Biochim. Biophys. Acta* **166,** 714.
Zimmerman, T.P., Robinson, B., and Gibbs, J. H. (1970). *Biochim. Biophys. Acta* **209,** 151.

# 4 Pyrimidine Photohydrates

*G. J. Fisher and H. E. Johns*



## A. Introduction

### 1. Discovery

As early as 1931, Heyroth and Loofbourow noticed that UV irradiation of Ura derivatives in aqueous solution results in almost total loss of the characteristic absorption peak near 260 nm. Sinsheimer and Hastings (1949) reported that this loss of optical density of Ura and Urd is largely reversible at room temperature at pH 1 or by boiling at neutral pH.

These changes in optical density are illustrated in Fig. 1. Curve a shows the absorption spectrum of UMP with its characteristic peak at 260 nm, while b is the spectrum of the solution after most of the parent material has been converted to the photoproduct. By reversing the photoproduct, > 90% of the parent absorption can be regenerated to give curve c. Cytidylic acids form even more unstable reversible

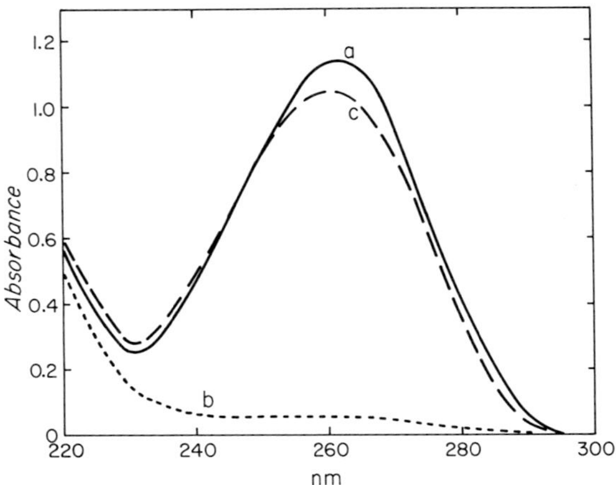


**Fig. 1.** *Spectra of UMP before irradiation (a), after irradiation (b), and after reversal of the hydration (c), adapted from Sinsheimer (1954). Curve b is essentially the spectrum of the hydrate.*

photoproducts (Sinsheimer, 1957; Wierzchowski and Shugar, 1957). In recent years the thermally reversible products of Pyr derivatives have been studied intensively and have been discussed in several reviews (McLaren and Shugar, 1964; Smith, 1966; Burr, 1968; Fahr, 1969; Smith and Hanawalt, 1969; Varghese, 1972). In this chapter we present a detailed review of the formation and properties of these photoproducts.

## 2. Nature of the Reversible Photoproducts

### *a. Uracil Derivatives*

From the loss of the UV absorption of the 5,6 double bond of 1,3-$Me_2$Ura, Moore and Thomson (1955) concluded that the bond is saturated in the reversible photoproduct. In addition, the product does not react with bromine water in the manner typical of unsaturated pyrimidines. They isolated the $Me_2$Ura photoproduct in low yield and found that its infrared spectrum in chloroform shows the 2.98 $\mu$m band characteristic of an OH group. Furthermore, the product has a molecular weight of 158.3, compared with 140 for $Me_2$Ura; the difference is one oxygen and two hydrogen atoms. These results led to the conclusion that the photoproduct was either **II** or **III. III** was ruled out when it was synthesized and shown to differ from the $Me_2$Ura photoproduct (Wang *et al.*, 1956). The product is, therefore, **II** ($ho^6Me_2hUra$). This compound was then synthesized and shown by

Wang *et al.* (1956) and by Moore and Thomson (1957) to be identical to the photoproduct (see Scheme 1).

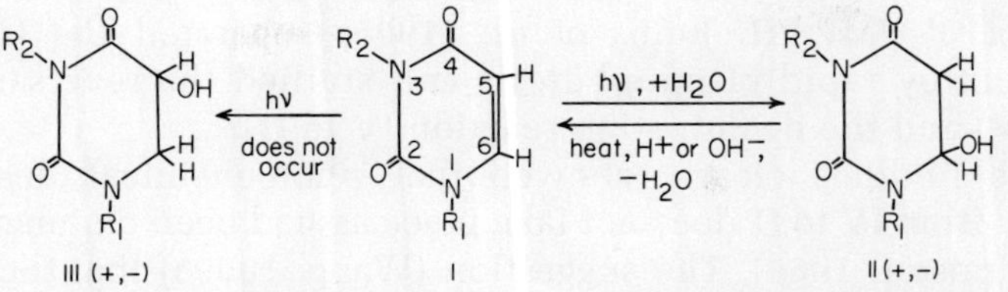

**Scheme 1.** **I** *is Ura when* $R_1 = R_2 = H$; **I** *is 1,3-Me$_2$Ura when* $R_1 = R_2 = CH_3$, and **I** *represents the corresponding nucleoside or nucleotide when* $R_1$ *is a Rib or Rib phosphate group.* **III** *is not formed photochemically.*

The ho$^6$hUra was synthesized and was shown to be identical to the Ura photoproduct (Moore, 1958; Gattner and Fahr, 1963). In 1964, Gattner and Fahr showed that reversible photoproducts of Urd and UMP are the same as synthetic compounds of form **II**. The NMR studies of Wechter and Smith (1968) indicated that the photoproduct of Urd exists in two enantiomeric forms of **II**, i.e., with the OH group in either of two orientations relative to the Rib.

In summary, all of the available evidence points to the water addition compound **II** as the structure of the reversible photoproduct, the so-called photohydrate of compounds of the uracil family.

*b. Cytosine Derivatives*

Irradiation of Cyt (**IV**) and its derivatives results in a reversible photoproduct which is much less stable than the corresponding product of Ura. Since this photoproduct is less stable there has been more difficulty in determining its structure. We now give some of the evidence which demonstrates that the photoproduct is **V** (see Scheme 2).

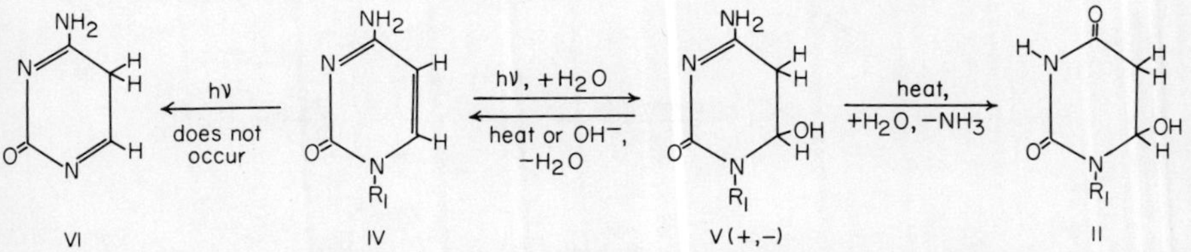

**Scheme 2.** *Structure* **IV** *represents Cyt when* $R_1 = H$; *Cyd, when* $R_1$ *is Rib; and CMP when* $R_1$ *is a Rib phosphate group.*

In 1961, Wierzchowski and Shugar showed that the photoreaction is first order, involving only one Cyt molecule, and that the products show an absorption peak near 240 nm and lack one at ~270 nm, which is partial evidence for 5,6-saturated cytosines. In 1964, Schuster showed that the photoproduct of 3′-CMP is largely reversible to the parent but that a small fraction deaminates to yield the photohydrate of 3′-UMP, **II**. Johns *et al.* (1965) separated the CMP photoproduct by rapid electrophoresis and studied the reversal reaction (**V** to **IV**) and the deamination reaction (**V** to **II**).

These findings clearly showed that photochemical deamination directly from **IV** to **II** does not take place as had been claimed (Daniels and Grimison, 1964). The suggestion (Wang, 1959a) that the unstable photoproduct of Cyt may be a tautomer **VI** was also ruled out, since blocking N(1) with a side group does not prevent the formation of a similar photoproduct, as shown by Wierzchowski and Shugar (1957) for 1-MeCyt and by Wang (1959a) for Cyd and CMP.

The suspicion from these experiments that the Cyt family forms a water addition compound **V** as a major unstable photoproduct was confirmed by Kleber *et al.* (1965). These investigators synthesized ho$^6$hCyd and ho$^6$-2′(3′)hCMP and found that these compounds have the same $R_f$ values and reversal rates as the corresponding photoproducts.

Finally, in 1968 Miller and Cerutti reduced the photoproduct of Cyd with borohydride, a reaction which leads to ring opening between C(6) and N(1) and to deamination to form **VII,** N(1)-($\alpha,\beta$-D-ribopyranosyl)-N(3)-($\gamma$-hydroxypropyl)urea (See Scheme 3.) **VII** has been identified by infrared and NMR spectra and by elemental analysis. This reduction product strongly suggests that the hydrate has the OH group on C(6) rather than on C(5) and that the photoproduct is ho$^6$hCyd. NMR studies of irradiated Cyd confirmed that such a hydrate is the principal product (DeBoer and Johns, 1970).

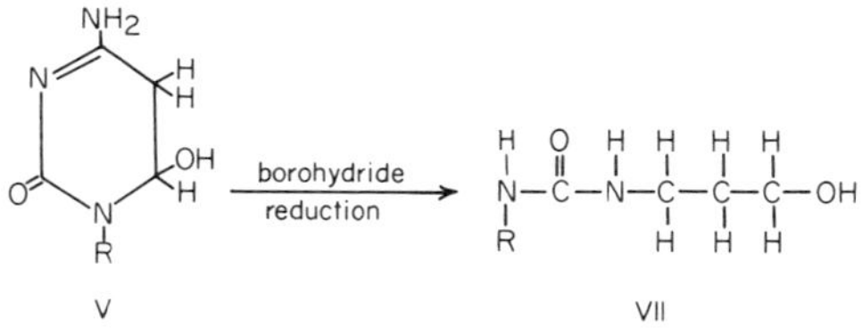

V —borohydride reduction→ VII

**Scheme 3**

To summarize: the evidence of synthesis, reduction, NMR data, and deamination to the hydrate of the corresponding Ura derivative leads to the conclusion that the major photoproduct of members of the Cyt family in aqueous solution is the 6-hydroxy-5,6-dihydro compound **V**.

*c. Thymine Derivatives*

The question of whether Thy and its derivatives form photohydrates analogous to those of Ura and Cyt has been the subject of many inconclusive discussions. Thy hydrates can be synthesized and are reasonably stable with lifetimes of hours (Nofre and Ogier, 1966), but most previous attempts to produce them photochemically and to detect them have failed. We now know the reason for the reported failures, since we have shown that the quantum yield for hydrate production is very small ($\sim 10^{-5}$). We have also shown (Fisher and Johns, 1973) that the Thy photohydrate is quite similar to the hydrates of Ura and Cyt and has the structure of chemically synthesized *cis*-$ho^6hThy$ (Cadet and Teoule, 1971b).

## B. Photohydrate Formation

### 1. Determination of Quantum Yields

*a. Cross Sections*

Detailed discussions of methods for determining photochemical quantum yields have been presented elsewhere (Johns, 1968; 1969, 1971). The approach used is to determine the photochemical cross section $\sigma$ for forming a product and the absorption cross section $\sigma_a$ for absorption of a photon by the parent molecule. The quantum yield for product formation is simply

$$\phi = \sigma / \sigma_a \tag{1}$$

The conversion cross section $\sigma$ is determined by the relation

$$\Delta N = \sigma N \Delta \bar{L} \tag{2}$$

where $\Delta N$ is the number of molecules of product formed when $N$ molecules of the parent are exposed to $\Delta \bar{L}$, an increment in the flux of photons per unit area. When the exposure $\Delta \bar{L}$ is expressed in $\mu E/cm^2$, then $\sigma$ must be given in $cm^2/\mu E$, so that the product $\sigma \Delta \bar{L}$ is dimensionless.* $\Delta N$ and $N$ are concentrations and may be expressed in moles/liter, molecules/$cm^3$, etc.

* A microeinstein ($\mu E$) is a micromole of photons ($6.02 \times 10^{17}$). Other useful units of exposure and cross section are photons/$cm^2$ and $cm^2$/photon, ergs/$mm^2$ and $mm^2$/erg, etc. The conversion factors from one set of units to another are found elsewhere (Johns, 1968, 1969, 1971).

The absorption cross section $\sigma_a$ is determined from the molar extinction coefficient of the parent, $\epsilon_\lambda$, at the wavelength of interest. The dimensions of $\sigma_a$ must be compatible with $\sigma$ and $\Delta\bar{L}$. When $\Delta\bar{L}$ is in $\mu E/cm^2$, then $\sigma_a$ is given by

$$\sigma_a = 2.303 \times 10^{-3}\epsilon_\lambda \tag{3}$$

*b. Average Exposure*

Since photochemical reactions require absorption of light, the intensity of the irradiation beam decreases as it passes through the solution. To obtain accurate quantum yields, then, it is necessary to keep the solution well stirred and to calculate the *average* photon fluence or exposure $\bar{L}$ *seen by all of the molecules in the solution*. This calculation requires a knowledge of the incident light flux $L_i$, the absorbance of the solution as a function of time during the irradiation, and the mean path length of the beam through the solution. The latter is difficult to determine if the irradiating beam is not parallel or if the solution is not in a rectangular vessel. Since the relation between $L_i$ and $\bar{L}$ depends on the absorbance, which is continually changing, it is necessary to calculate an increment in $\Delta\bar{L}$ after increments of incident flux $\Delta L_i$. The quantity $\bar{L}$ is simply the sum of the $\Delta\bar{L}$ increments. The methods for obtaining $\bar{L}$ from $L_i$ have been described in detail by Johns (1971).

The simplest method of monitoring many chemical reactions in solution is to observe changes in absorbance. This is a useful procedure when absorption spectra of the starting materials and the products differ, as they do, for example, in the case of Urd (Fig. 1). Figure 2a shows plots of the absorbance as a function of both $L_i$ and $\bar{L}$. The curves are quite different. The incident fluence is directly related to the "time of exposure" while $\bar{L}$ is not, since $\bar{L}$ is not a constant fraction of $L_i$ but varies as the irradiation progresses. The importance of distinguishing between $\bar{L}$ and $L_i$ can be seen by analyzing Fig. 2a as follows.

If irradiation leads to only one photoproduct which does not absorb at the wavelength of the irradiation, then a plot of absorbance $A$ *vs.* $\bar{L}$ will yield the photochemical cross section from Eq. (2). $A$ is proportional to $N$, the concentration of parent, and $\Delta A$ is proportional to $-\Delta N$. The minus sign occurs because the absorbance decreases as photoproducts are made. Hence

$$\Delta A = -\sigma A_0 \Delta\bar{L} \text{ or } \sigma = -\frac{1}{A_0}\frac{\Delta A}{\Delta\bar{L}} \tag{4}$$

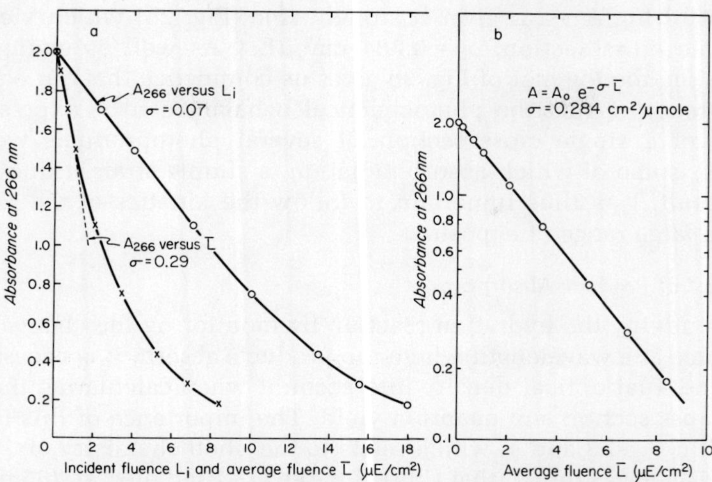


**Fig. 2.** *(a) Plot of absorbance at 266 nm as a function of $L_i$ and $\bar{L}$ for irradiation of $Me_2Ura$ at 265 nm, pH 7.0, 25° C. (b) Presentation of the same data showing how a straight line is obtained when log A is plotted against $\bar{L}$.*

in which $A_0$ is the initial absorbance. The initial slope gives $\sigma = 0.07$ cm²/$\mu$E when the absorbance is plotted against incident fluence $L_i$, but 0.29 cm²/$\mu$E when the results are correctly plotted against the average fluence, $\bar{L}$. The factor of four difference clearly demonstrates the need to distinguish between $L_i$ and $\bar{L}$ when valid quantitative information is desired.

In addition to being essential for a correct determination of the average dose, measurements of absorbance changes at small increments of exposure are also important in tracing the course of the reaction. It is evident from Fig. 2a that the absorbance does not decrease linearly with exposure, and clearly in the case of $Me_2Ura$ a single measurement of the absorbance drop would lead to an underestimation of the rate of change. Since the corrections for $\bar{L}$ are a function of absorbance and, therefore, of the irradiation wavelength, quantitative studies with polychromatic incident light are hopelessly complex. Methods for obtaining monochromatic light have been discussed elsewhere (Johns and Rauth, 1965; Muel and Malpiece, 1969).

The cross section $\sigma$ determined using Eq. (4) is, however, only approximate, since it depends on the operator's skills in drawing the best line for the initial slope. To obtain an accurate cross section it is convenient to manipulate the data to give a straight line. Equation (4) is really a differential equation which may be integrated to yield

$$A = A_0 e^{-\sigma \bar{L}} \tag{5}$$

A plot of log $A$ vs. $\bar{L}$ gives a straight line (Fig. 2b) which yields an accurate cross section, $\sigma = 0.284$ cm$^2$/$\mu$E.* As well as giving more precision, the log plot of Fig. 2b gives us confidence that our equation is correct and that the photochemical behavior can be expressed in terms of a single cross section. If several photoproducts were involved, some of which absorb radiation, a simple straight line would not result. It is thus important to follow the kinetics of the reaction over a large range of exposures.

*c. Effect of Product Absorption*

In studying the hydration reaction by monitoring the change in absorbance at a wavelength where the products absorb, it is necessary to take the final optical density into account when calculating the reaction cross section and quantum yield. The importance of this is illustrated in Figs. 3 and 4, which relate to the photochemistry of 3′-CMP. Suppose, for example, that Cp (pH 8.4) were irradiated at 265 nm and the absorbance were monitored at 258 nm. At this wavelength the ho$^6$hCp absorbs, so that even if no other reaction occurred, the absorbance would at first fall rapidly and then tend to a final value quite

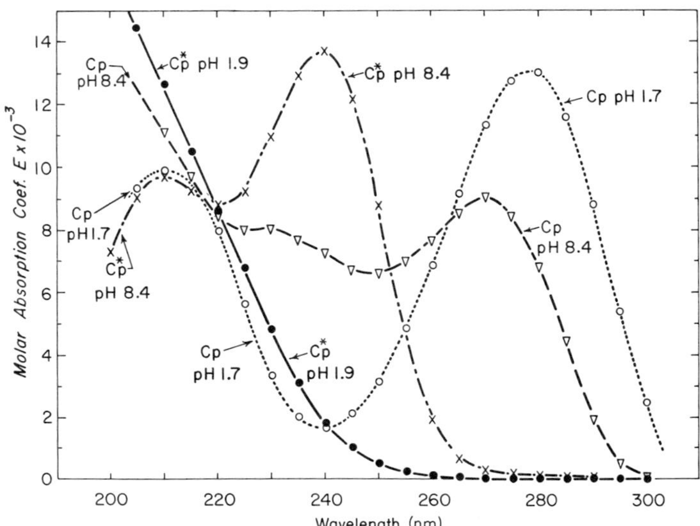


**Fig. 3.** *Molar extinction coefficient of ho$^6$hCp(C*p) and 3′-CMP(Cp) as a function of wavelength. The measured spectra for C*p were corrected for the amount of C*p converted back to Cp during the experiment (From Johns* et al., *1965).*

* From the log plot $\sigma$ is most easily determined by $\sigma = 2.303/\bar{L}_{10}$, where $\bar{L}_{10}$ is the exposure required to reduce the absorbance to one-tenth of its initial value.

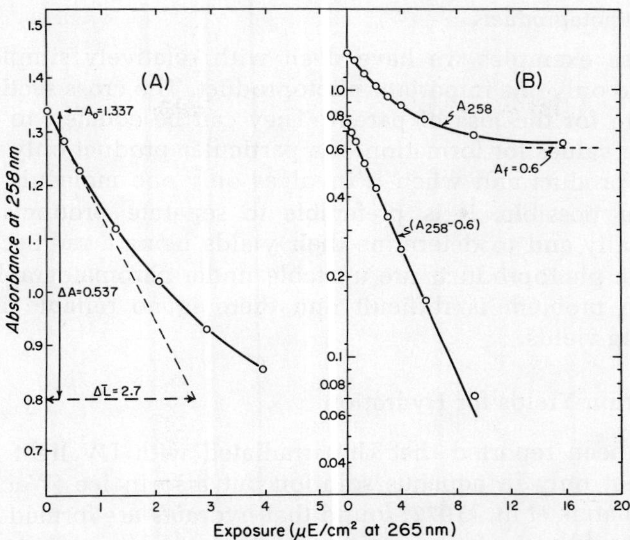


**Fig. 4.** *Irradiation at 265 nm of 3'-CMP at pH 8.37 in 0.01* M *Tris buffer at 0°C; initial absorbance,* $A_{258} = 1.337$. *(A) Plot of* $A_{258}$ *on linear scale vs. exposure. (B) Plot of* $A_{258}$ *and* $(A_{258} - 0.6)$ *on a log scale vs. exposure (from Johns, 1971).*

different from zero. From the linear plot (Fig. 4A) one would obtain from the initial slope the incorrect cross section as

$$\sigma = \frac{1}{A_0}\frac{\Delta A}{\Delta \bar{L}} = \frac{1}{1.337}\frac{0.537}{2.7} = 0.15 \text{ cm}^2/\mu\text{E}$$

The correct equation describing this reaction is obtained from Eq. (4) by integration and by assuming that after large exposures the final optical density is $A_f$. We now obtain the relation

$$(A - A_f) = (A_0 - A_f)\, e^{-\sigma \bar{L}} \tag{6}$$

Thus a plot of $\log(A - A_f)$ against $\bar{L}$ should give a straight line, and it does (Fig. 4B), yielding the correct cross section:

$$\sigma = \frac{2.303}{\bar{L}_{10}} = 0.274 \text{ cm}^2/\mu\text{E}$$

This differs by a factor of 2 from the value obtained from the linear plot. The importance of manipulating the data to yield a straight line is again evident.

*d. Many Photoproducts*

In these examples we have dealt with relatively simple systems leading to only one important photoproduct. The cross sections so obtained are for the loss of parent. They can be equated to the corresponding values for formation of a particular product only when it is the sole product and when it involves only one molecule of parent. Wherever possible, it is preferable to separate products chromatographically and to determine their yields using a radioactive label. When the photoproducts are unstable under chromatographic conditions the problem is difficult and there is no reliable method of measuring yields.

## 2. Quantum Yields for Hydration

It has been reported that Ura irradiated with UV light forms hydrates, not only in aqueous solution but also in ice (Wacker *et al.*, 1961). Khattak *et al.* (1972) found that hydrates are formed efficiently in acid puddles in frozen acidic solutions of Ura and its *N*-methyl derivatives. However, most of the experimental and theoretical interest in the last decade has centered on hydration in fluid solution, and we will deal only with this case. The many determinations of the quantum yields for photohydration, ($\phi_H$) of pyrimidines which have been reported are summarized in Table 1. In addition to the results quoted in the table, there are many reports in which the photochemistry is discussed in terms of relative quantum yields or rates of reaction in units of (min of irradiation)$^{-1}$. Since such results cannot be compared with the work of other laboratories, these values are not included.

*a. Precautions in Measuring Quantum Yields*

As has been discussed, incorrect dosimetry can be a major factor in reducing the apparent quantum yield below the actual value. Figure 2a illustrates how the failure to correct the incident intensity to the average can result in an error of a factor of four in estimating the reaction efficiency. For this reason we favor higher values of the quantum yield for loss of parent.

Failure to correct for finite absorption of photoproducts can also lead to an incorrect low value for the quantum yield, as has just been shown in connection with Fig. 4. This helps to account for the range of values reported for 3′-CMP (Table 1).

The instability of the hydrate likewise reduces the apparent quantum efficiency, since reversal of hydrates during prolonged irradiation means that not all of the product molecules are counted. For example, we believe that the lower values reported for the quantum yield of hydration of CMP (Table 1) are incorrect largely because of the rapid reversal of the hydrate and that the higher value obtained by Becker *et al.* (1967) is more nearly correct. This difficulty can best be overcome by using a high-intensity source to reduce the elapsed time of the experiment and by correcting for reversal during the experiment. Cooling the solutions to minimize reversal has also been advocated (DeBoer *et al.,* 1970), although temperature may affect quantum yield (Burr *et al.,* 1972).

Ignoring the possible formation of other photoproducts has the opposite effect of inflating the quantum yield for hydration, since some of the loss of parent absorbance is caused by other products. The possible importance of side reactions has been known for many years. For example, as the reaction proceeds, the degree of reversibility of Cyt decreases from 60% recovery of absorbance after 25% photolysis to no recovery after 80% photolysis. This indicates that irreversible products accumulate after prolonged irradiation, but it is not known whether they are direct photoproducts of Cyt or of its hydrate. Addition of 0.02 *M* phosphate buffer greatly enhances the formation of irreversible products and there is very little recovery of absorbance after 10% photolysis under these conditions (Wierzchowski and Shugar, 1961). Ionic strength (Moore, 1963) and temperature (DeBoer *et al.,* 1970) also affect the relative importance of side reactions. The conclusion is that hydration quantum yields should be determined at small exposures under conditions which give a high degree of reversibility. But even then, there exists the possibility of the formation of thermally reversible dimers. This process can probably be neglected for Cyt derivatives, since studies of their reversal kinetics generally follow a single exponential, suggesting that only one major heat reversible product is formed (DeBoer *et al.,* 1970). However, some dimers of Thy and Ura are known to be thermally reversible (Herbert *et al.,* 1969; Varghese, 1971).

Finally, it must be emphasized that even when quantum yields are correctly determined they apply only to the irradiation conditions used. Solvent, viscosity, temperature, wavelength, concentration, ions, and sensitizer or quencher molecules, including oxygen in inadequately degassed solutions, are among the factors which can affect photochemical quantum yields. They all should be specified.

To summarize, meaningful quantum yield determinations require adequate mixing, calculation of the average exposure in the solution, measurements after several increments of exposure, and knowledge of

**Table 1** Quantum Yields for Photohydration

| Material | pH | $\phi_H$ | Comments | Precautions[a] | Reference |
|---|---|---|---|---|---|
| Uracil Derivatives | | | | | |
| Ura | 2.6 | 0.0090 | Products separated; careful dosimetry | 1,2 | Brown and Johns (1968) |
| | 4.5 | 0.0019 | Independent of $O_2$, concentration, and λ from 230 to 280 nm | 1,2 | |
| | 6.2 | 0.0007 | | | |
| 1-MeUra | – | 0.0125 | $H_2O$ | | Shugar and Wierzchowski (1957) |
| | | 0.0057 | $D_2O$ | | |
| 3-MeUra | – | – | Yield at pH 3 is 9 times the value at pH 6 | | Burr *et al.* (1972) |
| | | – | | | |
| $Me_2Ura$ | – | 0.01 | | | Moore and Thomson (1956) |
| | 5–8 | 0.015(0.5–10 m*M*; 28–40°C) | | | Wang (1962b) |
| | – | 0.0039(0.1 m*M*) | | | Burr *et al.* (1968a) |
| | 7.0 | 0.014(1 m*M*) | | 1 | Johns (1971) |
| | 5.8 | 0.013(0.1 m*M*) | | 1 | J. C. LeBlanc (unpublished data, 1972) |
| | 5.8 | 0.013(10 m*M*) | | 1 | |
| Urd | – | 0.016 | | | Guschlbauer *et al.* (1965) |
| | – | 0.017 | Independent of λ from 238 to 280 nm | | Swenson and Setlow (1963) |
| 2′-UMP and 3′-UMP | – | 0.022 | | 1 | Sinsheimer (1954) |
| 2′(3′)-UMP (mixed isomers) | 7.0 | 0.02 | | | Wierzchowski and Shugar (1961) |
| | – | 0.022 | | | Rushizky and Pardee (1962) |
| | 2.0 | 0.02 | Absorbance study (2.0–UTP) | | Guschlbauer *et al.* (1965) |
| | 6 | 0.02 | | | |
| | 11 | 0.012 | | | |
| UDP | 2–11 | 0.016–0.019 | | | |
| UTP | – | 0.014 | | | |
| Cytosine derivatives | | | | | |
| Cyt | 7.2 | 0.002 | Absorbance study (Cyt–1,4-$Me_2$Cyt) | | Fikus *et al.* (1962) |
| 1-MeCyt | 7.2 | 0.002 | | | |
| 1,4-$Me_2$Cyt | 7.2 | 0.00001 | | | |

| | | | | | |
|---|---|---|---|---|---|
| Cyt | 2,13 | – | Irreversible absorbance loss | | Wierzchowski and Shugar (1957) |
| 1-MeCyt | 2,13 | – | | | |
| Cyd | 2 | 0.0016 | Not corrected for reversal during irradiation at 254 nm; dosimetry doubtful | | Wierzchowski and Shugar (1957) |
| | 5.6 | 0.009 | | | |
| | 7.1 | 0.010 | | | |
| dCyd | 7.1 | 0.0086 | | | |
| 2′(3′)-CMP (mixed isomers) | 1 | 0.0011 | | | |
| | 7.1 | 0.0115 | | | |
| 2′-CMP and 3′-CMP | 7 | 0.017 | 254 nm | 1,4 | Sinsheimer (1957) |
| 3′-CMP | 1–3 | 0.0013 | 280 nm | 1,2,3 | Becker *et al.* (1967) |
| | 4 | 0.0052 | | 1,2,3 | |
| | 5 | 0.010 | | 1,2,3 | |
| | 6–9 | 0.013 | | 1,2,3 | |
| | 6–9 | 0.011 | 240 nm | 1,2,3 | |
| | 8.37 | 0.016 | 265 nm; 0.01 *M* Tris buffer; 0°C | 1,4 | Johns (1971) |
| CMP | 7.1 | 0.0025 | Doubtful dosimetry; *not* corrected for reversal | | Rushizky and Pardee (1962) |
| | 7.1 | 0.0029 | | | Shugar and Wierzchowski (1958) |
| | 6.5 | 0.0086 | 265 nm | 1,3 | Becker *et al.* (1967) |
| 5- and 6-substituted uracils | | | | | |
| Thy | 3 | $2 \times 10^{-5}$ | Hydrate separated chromatographically | 1,2 | Fisher and Johns (1973) |
| | 6 | $3 \times 10^{-6}$ | | 1,2 | |
| FlUra | 5.6 | 0.017 | Product isolated | | Fikus *et al.* (1965) |
| 3-MeFlUra | 5.6 | 0.016 | | | |
| 1,3-$Me_2$FlUra | 5.6 | 0.023 | | | |
| FlUrd | 5.6 | 0.047 | | | |
| FlUMP | 5.6 | 0.045 | | | |
| FlUrd | 2 | 0.050 | Absorbance study only | | Guschlbauer *et al.* (1965) |
| | 6 | 0.040 | | | |
| $ho^5$Ura | 4.75 | $2 \times 10^{-4}$ | Absorbance study only | 1 | E. Lo, G. J. Fisher, and H. E. Johns (unpublished data, 1971) |
| 5-$NO_2$Ura | Unbuffered | $2 \times 10^{-5}$ | | | |
| $Am^5$Ura | Unbuffered | $1.0 \times 10^{-2}$ | | | |
| $Am^6$Ura | Unbuffered | 0 | | | |
| Ura-6-sulfonamide | Unbuffered | $1.0 \times 10^{-2}$ | | | |

(*Continued*)

**Table 1** *(Continued)*

| Material | pH | $\phi_H$ | Comments | Precautions[a] | Reference |
|---|---|---|---|---|---|
| Oligo- and Polynucleotides[b] | | | | | |
| UpU | | 0.0091 | 280 nm; $\phi$ for first hydration | 1,2 | Brown *et al.* (1966) |
| dUpU | | 0.018 | 280 nm | 1 | Helleiner *et al.* (1963) |
| UpGpA | | 0.0105 | | | |
| GpUpA | | 0.0070 | 254 nm | | Wierzchowski and Shugar (1962) |
| GpApU | | 0.0045 | | | |
| CpC | | 0.0064 | 254 nm | 1,2 | Hariharan and Johns (1968) |
| CpCp | | 0.0075 | | | |
| ApCp | | 0.0082 | 254 nm | | Wierzchowski and Shugar (1962) |
| GpCp | | 0.0055 | | | |
| TpC | | 0.006 | pH 7; independent of λ | | Haug (1964) |
| poly(rU) | | 0.012 | 254 nm | 1,2 | Pearson *et al.* (1966) |
| poly(rA-rU) | | 0.0009 | 280 nm; isolated hydrate | 1,2 | Pearson and Johns (1966b) |
| poly(rA-rU) | | 0.0068 | 280 nm; adjacent to a dimer | | |
| poly(rA-2rU) | | 0.0012 | 280 nm | | DeBoer *et al.* (1967) |
| poly(rC) | | 0.0117 | 254 nm | | Rhoades and Wang (1971) |

[a] Precautions: 1, reliable dosimetry; 2, hydrate separated; 3, corrected for reversal; 4, corrected for product absorbance.
[b] $\phi_H$ is calculated on the basis of number of hydrates produced per photon absorbed per Pyr. We neglect energy transfer and hypochromicity.

the relative amounts and stabilities of the products. If carried out carefully, such experiments give accurate values of quantum yields for particular photoproducts under the particular conditions of the irradiation. Results from different laboratories should then be directly comparable, which is not the case when reaction kinetics are carelessly described in terms of minutes of irradiation.

That the determination of an accurate quantum yield is not easy is illustrated by the differences in the values reported in Table 1. We attempted to evaluate the reliability of these results on the basis of precautions (Table 1, column 5) taken by the various workers to eliminate these sources of error.

*b. Effects of pH*

A number of workers studied $\phi_H$ as a function of pH, and some of their results are shown in Fig. 5. For Ura and its derivatives which are not substituted at N(1), the trend is clear. The quantum yield drops by a factor of ~10 as the pH is shifted from 3 to 5 (Brown and Johns, 1968). This conclusion is supported by the work of Wang and Nnadi (1968) and Burr *et al.* (1972) who determined relative yields or rates of hydration for several Ura derivatives but who, unfortunately, did not determine absolute quantum yields. For Ura derivatives which are substituted at N(1), the effect of pH on hydration is much smaller with a slight reduction in yield occurring at about pH 6 (Wang and Nnadi, 1968; Burr *et al.*, 1972). These findings may be interpreted in terms of

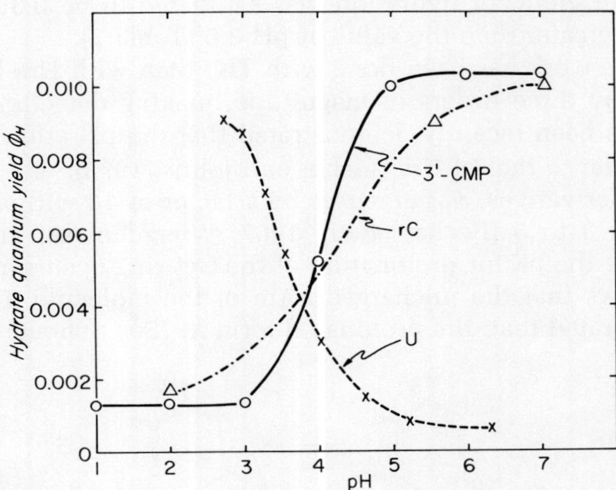


**Fig. 5.** *Dependence on pH of quantum yields for the photohydration of Ura (Brown and Johns, 1968), Cyd (Wierzchowski and Shugar, 1961), and 3′-CMP (Becker et al., 1967).*

the structures of the singly charged molecules **VIII** and **IX**. The p*K* for loss of a proton from the N(1) position of Ura occurs in the ground state above pH 9. Fluorescence data from many laboratories summarized by Burr *et al.* (1972) show that in the excited singlet state this ionization occurs at about pH 4. Delocalization of the negative charge affects the 5,6-bond, as shown in structure **VIII,** and the change in

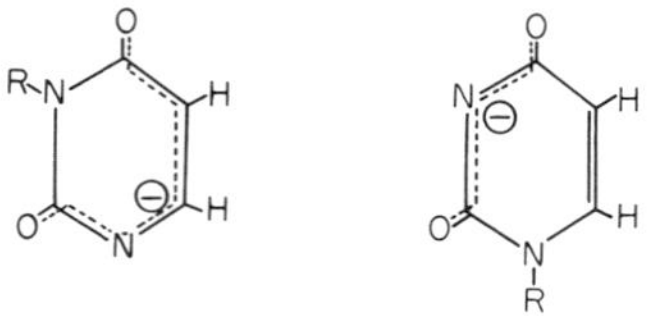

VIII IX

electron distribution greatly reduces the efficiency of the nucleophilic addition of water to this bond. On the other hand, derivatives which are substituted at N(1) can lose only the N(3) proton, giving an excited ion of type **IX** above pH 6. In this species the charge delocalization does not involve the 5,6-bond, and the hydrate yield differs only slightly from the yield for the neutral molecule as observed by Burr *et al.* (1972). The same authors reported that in highly acidic solutions ($pH < 2$) of Ura derivatives at room temperature the yield of hydrates is again decreased; however this may be due to the instability of hydrates in acid, leading to an underestimate of the yield. Khattak *et al.* (1972) irradiated Ura in 6.4 *M* $H_2SO_4$ solutions at $-30$°C. At this temperature the hydrates are reasonably stable despite the acid, and the quantum yield for hydration was estimated to be 0.35, which is forty times greater than the value at pH 2.6 (Table 1).

Much less work has been done with Thy than with Ura because $\phi_H$ is smaller by three orders of magnitude, making detection very difficult. It has been recently demonstrated that the pH effect on hydration is similar to that of Ura (Fisher and Johns, 1973).

For Cyt derivatives, $\phi_H$ *increases* by a factor of 10 with an increase in pH from 3 to 5 (Becker *et al.*, 1967; Wierzchowski and Shugar, 1961). Since the p*K* for protonation of the Cyt ring occurs at about pH 4, this shows that the uncharged form of the molecule (**IV**) is more readily hydrated than the protonated form **X.** (See Scheme 4.)

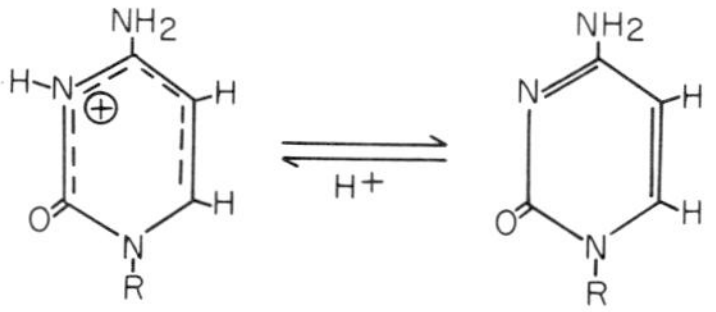

X IV

**Scheme 4**

At extremely basic pH, photolysis results in the formation of irreversible products because of ring opening of the hydrates or because of different photochemical reactions of the ionized molecules. Cyt and 1-MeCyt also undergo irreversible changes at pH 2 (Wierzchowski and Shugar, 1957).

*c. Effect of Wavelength*

Study of the effect of wavelength on $\phi_H$ requires a monochromatic source of light. Since such a source is not available in many photochemical laboratories, little work has been done on the problem. Most of the available results are summarized in Fig. 6. $\phi_H$ is independent of wavelength from 230 to 280 nm for Ura and is nearly so for Urd. In contrast, in dilute solution the quantum yield for dimerization (see Chapter 5) decreases by a factor of $\sim 3$ with increasing wavelength over the same range. This difference in behavior is strong evidence that hydrates and dimers come from different excited states: the hydrate from a singlet and the dimer from a triplet (see Section B,3).

Unlike Ura, 3′-CMP seems to show a small dependence of $\phi_H$ on wavelength. Hélène and co-workers (1964a, 1967, and references cited therein) reported on the existence of tautomeric forms of Cyt, which are found in aqueous solution in low proportions. They believe that one tautomer is responsible for the absorption shoulder at 240 nm

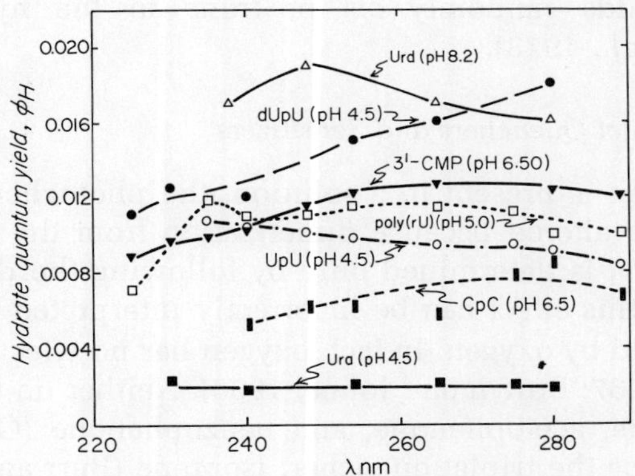


**Fig. 6.** *Quantum yield as a function of wavelength for Ura (Brown and Johns, 1968), Urd (Swenson and Setlow, 1963), 3′-CMP (Becker* et al.*, 1967) the dinucleotides CpC (Hariharan and Johns, 1968), UpU (Brown* et al.*, 1966) and dUpU (Helleiner* et al.*, 1963), and poly(rU) (Pearson* et al.*, 1966). For the last three cases, $\phi_H$ was calculated from the published values for the reaction cross section assuming the absorption cross section to be the same as that of UMP.*

(Fig. 3) and that it is the precursor for ethanol photoaddition (Hélène and Douzou, 1964). On the other hand, Johnson *et al.* (1971) maintain that the shoulder is due to another electronic transition. Regardless of which of these interpretations is correct, if the hydrate does not arise equally from all the absorbing species the quantum yields near 240 nm would have to be increased and the small wavelength dependence might disappear. The marked depression of quantum yields by methylation of the amino group (Table 1) might also reflect the suppression of a tautomeric form which is responsible for most of the hydration. Clearly this subject still requires a great deal of work.

*d. Isotope Effect*

When the hydration reaction is carried out in heavy water, the yield of hydrates is reduced by a factor of ~2. This finding applies to the derivatives of Ura (Wechter and Smith, 1968; Wierzchowski and Shugar, 1961) and Cyt (Wierzchowski and Shugar, 1957), although Ura itself shows a much smaller reduction in yield (Burr and Park, 1968). As a result of this reduced hydrate yield, dimers constitute a larger fraction of the photoproducts in $D_2O$ than in $H_2O$ (Nnadi and Wang, 1969). There is one report that hydration is enhanced when C(5) and C(6) of Ura are tritiated (Wacker *et al.*, 1964). NMR studies indicate that the addition of deuterium at C(5) in heavy water is not stereospecific; that is, the deuterium may add either *cis* or *trans* to the OD group on C(6) (Wechter and Smith, 1968). A similar study, of the addition of light water to 5,6-deuterated 1-EtUra, has confirmed that the proton adds randomly *cis* or *trans* to the hydroxyl group (Summers, *et al.*, 1973).

*e. Effect of Triplet Quenchers and Sensitizers*

When oxygen is present in a solution, the photochemistry of most pyrimidines is altered because dimerization from the triplet state is quenched. If $\phi_H$ is determined only by following the decrease in optical density, this effect can be incorrectly interpreted as a reduction of hydrate yield by oxygen. In fact, oxygen has no effect on $\phi_H$ (Greenstock *et al.*, 1967; Brown and Johns, 1968). Neither do the triplet sensitizers acetone, acetophenone, and benzophenone (Greenstock and Johns, 1968), or the triplet quencher, isoprene (Burr and Park, 1967), change the hydrate yield, although they have marked effects on dimer formation. These results together with the lack of a significant wavelength effect demonstrate that hydrates are formed from an excited precursor state in the singlet manifold.

*f. Effects of Other Components of the Solution*

The hydration reactions of $Me_2Ura$ (Wierzchowski and Shugar, 1959) and Ura (Burr *et al.*, 1968a) are unaffected by neutral salts, such as NaCl, at concentrations of up to 1 *M*. This implies that the rate-limiting step in the hydration of Ura derivatives does not involve two charged species such as an excited Ura ion and $H^+$ or $OH^-$.

In mixtures of acetonitrile or dioxane and water, the quantum yield for hydration of Ura increases linearly with the molar concentration of water, while for $Me_2Ura$ $\phi_H$ depends on the square of the water concentration (Burr and Park, 1968). This may indicate that different reaction mechanisms are involved for the two compounds. On the other hand, the linear and quadratic relationships may be fortuitous since organic solvents have large effects on intersystem crossing (Wagner and Bucheck, 1970), solute aggregation (Lamola and Mittal, 1966), excited-state lifetimes (Szabo *et al.*, 1970; Whillans and Johns, 1971), diffusion rates, dielectric constants, and other properties which can affect reaction efficiencies. When the relative concentrations of water and acetonitrile are varied, all of these properties may be expected to change and to affect the reaction efficiency. Thus, changes in rate depend not merely on the number of water molecules available to react but also on the influence of this water on other solution properties.

*g. Effects of Concentration*

The quantum yield for hydration of Ura is independent of the concentration of Ura between 15 $\mu M$ and 1 m*M* (Brown and Johns, 1968), as would be expected for a reaction involving only one molecule of the base. Similarly, concentration has no effect on $\phi_H$ for $Me_2Ura$, over the range 0.1 m*M*–0.1 *M* (Wang, 1962b; Stepien *et al.*, 1973; J. C. LeBlanc, unpublished data, 1972; see Table 1). However, it has been observed that the quantum yield for loss of $Me_2Ura$ increases at high concentrations (Wierzchowski and Shugar, 1959; Burr *et al.*, 1968a). This implies that other products besides the hydrate are formed on photolysis of concentrated solutions, and in fact it has been shown that photodimers of $Me_2Ura$ are produced (Nnadi and Wang, 1969). In 10 m*M* solution dimers account for 20% of the products, while at 0.1 *M* they comprise $>55\%$ of the total (Stepien *et al.*, 1973). At such high concentrations the dimers are formed from the photoexcitation of molecular aggregates (Fisher and Johns, 1970; see Chapter 5).

Wierzchowski and Shugar (1959) observed that the quantum yield for loss of absorbance of Urd increases with concentration. As with $Me_2Ura$, this can be interpreted in terms of the greater importance of dimerizations at high concentrations. However, Wang (1962b) found no concentration dependence of yield for Urd up to 10 m*M*. Wacker *et al.* (1961) reported that irradiation of 0.5 m*M* Urd leads, at low exposures, to roughly equal amounts of dimer and hydrate. As the photolysis proceeds, hydrates continue to form while dimers reach a low photostationary concentration and eventually decrease. This illustrates the importance of repeatedly monitoring the concentration of each product during the irradiation.

*h. Effects of N-Substitution*

Substituents at N(1) tend to increase $\phi_H$ for Ura derivatives, the amount of increase being greater for the more electronegative groups. On the other hand, pH no longer has much effect on the yield. Wang and Nnadi (1968) reported hydration rates in units of $min^{-1}$ for a number of derivatives. Relative to the rate for Ura at neutral pH, they gave rates approximately 4.5, 5.5, and 8 times larger for 1-MeUra, Ura-1-acetic acid, and Ura-1-acetamide derivatives, respectively. Others have reported rates of hydration which are approximately 14 and 34 times larger than the rate for Ura, for the 1-ethyl and 1-cyclohexyl derivatives (Burr *et al.*, 1968a,b). It is difficult to imagine that both sets of data are quantitatively correct, but the trend seems clear: N(1)-substituted uracils have a quantum yield for hydration, at all values of pH, comparable to or greater than the yield for Ura at acidic pH, and much greater than the yield for Ura at neutral pH.

Urd has been reported to be sixteen times as photoreactive as Ura at pH 7 (Sinsheimer and Hastings, 1949). Later estimates for the ratio of photolysis rates of Urd and Ura at neutral pH were 6.3:1 (Moore and Thomson, 1956), 11:1 (Wang and Nnadi, 1968), and 24:1 (Burr *et al.*, 1968a,b). The photohydration quantum yields of 0.016 for Urd (Guschlbauer *et al.*, 1965) and 0.0007 for Ura at pH 7 (Brown and Johns, 1968) are in the ratio 23:1. We think that the larger ratios are more likely to be correct, because failure to control the pH carefully may lead to a high value for the Ura hydrate yield below pH 5, giving a reduced, incorrect ratio.

2′- and 3′-UMP have the same value for $\phi_H$, 0.022 (Sinsheimer, 1954). This value has been confirmed by other workers (Table 1). The diphosphate, UDP, has almost as high a quantum yield, while UTP, perhaps because of its bulky charged group, forms a hydrate with a somewhat lower efficiency (Guschlbauer *et al.*, 1965; Table 1).

Methylation of Ura at N(3) reduces the hydrate yield by a factor of 2–3 (Wang and Nnadi, 1968; Burr *et al.*, 1972). For the disubstituted derivative, 1,3-$Me_2$Ura, the absolute value for the quantum yield of hydration in aqueous solution is well established. Johns (1971) presented the results shown in Fig. 2 and calculated a quantum yield of 0.014. This is in agreement with the earlier values of 0.01 (Moore and Thomson, 1956), 0.015 (Wang, 1962b), and the recent value of 0.013 (J. C. LeBlanc, unpublished data, 1972). However, it disagrees with the value of Burr *et al.* (1968a) who cited a yield of 0.00386 and of Stepien *et al.* (1973) who reported $\phi_H = 0.005$. These values are almost certainly too small.

For Cyt nucleosides and nucleotides, as for Ura derivatives, hydration quantum yields are considerably higher than for the parent Pyr (Table 1). This difference in yield has been attributed to participation of sugar hydroxyls in hydration (Shugar and Wierzchowski, 1957), although the N(1) glycosyl bond may also affect electron densities in the Pyr ring. As already mentioned, methylation of the amino group sharply reduces the hydration quantum yield either by suppressing a tautomer or by increasing the electron density of the ring, thereby inhibiting nucleophilic attack by water.

*i. Effects of 5- and 6-Substitution*

Ura and Cyt substituted on the ring nitrogens form hydrates quite efficiently. Thy (5-methyluracil) forms photohydrates in very low yield (Fisher and Johns, 1973) and no such reaction has yet been observed for Oro, **XI** (Sztumpf and Shugar, 1965; many experiments in our laboratory), for 5-EtdUrd (Pietrzykowska and Shugar, 1970), for 5-PrUrd (Krajewska and Shugar, 1971), or for ψrd (Lis and Allen, 1961). From a chemical viewpoint it is of interest to inquire how substitution at the 5,6-bond affects the efficiency of water addition to that bond.

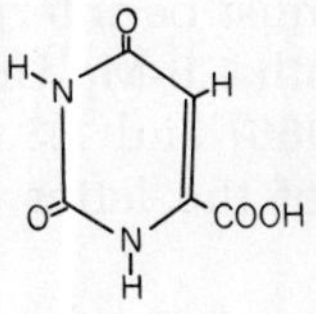

One of the most carefully studied classes of derivatives is that of 5-FlUra and its N-substituted analogs, **XII.** Its photohydrate is identical to the chemically synthesized product, **XIII** (Scheme **5**) (Lozeron *et al.*, 1964). Values of $\phi_H$ for FlUra and some related compounds are presented in Table 1.

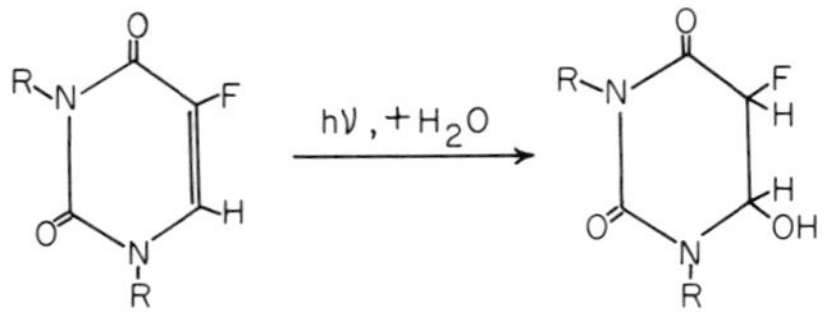

**Scheme 5**

In the 5-chloro and other halogen derivatives of Ura, photoproduction of radicals by cleavage of the carbon–halogen bond and subsequent dark reactions are more important than hydration (Langmuir and Hayon, 1969). The photochemical conversion of 6-ClUra to barbituric acid, on the other hand, is believed to proceed by photohydration followed by the elimination of HCl (Kazimierczuk and Shugar, 1971).

A number of other Ura derivatives have been subjected to cursory examinations. The definition of photohydrate has generally been "heat reversible product" as monitored by absorbance changes, but the fact that at least some isomers of dimers may be thermally unstable (Herbert *et al.*, 1969) was ignored. Wang and Nnadi (1968) attributed the loss of absorbance of 6-MeUra in water entirely to hydration. E. Lo, G. J. Fisher, and H. E. Johns (unpublished data, 1971 summarized by Fisher, 1973) found that ~10% of the absorbance loss of 6-MeUra on irradiation in deoxygenated unbuffered 0.1 m*M* solution at 265 nm is reversible on heating but virtually all of the loss is restored by irradiating the solution at 235 nm. This suggests that the products of 6-MeUra in solution are almost all dimers, including at least one thermally unstable isomer. Therefore, it is worth stressing that, in the absence of structural evidence concerning the nature of the products, absorbance data must be interpreted with great care.

In contrast to 6-MeUra, both 1,6-$Me_2$Ura (Wierzchowski and Shugar, 1960; Fikus and Shugar, 1966) and 1,3,6-$Me_3$Ura (Nnadi, 1968) form hydrates, and the hydrate of the latter compound has been isolated (Nnadi, 1968).

Table 1 presents a number of quantum yields for reversible absorbance losses of substituted uracils. Little can be concluded about the mechanistic significance of these results, but the yields are quite high for the derivatives with the electron-donating amino group at C(5) and with the electron-withdrawing sulfonamide group at C(6).

In summary, substituents at C(5) and C(6) strongly affect the susceptibility of Ura derivatives to photohydration. These substituents could alter singlet lifetimes, the pK of the excited state, the position of tautomeric equilibrium, bond strengths, steric repulsion of the attacking species, and the solvent orientation around the ground-state molecules. The importance of these factors can be determined only by detailed investigations.

*j. Effects of Neighboring Bases*

The photochemistry of di- and polynucleotides is complex and hence information is limited. Mathematical treatments of the formation of several products in di- and polynucleotides have been presented elsewhere (Brown and Johns, 1967; Johns *et al.*, 1966). We will discuss some of the problems of quantitative studies and some of the implications of the available results.

When bases are joined together in a polymer, their excited-state properties are affected and the extinction coefficients are less per base than for monomeric nucleotides. In single-stranded polymers this hypochromism is small, but in ordered structures such as the double-helical complex of poly(rA) and poly(rU) the absorption can be reduced by as much as 30% (DeBoer *et al.*, 1967). This leads to difficulty in choosing the appropriate absorption cross section and in determining quantum yields in ordered polynucleotides by observing absorbance changes. The formation of a product causes not only a loss of the absorption of the base which has reacted, but also an increase in the absorption of neighboring bases because of local denaturation. In addition to hydrates, dimers and other products are readily formed in polynucleotides. Hydrates are unstable to acid hydrolysis of the polymer and an adequate assay is lacking, although a number of assays have been proposed (Section C.1.b, C. 2.d and C.2.e). A reliable method for estimating the amount of hydrate consists of enzymatic hydrolysis, under mild conditions in which the hydrates are known to be stable, followed by chromatographic analysis. Such a procedure has been developed for polyribonucleotides (Pearson and Johns, 1966a) but not for polydeoxyribonucleotides.

In addition to the difficulties in measuring the amount of hydrate produced, energy transfer is also a problem in quantitative work. Consider, for example, the formation of Ura hydrates in the relatively simple oligonucleotide UpGpA. Purines are quite unreactive and dimers cannot form so the hydrate is the only possible photoproduct. Furthermore, hypochromism is negligible in solution. As a result, the cross section for hydration can be determined in this case by monitoring absorbance losses, and a quantum yield $\phi_H$ may be determined by dividing this loss by the absorption cross section of UMP. However, it is known that singlet excitations can be transferred along polynucleotide chains over a distance of several bases (Sutherland and Sutherland, 1970). Hence, some of the hydrates formed in UpGpA may derive from photons absorbed in Gua or Ade, and the calculated quantum yield may be too high. The values given in Table 1 assume no energy transfer, so they are probably overestimates. Despite these reservations, many of the values in Table 1 are close to the values for the monomers.

In ordered structures such as poly(rA-rU) in which a poly(rU) chain is hydrogen bonded to a poly(rA) chain, the yield of hydrates is very small (0.0009) because of the exclusion of water from the interior of the double helix. If the ordered structure is denatured by heating, the quantum yield increases to 0.012 for poly(rU) since the bases are now more exposed to water. Also, the local disruption of the helix by the formation of a dimer leads to an increase in the efficiency of hydration of a neighboring base (Pearson and Johns, 1966b); that is, the quantum yield for one product is altered by the presence of another product, which changes the local molecular environment.

Similar observations have been made for single-stranded poly(rC), which has some helical structure in solution (Lomant and Fresco, 1972; see also Rhoades and Wang, 1971).

Other changes in the structure, such as complexing with polyamines or with a second strand of poly(rU), also affect the hydrate yield (DeBoer *et al.*, 1967). It has recently been shown that base stacking reduces the efficiency of solvent photoaddition to pyrimidines (Leonov and Elad, 1974).

### 3. Mechanism of Hydration

#### *a. Electronic Precursor State*

Upon absorption of a photon, a Pyr molecule is excited from the ground state ($^1P_0$) to a higher singlet state ($^1P_1$). This excited state has a lifetime at room temperature of only a few picoseconds (Eisinger and Shulman, 1968; Hauswirth and Daniels, 1971), during which it can react chemically (path a in Fig. 7) perhaps via an excited complex with a water molecule (Summers and Burr, 1972); it can lose its energy by a nonradiative transition to one of the upper vibrational levels of the ground state (path b), which might in turn react with water to form the hydrate (path d), or it can undergo a spin inversion (path c), leading to a triplet state ($^3P_1$). The triplet state has a lifetime of several microseconds, during which it can interact with a ground-state molecule to form a dimer (see Chapter 5). The triplet may also be populated by energy transfer from a donor molecule, such as acetone

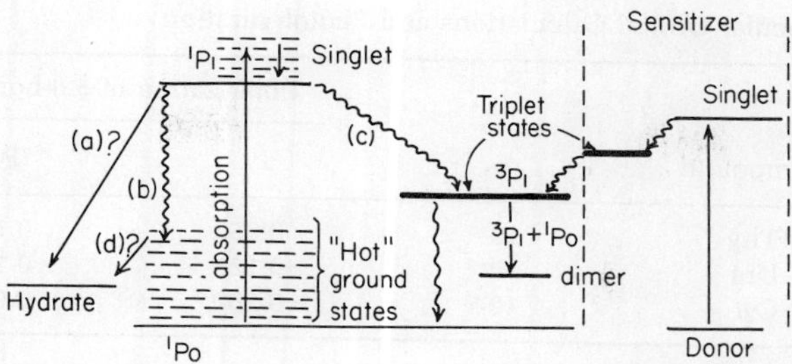


**Fig. 7.** *Energy level scheme for the pyrimidines indicating the precursor states responsible for dimers and hydrates.*

excited to its triplet. When this is done triplet states are produced without any involvement of the Pyr singlet. Under these circumstances, Greenstock and Johns (1968) showed that no hydrates result. *Hence, hydrates do not arise from the triplet state.* [The 6-azapyrimidines are exceptions to this rule (Kittler and Löber, 1969; Kittler, 1972); however, they are not true pyrimidines.] This leaves paths a and d as possible precursors for the hydrate, along with such suggestions as an alternate singlet-state or an excited-state tautomer (for a discussion, see Summers and Burr, 1972).

EVIDENCE FOR AND AGAINST THE HOT GROUND STATE. We first consider the dependence of hydration on pH, illustrated in Fig. 5 which shows that hydrates are formed by interactions of water with a neutral Pyr. For example, with Cyt and its derivatives, the hydrate yield increases markedly above pH 4 when the base becomes uncharged in its ground state. The simplest explanation is that hydration occurs from a hot uncharged ground state.

For Ura, which is neutral up to pH 9.3 in the ground state (Shugar and Fox, 1952), a sudden reduction in hydrate yield occurs at about pH 4. Fluorescence data, summarized by Burr *et al.* (1972), indicate that this is the pH at which the excited state becomes negatively charged. The simplest explanation in this case is that hydration of Ura takes place mainly from an uncharged excited singlet state. However, the involvement of an uncharged hot ground state cannot be ruled out. For example, if a neutral Ura molecule at pH 7 were excited, it would quickly deprotonate to yield a negatively charged singlet. If this molecule underwent internal conversion in less time than that required to regain a proton by diffusion, it would then be uncharged in a hot ground state and so available for hydration. For Ura the pH evidence alone cannot allow us to choose between the two models, pathways a and d of Fig. 7. Therefore, other lines of evidence must be considered.

**Table 2** Molecular Orbital Calculations and Photohydration

| Compound | $\phi_H$ | Polarization of 5,6-bond | |
|---|---|---|---|
| | | $^1P_0$ | $^1P_1$ |
| Thy | $10^{-5}$ | 0.052 | 0.158 |
| Ura | $10^{-2}$ | −0.122 | −0.193 |
| Cyt | $10^{-2}$ | −0.189 | 0.083 |

Molecular orbital calculations have been carried out by Danilov (1967) and Malrieu (1967), and some of their results are reproduced in Table 2. The polarization of the 5,6-bond is defined as the difference between the calculated charge densities on C(5) and C(6). A positive value indicates that C(6) is relatively negative and, therefore, that it has a lower probability for nucleophilic addition of a water molecule. Hence a positive polarization would be expected to lead to a low hydrate yield. The values for $\phi_H$ are approximate values for the neutral molecules taken from Table 1. The yield of hydrates correlates with the polarization of the 5,6-bond for the electronic ground state much better than with the values for the first excited singlet, $^1P_1$.

Whitten *et al.* (1970) reported that the quenching of fluorescence, and thus of the electronic singlet, of $Me_2$Ura by a variety of nucleophiles, including water, does not correlate well with the rate of photochemical addition to the Pyr. This suggests that the addition is not to the fluorescent singlet state but rather to a vibrationally excited ground state. Wang and Nnadi (1968) and Wang *et al.* (1968, 1970) also interpreted substituent, pH, and isotope effects in terms of a hot ground state intermediate.

The nucleophilic attack of bisulfite, $HSO_3^-$, on Ura and Cyt derivatives to give the 6-sulfite product analogous to the hydrate has been reported (Shapiro *et al.*, 1970; Hayatsu *et al.*, 1970). Since this reaction is a thermal addition which cannot involve the electronically excited singlet state, it provides yet another suggestion that hydration may occur via a hot ground state.

The thermal deamination of Cyd at 95°C may also involve the formation of a hydrate intermediate with a saturated 5,6-bond which deaminates and then rapidly eliminates water (Wechter and Kelly, 1970). Such a thermal hydrate could be formed only from a vibrationally excited ground state, since the electronic singlet is not produced by heating.

Although some evidence can be interpreted as showing that Pyr photohydrates are formed from the electronically excited singlet state, all of the reported observations and calculations are compatible with hydration of an uncharged, vibrationally excited ground-state molecule.

*b. Reaction Scheme*

The lack of a neutral salt effect on the hydrate yield of Ura indicates that the reaction does not occur between two ionic species (Burr *et al.*, 1968a). Furthermore, the pH dependence of photohydration is not compatible with an attack by hydroxyl or hydronium ions. That is, the rate-limiting step in hydrate formation is an interaction between the excited Pyr and a neutral water molecule. It has been observed by NMR that the addition of OH to C(6) and H to C(5) of Ura derivatives is not stereospecific but is almost completely random, showing that hydration is not a concerted process (Wechter and Smith, 1968; Summers, *et al.*, 1973). Rather, the C—OH and C—H bonds must be formed in separate steps. Since viscosity has little effect on yield unless the water is diluted with acetonitrile (Summers and Burr, 1972), it is apparent that the rate-limiting step involves a water molecule adjacent to the excited base.

The excited states of compounds such as aromatic ketones are zwitterionic, with the negative charge on the oxygen and the positive charge on a carbon at the opposite end of the conjugated system (Pickett, *et al.*, 1953; Braude *et al.*, 1954). This concept has been applied to the photohydration of pyrimidines (Wang *et al.*, 1956; Wang, 1958a; Moore, 1959; Wacker *et al.*, 1964). Recent work with N-substituted derivatives of Ura led to the suggestion that the intermediate zwitterion may have a portion of its positive charge on N(3) (Wang and Nnadi, 1968; Khattak *et al.*, 1972). However, the simplified reaction Scheme 6 accounts reasonably well for most observations.

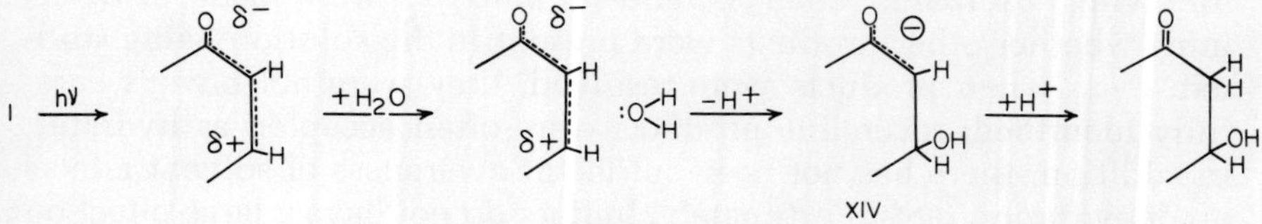

**Scheme 6**

Upon absorption of a photon, the neutral Pyr molecule is electronically excited to the neutral singlet and undergoes internal conversion to the vibrationally hot ground state, in which C(6) is relatively positive. A water molecule attacks C(6) and releases a proton, possibly by a concerted process involving a second water molecule. Finally, a proton adds to C(5) attacking the intermediate **XIV** from either side of the ring. When the excited Pyr is negatively ionized, above pH 4, the reaction proceeds less efficiently, while under extremely acidic conditions the 5,6-bond of the excited Ura cation is even more positive and hence more susceptible to nucleophilic attack (Khattak *et al.*, 1972). For neutral Cyt derivatives the mechanism is essentially the same as for neutral uracils, although the charge delocalization will be somewhat different than that shown, and the details are less well understood.

Work is continuing on the details of the reaction mechanism, using such tools as NMR to investigate configurations of intermediates and to confirm that hydration proceeds mainly through a hot ground state (Hauswirth *et al.*, 1972). Studies with fluorescent derivatives suggest that the excited singlet also plays a role in the process of hydrate formation (Burr *et al.*, 1975).

## C. Properties of the Photohydrate

### 1. Dehydration and Deamination

#### *a. Introduction*

Reversibility of a photoproduct to parent, upon heating in acidic or in basic solution, has long been believed to be diagnostic of the hydrate. Indeed this reversibility is characteristic of hydrates but not exclusively so. As is discussed in Chapter 5, some photodimers may also revert to parent in acid or base. Since this possibility has generally been neglected, the quantitative aspects of many dehydration experiments reported in the literature are of doubtful validity. Most of the hydrate reversal rates discussed in this section were obtained by following absorbance changes, and no attempts were made to determine whether other products were present in the solution being studied. Even when products were resolved, they were not always carefully identified; reversible products were often accepted as hydrates. In addition, there has not been sufficient awareness of solvent effects on dehydration rates. Fortunately, buffers do not have a large effect on the stability of hydrates of Ura derivatives, but for Cyd hydrate at pH 4 the rate for elimination of water is determined almost entirely by buffers (DeBoer *et al.*, 1970). Nevertheless, published results on the dehydration of Pyr hydrates present a remarkably coherent picture.

#### *b. High pH*

Regeneration of $Me_2Ura$ from hydrate is almost quantitative in 1 *M* $NH_4OH$ (Wang, 1962b). On the other hand, Wang (1962a) found that

Urd is not fully regenerated from its hydrate, and Schuster (1964) reported that only 65% of the hydrate of UMP reverses to the parent in alkaline solution. The other 35% undergoes ring opening at the 3,4-bond in the manner typical of 5,6-saturated pyrimidines (Batt *et al.*, 1954; Fink *et al.*, 1956). In spite of this Schuster (1964) used ring opening as an assay for hydrates in irradiated RNA, although it seems probable that the fractions of hydrates reversed and degraded in RNA are quite different from the corresponding values for monomers in solution. The apparent discrepancy between the various observations on the extent of reversibility was studied in detail by Fikus and Shugar (1966), who found three competing reactions of hydrates in basic solutions. The hydrates of 1,6-$Me_2$Ura, Urd, and other glycosides undergo direct base-catalyzed elimination of water, and also reversal by way of ring opening at the 1,6-bond (**XV**), dehydration to give an unstable intermediate absorbing at 290 nm, and ring closure to regenerate parent. Scheme 7 is not unreasonable in view of the fact that the 1,6-bond of Urd hydrate is also susceptible to cleavage by borohydride

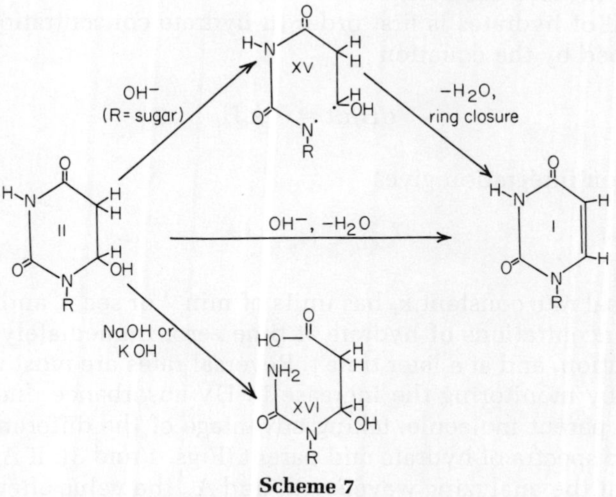

**Scheme 7**

(Miller and Cerutti, 1968). A third reaction, which occurs if the solution is made basic with NaOH or KOH, involves an irreversible ring opening at the 3,4-bond to give **XVI** (Fink *et al.*, 1956). This reaction accounts for the different results in $NH_4OH$ and in KOH since it does not occur in $NH_4OH$. In contrast to the glycosides, 1-MeUra hydrate does not undergo the second reaction, presumably because of the different effects of sugars and alkyl groups on the strength of the 1,6-bond; base-catalyzed reversal to 1-MeUra occurs only by the direct pathway. The hydrates of 5-FlUra and $Fl^5Me_2{}^{1,3}$Ura undergo similar reactions, but in 1 *N* KOH the ring cleaves at both the 3,4- and the 1,6-bonds to give two fragments (Lozeron *et al.*, 1964; Fikus and Shugar, 1966).

The hydrate of 2′(3′)-CMP undergoes both ring opening and deamination at extremely basic pH, in the presence of $Na^+$, to give the same product that was observed for Ura derivatives (Schuster, 1964).

In general, hydrates are too unstable above pH 12 to permit careful studies of their reversal rates, and the other reactions just described make studies more complicated. As a result, the remainder of this section deals only with work which has been carried out between pH 2 and pH 11.

*c. Determination of Reversal Rates*

"Unstable" or "rapidly reversible" no longer constitutes a satisfactory description of the stability of a hydrate. It has become important to discuss dehydration rates quantitatively and to specify the experimental conditions carefully.

Reversal of hydrates is first order in hydrate concentration and can be described by the equation

$$dH/dt = -k_r H_t \tag{7a}$$

which upon integration gives

$$H_t = H_0\, e^{-k_r t} \tag{7b}$$

The reversal rate constant $k_r$ has units of $min^{-1}$ or $sec^{-1}$, and $H_0$ and $H_t$ are the concentrations of hydrate at time zero, immediately following the irradiation, and at a later time $t$. Reversal rates are most readily determined by monitoring the increase in UV absorbance due to recovery of the parent molecule, taking advantage of the differences in the absorption spectra of hydrate and parent (Figs. 1 and 3). If $A_t$ is the absorbance at the analyzing wavelength and $A_\infty$ the value after complete dehydration, then $(A_\infty - A_t)$ is proportional to the hydrate concentration at time $t$. Therefore, Eq. (7) can be rewritten as

$$(A_\infty - A_t) = (A_\infty - A_0)\, e^{-k_r t} \tag{8}$$

Figure 8 shows a typical reversal curve for Cyd hydrate. The straight line on semilog paper verifies the applicability of Eq. (8) and allows an accurate determination of the rate constant for reversal. If there are

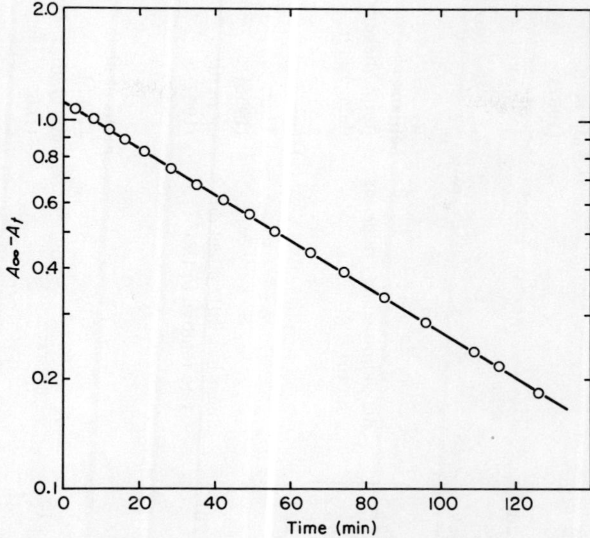


**Fig. 8.** *Dehydration curve for Cyd hydrate in 0.004* M *cacodylic acid at 25°C and pH 4.0. The analyzing wavelength was 271 nm* ($t_{1/2} = 48$ *min;* $k_r = 0.693/48 = 1.44 \times 10^{-2}$ *min*$^{-1}$) (*DeBoer* et al., *1970*).

other thermally labile photoproducts such as dimers in the solution, the absorbance varies with time in a more complicated manner as the sum of exponential terms. The pH has an extremely large effect on the reversal rates of Pyr hydrates; accordingly, rate constants are meaningless unless the pH is specified. Rates also increase rapidly with temperature, which, therefore, must be kept constant.

A number of important dehydration rates from the literature are compiled in Table 3. Figure 9c shows the effect of pH and temperature on the reversal of some Ura derivatives, while Fig. 11c presents the effects of buffers and pH on dehydration of the Cyd hydrate at 25°C. It is apparent that reversal is catalyzed in basic solutions but that the behavior of derivatives of Ura and of Cyt differs in acid. For this reason we will discuss the two classes of compounds separately.

### *d. Reversal of Hydrates of Uracil Derivatives*

i. REACTION SCHEME. Hydrates of Ura derivatives have maximum stability near pH 5, and their reversal rates increase rapidly in both acid and base. The reaction model of Fig. 9a has been proposed to account for this behavior (DeBoer, 1970). By this model, hydrate reversal may proceed slowly by interaction with neutral water molecules and may be catalyzed by $H^+$ or $OH^-$ ions and by neutral molecules of some

**Table 3** Representative Reversal Rate Constants

| Parent | pH | T (°C) | $10^4 k_r (min^{-1})$ | $E_a$(kcal/mole) | Comments | Reference |
|---|---|---|---|---|---|---|
| Uracil Derivatives | | | | | | |
| Ura | Unbuffered | 20 | 0.13 | | | Fahr *et al.* (1967) |
| | Unbuffered | 50 | 3.18 | | | |
| | 2.0 | 20 | 19.4 | 20 | | |
| | 2.0 | 50 | 486 | | | |
| | 7.0 | 20 | 0.69[a] | | | |
| | 7.0 | 50 | 16.8 | | | |
| | 9.5 | 20 | 87.3 | | | |
| | 2.0 | 27 | 74 | 21.1 | Activation energy depends on pH | LeBlanc (unpublished data, 1969) |
| | 2.0 | 50 | 850 | | | |
| | 7.0 | 27 | 1.45 | 25.2 | | |
| | 7.0 | 50 | 26 | | | |
| $Me_2$Ura | 4.76 | 20 | 0.17 | 22.6 | | Moore and Thomson (1956) |
| Urd | 2.0 | 20 | 0.48 | 20 | Sugar has little effect on rate relative to Ura | Fahr *et al.* (1967) |
| | 2.0 | 50 | 11.3 | | | |
| | 7.0 | 20 | 0.56[a] | | | |
| | 7.0 | 50 | 13.0 | | | |
| | 2.0 | 60 | 80 | | | Fikus *et al.* (1962) |
| | 7.0 | 60 | 85 | | | |
| 2′(3′)-UMP (mixed isomers) | 2.0 | 20 | 0.37 | 21 | | Fahr *et al.* (1967) |
| | 2.0 | 50 | 10.2 | | | |
| | 7.0 | 20 | 0.44[a] | | | |
| | 7.0 | 50 | 10.8 | | | |
| UMP | 2.0 | 60 | 600 | | | Fikus *et al.* (1962) |
| | 7.0 | 60 | 26 | | | |
| | 2.0 | 20 | 1.14 | | 5′ isomer reverses faster than 2′,3′ isomers | Fahr *et al.* (1967) |
| | 2.0 | 50 | 56.8 | | | |
| | Unbuffered | 50 | 8.8 | 25 | | |

| | | | | | | |
|---|---|---|---|---|---|---|
| | 7.0 | 20 | 0.44[a] | | | |
| | 7.0 | 50 | 10.7 | | | |
| Cytosine Derivatives | | | | | | |
| Cyt | Unbuffered | 20 | 49 | 21 | | Fahr *et al.* (1966) |
| | 7.4, phosphate | 20 | 110 | | | |
| Cyd | Unbuffered | 20 | 20 | 13 | Sugar decreases rate, relative to Cyt | Fahr *et al.* (1966) |
| | 7.4, phosphate | 20 | 50 | | | |
| | 7, 0.02 *M* phosphate | 25 | 66 | | | Wierzchowski and Shugar (1961) |
| | 4, 0.01 *M* cacodylate | 25 | 250 | | Carefully done; hydrate nearly pure; strongly dependent on buffer. See Fig. 11. | DeBoer *et al.* (1970) |
| | 4, 0.01 *M* acetate | 25 | 210 | | | |
| | 4, 0.01 *M* formate | 25 | 150 | | | |
| | 4, unbuffered | 25 | 55 | | | |
| dCyd | 7, 0.02 *M* phosphate | 25 | 102 | | Deoxy sugar increases rate | Wierzchowski and Shugar (1961) |
| 2′-CMP | 7, 0.02 *M* phosphate | 25 | 240 | | | |
| | 7, 0.01 *M* phosphate | 30 | 455 | | | Sinsheimer (1957) |
| 3′-dCMP | 7, 0.01 *M* phosphate | 30 | "similar to 2′-CMP" | | Deoxy sugar increases rate | Sinsheimer (1957) |
| 3′-CMP | 7, 0.01 *M* phosphate | 30 | 104 | | Slower than 2′-CMP | Sinsheimer (1957) |
| | 7, 0.02 *M* phosphate | 25 | 60 | | | Wierzchowski and Shugar (1961) |
| | 7.3, 0.01 *M* phosphate | 25 | 37 | 16 | | Johns *et al.* (1965) |

*(Continued)*

**Table 3** *(Continued)*

| Parent | pH | T (°C) | $10^4 k_r$(min$^{-1}$) | $E_a$(kcal/mole) | Comments | Reference |
|---|---|---|---|---|---|---|
| CMP | 7, 0.02 *M* phosphate | 25 | 390 | 15.5 | | Wierzchowski and Shugar (1961) |
| | 6, 0.02 *M* phosphate | 25 | 1320 | | High | Fikus *et al.* (1962) |
| | 4 | 25 | 430 | | Carefully done; hydrate nearly pure; independent of buffer; depends on secondary ionization of the 5′ phosphate | DeBoer *et al.* (1970) |
| | 6 | 25 | 860 | | | |
| | 7 | 25 | 480 | | | |
| | 8 | 25 | 100 | | | |
| | 9 | 25 | 35 | | | |
| | 11 | 25 | 120 | | | |
| 5′-MeCMP | 6 | 25 | 190 | | Methylation eliminates the phosphate ionization | Fikus *et al.* (1962) |
| | 7 | 25 | 140 | | | |
| | 8 | 25 | 120 | | | |
| 5-Substituted Uracils | | | | | | |
| Thy | 4.0 | 60 | 47 | 26.9 | | |
| | 4.0 0.01 *M* citrate or phosphate | 70 | 150 | | *cis* isomer | Fisher and Johns (1973a) |
| | 5.5 | 70 | 15 | | | |
| | 7.0 | 70 | 120 | | | |
| | 7.2, 0.067 *M* phosphate | 37 | 14.5 | 22.6 | *cis* isomer | Nofre and Ogier (1966) |
| 1,3-$Me_2$FlUra | 1 N HCl | 90 | 1800 | 37 | | Fikus *et al.* (1965) |

[a] The values reported by Fahr at pH 7.0 and 20°C were calculated from the measured rate at 50°C using an (incorrect) activation energy of 20 kcal/mole.

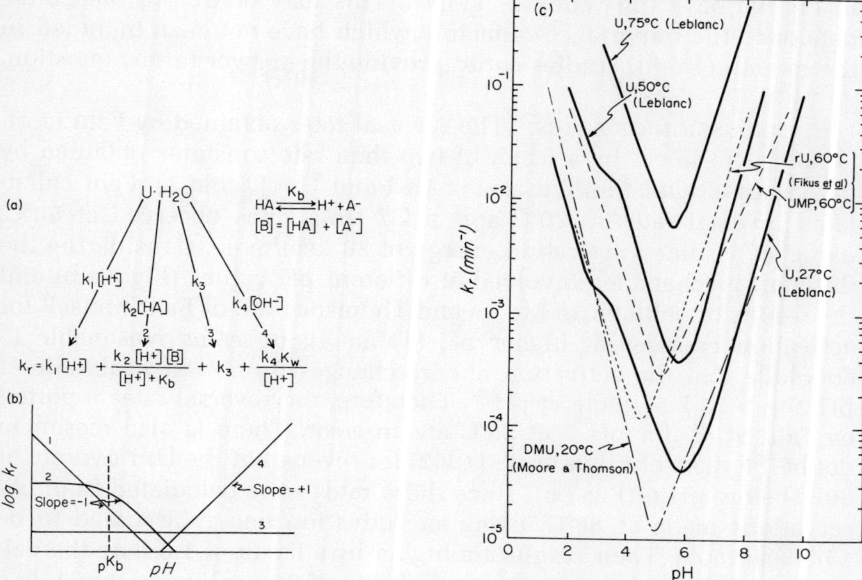


**Fig. 9** *(a) Proposed model for the reversal of Ura hydrate (DeBoer, 1970). (b) The observed reversal rate $k_r$ by this model is the sum of the contributions of paths 1 (specific acid catalysis, with a rate $k_1[H^+]$), 2 (general acid catalysis by the neutral form of the buffer, with rate $k_2[HA]$), 3 (spontaneous or water catalyzed reversal, with a rate $k_3$), and 4, (specific base catalysis, with a rate $k_4[OH^-]$). (c) Observed rates of reversal of hydrates of Ura (J. C. LeBlanc, unpublished data, 1969); $Me_2$Ura (Moore and Thomson, 1956), Urd, and UMP (Fikus* et al., *1962).*

buffers. At high pH, specific base catalysis by $OH^-$ predominates, and the rate ($k_4[OH^-]$) increases by an order of magnitude with each unit increase of pH, resulting in a slope of $+1$ on a log plot of $k_r$ against pH (Fig. 9b,c). At low pH, specific acid catalysis by $H^+$ is the major process and leads to a slope of $-1$. General acid catalysis by neutral buffer molecules, HA, leads to a shoulder on the low pH section of the curve. The position of this shoulder is determined by the pK of the buffer. Our results for reversal of Ura hydrates in McIlvaine's citrate and phosphate buffer (J. C. LeBlanc, unpublished data, 1969; Fig. 9c) exhibit the pH dependence predicted by the model, including the low pH shoulder. These results lead us to discount the importance of catalysis by the buffer anion, $A^-$, since there is no shoulder at high pH. Catalysis by neutral water may be an insignificant mechanism for dehydration. Only a detailed study of the sharpness of the minimum could reveal this. It should be noted that not all of the results shown in Fig. 9c have the "correct" slopes. This may be due to inaccurate data or to the importance of factors which have not been included in the model. Careful studies could provide the answer to this question.

ii. DISCUSSION OF RATES. The reversal rates obtained by Fahr *et al.* (1967) seem lower by a factor of two than rate constants obtained by other workers for the hydrates of Urd and Ura (Table 3). Fahr calculated reversal rates at 20°C and pH 7 from rates observed at 50°C, using an assumed activation energy of 20 kcal/mole. Considering the different mechanisms involved at different pH values (Fig. 9a,b) and the results of LeBlanc, of Moore and Thomson, and of Fahr himself for activation energies at higher pH (Table 3), it seems reasonable to conclude that the activation energy changes from $\sim 20$ kcal/mole at pH 2 to $>25$ kcal/mole at pH 7. Therefore, the reversal rates reported by Fahr *et al.* for pH 7 at 20°C are in error. There is also reason to doubt the rates of Fikus *et al.* (1962) for reversal of the Urd hydrate at 60°C below pH 6 (Fig. 9c), since these rates were calculated from observations made at 80°C, using an activation energy assumed to be 15.5 kcal/mole. Their results are higher by a factor of 1.6 than the values calculated using an activation energy of 21 kcal/mole, which is a better estimate in this pH range.

For Urd hydrate, the reversal rate in $D_2O$ at pD 1.6 is almost twice as great as the rate in $H_2O$ at pH 1.6 (Wierzchowski and Shugar, 1961). For $Fl^5Me_2{}^{1,3}$Ura in 1 *N* acid, hydrate reversal is $\sim 2.6$ times as rapid in $D_2O$ as in $H_2O$ (Fikus *et al.*, 1965).

The reversal rates of 2′-UMP and 3′-UMP hydrates differ by $\sim 50\%$ in 0.05 *M* Tris buffer at pH 8.4 and 86°C (Logan and Whitmore, 1966). However, whether this represents a major difference in stability, a slight shift in the stability curves with pH (as for Urd and Ura in Fig. 9c), the influence of the buffer or of a hydrate pK, or a distortion due to other photoproducts such as thermally reversible dimers is not clear. On the other hand, there is a marked and real difference in the behavior of the hydrates of 2′(3′)-UMP and UMP (Table 3), the latter being much less stable. Such a result was also found for hydrates of CMP, and the high rate of reversal for the 5′ derivative is attributable to catalysis of dehydration by the phosphate (DeBoer *et al.*, 1970).

iii. ISOMERIC HYDRATES. Water could add to the 5,6-bond of Urd or dUrd in two orientations relative to the sugar, leading to two isomeric forms of the hydrate, **XVII** and **XVIII.** The two isomers of the dUrd hydrate have reportedly been resolved by chromatography, but they have not been unambiguously identified (Pietrzykowska and Shugar,

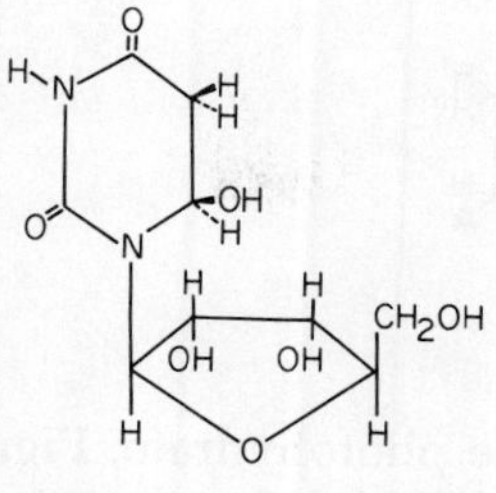

XVII 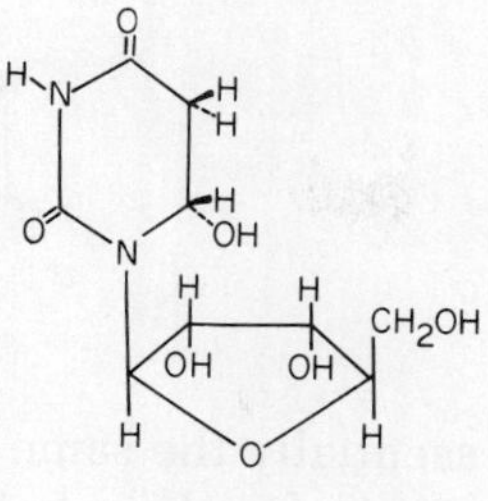 XVIII

1969). The two products showed such different behavior, in both acid and base, that it is difficult to believe they are simply steric isomers. In attempting to repeat these experiments in our laboratory (O. Klinghoffer, unpublished data, 1970), we found that one of the products is not reversible to the parent. S. Y. Wang (personal communication, 1972) confirmed that this photoproduct is not a hydrate. Efforts should be made to determine the structures of these two products in order to clarify their behavior. NMR should also be used as an alternate and highly specific method of monitoring isomerization and dehydration, following the lead of Wechter and Smith (1968). Recently, crystallization of the two diastomeric hydrates has been reported (Pietrzykowska and Shugar, 1974).

*e. Deamination and Reversal of Hydrates of Cytosine Derivatives*

i. DEAMINATION. The main dark reaction of the hydrates of Cyt derivatives is the elimination of water to give the parent compound, with a rate constant $k_r$, but a finite number of the hydrates deaminate with a rate constant $k_d$ and then eliminate water at the slower rate characteristic of Ura derivatives, according to Scheme 8 (in which the

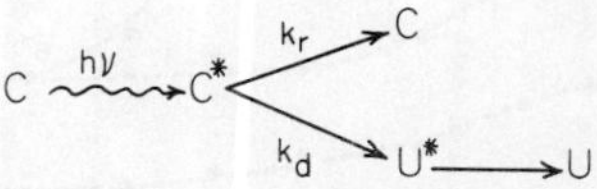

$$C \xrightarrow{h\nu} C^* \begin{cases} \xrightarrow{k_r} C \\ \xrightarrow{k_d} U^* \longrightarrow U \end{cases}$$

**Scheme 8**

asterisk denotes the water addition product). Deamination accounts for the common observation (for example, by Daniels and Grimison, 1964) that Ura is a photoproduct of Cyt. On the contrary, Ura results from dark reactions of the Cyt photohydrate. In fact, Wechter and Kelly (1970) reported the thermal deamination of Cyd entirely in the dark at 95°C, by way of the postulated acid addition intermediate **XIX,**

XIX

a structure essentially the same as the photohydrate. Figure 10 shows the rate constants for dehydration ($k_r$) and for deamination ($k_d$) of 3′-CMP hydrate as a function of pH. The rates were measured by rapid electrophoretic separation of 3′-CMP, its hydrate, and the hydrate of 3′-UMP after a series of incubation times, followed by quantitation using a radioactive label (Becker *et al.*, 1967).

In contrast to reversal, deamination is relatively independent of buffer or pH and involves less than 10% of the hydrates. Although a relatively minor occurrence, deamination has been the subject of much interest because of its possible role in mutagenesis, resulting in conversion of Cyt to Ura by way of the photohydrate (Grossman and Rodgers, 1968). However, it is not clear what proportions of hydrates deaminate and dehydrate in the environment of a polynucleotide, in which steric and catalytic effects are likely to influence both processes.

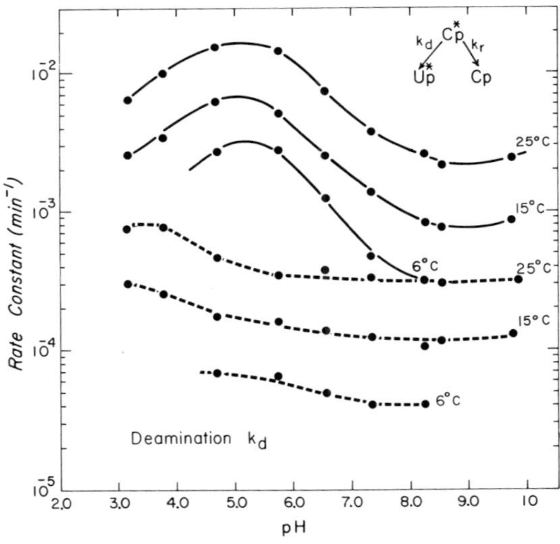


**Fig. 10.** *Rate constants for reversal of the hydrate of 3′-CMP to the parent (solid lines) and for deamination to the hydrate of 3′-UMP (dashed lines) as a function of pH, at phosphate buffer concentrations of 0.01* M *(Becker* et al., *1967).*

ii. FACTORS AFFECTING ACCURACY OF REVERSAL RATES. Table 3 summarizes several reversal rate studies on the photohydrates of Cyt derivatives. In determining rate constants for the elimination of water from these compounds, a number of important considerations are often neglected.

The nature and concentration of buffer have a large effect on the reversal rate between pH 3 and 7 for all Cyt derivatives except CMP (Fig. 11c; DeBoer et al., 1970). This effect will be discussed in connection with the mechanism of dehydration. Whereas most of the errors in determining the efficiency of forming hydrates lead to low values of $\phi_H$, buffers act as catalysts to increase reversal rates.

Often, insufficient care has been taken to ensure that hydrates are the principal products and that they are present in good yield. Failure to optimize conditions of irradiation may result in very small yields of hydrate and, therefore, only small increases in absorbance as the hydrate reverses, leading to inaccurate determination of the rate. When other products are formed, they may reverse at a very different rate. For example, a very unstable, reversible photoproduct is formed by 3′-CMP and CMP in ~10% of the yield of hydrate (Becker *et al.*, 1967). If it were not recognized that the recovery curve is a double exponential in such a case, the rapid initial rate of recovery of the parent might lead to an overestimate for the dehydration rate constant.

In their studies of Cyd and CMP hydrates, DeBoer *et al.* (1970) optimized irradiation conditions so as to produce large yields of products which were almost totally reversible to the parent. These precautions included lowering the temperature to 0°C to minimize the back reaction during photolysis, exclusion of buffers which catalyze reversal, titration to pH 8 or 9 to give maximum quantum yield (Fig. 5) and maximum stability (Fig. 11c), and use of an intense lamp to reduce the required irradiation time. All of this was necessary because the degree of reversibility of irradiated derivatives varies greatly with irradiation conditions. Under these conditions, reversibility of >90% can be achieved after 90% photolysis.

A third factor which is usually neglected is deamination, since it is often assumed that water elimination is the only reaction which hydrates undergo. As has been shown, as much as 10% of the hydrates of 3′-CMP deaminate to the more stable hydrate of 3′-UMP. The deaminating fractions of the hydrates of other derivatives are not known. Although this is a relatively small perturbation it should be kept in mind.

iii. DISCUSSION OF REVERSAL RATES. From Table 3 it is clear that the nature of the sugar residue has a relatively small effect on dehy-

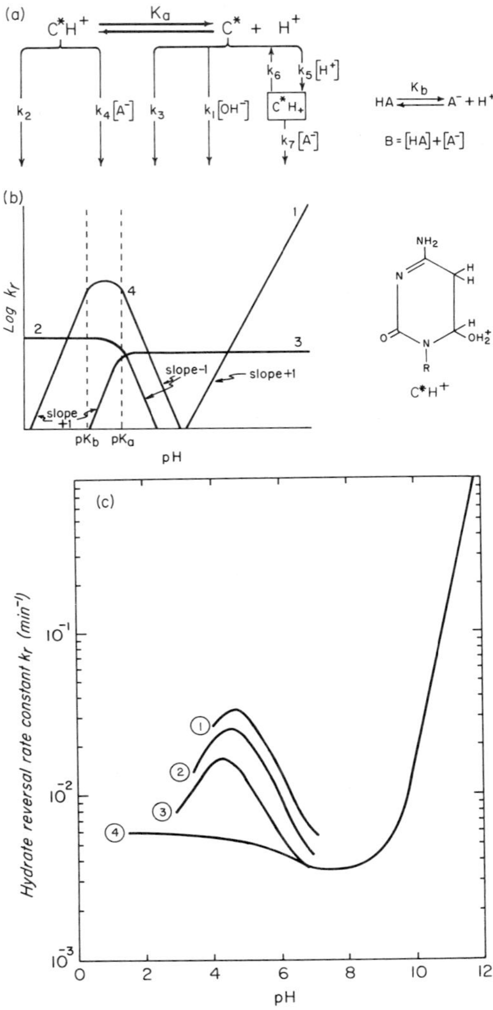


**Fig. 11.** *(a) Model for catalysis of the reversal of Cyt and Cyd hydrates as proposed by DeBoer (1970). (b) The observed reversal rate $k_r$ is the sum of contributions of processes 1 (specific base catalysis of the neutral hydrate, with rate $k_1$ $[OH^-]$), 2 and 3 (spontaneous or water-catalyzed reversal of the protonated and neutral forms of the hydrate, with rates $k_2$ and $k_3$), and 4 (general base catalysis of the protonated hydrate by the buffer anion, with rate $k_4[A^-]$). In addition to the ring nitrogen, the hydroxyl can also protonate, and processes 5–7 are a refinement allowing for this and slightly affecting the position of the maximum (DeBoer* et al., *1970). (c) Reversal rate constants for Cyd hydrate at 25°C as a function of pH, showing the effect of buffers (curve 1:0.01* M *cacodylic acid, pK = 6.19; curve 2:0.01* M *acetic acid, pK = 4.75; curve 3:0.01* M *formic acid, pK = 3.75; curve 4:unbuffered) (data from DeBoer* et al., *1970).*

dration rate. The hydrate of dCyd reverses a little less than twice as fast as Cyd hydrate. On the other hand, the position of the phosphate has a large effect. CMP hydrate reverses six times as fast as 3′-CMP hydrate under identical conditions. This has been attributed to catalysis of dehydration by the phosphate group of the nucleotide (Wierzchowski and Shugar, 1961). In 3′-CMP the phosphate is sterically prevented from influencing the reaction. These results are consistent with what is known about the conformation of nucleotides in solution: the 5′ phosphate is close to the 5,6-bond of the base, in the so-called *anti* conformation, **XX** (Ts'o, 1970). The importance of the

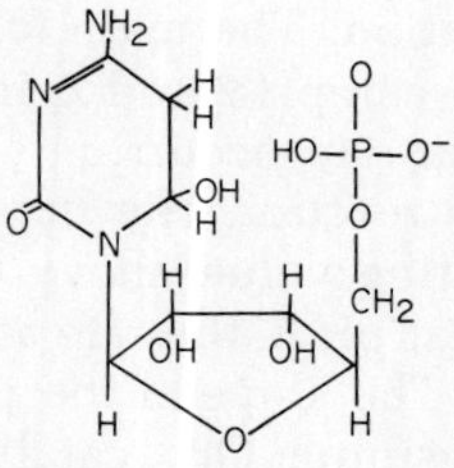

XX

ionization of the phosphate is indicated by the different effects of pH on the reversal rates of the hydrates of CMP and 5′-MeCMP (Table 3). The latter compound lacks the secondary pK near pH 6 and, therefore, cannot have a doubly charged phosphate group. In polynucleotides this pK is also lacking. Phosphate in the form of added buffer also enhances the reversal rates of many derivatives, including the hydrates of Cyt, Cyd, and 3′-CMP. The effect of pH on reversal (Fig. 11c is typical) may be attributed in part to ionization of the buffer (DeBoer *et al.*, 1970). Interestingly, the hydrate of CMP reverses several times more rapidly in the dry state than in water (DeBoer and Johns, 1970). This unusual behavior may be due to the arrangement of phosphate groups in the solid.

Reversal rates of the two optical isomers of the hydrates of Cyd, analogous to **XVII** and **XVIII,** have been shown by NMR techniques to be very similar at pH 9 (DeBoer, 1970), in contrast to the results for the reported isomeric hydrates of dUrd.

iv. ISOTOPE EFFECTS. Wierzchowski and Shugar (1961) studied the deuterium isotope effect on dehydration of Cyd hydrate. "Hydrates" formed in $D_2O$ reverse ~30% more slowly than $H_2O$ adducts. Comparing water elimination rates in $H_2O$ and $D_2O$, at a pH or pD of 1.6, they observed much slower elimination in $D_2O$: $k_{D_2O}/k_{H_2O} = 0.3$. This is quite different from the corresponding result for Urd hydrate, which reverses more rapidly in $D_2O$.

V. BUFFER EFFECTS. The most detailed and systematic study of photohydrate reversal to date has been carried out by DeBoer *et al.* (1970). Having optimized the irradiation conditions, they studied reversibility of the photohydrate by following the absorbance as a function of time $t$, using the relation given in Eq. (8). Figure 11c summarizes some of the results of these experiments for Cyd with various buffers. Cyt behaves similarly. However, CMP is not influenced by external buffers because its intramolecular catalyst (the phosphate group) is effectively present in high concentration. The main feature of the Cyd results is that the reversal rate, between pH 2 and 7, increases in the presence of buffer, with the maximum rate occurring at a pH which depends on the pK of the buffer. The reaction rate depends on buffer concentration, approaching a limiting value above 0.05 *M*. Beyond the minimum rate, which occurs at pH 8, the rate again increases with pH but is independent of buffer. The slope of the plot of log $k_r$ *vs.* pH is 1.0, as would be expected for simple $OH^-$ catalysis (Fig. 11b).

VI. REACTION SCHEME. To account for the observations on elimination of water from Cyt and Cyd hydrates, DeBoer presented the model reaction scheme shown in Fig. 11a. The main features of this model include specific base catalysis by $OH^-$, which predominates at high pH ($k_1$); uncatalyzed or water-catalyzed dehydration of the positive and neutral forms of the hydrate ($k_2$ and $k_3$); and general base catalysis of the protonated form of the hydrate by the anionic form of the buffer ($k_4$). The latter pathway is important only between the pKs of the hydrate and the buffer. This model, with a refinement allowing for protonation of the hydroxyl group, accounts for the observed effects of buffer and pH on dehydration rates (DeBoer *et al.*, 1970).

*f. Reversal of Hydrates of Thymine and Thymidine*

Studies of the chemically synthesized water addition products of Thy and Thd have shown that these compounds are comparable to Ura hydrates in their stability, having a lifetime of several hours in neutral solution at room temperature (Nofre and Ogier, 1966; Cadet and Téoule, 1971a; Table 3). In acidic or basic solution and on heating, the hydrates reverse readily to parent molecules (Cadet and Téoule, 1971b).

Because of the very small quantum yield for hydration of Thy (see Table 1), photohydration was not observed in early studies and investigators were led to the erroneous conclusion that the Thy hydrate must be too unstable to be detected (Moore, 1959; Wang, 1959b; Witkop, 1968). We measured the reversal rate of the photohydrate of Thy at 60°C and 70°C and showed (Fisher and Johns, 1973) that it has the same properties as the synthesized *cis* water addition product (**XXI**). We have not been able to detect the *trans* hydrate (**XXII**) by UV

XXI

*cis* hydrate of thymine

XXII

*trans* hydrate of thymine

irradiation, although it is produced by x-rays (Cadet and Téoule, 1971b). The reversal rate of the Thy photohydrate depends on pH in the same manner as the rate shown for the Ura hydrate in Fig. 9c.

## 2. Characterization of Hydrates

### *a. Synthesis*

The initial step in the chemical synthesis of hydrates of Ura and its N-substituted derivatives is bromination. A single equivalent of bromine water adds to the 5,6-bond of Ura, $Me_2Ura$, Urd, and UMP (Wang *et al.,* 1956; Moore and Anderson, 1959; Wang, 1959c; 1962a). The hydrate is then formed by catalytic reduction of the brominated compound under hydrogen. (see Scheme 9.) Platinum and aluminum

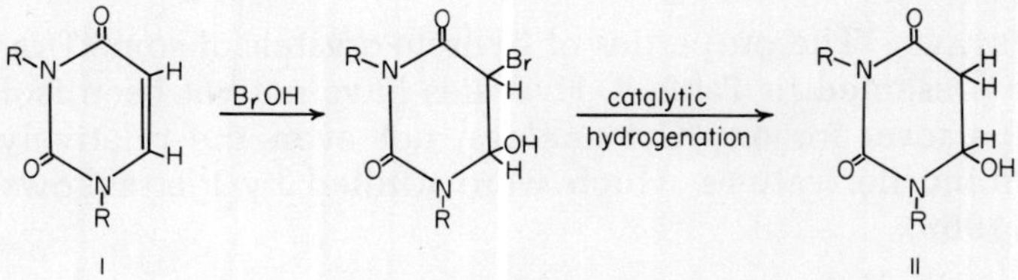

I

BrOH

catalytic hydrogenation

II

**Scheme 9**

oxide are unsatisfactory catalysts, since they also catalyze the dehydration of the hydrate before it can be isolated (Moore and Thomson, 1957). Zinc dust in acetic acid at pH 4 (Moore, 1958), charcoal impregnated with palladium and buffered at pH 7 (Wang, 1958a), and Raney nickel (Gattner and Fahr, 1963) are satisfactory for the production of hydrates in good yield.

Syntheses of the hydrates of Cyd and 2′(3′)-CMP have been reported (Kleber *et al.*, 1965; Fahr *et al.*, 1966). The procedure is essentially the same as that used for Ura derivatives, involving the addition of bromine water followed by hydrogenation with a Raney nickel catalyst. The reversal rates of the synthetic compounds were identical to those of the photohydrates. In addition, the hydrate of 2′(3′)-CMP was resolved from the parent by thin layer chromatography with NaCl solution at 0°C. On elution of the hydrate in dilute HCl, the parent was regenerated. These findings clearly show that the synthetic hydrate and the photoproduct are the same.

Attempting to identify some of the products of $\gamma$-radiolysis of Thy solutions, Nofre and Ogier (1966) synthesized and characterized the *cis* hydrate of Thy (**XXI**). They used the common synthesis, catalyzing the hydrogenation with zinc in acetate buffer at pH 4. Palladium–charcoal had previously been shown to be an unsatisfactory catalyst for $Me_2$Thy (Wang, 1959b). More recently, other investigators have synthesized both of the optical isomers of the *cis* hydrate of Thd (Cadet and Teoule, 1971a) and the *trans* hydrate of Thy (Cadet and Teoule, 1971b). The latter synthesis involves the initial preparation of the *cis* hydrate, followed by peroxidation, gentle iodide reduction, and chromatographic isolation.

$ho^5$hThy also has been synthesized (Nofre *et al.*, 1965) by an adaptation of the method of Fourneau (1909) and is even more stable than $ho^6$hThy. The corresponding 5-hydroxy derivative of $Me_2$Ura has also been synthesized and was shown to have no relation to the photoproduct (Wang *et al.*, 1956). $ho^6Fl^5$hUra has been synthesized and is identical to the photohydrate (Lozeron et al., 1964).

*b. Physical Properties*

i. CRYSTALS. The properties of hydrate crystals of some Ura derivatives are presented in Table 4. Hydrates have not yet been isolated in crystalline form for any Cyt analogs, not even the relatively stable methylamino derivatives which were studied by Pietrzykowska and Shugar (1969).

**Table 4** Properties of Hydrate Crystals

| Parent | Crystal form | Melting point (°C) | Reference |
|---|---|---|---|
| $Me_2$Ura | White needles | 104°–105° | Moore and Thomson (1955) |
| Urd | White powder | 95° (dec.) | Wang (1962a) |
| dUrd (one isomer) | Hygroscopic platelets | 53° (eliminates $H_2O$) | Pietrzykowska and Shugar (1969) |
| FlUra | Off-white | 182° | Lozeron *et al.* (1964) |

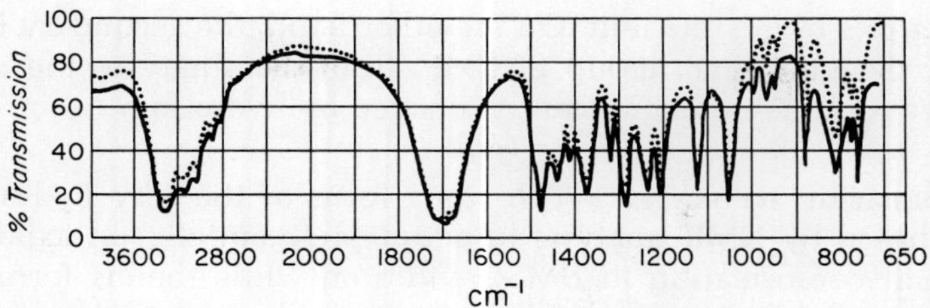


**Fig. 12.** *Infrared spectra, in KBr pellets, of the Ura hydrate obtained by UV irradiation (...) and by chemical synthesis (—), showing the characteristic O—H absorption at 2.98 $\mu$ or 3.36 $\times$ $10^3$ $cm^{-1}$ (Gattner and Fahr, 1963).*

ii. ABSORPTION SPECTRA. The UV absorption spectra of hydrates of Ura derivatives are similar to those for Urd (Fig. 1). It should be noted that published spectra are not for the pure hydrate but for photolyzed solutions of the parent. The hydrate of 3′-CMP (Fig. 3) shows the absorption peak near 240 nm which is considered to be typical of 5,6-saturated cytosines (Janion and Shugar, 1960; Brown and Hewlins, 1968).

The only complete infrared spectrum of a pyrimidine photohydrate in the open literature is for ho$^6$hUra. The synthesized and photochemical hydrates have identical spectra (Fig. 12) showing the 2.98 $\mu$ band (3.36 $\times$ $10^3$ $cm^{-1}$) of O—H in KBr pellets (Gattner and Fahr, 1963). The same band was also found in the spectrum of Ura hydrate in dry chloroform (Moore and Thomson, 1955). In his thesis Nnadi (1968) presented a similar spectrum for the Ura hydrate and a spectrum for the hydrate of 1,3,6-$Me_3$Ura.

iii. p*K*. Thermodynamic p*K*'s of some hydrates and parent compounds have been determined by DeBoer *et al.* (1970) by potentiometric titration with corrections made for reversal of hydrates during the experiment. These values are presented in Table 5. Hydration raises

**Table 5** Thermodynamic Ionization Constants at 25°C

| Compound | pK | Reference |
|---|---|---|
| Cyt | 4.60 | DeBoer *et al.* (1970) |
| Cyt hydrate, ho$^6$hCyt | 5.70 | |
| Cyd | 4.09 | |
| Cyd hydrate, ho$^6$hCyd | 5.12 | |
| CMP | 4.45, 6.58 | |
| CMP hydrate, ho$^6$hCMP | 5.35, 6.56 | |
| 3′-CMP | 4.28 | Cavalieri (1952) |
| 3′-CMP hydrate, ho$^6$hCp | 5.57 | Johns *et al.* (1965) |

the ring pKs by ~1 pH unit. On the other hand, the secondary ionization of the phosphate group of CMP, from the singly to the doubly negative form, is not significantly affected by hydration.

iv. ISOMERIC HYDRATES. The *trans* form of the Thy hydrate has been shown by NMR analysis to adopt preferentially an axial–axial half-chair conformation in DMSO solution, while the *cis* form, **XXI,** has the methyl in an equatorial position and the hydroxyl group axial (Cadet and Téoule, 1971b).

NMR methods have been applied to the question of stereoisomers of the form of **XVII** and **XVIII,** in which the OH can have either of two orientations relative to the sugar. Evidence has been obtained in this manner for the formation of two isomers of the photohydrates of UMP (Chambers, 1968), Urd (Wechter and Smith, 1968), Cyd, and 3′-CMP (DeBoer, 1970), and of the synthesized hydrate of dThd (Cadet and Téoule, 1971a).

Two reversible products of Urd, tentatively identified as isomeric hydrates, have been resolved by column chromatography (Fürst, 1968, cited by Fahr, 1969). Thin layer chromatographic resolution of two reversible products, said to be hydrate isomers, has also been reported for Urd, 3-MeUrd, and dUrd (Pietrzykowska and Shugar, 1969; Table 6). The large difference in chromatographic mobilities for the dUrd products could not be repeated in our laboratory (O. Klinghoffer, unpublished data, 1970), and stands in contrast to the similar properties of the two isomeric hydrates of dThd (Cadet and Téoule, 1971a; Table 6). S. Y. Wang (personal communication, 1972) maintains that both isomers are formed and have the same $R_f$ (0.22), while the material with $R_f$ 0.67 is not a hydrate.

v. CHROMATOGRAPHY. Some chromatographic properties of the hydrates of pyrimidines are presented in Table 6. No results are given for Cyt derivatives because they are too unstable for chromatography. However, they may readily be separated by rapid high-voltage electrophoresis (Becker *et al.*, 1967).

Column separation methods have also been found to be useful, especially by Fahr *et al.* (e.g., 1967). A further means of separation involves the partition of $Me_2$Ura and its hydrate between chloroform and water (Moore and Thomson, 1957).

*c. Photolysis of Hydrates*

When $Me_2$Ura photohydrate itself is irradiated, oxidation at C(6) leads to 1,3-dimethylbarbituric acid (**XXIII**) and ultimately to N,N′-

**Table 6** Chromatographic Properties of Photohydrates

| Parent | Solvent | Medium | $R_f$ of hydrate | $R_p$[a] | Reference |
|---|---|---|---|---|---|
| Ura | *n*-PrOH/$H_2O$ (7:3) | Cellulose thin layer | | 0.90 | Brown and Johns (1968) |
| | *n*-PrOH/$H_2O$ (7:3) | Paper | 0.42 | 0.84 | Moore (1958) |
| | i-PrOH/$H_2O$ (7:5) | Paper | 0.68 | 1.00 | |
| | BuOH saturated with $H_2O$ | Paper | 0.22 | 0.63 | |
| | $H_2O$ saturated with BuOH | Paper | 0.78 | 1.07 | |
| dUrd (2 isomers ?) | Chloroform/MeOH (85:15) run three times in succession | Silica gel | 0.22 | 0.47 | Pietrzykowska and Shugar (1969) |
| | | | 0.67 | 1.43 | |
| UMP | *n*-BuOH/HAc/$H_2O$ (5:3:2) | Whatman No. 1 | 0.20 | 0.80 | Fahr *et al.* (1967) |
| Thy | *n*-PrOH/$H_2O$ (10:3) | Cellulose thin layer | 0.65 (*cis*) | 0.88 | Cadet and Téoule (1971b) |
| | | | 0.66 (*trans*) | 0.89 | |
| | EtAc/*i*-PrOH/$H_2O$ (75:16:9) | Cellulose thin layer | 0.37 (*cis*) | 0.88 | |
| | | | 0.41 (*trans*) | 0.98 | |
| dThd (2 optical isomers of *cis* hydrate) | Chloroform/MeOH/$H_2O$ (4:2:1) | Silica gel thin layer | 0.26 (*d*) | 0.70 | Cadet and Téoule (1971a) |
| | | | 0.21 (*l*) | 0.57 | |

[a] $R_p$ = ratio of distances traveled by hydrate and the parent.

dimethylmalonamide (**XXIV**) (Scheme 10) (Wang, 1958b). This illustrates how prolonged irradiation of a given Pyr can lead through several intermediate steps to materials which are not produced directly from the parent molecule and shows again the need for frequent monitoring of a reaction, so that **XXIV,** for example, is not mistaken for a direct photoproduct of $Me_2Ura$.

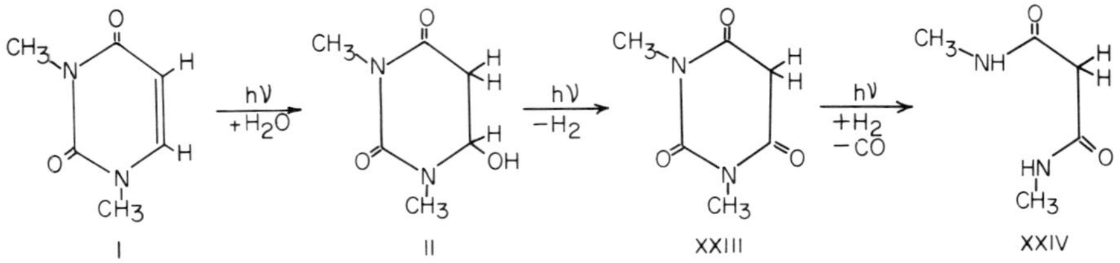

Scheme 10

The formation of the methylated analog of **XXIV,** N,N′-dimethylmethylmalonamide, by prolonged irradiation of $Me_2Thy$, has been interpreted in terms of the initial production of $ho^6Me_2hThy$ and its subsequent photolysis by analogy with the results for $ho^6Me_2hUra$ (Wang, 1959b). Proof of this must await more detailed studies on the course and yield of the photoreaction and the isolation of intermediates. If the Thy hydrate is involved, the overall efficiency of the reaction must be very low.

*d. Borohydride Reduction*

Sodium borohydride reduces the photohydrate of Urd to a mixture of the $\alpha$ and $\beta$ isomers of D-ribopyranosylurea (**XXV**) by a process involving cleavage of the Pyr ring at the 3,4- and 1,6-bonds and rearrangement of the sugar (Scheme 11) (Miller and Cerutti, 1968). In

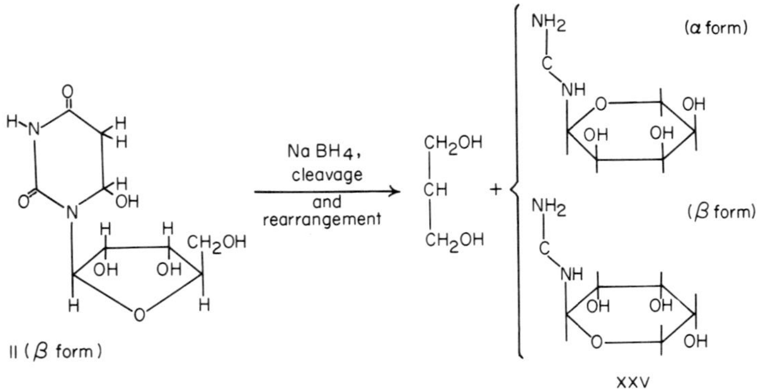

Scheme 11

the absence of an OH at C(6), as in the case of hUrd, the 1,6-bond does not break (Witkop, 1968). The small fragment of hydrate cleavage, 1,3-propanediol, contains three protons from borohydride. This degradation of the Urd hydrate by borohydride (or rather, by borotritiide) has been used as an assay for photohydration in irradiated RNA of various conformations, on the assumption that the reduction of hydrates and the labeling of the diol proceed with the same efficiency in solution and in polymers (Cerutti *et al.*, 1969). We believe that this will be a doubtful assay for the hydrate in RNA until the possible effects of the polymer environment can be taken into account. In contrast to the cleavage of the Urd hydrate at two positions in the ring, the Cyd photohydrate opens only at the 1,6-bond on borohydride reduction to give product **VII** (Miller and Cerutti, 1968).

The formation of N-deoxyribosylurea, the deoxy analogue of **XXV**, by the irradiation of dThd in solution with borohydride is thought to indicate the initial formation of a photohydrate, which then undergoes ring cleavage at the 1,6- and 3,4-bonds (Witkop, 1968; Kondo and Witkop, 1968).

*e. Hydrogen Exchange*

The exchange of hydrogen atoms attached to C(5) of Pyr photohydrates has been the subject of a number of investigations. However, hydrogen exchange is a very complex phenomenon, and frequently there has not been enough care taken to provide the quantitative data necessary to understand the problem.

One thorough study of hydrogen exchange by Cyd and CMP hydrates has been reported (DeBoer and Johns, 1970). DeBoer irradiated 5-tritiated Cyd or CMP in water under carefully controlled conditions which led to 95% conversion to hydrate within one minute. Immediately after the irradiation, ~15–20% of the bound tritium ($T_b$) was lost (Fig. 13), indicating that there is exchange in an excited state prior to formation of a stable hydrate. A slow loss of another 30% of the activity occurred in a period of hours during hydrate reversal (Fig. 13). Detailed analysis suggested that most of this loss takes place by true exchange, rather than by elimination of tritiated water. Loss of all of the activity occurred only after several cycles of photohydration and reversal. DeBoer and Johns (1970) confirmed their findings using proton magnetic resonance. The report that loss of tritium is quantitative when labeled Cyd is irradiated (Grossman and Rodgers, 1968) is, therefore, incorrect; the experiments were probably carried out under conditions which led to repeated formation and reversal of hydrates.

Conflicting findings have been reported for hydrates of Ura derivatives. In acidic $D_2O$, $D^5Do^6$hUrd dehydrates at only half the rate of

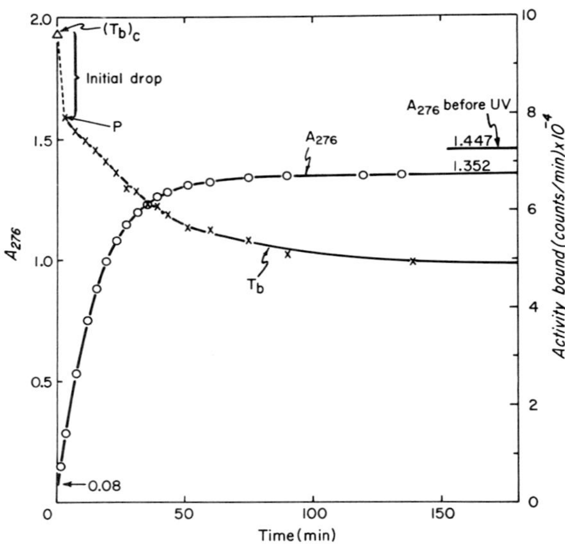


**Fig. 13.** *Absorbance at 276 nm* ($A_{276}$) *and bound activity* ($T_b$) vs. *time after a 1-min irradiation of* [$^3$H]5′-*CMP. The solution was buffered at pH 4.45 and kept at room temperature. A rapid drop in bound tritium occurs between the control level* $(T_b)_c$ *and the first post-irradiation point* P *(DeBoer and Johns, 1970).*

ho$^6$hUrd (Wierzchowski and Shugar, 1961; D. Shugar, personal communication, 1972). If exchange were rapid, this difference would not persist. Dehydration of T$^5$ho$^6$hUra in water leads to a loss of only 20% of the tritium (Wacker *et al.*, 1964). Again, if exchange were rapid in comparison with reversal, most of the tritium would be lost. Furthermore, incubation of ho$^6$hUMP in tritiated water leads to the tritiation of approximately half of the nucleotides (Chambers, 1968). On the other hand, NMR studies have suggested that a proton attached to the C(5) of Urd hydrate exchanges rapidly with the solvent at a rate much faster than the dehydration rate (Chambers, 1968; Wechter and Smith, 1968). This conflicts with the other evidence and is probably less reliable.

We believe that hydrogen exchange does occur for hydrates of Ura derivatives. However, this exchange is somewhat slower than reversal, accounting for the small effects on tritium labeling and dehydration. In addition, the situation is doubtless made more complicated by isotope effects: a C—T bond should be less labile than a C—H bond. For a complete understanding of these phenomena, detailed quantitative studies, such as those for Cyd (DeBoer and Johns, 1970), must be carried out for Ura derivatives, with carefully controlled irradiation and incubation conditions.

It has been proposed that hydrogen exchange might be a useful assay for photohydrates in DNA and RNA (Chambers, 1968; Grossman and Rodgers, 1968). Such an experiment would consist of either irradiation of a tritiated polymer followed by measurement of the loss of activity, or incubation of unlabeled irradiated polynucleotide in tritiated water followed by measurement of the uptake of activity. Clearly such an assay is not quantitative even for monomers. The efficiency of exchange is probably altered further by the polymer environment; such effects have not been studied. Finally, pyrimidines can exchange the hydrogen at C(5) by a thermal reaction even when the photohydrate is not formed (Wechter and Kelly, 1970; Heller, 1968). In fact, this procedure has been used by Wechter (1970) to produce Cyt labeled with deuterium at C(5), by heating Cyt in heavy water. Exchange is, therefore, an unsatisfactory assay for hydrates in nucleic acids, since it is neither quantitative nor specific for hydrates.

## D. Summary

### 1. Formation of Hydrates

(a) The photohydrates of Ura and its derivatives have the basic structure of ho$^6$hUra, as shown by UV, IR, and NMR spectra, molecular weight determinations, elemental analysis, chemical synthesis, and reaction with sodium borohydride.

(b) The photohydrates of Cyt and its derivatives have the basic structure of ho$^6$hCyt, as shown by UV, IR, and NMR spectra, chemical synthesis, borohydride reduction, and deamination to the hydrate of the corresponding Ura derivative.

(c) The photohydrate of Thy (and its derivatives) has the same structure as the chemically synthesized hydrate, ho$^6$hThy; chromatography, thermal reversal rates, photodegradation, and borohydride reduction support this assignment.

(d) Quantum yields for photohydration are unaffected by triplet sensitizers or quenchers or by neutral salts. They are independent of Pyr concentration and are affected only slightly by wavelengths shorter than 230 nm.

(e) For Ura and its N(3)-substituted derivatives, the quantum yield for hydration decreases by almost an order of magnitude between pH 3 and pH 5.

(f) For N(1)-substituted Ura derivatives, the quantum yield is almost independent of pH, dropping by only 10% near pH 6.

(g) Substitution on C(5) or C(6) decreases the hydrate yield. The yield is three orders of magnitude smaller for Thy than for Ura, but the pH dependence is similar.

(h) For Cyt derivatives the hydration quantum yield increases by an order of magnitude between pH 3 and pH 5.

(i) In oligo- and polynucleotides, the efficiency of hydrate formation is less than that for the corresponding monomers because the bases are shielded from water.

(j) There is some evidence that photohydration results mainly from the addition of neutral water to an uncharged Pyr molecule in a vibrationally excited electronic ground state.

### 2. Properties of Hydrates

(a) Dehydration of the photohydrates of Ura derivatives is a thermal reaction subject to specific base catalysis and specific or (perhaps) general acid catalysis. Reversal to the parent is almost quantitative below pH 9. Maximum stability occurs at approximately pH 5 or 6 and is enhanced by substitution at N(1). The activation energy for dehydration is ~20 kcal/mole at pH 2 and 25 kcal/mole at pH 7. Under extremely basic conditions some irreversible ring opening takes place.

(b) Reversal of the photohydrate of Cyt derivatives to the parent compound by elimination of water is nearly quantitative and is subject to general base catalysis by buffers or by internal phosphate groups in the case of nucleotides. Maximum stability occurs at pH 8. The activation energy for dehydration is ~15 kcal/mole.

(c) For the photohydrates of Ura derivatives, dehydration is almost twice as rapid in $D_2O$ as in $H_2O$. In contrast to this behavior, Cyt hydrates reverse more slowly in heavy water.

(d) A small fraction of the hydrates of Cyt derivatives deaminate to give, after subsequent dehydration, the corresponding Ura derivatives.

(e) The photohydrates of Ura and Cyt derivatives have slowly exchangeable hydrogens at C(5).

(f) Pyr photohydrates have labile 1,6-bonds which open on reduction with borohydride; Ura derivatives also undergo ring cleavage at the 3,4-bond.

(g) Continued irradiation of photohydrates of Ura analogs leads to barbituric acid derivatives and eventually to ring opening.

(h) The best available assay for hydrates in polynucleotides consists of enzymatic hydrolysis under mild conditions followed by chromatographic analysis of the fragments. Such alternatives as borohydride reduction, deamination, ring opening at high pH, hydrogen exchange, and heat-reversible absorbance changes are too fraught with difficulties, involving large and still unknown correction factors.

### 3. Remaining Problems

If the biological effects of hydrates are to be fully understood, additional work must be done to standardize more assays for hydrates in polynucleotides.

Much work remains to be done, by NMR and isotope studies, on the process of hydrogen exchange by excited pyrimidines and by hydrates.

A number of aspects of the hydration reaction remain to be studied. For example, the order of the reaction in mixed organic–aqueous solvents may shed light on the number of water molecules involved in hydrate formation. Mode-locked lasers should help to resolve some of the details of the reaction on a picosecond time scale. However, a complete understanding of photohydration probably must await detailed spectral analysis of the absorption and emission bands and much more sophisticated and reliable molecular orbital calculations, so that the nature of the electronically and vibrationally excited precursors of hydrates can be learned.

The influence of substituents on the quantum yield and on the stability and fate of the products is not clear. Why, for example, is the yield of hydrates in Thy three orders of magnitude smaller than in Ura? Does 6-MeUra or Oro form photohydrates? A systematic study of the effects of substituents on the photochemistry of pyrimidines of biological interest could lead to a better theoretical understanding of the behavior of excited molecules in general.

## References

Batt, R. D., Martin, J. K., Ploeser, J. M., and Murray, J. (1954). *J. Amer. Chem. Soc.* **76,** 3663.

Becker, H., LeBlanc, J. C., and Johns, H. E. (1967). *Photochem. & Photobiol.* **6,** 733.

Braude, E. A., Sondheimer, F., and Forbes, W. F. (1954). *Nature (London)* **173,** 117.

Brown, D. M., and Hewlins, M. J. E. (1968). *J. Chem. Soc., C,* p. 2050.

Brown, I. H., and Johns, H. E. (1967). *Photochem. & Photobiol.* **6,** 469.

Brown, I. H., and Johns, H. E. (1968). *Photochem. & Photobiol.* **8,** 273.

Brown, I. H., Freeman, K. B., and Johns, H. E. (1966). *J. Mol. Biol.* **15,** 640.

Burr, J. G. (1968). *Advan. Photochem.* **6,** 193.

Burr, J. G., and Park, E. H. (1967). *Radiat. Res.* **31,** 547.

Burr, J. G., and Park, E. H. (1968). *Advan. Chem. Ser.* **81,** 435.

Burr, J. G., Gordon, B. R., and Park, E. H. (1968a). *Advan. Chem. Ser.* **81,** 418.

Burr, J. G., Gordon, B. R., and Park, E. H. (1968b). *Photochem. & Photobiol.* **8,** 73.

Burr, J. G., Park, E. H., and Chan, A. (1972). *J. Amer. Chem. Soc.* **94,** 5866.
Burr, J. G., Summers, W. A., and Lee, Y. J. (1975). *J. Amer. Chem. Soc.* **97,** 246.
Cadet, J., and Téoule, R. (1971a). *C. R. Acad. Sci, Ser. D* **272,** 2254.
Cadet, J., and Téoule, R. (1971b). *Int. J. Appl. Radiat. Isotop.* **22,** 273.
Cavalieri, L. F. (1952). *J. Amer. Chem. Soc.* **74,** 5804.
Cerutti, P. A., Miller, N., Pleiss, M. G., Remsen, J. F., and Ramsay, W. J. (1969). *Proc. Nat. Acad. Sci. U.S.* **64,** 731.
Chambers, R. W. (1968). *J. Amer. Chem. Soc.* **90,** 2192.
Daniels, M., and Grimison, A. (1964). *Biochem. Biophys. Res. Commun.* **16,** 428.
Danilov, V. I. (1967). *Photochem. & Photobiol.* **6,** 233.
DeBoer, G. (1970). Ph.D. Thesis, Department of Medical Biophysics, University of Toronto.
DeBoer, G., and Johns, H. E. (1970). *Biochim. Biophys. Acta* **204,** 18.
DeBoer, G., Pearson, M. L., and Johns, H. E. (1967). *J. Mol. Biol.* **27,** 131.
DeBoer, G., Klinghoffer, O., and Johns, H. E. (1970). *Biochim. Biophys. Acta* **213,** 253.
Eisinger, J., and Shulman, R. G. (1968). *Science* **161,** 1311.
Fahr, E. (1969). *Angew. Chem., Int. Ed. Engl.* **8,** 578.
Fahr, E., Kleber, R., and Boebinger, E. (1966). *Z. Naturforsch. B* **21,** 219.
Fahr, E., Gattner, H., Dörhöfer, G., Kleber, R., and Popp, H. (1967). *Z. Naturforsch. B* **22,** 1256.
Fikus, M., and Shugar, D. (1966). *Acta Biochim. Pol.* **13,** 39.
Fikus, M., Wierzchowski, K. L., and Shugar, D. (1962). *Photochem. & Photobiol.* **1,** 325.
Fikus, M., Wierzchowski, K. L., and Shugar, D. (1965). *Photochem. & Photobiol.* **4,** 521.
Fink, R. M., Cline, R. E., McGaughey, C., and Fink, K. (1956). *Anal. Chem.* **28,** 4.
Fisher, G. J. (1973). Ph.D. Thesis, Dept. of Medical Biophys., University of Toronto.
Fisher, G. J., and Johns, H. E. (1970). *Photochem. & Photobiol.* **11,** 429.
Fisher, G. J., and Johns, H. E. (1973). *Photochem. & Photobiol.* **18,** 23.
Fourneau, E. (1909). *Bull. Soc. Chim. Fr.* **5,** 229.
Fürst, G. (1968). Dissertation, University of Würzburg.
Gattner, H., and Fahr, E. (1963). *Justus Liebigs Ann. Chem.* **670,** 84.
Gattner, H., and Fahr, E. (1964). *Z. Naturforsch. B* **19,** 74.
Greenstock, C. L., and Johns, H. E. (1968). *Biochem. Biophys. Res. Commun.* **30,** 21.
Greenstock, C. L., Brown, I. H., Hunt, J. W., and Johns, H. E. (1967). *Biochem. Biophys. Res. Commun.* **27,** 431.
Grossman, L., and Rodgers, E. (1968). *Biochem. Biophys. Res. Commun.* **33,** 975.
Guschlbauer, W., Favre, A., and Michelson, A. M. (1965). *Z. Naturforsch. B* **20,** 1141.
Hariharan, P. V., and Johns, H. E. (1968). *Can. J. Biochem.* **46,** 911.
Haug, A. (1964). *Photochem. & Photobiol.* **3,** 207.
Hauswirth, W., and Daniels, M. (1971). *Chem. Phys. Lett.* **10,** 140.
Hauswirth, W., Hahn, B. S., and Wang, S. Y. (1972). *Biochem. Biophys. Res. Commun.* **48,** 1614.
Hayatsu, H., Wataya, Y., and Kai, K. (1970). *J. Amer. Chem. Soc.* **92,** 724.
Hélène, C., and Douzou, P. (1964a). *C. R. Acad. Sci.* **259,** 4853.
Hélène, C., Haug, A., Delbruck, M., and Douzou, P. (1964b). *C. R. Acad. Sci.* **259,** 3385.
Hélène, C., Rivail, J.-L., and Pochon, F. (1967). *C. R. Acad. Sci., Ser. D* **264,** 861.
Helleiner, C. W., Pearson, M. L., and Johns, H. E. (1963). *Proc. Nat. Acad. Sci. U.S.* **50,** 761.
Heller, S. R. (1968). *Biochem. Biophys. Res. Commun.* **32,** 998.
Herbert, M. A., LeBlanc, J. C., Weinblum, D., and Johns, H. E. (1969). *Photochem. & Photobiol.* **9,** 33.
Heyroth, F. F., and Loofbourow, J. R. (1931). *J. Amer. Chem. Soc.* **53,** 3441.

Janion, C., and Shugar, D. (1960). *Acta Biochim. Pol.* **7,** 309.
Johns, H. E. (1968). *Photochem. & Photobiol.* **8,** 547.
Johns, H. E. (1969). *In* "Methods in Enzymology" (K. Kustin, ed.), Vol. 16, p. 257. Academic Press, New York.
Johns, H. E. (1971). *In* "Creation and Detection of the Excited State" (A. A. Lamola, ed.), p. 123. Dekker, New York.
Johns, H. E., and Rauth, A. M. (1965). *Photochem. & Photobiol.* **4,** 673 and 693.
Johns, H. E., LeBlanc, J. C., and Freeman, K. B. (1965). *J. Mol. Biol.* **13,** 849.
Johns, H. E., Pearson, M. L., and Brown, I. H. (1966). *J. Mol. Biol.* **20,** 231.
Johnson, W. C., Vipond, P. M., and Girod, J. C. (1971). *Biopolymers* **10,** 923.
Kazimierczuk, Z., and Shugar, D. (1971). *Biochim. Biophys. Acta* **254,** 157.
Khattak, M. N., Hauswirth, W., and Wang, S. Y. (1972). *Biochem. Biophys. Res. Commun.* **48,** 1622.
Kittler, L. (1972). *Photochem. & Photobiol.* **16,** 39.
Kittler, L., and Löber, G. (1969). *Photochem. & Photobiol.* **10,** 35.
Kleber, R., Fahr, E., and Boebinger, E. (1965). *Naturwissenschaften* **52,** 513.
Kondo, Y., and Witkop, B. (1968). *J. Amer. Chem. Soc.* **90,** 764.
Krajewska, E., and Shugar, D. (1971). *Science* **173,** 435.
Lamola, A. A., and Mittal, J. P. (1966). *Science* **154,** 1560.
Langmuir, M. E., and Hayon, E. (1969). *J. Chem. Phys.* **51,** 4893.
Leonov, D., and Elad, D. (1974). *J. Amer. Chem. Soc.* **96,** 5635.
Lis, A. W., and Allen, F. W. (1961). *Biochem. Biophys. Acta* **49,** 190.
Logan, D. M., and Whitmore, J. F. (1966). *Photochem. & Photobiol.* **5,** 143.
Lomant, A. J., and Fresco, J. R. (1972). *J. Mol. Biol.* **66,** 49.
Lozeron, H. A., Gordon, M. P., Gabriel, T., Tautz, W., and Duschinsky, R. (1964). *Biochemistry* **3,** 1844.
McLaren, A. D., and Shugar, D. (1964). "Photochemistry of Proteins and Nucleic Acids." Macmillan, New York.
Malrieu, J.-P. (1967). *C. R. Acad. Sci., Ser. D* **264,** 662.
Miller, N., and Cerutti, P. (1968). *Proc. Nat. Acad. Sci. U. S.* **59,** 34.
Moore, A. M. (1958). *Can. J. Chem.* **36,** 281.
Moore, A. M. (1959). *Can. J. Chem.* **37,** 1281.
Moore, A. M. (1963). *Can. J. Chem.* **41,** 1937.
Moore, A. M., and Anderson, S. M. (1959). *Can. J. Chem.* **37,** 590.
Moore, A. M., and Thomson, C. H. (1955). *Science* **122,** 594.
Moore, A. M., and Thomson, C. H. (1956). *Progr. Radiobiol.* **4,** 75.
Moore, A. M., and Thomson, C. H. (1957). *Can. J. Chem.* **35,** 163.
Muel, B. and Malpiece, C. (1969). *Photochem. & Photobiol.* **10,** 283.
Nnadi, J. C. (1968). Ph.D. Thesis, Johns Hopkins University, Baltimore, Maryland.
Nnadi, J. C., and Wang, S. Y. (1969). *Tetrahedron Lett.* p. 2211.
Nofre, C., and Ogier, M.-H. (1966). *C. R. Acad. Sci., Ser. C* **263,** 1401.
Nofre, C., Cier, A., Chapurlat, R., and Mareschi, J. -P. (1965). *Bull. Soc. Chim. Fr.* p. 332.
Pearson, M. L., and Johns, H. E. (1966a). *J. Mol. Biol.* **19,** 303.
Pearson, M. L., and Johns, H. E. (1966b). *J. Mol. Biol.* **20,** 215.
Pearson, M. L., Whillans, D. W., LeBlanc, J. C., and Johns, H. E. (1966). *J. Mol. Biol.* **20,** 245.
Pickett, L. W., Muller, N., and Mulliken, R. S. (1953). *J. Chem. Phys.* **21,** 1400.
Pietrzykowska, I., and Shugar, D. (1969). *Biochem. Biophys. Res. Commun.* **37,** 225.
Pietrzykowska, I., and Shugar, D. (1970). *Acta Biochim. Pol.* **17,** 361.
Pietrzykowska, I., and Shugar, D. (1974). *Acta Biochim. Pol.* **21,** 187.
Rhoades, D. F., and Wang, S. Y. (1971). *Biochemistry* **10,** 4603.

Rushizky, G. W., and Pardee, A. B. (1962). *Photochem. & Photobiol.* **1,** 15.
Schuster, H. (1964). *Z. Naturforsch. B* **19,** 815.
Shapiro, R., Servis, R. E., and Welcher, M. (1970). *J. Amer. Chem. Soc.* **92,** 422.
Shugar, D., and Fox, J. J. (1952). *Biochim. Biophys. Acta* **9,** 199.
Shugar, D., and Wierzchowski, K. L. (1957). *Biochim. Biophys. Acta* **23,** 657.
Shugar, D., and Wierzchowski, K. L. (1958). *J. Polym. Sci.* **31,** 269.
Sinsheimer, R. L. (1954). *Radiat. Res.* **1,** 505.
Sinsheimer, R. L. (1957). *Radiat. Res.* **6,** 121.
Sinsheimer, R. L., and Hastings, R. (1949). *Science* **110,** 525.
Smith, K. C. (1966). *Radiat. Res., Suppl.* **6,** 54.
Smith, K. C., and Hanawalt, P. C. (1969). "Molecular Photobiology:Inactivation and Recovery." Academic Press, New York.
Stepien, E., Lisewski, R., and Wierzchowski, K. L. (1973). *Acta Biochim. Pol.* **20,** 313.
Summers, W. A., and Burr, J. G. (1972). *J. Phys. Chem.* **76,** 3137.
Summers, W. A., Enwall, C., Burr, J. G., and Letsinger, R. L. (1973). *Photochem. & Photobiol.* **17,** 295.
Sutherland, J. C., and Sutherland, B. M. (1970). *Biopolymers* **9,** 639.
Swenson, P. A., and Setlow, R. B. (1963). *Photochem. & Photobiol.* **2,** 419.
Szabo, A. G., Riddell, W. D., and Yip, R. W. (1970). *Can. J. Chem.* **48,** 694.
Sztumpf, E., and Shugar, D. (1965). *Photochem. & Photobiol.* **4,** 719.
Ts'o, P. O. P. (1970). *In* "Fine Structure of Proteins and Nucleic Acids" G. D. Fasman and S. N. Timasheff, eds.), p. 49. Dekker, New York.
Varghese, A. J. (1971). *Biochemistry* **10,** 4283.
Varghese, A. J. (1972). *Photophysiology* **7,** 207.
Wacker, A., Weinblum, D., Träger, L., and Moustafa, Z. H. (1961). *J. Mol. Biol.* **3,** 790.
Wacker, A., Dellweg, H., Träger, L., Kornhauser, A., Lodemann, E., Türck, G., Selzer, R., Chandra, P., and Ishimoto, M. (1964). *Photochem. & Photobiol.* **3,** 369.
Wagner, P., and Bucheck, D. (1970). *J. Amer. Chem. Soc.* **92,** 181.
Wang, S. Y. (1958a). *J. Amer. Chem. Soc.* **80,** 6196.
Wang, S. Y. (1958b). *J. Amer. Chem. Soc.* **80,** 6199.
Wang, S. Y. (1959a). *Nature (London)* **184,** 184.
Wang, S. Y. (1959b). *Nature (London)* **184,** B. A. 59.
Wang, S. Y. (1959c). *J. Org. Chem.* **24,** 11.
Wang, S. Y. (1962a). *Photochem. & Photobiol.* **1,** 37.
Wang, S. Y. (1962b). *Photochem. & Photobiol.* **1,** 135.
Wang, S. Y., and Nnadi, J. C. (1968). *Chem. Commun.* p. 1160.
Wang, S. Y., Apicella, M., and Stone, B. R. (1956). *J. Amer. Chem. Soc.* **78,** 4180.
Wang, S. Y., Nnadi, J. C., and Greenfeld, D. (1968). *Chem. Commun.* p. 1162.
Wang, S. Y., Nnadi, J. C., and Greenfeld, D. (1970). *Tetrahedron* **26,** 5913.
Wechter, W. J. (1970). *Collect. Czech. Chem. Commun.* **35,** 2003.
Wechter, W. J., and Kelly, R. C. (1970). *Collect. Czech. Chem. Commun.* **35,** 1991.
Wechter, W. J., and Smith, K. C. (1968). *Biochemistry* **7,** 4064.
Whillans, D. W., and Johns, H. E. (1971). *J. Amer. Chem. Soc.* **93,** 1358.
Whitten, D. G., Happ, J. W., Carlson, G. L. B., and McCall, M. T. (1970). *J. Amer. Chem. Soc.* **92,** 3499.
Wierzchowski, K. L., and Shugar, D. (1957). *Biochim. Biophys. Acta* **25,** 355.
Wierzchowski, K. L., and Shugar, D. (1959). *Acta Biochim. Pol.* **6,** 313.
Wierzchowski, K. L., and Shugar, D. (1960). *Acta Biochim. Pol.* **7,** 63.
Wierzchowski, K. L., and Shugar, D. (1961). *Acta Biochim. Pol.* **8,** 219.
Wierzchowski, K. L., and Shugar, D. (1962). *Photochem. & Photobiol.* **1,** 21.
Witkop, B. (1968). *Photochem. & Photobiol.* **7,** 813.

# 5 Pyrimidine Photodimers

*G. J. Fisher and H. E. Johns*

## A. Introduction*

In the preceding chapter the heat reversible water addition photoproduct of pyrimidines was discussed. This chapter deals with a second major photoproduct, the cyclobutane-type dimer, in which two bases are covalently bonded by a four-carbon ring (Fig. 1a). Early workers noted that neither heating nor acidification of UV-irradiated Ura solutions results in complete recovery of the parent (Sinsheimer and Hastings, 1949), thus suggesting that hydrates are not the sole product. In 1954, Sinsheimer observed a concentration-dependent reaction in UMP, implying the involvement of two UMP molecules to form the product. Beukers and co-workers made major advances in their studies of this bimolecular process and found that the thermally stable product of this reaction reached a photostationary state (Rörsch *et al.*, 1958) with the equilibrium concentration of the product being reduced by oxygen (Beukers *et al.*, 1959a) and by paramagnetic ions (Beukers and Berends, 1960a) in liquid solution. These observations were interpreted in terms of an excited triplet-state precursor which was quenched by paramagnetic solutes. Oro, **I**, was shown to be more reactive than Ura, **II**, which in turn is more reactive than Thy, **III** (Beukers *et al.*, 1959a). In frozen solution, however, Thy was found to react most efficiently. The equilibrium concentration of the Thy photoproduct increases from a very low value in water to $>40\%$ in ice (Beukers *et al.*, 1958). The product was first isolated from irradiated ice by Wang (1960) and by Beukers and Berends (1960b). Elementary

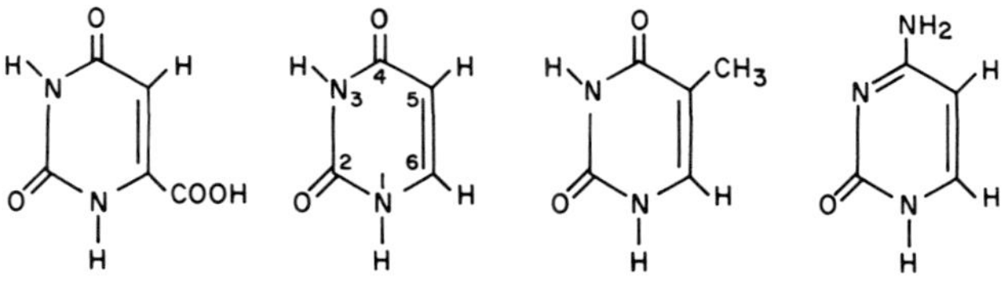

I II III IV

* Editor's notes: It is possible to advance alternative interpretations for historical events leading to the discovery of the thymine dimer. Generally, it is agreed that the most significant breakthrough in DNA photochemistry was the isolation and identification of the thymine dimer and it was made simultaneously and independently by Beukers and Berends and by Wang [see Vol. II: p. 54 (Chapter 2) and p. 235 (Chapter 6)]. The cited studies before 1960 had no notion of dealing with a photodimerization process. These observations may be interpreted in retrospect in terms of what we now know as photodimerization.

Also, a "molecular aggregation-puddle formation" hypothesis was proposed by Wang

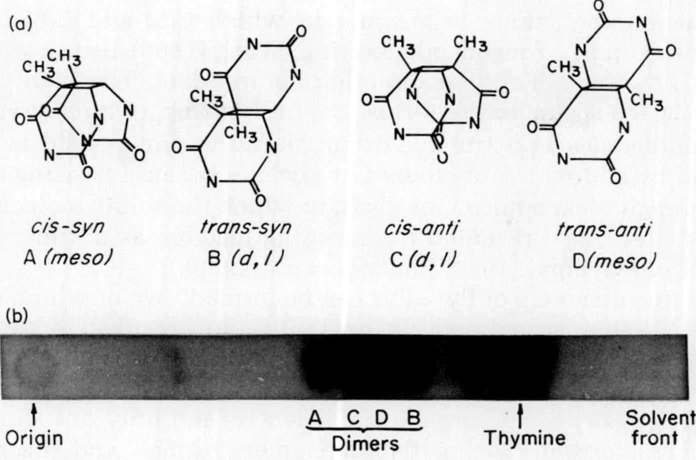


**Fig. 1.** *(a) Four isomeric cyclobutane-type photodimers of Thy; the letters A–D refer mainly to publications from our group (Weinblum and Johns, 1966; Herbert* et al.*, 1969). Dimers B and C exist in optically active forms. The nomenclature, which will be used throughout this review, is that generally used:* cis *and* trans *refer to the proximity of the bases, while* syn *and* anti *indicate whether their orientation is head-to-head, or head-to-tail. For example this convention has been used by Wang (1965), Blackburn and Davies (1966a,b), Jordan and Pullman (1968), Richter and Fahr (1969), Pietrzykowska and Shugar (1970), Elad* et al. *(1971), Birnbaum* et al. *(1971), Varghese (1971b), Konnert and Karle (1971), and Khattak and Wang (1972). Alternate systems use* syn *and* anti *to describe the nearness of the bases, while head-to-head and head-to-tail, or simply h-h and h-t, give the orientations of the bases (Camerman* et al.*, 1969; Morrison and Kleopfer, 1968; Jennings* et al.*, 1972). More confusing is a complete reversal of the common terms, so that dimers B and C are described as* cis-anti *and* trans-syn *(Kunieda and Witkop, 1971; Miskew, 1970). For the purpose of this review, we will use the standard nomenclature. For conciseness, the initials alone are sometimes used (c,s to represent* cis-syn*, etc.). (b) Separation, by thin layer chromatography on cellulose, of the four Thy◇Thy labelled with $^{14}C$. The solvent is* n*-butanol/acetic acid/water, 80:12:30 (Fisher and Johns, 1970).*

analysis showed the product to have the same composition as Thy, while a molecular weight determination and x-ray analysis indicated that the ice product is a dimer. The infrared spectrum showed the 870 $cm^{-1}$ peaks characteristic of a cyclobutane ring, in agreement with nuclear magnetic resonance findings and with the lack of the UV absorption due to the 5,6-double bond (Beukers and Berends, 1960b; Fig.

---

(1961; 1965) to interpret the reactions in a frozen state. This hypothesis differs from that advanced by Szent-Györgyi (1957) or Beukers *et al.* (1959b) [see Chapter 8]. The mechanistic understanding of the irradiation in frozen solutions is necessary for further understanding of the photodimerization processes.

2a). The product, then, is a dimer in which C(5) and C(6) of each monomer join in a four-membered ring. Wang (1960) also showed that 1,3-$Me_2$Thy forms a similar dimeric photoproduct. The high yield of dimers in ice is due to the formation, on freezing, of microcrystals of solute molecules in a suitable orientation; the dimer yield is greatly reduced by adding 5% of ethanol or glycerol because freezing of such a mixture gives an amorphous glass, in which the solute molecules are isolated and thus prevented from joining together as a dimer (Wang, 1961; Szent-Györgyi, 1957; Beukers *et al.*, 1959b).

Four stereoisomers of Pyr◇Pyr can be formed, two of which may be optically active, as pointed out by Wulff and Fraenkel in 1961 (Fig. 1a). All four isomers of the dimers of Thy and Ura have been isolated along with two for Oro and four for 6-MeUra ("*iso*-Thy"). The derivatives of Cyt (**IV**) dimerize very inefficiently, and only one dimer has been characterized. The particular isomers formed and the relative yields of each depend strongly on the irradiation conditions. For example, Thy in ice or in crystals is oriented in such a way as to allow formation of only one isomer, the *cis-syn* form (Gerdil, 1961; Wang, 1963), which is also the principal dimer formed in native DNA (Wacker *et al.*, 1960). On the other hand, all four isomers are formed in solutions in ratios that depend on concentration, temperature, pH, and the dielectric constant of the solvent. These effects and their mechanistic implications will be discussed in more detail. The properties of the isomeric dimers have also been studied extensively, and many of these results will be presented. Details of the NMR studies and x-ray structure determinations are reviewed in Chapters 10 and 11.

Wang (1961) pointed out that such dimers are not unique to pyrimidines. Rather, they are common photoproducts of $\alpha,\beta$-unsaturated ketones (Mustafa, 1951) such as chalcones (Montaudo and Caccamese, 1973) and coumarins (Anet, 1962; Schenk *et al.*, 1962; Rao *et al.*, 1973) and of simple alkenes such as cyclooctatetraene (Schröder and Martin, 1966). Nor are dimers the only cyclobutane-type photoproducts of pyrimidines: coumarins (Song *et al.*, 1971; Pathak and Krämer, 1969), psoralens (Bevilacqua and Bordin, 1973), vinyl carbonate (Beuglemans *et al.*, 1972), vinyl acetate (Hyatt and Swenton, 1972), and propylene (Krajewska and Shugar, 1972) all add to Pyr bases to give cyclobutane products. 5-PrUra even undergoes intramolecular cyclization to form a cyclobutane ring (Krajewska and Shugar, 1971). However, this chapter will be limited to the discussion of Pyr◇Pyr.

In the preparation of this review, the authors have had frequent recourse to the monograph of McLaren and Shugar (1964); to the general reviews by Smith (1964, 1966a), Burr (1968), and Varghese (1972b); and to the summary of chemical properties by Fahr (1969).

## B. Photodimerization of Monomers in Solution

### 1. Measurement of Quantum Yield

#### *a. Absorption Spectra of the Dimers*

The absorption spectra of pyrimidines show a strong peak at ~260 nm, due to the 5,6-double bond. Typical spectra are shown in Fig. 2a. When a Pyr is converted to dimer, the 5,6-bond is saturated and the absorption is lost (Fig. 2a).

#### *b. Simple Test for Dimer*

When a deoxygenated solution of a Pyr is irradiated near $\lambda_{max}$ (the wavelength of its maximum absorption), monomers are converted to dimers and the absorbance decreases, as shown in Fig. 2b. If the solution is then irradiated at a shorter wavelength at which the dimer absorbs more strongly, the optical density increases at the $\lambda_{max}$ of the

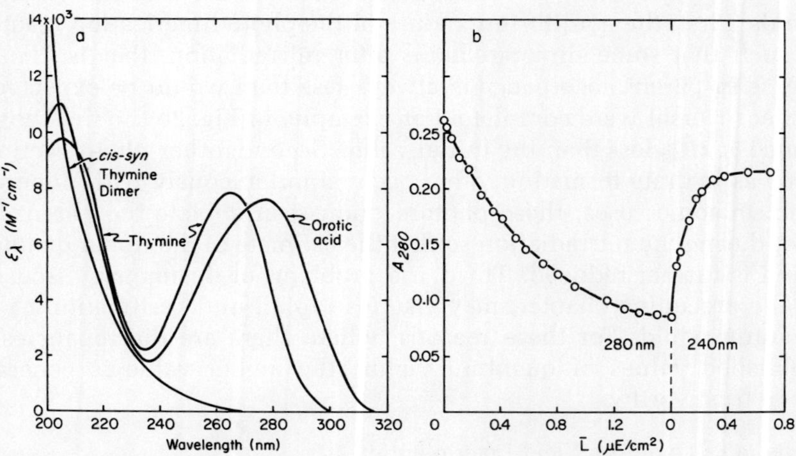


**Fig. 2.** *(a) Ultraviolet absorption spectra of Oro, Thy, and Thy◇Thy(c,s) measured in neutral aqueous solution. (Dimer spectrum from Herbert et al., 1969.) (b) Decrease in the absorbance of Oro at 280 nm ($3.6 \times 10^{-5}$ M, pH 3.2, $N_2$) as dimers are formed during irradiation at 280 nm; and approach to a new photostationary state during irradiation at 240 nm, as dimers are reversed.*

original monomer. This *increase* in optical density is due to the photoreversal of dimers to monomers. These reactions are represented by Eq. (1)*

$$P + P \underset{\text{favored at short wavelength}}{\overset{\text{favored at long wavelength}}{\rightleftharpoons}} P \diamond P \tag{1}$$

Upon continuous irradiation, the forward and backward reactions lead to a photosteady state with the rate of production of dimers balanced by their rate of reversal. The equilibrium concentration of dimers strongly depends on the wavelength of the radiation, which necessitates a monoenergetic source of UV light for such studies (Johns and Rauth, 1965). The quantum yield for the forward reaction is small ($<0.1$) while the reversal quantum yield is $\sim 1.0$ (Setlow, 1961; Herbert *et al.*, 1969).

The reversal properties of dimers can be used as a simple test for their presence. If a solution has been irradiated to equilibrium at a long wavelength, causing a decrease in absorbance, and if a dimer is the principal product, reirradiation at a short wavelength will lead to an increase in absorbance at $\lambda_{max}$ as the monomer concentration increases. This procedure is often used to indicate whether dimers are formed, and the amount of recovery of parent absorbance is occasionally equated to the amount of dimer which was present. Several difficulties make this an inadequate method of determining quantum yields. First, the equilibrium nature of the photodimerization reaction is such that some dimer remains after reirradiation; that is, the increase in parent absorbance is always less than would be expected if dimer reversal were complete. For example, in Fig. 2b the final absorbance is 20% less than the initial value. Second, other photoreactions, such as hydrate formation, often occur simultaneously with dimerization. In some cases, these photoreactions can deplete the parent further during the reirradiation so that the estimate of the dimer quantum yield is further reduced. Third, the problems of dosimetry, discussed in the preceding chapter, may also lead to an underestimation of the quantum yield. For these reasons, where there are discrepancies in published values of quantum yields, the higher value is generally more trustworthy.

*c. Reliable Quantum Yield Determination*

A more reliable method of determining quantum yields involves chromatographic separation of products at several increments of ex-

* P represents the pyrimidine monomer, P$\diamond$P or D represent cyclobutane photodimers, and $\underset{\smile}{PP}$ indicates an associated pair of monomers.

posure and quantitation using a radioactive label, as described by Johns (1971). This avoids the two problems of the absorbance assay, and it also permits kinetic studies of *each* photoproduct individually. This is important for a full understanding of dimerization, since four isomeric dimers may be formed in varying yields under different conditions (Fig. 1b). It also circumvents the problem that photoproducts other than dimers may be photoreversible.

### 2. General Model for the Reaction

Absorption of a photon by a molecule leads to the production of an electronically excited singlet state. Most of these revert to the ground state, with a lifetime of only $\sim 10^{-12}$ sec (Hauswirth and Daniels, 1971a), but a few may undergo a spin inversion leading to a long-lived triplet state. In general both singlets and triplets may give rise to photochemical reactions, the state responsible depending upon the molecular environment and the photoproduct. Formation of a hydrate requires an excited Pyr plus a water molecule, and since the water concentration is high (55 *M*), there should be ample time for a singlet state to react with water. We then expect hydrates to arise from the singlet manifold, as discussed in Chapter 4. On the other hand, dimer formation requires the proximity of an excited Pyr to another Pyr with the correct orientation. Since in dilute solution such a chance collision requires time, dimers in solution are expected to arise from a long-lived triplet. If, however, two pyrimidines are bonded by sugar–phosphate linkages as in the cases of TpT, UpU, or a longer polymer such as poly(rU), then the dimer might arise from the singlet as well as the triplet. Pyr derivatives may also be found in aggregates in concentrated solutions (Ts'o, 1970). These aggregates involve two or more pyrimidines in loose association, which will be represented by $\underset{\smile}{PP}$. Dimers from $\underset{\smile}{PP}$ could arise from singlets or triplets. We will now discuss the possible dimerization pathways in terms of the basic rate constants shown in Fig. 3.

#### a. *Rate Constants*

Figure 3 shows the relevant rate constants required for an analysis of dimerization yields. The absorption cross section leading to an excited singlet level is $\sigma_a$. When it is expressed in $cm^2/\mu E$,* it is related to the molar extinction coefficient, $\epsilon_\lambda$, by

$$\sigma_a = 2.303 \times 10^{-3}\,\epsilon_\lambda \tag{2}$$

* One micro-Einstein ($\mu$E) is $6.023 \times 10^{17}$ photons. At 265 nm this is equivalent to 0.452 J, 0.108 cal, and to $2.82 \times 10^{18}$ eV.

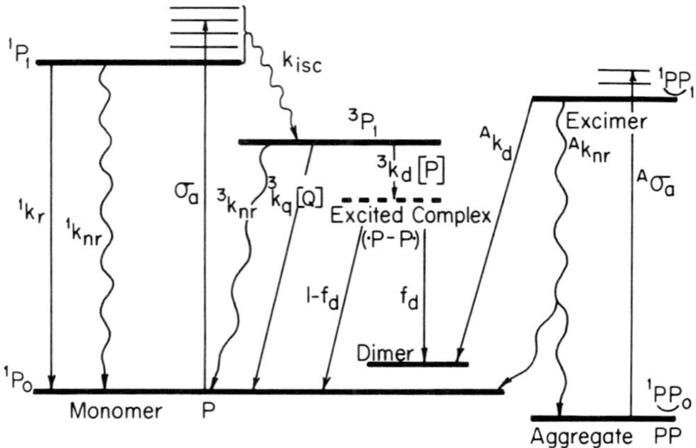


**Fig. 3.** *Schematic diagram showing excited states and rate constants for their deactivation in solution. The triplet state and aggregates are indicated by the superscripts 3 and A, respectively.*

The efficiency of intersystem crossing, $\phi_{isc}$, is the fraction of singlet states which undergoes a spin inversion leading to levels in the triplet manifold. The triplet may lose its energy by a nonradiative transition to the solvent with rate constant $^3k_{nr}$ ($sec^{-1}$); be quenched with quencher Q with a rate $k_q$ [Q]; or interact with a ground-state molecule P at rate $^3k_d$ [P]. A fraction $f_d$ of the resulting excited dimers may then relax to give stable Pyr◇Pyr, while the remainder, $1 - f_d$, revert spontaneously to two ground-state molecules. The latter process, known as self-quenching, leads to no detectable change in the solution but causes the quantum yield for dimer formation, $\phi_d$, to be less than $\phi_{isc}$.

We will now analyze Pyr◇Pyr yields in terms of these constants by a method initially proposed by Sztumpf-Kulikowska *et al.* (1967) which has been extended in our laboratory (Brown and Johns, 1968; Whillans and Johns, 1969, 1973; Fisher and Johns, 1970). Throughout this treatment, the average exposure, fluence, or "dose" is expressed in $\mu E/cm^2$ and is determined as described elsewhere (Johns, 1969, 1971).

*b. Dimers in Dilute Solution*

After an increment of exposure $\Delta\bar{L}$, the number of excited singlet states produced is $\sigma_a[P]\,\Delta\bar{L}$, and the number of triplet states formed is $\sigma_a[P]\phi_{isc}\,\Delta\bar{L}$. Of these a fraction,

$$f = \frac{^3k_d[P]f_d}{^3k_d[P] + {}^3k_q[Q] + {}^3k_{nr}} \tag{3}$$

will lead to dimers. Hence the change in dimer concentration, $\Delta D$, after exposure $\Delta\bar{L}$, (neglecting reversal) is

$$\Delta D = \sigma_a[\mathrm{P}]\phi_{\mathrm{isc}}f\,\Delta\bar{L} \tag{4}$$

The *initial* rate of dimer production from the triplet, $^3(\Delta D/\Delta L)_0$, may be written as

$$^3\left(\frac{\Delta D}{\Delta\bar{L}}\right)_0 = \sigma_a[\mathrm{P}]_0\phi_{\mathrm{isc}}f \tag{5}$$

The quantum yield for dimerization, $^3\phi_d$, can be obtained from the initial rate of dimer formation. It is the ratio of dimers formed via the triplet to photons absorbed and is given by

$$^3\phi_d = \frac{1}{\sigma_a[\mathrm{P}]_0}\,{}^3\left(\frac{\Delta D}{\Delta\bar{L}}\right)_0 \tag{6}$$

Combining this relation with Eq. (5), one obtains

$$^3\phi_d = \phi_{\mathrm{isc}}f \tag{7}$$

At high concentrations of parent and low concentrations of Q, $f$ approaches $f_d$, and the limiting value of the dimer yield from the triplet is

$$^3\phi_d = \phi_{\mathrm{isc}}f_d \tag{8}$$

*c. Dimers from Aggregates*

In concentrated aqueous solutions, purines, pyrimidines, and their nucleosides form aggregates or associated groups of stacked ground-state molecules (Ts'o, 1970; Solie and Schellman, 1968). At concentrations up to the solubility limit of most pyrimidines (~0.3 *M*) only pairwise associations of molecules need to be considered. These aggregates are in an equilibrium characterized by an association constant $K$,

$$\mathrm{P} + \mathrm{P} \overset{K}{\rightleftharpoons} \underset{\smile}{\mathrm{P\,P}}$$

where $[\underset{\smile}{\mathrm{PP}}] = K[\mathrm{P}]^2$.

The number of aggregates excited during an exposure $\Delta\bar{L}$ is $2\,{}^A\sigma_a[\underset{\smile}{\mathrm{PP}}]\,\Delta\bar{L}$ where $^A\sigma_a$ is the absorption cross section of a base in an aggregate. Since aggregates involve a loose association of bases, hyperchromic effects are small and ${}^{A}\sigma_a \cong \sigma_a$. The initial rate of formation of dimers from such aggregates is

$$^{A}\left(\frac{\Delta D}{\Delta \bar{L}}\right)_0 = 2\sigma_a K[P]^2\phi^A \tag{9}$$

From Fig. 3 the quantum efficiency $\phi^A$ is simply ${}^{A}k_d/({}^{A}k_d + {}^{A}k_{nr})$.

*d. Total Dimer Yield*

The total yield of dimers from both pathways is then obtained from Eqs. (3), (5), and (9):

$$\left(\frac{\Delta D}{\Delta \bar{L}}\right)_0 = \sigma_a[P]\left(\frac{{}^3k_d[P]f_d\phi_{isc}}{{}^3k_d[P] + {}^3k_q[Q] + {}^3k_{nr}} + 2K[P]\phi^A\right) \tag{10}$$

The quantum yield for dimerization is given by Eq. (11):

$$\begin{aligned}\phi_d &= \frac{1}{\sigma_a[P]}\left(\frac{\Delta D}{\Delta \bar{L}}\right)_0 \\ &= \phi_{isc}\frac{{}^3k_d[P]f_d}{{}^3k_d[P] + {}^3k_q[Q] + {}^3k_{nr}} + 2K\phi^A[P] \\ &= {}^3\phi_d + 2K\phi^A[P]\end{aligned} \tag{11}$$

At low concentrations, dimerization from aggregates is negligible because of the low number of pairs. The triplet mechanism predominates, and Eq. (11) may be rearranged to the form

$$\frac{1}{\phi_d} = \frac{1}{\phi_{isc}f_d}\left(1 + \frac{{}^3k_q[Q]}{{}^3k_d[P]} + \frac{{}^3k_{nr}}{{}^3k_d[P]}\right) \tag{12}$$

In the absence of quenchers, a plot of $\phi_d^{-1}$ *vs.* $[P]^{-1}$ would be expected to give a straight line with intercept $1/\phi_{isc}f_d$ and slope $(1/\phi_{isc}f_d)$ $({}^3k_{nr}/{}^3k_d)$. Similarly, with fixed Pyr concentration, a plot of $\phi_d^{-1}$ *vs.* quencher concentration should give a straight line with slope $(1/\phi_{isc}f_d)$ $({}^3k_q/{}^3k_d[P])$. These methods of treating the experimental results have in fact been used for several Pyr bases with good results, some of which are shown in Fig. 4. Unfortunately, manipulation of the data can yield only ratios of rate constants rather than individual values, which must be measured by flash photolysis techniques.

*e. Self-Quenching*

The pathway in which a triplet state is quenched by a ground-state molecule without reacting causes difficulty. The possible presence

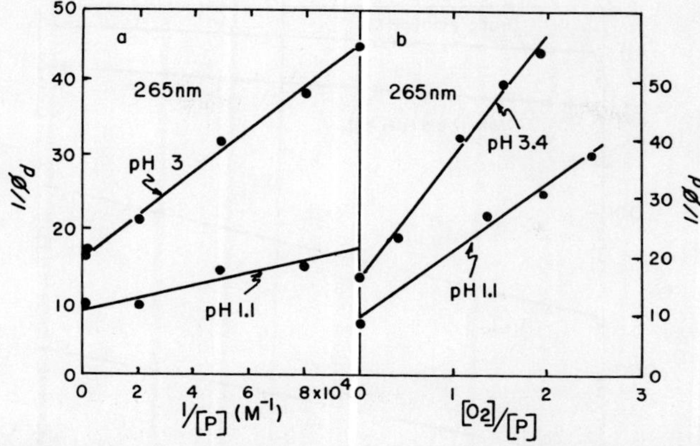


**Fig. 4.** *(a) Plot of $1/\phi_d$ vs. $1/[P]$ for Oro ($[O_2] = 0$). The straight line is predicted from Eq. (12). (b) Plot of $1/\phi_d$ vs. the concentration of oxygen, a triplet quencher (Whillans and Johns, 1969).*

of this self-quenching prevents the determination of an absolute value for $\phi_{isc}$. All that can be determined is $\phi_{isc}f_d$ where $f_d$ is the fraction of the triplets which react with parent molecules to give stable dimers. If we assume that *all* the triplets which are quenched by the parent lead to stable dimers, then $f_d = 1.0$ and we obtain a lower limit to $\phi_{isc}$. If, on the other hand, $f_d$ were only 0.1, then 9 out of 10 of the triplets which were quenched by the parent would not lead to dimers, and by counting the dimers one would underestimate $\phi_{isc}$ by a factor of 10. To resolve this problem independent methods for counting triplets are necessary. One such method was developed at the Bell Labs by Eisinger and Lamola (1971) who used $Eu^{3+}$ as a triplet counter. The reaction of cysteine with excited Ura (Jellinek and Johns, 1970) and with Thy (Fisher *et al.*, 1974), and the isomerization of organic dienes (Wagner and Bucheck, 1970) have been used to estimate triplet yields. Such studies show that $f_d$ may be as small as 0.25 for Ura and 0.07 for Thy in water, but reliable values for $\phi_{isc}$ are still not available for pyrimidines.

*f. Determination of $\phi_{isc}f_d$*

Figure 5 shows the dependence of $\phi_{isc}f_d$ on the energy of the exciting radiation. For Oro $\phi_{isc}f_d$ is larger by a factor of $\sim 10$ than it is for Ura, while for Ura the yield of dimers is another order of magnitude greater than for Thy. These differences are due mainly to differences in $f_d$ (Fisher and Johns, 1974).

For all three bases, Fig. 5 shows that $\phi_{isc}f_d$ increases with the energy of excitation. This suggests that intersystem crossing proceeds from the higher vibrational levels of the singlet state. For most organic com-

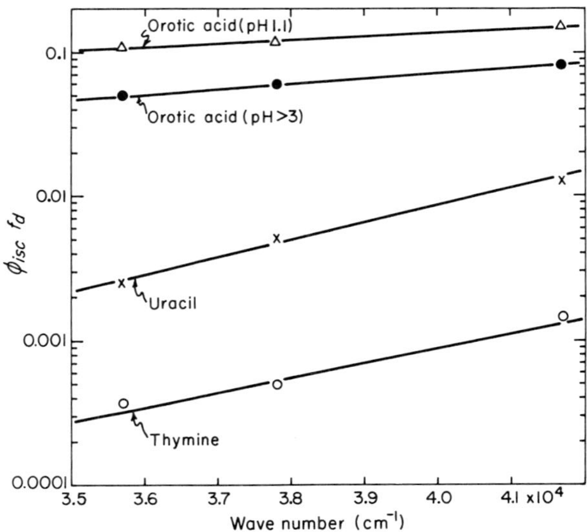


**Fig. 5.** *Plot of $\phi_{isc}f_d$ as a function of the energy of excitation. Results are given for the charged and uncharged forms of Oro (Whillans and Johns, 1969) and for neutral Ura (Brown and Johns, 1968) and Thy (Fisher and Johns, 1970). Later it will be shown that for Oro $f_d = 1.0$ so that the figure shows $\phi_{isc}$ as a function of energy for this base. The wave number range shown in the figure corresponds to wavelengths ranging from 280 to 240 nm.*

pounds, vibrational levels are so short lived ($\sim 10^{-13}$ sec) relative to the lowest level of the singlet ($> 10^{-9}$ sec) that their contribution to intersystem crossing is negligible. However, for pyrimidines at room temperature, the singlet lifetime is much shorter ($10^{-12}$ sec, Hauswirth and Daniels, 1971b), so that the excited molecule spends relatively more time in higher vibrational levels, and excitation by higher energy photons to these levels increases intersystem crossing. The increase of $\phi_{isc}$ with photon energy has been corroborated by Eisinger and Lamola (1971), using europium as a counter of triplets.

g. *Determination of $K\phi^A$*

At concentrations $>1$ mM almost all of the triplet states are quenched by ground-state molecules, and Eq. (11) reduces to

$$\phi_d = \phi_{isc} f_d + 2K\phi^A[P] \tag{13}$$

Under these circumstances a plot of $\phi_d$ *vs.* [P] should give a straight line with slope $2K\phi^A$ and an intercept of $\phi_{isc}f_d$ (Fig. 6). It is impossible to determine both $K$ and $\phi^A$ but their product can be estimated. For Thy at 24°C, $K\phi^A = 0.05$ (Fisher and Johns, 1970). Association constants have been measured by osmometric techniques for several nucleosides, including Urd ($K = 0.61$–$0.70$) and Thd ($K = 0.91$–$1.2$) (Ts'o *et al.*, 1963; Solie and Schellman, 1968). Most Pyr bases are too insoluble to study by osmometry, but their association constants are probably comparable to those of the nucleosides. From this assumption, one can obtain a value for $\phi^A$ of 0.05 for Thy at 24°C. The yield of dimers from aggregates decreases with increase in temperature, as would be expected since the aggregates fall apart on heating.

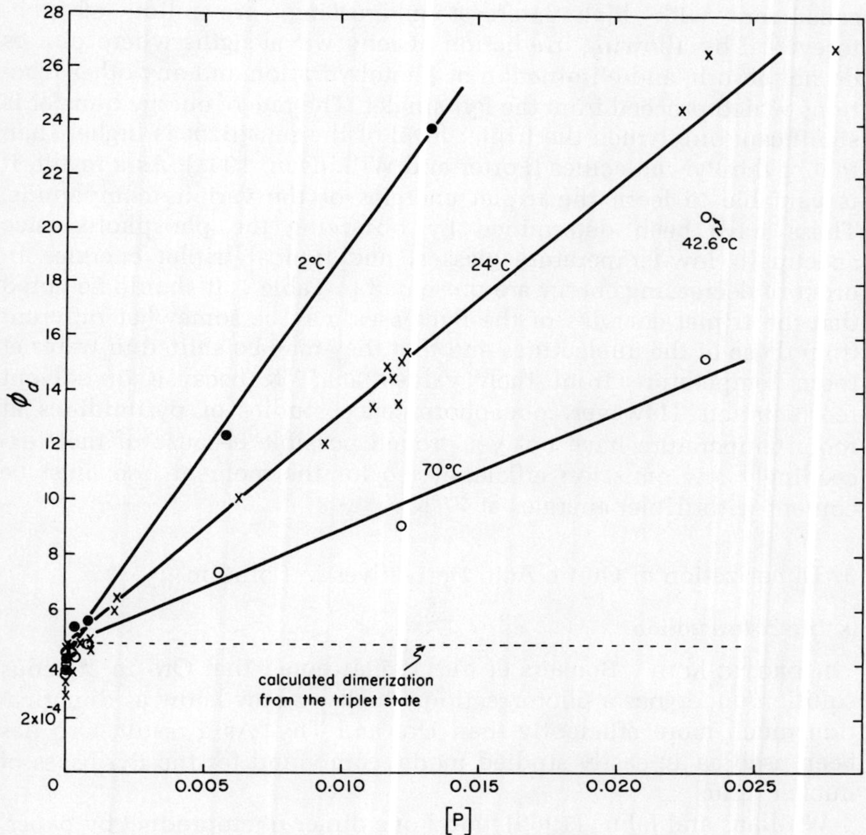


**Fig. 6.** *Yields of Thy◇Thy (deoxygenated solution, 265 nm) at four temperatures, plotted according to Eq. (13) to emphasize the behavior at high concentrations, and the concentration dependence calculated for dimerization from the triplet state. The difference between the observed and calculated values is attributed to dimerization from aggregates. The straight lines for* $[P] > 1$ *mM have slopes* $2K\phi^A$. *Increasing the temperature reduces dimerization from aggregates by shaking apart or "melting" the pairs of molecules (Fisher and Johns, 1970).*

*h. Sensitization*

In Chapter 4 we showed how reactions which proceed by way of the triplet state can be induced by photosensitizers such as acetone, acetophenone, and benzophenone. The sensitizer is excited to its triplet state by long-wavelength radiation which is *not* absorbed by the Pyr molecule. Energy transfer then takes place from the sensitizer triplet to the triplet state of the Pyr. Thus the Pyr is excited to its triplet state without having its singlet state populated. Advantages of triplet sensitization include more efficient dimer production through the use of sensitizers with high intersystem crossing, prevention of photoreversal by allowing irradiation at long wavelengths where dimers do not absorb, and elimination of photohydration and any other reactions which proceed from the Pyr singlet. The rate of energy transfer is significant only when the triplet level of the sensitizer is higher than that of the Pyr molecules (Porter and Wilkinson, 1961). As a result, it is valuable to learn the triplet energies of the various compounds. These have been determined by observing the phosphorescence spectra in low-temperature glasses, and typical triplet energies in order of decreasing energy are presented in Table 1. It should be noted that the triplet energies of the Pyr bases may be somewhat different from those of the nucleotides and that they may be shifted in water at room temperature from their values at 77°K because of solvent reorientation. However, phosphorescence studies of pyrimidines at room temperature have not yet proved possible because of their exceedingly low emission efficiency, so for the moment one must be content with triplet energies at 77°K.

## 3. Dimerization of Orotic Acid Derivatives in Solution

*a. Direct Irradiation*

i. OROTIC ACID. Beukers *et al.* (1959a) noted that Oro in aqueous solution undergoes a photoreaction which we now know as dimerization much more efficiently than Ura and Thy. As a result, Oro has been used as an easily studied model compound for the Pyr bases of nucleic acids.

Whillans and Johns (1969) found one dimer photoproduct by paper, thin layer, and column chromatography, and by electrophoresis. Charlier *et al.* (1969) also reported only one photodimer by direct and by sensitized irradiation. However, there have been reports of the existence of a second dimer in low yield (Sztumpf and Shugar, 1965; Birnbaum *et al.*, 1971). Birnbaum *et al.* showed by x-ray analysis that the principal Oro◇Oro is the *trans-syn* isomer.

Since only one product could consistently be detected by our group,

**Table 1** Triplet Energies at 77°K

| Material | $E_T$ (cm$^{-1}$) × $10^{-4}$ | $E_T$ (eV) | Reference |
|---|---|---|---|
| Acetone | 2.82 | 3.48 | Borkman and Kearns (1966) |
| CMP | 2.79 | 3.45 | Lamola *et al.* (1967) |
| UMP | 2.75 | 3.40 | Lamola *et al.* (1967) |
| GMP | 2.72 | 3.36 | Guéron *et al.* (1967) |
| AMP | 2.67 | 3.30 | Guéron *et al.* (1967) |
| Acetophenone | 2.65 | 3.28 | Kearns and Case (1966) |
| TMP | 2.63 | 3.25 | Lamola *et al.* (1967) |
| Benzophenone | 2.48 | 3.06 | Kearns and Case (1966) |
| Orotic acid | 2.13 | 2.64 | Charlier and Hélène (1967) |

the dimer yield can be measured just as well by monitoring the absorbance peak of Oro (Fig. 2) as by separating the photoproduct. Whillans and Johns (1969) obtained the same results using both methods and some of these results are plotted in Fig. 4. Apparently, $1/\phi_d$ depends linearly on $[P]^{-1}$ and $[O_2]$ as would be expected from the model for dimerization described by Eq. 12. The reduction of dimer yield by oxygen agrees with earlier observations (e.g., Beukers *et al.*, 1959a; Sztumpf and Shugar, 1965; Sztumpf-Kulikowska *et al.*, 1967; Greenstock *et al.*, 1967) and the ratio of rate constants ${}^3k_q/{}^3k_d$ agrees with the values obtained in flash photolysis studies of the Oro triplet (Herbert and Johns, 1971; Table 2).

Unfortunately, the concentration dependence studies of Whillans and Johns (1969) were carried out with unbuffered solution, and the pH varied from 3 to 6 as the concentration of Oro was altered. The importance of pH in such a study was later realized and the results (Herbert and Johns, 1971) obtained by flash photolysis are more accurate. In these later studies, ${}^3k_d$ was found to decrease by a factor of $> 2.0$ as one increases the pH from below to above the pK at 4.6. Below pH 4.6 only the triplet molecule is charged so that no repulsion occurs between the triplet and ground state. From pH 4.6 to 9.4 both species are charged so that repulsion reduces ${}^3k_d$, and finally, above 9.4 both species are doubly charged, greatly reducing ${}^3k_d$. For a detailed discussion of the influence of pH, see Herbert and Johns (1971). As a result of these effects, the dimerization efficiency is very low in alkaline solution and the photostationary equilibrium is shifted in favor of monomers (Sztumpf and Shugar, 1965; J. C. LeBlanc, unpublished data, 1971).

The nonradiative decay rate of the Oro triplet, ${}^3k_{nr}$, can change by a factor of 2 or more depending on the purity of the water (Herbert and Johns, 1971). The greater the concentration of impurities, the more rapidly the triplet decays. Since ordinary distilled water was used in

**Table 2** Summary of Pyrimidine Dimerization and Triplet-State Properties in Aqueous Solution

| | Oro | | | |
|---|---|---|---|---|
| | pH 1 | pH > 3 | Ura | Thy |
| Dimerization results[a] | | | | |
| $\phi_{isc}f_d$ | 0.12 | 0.060 | 0.0050 | 0.00047 |
| $^3k_q/^3k_d$ | | | | |
| Oxygen | 1.1 | 1.2 | 1.1 | 1.1 |
| $Mn^{2+}$ | | | | $10^{-2}$ |
| $^3k_{nr}/^3k_d$ | $0.5 \times 10^{-5}$ | $1.8 \times 10^{-5}$ (pH 3–6) | $5.9 \times 10^{-5}$ | $0.5 \times 10^{-5}$ |
| Flash photolysis results[b] | | | | |
| $^3k_d$ ($M^{-1}$ sec$^{-1}$) | $2.0 \times 10^9$ | $1.6 \times 10^9$ (pH 3)<br>$8.0 \times 10^8$ (pH 6) | $2.9 \times 10^9$ | $2.3 \times 10^9$ |
| $^3k_q$ ($M^{-1}$ sec$^{-1}$) | | | | |
| Oxygen | $2.2 \times 10^9$ | $3.0 \times 10^9$ | $3.9 \times 10^9$ | $3.4 \times 10^9$ |
| $Mn^{2+}$ [c] | | | | $4 \times 10^7$ |
| $^3k_{nr}$ (sec$^{-1}$) | $7.3 \times 10^3$ | $7.3 \times 10^3$ | $166 \times 10^3$ | $8 \times 10^3$ |

[a] Dimerization results from Whillans and Johns (1969), Brown and Johns (1968), and Fisher and Johns (1970), at 265 nm.
[b] Flash photolysis data from Herbert and Johns (1971) and Whillans and Johns (1971).
[c] Manganese quenching rate from Whillans (1972).

the dimerization study, rather than the multiply quartz-distilled water which was used in the flash photolysis work, it is not surprising that the ratio of $^3k_{nr}/^3k_d$ obtained by Whillans and Johns (1969) is somewhat larger than the ratio obtained by Herbert and Johns (1971; Table 2).

As might be expected for a triplet-state reaction, dimerization of Oro is inhibited by conjugated dienes, such as hexadienol, which quench triplets (Yip *et al.*, 1970) and which may also add to the excited Oro (Charlier *et al.*, 1969). In addition, the paramagnetic ions $Ni^{2+}$, $Cu^{2+}$, $Co^{2+}$, $Cr^{3+}$, $Mn^{2+}$, and $Fe^{2+}$, which form complexes with Oro (Haug, 1964), reduce the dimer yield, in keeping with their known triplet quenching ability whereas the diamagnetic ions $K^+$, $Al^{3+}$, $Ca^{2+}$, $Mg^{2+}$, $Cd^{2+}$, and $Zn^{2+}$ have no effect on dimer formation (Beukers and Berends, 1960a).

As yet unexplained is the finding that Oro dimerization is also reduced in the presence of Thy or Guo (Brown, 1968). From the energies of the triplets of Oro, TMP, and GMP in glasses at 77°K (Table 1), one would not expect triplet transfer from Oro to a Thy or Guo moiety.

With the europium counter, Eisinger and Lamola (1971) determined the absolute value of $\phi_{isc}$ and showed that it equals the dimer yield measured by Whillans and Johns (1969) for Oro. This shows that for Oro self-quenching is not important and $f_d = 1.0$.

Dimerization of Oro from aggregates has not been reported for three reasons: (1) no studies have been carried out at high enough concentrations (> 1 m*M*), (2) Oro does not aggregate as strongly as less polar pyrimidines, and (3) a small increase in the already high quantum yield would scarcely be detectable.

ii. DERIVATIVES OF OROTIC ACID. Sztumpf and Shugar (1965) studied the photochemical behavior of several methyl derivatives of Oro. 3-MeOro has a dimer yield and oxygen effect comparable to Oro while 1-MeOro and 1,3-$Me_2$Oro are somewhat more resistant to radiation. At neutral pH, 5-MeOro is at least an order of magnitude more resistant to UV light than is Oro, as has been confirmed by a recent survey of pyrimidines in our laboratory. The methyl esters of Oro and 5-MeOro behave photochemically in much the same way as their parent compounds. The nucleoside, Ord, is photolyzed readily to at least two products, only one of which, accounting for ~20% of the total, is a dimer.

*b. Photosensitized Dimerization of Orotic Acid*

The triplet energy level for Oro (Table 1) is lower than the levels for acetone, acetophenone, and benzophenone, allowing efficient energy transfer from these sensitizers. Charlier and Hélène (1967) fitted their data for sensitized dimerization to a triplet transfer model. They separated only one isomer with chromatographic properties identical to the Oro◇Oro (*trans, syn*) formed by direct irradiation. Greenstock and Johns (1968) also reported the sensitized formation of only one dimer. At an irradiation wavelength of 334 nm, they found that 1 m*M* Oro in anoxic solution dimerizes with an overall quantum yield of 0.035 in 0.1 *M* aqueous acetone, 0.022 in 1 m*M* acetophenone, and 0.11 in 0.5 m*M* benzophenone. Charlier *et al.* (1969) used flash photolysis techniques to demonstrate that sensitization of Oro◇Oro formation in fact does involve the interaction of benzophenone triplet with a ground-state Oro molecule, and Herbert and Johns (1971) showed that acetone acts in the same way.

### 4. Dimerization of Thymine and Its Derivatives in Solution

*a. Direct Irradiation*

i. THYMINE IN WATER. In comparison with Oro, the derivatives of Thy are very resistant to UV irradiation in dilute aqueous solutions. The insensitivity of Thy has been noted by many workers (Lis and

Allen, 1961; Moore and Thomson, 1956; Wierzchowski and Shugar, 1960). Wacker *et al.* (1961a) showed that by extensive irradiation of 0.1 m*M* aerated solution ~3% of Thy could be converted to dimer, as determined by column separation. Using thin layer chromatography, Greenstock *et al.* (1967) showed that Thy, Thd, and TMP form photodimers in low yield, and that the reactions are quenched by oxygen.

In the first detailed study of Thy◇Thy formation in deoxygenated solution, it was found that the triplet state is the precursor of virtually all of the dimers formed in solutions of $<1$ m*M* (Fisher and Johns, 1970). The dependence of the quantum yield $\phi_d$ on Thy concentration is shown in Fig. 7. In agreement with the expectations of the reaction model, $\phi_d^{-1}$ is a linear function of $[P]^{-1}$ except at high concentrations, in which the quantum yield is increased by dimerization from aggregates. Extrapolation of the straight-line portion of the graph gives a value for $\phi_d$ which is two orders of magnitude smaller than the dimer yield for Oro. This low yield accounts for past difficulties in observing the reaction for Thy. It was found too that $\phi_d$ increases with excitation energy (Fig. 5) in a manner similar to that of Oro. Analysis of data similar to that in Fig. 7 gives ratios of rate constants for quenching by oxygen, nonradiative decay, and bimolecular interaction which agree reasonably well with the values obtained by Whillans and Johns

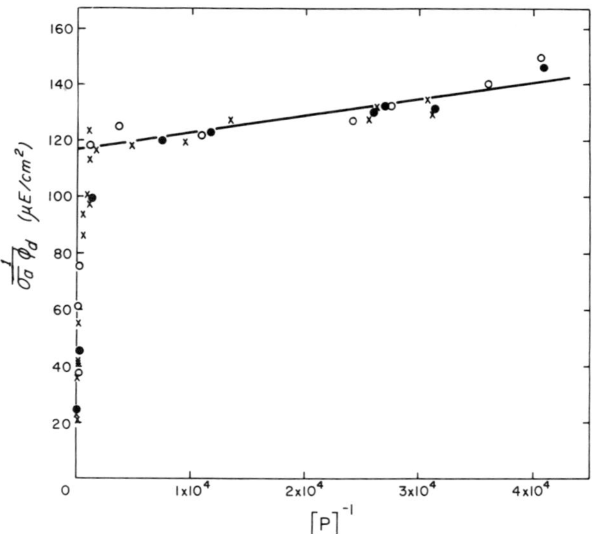


**Fig. 7.** *Plot of $1/\sigma_a\phi_d$ as a function of the reciprocal of Thy concentration. Irradiations were carried out in neutral deoxygenated solutions at 265 nm at temperatures of 2°C (●), 24°C (×), and 70°C (○) (Fisher and Johns, 1970).*

(1971), who used flash photolysis techniques to study the triplet state (Table 2).

By thin layer chromatography, it has been possible to show that all four of the possible diastereomers of the Thy◇Thy are formed in aqueous solution (Fig. 1b). It might be expected that changes in the temperature of the solution would influence intersystem crossing efficiencies, nonradiative decay rates, and diffusion rates, and recent flash photolysis studies of Thy show this to be the case (Fenster and Johns, 1973). Nevertheless, the total quantum yield of dimerization from the triplet does not change between 2°C and 70°C (Fig. 7; Fisher and Johns, 1970). What does change with temperature is the distribution among the various isomers. As the temperature increases, dimers B, C, and D all decrease, while the *cis-syn* dimer, A, increases (Table 3). The reason for this behavior is still unknown.

At concentrations of $>1$ m*M*, while dimerization from the triplet state of Thy occurs with a fixed quantum yield, the overall yield of dimers continues to increase linearly with concentration (Fig. 6) because of the increasing number of associated pairs of molecules. The linear dependence is consistent with the aggregation model, as formulated in Eq. (9). Four isomeric dimers may be formed by this mechanism, presumably from four types of aggregates in which the orientation of the molecules prefigures the resulting dimer. At elevated temperatures, the amount of dimerization from aggregates is reduced (Fig. 6) due to a decrease in the stability of the associated pairs.

ii. THYMINE IN ACETONITRILE. The photodimerization of Thy in dilute solution in acetonitrile was studied by Lamola and Mittal (1966). At a concentration of 1 m*M*, the reaction may be quenched completely by an equal concentration of the triplet quencher isoprene, indicating that dimers are formed from a triplet-state precursor in acetonitrile.

The rate constant for the bimolecular triplet–ground state interaction is $\sim 7 \times 10^8\ M^{-1}\ \mathrm{sec}^{-1}$ which is $<10\%$ of the diffusion rate in acetonitrile (Wagner and Bucheck, 1968). The flash photolysis results of Szabo *et al.* (1970) confirm this. The quantum yield for dimerization in acetonitrile has been reported to be 0.0025 for 0.7 m*M* Thy at 254 nm (Lamola and Mittal, 1966) and variously as 0.005 (Wagner and Bucheck, 1968) and 0.0025 (Wagner and Bucheck, 1970) for 0.6 m*M* Thy at wavelengths longer than 275 nm. This apparent lack of wavelength dependence stands in sharp contrast to the behavior of Thy in water and should be checked more carefully. By using 1,3-pentadiene to count the Thy triplets, Lamola and Mittal (1966) showed

**Table 3** Distribution of Isomeric Dimers Formed in Solution

| Material | Solvent (dielectric constant) | Precursor | Quantum yield[a] | % Yield of each isomer[a] | | | | References[b] |
|---|---|---|---|---|---|---|---|---|
| | | | | *cis-syn* | *trans-syn* | *cis-anti* | *trans-anti* | |
| Thy (2°C) | Water (79) | Triplet | 0.00047 (265 nm) | 15 | 22 | 34 | 29 | 1 |
| Thy (24°C) | Water (79) | Triplet | 0.00047 (265 nm) | 24 | 18 | 32 | 26 | 1 |
| Thy (70°C) | Water (79) | Triplet | 0.00047 (265 nm) | 43 | 13 | 24 | 20 | 1 |
| Thy | | Aggregates | 0.041 ($\phi^A$) | 33 | 10 | 31 | 26 | 1 |
| Thy (25°C) | 0.14 *M* Propiophenone in water (79) | Sensitized | | 24 | 20 | 33 | 23 | 2 |
| | 0.68 *M* Acetone in water (78) | Sensitized | | 22 | 21 | 34 | 23 | 2 |
| | 11 *M* Acetone and water (30) | Sensitized | | 20 | 21 | 20 | 39 | 2 |
| Thy | Acetonitrile (38) | Triplet | 0.005 | — | — | — | + | 3 |
| Thd | Water (79) | Triplet | 0.00056 (265 nm) | + | + | + | + | 1 |
| | Water (79) | Aggregates | 0.038 ($\phi^A$) | + | + | + | + | 1 |
| | Acetone/*t*-butanol 7:4 | Sensitized | | 26 | 6 | 13 | 55 | 4 |
| $Me_2Thy$ | Toluene (2) | Triplet | 0.008 | 25 | 4 | 59 | 12 | 5 |
| | Ethyl acetate (6) | Triplet | 0.009 | 46 | 5 | 41 | 7 | 5 |
| | Methanol (33) | Triplet | 0.0001 | 72 | 5 | 22 | 2 | 5 |
| | Acetonitrile (38) | Triplet | 0.0002 | 74 | 8 | 16 | 2 | 5 |
| | Water (79) | Triplet | 0.00004 | | | | | 6 |
| | Toluene (2) | Aggregates | 0.0017 (0.10 *M*) | 11 | 0 | 66 | 24 | 5 |
| | Ethyl acetate (6) | Aggregates | 0.0016 (0.10 *M*) | 22 | 1 | 59 | 19 | 5 |
| | Methanol (33) | Aggregates | 0.0019 (0.10 *M*) | 36 | 3 | 43 | 18 | 5 |
| | Acetonitrile (38) | Aggregates | 0.0013 (0.10 *M*) | 46 | 3 | 39 | 12 | 5 |
| | Water (79) | Aggregates | 0.014 (0.10 *M*) | 39 | 6 | 38 | 18 | 5 |
| | Water (79) | Aggregates | 0.18 ($\phi^A$) | + | + | + | + | 6 |
| | Water (79) | Aggregates | 0.065 ($\phi^A$) | | | | | 15 |
| 5-EtdUrd | Water (79) | Triplet | | 30 | 5 | 50 | 15 | 7 |
| 6-MeUra | Water (79) | Triplet | | 0 | 30 | 20 | 50 | 8 |
| Ura | Water (79) | Triplet | 0.0050 (265 nm) | 35 | 0 | 15 | 50 | 9,10 |
| | Water (79) | Triplet | | 19 | 3 | 76 | 2 | 16 |

| | | | | | | | | |
|---|---|---|---|---|---|---|---|---|
| Ura | 50% Aqueous acetone | Sensitized | | 68 | − | 29 | 3 | 17 |
| Urd | 50% Aqueous acetone | Sensitized | | + | − | + | 55 | 17 |
| $Me_2Ura$ | Water (79) | Triplet | < 0.0003 | | | | | 12 |
| | Water (79) | Aggregates | 0.03 ($\phi^A$) | 25 | 60 | 8 | 8 | 15 |
| | Benzene | Triplet | 0.0014 (0.2 *M*) | 20–23 | 15 | 39–40 | 22–26 | 11 |
| | Dioxane | Triplet | 0.004 (0.2 *M*) | 16–32 | 17–19 | 24–30 | 27–35 | 11 |
| | Acetonitrile (38) | Triplet | 0.0026 (0.2 *M*) | 35–36 | 30–31 | 21–27 | 8–12 | 11 |
| | 50% Aqueous acetone | Sensitized | 0.048 (0.2 *M*) | 49–50 | 21–24 | 15–16 | 11–14 | 11 |
| Cyt | Water (79) | Aggregates | 0.005 ($\phi^A$) | | | | | 13 |
| | 50% Aqueous acetone | Sensitized | | 13 | 0 | 5 | 32 | 14 |
| Cyd | 95% Aqueous acetone | Sensitized | | − | − | − | + | 14 |
| | 50% Aqueous acetone | Sensitized | | + | − | + | >60 | 14 |

[a] Detected but not quantitated: +; not detected: −. A blank indicates that no information was presented.

[b] References:

1. Fisher and Johns (1970).
2. Jennings *et al.* (1970).
3. Lamola and Mittal (1966).
4. Ben-Hur *et al.* (1967).
5. Kleopfer and Morrison (1972).
6. Lisewski and Wierzchowski (1971).
7. Pietrzykowska and Shugar (1970).
8. Khattak and Wang (1972).
9. Brown and Johns (1968).
10. A. J. Varghese, unpublished data (1972).
11. Elad *et al.* (1971).
12. Greenstock *et al.* (1967).
13. Calculated from I. H. Brown (1968).
14. Varghese (1971a).
15. Stepien *et al.* (1973b).
16. Fendler and Bogan (1974).
17. Varghese (1971b).

that $\phi_{isc}$ at 254 nm is 0.18. This is 72 times larger than the dimerization quantum yield, implying that the bimolecular interaction between a triplet and a ground-state molecule in acetonitrile is highly inefficient, with self-quenching a more common result than production of a stable dimer.

iii. 1,3-DIMETHYLTHYMINE. 1,3-$Me_2$Thy was studied in some detail by Morrison *et al.* (1968), who obtained all four dimers by irradiating 0.1 *M* aqueous solutions with wavelengths longer than 280 nm. The overall quantum yield for dimer formation under these conditions was originally presented as 0.05 but was later revised to 0.0141 (Kleopfer and Morrison, 1972). Lisewski and Wierzchowski (1971) showed that at 0.1 *M* most of the dimers come from aggregates, with a quantum yield $\phi^A$ of 0.165 per photon absorbed in aggregates. At lower concentrations they also showed that dimerization occurs from the triplet, with $\phi_{isc}f_d \cong 4 \times 10^{-5}$. In nonaqueous solvents $Me_2$Thy does not stack to the extent that it does in water and the triplet-state precursor is, therefore, relatively more important even at high concentrations of $Me_2$Thy.

Kleopfer and Morrison (1972) determined quantum efficiencies and relative yields of the four isomeric dimers of $Me_2$Thy upon direct irradiation, triplet sensitization, and with triplet quenching in several organic solvents. Their results are summarized in Table 3. The solvents are listed in order of increasing dielectric constant, and dimer yields were measured from the triplet state and from aggregates. By both mechanisms the relative yields of the *syn*, or head-to-head, isomers increase with increasing polarity of the solvent, while the *anti*, or head-to-tail, isomers decrease. While the quantum yield for dimerization via the triplet decreases by two orders of magnitude between toluene and methanol, the intersystem crossing efficiency decreases by a factor of only 5, suggesting that $f_d$ is 20 times smaller in methanol than in toluene.

iv. THYMINE NUCLEOSIDES AND NUCLEOTIDES. The photochemical properties of Thd and TMP have not been examined as thoroughly as those of Thy itself. Thd forms four isomeric dimers in water (Fisher and Johns, 1970), but structural assignments have not been made. The overall quantum yield ($\phi_{isc}f_d$) for Thd$\diamond$Thd from the triplet state in neutral aqueous solution is $5.6 \times 10^{-4}$ at 265 nm (Table 4). Other reported values are also presented in Table 4; some of these were calculated by us from published "rates" or relative yields by comparison with the reported values for Thy and the quantum yield of Fisher and Johns (1970) for Thy. The work of Guschlbauer *et al.* (1965)

**Table 4** Quantum Yields for Dimerization of Thymine Derivatives

| Material | Quantum yield × $10^4$ | Reference[a] | Comments |
|---|---|---|---|
| Thy | 4.7 | 1 | 265 nm, limiting yield from triplets |
| Thd | 5.6 | 1 | 265 nm, limiting yield from triplets |
| Thd | 4.0 | 3 | Calculated from rate data by comparison with the rate of Thy and the Thy quantum yield of Fisher and Johns (1970) |
| x-TMP (unspecified isomer) | 4.5 | 3 | |
| TMP | 1.0 | 4 | Aerated solution |
| rThd (pH 6) | 4.5 | 2 | Concentration unspecified; reported yield for loss of parent divided by 2 on the assumption that products are dimers |
| 3-MeThd | 1.5 | 5 | |

[a] References:
1. Fisher and Johns (1970)
2. Guschlbauer *et al.* (1965)
3. Moore and Thomson (1956)
4. Rushizky and Pardee (1962)
5. Wierzchowski and Shugar (1960)

suggested that rThd has approximately the same dimerization quantum yield as Thd, while methylation at the ring nitrogen, N(3), reduces the dimerization yield (Wierzchowski and Shugar, 1960).

At concentrations >1 m*M*, Thd forms dimers from aggregates as well as from triplets, with an efficiency $\phi^A$ of 0.04 per photon absorbed in an aggregated pair (Fisher and Johns, 1970).

Thymidylic acids have been studied even less than Thd. The dearth of reliable information on the photolysis of Thy nucleotides is unfortunate; x-TMP (unspecified isomer) is the only Thy derivative for which an independent estimate of the intersystem crossing efficiency in water is available ($\phi_{isc} = 0.008$ at 265 nm; Lamola and Eisinger, 1971).

*b. Photosensitized Dimerization*

i. THYMINE. A number of studies of the dimerization of Thy derivatives have been made using triplet sensitizers such as acetone and acetophenone. Early reports suggested that Thy forms either two (von Wilucki *et al.*, 1967) or three (Greenstock and Johns, 1968) isomeric dimers upon sensitized irradiation. We now know that four dimers are formed (Jennings *et al.*, 1970). The four structures were confirmed and the formation of each was monitored using NMR spectroscopy. Acetophenone was a less effective sensitizer than acetone or propiophenone. The yields of the isomers were found to be in the same ratio, in dilute acetone or propiophenone solution (Table 3), as the yields of dimer formed from the triplet state by direct irradiation in water (Fisher and Johns, 1970). On the other hand, when high concentrations of acetone were present (Table 3), the *trans-anti* dimer was formed in higher yield, at the expense of the *cis-anti* isomer, while the proportions of the *syn* dimers remained unchanged. The change may be attributed in part to the change in dielectric constant from 78 to 30 on going from a predominantly aqueous solvent to 11 *M* acetone.

ii. *N*-METHYLATED THYMINE. The relative yields of the four isomeric dimers of $Me_2Thy$ are the same by direct formation via the triplet and by sensitization by benzophenone in several organic solvents (Kleopfer and Morrison, 1972). No information is provided on the efficiency of the sensitized reaction or on the triplet energy levels of $Me_2Thy$ and benzophenone in the solvents used. Interestingly, the distribution of dimers is reported to vary with dielectric constant in a way which is quite different from the case for Thy: as the dielectric constant increases, the *syn* dimers of $Me_2Thy$ increase at the expense of the *anti* dimers (Table 3; Kleopfer and Morrison, 1972). The contrast in behavior of Thy and $Me_2Thy$ may reflect the importance of the *N*-methyl groups in changing the properties of the molecule, but it also suggests that the system should be reexamined, and the direct irradiation of Thy should be carried out in solvents of different dielectric constant.

Upon irradiation of $Me_2Thy$ in neat acetone, only two dimers have been isolated, with traces of others also reported (Elad *et al.*, 1967). The major dimers are identical to the two isomers formed by irradiation of frozen solutions of $Me_2Thy$: *cis-syn* (Blackburn and Davies, 1966a) and *cis-anti* (Weinblum and Johns, 1966). These are indeed the two isomers which are produced in the highest yields from the triplet state (Table 3; Kleopfer and Morrison, 1972). Acetone-sensitized irradiation of 1-MeThy leads to roughly equal yields of the *c,s*, *c,a*, and *t,s* dimers, but to virtually no *t,a* (Stepien *et al.*, 1973a).

iii. THYMIDINE. The sensitized dimerization of Thd in an acetone/*t*-butanol matrix has been reported (Ben-Hur *et al.*, 1967). Four dimers were detected, one of which was resolved into two optical isomers. The *trans-anti* isomer comprised 55% of the total product while the *cis-anti* and *trans-syn* dimers formed relatively small fractions of the total (Table 3).

### 5. Dimerization of Uracil and Its Derivatives in Solution

#### *a. Direct Irradiation*

i. URACIL. Early experiments by Rörsch *et al.* (1958), Beukers *et al.* (1959a), and Beukers and Berends (1961) demonstrated that Ura undergoes photochemical reactions suggestive of dimerization. The dimer was first separated by paper chromatography by Wacker *et al.* (1961b). During the course of an irradiation, the amount of dimer increases with increasing exposure to an equilibrium level, then declines as Ura is converted to other products (Wacker *et al.*, 1964).

Brown and Johns (1968) examined the kinetics of photodimerization of Ura and found that the quantum yield for the reaction depends on concentration in the way predicted by the triplet mechanism, Eq. (5), reaching a limiting value at Ura concentrations $>0.1$ m*M*. This limiting yield depends on the irradiation wavelength, increasing at higher photon energies (Fig. 5), and at all wavelengths is approximately an order of magnitude larger than the corresponding value for Thy. The ratios of rate constants for quenching by oxygen, nonradiative decay, and bimolecular interaction which were obtained in this work are compatible with the absolute values for the rate constants of the Ura triplet state as obtained by flash photolysis studies (Table 2; Whillans and Johns, 1971). The quantum yield is not affected by pH over the range 2–8. In contrast with Thy, Ura has not yet been shown to dimerize from aggregates in water because it associates to a lesser extent than Thy (Solie and Schellman, 1968), because it has not been studied at high enough concentrations ($>10$ m*M*), and because a small change would scarcely be noticed in the quantum yield, which is much larger than the yield for Thy. It has been reported that four photoreversible Ura◇Ura dimers may be resolved by thin layer chromatography and detected using a $^{14}C$ label (Brown and Johns, 1968), but no estimate was given of the relative yields of the various isomers. Recent experiments have suggested that the three major dimers are the *trans-anti* (50%), *cis-syn* (35%), and *cis-anti* (15%) isomers (A. J. Varghese, unpublished data) but the published report of Fendler and Bogan (1974) gives very different results (see Table 3). These large differences may be due to mistaken identification of the products, to concentration effects on product ratios, or to changes in dimer ratios with increasing irradiation. Fendler and Bogan (1974) demonstrated such a dose dependence, and presented initial yields of the four dimers.

Ura also dimerizes when irradiated in degassed acetonitrile, with a quantum yield of 0.019 at 0.39 m*M* and wavelengths longer than 275 nm (Wagner and Bucheck, 1970). It is estimated that at this concentration 83% of the triplets react with ground-state Ura molecules, but unfortunately this has not been demonstrated. At 254 nm, the intersystem crossing efficiency of Ura has been measured as 0.40 in acetonitrile, using 1,3-pentadiene as a triplet counter (Lamola and Mittal, 1966). The 20-fold difference in these values for the yields of triplets and of dimers in acetonitrile may reflect the importance of the self-quenching process. However, it may also be due in part to a wavelength dependence in intersystem crossing efficiency: in water, $\phi_{isc}$ is more than twice as large as 254 nm as at 280 nm (Fig. 5; Brown and Johns, 1968). Further, the fraction of triplets which react at 0.39 m*M* may be less than 83%. It is unfortunate that the quantum yield for dimerization was not measured at several concentrations.

There is evidence that formation of Ura◇Ura in acetonitrile may not be entirely by way of a triplet-state precursor. Only ~90% of the loss of Ura can be quenched by a triplet quencher (Lamola and Mittal, 1966). Whether the remaining 10% of the photolysis involves dimerization from aggregates or a reaction with the solvent analogous to photohydration in water is not clear.

ii. 1,3-DIMETHYLURACIL. The first evidence for dimerization of 1,3-$Me_2$Ura was provided when the irradiation of 0.2 *M* aqueous solutions was shown to give several products besides the well-known photohydrate (Moore and Thomson, 1956). At 10 m*M* the photolysis is not a first-order process, a fact which also suggests that hydration is not the only reaction, whereas at concentrations $<1$ m*M* first-order kinetics, characteristic of hydrate production, do prevail (Wierzchowski and Shugar, 1959).

At 0.1 m*M* dimers comprise less than 2% of the total photoproducts, the remainder being the hydrate (Greenstock *et al.*, 1967). From the known quantum yield (0.014) for photolysis of $Me_2$Ura (Wang, 1962; Johns, 1971; see also the discussion in Chapter 4), the dimerization quantum yield at 0.1 m*M* may be estimated as $<0.0003$. This is two orders of magnitude smaller than the corresponding figure for Ura. At high concentrations ($>10$ m*M*) dimers may be formed more efficiently (Nnadi and Wang, 1969), probably from aggregates. Detailed studies of the photochemical behavior of $Me_2$Ura at various concentrations and pHs are required in order to determine the quantum yields of hydrate formation and of dimerization from the triplet and from aggregates in water.

An important step in this direction was recently made by Stepien *et al.* (1973b), who found a quantum yield for dimerization in excited aggregates of ~0.03, with the principal isomer being the *trans-syn* form (see Table 3).

In organic solvents, the relative amount of *anti-* $Me_2$Ura◇$Me_2$Ura decreases with increasing solvent polarity (Table 3; Elad *et al.*, 1971). This is analogous to the behavior noted for $Me_2$Thy. However, unlike $Me_2$Thy, it seems that dimerization of $Me_2$Ura in organic solvents proceeds entirely through the triplet state, without involving aggregates to any significant extent.

iii. URIDINE. The photodimerization of Urd in aqueous solution has been demonstrated using radioactive label and by separating products by paper chromatography (Wacker *et al.*, 1961b). Absorbance studies have shown that dimerization proceeds somewhat more efficiently for Urd than for Ura and that it may be quenched by oxygen at 0.1 m*M* (Greenstock *et al.*, 1967). Further, the efficiency of dimer formation in aerated solution increases with concentration, in the manner characteristic of dimerization from aggregates, in $H_2O$ and in $D_2O$ (Nnadi and Wang, 1969). Interestingly, this is all that is known about the production of Urd◇Urd by direct irradiation. No quantum yield measurements have been made, and no study has been done on the relative yields of the various isomers. This important compound certainly deserves further study.

iv. URIDINE MONOPHOSPHATE. The photolysis of 2′(3′)-UMP, like that of Urd, seems to be increasingly efficient at high concentrations (Sinsheimer, 1954; Wierzchowski and Shugar, 1959). The quantum yield for loss of parent absorption at pH 7.0 has been reported to be 0.019 at 1 m*M* and 0.037 at 10 m*M* for an irradiation wavelength at 254 nm (Moore and Thomson, 1956). At low concentrations the principal photoproduct is the water addition product or hydrate (Sinsheimer, 1954; see Chapter 4); the increase in photolysis at high concentrations may be interpreted as being dimerization. A photoproduct resolved by chromatography on DEAE-cellulose was shown to be converted to parent on further irradiation (Zachau, 1964), but more rigorous characterization of the product as a dimer has not been attempted.

At 0.1 m*M*, both 2′(3′)-UMP and UMP dimerize only slightly, according to absorbance studies, and the quenching of the reaction by oxygen implicates the triplet state as a precursor at this concentration (Greenstock *et al.*, 1967).

The photodimerization of UMP merits more detailed study in order to determine the relative importance of triplets and aggregates as a function of concentration, the effect of phosphate ionization on dimer yield, and the relative yields of the isomeric dimers.

*b. Photosensitized Dimerization of Uracil Derivatives*

The irradiation of 1 m*M* aqueous solutions of Ura in 25% acetone with 315 nm light was reported by Krauch and co-workers (1967) to yield four products, of which only one, Ura◇Ura(*c*,*s*), was identified. Greenstock and Johns (1968) also showed that four products are formed using acetone, acetophenone, and benzophenone as sensitizers. On the other hand, Charlier and Hélène (1972) showed that benzophenone gives addition products to Ura, as well as dimers, which may have been confused with dimers in earlier work. Subsequently, Varghese (1971b) showed that irradiation of Ura in 50% acetone leads primarily to the formation of Ura◇Ura(*t*,*a*), with lesser amounts of the *c*,*s* and *c*,*a* isomers, and hUra, while the *t*,*s* dimer could not be detected. There is a conflicting report by Jennings and co-workers (1972) that the *syn* isomers are produced in largest yield. Detailed x-ray diffraction studies are required to clarify this inconsistency. (See the discussion of IR spectra in Section E,5,c.) It is unclear how acetophenone and benzophenone, which have low triplet energies (Table 1), can excite the Ura triplet; the efficiency for these sensitizers may be low (Jennings *et al.*, 1972). Definitive experiments on the quantum yields for the sensitized formation of Ura◇Ura have not been carried out.

Irradiation of Urd in 40% *t*-butanol in acetone has been reported to give a single isomeric dimer of undetermined structure (Rosenthal and Elad, 1968). In 50% aqueous acetone Urd◇Urd dimers have been detected, with the *t*,*a* isomer comprising 55% of the total. Both the *c*,*s* isomer and the *c*,*a* dimer (in two optically active forms) have been isolated in lesser yield (Varghese, 1971b). Because of the conflicting report of Jennings *et al.* (1972), doubt remains as to whether the *t*,*a* dimer might really be *t*,*s*.

In neat acetone, $Me_2$Ura forms four isomeric dimers (Elad *et al.*, 1967). The relative yields of the isomers on irradiation in organic solvents are independent of the sensitizer and are the same by direct and by sensitized irradiation (Table 3; Elad *et al.*, 1971). In more polar solvents, the *c*,*s* dimer predominates.

### 6. Dimerization of Cytosine and Its Derivatives

#### *a. Direct Irradiation.*

Available information on the dimerization of Cyt derivatives in solution is surprisingly limited in view of the large number of photochemical studies performed on these compounds. Studies based on absorbance measurements using dilute solutions ($< 1$ m*M*) generally lead to the conclusion that Cyt, Cyd, 3′-CMP, and CMP form no dimers but only hydrates by direct irradiation in water (Wierzchowski and Shugar, 1961). On the other hand, after irradiating 20 m*M* solutions of Cyt, Brown (1968) detected Ura◇Ura by thin layer chromatography. These dimers result from deamination of Cyt◇Cyt. The latter almost certainly arise from aggregates, since no dimers were produced in dilute solution. Assuming that dimerization is from aggregates and that the association constant of 0.9 for Cyd (Solie and

Schellman, 1968) also applies to Cyt, the quantum efficiency $\phi^A$ for dimer formation from excited aggregates is calculated to be $5 \times 10^{-3}$, an order of magnitude smaller than the corresponding value for Thy (Table 3). No information is available on the isomers which are formed.

Careful attempts by our group to observe a triplet species by flash photolysis of Cyt have proved inconclusive, suggesting that intersystem crossing is at least a factor of 10 smaller in Cyt than in Thy.

*b. Sensitized Dimerization*

Krauch *et al.* (1967) found that 1 m*M* Cyt forms dimers in low yield when irradiated in 25% aqueous acetone. Varghese (1971a) studied the reaction in greater detail using 50% acetone, and found three isomeric dimers, of which the *t,a* isomer is the major one (Table 3). When Cyd was irradiated in the same way, only the *t,a* isomer was detected. Wang's recent results, as yet unpublished, do not support all of Varghese's findings, and further work is required.

Since Cyt is one of the two main pyrimidines found in DNA, its photodimerization merits further investigation.

## 7. Dimerization of 5- and 6-Substituted Pyrimidine Derivatives

The variation of dimerization efficiencies over more than two orders of magnitude (Fig. 5) with only small structural changes from Oro to Thy has led to several studies with other derivatives in an attempt to clarify how the substituents affect the reaction. It was hoped that a quantitative correlation could be made between photodimer yield and the site and electronegativity of the substituent, analogous to the Hammett relationship for reactions of substituted benzoic acid. However, it has become apparent that such a simple relationship will not be found. Following is a summary of our present knowledge.

*a. 6-Methyluracil*

The dimerization of 6-MeUra ("isothymine," **V**) has recently been

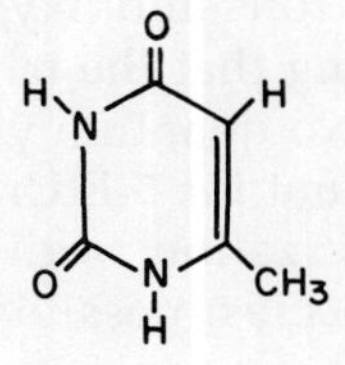

O
H N H
O N CH3
H

**V**

described (Khattak and Wang, 1972). By direct irradiation of 2 m*M* solutions of 6-MeUra, mainly the *trans* dimers are formed, along with a trace of the *c,a* isomer.

Recent unpublished work in our laboratory, based on simple absorbance studies rather than on quantitation of products, suggests that the quantum yield for dimerization is $\sim 1 \times 10^{-3}$ in deoxygenated solution (Table 5). This is twice the value for Thy (**III**) in which the methyl group is on C(5) but only one-fifth of the yield for unsubstituted Ura.

By acetone-sensitized irradiation in solution, only the two *trans* isomers, and not the *cis* isomers, are formed (Khattak and Wang, 1972). As the solvent is changed from dioxane to less polar media, the relative yield of the *t,a* dimer decreases from almost 50% of the total to zero, while the *t,s* dimer increases. This is in direct contrast to the behavior shown by Thy, $Me_2$Thy, and $Me_2$Ura (Table 3). The *t,s* dimer also becomes relatively more important at high pH.

*b. 5-Alkyl Analogs of Thymine*

When either 5-EtUra (**VI**), its riboside, or its deoxyriboside is irradiated in water at wavelengths shorter than 265 nm, some dimer is formed, but the major product detected is an intramolecular cyclobutane product, **VII** (see Scheme 1) (Pietrzykowska and Shugar, 1970).

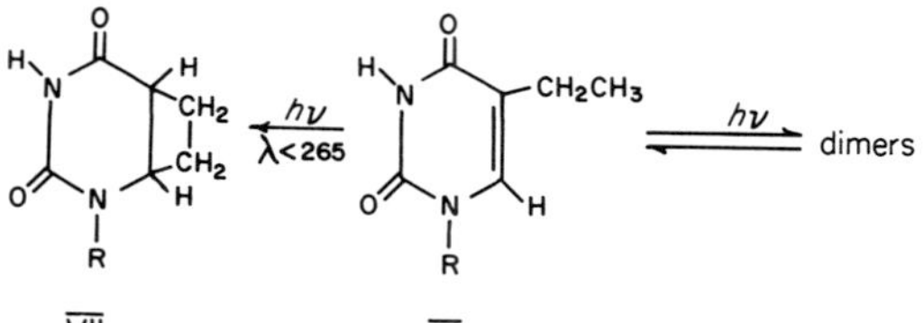

**Scheme 1**

At longer wavelengths, however, dimerization predominates. Four isomers have been detected, but the *cis* forms comprise $\sim 80\%$ of the total yield. Both manganese ions and oxygen quench dimer formation of 0.1 m*M* 5-EtUrd, indicating that the triplet state is the precursor of dimers in dilute solutions. No quantum yields are given, but the reaction is $\sim 3$ times more efficient for 5-EtUrd than for Thd. Upon sensitization with acetone, dimerization readily occurs, but $> 20\%$ of the product is a nondimeric species, possibly an oxetane adduct of acetone to 5-EtUra.

Direct irradiation of 5-PrUrd and 5-*i*PrUrd results only in a very slow conversion to intramolecular cyclobutane products analogous to **VII** (Krajewska and Shugar, 1972).

*c. Other 5- and 6-Substituted Pyrimidines*

We irradiated 0.1 m*M* aqueous solutions of a number of Ura derivatives under a nitrogen atmosphere, at 265 nm, and determined quantum yields for loss of parent, $\phi_{-P}$, by monitoring absorption. We estimated the dimer yield, $\phi_d$, in each case by observing the amount of recovery of absorption following reirradiation at 235 nm. As discussed in Section B,1, there are limitations to the use of this method, and quantum yields can only be estimated. These results and others are summarized in Table 5.

Evidently, Ura derivatives substituted at C(5), whether by electron-withdrawing (e.g., COOH, $NO_2$) or electron-donating (e.g., $CH_3$) groups, have dimer yields considerably lower than those of unsubsti-

**Table 5** Quantum Yields for Loss of Parent ($\phi_{-P}$) and for Dimerization ($\phi_d$) for Various Pyrimidines Irradiated at 265 nm in 0.1 m*M* Aqueous Solution

| Compound | $\phi_{-P}$ | $\phi_d$ | Reference[a] |
|---|---|---|---|
| Ura | 0.011 | 0.005 | 1 |
| Thy (5-MeUra) | 0.001 | 0.0005 | 2 |
| 5-AmUra | 0.01 | 0 | 3 |
| 5-hmUra | 0.006 | 0.001 | 3 |
| | 0.0016[b] | | 5 |
| 5-$NO_2$-Ura | 0.0002 | 0.0001 | 3 |
| 5-$CF_3$-Ura | 0.001 | 0.0005 | 3 |
| 5-Et-*iso*-Oro | 0.0026 | 0.0013 | 3 |
| *iso*-Oro | 0.0005 | $\sim 10^{-4}$ | 3 |
| Urd | 0.0023[b] | | 5 |
| 5-hoUra | 0.0005 | 0.0002 | 3 |
| 6-MeUra | 0.002 | 0.001 | 3 |
| 6-AmUra | 0.001 | 0.0005 | 3 |
| Oro | 0.10 (pH > 3) | 0.05 | 4 |
| Ura-6-sulfonamide | 0.03 | 0.01 | 3 |
| Ura-6-Ac | 0.03 | 0.01 | 3 |
| 5,6-$Am_2$Ura | 0.03 | 0 | 3 |
| 5,6-$Me_2$Ura | $10^{-4}$ | $5 \times 10^{-5}$ | 3 |
| 5-AmOro | 0.05 | 0 | 3 |
| 5-MeOro | $<10^{-5}$ | $<10^{-5}$ | 3 |
| 6-AmThy | 0.15[b] | 0 | 5 |

[a] References:
1. Brown and Johns (1968)
2. Fisher and Johns (1970)
3. Fisher (1973)
4. Whillans and Johns (1969)
5. Wierzchowski and Shugar (1960)

[b] Aerated, 254 nm.

tuted Ura. This suggests that groups at C(5) influence dimerization not so much by an electronic effect as by sterically hindering the reaction. Substitution of electron-withdrawing groups at C(6) seems to increase the dimer yield, while electron-donating groups reduce it. Disubstituted derivatives have very low dimer yields, probably because of steric hindrance.

Much more work will be required for an understanding of whether substituents affect dimerization by altering intersystem crossing or self-quenching.

## C. Dimerization in Solids

In solution, the yields of Pyr<>Pyr and their isomeric distributions are determined largely by the properties of the electronic excited states of parent compounds. In a solid matrix such as a dry film or a microcrystalline ice, on the other hand, the yields and isomers depend primarily on the spacing and orientation of the monomers in their ground state. Knowledge of crystal geometry, therefore, is important to an understanding of dimer formation. This information is not always available even after x-ray analysis, since often only the conformation of individual molecules is given, not the relationship between neighboring molecules.

### 1. Orotic Acid

Sztumpf and Shugar (1965) found that Oro is resistant to radiation in frozen solution, although it dimerizes efficiently in water. On the other hand, they found that 5-MeOro dimerizes much more readily in ice than in water. It forms two isomers, of which the major one has been shown, by x-ray analysis of the barium salt, to have the *c,a* configuration (P. T. Cheng, V. Miskew, S. C. Nyburg, and Weinblum, unpublished data; Miskew, 1970). The difference in behavior of Oro and 5-MeOro is due to differences in the crystal structure of the two pyrimidines; these structures have not yet been described. The methyl esters of both Oro and 5-MeOro are more reactive in ice than the parent acids (Sztumpf and Shugar, 1965).

Lisewski and Wierzchowski (1970) prepared crystals of Oro and 5-MeOro in a transparent KBr matrix to allow quantum yield determinations which are difficult in ice because of cracking and light scattering. Both bases rapidly reach an equilibrium level when conversion to the photodimer is balanced by photoreversal to monomers. At 275 nm, $>80\%$ of the parent may be converted to dimer, while at 254 nm, because of increased absorption by the dimer at shorter wavelengths, the equilibrium yield is only ~ 50%. One main isomer of Oro◇Oro is formed, with an infrared spectrum which differs from that of the *t,s* dimer produced in solution. The quantum efficiency of dimerization of Oro in the KBr pellet is 0.48; the corresponding value of 5-MeOro is only 0.16. These yields are different from the qualitative results obtained in ice, perhaps because in ice the crystal structures are affected by water of hydration.

2. Thymine and Its Derivatives

Published data on the dimerization of Thy and its N-substituted derivatives in ice, dry films, and KBr pellets, are summarized in Table 6. The influence of the irradiation wavelength on the equilibrium yield of dimers is apparent from this table.

Gerdil (1961) showed that Thy crystallizes from water as a monohydrate, with adjacent Thy molecules stacked in a position which is a suitable precursor for the *c,s* dimer. This in fact is the only isomer formed by Thy in the solid state (Wang, 1963; Weinblum and Johns, 1966; Lisewski and Wierzchowski, 1970). As water is removed from the Thy crystal, the ordered structure collapses, and dimerization is inhibited (Wang, 1963). The range of values for the extent of the reaction in ice (Table 6) probably reflects differences in the amount of water frozen into the Thy microcrystals and in the fraction of molecules which are trapped in sites unsuitable for dimerization.

When any of the four Thy◇Thy isomers is frozen in a glass and split by irradiation at 248 nm, the bases remain fixed in an orientation suitable for reforming the original dimer, and are closer together than monomers in a crystal (~ 2.8 Å apart, as compared with ~ 3.7 Å in crystals). The quantum efficiency for such a remaking of dimers with 280 nm light is close to unity (Eisinger and Lamola, 1969).

At small exposures only the *t,s* dimer is formed in 1-MeThy in the dry state, as would be expected on the basis of its crystal structure (Stewart, 1963; Lisewski and Wierzchowski, 1970). As the irradiation continues, however, the crystal structure breaks down, and *t,a* dimer is also formed (Lisewski and Wierzchowski, 1970), with small amounts of *c,s* and *c,a* (Stepien *et al.*, 1973a). Irradiation of 1-MeThy in frozen solution leads to the formation of both the *c,s* and *t,s* dimers (Blackburn and Davies, 1966a) and eventually the *t,a* isomer (Einstein *et al.*, 1967). The contrast between this result and the findings with the dry solid suggests that some of the microcrystals of 1-MeThy which form in ice have a structure which differs from that of the anhydrous crystals. A second type of structure may be formed by crys-

**Table 6** Dimerization of Thymine Derivatives in the Solid State

| Compound | State | Maximum conversion to dimer (%)[a] | $\phi_d$[a] | Isomers | References |
|---|---|---|---|---|---|
| Thy | KBr | 30 | 0.16 | *cis-syn* | Lisewski and Wierzchowski; 1970 |
| | Ice | 48 | ~0.5–1 | — | Fuchtbauer and Mazur, 1966 |
| | Ice | 60 | — | — | Wulff and Fraenkel, 1961 |
| | Ice | 68 | — | — | Gauri, 1967 |
| | Ice | 85 | — | — | Smietanowska and Shugar, 1961; Smith, 1964 |
| | Ice | 86 | ~0.23 | — | Wang, 1961 |
| | Ice | — | — | *cis-syn* | Weinblum and Johns, 1966 |
| | Solid film | 55 (98% humidity) | — | — | Wang, 1963 |
| | Solid film | 27 (30% humidity) | — | — | Wang, 1963 |
| | Split dimers in glass | — | 1 (280 nm) | — | Eisinger and Lamola, 1969 |
| 3-MeThy | Ice | 5–10 | — | *cis-syn* | Blackburn and Davies, 1966a |
| | KBr | 10 | 0.11 | | Lisewski and Wierzchowski, 1970 |
| 1-MeThy | KBr | 70 (254 nm) | 0.42 | *trans-syn* | Lisewski and Wierzchowski, 1970 |
| | KBr | 90 (>275 nm) | — | | Lisewski and Wierzchowski, 1970 |
| | Ice | 50 | — | *cis-syn*, ~70% *trans-syn*, ~30% | Blackburn and Davies, 1966a |
| | Ice | — | — | *trans-anti* | Einstein *et al.*, 1967 |
| | Solid | — | — | *trans-syn* | Stewart, 1963 |
| | Solid | — | — | *trans-syn* | Weinblum *et al.*, 1968 |
| $Me_2Thy$ | KBr | 40 (254 nm) | 0.16 | *cis-syn* only | Lisewski and Wierzchowski, 1970 |
| | | 90 (>275 nm) | — | | Lisewski and Wierzchowski, 1970 |
| | Ice | 60 | — | 2 isomers | Wulff and Fraenkel, 1961; Wang, 1961 |
| | Frozen cyclohexane | 75 | ~0.1 | 1 isomer only (the higher melting of the ice dimers) | Wang, 1965 |
| | Split dimers in glass | — | ~1 (280 nm) | — | Eisinger and Lamola, 1969 |

| | | | | | |
|---|---|---|---|---|---|
| dThd | Ice | 49 | ~0.16 | — | Wang, 1961 |
| | Ice | 50 | — | — | Gauri *et al.*, 1971 |
| | Ice | — | — | *cis-syn*, 29%<br>*cis-anti* (+), 21%<br>*cis-anti* (−), 43%<br>*trans-anti*, 7% | Weinblum and Johns, 1966 |
| | Ice | — | — | *cis-syn*, 38%<br>*cis-anti* (+), 14%<br>*cis-anti* (−), 34%<br>*trans-anti*, 14% | Ben-Hur *et al.*, 1967 |
| | Ice | 70 | — | *cis-syn*, 21%<br>*cis-anti* (+), 10%<br>*cis-anti* (−), 30%<br>*trans-anti*, 10%<br>Other products, 29% | Varghese, 1971c |
| | Dry film | 49 | — | *cis-syn*, 13%<br>*cis-anti* (+), 12%<br>*cis-anti* (−), 15%<br>*trans-anti*, 9%<br>Other products, 48% | Varghese, 1971c |
| TMP | Ice | 41 | ~0.003 | — | Wang, 1961 |
| Thy + 6-aza-Thy | Ice | 16 | — | — | Günther and Prusoff, 1967 |
| Thy | Ice | 75 | — | *cis-syn* only | Smith, 1966b |
| Thy + Ade (1:1) | Ice | 19 | — | *cis-syn*, 70%<br>*cis-anti*, 30% | |
| 1-MeThy + 9-MeAde (1:1) | KBr | 46 (254 nm)<br>63 (>275 nm) | ~0.15 | *trans-anti* | Lisewski and Wierzchowski, 1970 |

[a] Yields at 254 nm unless noted. Dashes indicate no data.

tallization of 1-MeThy from water (Johnson and Clapp, 1908; Hoogsteen, 1963), and this may give rise to the *c,s* dimer in ice. Unfortunately, the crystal structure of this second form has not been described.

Two isomeric dimers are also formed by the irradiation of $Me_2$Thy in ice (Wang, 1961; Wulff and Fraenkel, 1961), whereas only one is formed in the dry state (Lisewski and Wierzchowski, 1970) and in frozen cyclohexane solution (Wang, 1965). Reformation of split dimers of $Me_2$Thy in glasses is highly efficient (Eisinger and Lamola, 1969).

Thd forms four dimers, including two optical isomers of the *c,a* dimer, in both dry films and ice, but the relative yields are somewhat different in the two cases (Weinblum and Johns, 1966; Varghese, 1971c). The low efficiency of TMP dimerization in ice (Wang, 1961), is consistent with the crystal structure of TMP (Trueblood *et al.*, 1961), in which the orientation of the bases is not suitable for forming dimers.

Knowledge of the second crystal structure of 1-MeThy, the $Me_2$Thy structure in ice, and the possible crystal structures of Thd is necessary for a full interpretation of these results.

Hydrogen-bonded complexes of Thy and Ade, in ice, are oriented in such a way as to give the *c,a* dimer of Thy as well as *c,s* (Smith, 1966b), and 1-MeThy:9-MeAde complexes yield only the *t,a* dimer of 1-MeThy (Lisewski and Wierzchowski, 1970) in agreement with the known antiparallel stacking in such crystals (Stewart and Davidson, 1963). The reduction in quantum yield on the formation of these complexes (Table 6) is due to an increase in the separation of the 1-MeThy residues, from 3.7 Å in crystals of 1-MeThy alone to 4.8 Å in crystals of 1-MeThy:9-MeAde (Stewart, 1963). An addition of the radioprotector 6-aza-Thy to frozen Thy solutions also leads to a reduction in the Thy◇Thy yield. This may be due to a disruption of the crystal structure or to energy transfer from Thy to 6-aza-Thy (Günther and Prusoff, 1967).

Dimerization may be completely inhibited by the addition of 5% ethanol to 0.1 m*M* Thy solutions before freezing, because ethanol prevents the formation of crystals, leaving the Thy molecules isolated and unable to dimerize (Beukers *et al.*, 1959b; Smith, 1963).

### 3. Uracil and Its Derivatives

Table 7 summarizes available data on the dimerization of Ura derivatives in the solid state.

**Table 7** Dimerization of Uracil Derivatives in the Solid State

| Compound | State | Maximum conversion to dimer (%)[a] | $\phi_d$[a] | Isomers[a,b] | References |
|---|---|---|---|---|---|
| Ura | Ice | 68 | ~0.005 | — | Wang, 1961 |
| | Ice | 58 | — | — | Smietanowska and Shugar, 1961 |
| | Ice | 69 | — | 2 isomers | Smith, 1963 |
| | Ice | 64 | — | — | Gauri, 1967 |
| | Ice | — | — | *cis-syn* | Adman *et al.*, 1968<br>Dönges and Fahr, 1966 |
| | Ice | — | — | *cis-syn* and a small amount of *cis-anti* | Sasson *et al.*, 1970 |
| | Ice | — | — | *cis-syn*, 90%<br>*cis-anti*, 10% | Varghese, 1971b |
| | Dry film | — | — | *cis-syn* | Wang, 1963 |
| $Me_2$Ura | Ice | 25 | ~0.013 | 2 isomers | Wang, 1961 |
| | Ice | — | — | 4 isomers | Sasson *et al.*, 1970 |
| | Ice | — | — | *cis-syn*, 24%<br>*cis-anti*, 61%<br>*trans-syn*, 6%<br>*trans-anti*, 9% | Fürst *et al.*, 1967<br>Fahr *et al.*, 1972 |
| Urd | Ice | 21 | ~0.0019 | — | Wang, 1961 |
| | Ice | — | — | *cis-syn*, *cis-anti* (+) | Fahr *et al.*, 1967 |
| | Ice | 15 (and 3% other products) | — | *cis-syn*, *cis-anti* (+) } 71%<br>*cis-anti* (+), *cis-anti* (−) } 9%<br>*trans-anti*, 20% | Varghese, 1971b |
| | Dry film | 6 (and 6% other products) | — | *cis-syn*, *cis-anti* (+) } 88%<br>*cis-anti* (−), 12%<br>*trans-anti* 0 | Varghese, 1971b |
| 3′-UMP | Ice | 40 | ~0.0022 | — | Wang, 1961 |
| | Ice | — | — | — | Schuster, 1964 |

[a] Irradiation at 254 nm. Dashes indicate no data.
[b] (+) and (−) refer to optical rotation.

Ura itself forms two isomeric dimers in ice, with the *c,s* form accounting for 90% of the total (Varghese, 1971b). The remaining 10% is made up of the *c,a* dimer, which generally escaped detection in the earlier work because of its low yield and its instability to heat and extremes of pH. From the crystal structure of Ura (Parry, 1954), one would expect to obtain the *c,s* dimer; the second isomer may arise from imperfections, from crystal collapse during the course of the irradiation, from a second form of crystal, or by way of an intermediate photoproduct. Until detailed studies are made of wavelength and dose effects on dimer yields, no firm conclusions can be drawn. Since Ura, unlike Thy, does not include water in its crystals, dimerization in solid films is not affected by humidity (Wang, 1963). On the other hand, the results for Urd in ice and in dry films suggest that water of hydration may be essential for forming the *t,a* dimer of the nucleoside. The predominance of $Me_2$Ura◇$Me_2$Ura(*c,a*) in ice (Table 7) is in marked contrast to the low yield of this isomer in Ura and Urd. Information on the nucleotides is woefully lacking.

### 4. Cytosine and Its Derivatives

Compared to Ura and Thy, the derivatives of Cyt dimerize very inefficiently in the solid state (Günther and Prusoff, 1967; Fahr, 1969). However, Weinblum (1961) observed Ura◇Ura after the photolysis of Cyt in ice; it resulted from the spontaneous deamination of Cyt◇Cyt. Varghese and Rupert (1971) showed that 7% of Cyd and dCyd can be converted to dimers upon irradiation in ice at 254 nm.

Following deamination and acid hydrolysis to remove the sugars, the Cyd◇Cyd gives a product identical to Ura◇Ura(*c,s*) (Varghese, 1971a), suggesting that the Cyd◇Cyd itself has the *c,s* conformation. However, it may be that other, less stable dimers are lost during the hydrolysis and, in the light of different findings from Wang's laboratory, more work is required in this area.

### 5. Other Pyrimidine Derivatives

Table 8 summarizes the dimerization results obtained with derivatives of Ura in ice. Both 5-EtUra and 5-PrUra seem to be more radiation resistant than their 1-methylated analogs, probably because of differences in crystal structure. The 6-chloro derivatives of 5-alkyluracils do not dimerize in ice (Gauri, 1967). 6-MeUra forms exclusively the *c,s* dimer in the solid state, whereas in solution it forms only the other three isomers (Khattak and Wang, 1972).

**Table 8** Dimerization of Substituted Pyrimidines in Ice

| Compound | Maximum conversion to dimers (%) | Isomers | References |
|---|---|---|---|
| 5,6-$Me_2$Ura | 70 | — | Smietanowska and Shugar, 1961 |
| 1,3,5,6-$Me_4$Ura | 54 | — | Smietanowska and Shugar, 1961 |
| 1-Thia-Ura | 11 | *cis-syn* | Bremner *et al.*, 1971 |
| 6-MeUra | — | *cis-syn* | Konnert *et al.*, 1970 |
| | 60 | — | Wang, 1961 |
| 5-EtdUrd | $\frac{1}{5}$ the yield of Thd | — | Gauri *et al.*, 1971 |
| | Same rate and yield as Thd | — | Pietrzykowska and Shugar, 1970 |
| 5-EtUra | $\frac{1}{8}$ the rate of Thy | — | Pietrzykowska and Shugar, 1970 |
| 1-Me-5-EtUra | 30 | 3 isomers | Pietrzykowska and Shugar, 1970 |
| | 36 | — | Gauri, 1967 |
| 1-Me-5-PrUra | 38 | — | Gauri, 1967 |
| 5-PrUra | 0 | — | Gauri, 1967 |
| 5-PrdUrd | 0 | — | Gauri *et al.*, 1971 |
| 5-hmUra | $\frac{1}{6}$ the efficiency of Thy | — | Kaláb, 1966 |

### 6. Heterodimers

The irradiation of a frozen solution of Thy and Ura in equal concentrations leads to the production of the *c,s* dimers of Thy and Ura, and one isomer of Ura$\diamond$Thy, probably also the *c,s* form (Beukers and Berends, 1960b; Weinblum, 1967, Fahr *et al.*, 1974). Four isomers of Ura$\diamond$Thy may be detected after hydrolysis of Urd$\diamond$Thd dimers from ice (Fahr, 1969). One heterodimer of $Me_2$Ura$\diamond$$Me_2$Thy has been obtained in purified form, and additional products made in irradiated frozen solutions of $Me_2$Ura and $Me_2$Thy may be other isomeric dimers (Wang, 1965). At least one Thy$\diamond$Ura is observed after the photolysis of frozen mixed solutions of Thy and Cyt (Smith, 1963) and from hydrolysis of the ice photoproducts of Thy and Cyd (Varghese, 1971a).

## D. Dimerization in Oligo- and Polynucleotides

Absorbance changes are an unreliable measure of photoproduct yields in polynucleotides, particularly in double-stranded polymers, because of hyperchromic effects. The absorption per base may be as much as 30% less, at the peak wavelength, in a double-stranded helix than in denatured single strands, while at the long-wavelength edge, the absorbance may be slightly higher (Szer *et al.*, 1963). Since dimer formation disrupts the helix, causing denaturation of ~4.3 base pairs (Hayes *et al.*, 1971), the decrease in peak absorbance from the loss of two bases will be accompanied by a 30% increase in the absorption of $2(4.3 - 1) = 6.6$ bases. Thus, irradiation of a polymer may cause an initial increase in absorbance or a lag in loss of absorbance (DeBoer *et al.*, 1967; Tramer *et al.*, 1969). The problem is less serious in single-stranded helices. The difficulty may be partially circumvented by denaturing the helix completely before monitoring the absorbance, but when possible it is preferable to analyze for products more directly. The mild enzymatic hydrolysis of polyribonucleotides is ideal for this purpose (Pearson and Johns, 1966), but a comparable procedure has not been developed for polydeoxyribonucleotides.

Apart from the effect on assaying for products, the spectral changes caused by base stacking may lead to difficulties in determining quantum yields. In dinucleotides such as UpU, only a small percentage of the bases may be stacked. In these molecules the long-wavelength absorption will be greater than the absorption of the more stretched dinucleotides but since their numbers are small, the overall effect on absorbance of the solution will be negligible. Nevertheless, these stacked dinucleotides have a higher probability of forming dimers. By using the smaller average absorption cross section rather than the larger cross section of the stacked bases, the quantum yield at long wavelengths will be overestimated. This may account for the apparent increase in quantum yield in polymers and dinucleotides at long wavelengths (Fig. 8a).

Energy transfer is important in polymers and affects the calculations of quantum yields. Sutherland and Sutherland (1970) demonstrated that singlet excitation energy from photon absorption by one base in DNA may be transferred four bases along the chain, while triplet excitation may be transferred more than 15 bases. In either case, adjacent thymines may dimerize even when neither of them absorbs light directly. Neglecting energy transfer will cause an overestimate of quantum yields.

Despite these reservations, a number of the reported photochemical studies in the literature have provided useful information about dimerization in oligo- and polynucleotides. Many of these findings are summarized in Table 9 and Fig. 8.

Mathematical treatments of dimer formation in dinucleotides (Brown and Johns, 1967) and polynucleotides (Johns *et al.*, 1966) are presented elsewhere.

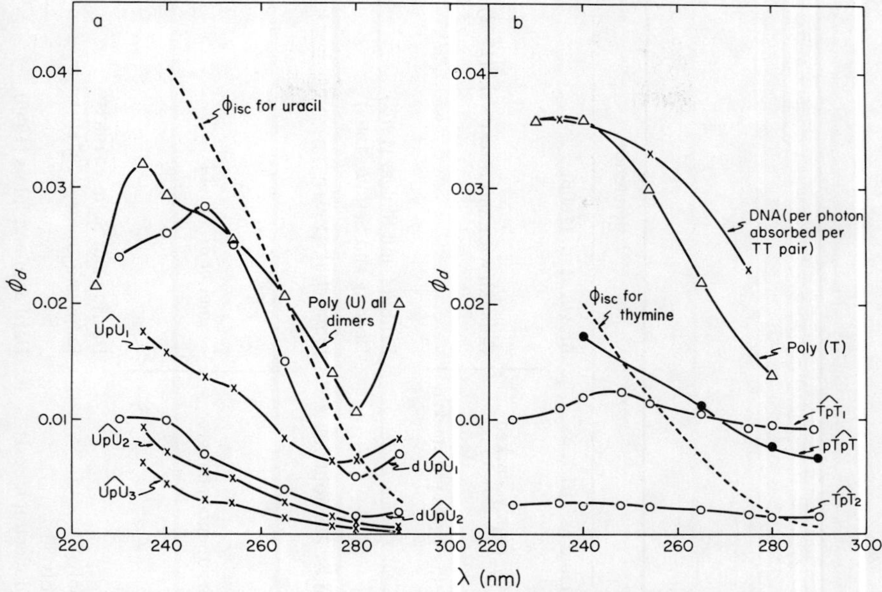


**Fig. 8.** *(a) Quantum yields plotted as a function of wavelength for two dimers of dUpU* (Helleiner et al., *1963), three dimers of UpU* (Brown et al., *1966), and poly(U)* (Pearson et al., *1966). The intersystem crossing efficiency of Ura, as determined by reactions with cysteine, is shown as a dashed line* (Fisher et al., *1974). (b) Quantum yields for dimer formation of TpT* (Johns et al., *1964), pTpT, poly(T) (Deering and Setlow, 1963), and Thy in DNA (Wulff, 1963), with $\phi_{isc}$ for Thy shown as a dashed line* (Fisher et al., *1974).*

In dinucleotides such as TpT and dUpU only two isomeric dimers, *c,s* and *t,s* are formed. In UpU, *three* dimeric products are resolved (Fig. 8a, Table 9). Two of these may be optical isomers of the *t,s* dimer, but the structures have not been determined. "Abbreviated" dinucleosides, consisting of two bases joined to the same sugar, form primarily the *c,s* dimer with a good yield of *t,s* in the dinucleoside containing Thy and Ura (Logue and Leonard, 1972).

In synthetic dinucleotides joined by three methylene groups instead of by a sugar–phosphate linkage, the *c,s* dimer predominates (Leonard *et al.*, 1969). The bases in such methylene-bridged compounds are stacked, and they dimerize quite efficiently with no dependence of yield on wavelength or $O_2$. When the bridge consists of 5 methylene groups, the bases are too far apart to stack, so that dimerization takes place with a lower efficiency, is quenched by oxygen, and is wavelength dependent (Table 9; Golankiewicz and Strekowski, 1972).

**Table 9** Dimer Yields in Oligo- and Polynucleotides

| Material | λ (nm) | $\phi_d$ | Comments | Reference |
|---|---|---|---|---|
| Thy-$(CH_2)_3$-Thy | — | — | "3.5 times as efficient as TpT" | Brown *et al.* (1968) |
| | 254 | 0.036 | No effect of λ or $O_2$; highly stacked | Golankiewicz and Strekowski (1972) |
| | 275 | 0.040 | | |
| Thy-$(CH_2)_3$-Ura | 254 | 0.022 | | |
| | 275 | 0.024 | | |
| Thy-$(CH_2)_5$-Thy | 254 | 0.004 | $O_2$ and λ affect yields; bases not stacked | |
| | 275 | 0.002 | | |
| pTpT | 254 | 0.01 | $\phi_d$ same at pH 2 and 7 | Sztumpf and Shugar (1962) |
| | 265 | 0.011 | | Deering and Setlow (1963) |
| TpT | 265 | 0.010, *cis-syn* | Products resolved by electrophoresis | Johns *et al.* (1964) |
| | | 0.002, *trans-syn* | | |
| TpTp | 254 | 0.011 (16°C) | Determined by absorbance changes | Tramer *et al.* (1969) |
| | | 0.008 (40°C) | | |
| Poly(rT) | 254 | 0.044 (16°C) | | |
| | | 0.022 (40°C) | | |
| Poly(T) | 254 | 0.033 (16°C) | | |
| | | 0.024 (40°C) | | |
| Oligo(rT) | 254 | 0.024 (16°C) | | |
| Poly(T) | 265 | 0.022 | Absorbance | Deering and Setlow (1963) |
| Poly(d-dT) | 254 | — | $\frac{1}{5}$ the "rate" for poly(T) | Rahn and Landry (1971) |
| | >300, acetone | — | Same "rate" as poly(T) | |
| CpC | 265 | 0.015, $\widehat{CpC}_1$ | Resolved by rapid electrophoresis | Hariharan and Johns (1968) |

| | | | | |
|---|---|---|---|---|
| dUpU | 265 | 0.015, $\widehat{dUpU}_1$ | Two dimers resolved | Helleiner *et al.* (1963) |
| | | 0.004, $\widehat{dUpU}_2$ | | |
| UpU | 265 | 0.0083, $\widehat{UpU}_1$ | Three dimers resolved | Brown *et al.* (1966) |
| | | 0.0027, $\widehat{UpU}_2$ | | |
| | | 0.0013, $\widehat{UpU}_3$ | | |
| Poly(U) | 265 | 0.006 | | Swenson and Setlow (1963) |
| | | 0.0056 | Isolated dimers | Pearson *et al.* (1966) |
| | | 0.0205 | Dimers in sequences | Pearson *et al.* (1966) |
| Poly(U) + spermine | 280 | 0.0083 (1°C) | | DeBoer *et al.* (1967) |
| | | 0.0067 (35°C) | | |
| | 235 | 0.0098 (1°C) | | DeBoer *et al.* (1967) |
| | | 0.0220 (35°C) | | |
| Poly(A + U) | 280 | 0.0009 | Isolated dimers | Pearson *et al.* (1966) |
| Poly(A + 2U) | | 0.0016 | | DeBoer *et al.* (1967) |

These findings, which illustrate the importance of conformation on product yields and on the mechanism of formation, may be interpreted in terms of a singlet precursor when bases are stacked and in terms of a triplet precursor when they are not.

Work with poly(d-dT) and poly(T) confirms this conclusion. Poly(d-dT) is formed from the alternating copolymer poly(dAT) by depurination. The Thy residues are then relatively isolated, and they dimerize much less efficiently than the bases in poly(T) since they cannot interact with a neighbor in the singlet lifetime ($10^{-12}$ sec). When the long-lived triplet state is populated by sensitization, the bases can interact and dimerize as efficiently in poly(d-dT) as in poly(T) (Table 9; Rahn and Landry, 1971).

In polynucleotides, secondary structure has marked effects on dimer yields. Single-stranded poly(U), when ordered into a helix by spermine at 1°C, shows no wavelength dependence of quantum yield. When converted to a random coil at 35°C, the yield is much greater at 235 nm than at 280 nm, again suggesting the importance of a triplet precursor in unstacked configurations (Table 9; DeBoer *et al.*, 1967). Interestingly, the quantum yield in more rigid double and triple helices is reduced by an order of magnitude below that in the more flexible single helix because the bases are not free to reorient themselves to form dimers (Table 9; Pearson *et al.*, 1966). These general features are clear, although the precise interpretation of the results with double and triple helices remains in some doubt because of problems in forming homogeneous populations of the copolymers (for a review, see Lomant and Fresco, 1972).

Apparently, the formation of a dimer causes local denaturation of the rigid helix (Zavil'gel'skii *et al.*, 1965) and enhances the probability that a second dimer will form next to the first (Pearson and Johns, 1967). Besides the increase in yield when the helix relaxes, this enhancement may be related to a mechanism of energy transfer along the polynucleotide chain: the presence of a dimer prevents further transfer, leaving the adjacent base excited and likely to dimerize (Pearson and Johns, 1967).

By the use of the triplet-state sensitizer, acetophenone, and near-UV light (> 300 nm), it is possible to populate exclusively the Thy triplet in DNA; Thy◇Thy(*c*,*s*) is the major product, along with small amounts (< 3%) of Cyt◇Thy, hThy, and Thy–acetophenone addition products (Lamola and Yamane, 1967; Lamola, 1969; Meistrich *et al.*, 1970). When acetone is used as a sensitizer, the Cyt triplet may also be excited, and some Cyt◇Cyt is formed (Ben Ishai *et al.*, 1968). Pyr◇Pyr can even be produced in DNA in the dark by the thermal decomposition of trimethyl-1,2-dioxetane to give triplet acetone, which in turn sensitizes dimer formation in the same way as UV-excited acetone (Lamola, 1971). Benzophenone sensitizes the formation of Thy◇Thy and strand breaks in DNA (Charlier *et al.*, 1972).

## E. Properties of Pyrimidine Photodimers

### 1. Crystal Properties

Table 10 summarizes much of the available information on crystal form, melting point, and density for Pyr◇Pyr. The literature on the subject is extensive; there are, for example at least nineteen published values for melting points of the various $Me_2$Thy◇$Me_2$Thy. In many cases, the isomeric form of the dimer is unknown or unspecified; however, the numerical values correspond well to the melting points quoted for particular isomers in later papers. As a result, only representative values are presented in Table 10. An exception is made for the *c,s*- and *t,a* forms of Thy◇Thy, for which large discrepancies exist.

"Crystal form" refers only to gross appearance; x-ray determinations of crystal and molecular structure are reviewed in Chapter 11. Densities, however, are generally calculated from the unit cell dimensions obtained using x-rays, rather than measured directly. The higher density of Ura◇Ura(*c,a*) is attributed to the formation of eight hydrogen bonds by each dimer molecule (Konnert *et al.*, 1970).

An interesting example of the effect of methylation on crystal stability is afforded by the data for dimers of Ura, Thy, $Me_2$Ura, and $Me_2$Thy. In general, the N-methylated dimers melt at lower temperatures than the corresponding dimers of Ura and Thy, with the single exception of the Thy◇Thy(*t,s*). Methylation of C(5) also has an effect on crystal stability: of the $Me_2$Ura◇$Me_2$Ura dimers the *c,s* isomer has the lowest melting point, while for $Me_2$Thy◇$Me_2$Thy, the *c,a* isomer forms the least stable crystals.

### 2. Dipole Moments

Dipole moments of dimers have been measured for $Me_2$Thy and $Me_2$Ura, and have been calculated for $Me_2$Thy and Thy. The calculations are based on the assumptions that the cyclobutane ring is a square and that the bases are planar and are oriented with their N(1)–C(4) axes parallel. In fact the rings are not flat, and the bases are typically rotated by $\sim 28°$ with respect to each other (Camerman and

**Table 10** Crystal Properties of Pyrimidine Dimers

| Dimer | Crystal form | Melting point (°C) | Density (gm cm$^{-3}$) | References |
|---|---|---|---|---|
| Thy◇Thy | | | | |
| Ice dimer(*c,s*) | White needles | 320° (sublimes, reverts to T) | 1.56 | Wulff and Fraenkel (1961); Beukers and Berends (1961) |
| From dry film (IR spectrum same as ice dimer) | | 242° (decomposes) | — | Ishihara (1963) |
| *trans-anti* | | 240° (reverts to T) | — | von Wilucki *et al.* (1967) |
| *trans-anti* | Colorless prisms | >300° | — | Kunieda and Witkop (1971) |
| *trans-syn* | | 190° (sublimes) | — | von Wilucki *et al.* (1967) |
| 1-MeThy◇1-MeThy | | | | |
| *cis-syn* | White | 330° (decomposes) | — | Blackburn and Davies (1966a) |
| *trans-syn* | White | 330° (decomposes) | — | Blackburn and Davies (1966a) |
| *trans-syn* | Elongated prisms | — | — | Einstein *et al.* (1967) |
| 3-MeThy◇3-MeThy | | | | |
| *cis-syn* | White | 320° (sublimes) | — | Blackburn and Davies (1966a) |
| $Me_2$Thy◇$Me_2$Thy | | | | |
| *cis-syn* | White hexagonal right prisms | 256° | — | Wulff and Fraenkel (1961); Kleopfer and Morrison (1972) |
| *trans-syn* | White | 255° | — | Blackburn and Davies (1966a); Kleopfer and Morrison (1972) |
| *cis-anti* | Parallelepipeds | 229° | 1.294 | Wulff and Fraenkel (1961); Kleopfer and Morrison (1972); Camerman *et al.* (1969) |

| | | | | |
|---|---|---|---|---|
| *trans-anti* | White | 261°–262° | — | Kleopfer and Morrison (1972) |
| Thd◇Thd | | | | |
| *cis-syn* | Colorless | 184°–188° | — | Kunieda and Witkop (1971) |
| *cis-anti* (−) | — | 150° | — | Witkop (1968) |
| 6-MeUra◇6-MeUra | — | — | 1.462 | Konnert *et al.* (1970) |
| Ura◇Ura | | | | |
| *cis-syn* | Two crystal forms with differing IR spectra (prisms and needles) | 308° (decomposes) | 1.605 | Varghese (1971b); Jennings *et al.* (1972); Smietanowska and Shugar (1961); Adman and Jensen (1970) |
| *trans-syn* | — | 320° (decomposes) | — | Richter and Fahr (1969) |
| *cis-anti* | — | — | 1.704 | Konnert *et al.* (1970) |
| *trans-anti* | — | 300° (decomposes) | — | Richter and Fahr (1969) |
| $Me_2$Ura◇$Me_2$Ura | | | | Elad *et al.* (1971) |
| *cis-syn* | — | 242°–244° | — | |
| *trans-syn* | — | 256°–258° | — | |
| *cis-anti* | — | 260°–261° | — | |
| *trans-anti* | — | 270°–272° | — | |
| Urd◇Urd | | | | |
| *cis-syn* | Colorless | 178°–179° | — | Fahr *et al.* (1967) |
| Unspecified isomer | — | 247°–250° | — | Rosenthal and Elad (1968a) |
| Cyd◇Cyd | | | | |
| *trans-anti* | Hydrated crystal | 150°–155° | — | Varghese (1972a) |
| MeOrotate | | | | |
| *trans-syn* | — | 195–210° (reverts to parent) | — | Birnbaum *et al.* (1971) |
| Oro◇Oro | | | | |
| *trans-syn* | White | 180° (reverts to OA) | — | Birnbaum *et al.* (1971); Sztumpf and Shugar (1965) |
| 5-MeOro◇5-MeOro | | | | |
| *cis-anti* | Hydrated crystal | 260°–265° (decomposes to T and $CO_2$) | — | Miskew (1970); Sztumpf and Shugar (1965) |

**Table 11** Dipole Moments of Dimers

| Parent | Comments | Dipole moment (debye) | | | | Reference |
|---|---|---|---|---|---|---|
| | | *cis-syn* | *trans-syn* | *cis-anti* | *trans-anti* | |
| $Me_2Thy$ | Measured in benzene | 6.04 | 2.79 | 5.75 | — | Weinblum *et al.* (1968) |
| $Me_2Thy$ | Calculated | 6.1–6.8 | 3.1–3.5 | 5.1–5.6 | 0.0 | Weinblum *et al.* (1968) |
| Thy | Calculated | 6.34 | 3.25 | 5.33 | 0.00 | Jordan and Pullman (1968) |
| $Me_2Ura$ | Measured in dioxane | 6.33 | 3.35 | 6.07 | 1.19 | Sasson *et al.* (1970) |

Camerman, 1968). Measured and calculated values are nevertheless in reasonable agreement, as shown in Table 11. No value is available for $Me_2Thy$◇$Me_2Thy$(*t*,*a*) because of difficulties in obtaining a pure product; its dipole moment is expected to be zero because the molecule possesses a center of symmetry.

### 3. Separation of Dimers

#### *a. Column Chromatography*

The literature contains many descriptions of the separation of dimers by column chromatography. Representative of these are a DEAE cellulose column for 2′(3′)-UMP (Zachau, 1964) and Urd (Fahr *et al.*, 1967), and Dowex columns for 3′-UMP (Schuster, 1964), dThd (Weinblum and Johns, 1966; Ben-Hur *et al.*, 1967), Thy (Jennings *et al.*, 1970), and Ura and Urd (Varghese, 1971b). Using column techniques it has been possible to resolve the optical isomers of dThd◇dThd(*c*,*a*) (Weinblum and Johns, 1966).

#### *b. Thin Layer and Paper Chromatography*

Reports of paper and thin layer chromatographic separations are extensive. For example, $R_f$ values have been presented for the dimers of Oro and five derivatives in eight solvent systems (Sztumpf and Shugar, 1965). A few of these results are summarized in Table 12; $R_f$ values of the parent molecules are given in parentheses.

In general, dimers are chromatographically less mobile than monomers. An indication of the degree of reproducibility which can be obtained in different laboratories is provided by the results for Ura and its *c*,*s* dimer on cellulose thin layers; the agreement is fairly good, but is not exact. The medium has an important influence on mobilities, as was shown for the (*c*,*s*)- and (*c*,*a*) forms of 6-MeUra◇6-MeUra by Khattak and Wang (1972). Using the solvent *sec*-butanol/water, 50:20, $R_f$ values were obtained for the two dimers of 0.38 and 0.48 on cellulose thin layers, and 0.23 and 0.32 on Whatman paper using the descending technique. From this it is clear that $R_f$ values are meaningless unless the conditions are specified.

An unusual feature of the results in Table 12 is that Ura◇Ura(*t*,*a*) is reported to have the smallest $R_f$ of the Ura◇Ura in several solvents (Varghese, 1971b). In contrast the Thy◇Thy(*t*,*a*) and the dimers of 5-EtdUrd and 6-MeUra have large $R_f$ values. The Ura◇Ura may be peculiar in their chromatographic behavior, or the identification of the dimers may be incorrect.

## 4. Solubility

Almost no quantitative information is available concerning the solubilities of Pyr◇Pyr. The *c*,*s* dimers of Ura and Thy are less soluble in water than are the parent compounds (Wang, 1961). The limit of solubility of the Ura◇Ura in water is 0.1 g/liter at 100°C (Smietanowska and Shugar, 1961), while several grams of Ura will dissolve under the same conditions. The Thy◇Thy(*t*,*a*) is insoluble in acetone (Jennings *et al.*, 1970). The dimer formed by 3-MeOro is soluble in water and alcohols but insoluble in ether, chloroform, and other nonpolar solvents (Sztumpf and Shugar, 1965). At pH 11.8, the Thy◇Thy(*c*,*s*) is soluble in water to the extent of 0.04 *M*, or 10 g/liter, at room temperature (Herbert *et al.*, 1969).

## 5. Optical Properties

### a. *Optically Active Forms*

Two of the photodimers of Thy, the *c*,*a* and *t*,*s* isomers, exist in optically active forms (Fig. 1a). For Thy itself it is not possible to separate these optical isomers from a racemic mixture, but for the nucleoside they can be resolved chromatographically. After mild acid hydrolysis of the glycosyl bonds, the corresponding Thy◇Thy is isolated. Weinblum *et al.* (1968) methylated one enantiomer of Thy◇Thy(*t*,*s*) and measured its optical rotation, obtaining a value of $[\alpha]_{D,20} = -136°$. Kunieda and Witkop (1971) showed that the two optical isomers of the Thy◇Thy(*c*,*a*) have $[\alpha]_{D,25} = +94°$ and $-92°$.

### b. *Ultraviolet Absorption Spectra*

UV absorption spectra of Thy◇Thy have been reported (Wulff and Fraenkel, 1961; Setlow, 1961; Johns *et al.*, 1962; Ishihara, 1963).

**Table 12** Chromatographic Mobilities of Pyrimidine Dimers in Solvents A, B, C, D, and E[a,b]

| Parent | Medium | Dimer | A | B | C | D | E | References |
|---|---|---|---|---|---|---|---|---|
| Oro | Whatman #1, ascending | *trans-syn* | | 0.24 (0.40)[c] | 0.0 (0.24) | | | Sztumpf and Shugar (1965) |
| 3-MeOro | | – | | 0.46 (0.57) | 0.37 (0.50) | | | |
| 5-MeOro | | Ice dimer (*cis-anti*) | | 0.34 (0.43) | 0.10 (0.33) | | | |
| Ord | | – | | 0.12 (0.27) | 0.12 (0.32) | | | |
| dThd | Whatman 3 MM | *cis-syn* | | 0.56[d] | | | | Ben-Hur *et al.* (1967) |
| | | *cis-anti* | | 0.48 | | | | |
| | | *trans-syn* | | 0.58 | | | | |
| | | *trans-anti* | | 0.53 | | | | |
| Thy | Whatman #1, decending | *cis-syn* | 0.14 | 0.24 (.59) | | | | Wacker *et al.* (1961b); Smith (1963) |
| Thy◇Ura | | *cis-syn* | 0.08 | 0.19 | | | | |
| Ura | | *cis-syn* | 0.02 (0.36) | 0.12 (.47) | | | | |
| Thy | Whatman 3 HR | *cis-syn* | 0.12 | | 0.49 | | | Weinblum and Johns (1966) |
| | | *cis-anti* | 0.17 | | 0.56 | | | |
| | | *trans-syn* | 0.24 | | 0.62 | | | |
| | | *trans-anti* | 0.24 | | 0.62 | | | |
| 1-MeThy | Cellulose thin layer | *cis-syn* | | 0.47 (0.74) | 0.64 (0.77) | 0.63 | | Stepien *et al.* (1973a) |
| | | *cis-anti* | | 0.62 | 0.80 | 0.75 | | |
| | | *trans-syn* | | 0.56 | 0.76 | 0.70 | | |
| | | *trans-anti* | | 0.70 | 0.72 | 0.73 | | |
| Thy | Cellulose thin layer | *cis-syn* | | 0.46 (.77) | | | | Fisher and Johns (1970) |
| | | *cis-anti* | | 0.51 | | | | |
| | | *trans-syn* | | 0.60 | | | | |
| | | *trans-anti* | | 0.55 | | | | |
| 5-EtdUrd | Whatman #1 | *cis-syn* | 0.30 (0.70)[e] | | | | | Pietrzykowska and Shugar (1970) |
| | | *cis-anti* | 0.40 | | | | | |
| | | *trans-syn* | 0.73 | | | | | |
| | | *trans-anti* | 0.50 | | | | | |

| | | | | | | | | |
|---|---|---|---|---|---|---|---|---|
| 6-MeUra | Whatman paper, descending | *cis-syn* | | 0.25 (.54) | | | | Khattak and Wang (1972) |
| | | *cis-anti* | | 0.29 | | | | |
| | | *trans-syn* | | 0.33 | | | | |
| | | *trans-anti* | | 0.33 | | | | |
| Ura | Cellulose thin layer | *cis-syn* | | 0.22 (.50) | 0.21 (0.54) | 0.31 (0.65) | 0.24 (0.62) | Varghese (1971b) |
| | | *cis-anti* | | 0.24 | 0.25 | 0.34 | 0.29 | |
| | | *trans-anti* | | 0.20 | 0.20 | 0.25 | 0.20 | |
| Ura | Cellulose thin layer | *cis-syn* | | | | 0.24 (0.59) | 0.20 (0.56)[f] | Greenstock and Johns (1968) |
| Ura | Cellulose thin layer | *cis-syn* | | | | 0.30 (0.66) | 0.20 (0.58) | Adman *et al.* (1968) |
| Ura | Silica gel thin layer | *cis-syn* | 0.18[e] | | | 0.47 | | Fahr *et al.* (1967) |

[a] Solvents: A, *n*-butanol/water (86:14); B, *n*-butanol/acetic acid/water (80:12:30); C, isopropanol/ammonium hydroxide/water (7:1:2); D, *n*-propanol/water (7:3); E, isopropanol/water (7:3).

[b] Numbers in parentheses give the $R_f$ of the parent.

[c] Solvent proportions 5:2:3.

[d] Solvent proportions 74:19:50.

[e] *n*-Butanol saturated with water.

[f] Solvent proportions 3:1.

The most careful study was carried out by Herbert and co-workers (1969), who measured extinction coefficients for the four isomeric Thy◇Thy and corrected for the thermal instability of the *t,a* dimer. One of their spectra, for the Thy◇Thy(*c,s*), is replotted in Fig. 2a. Their results for all four isomers are presented in Table 13. From these values it is evident that the four structurally similar species have spectra which are similar but not identical. The extinction coefficient of the *t,s* dimer is the greatest at wavelengths above 240 nm, but not at short wavelengths.

The same general end-absorption spectrum also applies to dimers of Ura (Smietanoswka and Shugar, 1961; Swenson and Setlow, 1963) and to the dimers of 2′(3′)-UMP (Zachau, 1964), 1-MeThy (Stepien *et al.*, 1973a), $Me_2$Ura (Fahr *et al.*, 1972), and several derivatives of Oro (Sztumpf and Shugar, 1965), as would be expected for 5,6-saturated pyrimidines, **VIII.** The Cyd◇Cyd(*c,s*), on the other hand, has the absorption shoulder at 240 nm (Varghese, 1971a) which is typical of 5,6-saturated Cyt derivatives (Janion and Shugar, 1960; Brown and Hewlins, 1968) because of the conjugated double bonds (**IX**) (Scheme 2). When N(3) is protonated at low pH, this absorption disappears (Varghese and Rupert, 1971).

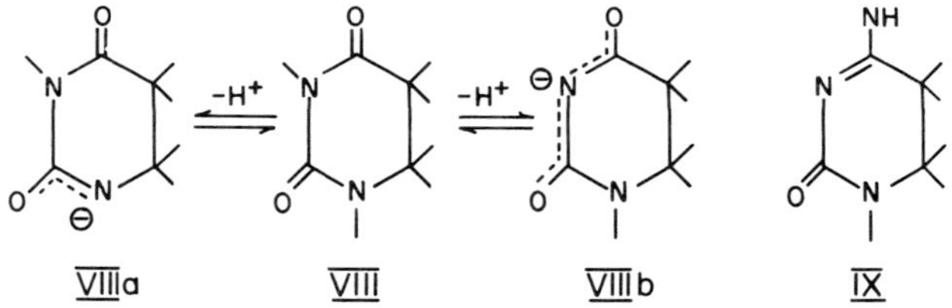

**Scheme 2**

In alkali, dimers of Ura◇Ura and Thy◇Thy (Smietanowska and Shugar, 1961; Ishihara, 1963), of Oro, Ord, and 5-MeOro (Sztumpf and Shugar, 1965), and of 1-MeThy (Blackburn and Davies, 1966a) exhibit a strong absorption band between 230 and 240 nm because of ionization at N(3) to give the conjugated species **VIIIb.** In contrast, the dimers of 3-MeThy (Blackburn and Davies, 1966a) and 3-MeOro (Sztumpf and Shugar, 1965) lack this absorption in alkali since they form, on ionization at N(1), the species **VIIIa,** which has much less conjugation.

*c. Infrared Spectra*

Infrared (IR) spectra are frequently used to distinguish among the various dimers. Numerous workers have published incomplete data,

**Table 13** Extinction Coefficients of Thy◇Thy at pH 7[a]

| λ (nm) | *cis-syn* | *trans-syn* | *cis-anti* | *trans-anti* |
|---|---|---|---|---|
| 200 | $1.9 \times 10^4$ | $2.0 \times 10^4$ | $1.7 \times 10^4$ | $2.2 \times 10^4$ |
| 210 | $7.6 \times 10^3$ | $8.5 \times 10^3$ | $7.9 \times 10^3$ | $9.8 \times 10^3$ |
| 225 | $3.0 \times 10^3$ | $3.6 \times 10^3$ | $4.2 \times 10^3$ | $5.2 \times 10^3$ |
| 230 | $2.1 \times 10^3$ | $2.9 \times 10^3$ | $3.2 \times 10^3$ | $3.4 \times 10^3$ |
| 235 | $1.5 \times 10^3$ | $2.6 \times 10^3$ | $2.3 \times 10^3$ | $2.1 \times 10^3$ |
| 240 | $1.0 \times 10^3$ | $2.2 \times 10^3$ | $1.5 \times 10^3$ | $1.3 \times 10^3$ |
| 248 | $5.3 \times 10^2$ | $1.4 \times 10^3$ | $8.0 \times 10^2$ | $6.3 \times 10^2$ |
| 254 | $3.0 \times 10^2$ | $9.2 \times 10^2$ | $4.8 \times 10^2$ | $3.3 \times 10^2$ |
| 265 | 87 | $4.1 \times 10^2$ | $1.2 \times 10^2$ | 48 |
| 275 | 18 | $1.2 \times 10^2$ | 16 | 4.4 |
| 280 | 8.0 | 51 | 4.3 | 1.2 |
| 289 | 1.4 | 12 | 0.17 | 0.15 |

[a] Data in $M^{-1}$ $cm^{-1}$.

but the first complete set of spectra (1800–1200 $cm^{-1}$) for all four Thy◇Thy was presented by Weinblum and Johns (1966). Since that time four spectra have been reported for Ura◇Ura (Varghese, 1971b) and are shown in Fig. 9. Ura◇Ura(*c*,*s*) exists in two crystalline forms with quite different spectra (compare Figs. 9a and 9b). Jennings *et al.* (1972) published almost identical spectra for these dimers, but with different assignments. The dimer identified by Varghese as *trans-anti* is referred to by Jennings as *trans-syn*. This inconsistency should be resolved by further work. Khattak and Wang (1972) presented spectra for the four dimers of 6-MeU, Fahr *et al.* (1972) for the 1,3-$Me_2$Ura dimers, and Stepien *et al.* (1973a) for the dimers of 1-MeThy and 1,3-$Me_2$Thy.

There is no apparent correlation between the type of dimer and the details of the IR spectra, so that it is impossible to predict the structure of an unknown dimer from its IR spectrum alone.

## 6. Synthesis of Dimers

The chemical synthesis of pyrimidine cyclobutane dimers has been reported only for uracil. The initial steps involve the preparation of the appropriate *trans* or *cis* anhydrides, **X** and **XI**, followed by boiling in anhydrous methanol and isolation of the isomeric half-esters **XII–XV.** These products are then converted via acid chlorides and azides to isocyanates, which react with ammonia to form the diureido derivatives **XVI–XIX.** These in turn undergo ring closure in acid with loss of methanol to give finally the 4 isomers of Ura◇Ura. Details of

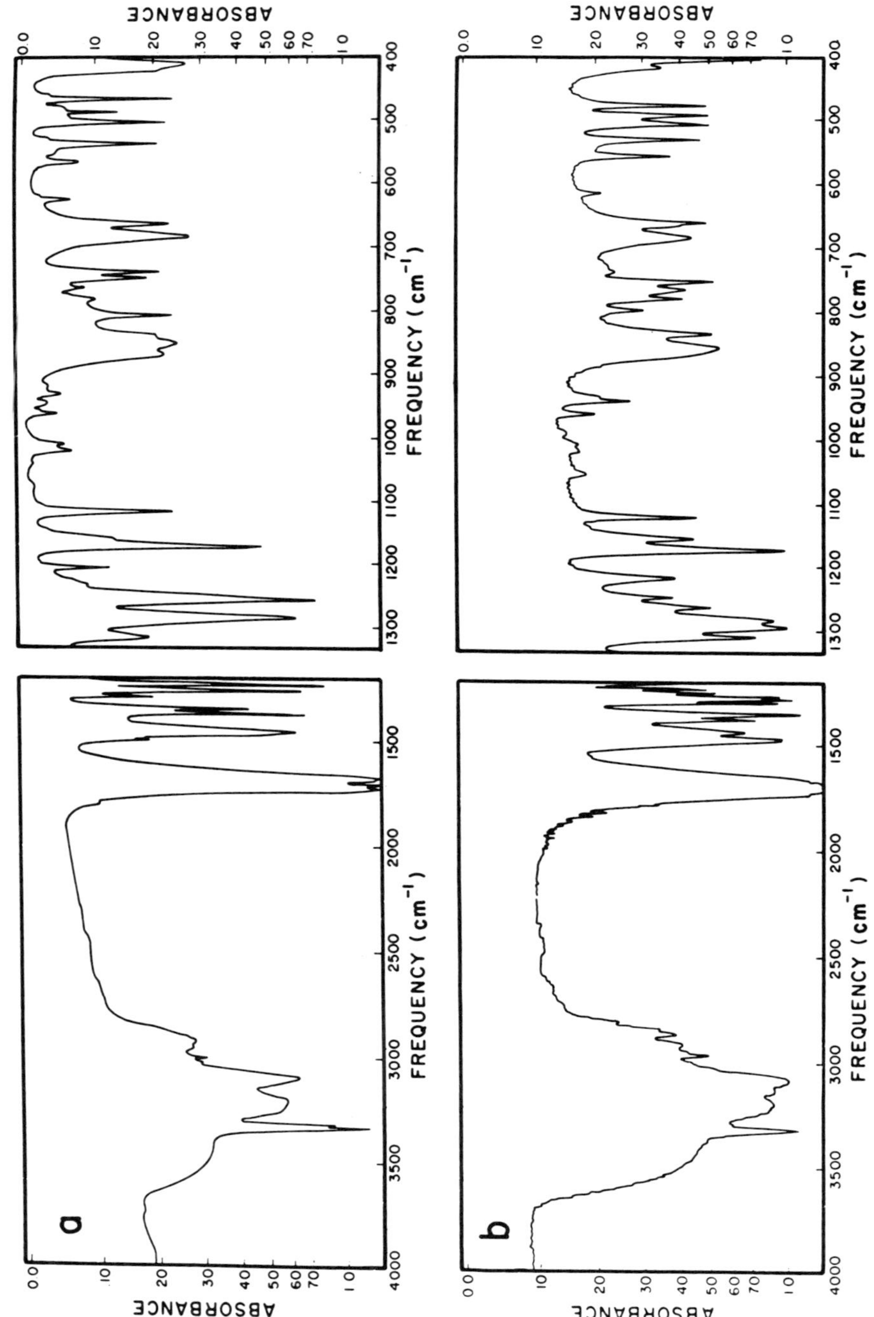
a
b
ABSORBANCE
FREQUENCY (cm$^{-1}$)

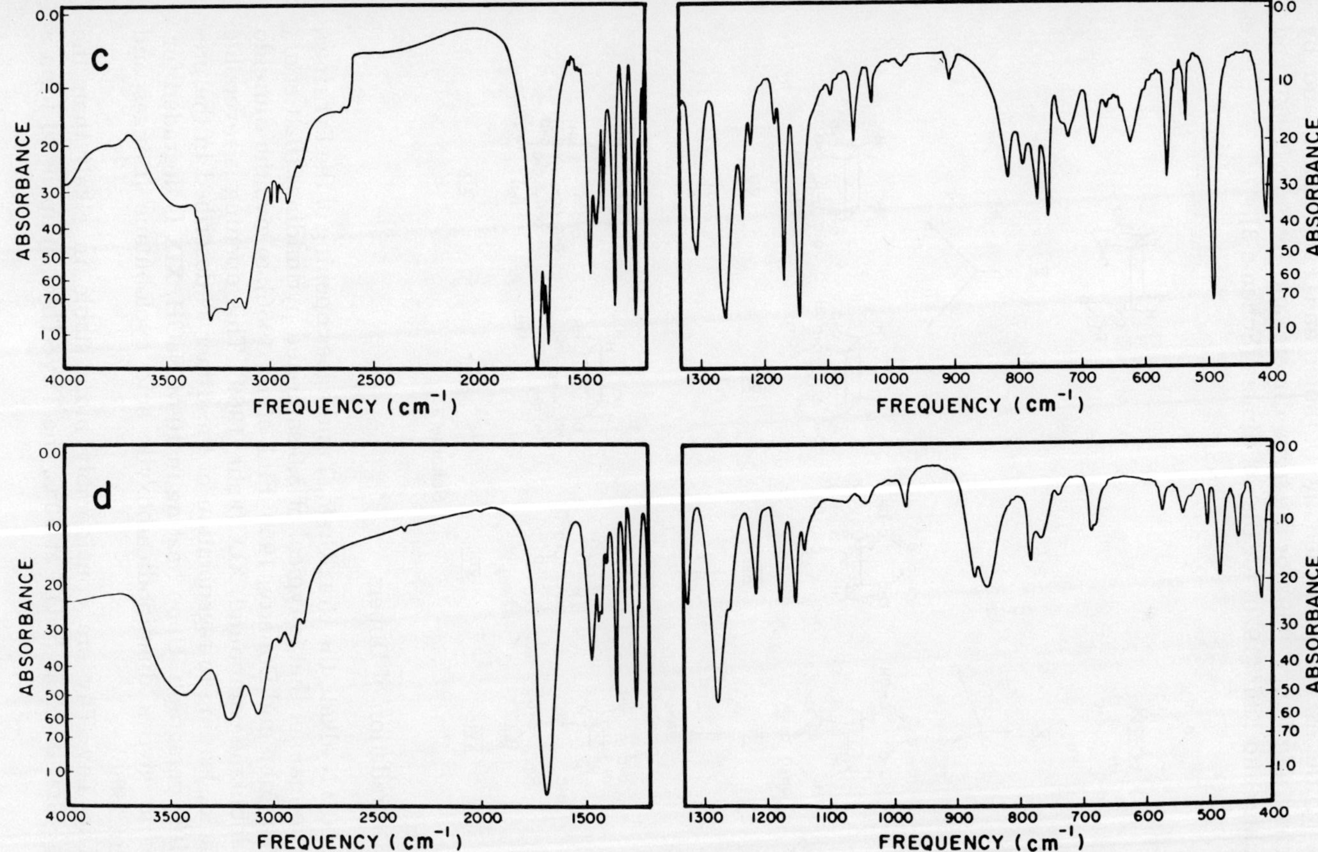


**Fig. 9.** *Infrared absorption spectra in KBr of (a) stable α form of cis-syn uracil dimer, (b) β form of cis-syn uracil dimer which was obtained from uridine dimer after hydrolysis,* (c) cis-anti *uracil dimer, and* (d) trans-anti *uracil dimer* (*from Varghese, 1971b*).

the syntheses are presented by Dörhöfer and Fahr (1966) and by Richter and Fahr (1969) for the *trans* dimers, and by Fahr (1969) and Richter and Fahr (1970) for the *cis* dimers (Scheme 3).

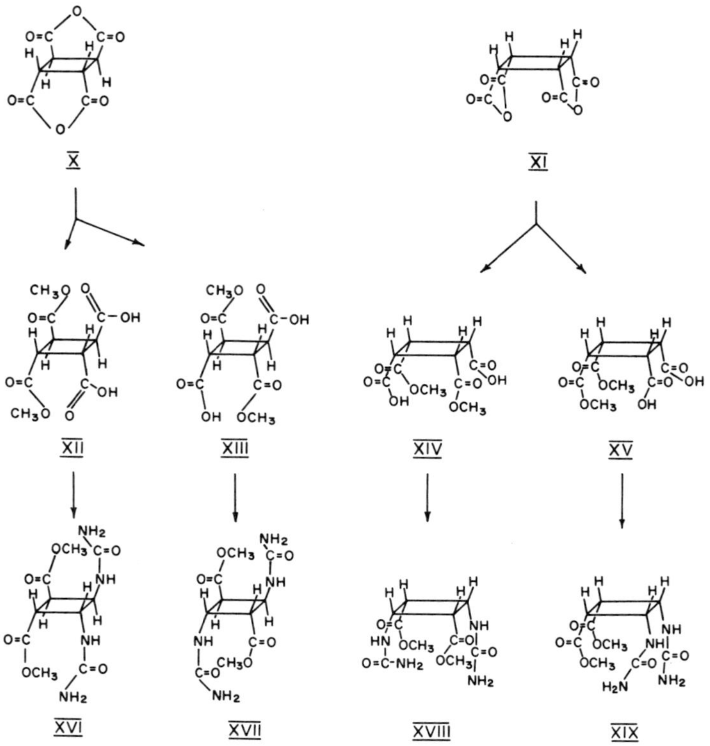

X XI

XII XIII XIV XV

XVI XVII XVIII XIX

**Scheme 3**

7. Degradation of Dimers

In 0.5 *N* alkali, the Ura◇Ura(*c,s*) undergoes opening of the Pyr rings, in the manner that is typical of 5,6-saturated pyrimidines (Batt *et al.*, 1954; Cohn and Doherty, 1956; Fink *et al.*, 1956), to give the diureido cyclobutane compound, **XIX** (Fahr, 1969). This opening is reversible in acid, leading to regeneration of the dimer as described in the preceding paragraph. Upon heating in 10 *M* NaOH, **XIX** is degraded further to give a diaminodicarboxylic acid cyclobutane (Dönges and Fahr, 1966).

The Thy◇Thy are considerably more stable in alkali than the Ura◇Ura. In 10 *M* NaOH at 50°C, the Thy◇Thy(*c,s*) opens to form a diureido product analogous to **XIX** (Blackburn and Davies, 1965). In acid this recyclizes to form dimer, while treatment with bromine leads to the triazatricyclodecane derivative, **XX**, which on pyrolysis gives 2,3-dimethylmaleimide (**XXI**) and 2-imidazolone (**XXII**) (Blackburn and Davies, 1966b). (See Scheme 4.) In fact, the presence of the two

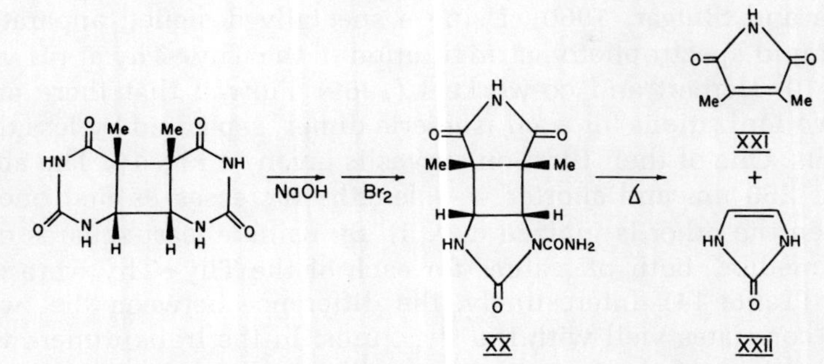

**Scheme 4**

methyl groups on adjacent carbons in **XXI** was among the first proofs that the Thy◇Thy from ice has the *c,s* configuration. Alkaline hydrolysis of the Thy◇Thy(*t,a*) has also been shown to lead to a diureido compound (von Wilucki *et al.*, 1967) under milder conditions of 0.01 *N* NaOH at room temperature (Kunieda and Witkop, 1971).

In the presence of excess sodium borohydride, Thy◇Thy(*c,s*) is reduced, with ring opening, to the ureido alcohols **XXIII** and **XXIV**. Permanganate oxidation of the latter compound leads to the same product that was obtained by alkaline degradation, **XXV** (Scheme 5) (Kunieda and Witkop, 1969, 1971; Witkop, 1968).

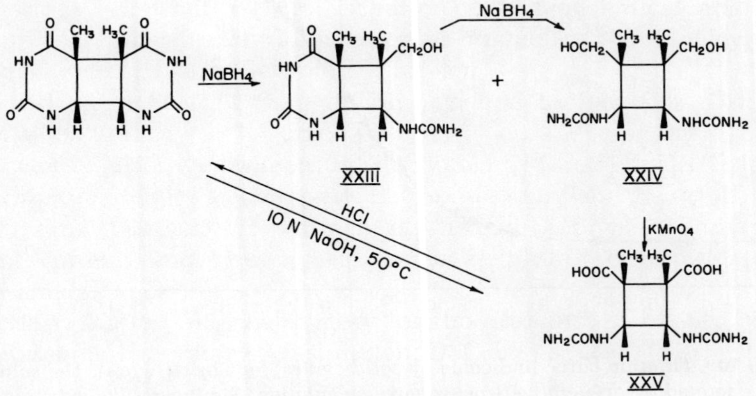

**Scheme 5**

## 8. Dissociation Constants for Dimers

Ura and Thy have pKs of 9.5 and 9.9, respectively, for deprotonation of one ring nitrogen (Shugar and Fox, 1952). When the 5,6-bond is saturated, the first ionization is displaced to pH 11 or higher, and alkaline instability makes accurate determinations of the pK difficult (Janion and Shugar, 1960). Using a specially designed apparatus to allow rapid spectrophotometric titration of the Thy◇Thy at pH values above 10, Herbert and co-workers (1969) showed that there are, in fact, two ionizations for *each* isomeric dimer, separated by less than 2 pH units. One of their titration curves is given in Fig. 10. The absorbance at 250 nm and shorter wavelengths increases as first one ring and then the other is ionized at N(3). By using a least-squares curve-fitting method, both pK values for each of the Thy◇Thy were determined (Table 14). Interestingly, the difference between the two pK values correlates well with the structures; in the *trans* dimers where N(3) positions of the two rings are well separated, ionization of one has a relatively small effect on the ionization of the second and the

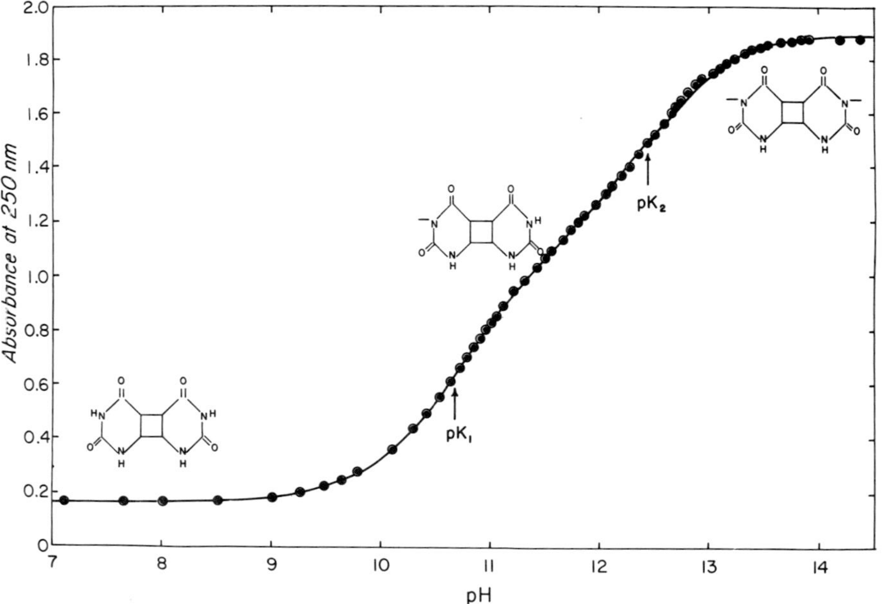


**Fig. 10.** *Titration curve (indicated by solid circles) for Thy◇Thy(c,s). The solid line was generated by an iterative least-squares calculation. Similar results were obtained by monitoring at a number of wavelengths shorter than 250 nm (Herbert et al., 1969).*

**Table 14** p*K* Values for Thy◇Thy[a,b]

| Dimer | p$K_1$ | p$K_2$ | p$K_2$ − p$K_1$ |
|---|---|---|---|
| *cis-syn* | 10.65 | 12.45 | 1.80 |
| *trans-syn* | 10.89 | 12.30 | 1.41 |
| *cis-anti* | 10.50 | 12.21 | 1.71 |
| *trans-anti* | 10.22 | 11.64 | 1.42 |

[a] Data from Herbert *et al.* (1969).
[b] Analyzing wavelength, 235 nm.

pKs differ by 1.4 units. When the rings are close together (*cis* dimers), ionization of one influences the other and the pKs differ by 1.8 units.

Protonation at N(3) of Cyd occurs below its p*K* of 4.09 (DeBoer *et al.*, 1970). This is displaced to higher pH values when the 5,6-bond is saturated, as in hCyd (p*K* 6.1, Ono *et al.*, 1965) and ho$^6$hCyd (p*K* 5.57, DeBoer *et al.*, 1970). The Cyd◇Cyd from frozen solutions, which is largely the *c*,*s* isomer, presumably has two p*K* values, for ionization of first one ring and then the other. From a spectrophotometric titration curve analogous to that in Fig. 10 only the midpoint, pH 5.4, has been reported (Varghese, 1971a). Probably, one p*K* value is greater than this, and one is smaller.

## 9. Methylation of Ring Nitrogens

Because the N-protons of pyrimidines are exchangeable, they can be replaced with methyl groups. The Thy◇Thy(*c*,*s*), for example, may be converted to the corresponding $Me_2$Thy◇$Me_2$Thy by room-temperature incubation with alkaline dimethyl sulfate (Wulff and Fraenkel, 1961; Beukers and Berends, 1961). The two ice dimers of 1-MeThy are converted, under similar conditions, to the (*c*,*s*)- and (*t*,*s*)$Me_2$Thy◇$Me_2$Thy (Blackburn and Davies, 1966a). The *anti*-Thy◇Thy have also been fully methylated (Weinblum and Johns, 1966; Weinblum *et al.*, 1968). The yields, however, are low—only 1% for the *t*,*a* isomer because of the alkaline instability of the dimers. When a large excess of dimethyl sulfate is not present, the major product is the partially methylated dimer of 3-MeThy (Blackburn and Davies, 1966a). This is consistent with the fact that N(3) is more labile, and hence more readily replaced, than the N(1)-proton.

The Ura◇Ura are much more resistant to full methylation, giving primarily the corresponding 3-MeUra◇3-MeUra (Blackburn and Davies, 1966c). Even with diazomethane it is not possible to replace more than three of the protons (Fahr, 1969). However, Elad *et al.*

(1971) reported complete methylation of the (*c*,*s*)- and (*c*,*a*)Ura◇Ura by prolonged stirring with silver oxide and methyl iodide in dimethylformamide.

## 10. Hydrolysis of Nucleoside Dimers

The N(1)–glycosyl bond which joins bases to the backbone of DNA or RNA may be cleaved by acid hydrolysis, for example by heating to 100°C in 70% perchloric acid for one hour. The *syn* dimers of Thy are stable under these conditions, but the *anti* dimers revert quantitatively to Thy. If DNA is hydrolyzed in an attempt to isolate photoproducts, the *anti* dimers will necessarily escape detection. However, dimerization weakens the glycosyl bond in nucleosides (Wacker and Träger, 1963). Therefore, it has been possible to obtain the (*c*,*a*)- and (*t*,*a*)Thy◇Thy in good yield by mild hydrolysis of the corresponding Thd◇Thd in 6 *N* HCl at 85°C for 15 minutes (Weinblum and Johns, 1966; Kunieda and Witkop, 1971). On boiling in 4 *N* HCl for 30 minutes, 20% of the Urd◇Urd(*c*,*a*) dimer, ~ 50% of the *t*,*a* dimer, and > 80% of the *c*,*s* dimer is converted to Ura◇Ura (Varghese, 1971b).

The Cyd◇Cyd(*c*,*s*), like many 5,6-saturated Cyt derivatives, may undergo deamination. Thus, after 90 minutes in trifluoroacetic acid at 165°C, the products include the Ura◇Ura in which the amino groups and the sugars have been lost (Varghese, 1971a; Varghese and Rupert, 1971).

## 11. Thermal and pH-Dependent Reversal

In strongly basic solution Pyr◇Pyr undergo ring opening. Under neutral or acidic conditions and in the solid state, the dimers are commonly believed to be very stable, even at elevated temperatures. This is the basis of the fallacious thermal-reversibility test for photohydrates, as discussed in Chapter 4.

In fact, even in the solid state the Thy◇Thy(*c*,*s*) reverts to parent at 320°C (Beukers and Berends, 1961) and the *t*,*a* dimer reverses at 240°C (von Wilucki *et al.*, 1967). The Oro◇Oro(*t*,*s*) and 5-MeOro◇5-MeOro (*c*,*a*) split at 180°C and 260°C, respectively (Sztumpf and Shugar, 1965), while the *trans-syn* methyl orotate dimer breaks down at 195°C (Birnbaum *et al.*, 1971). In the latter case, the lability of the cyclobutane ring was attributed to the long (1.622 Å), and hence weak, bond between C(6) of the bases.

There have been numerous qualitative and often conflicting reports on the stability of various Pyr◇Pyr in solution. One general conclusion which can be made is that the *anti* dimers are much less stable than the *syn* isomers, in both acid and base, for Thy and dThd (Weinblum and Johns, 1966; Ben-Hur *et al.*, 1967), 5-EtUra (Pietrzykowska and Shugar, 1970), 6-MeUra (Khattak and Wang, 1972), and Ura (Varghese, 1971b; Jennings *et al.*, 1972). Reported differences between the stabilities of the Ura◇Ura(*t,a*) in neutral solution by the last two references may be due to a mistaken structure assignment.

Further evidence for this generalization is provided by the detailed study of Herbert *et al.* (1969) on Thy◇Thy. The Thy◇Thy(*c,s*) is completely stable to acid hydrolysis under a variety of conditions, but it is destroyed in 2 *N* KOH at 100°C, with a half-life of 100 min, giving back Thy and a ring-opening product. The *t,s* dimer is similarly stable in acid but has a half-life of only 10 min in alkali. Both of the *anti* dimers have their maximum stability between pH 1 and 6. Figure 11 shows the pH dependence of the half-life for these two dimers; both revert to Thy.

Quantitative results have also been presented for the isomers of $Me_2$Ura◇$Me_2$Ura in 6 *N* HCl at 20°C: while the *c,s* and *t,s* forms are quite stable, the *c,a* and *t,a* revert to parent with respective half-lives of $37 \times 10^3$ min and $7.3 \times 10^3$ min (Fahr *et al.*, 1972).

Attempts to predict stabilities of Thy dimers by molecular orbital

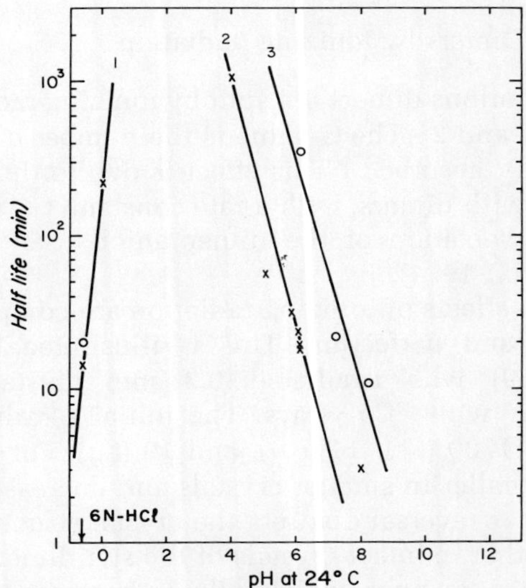


**Fig. 11.** *Effect of pH (measured at 24°) on stability of the* cis-anti *(C) and* trans-anti *(D) dimers of thymine at 100°C (Herbert* et al., *1969). All decay to thymine; curve 1 shows dimers C and D, curve 2 shows dimer D, and curve 3 shows dimer C.*

calculations (Jordan and Pullman, 1968; improved by Camerman and Nyburg, 1969) led to the erroneous conclusion that the *trans-anti* dimer is more stable than the *cis-anti* isomer. Steric arguments have been more successful, suggesting that greater steric shielding of the cyclobutane ring in *syn* dimers should make them more resistant to attack in acid or base and, hence, more stable than the *anti* dimers (Camerman and Camerman, 1970). A more reliable method of predicting stabilities may involve organic chemical considerations, as described by Khattak and Wang (1972).

The Cyd◇Cyd from frozen solution is peculiar in that it may deaminate in the manner typical of 5,6-saturated Cyt derivatives (Green and Cohen, 1957; Becker *et al.*, 1967), reverse to parent, undergo acid-catalyzed hydrolysis, or undergo base-induced ring opening. The relative importance of these various processes is a function of pH (DeBoer *et al.*, 1970; Varghese, 1971a).

The importance of understanding product stabilities before attempting quantitative photochemistry is illustrated by the following. When dThd is irradiated, at least seven products are formed, but only the *cis-syn* dimer and one adduct survive the conditions normally used to hydrolyze DNA (Varghese, 1971c). When one realizes that Cyt products are even less stable, the difficulty of isolating photoproducts from DNA is evident.

## 12. Splitting of Dimers by Ionizing Radiation

In aqueous solutions dimers are split by ionizing radiation with a *G*-value between 1 and 2. (The *G*-value is the number of molecules converted per 100 eV absorbed.) This effect is due to the reaction of solvated electrons with dimers, with a rate constant $1.5 \times 10^{10}$ $M^{-1}$ $sec^{-1}$, followed by dissociation of the dimer anion (Santus *et al.*, 1972; Grossweiner *et al.*, 1974).

In crystals the effects of ionizing radiation are complex and depend on crystal size and perfection. This is illustrated by the work of Weinblum (1969) who irradiated 0.3 mm crystals of the four $Me_2Thy$◇$Me_2Thy$ with $^{60}Co$ $\gamma$-rays. The initial *G*-values for splitting were 6300 (*c*,*s*), 1700 (*t*,*s*), 14 (*c*,*a*), and 19 (*t*,*a*). For the *syn* dimers, the *G*-value is smaller in smaller crystals and decreases during the irradiation as dimer reversal disrupts the crystal structure. These large *G*-values imply that in intact crystals of the *syn* dimers the molecules are arranged in a manner which allows a chain reaction so that reversal of one by the absorption of energy initiates the reversal of many more. That the reaction does not involve an excited state of a dimer has been demonstrated by Weinblum (1970), who showed that only one molecule of $Me_2Thy{\diamond}Me_2Thy(c,s)$ is split per UV photon absorbed in the solid. The mechanism may involve a dimer anion or radical.

The instability of dimers to ionizing radiation creates problems during x-ray analysis of crystal structures, because the crystals are destroyed during the study (see Birnbaum *et al.*, 1971; Camerman and Camerman, 1970; Leonard *et al.*, 1969).

## 13. Photoreversal of Dimers

### *a. Direct Photolysis*

The efficient splitting of Pyr◇Pyr when it absorbs a photon is one of the best-known properties of dimers. For Thy◇Thy dimers, careful determinations of quantum yields in neutral water showed that the efficiency is almost constant, between 0.6 and 0.8, for splitting four isomeric dimers over the wavelength range 225–289 nm (Herbert *et al.*, 1969).

For other pyrimidines, studies are less complete, but in all cases the quantum yields are in the range 0.2–0.8 (Swenson and Setlow, 1963; Sztumpf and Shugar, 1965; Pietrzykowska and Shugar, 1970; Varghese and Rupert, 1971). It should be noted that dimer reversal is orders of magnitude more efficient than dimer formation in solution.

UV irradiation of dimers in the crystalline state (Weinblum, 1970; Lisewski and Wierzchowski, 1970) or in frozen glasses (Eisinger and Lamola, 1969) also leads to efficient reversal, with quantum yields close to 1.0.

### *b. Photosensitized Reversal*

When dimers are irradiated in the presence of certain photosensitizers, splitting occurs even though the dimers absorb no light. For example, the following sensitizers have been used: uranyl acetate (Wacker *et al.*, 1964), 2,6-naphthalenedisulfonate and 2-triphenylene sulfonate (Lamola, 1966), chloranil (Rosenthal and Elad, 1968), a number of anthraquinonesulfonates (Ben-Hur and Rosenthal, 1970), and indole derivatives such as tryptophan (Hélène and Charlier, 1971; Charlier and Hélène, 1975). The quantum efficiency for splitting $Me_2Thy$ dimers is $\leq 10^{-3}$ for excitation of most of these sensitizers with light of wavelengths longer than 300 nm, but for 2-anthraquinone sulfonate the quantum yield may be as high as 1.0 (Lamola, 1972).

When photosensitizers such as acetone are used to form dimers, the mechanism proceeds via energy transfer to the Pyr triplet state. When sensitizers are used to reverse dimers, the situation is much more complicated. There seem to be at least three possible mechanisms for sensitized splitting of dimers, with the actual pathway in a particular case depending on the sensitizer and the irradiation conditions. (1) Photoelectrons ejected by excited indoles may attack the dimer, leading to dissociation of the negatively ionized molecule (Hélène and Charlier, 1971). (2) In concentrated or frozen solutions, the singlet state of the indole may interact with the dimer to transfer an electron directly (Hélène and Charlier, 1971). (3) The triplet state of the aromatic sulfonates or an excited state of the uranyl ion, may abstract an electron from the dimer, resulting in a positively ionized dimer which is split (Roth and Lamola, 1972; Rosenthal *et al.*, 1975).

c. *Photoenzymatic Splitting*

A number of cellular systems contain enzymes which form a complex with the Thy◇Thy(*c*,*s*) in UV-irradiated DNA, absorb visible light, and cause cleavage of the dimer back to monomers, thereby repairing some of the UV damage. This important repair process, summarized by Rupert in 1964, is dealt with in Chapter 6 of Vol. II.

## F. Summary

1. Dimers can exist in four isomeric forms, the *c*,*s*, *t*,*s*, *c*,*a*, and *t*,*a* isomers (see Fig. 1).

2. In general, dimers show only end absorption and may be reversed by short-wavelength light with a quantum yield near 1.

3. In dilute solution, dimers arise through the triplet state. In concentrated solutions where aggregates are present, they probably arise from the singlet state, although the triplet process cannot be completely ruled out.

4. In solution and polynucleotides, photosensitizers are useful in producing dimers. Energy is transferred from the sensitizer triplet state to the Pyr triplet, leading to dimer formation.

5. The yield of dimers in dilute solution gives the product of intersystem crossing efficiency, $\phi_{isc}$, and $f_d$, the fraction of triplet–ground state complexes which form stable dimers. Values for $f_d$ are 1.0 for Oro, 0.25 for Ura, and 0.07 for Thy.

6. Intersystem crossing increases with the energy of excitation, showing that crossover to the triplet takes place from the higher vibrational levels of the singlet state. At 265 nm, $\phi_{isc}$ is 0.12 in Oro (pH 1) 0.02 in Ura, and 0.0067 in Thy.

7. Quantum yields of the various isomeric dimers in solution are summarized in Table 3 for derivatives of Thy, Ura, and Cyt. In solution, steric factors and solvent effects determine the relative yields of the isomers.

8. Dimerization in the solid state is summarized in Tables 6 and 7. The particular isomer formed depends on the crystal structure.

9. Dimerization in polymers (see Table 9) is poorly understood because of the difficulty in isolating dimers and the complication of energy transfer along the chain.

10. Properties of dimer crystals are summarized in Table 10.

11. Chromatographic properties of dimers are summarized in Table 12.

12. Titration of dimers shows that the ionization of one ring influences the pH at which the second ring ionizes. Thus, each dimer has two p*K*s, separated by $\sim 1.5$ pH units.

13. The *anti* dimers, in which the bases are oriented head-to-tail, are more unstable, in both acid and base, than the *syn* or head to-head dimers.

## Acknowledgments

The authors are happy to acknowledge the financial support of the National Cancer Institute of Canada, and the Medical Research Council of Canada.

## References

Adman, E., and Jensen, L. H. (1970). *Acta Crystallogr., Sect. B* **26,** 1326.
Adman, E., Gordon, M. P., and Jensen, L. H. (1968). *Chem. Commun.* p. 1019.
Anet, R. (1962). *Can. J. Chem.* **40,** 1249.
Batt, R. D., Martin, J. K., Ploeser, J. M., and Murray, J. (1954). *J. Amer. Chem. Soc.* **76,** 3663.
Becker, H., LeBlanc, J. C., and Johns, H. E. (1967). *Photochem. & Photobiol.* **6,** 733.
Ben-Hur, E., and Rosenthal, I. (1970). *Photochem. & Photobiol.* **11,** 163.
Ben-Hur, E., Elad, D., and Ben-Ishai, R. (1967). *Biochim. Biophys. Acta* **149,** 355.
Ben-Ishai, R., Ben-Hur, E., and Hornfeld, Y. (1968). *Isr. J. Chem.* **6,** 769.
Beugelmans, R., Fourrey, J. L., Gero, S. D., LeGoff, M. T., Mercier, D., Ratovelomanana, V., and Janot, M. (1972). *C. R. Acad. Sci., Ser. C* **274,** 882.
Beukers, R., and Berends, W. (1960a). *Biochim. Biophys. Acta* **38,** 573.
Beukers, R., and Berends, W. (1960b). *Biochim. Biophys. Acta* **41,** 550.
Beukers, R., and Berends, W. (1961). *Biochim. Biophys. Acta* **49,** 181.
Beukers, R., Ijlstra, J., and Berends, W. (1958). *Rec. Trav. Chim. Pays-Bas* **77,** 729.
Beukers, R., Ijlstra, J., and Berends, W. (1959a). *Rec. Trav. Chim. Pays-Bas* **78,** 879.
Beukers, R., Ijlstra, J., and Berends, W. (1959b). *Rec. Trav. Chim. Pays-Bas* **78,** 883.
Bevilacqua, R., and Bordin, F. (1973). *Photochem. & Photobiol.* **17,** 191.
Birnbaum, G. I., Dunston, S. M., and Szabo, A. G. (1971). *Tetrahedron Lett.* p. 947.

Blackburn, G. M., and Davies, R. J. H. (1965). *Chem. Commun.* p. 215.
Blackburn, G. M., and Davies, R. J. H. (1966a). *J. Chem. Soc.*, C p. 1342.
Blackburn, G. M., and Davies, R. J. H. (1966b). *J. Chem. Soc.*, C p. 2239.
Blackburn, G. M., and Davies, R. J. H. (1966c). *Tetrahedron Lett.* p. 4471.
Borkman, R. F., and Kearns, D. R. (1966). *J. Chem. Phys.* **44,** 945.
Bremner, J. B., Warrener, R. N., Adman, E., and Jensen, L. H. (1971). *J. Amer. Chem. Soc.* **93,** 4574.
Brown, D. M., and Hewlins, M. J. E. (1968). *J. Chem. Soc., C* p. 2050.
Brown, D. T., Eisinger, J., and Leonard, N. J. (1968). *J. Amer. Chem. Soc.* **90,** 7302.
Brown, I. H. (1968). Ph.D. Thesis, Dept. of Medical Biophysics, University of Toronto.
Brown, I. H., and Johns, H. E. (1967). *Photochem. & Photobiol.* **6,** 469.
Brown, I. H.. and Johns, H. E. (1968). *Photochem. & Photobiol.* **8,** 273.
Brown, I. H., Freeman, K. B., and Johns, H. E. (1966). *J. Mol. Biol.* **15,** 640.
Burr, J. G. (1968). *Advan. Photochem.* **6,** 193.
Camerman, N., and Camerman, A. (1968). *Science* **160,** 1451.
Camerman, N., and Camerman, A. (1970). *J. Amer. Chem. Soc.* **92,** 2523.
Camerman, N., and Nyburg, S. C. (1969). *Theor. Chim. Acta* **13,** 162.
Camerman, N., Weinblum, D., and Nyburg, S. C. (1969). *J. Amer. Chem. Soc.* **91,** 982.
Charlier, M., and Hélène, C. (1967). *Photochem. & Photobiol.* **6,** 501.
Charlier, M., and Hélène, C. (1972). *Photochem. & Photobiol.* **21,** 31.
Charlier, M., and Hélène, C. (1975). *Photochem. & Photobiol.* **21,** 31.
Charlier, M., Hélène, C., and Dourlent, M. (1969). *J. Chim. Phys.* **66,** 700.
Charlier, M., Hélène, C., and Carrier, W. L. (1972). *Photochem. & Photobiol.* **15,** 527.
Cohn, W. E., and Doherty, D. G. (1956). *J. Amer. Chem. Soc.* **78,** 2863.
DeBoer, G., Pearson, M. L., and Johns, H. E. (1967). *J. Mol. Biol.* **27,** 131.
DeBoer, G., Klinghoffer, O., and Johns, H. E. (1970). *Biochim. Biophys. Acta* **213,** 253.
Deering, R. A., and Setlow, R. B. (1963). *Biochim. Biophys. Acta* **68,** 526.
Dönges, K.-H., and Fahr, E. (1966). *Z. Naturforsch. B* **21,** 87.
Dörhöfer, G., and Fahr, E. (1966). *Tetrahedron Lett.* p. 4511.
Einstein, J. R., Hosszu, J. L., Longworth, J. W., Rahn, R. O., and Wei, C. H. (1967). *Chem. Commun.* p. 1063.
Eisinger, J., and Lamola, A. A. (1969). *Mol. Photochem.* **1,** 209.
Eisinger, J., and Lamola, A. A. (1971). *Biochim. Biophys. Acta* **240,** 299.
Elad, D., Kruger, C., and Schmidt, G. M. J. (1967). *Photochem & Photobiol.* **6,** 495.
Elad, D., Rosenthal, I., and Sasson, S. (1971). *J. Chem. Soc., C* p. 2053.
Fahr, E. (1969). *Angew. Chem., Int. Ed. Engl.* **8,** 578.
Fahr, E., Fürst, G., Dörhöfer, G., and Popp, H. (1967). *Angew. Chem., Int. Ed. Engl.* **6,** 250.
Fahr, E., Fürst, G., Maul, P., and Wieser, H. (1972). *Z. Naturforsch. B.* **27,** 1475.
Fahr, E., Pastille, R., Pelz, N., and Scheutzow, D. (1974). *Z. Naturforsch. B* **29,** 410.
Fendler, J., and Bogan, G. (1974). *Photochem. & Photobiol.* **20,** 323.
Fenster, A., and Johns, H. E. (1973). *J. Phys. Chem.* **77,** 2246.
Fink, R. M., Cline, R. E., McGaughey, C., and Fink, E. (1956). *Anal. Chem.* **28,** 4.
Fisher, G. J. (1973). Ph.D. Thesis, University of Toronto.
Fisher, G. J., and Johns, H. E. (1970). *Photochem. & Photobiol.* **11,** 429.
Fisher, G. J., and Johns, H. E. (1974). *Photochem. & Photobiol.* **20,** 109.
Füchtbauer, W., and Mazur, P. (1966). *Phothochem. & Photobiol.* **5,** 323.
Fürst, G., Fahr, E., and Wieser, H. (1967). *Z. Naturforsch. B* **22,** 354.
Gauri, K. K. (1967). *Z. Naturforsch. B* **22,** 685.
Gauri, K. K., Reuger, W., and Wacker, A. (1971). *Z. Naturforsch. B* **26,** 167.

Gerdil, R. (1961). *Acta Crystallogr.* **14,** 333.
Golankiewicz, K., and Strekowski, L. (1972). *Mol. Photochem.* **4,** 189.
Green, M., and Cohen, S. S. (1957). *J. Biol. Chem.* **225,** 397.
Greenstock, C. L., and Johns, H. E. (1968). *Biochem. Biophys. Res. Commun.* **30,** 21.
Greenstock, C. L., Brown, I. H., Hunt, J. W., and Johns, H. E. (1967). *Biochem. Biophys. Res. Commun.* **27,** 431.
Grossweiner, L. I., Kepka, A. G., Santus, R., and Vigil, J. A. (1974). *Int. J. Radiat. Biol.* **25,** 521.
Guéron, M., Eisinger, J., and Shulman, R. G. (1967). *J. Chem. Phys.* **47,** 4077.
Günther, H. L., and Prusoff, W. H. (1967). *Biochim. Biophys. Acta* **149,** 361.
Guschlbauer, W., Favre, A., and Michelson, A. M. (1965). *Z. Naturforsch. B* **20,** 1141.
Hariharan, P. V., and Johns, H. E. (1968). *Can. J. Biochem.* **46,** 911.
Haug, A. (1964). *J. Amer. Chem. Soc.* **86,** 3381.
Hauswirth, W., and Daniels, M. (1971a). *Chem. Phys. Lett.* **10,** 140.
Hauswirth, W., and Daniels, M. (1971b). *Photochem. & Photobiol.* **13,** 157.
Hayes, F. N., Williams, D. L., Ratliff, R. L., Varghese, A. J., and Rupert, C. S. (1971). *J. Amer. Chem. Soc.* **93,** 4940.
Hélène, C., and Charlier, M. (1971). *Biochem. Biophys. Res. Commun.* **43,** 252.
Helleiner, C. W., Pearson, M. L., and Johns, H. E. (1963). *Proc. Nat. Acad. Sci. U.S.* **50,** 761.
Herbert, M. A., and Johns, H. E. (1971). *Photochem. & Photobiol.* **14,** 693.
Herbert, M. A., LeBlanc, J. C., Weinblum, D., and Johns, H. E. (1969). *Photochem. & Photobiol.* **9,** 33.
Hoogsteen, K. (1963). *Acta Crystallogr.* **16,** 28.
Hyatt, J. A., and Swenton, J. S. (1972). *J. Amer. Chem. Soc.* **94,** 7605.
Ishihara, H. (1963). *Photochem & Photobiol.* **2,** 455.
Janion, G., and Shugar, D. (1960). *Acta Biochim. Pol.* **7,** 309.
Jellinek, T., and Johns, R. B. (1970). *Photochem. & Photobiol.* **11,** 349.
Jennings, B. H., Pastra, S. C., and Wellington, J. L. (1970). *Photochem. & Photobiol.* **11,** 215.
Jennings, B. H., Pastra-Landis, S., and Lerman, J. W. (1972). *Photochem. & Photobiol.* **15,** 479.
Johns, H. E. (1969). *In* "Methods in Enzymology" (K. Kustin, ed.), Vol. 16, p. 257. Academic Press, New York.
Johns, H. E. (1971). *In* "Creation and Detection of the Triplet State" (A. A. Lamola, ed.), p. 123. Dekker, New York.
Johns, H. E., and Rauth, A. M. (1965). *Photochem. & Photobiol.* **4,** 673.
Johns, H. E., Rapaport, S. A., and Delbrück, M. (1962). *J. Mol. Biol.* **4,** 104.
Johns, H. E., Pearson, M. L., LeBlanc, J. C., and Helleiner, C. W. (1964). *J. Mol. Biol.* **9,** 503.
Johns, H. E., Pearson, M. L., and Brown, I. H. (1966). *J. Mol. Biol.* **20,** 231.
Johnson, T. B., and Clapp, S. H. (1908). *J. Biol. Chem.* **5,** 49.
Jordan, F., and Pullman, B. (1968). *Theor. Chim. Acta* **10,** 423.
Kaláb, D. (1966). *Experientia* **22,** 23.
Kearns, D. R., and Case, W. A. (1966). *J. Amer. Chem. Soc.* **88,** 5087.
Khattak, M. N., and Wang, S. Y. (1972). *Tetrahedron* **28,** 945.
Kleopfer, R., and Morrison, H. (1972). *J. Amer. Chem. Soc.* **94,** 255.
Konnert, J., and Karle, I. L. (1971). *J. Cryst. Mol. Struct.* **1,** 107.
Konnert, J., Gibson, J. W., Karle, I. L., Khattak, M. N., and Wang, S. Y. (1970). *Nature (London)* **227,** 953.

Krajewska, E., and Shugar, D. (1971). *Science* **173,** 435.
Krajewska, E., and Shugar, D. (1972). *Acta Biochim. Pol.* **19,** 207.
Krauch, C. H., Krämer, D. M., Chandra, P., Mildner, P., Feller, H., and Wacker, A. (1967). *Angew. Chem., Int. Ed. Engl.* **6,** 956.
Kunieda, T., and Witkop, B. (1969). *J. Amer. Chem. Soc.* **91,** 7751.
Kunieda, T., and Witkop, B. (1971). *J. Amer. Chem. Soc.* **93,** 3493.
Lamola, A. A. (1966). *J. Amer. Chem. Soc.* **88,** 813.
Lamola, A. A. (1969). *Photochem. & Photobiol.* **9,** 291.
Lamola, A. A. (1971). *Biochem. Biophys. Res. Commun.* **43,** 893.
Lamola, A. A. (1972). *Mol. Photochem.* **4,** 107.
Lamola, A. A., and Eisinger, J. (1971). *Biochim. Biophys. Acta* **240,** 313.
Lamola, A. A., and Mittal, J. P. (1966). *Science* **154,** 1560.
Lamola, A. A., and Yamane, T. (1967). *Proc. Nat. Acad. Sci. U.S.* **58,** 443.
Lamola, A. A., Guéron, M., Yamane, T., Eisinger, J., and Shulman, R. G. (1967). *J. Chem. Phys.* **47,** 2210.
Leonard, N. J., Golankiewicz, K., McCredie, R. S., Johnson, S. M., and Paul, I. C. (1969). *J. Amer. Chem. Soc.* **91,** 5855.
Lis, A. W., and Allen, F. W. (1961). *Biochim. Biophys. Acta* **49,** 190.
Lisewski, R., and Wierzchowski, K. L. (1970). *Photochem. & Photobiol.* **11,** 327.
Lisewski, R., and Wierzchowski, K. L. (1971). *Mol. Photochem.* **3,** 231.
Logue, M. W., and Leonard, N. J. (1972). *J. Amer. Chem. Soc.* **94,** 2842.
Lomant, A. J., and Fresco, J. R. (1972). *Progr. Nucl. Acid Res. Mol. Biol.* **12,** 1.
McLaren, A. D., and Shugar, D. (1964). "Photochemistry of Proteins and Nucleic Acids." Macmillan, New York.
Meistrich, M. L., Lamola, A. A., and Gabbay, E. (1970). *Photochem. & Photobiol.* **11,** 169.
Miskew, W. (1970). M.Sc. Thesis, Dept. of Chemistry, University of Toronto.
Montaudo, G., and Caccamese, S. (1973). *J. Org. Chem.* **38,** 710.
Moore, A. M., and Thomson, C. H. (1956). *Progr. Radiobiol.* **4,** 75.
Morrison, H., and Kleopfer, R. (1968). *J. Amer. Chem. Soc.* **90,** 5037.
Morrison, H., Feeley, A., and Kleopfer, R. (1968). *Chem. Commun.* p. 358.
Mustafa, A. (1951). *Chem. Rev.* **51,** 1.
Nnadi, J. C., and Wang, S. Y. (1969). *Tetrahedron Lett.* p. 2211.
Ono, J., Wilson, R. G., and Grossman, L. (1965). *J. Mol. Biol.* **11,** 600.
Parry, G. S. (1954). *Acta Crystallogr.* **7,** 313.
Pathak, M. A., and Krämer, D. M. (1969). *Biochim. Biophys. Acta* **195,** 197.
Pearson, M. L., and Johns, H. E. (1966). *J. Mol. Biol.* **19,** 303.
Pearson, M. L., and Johns, H. E. (1967). *Proc. Can. Cancer Res. Conf.* **7,** 353.
Pearson, M. L., Whillans, D. W., LeBlanc, J. C., and Johns, H. E. (1966). *J. Mol. Biol.* **20,** 245.
Pietrzykowska, I., and Shugar, D. (1970). *Acta Biochim. Pol.* **17,** 361.
Porter, G., and Wilkinson, F. (1961). *Proc. Roy. Soc. Ser. A* **264,** 1.
Rahn, R. O., and Landry, L. C. (1971). *Biochim. Biophys. Acta* **247,** 197.
Rao, D. V., Ulrich, H., Stuber, F. A., and Sayigh, A. A. R. (1973). *Chem. Ber.* **106,** 388.
Richter, P., and Fahr, E. (1969). *Angew. Chem. Int. Ed. Engl.* **8,** 208.
Richter, P., and Fahr, E. (1970). *Tetrahedron Lett.* p. 1921.
Rörsch, A., Beukers, R., Ijlstra, J., and Berends, W. (1958). *Rec. Trav. Chim. Pays-Bas* **77,** 423.
Rosenthal, I., and Elad, D. (1968). *Photochem. & Photobiol.* **8,** 145.
Rosenthal, I., and Elad, D. (1968b). *Biochem. Biophys. Res. Commun.* **32,** 599.

Rosenthal, I., Rao, M. M., and Salomon, J. (1975). *Biochim. Biophys. Acta,* **378,** 165.
Roth, H. D., and Lamola, A. A. (1972). *J. Amer. Chem. Soc.* **94,** 1013.
Rupert, C. S. (1964). *Photochem. & Photobiol.* **3,** 399.
Rushizky, G. W., and Pardee, A. B. (1962). *Photochem. & Photobiol.* 1, 15.
Santus, R., Hélène, C., Ovadia, J., and Grossweiner, L. I. (1972). *Photochem. & Photobiol.* **16,** 65.
Sasson, S., Rosenthal, I., and Elad, D. (1970). *Tetrahedron Lett.* p. 4513.
Schenk, G. O., von Wilucki, I., and Krauch, C. H. (1962). *Chem. Ber.* **95,** 1409.
Schröder, G., and Martin, W. (1966). *Angew. Chem.* **78,** 117.
Schuster, H. (1964). *Z. Naturforsch. B* **19,** 815.
Setlow, R. B. (1961). *Biochim. Biophys. Acta* **49,** 239.
Shugar, D., and Fox, J. J. (1952). *Biochim. Biophys. Acta* **9,** 199.
Sinsheimer, R. L. (1954). *Radiat. Res.* **1,** 505.
Sinsheimer, R. L., and Hastings, R. (1949). *Science* **110,** 525.
Smietanowska, A., and Shugar, D. (1961). *Bull. Acad. Pol. Sci., Cl. 2* **9,** 375.
Smith, K. C. (1963). *Photochem. & Photobiol.* **2,** 503.
Smith, K. C. (1964). *Photophysiology* **2,** 329.
Smith, K. C. (1966a). *Radiat. Res., Suppl.* **6,** 54.
Smith, K. C. (1966b). *Biochem. Biophys. Res. Commun.* **25,** 426.
Solie, T. N., and Schellman, J. A. (1968). *J. Mol. Biol.* **33,** 61.
Song, P.-S., Harter, M., Moore, T. A., and Herndon, W. C. (1971). *Photochem. & Photobiol.* **14,** 521.
Stepien, E., Lisewski, R., and Wierzchowski, K. L. (1973a). *Acta Biochim. Pol.* **20,** 297.
Stepien, E., Lisewski, R., and Wierzchowski, K. L. (1973b). *Acta Biochim. Pol.* **20,** 313.
Stewart, R. F. (1963). *Biochim. Biophys. Acta* **75,** 131.
Stewart, R. F., and Davidson, N. J. (1963). *J. Chem. Phys.* **39,** 255.
Sutherland, J. C., and Sutherland, B. M. (1970). *Biopolymers* **9,** 639.
Swenson, P. A., and Setlow, R. B. (1963). *Photochem. & Photobiol.* **2,** 419.
Szabo, A. G., Riddell, W. D., and Yip, R. W. (1970). *Can. J. Chem.* **48,** 694.
Szent-Györgyi, A. (1957). "Bioenergetics." Academic Press, New York.
Szer, W., Swierkowski, M., and Shugar, D. (1963). *Acta Biochim. Pol.* **10,** 87.
Sztumpf, E., and Shugar, D. (1962). *Biochim. Biophys. Acta* **61,** 555.
Sztumpf, E., and Shugar, D. (1965). *Photochem. & Photobiol.* **4,** 719.
Sztumpf-Kulikowska, E., Shugar, D., and Boag, J. (1967). *Photochem. & Photobiol.* **6,** 41.
Tramer, Z., Wierzchowski, K. L., and Shugar, D. (1969). *Acta Biochim. Pol.* **16,** 83.
Trueblood, K. N., Horn, P., and Luzzati, V. (1961). *Acta Crystallogr.* **14,** 965.
Ts'o, P. O. P. (1970). *In* "The Biological Macromolecules" (G. D. Fasman and S. N. Timasheff, eds.) p. 49. Dekker, New York.
Ts'o, P. O. P., Melvin, I., and Olson, A. (1963). *J. Amer. Chem. Soc.* **85,** 1289.
Varghese, A. J. (1971a). *Biochemistry* **10,** 2194.
Varghese, A. J. (1971b). *Biochemistry* **10,** 4283.
Varghese, A. J. (1971c). *Photochem. & Photobiol.* **13,** 357.
Varghese, A. J. (1972a). *Photochem. & Photobiol.* **15,** 113.
Varghese, A. J. (1972b). *Photophysiology* **7,** 208.
Varghese, A. J., and Rupert, C. S. (1971). *Photochem. & Photobiol.* **13,** 365.
von Wilucki, I., Matthäus, H., and Krauch, C. H. (1967). *Photochem. & Photobiol.* **6,** 497.
Wacker, A., and Träger, L. (1963). *Z. Naturforsch. B* **18,** 13.
Wacker, A., Dellweg, H., and Weinblum, D. (1960). *Naturwissenschaften* **47,** 477.
Wacker, A., Dellweg, H., and Lodemann, E. (1961a). *Angew. Chem.* **73,** 64.

Wacker, A., Weinblum, D., Träger, L., and Moustafa, Z. H. (1961b). *J. Mol. Biol.* **3,** 790.
Wacker, A., Dellweg, H., Träger, L., Kornhauser, A., Lodemann, E., Turck, G., Selzer, R., Chandra, P., and Ishimoto, M. (1964). *Photochem. & Photobiol.* **3,** 369.
Wagner, P. J., and Bucheck, D. J. (1968). *J. Amer. Chem. Soc.* **90,** 6530.
Wagner, P. J., and Bucheck, D. J. (1970). *J. Amer. Chem. Soc.* **92,** 181.
Wang, S. Y. (1960). *Nature (London)* **188,** 844.
Wang, S. Y. (1961). *Nature (London)* **190,** 690.
Wang, S. Y. (1962). *Photochem. & Photobiol.* **1,** 135.
Wang, S. Y. (1963). *Nature (London)* **200,** 879.
Wang, S. Y. (1965). *Fed. Proc., Amer. Soc. Exp. Biol.* **24,** Suppl. 15, 71.
Weinblum, D. (1961). Ph.D. Thesis, Technische Universität, Berlin.
Weinblum, D. (1967). *Biochem. Biophys. Res. Commun.* **27,** 384.
Weinblum, D. (1969). *Radiat. Res.* **39,** 731.
Weinblum, D. (1970). *Photochem. & Photobiol.* **12,** 509.
Weinblum, D., and Johns, H. E. (1966). *Biochim. Biophys. Acta* **114,** 450.
Weinblum, D., Ottensmeyer, F. P., and Wright, G. F. (1968). *Biochim. Biophys. Acta* **115,** 24.
Whillans, D. W. (1972). Ph.D. Thesis, Dept. of Medical Biophysics, University of Toronto.
Whillans, D. W., and Johns, H. E. (1969). *Photochem. & Photobiol.* **9,** 323.
Whillans, D. W., and Johns, H. E. (1971). *J. Amer. Chem. Soc.* **93,** 1358.
Whillans, D. W., and Johns, H. E. (1973). *Curr. Top. Radiat. Res.* **9,** 119.
Wierzchowski, K. L., and Shugar, D. (1959). *Acta Biochim. Pol.* **6,** 313.
Wierzchowski, K. L., and Shugar, D. (1960). *Acta Biochim. Pol.* **7,** 63.
Wierzchowski, K. L., and Shugar, D. (1961). *Acta Biochim. Pol.* **8,** 219.
Witkop, B. (1968). *Photochem. & Photobiol.* **7,** 813.
Wulff, D. L. (1963). *Biophys. J.* **3,** 355.
Wulff, D. L., and Fraenkel, G. (1961). *Biochim. Biophys. Acta* **51,** 332.
Yip, R. W., Riddell, W. D., and Szabo, A. G. (1970). *Can. J. Chem.* **48,** 987.
Zachau, H. G. (1964). *Hoppe-Seyler's Z. Physiol. Chem.* **336,** 176.
Zavil'gel'skii, G. B., Minchenkova, L. E., Minyat, E. E., and Savich, A. P. (1965). *Biokhimiya* **30,** 652.

# 6 Pyrimidine Bimolecular Photoproducts

*Shih Yi Wang*

This chapter deals with singly-bonded dipyrimidine photoproducts. On the bases of structural and photochemical mechanistic differences, these photoproducts can be grouped into three classes. As a result this chapter is divided into three parts: each is treated independently from the others since there is neither interconnection nor interdependency between them.

# PART A. PHOTOCHEMISTRY OF 5-HALOGENOURACIL DERIVATIVES

## A. 5-Bromo-, 5-Chloro-, and 5-Iodouracil Derivatives

The replacement of thymine (Thy) with 5-bromouracil (BrUra), 5-iodouracil (IoUra), and 5-chlorouracil (ClUra) in DNA increases the sensitivity of the transforming principle, viruses, bacteria, and mammalian cells to UV light as well as to $\gamma$- or x-rays (Greer and Zamenhof, 1957; Kozinski and Szybalski, 1959; Greer, 1960; Djordjevic and Szybalski, 1960; Kaplan *et al.*, 1961, 1962; Opara-Kubinska *et al.*, 1961; Stahl *et al.*, 1961; Sauerbier, 1961; Smith, 1962). [BrUra incorporation into DNA itself also induces mutations (Zamenhof *et al.*, 1958; Benzer and Freese, 1958; Litman and Pardee, 1960; Howard and Tessman, 1964; Puck and Kao, 1967; Rosner and Yagil, 1969); however, these mutations do not necessarily relate to photochemical changes.]

The photochemical mechanism responsible for this effect has been studied intensively and two possible molecular mechanisms have been suggested: (1) formation of a coupled product, 5,5′-diuracilyl; and (2) single-strand breaks in DNA.

A third alternative, that Ura itself is a lethal product, appears to be unlikely. This section will present the findings obtained so far and the pros and cons for each mechanistic interpretation. The discussion will be restricted to the chemical aspects since the *in vitro* and *in vivo* DNA studies are presented in Chapter 3, Volume II.

### 1. Formation of a Coupled Product, 5,5′-Diuracilyl

#### *a. Uracil as a Major Photoproduct*

BrUra, a structural analog of Thy, can be incorporated in place of Thy in bacteria (Weygand *et al.*, 1951, 1956). The resulting UV sensitivity of DNA (Greer and Zamenhof, 1957) was shown to be propor-

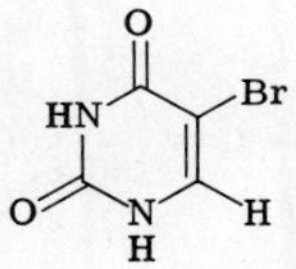

BrUra 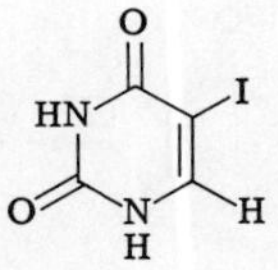 IoUra 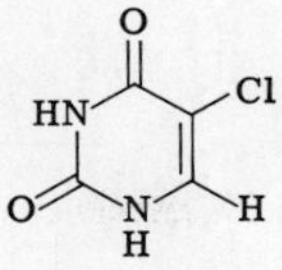 ClUra

tional to the amount of Thy replaced by BrUra (Greer, 1960). In order to understand the molecular basis of this phenomenon, several 5-halogenouracil derivatives have been studied.

When irradiated with 254 nm light ($1.5 \times 10^5$ ergs/mm$^2$) in an aqueous solution (0.1 m*M*), BrUra is only slightly altered (Wacker *et al.*, 1961), and only 3% BrUrd undergoes change. However, when [2-$^{14}$C] BrUra was incorporated into the DNA of *E. coli* 15T$^-$, a 56% change *in vivo* and a 72% change *in vitro* were observed under the same conditions, as shown by the results of paper chromatograms of the acid hydrolysates. Based on $R_f$ values, the main photoproduct was identified as Ura. Similar experiments performed with 313 and 366 nm light gave essentially identical results (Wacker *et al.*, 1962).

In contrast, irradiation of IoUra in aqueous solution at 290 nm resulted in nearly quantitative deiodination (Wacker *et al.*, 1964). The absorption spectrum of IoUra with a maximum at 284 nm gradually changes to the spectrum of Ura ($\lambda_{max} = 258$ nm). Dehalogenation to Ura was also assayed in DNA isolated from *Enterococcus stei* grown with radioactively labeled ClUra, BrUra, or IoUra by paper chromatography of the acid hydrolysates of the cells (Dellweg and Wacker, 1964). The latter two were found to dehalogenate at similar rates while ClUra was much slower. Dehalogenation was fast with native DNA, slower with heat-denatured DNA, and only slight with apurinic acid. These results, which are consistent with the formation of Ura from 5-halogenouracil by dehalogenation, led to the conclusion that UV sensitization of the cells upon 5-halogenouracil incorporation was due to photochemical formation of Ura (Wacker *et al.*, 1964).

*b. Uracil Radical Formation*

Subsequently, the photochemical reactions of IoUra were investigated in more detail (Rupp and Prusoff, 1965a). (See Scheme 1.) Irradiation of IoUra in an aqueous solution resulted in a decrease of UV absorption longer than 250 nm. The possibility of a photohydrate of either IoUra or Ura was ruled out since heating under acidic conditions resulted in no spectral changes. Based on qualitative tests, the major product was believed to be alloxanic acid (**I**). Since the conver-

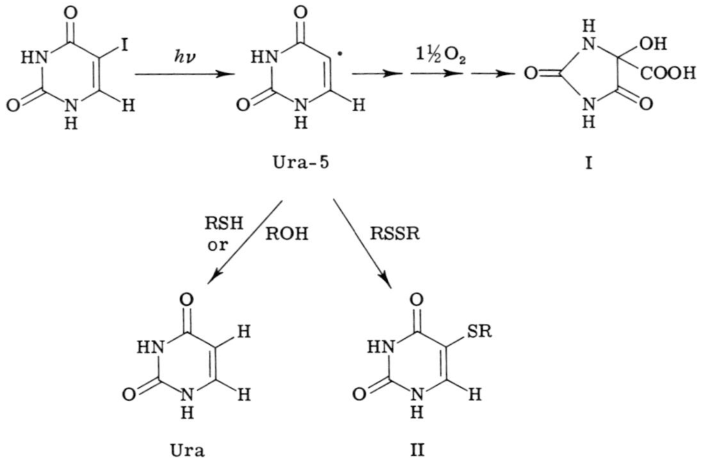

$h\nu$ — Ura-5· — $1\frac{1}{2}O_2$ — I

RSH or ROH — RSSR

Ura — II

**Scheme 1**

sion of IoUra to **I** requires the addition of oxygen atoms, the uptake of $O_2$ during irradiation was measured and found to be substantial. Also, the release of iodine during irradiation was found to be 0.73–0.99 mole equivalent (see also Gilbert and Schulte-Frohlinde, 1970). This evidence suggests that the loss of an I atom may result in the formation of a Ura-**5**· radical. Additional evidence for Ura-**5**· was gained by the irradiation of IoUra in ethanol or in aqueous solutions of mercaptoethanol or cysteamine which serve as radical scavengers (Rupp and Prusoff, 1965b). Both Ura and 5-CysUra (**II**, R = cysteinyl) were suggested as photoproducts. This radical scavenging may explain the specific protection afforded by cysteamine and cystamine against inactivation by UV irradiation in T1 phage substituted with BrUra (Hotz, 1963, 1968) or IoUra (Rupp and Prusoff, 1964). Additionally, the fact that the major function of ethanol or mercaptoethanol in UV protection is the formation of Ura argues against the earlier conclusion that Ura is a lethal product.

*c. Isolation of 5,5′-Diuracilyl*

In view of this conclusion which suggests the nonlethality of Ura in DNA and considering that *B. subtilis* phages PBS-1 and PBS-2 contain Ura in place of Thy (Takahashi and Marmur, 1963) and that the Kornberg DNA polymerase can use dUTP for dTTP (Bessman *et al.*, 1958), lethal photoproducts other than Ura should be considered. For such a purpose, a study was made involving a thorough investigation of the photochemistry of BrUra and its 1,3-dimethyl derivative, $Me_2BrUra$

**Table 1** Photoproducts from 10 mmoles (1.91 g) of BrUra[a]

| Method of separation | Photoproduct | Yield | |
|---|---|---|---|
| | | mg | mequiv |
| Dowex 1-X8($Cl^-$) column | Glyoxaldiureine | 48 | 0.34 |
| | Ura | 100 | 0.89 |
| | Barbituric acid[b] | | |
| | Oxalic acid | 23 | 0.18[c] |
| | Isoorotic acid[d] | | |
| | Ura(5–5)Ura (**V**) | 29 | 0.13 |
| | BrUra | 425 | |
| Ether extraction | Parabanic acid | 12 | 0.11 |
| Dowex 50-X8($H^+$) column | Urea | 35 | 0.09[e] |
| | Ammonia | 1.5 | 0.09 |
| 2,4-Dinitrophenylhydrazone | Glyoxal | 250 | 0.60 |
| | Formaldehyde | Not detected | |
| Others | 5-OHUra | Not detected | |
| | Oxamide | Not detected | |
| | Alloxanic acid | Not detected | |

[a] From Ishihara and Wang (1966c).
[b] By chromatography and ultraviolet spectrum.
[c] As Ca salt.
[d] By $R_f$ and ultraviolet spectrum.
[e] As dixanthylurea.

(Ishihara and Wang, 1966a,b,c). Irradiation in an aqueous solution at biological doses results in a slight but gradual increase at the ~240 nm region and no appreciable decrease in the absorbancy maximum at the ~280 nm region. These spectral changes are indicative of a photoproduct having high absorbancy in the 240 nm region, retaining the Pyr nucleus, and forming within the dose range used for biological studies. In order to identify this photoproduct, both compounds were subjected to extensive irradiation and the photoproducts were separated and identified by UV, IR, and, at times, NMR spectra and synthesis. The quantitative results are summarized in Tables 1 and 2.

It is apparent that BrUra and $Me_2$BrUra yield parallel results, and the main features may be easily perceived in Scheme 2. With the presence of oxygen in the reaction medium, Ura($O_2\cdot$)$^\alpha$ could be formed and give rise to the oxidative products (Ishihara and Wang, 1966a). Indeed the formation of Ura($O_2\cdot$)$^\alpha$ was found to be the predominant reaction in an aerated aqueous solution and Ura($O_2\cdot$)$^\alpha$ appeared to be quantitatively converted to isodialuric acid (**III**) (Gilbert *et al.*, 1971; Gilbert and Wagner, 1972). In the presence of potassium iodide, a reducing

**Table 2** Photoproducts from Irradiation of 10 mmoles (2.19 g) of 1,3-$Me_2$BrUra

| Fraction | Product | Yield mg | Yield mequiv |
|---|---|---|---|
| Ether | $Me_2$BrUra | 258.5 | 1.2 |
| | *sym*-Dimethyloxamide | 13.5 | 0.11 |
| | $Me_2$Ura | 135.2 | 0.97 |
| | $Me_2$Ura(5–5)$Me_2$Ura (**V**) | 75.1 | 0.27 |
| | *sym*-Dimethylurea | 25.2 | 0.29 |
| Distillate | Methylamine | 20.8 | 0.67 |
| | Ammonia (as $CH_3NH_2$) | | |
| Aqueous | $Cb^5Me_2$Ura | 18.2 | 0.12 |
| | Acetic acid (as $CH_3COONa{\cdot}3H_2O$) | 8.5 | 0.07 |
| | Aldehyde | No 2,4-dinitrophenylhydrazone precipitate | |
| | $ho^5Me_2$Ura | Not detected by ferric chloride test | |

[a] From Ishihara and Wang (1966b).

agent, isobarbituric acid (**IV**) and other unknown compounds may be formed at the expense of isodialuric acid at rates dependent on pH (Gilbert *et al.*, 1972). However, the most important finding is the isolation and identification of the coupled products (**V**), Ura(**5–5**)Ura and $Me_2$Ura(**5–5**)$Me_2$Ura (Ishihara and Wang, 1966a). The latter can be synthesized by the coupling of $Me_2$BrUra under Ullman reaction conditions. Another chemical synthesis to Ura(**5-5**)Ura has recently been accomplished by H. Ishihara (personal communication). This involves the reaction sequence: hydurilic acid ($R_2{=}R_4{=}R_6{=}OH$) → hexachloro-5,5′-dipyrimidine (m.p. 194–195°C) → 2,4,2′,4′-tetramethoxy-6,6′-dichloro-5,5′-dipyrimidine (m.p. 132–134°C) → 2,4,2′,4′-tetramethoxy-5,5′-dipyrimidine (m.p. 215–217°C) → 2,4,2′,4′-tetrahydroxy-5,5′-dipyrimidine i.e. Ura(**5-5**)Ura. Hydurilic acid was prepared according to the method of Biltz and Heyn (1919). The procedures for the conversion of this acid to Ura(**5-5**)Ura are analogous to those of barbituric acid to Ura (Baddiley and Topham, 1944; Fisher and Johnson, 1932).

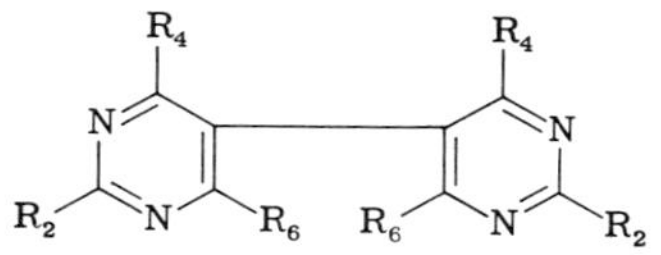

The following conclusions may be inferred:

1. Ura derivatives are formed in highest yields as previously shown.

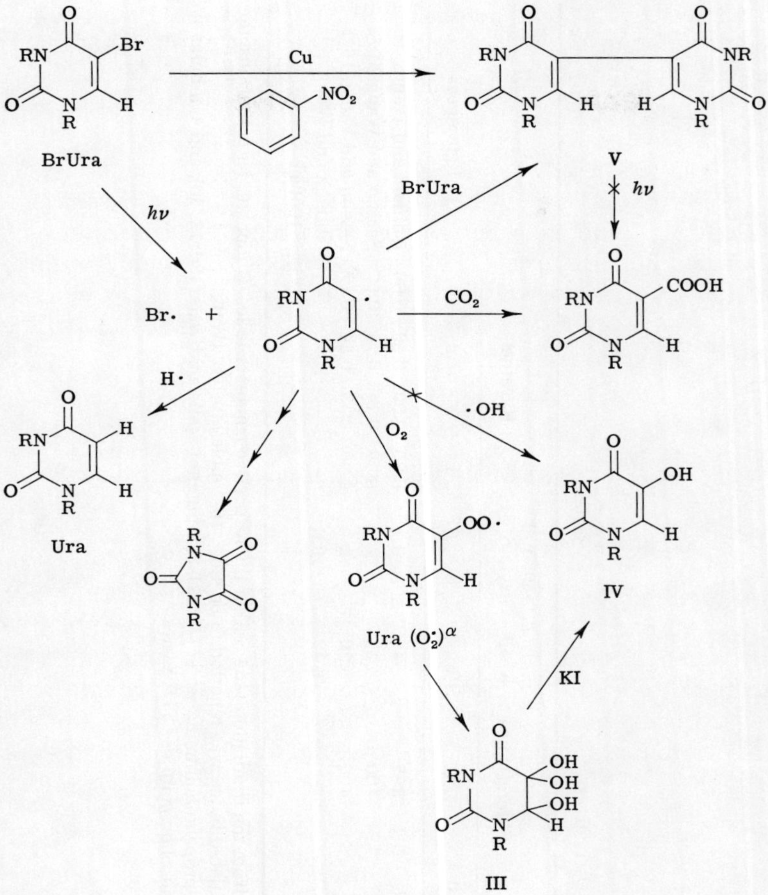

**Scheme 2**

2. The coupled product, Ura(5–5)Ura, is an end product which is unchanged upon further irradiation (Ishihara and Wang, 1966a,b,c).

3. UV spectral properties of Ura(5–5)Ura (Table 3) conform to the spectral changes observed when BrUra and $BrMe_2Ura$ are irradiated at biological UV doses, and, therefore, these products may be of importance in photobiology.

4. The formation of the coupled products probably requires the collision of a Ura-5· with Ura or two Ura-5· radicals.

5. Ura-5· radicals can also react with surrounding molecular species including HOH but will react preferentially with good H· donors such as ROH or RSH. The rate constants for the reaction of Ura-5· with H-donor solvents and various compounds have been stud-

**Table 3** Properties of 5,5′-Diuracilyl and Derivatives

| | UV Absorption | | Fluorescence | NMR | Quantum yield[a] | $R_f$ value[b] | | | | |
|---|---|---|---|---|---|---|---|---|---|---|
| Compound | $\lambda_{max}^{HOH}$ (nm) | $\epsilon_{max} \times 10^{-3}$ | $\lambda_{max}^{HOH}$ | H(6), $\delta$ | $\phi \times 10^3$ | 1 | 2 | 3 | 4 | References |
| Ura(5–5)Ura | 269 | 15.0 | 395 | — | ~0 | 0.00 | 0.00 | 0.00 | 0.00 | Ishihara and Wang, 1966a |
| Urd(5–5)Urd | 277 | 16.0 | 400 | 8.22 | 2 | 0.19 | 0.14 | 0.14 | 0.07 | Sasson and Wang, 1975 |
| dUrd(5–5)dUrd | 278 | 16.5 | 400 | 8.30 | 2 | 0.36 | 0.33 | 0.34 | 0.17 | Sasson and Wang, 1975 |
| $Me_2$Ura(5–5)$Me_2$Ura | 272 | 15.1 | — | 8.42 | ~0 | — | — | — | — | Ishihara and Wang, 1966a,b |

[a] Quantum yields given were estimated from the irradiation (254 nm) of the respective monomer in ice. Under similar conditions, the quantum yield of Thd◇Thd from Thd irradiated in ice was reported to be $4.0 \times 10^{-3}$ mole/Einstein (Wang, 1961).

[b] Eluent 1, *n*-propanol/water (10:3); 2, *s*-propanol/ammonia/water (7:1:2); 3, *n*-butanol/acetic acid/water (80:12:30); and 4, *n*-butanol/water (84:14) were used to obtain these values on cellulose TLC plates.

ied (Ebel and Kraljic, 1971; Gilbert and Wagner, 1972; Campbell *et al.*, 1974; Hutchinson, 1973).

Considering the stereoconfigurations of DNA, the formation of Ura(5–5)Ura by reaction of intrastrand neighboring bases should be favored, thus enhancing the possible importance of Ura(5–5)Ura in photobiology. It remains to be established whether or not Ura(5–5)Ura is actually formed in DNA *in vivo* and *in vitro*. The most direct approach would be to repeat the study of Wacker *et al.* (1961) and to look for Ura(5–5)Ura. This was done with *E. coli* 15T$^-$ cells with [2-$^{14}$C]BrUra incorporated into its DNA (H. Ishihara and S. Y. Wang, unpublished data). The cells after irradiation were hydrolyzed in trifluoracetic acid at 175°C and chromatographed on paper with *n*-butanol/acetic acid/water (80:12:30). Two main products were located at $R_f$ 0.49 (Ura) and $R_f \sim 0.00$; the ratio of the two was 50:3 at $2.88 \times 10^4$ ergs/mm$^2$. The radioactivity counts for the appearance of Ura increased with increasing UV doses but those of the $R_f$ 0.00 material were inconsistent. However, the spot at the origin showed the same $R_f$ value and fluorescent properties as those of Ura(5–5)Ura. As with Ura(5–5)Ura, many solvents failed to elute this material from the origin (a strong basic solvent such as 0.5 *N* NaOH caused too much trailing). While this is not conclusive evidence, it is indicative that Ura(5–5)Ura is formed in *in vivo* systems. Although our finding of two photoproducts is considerably different from that reported by Smith (1964), he also detected a photoproduct with $R_f \sim 0.01$; its radioactivity counts increased linearly with increasing UV doses.

In addition to the incorporation of BrUra into DNA in place of Thy, polyribobromouridylic acid [poly(rBrU)] can be prepared by the polymerization of BrUDP with polynucleotide phosphorylase (Riley and Paul, 1970). The photolysis of this synthetic polymer with 313 nm light has been studied (Ehrlich and Riley, 1972a,b). At a dose of $6.45 \times 10^6$ ergs/mm$^2$ at neutral pH, extensive debromination and a loss of 45% absorbancy at 280 nm was observed. Oligonucleotides found after RNase hydrolysis of these irradiated samples were isolated by DEAE-cellulose chromatography. The dinucleotide fraction was the most abundant and was not susceptible to alkaline hydrolysis or to snake venom phosphodiesterase. It was concluded that these dinucleotides contain a photoproduct since unirradiated poly(rBrU) solutions were degraded completely by RNase to mononucleotides. The UV spectrum of the major product of the dinucleotide fraction is characteristic of Ura(5–5)Ura and not of Pyr◇Pyr or of Pyr adducts.

Subsequently, irradiation of double-stranded synthetic polynucleotides containing BrUra and Ade have been investigated (Ehrlich and Riley, 1974). Homopolymer pairs and alternating copolymers composed of either ribo- or deoxyribonucleotides were irradiated (313 or ~285 nm). Similar changes in the UV absorption spectra, extensive debromination, strand separation, and a modest amount of strand breakage following irradiation of the homopolymer pairs were observed for rA · rBrU and dA · dBrU. Such irradiation brought about little if any change in the alternating copolymers. An RNase digest of irradiated rA · [$^{14}$C](rBrU), analyzed by column chromatography, showed components similar to those found upon irradiation of single-strand poly(rBrU), and the major product was in the dinucleotide fraction as described earlier.

Since the dinucleotide fraction was found to have properties in common with Ura(5–5)Ura, further hydrolysis of this fraction should yield Ura(5–5)Ura. However, the properties of Ura(5–5)Ura were found not to be suitable for identification purposes; therefore, Urd(5–5)Urd and dUrd(5–5)dUrd were prepared (S. Sasson and S. Y. Wang, unpublished data). In contrast to the inertness of BrUra itself, BrUrd and BrdUrd were found to form the respective coupled product in reasonably high quantum yields when irradiated (254 nm) in ice. Their properties together with those of Ura(5–5)Ura are provided in Table 3. Subsequently, Urd(5–5)Urd was found to be a major product when the dinucleotide fraction was further hydrolyzed in 3.5% $NH_4OH$ in a sealed tube at 145°–155°C for 3.5 hr (Sasson *et al.*, 1976). This hydrolyzed product was identified not only by $R_f$ values but also by its UV and fluorescence spectra as the authentic Urd(5–5)Urd.

Thus, it may be concluded that Ura(5–5)Ura is formed when BrUra containing single- and double-stranded homopolynucleotides are irradiated with 254–313 nm light. There are also indications that Ura(5–5)Ura is formed in cells under similar conditions as discussed above.

### *d. Heterocoupled Products of Bromouracil with Other Pyrimidines*

Photoreaction (254 nm) of BrUra with other bases has been studied with radioactively labeled compounds in frozen aqueous solutions or in dried films (Smith, 1963a). Irradiation of BrUra or BrUra and Thy formed no photoproducts, but when BrUra was irradiated in the presence of Ura, Urd, Cyt, or NaOH, it formed many photoproducts. Most products were devoid of bromide, but two still contained Br. One of these has been suggested to be a coupled product. The others are suspected to be the mixed dimer of Ura and BrUra and the dimer of BrUra. Since the formation of cyclobutyl dimers of BrUra derivatives has been questioned (Ishihara and Wang, 1966a), a detailed isolation and characterization of these products should be of value.

UV irradiation (280 nm) of two synthetic dinucleotides, BrdUpT and BrdUpdC has been studied (Haug, 1964; Peter and Drewer, 1970). Each gave a single photoproduct which in turn gave a single acid-hydrolyzed product. The structure for the photoproduct from BrdUpdC was not studied. The photoproduct from BrdUpT was recently suggested to have an eight-membered ring (Peter and Drewer, 1970), which would mean that the earlier suggestion (Haug, 1964) of a cyclobutene structure is incorrect. In order to demonstrate the existence of these photoproducts in UV-irradiated BrUra-DNA, the latter was hydrolyzed in trifluoracetic acid (Peter and Drewer, 1970). However, at 155°C both the decomposition of the photoproducts and incomplete hydrolysis may result, and thus the data would be difficult to interpret.

Since some difficulty has been encountered (Haug, 1964; Peter and Drewer, 1970) in assigning conventional structures for the photoproducts formed with dinucleotides containing BrUra, it seems to be appropriate here to discuss the photochemical mechanism of coupled products. In organic reactions, one of the possible modes for the formation of biaryl compounds is sometimes assumed to be the collision of two free radicals as shown in Scheme 3. However, since Ura-5· species could not be observed under flash photolysis conditions (Langmuir and Hayon, 1969), Ura-5· is probably short-lived. The steady-state concentrations of Ura-5· are, therefore, fairly low; thus the direct coupling of two radicals may become unimportant. An alternative is that a Ura-5· could react with a BrUra molecule to give **VI** which, upon elimination of Br·, would yield Ura(5–5)Ura (**V**). By a similar mechanism, Ura(5–5)Ura would be obtained when Ura-5· reacts with Ura to yield **VI** and then H· is eliminated. Likewise, Ura-5· could add to Cyt and the product would be Ura(5–5)Cyt (**VII**) via **VIII.** Although Ura-5· could also add to Thy yielding a similar intermediate **IX**, the elimination of $CH_3$ from **IX** is unlikely. Therefore, recombination of **IX** with a radical should be the termination step of this radical reaction. For dinucleotides containing BrUra, only Br· is readily available and the product would be Ura(5–5)$m^5Br^6$hUra (**X**). One may argue that H· addition is a process which is likely to yield Ura(5–5)$m^5$hUra (**XI**); however, HOH is not a good donor and will react in the absence of Br· or other better H· donors. It seems reasonable to assume that the Br-containing Ura(5–5)$m^5Br^6$hUra should be the major product. It should be noted that in the study of BrdUpT one report (Peter and Drewer, 1970) claimed that one Br atom is present per photoproduct molecule and another (Haug, 1964) found the photoproduct to be essentially free of Br. Not only could the present

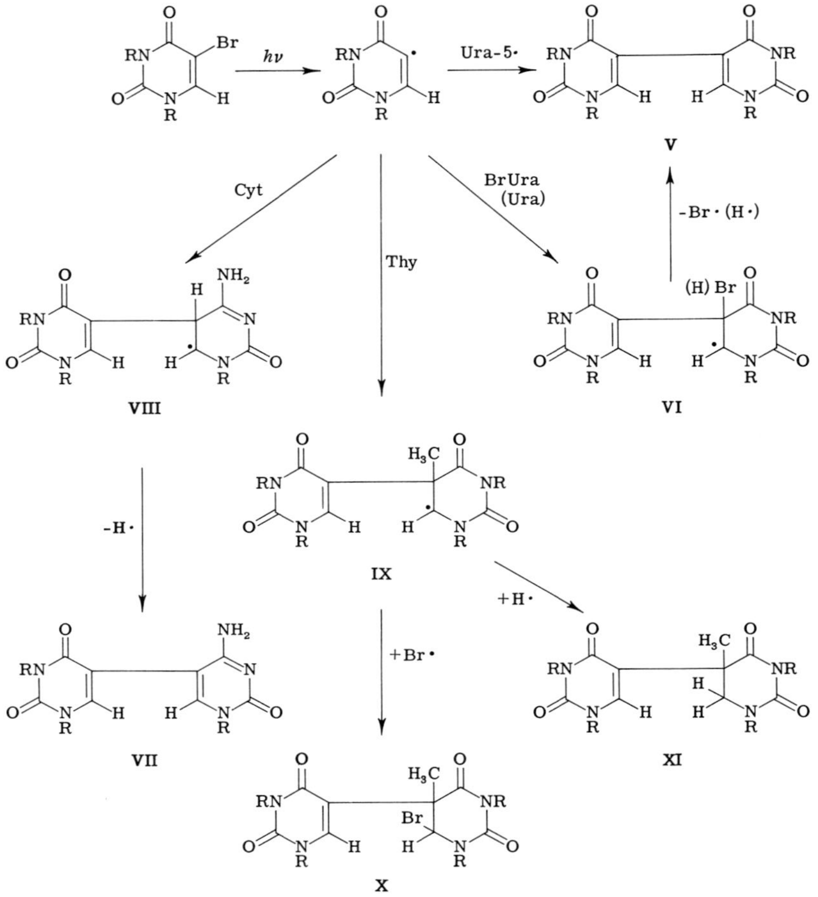

**Scheme 3**

mechanism satisfactorily account for these differences but also the molecular structure, Ura(**5–5**)m[5]Br[6]hUra, is consistent with the reported UV and NMR data [i.e., $CH_3$(s, δ 1.47); CH (s, δ 5.22); C=CH (s, δ 8.02) relative to $(CH_3)_4Si$ as external standard]. These products may, therefore, be formed in nucleic acids. Their detection should be carried out in milder conditions than acid hydrolysis in order to prevent decomposition or modification.

### e. *Excited-State Studies*

Data on the excited singlet states of these compounds are rather limited despite the availability of UV absorption spectra of halogen-

**Table 4** Ultraviolet Absorption Maxima of 5-Halogenouracil Derivatives

| Compound | Neutral | | | Monoanion | | |
|---|---|---|---|---|---|---|
| | λ nm | $\epsilon \times 10^{-3}$ | pH | λ nm | $\epsilon \times 10^{-3}$ | pH ($pK_1$) |
| Uracil (Ura) | 259.5 | 8.20 | <7 | 284 | 6.15 | 12 (9.50) |
| 5-Me | 264.5 | 7.90 | <7 | 291 | 5.44 | 12 (9.90) |
| 5-Fl[a] | 268 | 7.00 | <5 | 270 | 4.10 | 10–12 (8.00) |
| 5-Cl[a] | 275 | 7.06 | <5 | 300 | 6.30 | 10–12 (7.95) |
| 5-Br[a] | 277.5 | 7.14 | <5 | 302.5 | 6.90 | 10–12 (8.05) |
| 5-Io[a] | 285 | 6.15 | <5 | 305 | 7.10 | 10–12 (8.25) |
| Uridine (Urd) | 262 | 10.1 | <7 | 264.5 | 7.50 | 14 (9.25) |
| 5-Me | 267 | 9.85 | <7 | 268 | 7.50 | 13 (9.68) |
| 5-Fl[a] | 271 | 9.22 | <5 | 272 | 7.15 | 10–12 (7.75) |
| 5-Cl[a] | 278 | 9.11 | <5 | 278 | 6.73 | 10–12 (8.20) |
| 5-Br[a] | 279 | 9.30 | <5 | 278 | 6.40 | 10–12 (8.20) |
| 5-Io[a] | 291 | 7.66 | <5 | 282.5 | 5.45 | 10–12 (8.50) |
| Deoxyuridine (dUrd) | 262 | 10.2 | <7 | 262 | 7.60 | 12 (9.30) |
| 5-Me | 267 | 9.65 | <7 | 267 | 7.40 | 13 (9.80) |
| 5-Fl[a] | 271 | 9.17 | <5 | 271 | 7.30 | 10–12 (7.80) |
| 5-Cl[a] | 279 | 9.16 | <5 | 276.5 | 6.50 | 10–12 (7.90) |
| 5-Br[a] | 280 | 9.23 | <5 | 280 | 6.50 | 10–12 (7.90) |
| 5-Io[a] | 287.5 | 7.50 | <5 | 278.5 | 5.50 | 10–12 (8.20) |
| Uridine 5′-phosphate (pU) | 262 | 10.0 | 7 | 261 | 7.8 | 11 (9.50) |
| 5-Me | — | — | — | — | — | — |
| 5-Fl[b] | 270 | 6.0 | 1 | — | — | (7.75) |
| 5-Cl[c] | 276 | 7.8 | 2 | 274 | 6.2 | 12 (8.30) |
| 5-Br[c] | 279 | 8.7 | 2 | 275.5 | 6.5 | 12 (8.25) |
| 5-Io[c] | 289 | 8.0 | 2 | 279.5 | 5.7 | 12 (8.50) |
| Deoxyuridine 5′-phosphate (pdU) | 260 | 9.8 | 7 | 261 | 7.6 | 12 |
| 5-Me | 267 | 10.2 | <7 | — | — | (10.0) |
| 5-Fl | — | — | — | — | — | — |
| 5-Cl[d] | 276 | — | 1 | 274 | — | 13 |
| 5-Br[d] | 279 | — | — | — | — | — |
| 5-Io[d] | 285 | — | — | — | — | — |
| Poly(rU)[c] | 261 | 8.5 | 7 | 260 | 6.4 | 12 (9.81) |
| 5-Me[e] | 266 | 6.8 | 7 | — | — | — |
| 5-Fl[c] | 267 | 6.5 | 7 | — | — | (8.1) |
| 5-Cl[c] | 275 | 7.5 | 7 | 274 | — | 12 (8.35) |
| 5-Br[f] | 279 | 8.2 | 7 | 277 | 6.1 | 12 (8.43) |
| 5-Io[c] | 289 | 5.1 | 7 | 278 | — | 12 (8.64) |

[a] Berens and Shugar (1963).
[b] Kahan and Hurwitz (1962).
[c] Michelson *et al.* (1962).
[d] Dunn and Smith (1957).
[e] Griffin *et al.* (1958).
[f] Riley and Paul (1970).

ated uracils and their nucleosides (Berens and Shugar, 1963; see also Table 4). However, the excited triplet states of BrUra and IoUra have been more thoroughly examined. The usual method of estimating the energy of the lowest triplet state from the short-wavelength limit of phosphorescence emission is inapplicable because the un-ionized BrUra and IoUra do not measurably phosphoresce. Other approaches have other limitations, and the phosphorescence excitation method was resorted to for measuring the $S_0 \rightarrow T$ absorption spectra of crystalline samples of BrUra and IoUra (Rothman and Kearns, 1967). Each host crystal was doped with a small amount of naphthalene, which has a lower-energy triplet state than that of the host. The guest molecules then function as triplet energy traps for triplet halogenouracils, which results in sensitized naphthalene phosphorescence. The phosphorescence excitation spectra characteristic of BrUra and IoUra were obtained by monitoring the variation in the intensity of guest phosphorescence with the wavelength of excitation. From these spectra at 77°K, the triplet-state energy of BrUra was estimated to be ~400 nm and that of IoUra to be ~405 nm. However, the triplet states of these molecules in solution are anticipated to be ~10 nm higher in energy than in the crystal.

The triplet states of ionized BrUra and IoUra were studied in 9 *M* LiCl aqueous solution at pH 7 and 13 (NaOH added) at 77°K (Rothman and Kearns, 1967). IoUra did not appear to phosphoresce, but strong emission of BrUra anion from a pH 13 solution was observed with a lifetime of 6 $\mu$sec and $\lambda_{max}^{phos}$ at ~425 nm. However, Kleinwächter *et al.* (1966) reported that the triplet-state lifetime of BrUra anion is a factor of 100 times longer. This evidence suggests that the bromine atom in the BrUra anion does not perturb the lowest triplet state of this molecule.

In an effort to gain evidence for the formation of triplet states, Danziger *et al.* (1968) studied the photochemistry of aqueous solutions of BrUra using a fast-reaction technique of kinetic absorption spectroscopy. Under this experimental condition, no short-lived species which could be assigned to the triplet excited state was observed. Instead, evidence was obtained for the rupture of the C—Br bond and formation of Br·. Two transient species with $\lambda_{max}$ ~ 320 and 420 nm were observed to be simultaneous with the C—Br cleavage. In order to identify these two species, the flash photolysis of aqueous solutions of BrUra and its $Me^1$-,$Me^3$-, and $Me_2^{1,3}$-derivatives was studied (Langmuir and Hayon, 1969) as a function of pH in the presence and absence of oxygen and nitrous oxide. Although the yields varied, the same two intermediates were observed in all cases and were as-

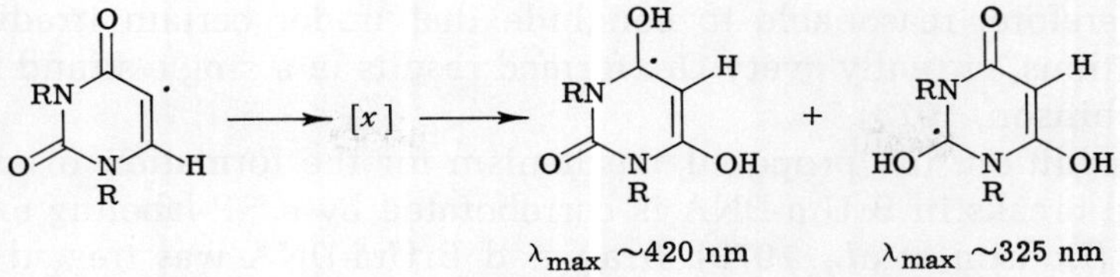

$\lambda_{max}$ ~420 nm $\lambda_{max}$ ~325 nm

**Scheme 4**

signed as ketyl radicals of barbituric acid as shown in Scheme 4. However, the intermediate steps, [$x$], leading to their formation are not obvious.

## 2. Single-Strand Breaks in DNA

Irradiation of BrUra-DNA with 254–280 nm light results in single-strand breaks as detected by denaturation with alkali (Lion, 1968; Hewitt *et al.*, 1969; Hotz and Walser, 1970; Hutchinson and Hales, 1970) or formamide (Hutchinson and Hales, 1970). Since the same sedimentation patterns were observed whether the irradiated BrUra-DNA was denatured with alkali or with neutral formamide, it may be concluded that incorporated BrUra induces actual single-strand breaks rather than alkali labile bonds in a DNA double helix. Also, the rate of production of these breaks is ~ 2–3 orders of magnitude greater than that in normal DNA (Hutchinson and Hales, 1970; Lion, 1970; Stephan *et al.*, 1970).

In order to account for these findings as well as for experimental observations, a possible mechanism was proposed (see a review by Hutchinson, 1973). In the normal B structure for DNA (see Chapter 1, Volume II), the Thy $CH_3$ group is nearest to the C(2′)H of the dRib moiety of the succeeding nucleotide. The Br atom of BrUra should occupy the same position as the Thy $CH_3$ group in DNA molecules. Upon UV irradiation, such as BrUra moiety forms a Ura-5· which may readily extract a H· from C(2′) of the nearest dRib to form Ura. The resulting sugar radicals, as determined by ESR (Koehnlien and Hutchinson, 1969), undergo further, as yet unknown reactions (Hotz and Reuschl, 1967) leading to the single-strand breaks. In the presence of radical scavenger, Ura-5· abstracts H· from the scavenger molecules rather than from dRib, thus preventing the chain breaks (Lion, 1968; Hotz and Walser, 1970; Stephan *et al.*, 1970).

It is interesting to note that the quantum yields for the formation of single-strand breaks are comparable to those for producing Ura from halogenated Ura in monomeric and polymeric states ($2–8 \times 10^{-3}$). It is, therefore, reasonable to conclude that under certain irradiation conditions, virtually every Ura formed results in a single-strand break (Hutchinson, 1973).

In addition, the proposed mechanism for the formation of single-strand breaks in BrUra-DNA is corroborated by a $^{32}$P-labeling experiment (Dodson *et al.*, 1972). Irradiated BrUra-DNA was treated with alkaline phosphatase to remove all terminal phosphate groups and the 5′-phosphate groups were restored using $\gamma$-$^{32}PO_4$-ATP with 5′-polynucleotide kinase. Subsequently, the labeled 5′-mononucleotides obtained by enzymatic degradation showed a UV dose-dependent formation of dUMP. This newly formed dUMP must have been derived from the BrUra moiety as anticipated from the model mechanism.

The coherence between the experimental findings and the proposed mechanism is exceedingly satisfactory, and the major photochemical lesions in BrUra-DNA may well be single-strand breaks. However, these conditions are insufficient to explain the biological sensitization produced by BrUra incorporation in DNA. Especially, single-strand breaks have been shown to be reparable (Smets and Cornelis, 1971; Benhur and Elkind, 1972; Ley and Setlow, 1972; Dennis and Hutchinson, 1972). It is possible that single-strand breaks may induce breaks on opposing strands thus resulting in lethal double-strand breaks. However, it has been pointed out (Hutchinson, 1973) that the occurrence of such double-strand breaks is rare.

It should be noted that the BrUra strand of both rBrU·rA and dBrU·dA was broken by irradiation as determined by the sedimentation profiles of the alkaline-denatured samples (Ehrlich and Riley, 1974). At 54% debromination (34 $\mu$E/cm$^2$, 313 nm light), rBrU·rA suffered one break per ~650 nucleotides; at 22% debromination (31 $\mu$E/cm$^2$, ~285 nm light), dBrU·dA suffered one break per ~2000 nucleotides. These experiments did not distinguish between double- and single-strand breaks in the double helices, but it is clear that the degree of breakage observed in the BrUra strand of the synthetic polymers was a great deal less than that found following irradiation of BrUra-DNA (Hutchinson and Hales, 1970; see also Hutchinson, 1973). The reason for this difference is not known.

This brief account does not include those aspects in the realm of photochemistry of DNA (see Chapter 3, Volume II) which relate to single-strand breaks in the irradiation of BrUra-DNA, but does indicate the necessity for further chemical studies of this sensitization phenomenon. Finally, it is interesting to note that many biological studies have made use of the increased radiation sensitivity of systems containing BrUra-DNA (Djordjevic and Szybalski, 1960;

Setlow and Boyce, 1963; Fox and Meselson, 1963; Denhardt and Sinsheimer, 1965; Puck and Kao, 1967; Menningmann, 1967; Cleaver, 1968; Eisenberg and Pardee, 1969), the formation of Ura-5· from IoUrd (Voytek *et al.*, 1972), and the production of single-strand breaks in DNA (Regan *et al.*, 1971; Buhl *et al.*, 1972).

## B. 5-Fluorouracil Derivatives

FlUra has been incorporated into the RNA of tobacco mosaic virus (Gordon and Staehelin, 1959). This incorporation sensitizes both the intact virus (Bećarević *et al.*, 1963; Lozeron and Gordon, 1964) and the infectious RNA (Lozeron and Gordon, 1964) to UV light. Sensitivity is proportional to the incorporation of FlUra in place of Ura. Also, the sensitivity of the 50% FlUra-substituted RNA was found to be wavelength dependent and to increase progressively as the wavelength of the irradiation increases from 254 to 297 nm. This was attributed to the absorbancy differences between FlUra and Ura. On the other hand, *E. coli* cells grown in the presence of FlUra were found to respond to UV irradiation differently depending on the concentration of FlUra present during preincubation. Low concentrations of FlUra enhanced resistance to UV light, whereas higher concentrations reduced resistance (Ben-Ishai *et al.*, 1962). FlUra-induced resistance to UV was found to be due to the incorporation of FlUra into messenger RNA and to the accumulation of this RNA fraction (Ben-Ishai *et al.*, 1965). Upon removal of FlUra, a rapid turnover of this FlUra-containing messenger RNA occurred and resulted in the loss of the induced resistance. This induced resistance may indicate that FlUra is less sensitive to UV than Ura.

Study of the photochemistry (254 nm) of FlUra in aqueous solution revealed that the major photoproduct when compared with the synthetic material (80%) is 5-fluoro-6-hydroxyhydrouracil ($Fl^5OH^6hUra$) (see Scheme 5) (Lozeron *et al.*, 1964). The structure and synthesis of this irradiation product are parallel to those of 1,3-dimethyl-6-hydroxyhydrouracil, a photohydrate (Wang *et al.*, 1956; Moore and Thomson, 1957; Wang, 1958a). This compound is stable in 1 *N* HCl but decomposes in 1 *N* NaOH to form urea and probably fluoromalonaldehydic acid (Lozeron *et al.*, 1964).

In a similar study (Fikus *et al.*, 1965), 1,3-$Me_2$FlUra was shown to give initially a photohydrate which upon further irradiation yields 1,3-dimethylbarbituric acid by elimination of HF. This latter step is again parallel to that observed in the photohydration of 1,3-$Me_2$Ura

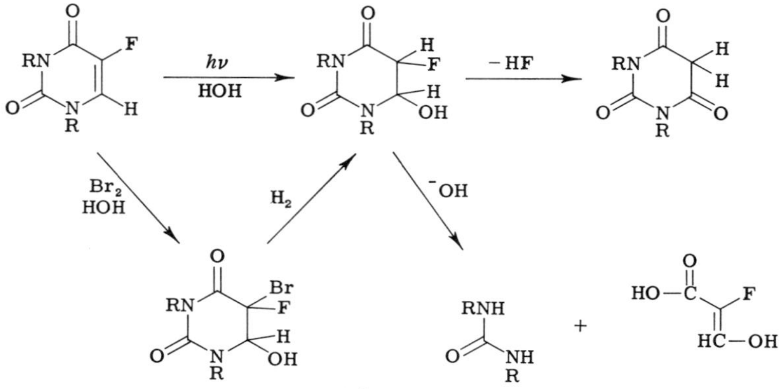

Scheme 5

(Wang, 1958b). This photohydrate of $Me_2FlUra$ is stable in neutral and acid medium at room temperature. In 1 $N$ HCl at 100°C, $Me_2FlUra$ is regenerated by dehydration. The dehydration reaction gives first-order rate constants and shows a deuterium isotope effect of 2.6 ($K_{D_2O}/K_{H_2O}$); its energy of activation is calculated to be 37 kcal/mole. Similar ring-opening products such as those of FlUra hydrate were obtained in an alkaline solution (pH > 10). Fikus *et al.* (1965) also studied the photochemical transformation of nucleosides and nucleotides of FlUra and found that their behavior is qualitatively similar to FlUra. Table 5 lists the quantum yields for these compounds in neutral and anionic forms.

It remains to be established whether or not FlUra incorporated in

**Table 5** Quantum Yields for Photochemical Transformation of $10^{-3}$–$10^{-4}$ $M$ Solutions of 5-Fluorouracil Derivatives[a]

| Derivative | $\phi_{H_2O} \times 10^2$ | | $\phi_{H_2O}/\phi_{D_2O}$ |
|---|---|---|---|
| | pH 5.6 | pH 9.5 | |
| FlUra | 1.7 | 1.0 | — |
| 3-MeFlUra | 1.6 | — | — |
| $Me_2FlUra$ | 2.3 (1.6)[b] | 2.1 | 1.43 |
| FlUrd | 4.7 | 4.0 | — |
| FldUrd | 4.4 (3.2) | 4.4 | 1.37 |
| 2′,3′-iprFlUrd | 4.1 | 3.3 | — |
| 3-MeFldUrd | 3.7 (2.7) | — | 1.37 |
| FU-5′-P | 4.5 | 3.3 | — |
| Poly(FU) | 6.0 (6.0) | — | 1.00 |

[a] In 0.02 $M$ phosphate, pH 5.6 (neutral forms) and in 0.02 $M$ borate pH 9.5 (anionic forms), irradiation was at 253.7 nm at incident intensity of $2 \times 10^{-7}$ E/min/cm².

[b] Values in parentheses refer to $D_2O$ solutions.

RNA undergoes photohydration as in the monomers. The most direct approach is to study the irradiation of poly(FU) (Fikus *et al.*, 1965). In a pH 5.6 phosphate-buffered solution, irradiation at wavelength >270 nm resulted in a first-order decrease of the poly(FU) absorption maximum. After the complete disappearance of the absorption maximum, the irradiated solution showed only a slight reversal when acidified to 1 *N* HCl and heated for 90 min. However, on raising the temperature to 125°C, a further 90 min heating largely restored the original optical density. From this the authors concluded that the major reaction in poly(FU) is photohydration. Although the rather drastic conditions required for dehydration leave such a conclusion open to question, the possibility that the dehydration rates for FlUra hydrate in polymers are slower than monomers should be considered. However, as seen in Table 5, the quantum yields determined for photohydration in light and heavy water indicate that the isotope effects observed for the monomers ($\phi_{H_2O}/\phi_{D_2O} = 1.37–1.43$) are absent in poly(FU). This difference implies that the spectral reversal may be due not to the dehydration of FlUra hydrate but to other reactions such as dimer reversal.

It would be appropriate to ask why the formation of FlUra dimer is now suggested while the earlier notion (Ishihara and Wang, 1966a) indicated that a BrUra homodimer, if formed, may be unstable, since the electronic effect of the Br atom would be expected to confer instability on the dimer, possibly causing decomposition to Ura(5–5)Ura according to Scheme 6. The major reason is probably that the strong C—F bond, (105 kcal/mole), differing from C—Br bonds (66 kcal/mole) gives stable dimers rather than homolysis or heterolysis. Thus, photohydration, photodimerization, etc., instead of the formation of Ura-5·, become important chemical processes for FlUra.

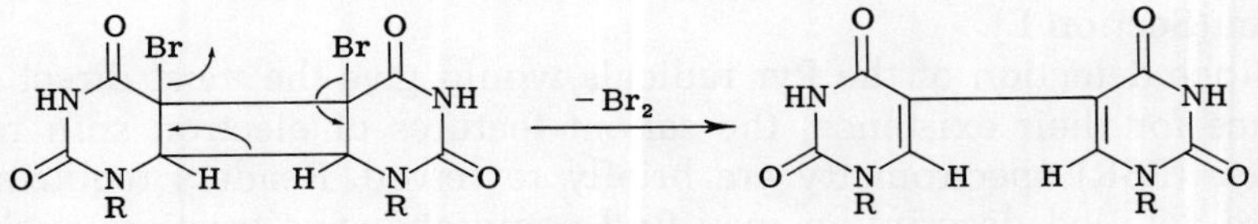

$$\xrightarrow{-Br_2}$$

Scheme 6

# PART B. BIPYRIMIDINE PHOTOPRODUCTS DERIVED FROM PYRIMIDINE RADICALS

Irradiation of Pyr in the presence of suitable hydrogen acceptors or hydrogen donors may lead to Pyr radicals which then yield stable products by radical combinations or proton abstractions. Two main classes of Pyr radicals have been characterized. One is composed of 5,6-dihydropyrimidine radicals which are derived from Ura and Thy and have an odd electron on the Pyr ring, such as hUra-5·, hUra-6·, hThy-5·, and hThy-6·. While similar radicals are expected to form

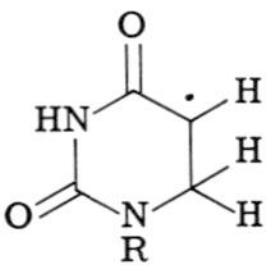

hUra-5· 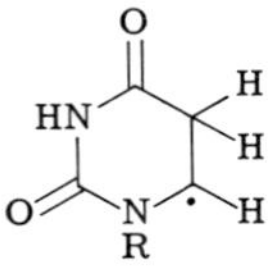 hUra-6· 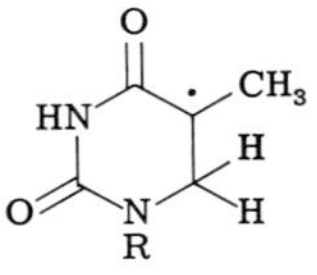 hThy-5·

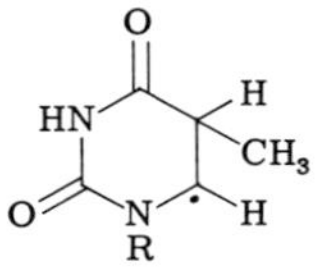

hThy-6· 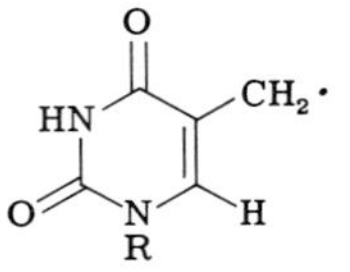 Thy$^{\alpha}$·

with other Pur and Pyr derivatives by the addition or abstraction of H·, Pur and Pyr radicals are also formed by the attack of a radical other than H· on the double bonds. The latter reactions, which involve the heteroaddition products of amino acids, alcohols, etc., are discussed in Chapters 7 and 12, Volume I, and Chapter 5, Volume II. The other class is the thyminyl radicals (Thy$^{\alpha}$·) derived from Thy by H· abstraction from the $\alpha$-methyl group. Only bipyrimidine photoproducts formed from these two classes of radicals will be discussed in this section. Since the bipyrimidine photoproducts reported to date concern only those derived from Thy, this section is further limited to the formation of hThy· and related radicals (Section C), the formation of Thy$^{\alpha}$· radicals (Section D), and the bipyrimidine products derived from them (Section E).

Since detection of the Pyr radicals would give the most direct evidence for their existence, the salient features of electron spin resonance (ESR) spectrometry are briefly reviewed. Readers requiring a more detailed description may find comprehensive treatises such as Gordy *et al.* (1953) and Ingram (1958) useful. A radical is distinguished from other chemical species by its possession of an odd electron which gives the molecules a net magnetic moment. However, radicals are extremely reactive and rapidly recombine to yield nonradical products. Recombination may be prevented by cooling to low temperatures or by evaporating to dryness so that the diffusion of radicals is effectively inhibited. Under such conditions, the existence of radicals can be readily observed by ESR spectrometry, which detects resonance absorption of electromagnetic microwave radiation by radicals, more generally paramagnetic molecules, placed in a magnetic field of several thousand gauss. This method not only furnishes information on the molecular structure but also offers the possibility of determining the radical concentrations. Due to these advantages, most information concerning radiation-produced radicals in nucleic acids has been obtained by ESR techniques.

## C. 5,6-Dihydropyrimidine Radicals

Gordy and associates, who introduced ESR spectrometry to the study of radiation-produced radicals in biological systems and related compounds, first published ESR spectra of $\gamma$-irradiated nucleic acids and their constituents (Shields and Gordy, 1959). Similar studies were also carried out by other laboratories (Shen *et al.*, 1959; Boag and Müller, 1959). Radical structures of other constituents in nucleic acids, mostly generated by $\gamma$-irradiation, have been sufficiently characterized and satisfactorily reviewed by Müller (1967). It may be seen that these ESR studies are carried out with high-energy rather than with UV radiation. This is to be expected since the interaction of high-energy radiation with a molecule is nonselective and is mainly a function of its fractional electron density. In dilute aqueous solutions, this interaction primarily induces radical products of water which in turn generate Pyr radicals. These reactions are reviewed in Chapter 12, Volume I and Chapter 5, Volume II.

Evidence for the photochemical generation of hThy-**5**· was obtained from the ESR spectral studies on UV-irradiated (240–280 nm) DNA and Thy (Eisinger and Shulman, 1963) which showed the same octet pattern as that obtained in $\gamma$-irradiated samples (Ehrenberg *et al.*, 1963; Salovey *et al.*, 1963; Müller, 1963). The observed hyperfine octet pattern of Thy, like that of dThd, is interpreted (Shields and Gordy, 1959) as a triplet of quartets arising from sets of two ($CH_aH_b$) and three ($CH_3$) equivalent coupling protons (Fig. 1). This should resolve into twelve lines; however, due to superposition of the four central pairs of lines, the structure has only eight well-resolved hyperfine lines. The hyperfine coupling constant of the C(5) odd electron with three methyl protons is 20.8 gauss, and that of the two methylene protons is 37.7 gauss, as shown. This assignment was confirmed in a study (Pershan *et al.*, 1964) of the hyperfine structure of $CD_3$-Thy which was incorporated into the DNA of *E. coli* $15T^-$. After UV irradiation of the DNA extract, the hyperfine structure exhibited only one satellite on each side of the main line, as in the $CD_3$-Thy. Apparently the methyl

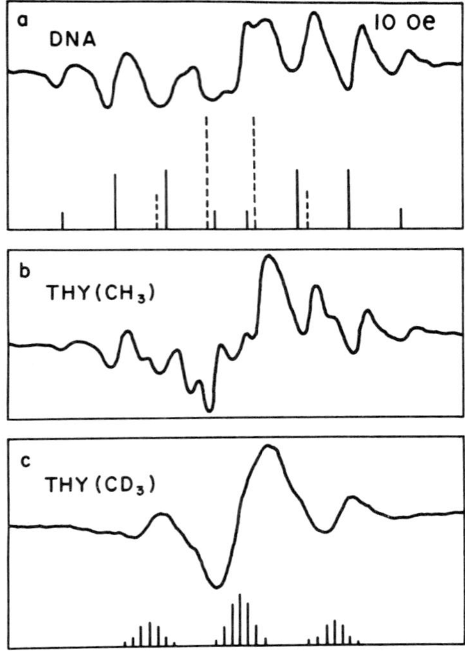


**Fig. 1.** *First derivative of ESR absorption in UV-irradiated (a) calf thymus DNA, (b) polycrystalline Thy, and (c) polycrystalline $CD_3$-Thy. The theoretical patterns expected from the* $—CH_2—\dot{C}(|)—CH_3$ *radical and from the* $—CH_2—\dot{C}(|)—CD_3$ *are shown as lines in (a) and (c), respectively. (Redrawn from Pershan* et al., *1964).*

quartet had collapsed into a broad single line because of the smaller hyperfine coupling (3.2 gauss) with the deuterons. Therefore, the Thy resonance observed in DNA after UV irradiation probably arises from an odd electron interacting with the methyl group. These radicals are produced by an addition of H· at C(6) to the C=C bond. This was tested (Pershan *et al.*, 1964; Shulman and Pershan, 1964) by means of calf thymus DNA equilibrated with $H_2O$ or $D_2O$. After UV irradiation at 77°K, the typical octet spectrum was obtained for the $H_2O$-equilibrated DNA. A different pattern emerged when $D_2O$ was used. From the observed modifications, the structure

$$—CHD—\dot{C}(|)—CH_3$$

can be assigned. Thus, ESR study has established the formation of hThy-**5**· in Thy and in DNA upon UV irradiation. Similar conclusions were also reached by other laboratories (Gordy *et al.*, 1965; Herak and

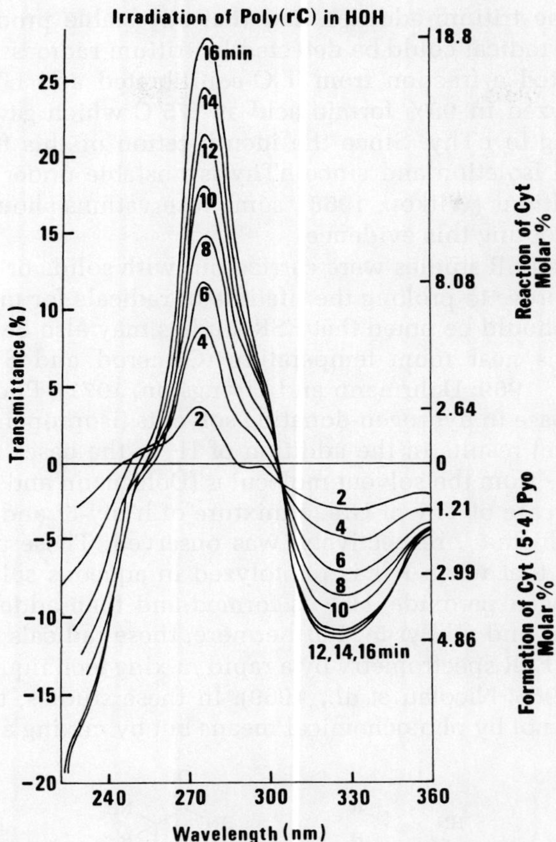


**Fig. 2.** *Difference transmittance spectra of poly(rC) irradiated at 1.17 μE/min in water at 2-min intervals (from Rhoades and Wang, 1971b).*

Gordy, 1965; Heller and Cole, 1965; LaCroix and Van de Vorst, 1967).

The identification of hThy from UV-irradiated DNA was undertaken in order to demonstrate that photoproducts derived from Thy radicals may be involved in photobiology (Yamane *et al.*, 1967). The formation of hThy was thought to occur with the addition of H· to hThy-5·(Scheme 7). It was reasoned that DNA equilibrated with $T_2O$ would

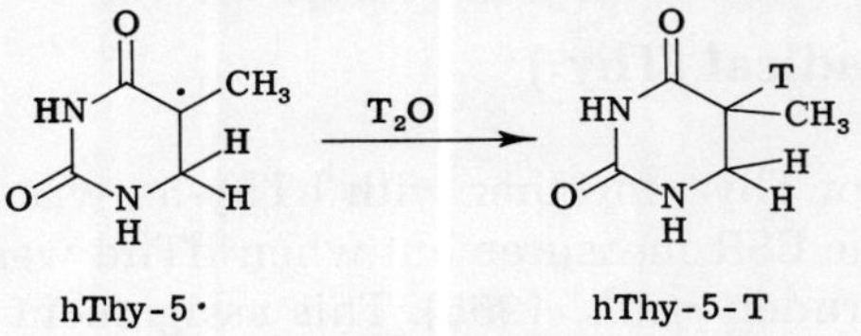

hThy-5· $\xrightarrow{T_2O}$ hThy-5-T

**Scheme 7**

result in some tritium addition and that any stable product formed from the free radical could be detected by tritium radioassay. Yamane *et al.* separated a fraction from $T_2O$-equilibrated and UV-irradiated DNA hydrolyzed in 90% formic acid at 175°C which gave $R_f$ values corresponding to hThy. Since the identification of this fraction was not by actual isolation and since hThy is unstable under the experimental conditions (Witkop, 1968), some reservations should be maintained in accepting this evidence.

While these ESR studies were carried out with solids or at low temperatures in order to prolong the lifetime of radicals for measurement purposes, it should be noted that ESR studies may also be carried out with solutions near room temperature (Ormerod and Singh, 1966; Nicolau *et al.*, 1969; Dohrmann and Livingston, 1971). Photolysis of a Pyr or a Pur base in hydrogen-donating solvents (isopropanol, ethanol, and *p*-dioxane) results in the addition of H· to the bases and the abstraction of H· from the solvent molecules (Dohrmann and Livingston, 1971). In the case of Thy or Ura, a mixture of hThy-5· and hThy-6· or hUra-5· and hUra-6·, respectively, was observed. These authors further reported that when Pyr is photolyzed in aqueous solutions containing hydrogen peroxide, HO· is formed and then added to Pyr to form $o^5$hPyr-6· and $o^6$hPyr-5·. Furthermore, these radicals can also be generated for ESR spectrometry by a rapid mixing technique (Ormerod and Singh, 1966; Nicolau *et al.*, 1969). In these studies, the radicals are generated not by photochemical means but by mixing a solution of

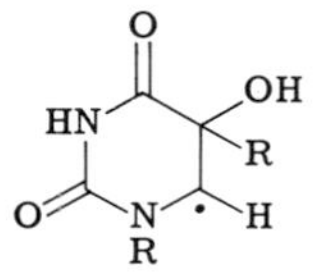

$o^5$hPyr-6· 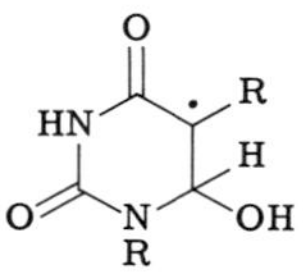 $o^6$hPyr-5· 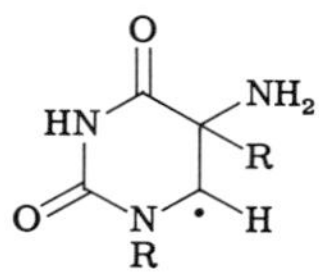 $a^5$hPyr-6· 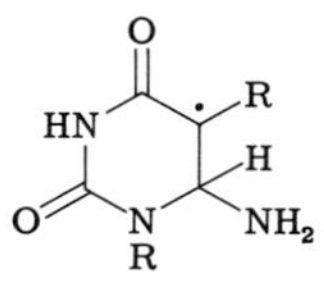 $a^6$hPyr-5·

the appropriate concentration of $TiCl_3$ in concentrated $H_2SO_4$ with $H_2O_2$ for HO· and with $HONH_2 \cdot HCl$ for $H_2N$·. These ESR findings are of importance in the study of the radiolysis of Pyr and nucleic acid-protein cross linking.

## D. Thyminyl Radical (Thy$^{\alpha}_{\cdot}$)

The existence of Thy$^{\alpha}_{\cdot}$, together with hThy-5·, was detected from a weak triplet in the ESR measurement when dThd was $\gamma$-irradiated in single crystals (Pruden *et al.*, 1965). This assignment was ambiguous since such a triplet might arise from a radical produced from the sugar moiety. However, upon, $\gamma$-irradiation of Thy monohydrate crystals at 77°K, both hThy-5· and $Thy^{\alpha}_{\cdot}$ have been identified (Hütterman, 1970) and are formed in approximately equal proportions. On the other hand, UV irradiation of Thy at pH 12 and −196°C gave two ESR resonances corresponding to anionic tautomers $Thy{:}^{1}$ and $Thy{:}^{3}$ (Rahn, 1969). Excitation at ~270 nm gave predominantly the low-field signal (1067 gauss) while excitation at ~300 nm produced only high-field resonance (1260 gauss). The former was attributed to $Thy{:}^{1}$ and the latter was assigned to $Thy{:}^{3}$, since UV absorptions by these tautomers are rather well separated with maxima at ~270 and ~290 nm, respectively (Shugar and Wierzchowski, 1965). Based on this information, an ESR study of Thy photolyzed in an alkaline glass (8 *N* NaOD at 77°K) and in an acid glass (6 *M* $D_3PO_4$ at 77°K) was carried out. ESR spectra corresponding to deprotonated $Thy^{\alpha}_{\cdot}$ and protonated $Thy^{\alpha}_{\cdot}$, respectively, were observed (Sevilla, 1971). Thus, under appropriate conditions, UV as well as $\gamma$-rays promote the cleavage of CH bonds of the 5-methyl in Thy.

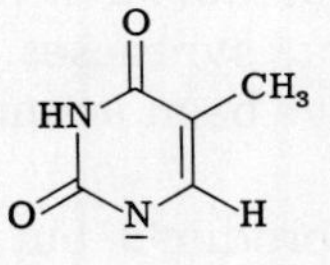

Thy: ¹ 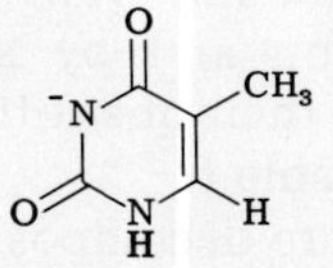 Thy: ³ 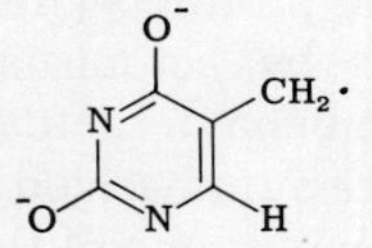 Deprotonated $Thy^{\alpha}_{\cdot}$

Although direct ESR detection of $Thy^{\alpha}_{\cdot}$ in UV-irradiated neutral solution has not been accomplished, convincing evidence for its existence during photolysis arises from photochemical studies of Thy derivatives (Alcantara and Wang, 1965a,b; Wang and Alcantara, 1965). In aqueous solutions, irradiation of Thy derivatives results in a slow decrease of the UV absorption spectra which shows no reversal by treatment with acid, alkali, or heat (Sinsheimer, 1957). This behavior is different from that of Ura derivatives. It was then assumed that this difference may be due to structural differences (Shugar and Wierzchowski, 1958). However, since the electronic configurations of Thy and Ura are similar, and since photochemical changes are closely related to their electronic states, the photohydration step should be the same for both groups of compounds (Wang, 1959). [Thy photohydration has been observed recently albeit in low quantum yield; see Chapter 4]. Spurred by the apparent lack of direct evidence of Thy photohydration, a detailed photochemical study of Thy and its 1,3-dimethyl derivative, $Me_2Thy$, following extensive irradiation in

**Table 6** Photoproducts from Irradiation (254 nm) of 10 mmoles of 1,3-Dimethylthymine (1.54 g)[a]

| Product | Yield mequiv | Yield mg |
|---|---|---|
| $CH_3NH_2$, $NH_3$ | 0.44 | 14 |
| 5-carboxy-1,3-$Me_2$Ura | 0.32 | 59 |
| $CH_3COOH$ | 0.70 | 42 |
| HCOOH | 0.91 | 42 |
| 5-formyl-1,3-$Me_2$Ura | 0.76 | 128 |
| HCHO | 1.12 | 34 |

[a] From Alcantara and Wang (1965a).

aqueous solution resulted in the demonstration of the formation of Thy$^{\alpha}$ and related photoproducts.

In the case of $Me_2$Thy, irradiation in an aqueous solution causing ~50% reduction in absorption at $\lambda_{max}$ resulted in the formation of a number of products (Alcantara and Wang, 1965a). Seven of these were identified by actual isolation and by spectra, syntheses, chromatographic behavior, etc., and their quantities have been estimated. The results are summarized in Table 6.

Obviously, some of these are decomposition products, but 5-formyl-$Me_2$Ura (a major product) and 5-carboxy-$Me_2$Ura must have resulted from the oxidation of the methyl group. If this oxidation should occur in a stepwise manner, the most likely reaction sequence would be shown in Scheme 8.

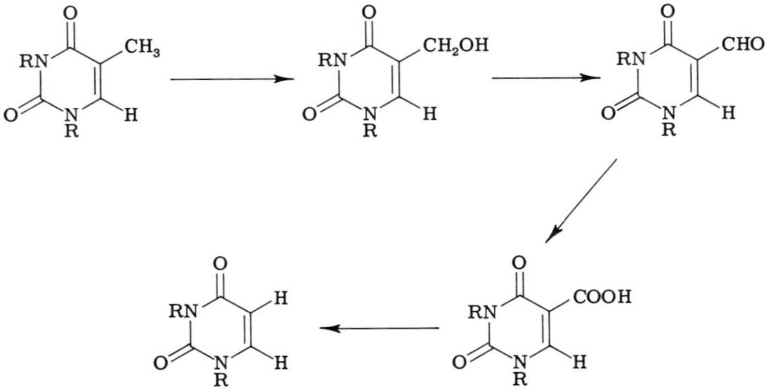

**Scheme 8**

In a similar study (Alcantara and Wang, 1965b), irradiation of Thy in an aqueous solution for ~16 hr decreased its absorption by ~50%. The products were identified by paper chromatography as 5-hydroxymethyl-Ura, 5-formyl-Ura, 5-carboxy-Ura, and Ura, among others.

In principle, the most efficient pathway for these oxidation products would be via a common intermediate formed by the absorption of one photon for each molecule. The most likely candidate would be a peroxide of Thy$^{\alpha}$· (Thy$^{\alpha}$—O—O—R where R represents H, Thy$^{\alpha}$·, or other radicals). In any event, it seems that the initial step must be the homolytic cleavage of a CH bond to yield Thy$^{\alpha}$·. This, in turn, reacts with $O_2$ to give Thy$^{\alpha}O_2$· which may lead to the eventual formation of derivatives of alcohol, aldehyde, etc. (Wang and Alcantara, 1965).

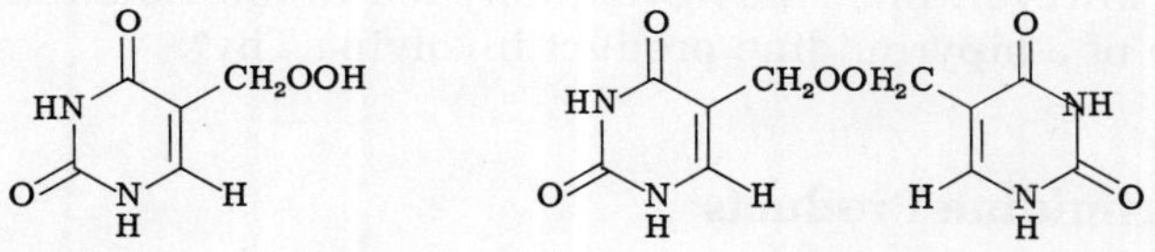

Thy$^{\alpha}O_2H$

Thy$^{\alpha}$—O—O—$^{\alpha}$Thy

Since low-pressure Hg germicidal lamps emit 185 nm as well as 254 nm light, and since both the cleavage of the CH bonds and the formation of peroxides are typical reactions for high-energy radiation, this reaction condition was designed to exclude 185 nm light (Alcantara and Wang, 1965b). Furthermore, irradiation of Thy in an aqueous solution with 185 nm light resulted in the formation of positional isomers of hydroxyhydroperoxy-Thy and of Thy glycol (Daniels and Grimison, 1967). This is understandable since a considerable fraction of 185 nm light is absorbed by water molecules to yield H· and HO·. These radicals add to the 5,6-double bond of Thy resulting in the above-mentioned products which are well known in radiation chemistry (Chapter 12). Thus, the products from the photolysis of Thy in aqueous solutions with 254 nm light are quite different from those with 185 nm light and are not due to initial water photolysis.

The oxidation of the 5-methyl group to form an alcohol, an aldehyde, and an acid which, in turn, forms Ura derivatives through decarboxylation appears to be a common reaction sequence for Thy derivatives. This photooxidation cannot occur with Ura or other bases which do not have a methyl moiety on C(5), unlike photoadduct formation, photodimerization, photohydration, etc., which occur for all natural Pyr. Therefore, this unique photoreaction of Thy may be important in the different photochemical behaviors of DNA and RNA (Wang and Alcantara, 1965).

In order to test for the removal of an H· from the C(5)-$CH_3$ of Thy, bacterial cells were doubly labeled with a mixture of Thy carrying tritium at the C(5)-$CH_3$ and $^{14}C$ at the C(2) of the ring, respectively. The labeled cells were irradiated and hydrolyzed, and the T/$^{14}C$ ratio in the chromatographically separated products was examined. The occurrence of a reaction involving the loss of one H on the C(5)-$CH_3$ would result in a reduction of the T/$^{14}C$ ratio. On the other hand, with reactions not involving C(5)-$CH_3$, the T/$^{14}C$ ratio would be unchanged. Although the originally reported results showed a reduction of the T/$^{14}C$ ratio in the photoproducts from the UV-irradiated cells (Wang *et al.*, 1967a), these findings could not be repeated and the T/$^{14}C$ ratio was found to be unchanged in the photoproducts *in vivo* (Wang *et al.*, 1967b). However, this study eventually led to the isolation and identification of a bipyrimidine product involving $Thy^{\alpha}_{\cdot}$.

## E. Bipyrimidine Products

In theory, there are three types of bithymine derivatives: Thy(α–α)Thy′, hThy(5–5)hThy′ or 5,6 isomers, and Thy(α–5)hThy. The heavy lines in the formulas below indicate the essential molecular features. The two Pyr moieties must be 5-methylpyrimidines such as Thy or Thy derivatives, but they do not have to be identical in one molecule. Thus far, the first type has been isolated from the UV irradiation of Thy(6–4)Pyo as a tetramer (Wang and Rhoades, 1971) (see Scheme 10). The third type has been isolated as a photoproduct from UV-irradiated DNA (Varghese, 1970a) and was originally separated on radiochromatograms of acid hydrolysates of DNA irradiated *in vitro*

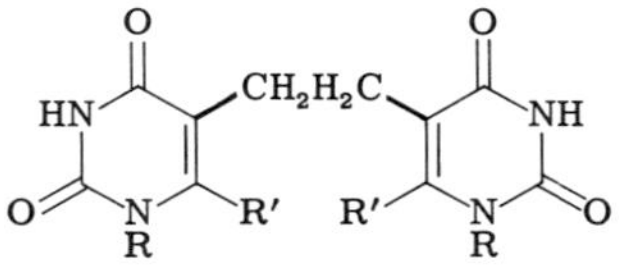

Thy(α-α)Thy′

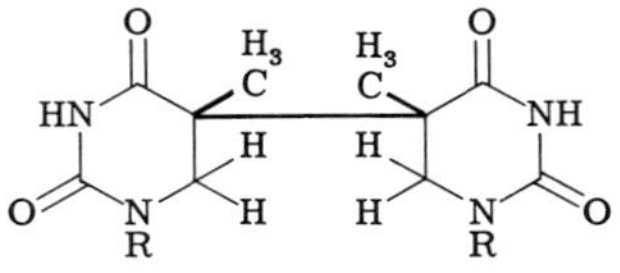

hThy(5-5)hThy′

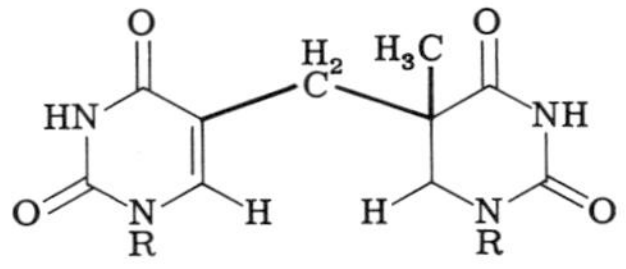

Thy(α-5)hThy

with UV light (Wang *et al.*, 1967a). A minor product is Thy($\alpha$–5)hThy ($P_3$)* which is separable from Thy◇Ura($P_1$), Thy◇Thy($P_2$A), Thy(6–4)Pyo ($P_2$B), and another as yet unidentified photoproduct ($P_4$) (Wang *et al.*, 1967a). It is interesting that $P_3$ and $P_4$ are photoproducts when DNA is irradiated *in vitro* but not *in vivo*. This finding indicates that identical DNA produces different photoproducts in different environments.

It has been known for some time that bacterial spores are very resistant to the deleterious effects of UV light, ionizing radiation, heat, etc. (Becquerel, 1910). Three photoproducts formed in spores were separated by paper chromatography from UV irradiated *Bacillus megaterium* spores and were shown to be different from the dimers (Donnellan and Setlow, 1965). A similar study was carried out with *B. subtilis* spores (Smith and Yoshikawa, 1966). The major product is apparently Thy($\alpha$–5)hThy (Varghese, 1970a). Also, dThd($\alpha$–5)hdThd was indicated as a photoproduct from irradiation of dThd in ice, (Varghese, 1970b) and as a solid film (Varghese, 1971c). This compound has a UV spectrum ($\lambda_{max} = 264$ nm and $\epsilon = 8{,}200$ at pH 2–6; $\lambda_{max} = 290$ nm and $\epsilon = 7500$ at pH 12) similar to Thy and is not reversed to Thy by irradiation in an aqueous solution. The mechanism leading to its formation should involve first the formation of a $Thy^{\alpha}\cdot$, as discussed above, which adds across the 5,6-double bond of a neighboring Thy and is followed by the H· addition at C(6) as shown in Scheme 9. The formation of Thy($\alpha$–5)hThy is a result of the coupling

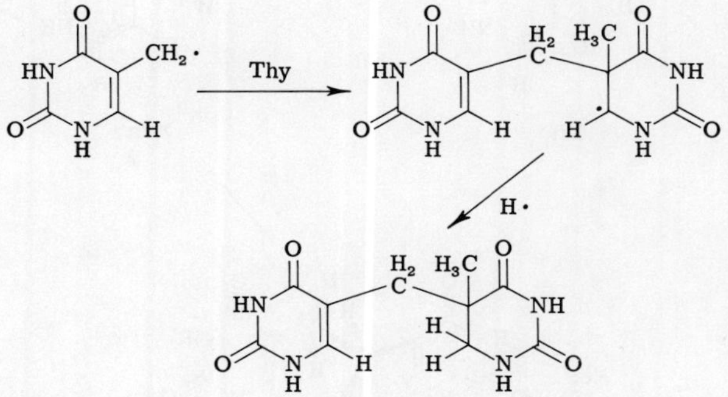

**Scheme 9**

* Two photoproducts separated from the acid hydrolysates of DNA irradiated *in vivo* are designated as $P_1$ and $P_2$ and have $R_f$ values of 0.20 and 0.29, respectively, in the solvent system *n*-butanol/acetic acid/water (80:12:30). However, four products are formed when DNA is irradiated *in vitro*. In addition to $P_1$ and $P_2$, $P_3$ and $P_4$ with $R_f$ values of 0.37 and 0.45, respectively, were found. $P_1$ is generally believed to be Thy◇Ura, $P_2$ is a mixture of $P_2$A [Thy◇Thy(*c,s*)] and $P_2$B [Thy(6–4)Pyo], $P_3$ is Thy($\alpha$–5)hThy, and $P_4$ is still unidentified.

of Thy$^{\alpha}$ and hThy-5· which has been suggested (Varghese, 1970a) to be a rare occurrence.

Since radicals generally are more stable at low temperatures or in a dry state, Thy($\alpha$–5)hThy formation should be favored under these conditions. At low temperatures, irradiation of bacteria and phages (Ashwood-Smith *et al.*, 1965; Smith and Yoshikawa, 1966) as well as isolated DNA (Rahn and Hosszu, 1968) resulted in the formation of Thy($\alpha$–5)hThy, the yields of which are a function of temperature, reaching a maximum at $\sim -100°C$ (Rahn and Hosszu, 1968). Also, after equilibration at various relative humidities, irradiation of DNA films resulted in the formation of Thy($\alpha$–5)hThy, and its yield increased sharply when the relative humidity was $<65\%$ (Rahn and Hosszu, 1969a). In contrast, the yield of Thy◇Thy is known to be in direct proportion to the relative humidity and reaches a maximum with humidity $>75\%$ (Wang, 1963). It is likely that conditions favoring dimer formation would result in lower yields of Thy($\alpha$–5)hThy and vice versa. These two competing processes probably operate both in Thy crystals and in DNA molecules.

Recently, it was found that irradiation of Thy(6–4)Pyo in an aqueous solution yields Pyr phototetramers (see Scheme 10) (Wang and Rhoades, 1971). These structures were established positively not

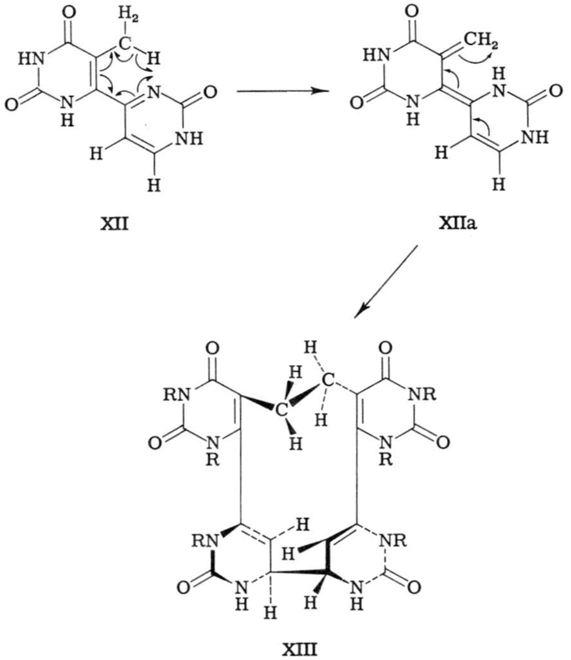

XII

XIIa

XIII

Scheme 10

only by UV, IR, mass, and NMR spectral evidence but also by x-ray diffraction analysis (Flippen *et al.*, 1971). Clearly, the $—CH_2—CH_2—$ moiety of the molecular structure of the product is conclusive evidence for the cleavage of a CH bond of a Thy derivative. A possible mechanism is given below. Also, since this photoreaction was induced by 313 and 360 nm light, 360 nm energy seems to be sufficient to form Thy$^{\alpha}$ derivatives. This leads to the suggestion that Thy$^{\alpha}$ derivative formation could occur in DNA if the longer wavelength light were absorbed by some other chromophoric groups acting as a sensitizer (Wang and Rhoades, 1971).

As a final note, it must be pointed out that the photoproduct from TpBU, according to the suggestion of Wang (see Scheme 3), may be

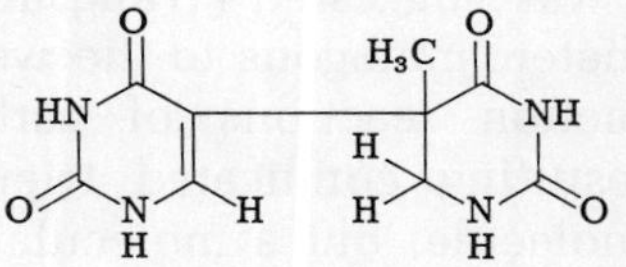

O H$_3$C O
HN NH
H
O N H H N O
H H

classified in this section since the photoproduct may be considered as the joining of Ura-5· with a Thy to give Ura(5–5)hThy as a final product. Apparently, the initial homolysis step for 5-halogenouracils to give Ura-5· radicals differs from that for Thy giving Thy$^{\alpha}$ or hThy-5·. However, the mechanism for the recombination step seems to be similar for both types of compounds.

## F. Photochemical Mechanisms

Mechanistic aspects of these photochemically produced radicals have not been reported. However, a recent study on Pyr–cysteine photoreaction (Fisher *et al.*, 1974) may be used as a model system. Charge-transfer interaction was proposed to account for the formation of hThy· with the rate-limiting step involving formation of a charge-transfer complex (Dack, 1973). The donor transfers a nonbonding electron to an excited acceptor Pyr across a hydrogen bridge (Riehl, 1968); this is followed by proton transfer before the complex can dissociate and escape the solvent cage (Lamola, 1972). Such reactions between ground-state electron donors and excited triplet acceptor molecules are well known (Loutfy and Loutfy, 1972). Furthermore, Fisher *et al.* (1974) ruled out triplet-state hydrogen abstraction as the rate-limiting step in the formation of hThy· by arguing that the rate constant for this reaction ($2–3 \times 10^8$ $M^{-1}$ sec$^{-1}$) is far too large relative to the rate constants of $10^7$ $M^{-1}$ sec$^{-1}$ generally observed for this type of process (Cohen, 1967). Additionally, no deuterium isotope effect was seen.

The formation of Thy$^{\alpha}_{\cdot}$ by the homolytic cleavage of an $\alpha$-methyl C—H bond was suggested to occur via one of two possible processes (Fisher *et al.*, 1974). It may arise from the abstraction of H· from a Thy molecule or its triplet by radicals or radical anions (Neta, 1972; Infante *et al.*, 1973). Alternatively, it may occur by a biphotonic process (Van de Vorst and Lion, 1973 ) in which the triplet–triplet absorption by a photosensitizer results in photoionization. This gives rise to an H· which leads to steps of addition and abstraction with the eventual formation of Thy$^{\alpha}_{\cdot}$.

The case of phototetramer formation via the apparent homolytic cleavage of the $\alpha$-methyl C—H bond of the photoadduct (XII, Scheme 10) is probably different. This appears to involve an intramolecular hydrogen transfer. It was suggested (Wang and Rhoades, 1971) that this process may be heteroanalogous to the well-known Norrish type II $\alpha$-hydrogen abstraction reactions of carbonyl compound (see Scheme 10). The resulting conjugated triene intermediate (XIIa) should be a planar molecule, but a molecular model shows the exocyclic methylene group forced out of the plane of the molecule, thus making it quite reactive. Dimerization in a concerted manner would yield *cis-syn* and *trans-syn* isomers of the phototetramers, XIII. Obviously, further work is required to substantiate this suggestion.

## PART C. BIPYRIMIDINE PHOTOADDUCTS

Pyrimidine derivatives form another type of UV-induced bipyrimidine product generally referred to as Pyr adducts. Homoadducts and heteroadducts formed with identical and nonidentical Pyr bases, respectively, have been studied. These adducts have fluorescent properties and UV absorbance maxima at $\geq 300$ nm which are unique among known Pyr photoproducts. It is these characteristics that led to their discovery, first from the acid hydrolysates of UV-irradiated DNA *in vivo* and *in vitro*.

### G. Heteroadduct of Thy and Cyt and of Thy and Ura and Related Compounds

#### 1. Isolation and Identification of Thy(6–4)Pyo *in Vivo* and *in Vitro*

When DNA was irradiated with 254 nm light at a dose rate of 28 ergs/mm$^2$ sec, there was a decrease in absorbance at 260 nm with a simultaneous increase in the absorbance at 320 nm. This increase seemed to vary directly with the increase in Ade–Thy content of the DNA as indicated by the irradiation of *Micrococcus lysodeikticus*, calf thymus, and *B. cereus* DNA (Wang, 1962, as cited by Smith, 1963b). This observation could not be accounted for by the formation of Pyr◇Pyr. Furthermore, the so-called Thy◇Thy band detected on paper chromatograms of the acid hydrolysates of UV-irradiated DNA shows strong fluorescence but purified Thy◇Thy does not. Apparently, this fluorescent material cochromatographs with Thy◇Thy(*c,s*) in various neutral and acidic eluents. Its identification was not determined until a suitable eluent was found for its separation from Thy◇Thy(*c,s*) (Varghese and Wang, 1967). It was found that its absorbance maxima are at 316, 316, and 304 nm in neutral, pH 2, and pH 11 aqueous solutions, respectively. Based on UV, IR, NMR, and mass spectra, a probable structure was suggested to be 6–4′-[pyrimidin-2′-one]-thymine, Thy(6–4)Pyo (referred to as $P_2B$). The original photolytic process for the formation of Thy(6–4)Pyo probably involves the addition of the C=NH group of Cyt to the 5,6-double bond of Thy to form a compound with a four-membered azetidine ring which opens spontaneously to produce a$^5$hThy(6–4)Pyo and, with the subsequent elimination of $NH_3$ during acid hydrolysis, becomes Thy(6–4)Pyo (Scheme 11) (Wang and Varghese, 1967). Scheme 11 is

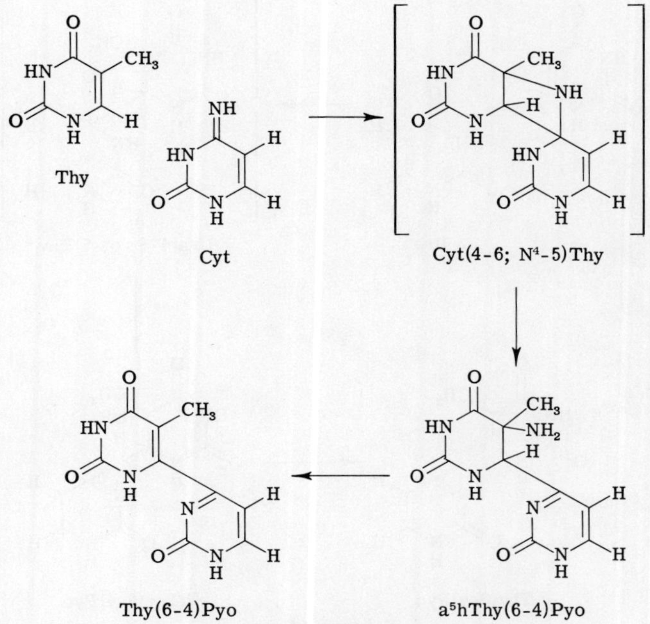

Thy

Cyt

Cyt(4-6; N$^4$-5)Thy

Thy(6-4)Pyo

a$^5$hThy(6-4)Pyo

**Scheme 11**

further supported by double-labeling experiments, since Thy(6–4)Pyo isolated from *E. coli* 15 $\overline{\text{TAU}}$ either *in vivo* or *in vitro* contained both $^3$H and $^{14}$C when $C^3H_3$-dThd and [2-$^{14}$C]Cyd were used as the precursors (Varghese and Patrick, 1969). These authors claimed that Thy(6–4)Pyo can be isolated from acid hydrolysates of equimolar mixtures of Cyd and dThd and of CMP and TMP. Since there is no direct evidence, not even a UV spectrum, for either the adduct of nucleosides or nucleotides, this finding requires confirmation, particularly since we detected no such products under a similar condition. However, the mixture of Cyt and Thy gives a barely detectable yield of Thy(6–4)Pyo. This is to be expected since the freezing process can bring two bases in a close proximity which facilitates such a bonding formation while substituents on N(1) such as sugar and phosphate moieties would prevent such a close contact. For example, quantum yields of photodimerization of dThd and TMP are ~100-fold less compared to that of Thy (Wang, 1961). (See also Chapter 2, Volume II for further discussion on this Thy–Cyt adduct in DNA.)

2. Isolation and Identification of o$^5$hThy(6–4)Pyo

Thy(6–4)Pyo may also be obtained from Thy and Ura according to Scheme 12. In this case, a four-membered oxetane ring must be the

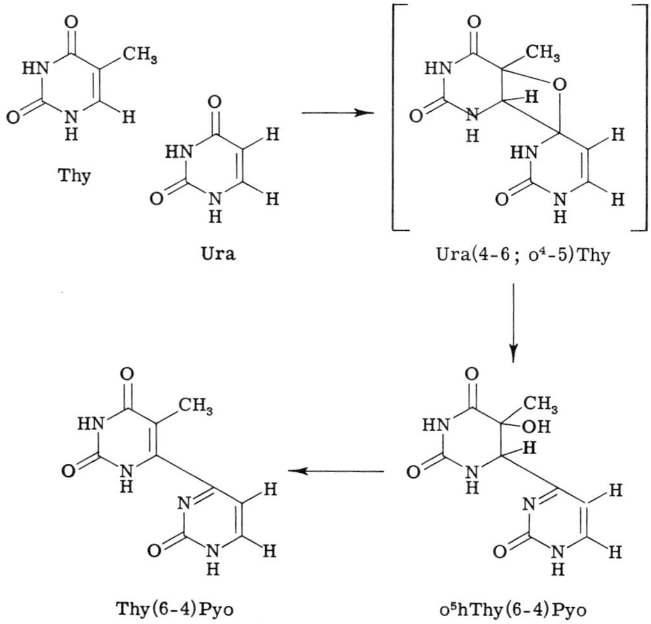

Thy + Ura → [Ura(4-6; o$^4$-5)Thy] → o$^5$hThy(6-4)Pyo → Thy(6-4)Pyo

**Scheme 12**

intermediate which will open to yield $o^5$hThy(6–4)Pyo and this compound was isolated from a Ura and Thy mixture irradiated in a frozen aqueous solution (Rhoades and Wang, 1970). Structural assignment was based on mass, UV, IR, and NMR spectra. The compound readily underwent dehydration with heat and acid to yield Thy(6–4)Pyo ($P_2B$) (Rhoades and Wang, 1970). The identification of a precursor of Thy(6–4)Pyo not only gives further credence to the reaction scheme but also provides additional knowledge of this type of compound. Elucidation of its chemical properties and structural configuration should be of help in ascertaining the role of photoadducts in photobiology.

## H. Homoadduct of Thy and Related Compounds

### 1. Thy(6–4)$m^5$Pyo and $o^5$hThy(6–4)$m^5$Pyo

Earlier, an absorbancy increase above 300 nm, comparable to that in DNA, was observed in the thawed aqueous solution of Thy irradiated in the frozen state (Setlow, cited by Smith, 1963b). Also, irradiation at 225 to 289 nm of dithymidylates (dT–dT, pdT–dT) in an aqueous solution yielded one product, among four, which has an absorbancy maximum at 325 nm (Johns *et al.*, 1964). This product is produced irreversibly at a quantum yield of $<0.001$. It fluoresces at 405 nm when excited at 325 nm, has a $pK_a$ at 10.75, and is unstable to prolonged treatment in extreme pH (Pearson *et al.*, 1965). Subsequently, the substance responsible for this increased absorbancy at $\sim 320$ nm and fluorescence at 400 nm was isolated and identified as $o^5$hThy(6–4)$m^5$Pyo by UV, IR, NMR, and mass spectra (Varghese and Wang, 1968a). (See Scheme 13.) The molecular structure was further confirmed by x-ray diffraction analysis (Karle *et al.*, 1969; Karle, 1969). This compound cochromatographs with Thy◇Thy(*c,s*) in many eluents and is responsible for the fluorescence observed in the crude dimer (Varghese and Wang, 1968a). A low-temperature study (Rahn and Hosszu, 1969b) showed that fluorescence does not appear directly after irradiation at liquid $N_2$ temperatures ($-196°C$) but does occur at temperatures above $-80°C$. This may be interpreted as possible evidence for a precursor of the adduct which is stable at low temperatures and rearranges to the adduct upon annealing. The precursor is probably the oxetane intermediate as suggested (Varghese and Wang, 1968a). $o^5$hThy(6–4)$m^5$Pyo dehydrates readily in 0.5 *N* HCl with refluxing to give Thy(6–4)$m^5$Pyo, a compound analogous to Thy(6–4)Pyo (Varghese and Wang, 1968a). The UV spectrum of the former has a maximum characteristic of a Pyo derivative, and the lat-

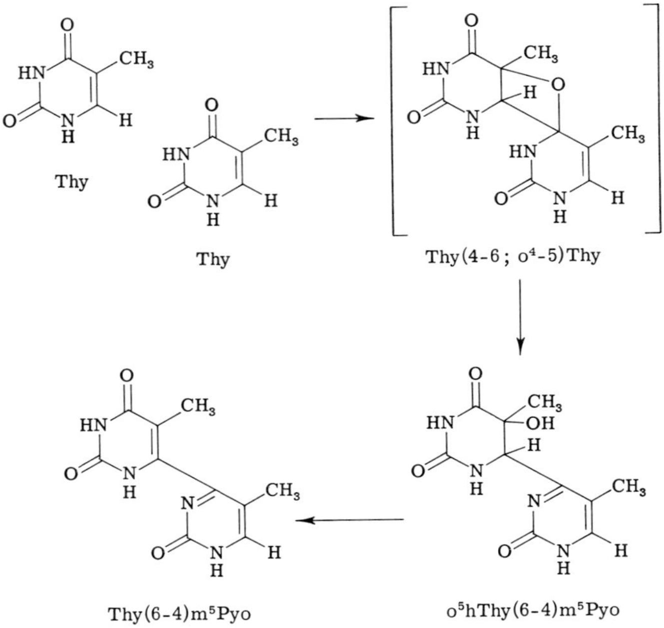

Thy Thy Thy(4-6; $o^4$-5)Thy

Thy(6-4)$m^5$Pyo $o^5$hThy(6-4)$m^5$Pyo

**Scheme 13**

ter has two maxima characteristics of Thy and Pyo derivatives, respectively. This indicates that the two chromophores may not interact and that the two rings of the molecule may be in a skew conformation (Varghese and Wang, 1968a, also see Section K). Corresponding adducts have been observed when dThd is irradiated in frozen solutions or as solid films (Varghese, 1970b; Kunieda and Witkop, 1971). Irradiation of 1-MeThy in a KBr matrix (Lisewski and Wierzchowski, 1970) was found to yield a product having properties similar to those of the Thy adduct.

## 2. Thymine Phototrimer

A trimeric product was isolated from Thy irradiated in a frozen aqueous solution (Varghese and Wang, 1968b). This compound has the same UV characteristics as one observed earlier (Smith, 1963b) except for a reported 282 nm peak which may be due to impurities since the UV spectra of the trimer has no absorption maximum in this region. The molecular structure of this trimer was established by x-ray diffraction analysis (Flippen *et al.*, 1970; Flippen and Karle, 1971), UV, IR, and NMR spectra and chemical reactivities (Wang, 1971), and mass spectrum (Fenselau *et al.*, 1970). It is composed of one water and three Thy molecules joined by a cyclobutyl ring and an adduct linkage (see Scheme 14). The diol structure gives strong evidence that the oxetane ring is in the photoproduct initially formed upon irradiation of Thy. Derivatives containing oxetane are particularly susceptible to at-

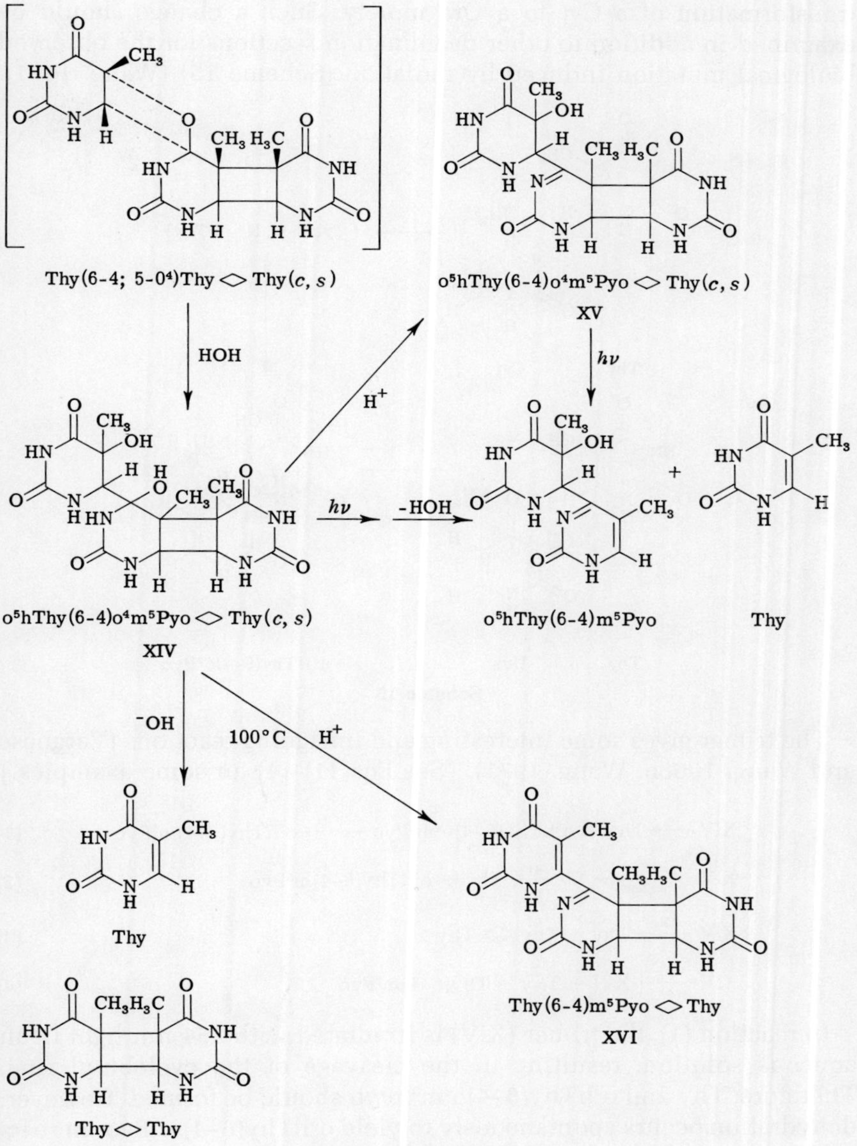

Thy(6-4; 5-0⁴)Thy ◇ Thy(*c*, *s*)

HOH

o⁵hThy(6-4)o⁴m⁵Pyo ◇ Thy(*c*, *s*)

XIV

H⁺

o⁵hThy(6-4)o⁴m⁵Pyo ◇ Thy(*c*, *s*)

XV

*hν*

*hν* - HOH

o⁵hThy(6-4)m⁵Pyo + Thy

⁻OH

Thy

Thy ◇ Thy

100° C H⁺

Thy(6-4)m⁵Pyo ◇ Thy

XVI

Scheme 14

tack by reagents such as $H_2O$ and in this case result in a ring opening to the diol moiety (Wang, 1971). This unique structure may be of interest in photobiology since it would conceivably allow the reversal of an adduct to two monomers. If such a reaction involves a Cyt moiety (as in Thy(6–4)Pyo), deamination may occur, resulting in the transformation of a Cyt to a Ura moiety. Such a change should be examined in addition to other deamination reactions for the observed biological mutation induced by radiation (Scheme 15) (Wang, 1971).

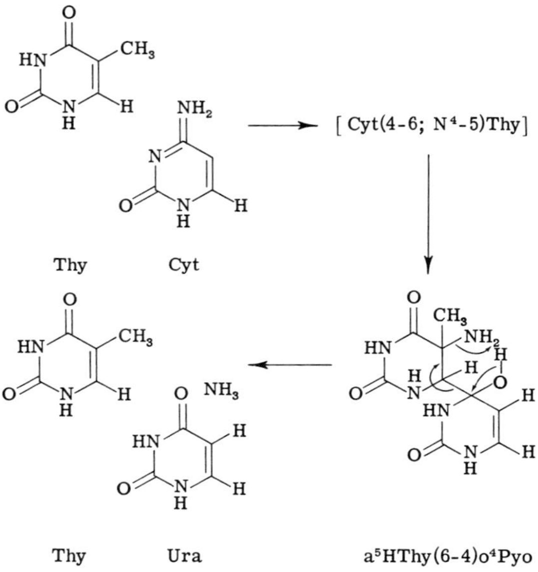

Thy + Cyt ⟶ [Cyt(4-6; $N^4$-5)Thy] ⟶ $a^5$HThy(6-4)$o^4$Pyo ⟶ Thy + Ura + $NH_3$

Thy Cyt

Thy Ura $a^5$HThy(6-4)$o^4$Pyo

**Scheme 15**

The trimer gives some interesting and intriguing reactions (Varghese and Wang, 1968b; Wang, 1971). [See Eqs. (1)–(4) for some examples.]

$$\text{XIV} \xrightarrow{h\nu} \text{Thy} + \text{o}^5\text{hThy(6–4)o}^4\text{m}^5\text{Pyo} \xrightarrow{-\text{HOH}} \text{o}^5\text{hThy(6–4)m}^5\text{Pyo} \quad (1)$$

$$\text{XIV} \xrightarrow[\text{room temp.}]{\text{H}^+} \text{XV} \xrightarrow{h\nu} \text{Thy} + \text{o}^5\text{hThy(6–4)m}^5\text{Pyo} \quad (2)$$

$$\text{XIV} \xrightarrow{-\text{OH}} \text{Thy} + \text{Thy} \diamond \text{Thy} \quad (3)$$

$$\text{XIV} \xrightarrow[100^\circ\text{C}]{\text{H}^+} \text{XVI} + \text{Thy} + \text{Thy(6–4)m}^5\text{Pyo} \quad (4)$$

In reaction (1), the trimer (XIV) is irradiated with 254 nm light in an aqueous solution, resulting in the cleavage of the cyclobutyl ring. Therefore, Thy and $o^5$hThy(6–4)$o^4m^5$hPyo should be formed. However, dehydration occurs spontaneously to yield $o^5$hThy(6–4)$m^5$Pyo. In reaction (2), acid-catalyzed dehydration converts the trimer to (XV) which again gives Thy and o$^5$hThy(6–4)m$^5$Pyo upon irradiation. In reaction (3), base-catalyzed decomposition of the trimer yields Thy◇Thy(*c,s*) and Thy. In reaction (4), three separate events may take place concurrently. First, the trimer is further dehydrated at 100°C to (XVI), which has a $\lambda_{max}$ at 267 nm. Second, the cyclobutyl ring undergoes cleavage to give Thy(6–4)m$^5$Pyo. Third, acid-catalyzed decomposition forms Thy (Varghese and Wang, 1968b; Wang, 1971).

The yield of trimer is a sigmoidal function of the UV doses (Rahn and Hosszu, 1969b). Apparently, the amount of adduct reaches a maximum of ~ 3%, and additional irradiation leads to a reduction with an increase in trimer. At high UV doses, the amount of trimer approaches 10%, and the trimer seems to form at the expense of the adduct. Since the formation of the trimer probably requires two photolytic steps for the covalent linking of three Thy molecules, this observation may indicate that the adduct is the precursor of the trimer as the dimer absorbs only weakly at 254 nm and is unlikely to absorb a second photon effectively. To substantiate this mechanism, the irradiation of an equimolar proportion of the radioactive labeled adduct and unlabeled Thy should yield the trimer with essentially unchanged specific activity, but such an experiment has not yet been carried out.

## I. Homoadduct of Ura and Ura Phototrimer

A similar adduct was isolated from Ura irradiated in a frozen aqueous solution (Khattak and Wang, 1969). UV, IR, NMR, and mass spectra indicate that it is probably 6–4′[pyrimidin-2′-one]-uracil, Ura(6–4)Pyo. Apparently, it is the dehydration product of a homoadduct of Ura. Since only Ura(6–4)Pyo can be isolated, the Ura adduct must undergo rapid spontaneous dehydration. UV spectra with $\lambda_{max}$ at 332 nm suggest that the two chromophores may be more coplanar, in contrast to those adducts containing a Thy moiety (see Section K,1). Urd, under similar conditions, also gives rise to Urd adducts (Varghese, 1971b), one of which is probably the nucleoside of Ura(6–4)Pyo because this compound gave Ura(6–4)Pyo upon acid hydrolysis. The possible biological importance of Ura(6–4)Pyo in the photochemistry of RNA should be considered (Chapter 7, Volume II).

There is only one report (Varghese, 1971a) that Ura(6–4)Pyo is a photoproduct which can be isolated from the acid hydrolysates of UV-irradiated DNA. Ura(6–4)Pyo presumably resulted from the acid-catalyzed deamination of a corresponding Cyt adduct, Cyt(6–4)Pyo.

This adduct has not been characterized by this author or reported by other laboratories. This claim seems to be inconsistent with the fact that once such a Cyt adduct is formed it should be stable under acid hydrolysis. Deamination, therefore, is unlikely to occur, since it does not occur in the case of Cyt(5–4)Pyo whose structure has been fully characterized (see Section J). This report must, for this reason, be viewed with reservation.

A Ura trimer has been observed (Khattak and Wang, 1969) in the irradiation of Ura in a frozen aqueous solution. Although UV, NMR, and IR spectra were reported, the structure of this trimer has not been established due to its extreme lability. This trimer has also been observed in the irradiation of Ura in solution (E. Fahr, personal communication).

## J. Homoadduct of Cyt and Heteroadduct of Cyt and 4-Sra and of Cyt and Ura

Since the same photoadduct, Cyt(5–4)Pyo, is obtained by reactions between Cyt and Cyt, Cyt and 4-Sra, and Cyt and Ura, these photoaddition reactions will be discussed together.

### 1. Isolation from Cyt Derivatives

Homoadduct formation for Cyt derivatives has been studied with poly(rC), Cyd, and dCyd (Rhoades and Wang, 1971a,b). Irradiation of poly(rC) in buffered pH 7 solutions at or in unbuffered aqueous solutions results in an absorbancy decrease in the 275 nm region with a concomitant increase in the 300 nm region (Fig. 2), similar to that in DNA (Wang, 1962). The compound responsible for this increase was isolated from the acid hydrolysates of irradiated poly(rC) in an aqueous solution (pH 4–7), and its structure was characterized as 5-4′-[pyrimidin-2′-one]-cytosine, Cyt(5–4)Pyo.

The isolation of Cyt(5–4)Pyo directly from the irradiation of Cyt in a frozen state is not as fruitful as isolation from its ribosyl derivatives, Cyd and dCyd. It was also found that the pH of the medium plays a critical role in the formation of Cyd and dCyd homoadducts. This is clearly shown in Fig. 3 where the yield rises slightly from pH 2.4 to pH 3.4, followed by a sharp rise with a maximum at pH 3.8. Further increase in pH sharply reduces the yield to the level observed at pH 2 to 3. For both Cyd and dCyd, the pH for the maximal yields are calculated to be at 0.5 molar equivalents in HCl. Thus, formation of

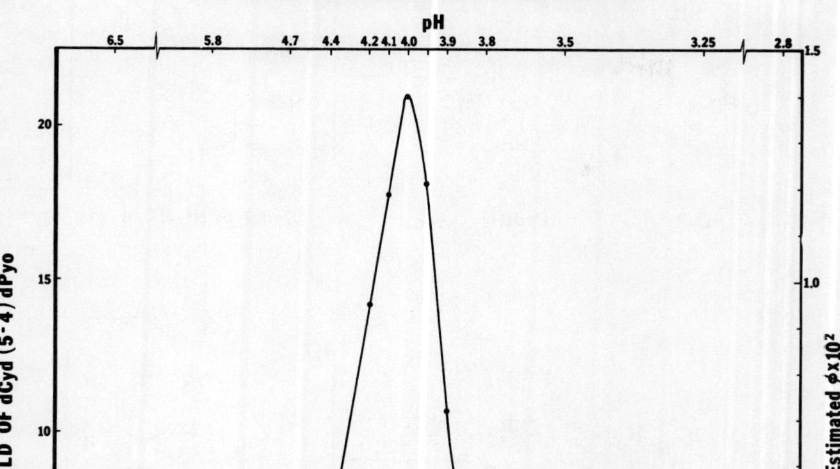


**Fig. 3.** *Yields and quantum yields of dCyd(5–4)dPdo from dCyd irradiated in the frozen state at various pHs or molar ratios of HCl/dCyd (from Rhoades and Wang, 1971b).*

Cyd(5–4)Pdo or dCyd(5–4)dPdo requires one molecule of HCl and two molecules of the nucleosides. As shown in Scheme 16, formation of azetidine derivatives involves one neutral and one protonated Cyt. Subsequently, this four-membered ring may be opened in solutions to give the intermediate, Cyd(5–4)a$^4$hPdo which deaminates readily to the final product, Cyd(5–4)Pdo.

A similar interaction between Cyd and its cationic species has been suggested for poly(rC), CpC, and Cyd aggregates in ice at 77°K or in glass (Montenay-Garestier and Hélène, 1970; see Chapter 8) because a long-wavelength fluorescent emission was observed having a pH-dependent intensity (see also Chapter 3). This pH dependency closely

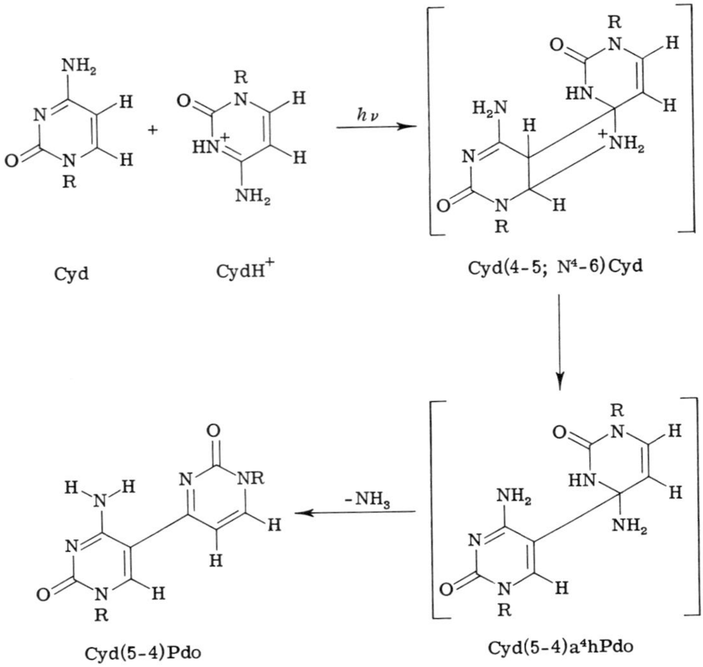

Cyd + CydH$^+$ $\xrightarrow{h\nu}$ [Cyd(4-5; N$^4$-6)Cyd] → [Cyd(5-4)a$^4$hPdo] $\xrightarrow{-NH_3}$ Cyd(5-4)Pdo

**Scheme 16**

resembles that observed in the formation of Cyd and dCyd adducts, with the maxima in both cases being close to the ground-state pK. Similar long-wavelength fluorescence emission has been detected in CppC at room temperature (Vigny and Favre, 1974) and is similar to those previously reported (Thiele *et al.*, 1972; Favre, 1972) for poly(rC·rC$^+$) and poly(rI·rC·rC$^+$).

### 2. Isolation from Cyt and 4-Sra or Ura *in Vitro* and *in Vivo*

It was found (Favre *et al.*, 1969; Yaniv *et al.*, 1969; Chaffin *et al.*, 1971) that irradiation of *E. coli* tRNA at 335 nm induced a covalent bond formation between the 4-Srd in position 8 and Cyd in position 13 from the 5′-terminal end. The photoproduct in the intact tRNA can subsequently be converted into a strongly fluorescent compound by reduction with $NaBH_4$ under mild conditions (Favre and Yaniv, 1971; Krauskopf *et al.*, 1972). The usefulness of this photoreaction for probing the conformation of tRNA and the fluorescence properties of the reduced product for quantitative estimation of tRNA will be covered in Chapter 7, Volume II. This section is concerned only with the identification of these two compounds.

Irradiation at 335 nm between 4-Sra and Cyt at 4°C in aqueous solution leads to the formation of Cyt(5–4)Pyo (Leonard *et al.*, 1971), which is identical to that obtained from Cyt derivatives irradiated in a frozen state (Rhoades and Wang, 1971a,b). This is to be expected if one postulates the formation of a thietane intermediate, followed by a ring opening to form s$^6$hCyt(5–4)Pyo, since subsequent elimination of $H_2S$ then gives Cyt(5–4)Pyo (Scheme 17). In this reaction a second

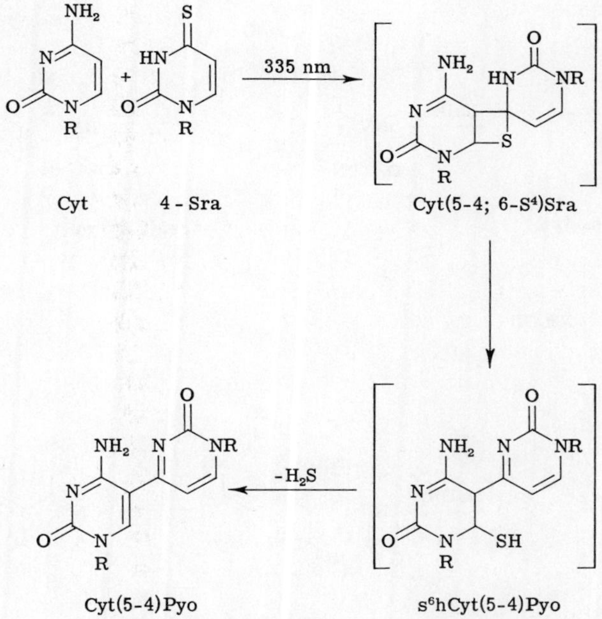

**Scheme 17**

photoproduct, Sra(5–4)Pyo was isolated (Bergstrom and Leonard *et al.*, 1971). This compound also precipitated in pure form on irradiation (335 nm) of an aqueous solution of Sra. In addition, irradiation (335 nm) of a mixture of Ura and Sra gave principally Ura(5–4)Pyo. Treatment of Sra(5–4)Pyo with concentrated ammonium hydroxide at 95°C gave Cyt(5–4)Pyo which, in turn, could be converted to Ura(5–4)Pyo by heating 6 *N* HCl at 140°C (Bergstrom and Leonard, 1971a).

The strong fluorescent compound formed by $NaBH_4$ reduction was identified as 5–4′-[pyrimidin-2′-one]-3,6-dihydrocytosine, 3,6-hCyt(5–4)Pyo (Bergstrom and Leonard, 1972a; Favre *et al.*, 1972). Its high fluorescence quantum yield is consistent with the general trend for adducts, in which adducts containing a saturated Pyr moiety are more fluorescent than their unsaturated derivatives (Hauswirth and Wang, 1973).

Catalytic oxygenation (Pt, $O_2$) converted 3,6-hCyt(5–4)Pyo back to Cyt(5–4)Pyo. Treatment of 3,6-hCyt(5–4)Pyo with aqueous acid gave 3,6-hUra(5–4)Pyo which could also be obtained by treatment of Ura(5–4)Pyo with $NaBH_4$. Sra(5–4)Pyo is reduced by $NaBH_4$ to 3,6-hSra(5–4)Pyo (Scheme 18) (Bergström and Leonard, 1972b).

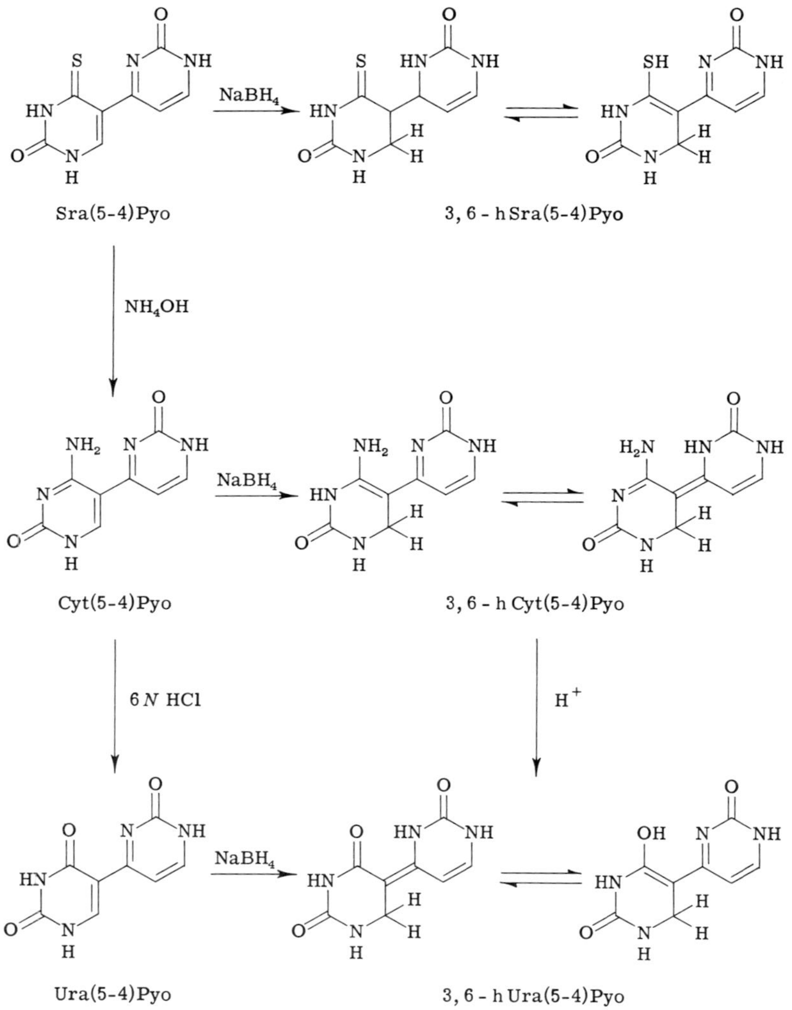

Sra(5-4)Pyo $\xrightarrow{NaBH_4}$ 3, 6 - hSra(5-4)Pyo

$NH_4OH$

Cyt(5-4)Pyo $\xrightarrow{NaBH_4}$ 3, 6 - hCyt(5-4)Pyo

6 *N* HCl

$H^+$

Ura(5-4)Pyo $\xrightarrow{NaBH_4}$ 3, 6 - hUra(5-4)Pyo

**Scheme 18**

### 3. Synthesis of Cyt (5–4)Pyo

In order to confirm the structure of Cyt(5–4)Pyo by means other than spectral data, Bergstrom *et al.* (1972) devised an unequivocal synthesis which also serves as a general method for the preparation of bipyrimidines with a 5–4′ or possibly a 6–4′ linkage. This multistep synthesis is rather elegant although the yields are quite low. Improved methods leading to higher yields are desirable in order to prepare various adducts in sufficient quantities for studying their possible biological importance.

## K. Properties of Photoadducts

### 1. UV Absorption and Optical Isomers

UV absorption spectra of the ten known bipyrimidine adducts are shown in Fig. 4. Phototrimers of Thy and Ura have only end absorp-

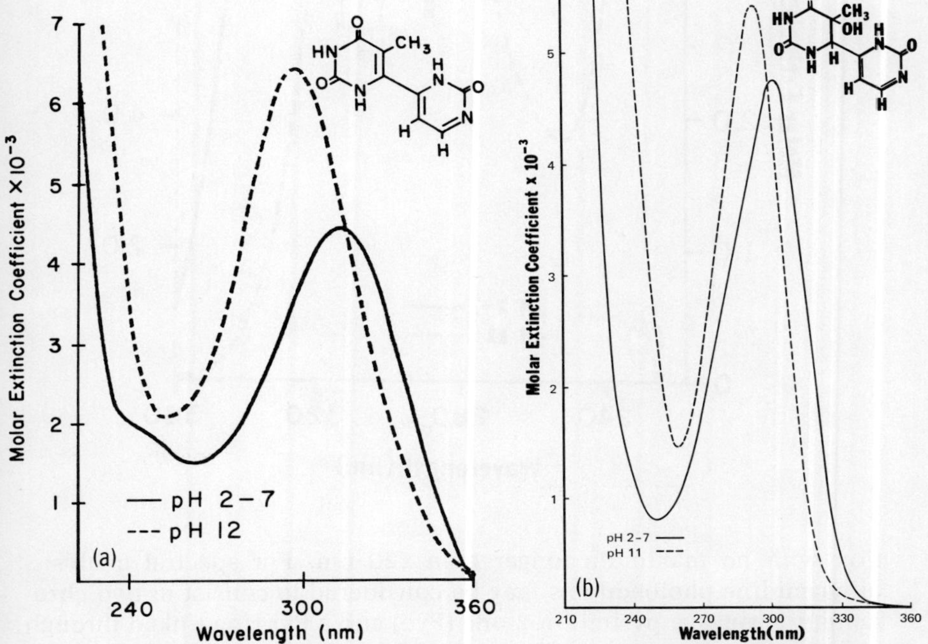


**Fig. 4.** *Ultraviolet absorption spectra of (a) Thy(6–4)Pyo, (b) o⁵hThy(6–4)Pyo, (c) o⁵hThy(6–4)m⁵Pyo, (d) Thy(6–4)m⁵Pyo, (e) Ura(6–4)Pyo, (f) Cyt(5–4)Pyo, (g) Cyd(5–4)Pdo, (h) dCyd(5–4)dPdo, (i) Ura(5–4)Pyo, and (j) Sra(5–4)Pyo in aqueous solutions (see Table 8 for references).*

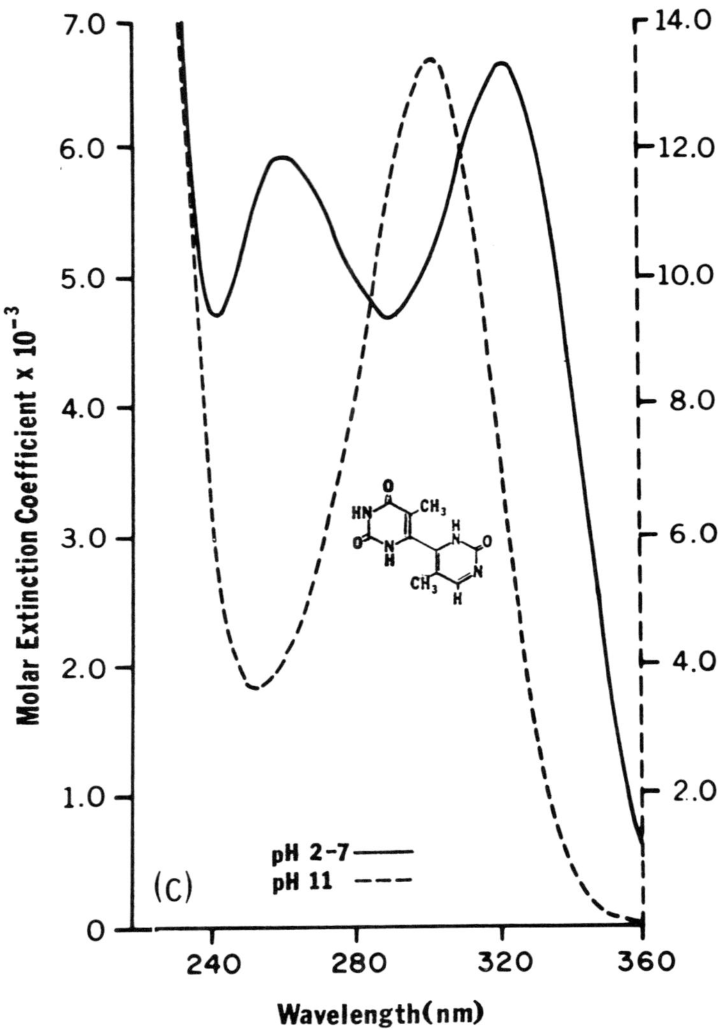


tion with no maximum longer than 220 nm. For spectral analysis, bipyrimidine photoadducts may be considered to consist of two chromophoric rings, a pyrimidin-2-one (Pyo) and a Pyr ring linked through either a 6–4′ or a 5–4′-bond. Additionally, the Pyr ring may be saturated at C(5)–C(6) giving hPyr, which in isolation does not absorb at wavelengths longer than 220 nm. Such saturation results in two asymmetric carbons and a pair of diastereoisomers should, therefore, exist.

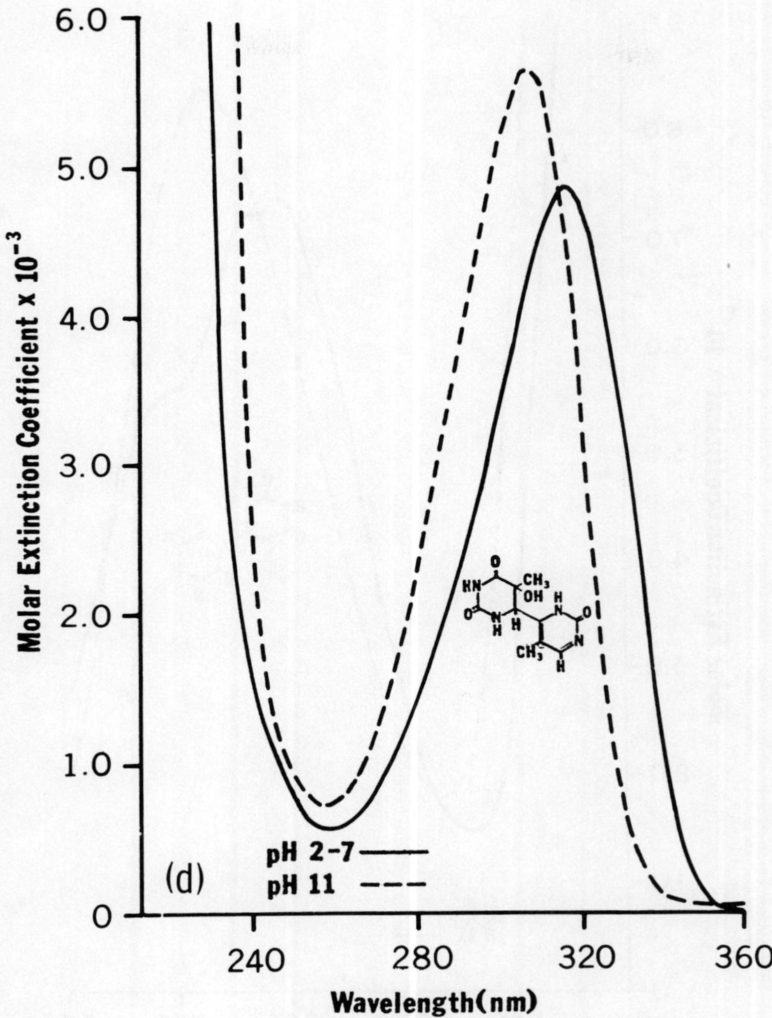


Indeed such isomers were detected by x-ray diffraction analysis of o$^5$hThy(6–4)m$^5$Pyo (Karle *et al.*, 1969; Karle, 1969); however, so far no attempt has been made to resolve them.

The absorption spectra of photoadducts (Fig. 5) are consistent in that those containing hPyr (**a, c, g**) have only a single absorption manifold longer than 240 nm characteristic of Pyo(~ 300 nm), while those with a Pyr (**b, d, e, f, h**) exhibit multicomponent spectra with the

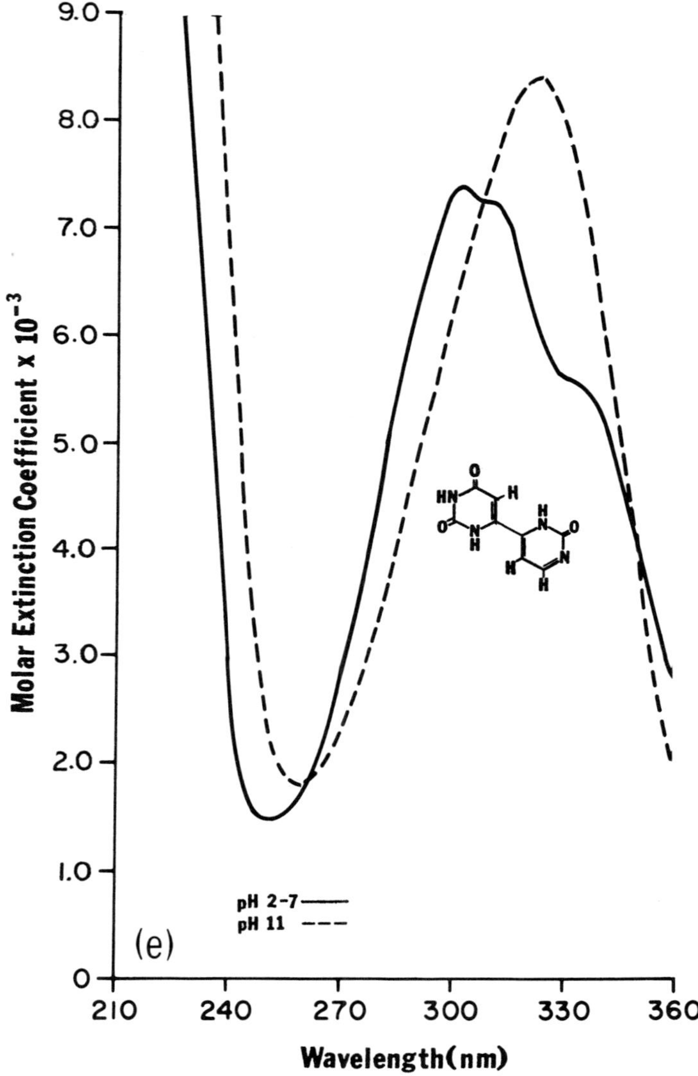


lowest energy component red-shifted 10–40 nm from isolated Pyo absorption, Fig. 5 and Table 7. Since an increase in conjugation of a $\pi$ system generally shifts that electronic transition to lower energy (Jaffe and Orchin, 1962), the simplest explanation for these shifts is that interaction between Pyo and Pyr rings expands $\pi$ conjugation to a

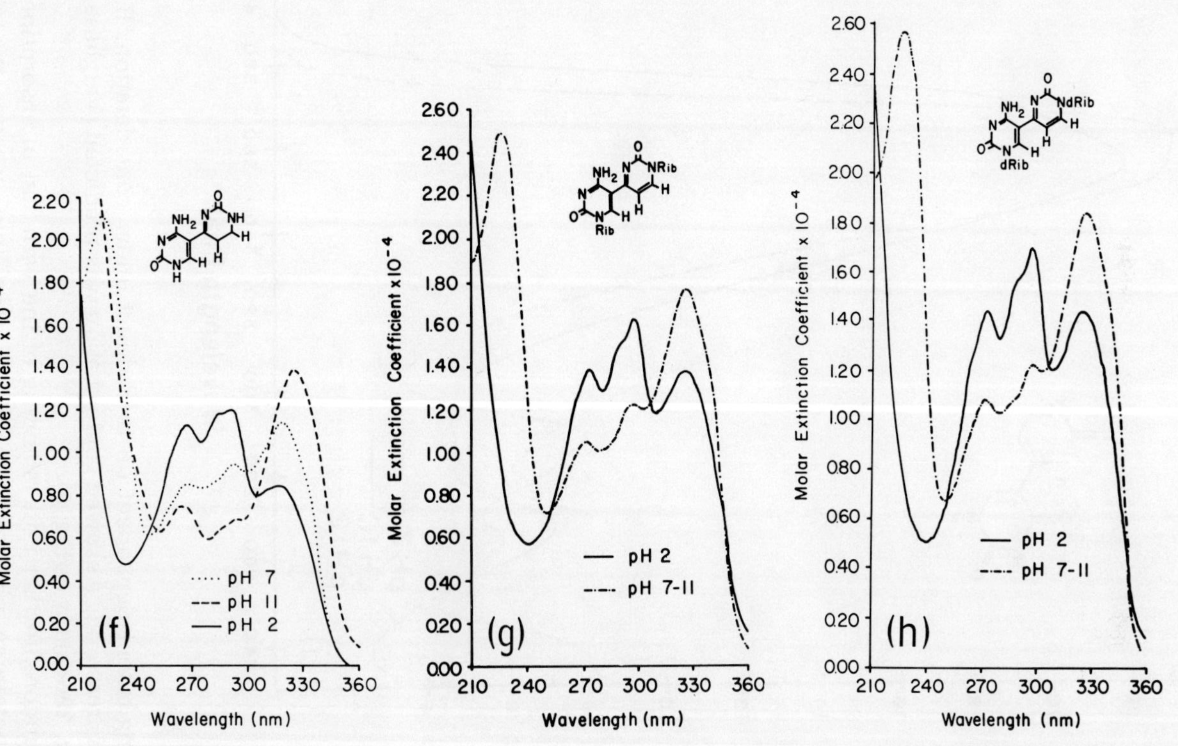
(f)
pH 7
pH 11
pH 2
Molar Extinction Coefficient x 10⁻⁴
Wavelength (nm)
(g)
pH 2
pH 7-11
Molar Extinction Coefficient x 10⁻⁴
Wavelength (nm)
(h)
pH 2
pH 7-11
Molar Extinction Coefficient x 10⁻⁴
Wavelength (nm)

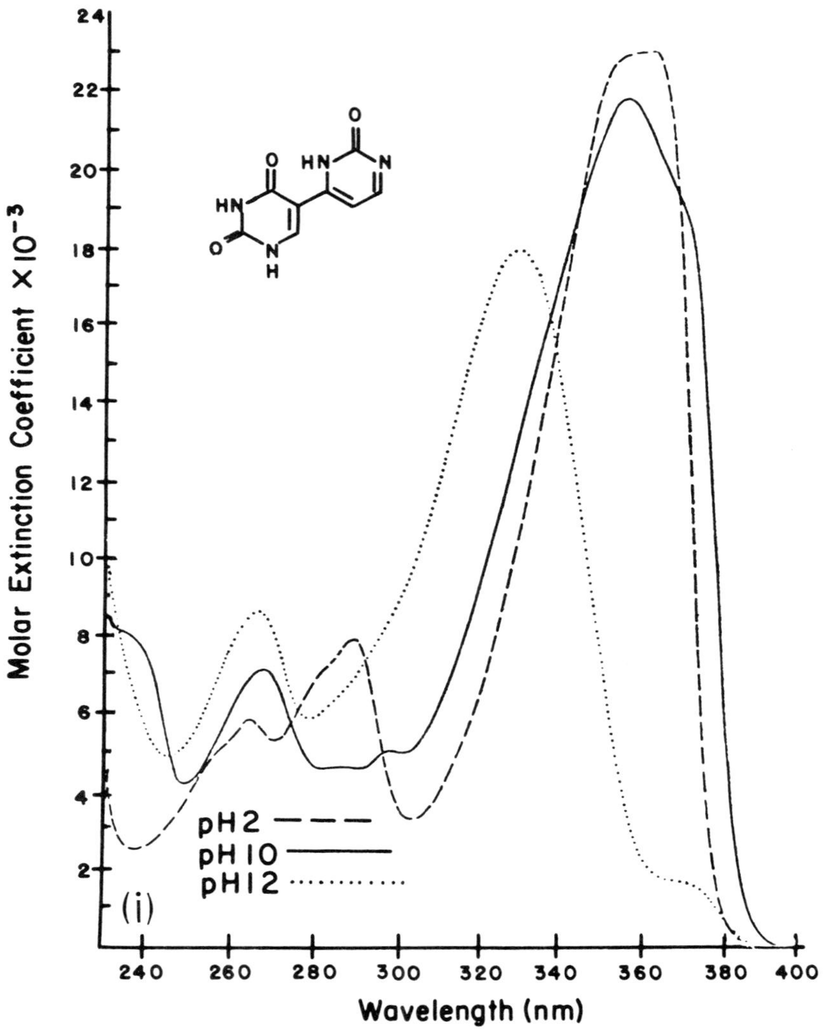


varying extent dependent on the adduct under consideration. Based on this reasoning a skewed conformation was predicted (Varghese and Wang, 1968a) for **b** (Fig. 5).

Therefore, further understanding of the variation in absorption for Pyr-containing photoadducts should take into account the steric factors inhibiting the coplanarity of the two rings. The two extremes, coplanarity (inter-ring torsional angle $\theta = 0°$) and perpendicularity

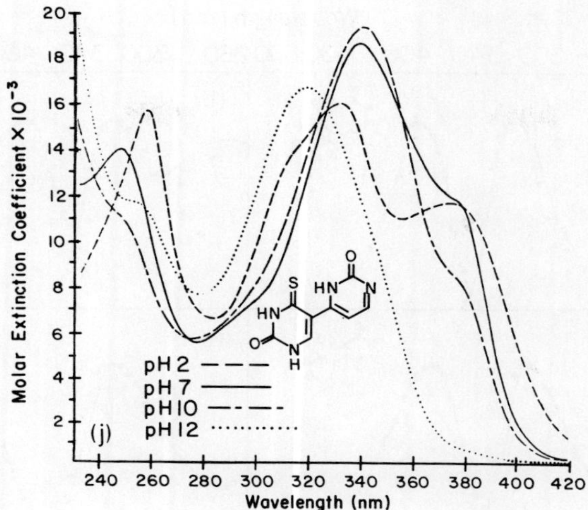


($\theta = 90°$) are expected and allow maximal and minimal overlaps, respectively. This would result in either red-shifts of both Pyr and Pyo bands with possible spectral overlap in the first case or Pyr and Pyo bands similar to those for isolated rings in the second case.

The major factor controlling $\theta$ is the bulkiness of the substituents at positions *ortho* to the single bond connecting the two rings; an analogy can be drawn with the absorption of *ortho*-substituted biphenyls (Suzuki, 1967). Thus, the most planar adduct should be **e** and the least planar one **b**. Within the middle group **f** and **h** are expected to be somewhat more planar than **d** due to a hydrogen bridge between Pyd-$NH_2$ and Pdo-N(3) detected by NMR (Rhoades and Wang, 1971a,b). Absorption spectra appear to confirm this analysis. **e** has the lowest-energy transition (340 nm) presumably due to a red-shifted Pyo and no band of energy higher than 300 nm, suggesting that the Pyr transition has shifted to the 300–340 nm region. At the other extreme, **b** exhibits a relatively unperturbed Pyo band at 325 nm and a Pyr band at 268 nm.

A more quantitative estimate of $\theta$ (Hauswirth and Wang, 1976) can be gained using the relation

$$E_r\ (\theta) \cong E_r\ (0) \cos^2\theta$$

in which $E_r$ (0) is the $\pi$-resonance energy in a coplanar "biphenyl-like" molecule, and $E_r$ ($\theta$) is the $\pi$-resonance energy in a molecule

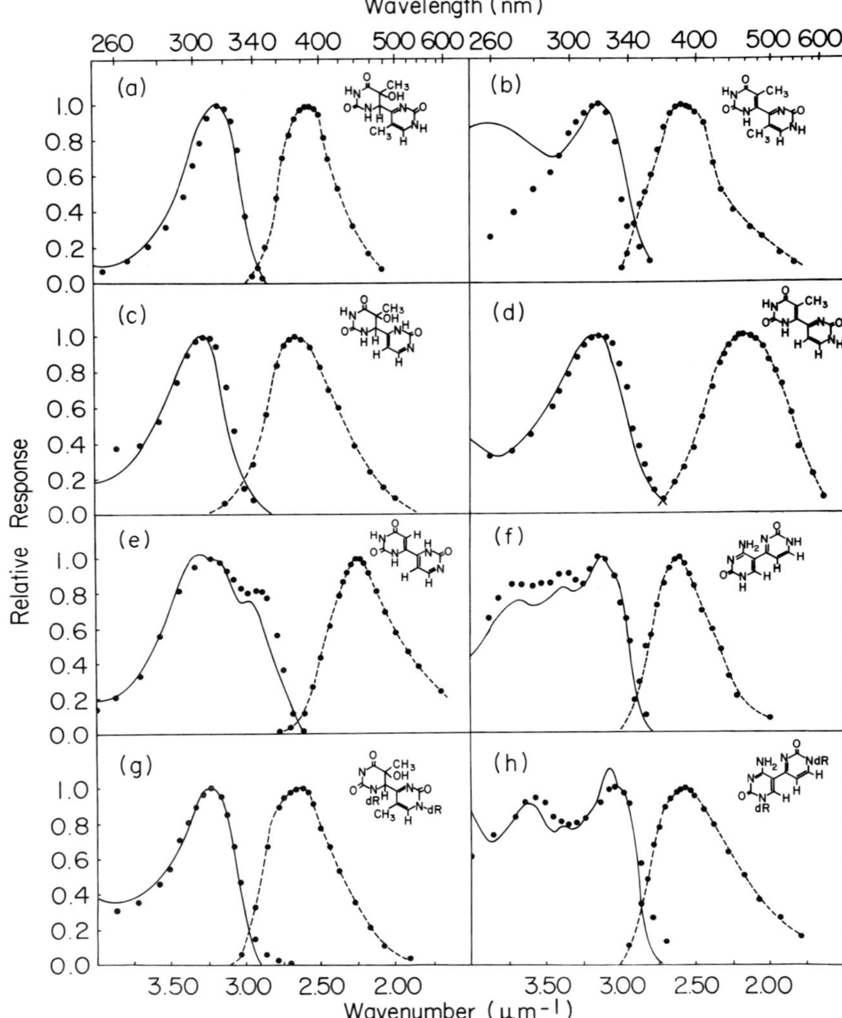


**Fig. 5.** *Absorption* (—), *corrected fluorescence excitation* (· · ·), *and corrected fluorescence emission* (·----) *spectra of Pyr adducts (W. Hauswirth and S. Y. Wang, unpublished data).*

whose rings are $\theta°$ out of coplanarity (Suzuki, 1967). As a first approximation it is assumed that

$$E_r\,(0) - E_r\,(\theta) \approx \Delta\,\bar{\nu}_{abs} = \bar{\nu}_{abs}\,(\theta) - \bar{\nu}_{abs}\,(0)$$

i.e., the difference in resonance energy between 0° and $\theta°$ is approxi-

**Table 7** Spectral Properties of Pyrimidines and Pyrimidine Adducts

| Compound | Absorption $\lambda_{max}$ (nm) 300°K | | Fluorescence $\lambda^{ex}_{max}$ (nm) 300°K | | Fluorescence $\lambda^{fl}_{max}$ (nm) 300°K | | Fluorescence $\lambda^{fl}_{max}$ (nm) 77°K[a] | | Phosphorescence $\lambda^{ph}_{max}$ (nm) 77°K[a] | | $\Delta\bar{\nu}_{abs}$ (kK) | $\theta$ (calculated) |
|---|---|---|---|---|---|---|---|---|---|---|---|---|
| | Pyr | Pyo | Pyr | Pyo | Pyr | Pyo | Pyr | Pyo | Pyr | Pyo | | |
| Pyo | — | 295 | — | 295 | — | 371 | — | 347 | — | 449 | — | — |
| $m^5$Pyo | — | 310 | — | 313 | — | 371 | — | 375 | — | 449 | — | — |
| Cyt | 267 | — | 274[b] | — | 303[b] | — | 315[c] | — | — | — | — | — |
| Thy | 265 | — | 269[b] | — | 337[b] | — | 316[c] | — | — | — | — | — |
| Ura | 259 | — | 267[b] | — | 308[b] | — | 315[c] | — | — | — | — | — |
| Thy(6–4)$m^5$Pyo, **b** | 257 | 319 | 260[d] | 315 | 333[d] | 385 | — | 360 | — | 445 | 0.95 | 49°–54° |
| Thy(6–4)Pyo, **d** | 250[d] | 313 | — | 318 | — | 457 | — | 392 | — | 445 | 1.90 | 61°–67° |
| Ura(6–4)Pyo, **e** | 300 | 338 | 308 | 338 | — | 445 | — | 379 | — | 435 | 4.0 | 23°–29° |
| Cyt(5–4)Pyo, **f** | 297 | 321 | 297 | 321 | — | 385 | — | 368 | — | 440 | 2.7 | 40°–45° |

[a] Uncorrected data.
[b] Daniels and Hauswirth, 1971.
[c] Longworth *et al.*, 1966.
[d] Very weak, maximum only approximate and uncorrected.

mately reflected in the absorption energy difference between compounds of 0° and $\theta$°. Adduct $\theta$'s are seen to be in the order **b** > **d** > **f** = **h** > **e**, the same as predicted qualitatively from steric repulsion at position *ortho* to the inter-ring bonding.

The foregoing discussed the restricted rotation of two-ring molecules which in actuality resulted in conformational isomerism, atropisomerism, and isomers which can also be resolved by the usual techniques (Bentley, 1969).

### 2. Emission

Absorption analysis leads to the general conclusion that photoadduct absorption can be partially resolved into transitions in either Pyr or Pyo. This resolution should carry over to emission as well. Adduct **b**, having noninteracting rings in the ground state, also exhibits two apparently independent fluorescence modes since excitation of Pyo (325 nm) results in Pyo-like emission while excitation of Pyr (260 nm) gives apparently Thy-like emission ($\sim$ 330 nm, $\phi_f << 10^{-3}$). These processes are shown schematically in Fig. 6a.

For the remaining adducts, excitation into any absorption band gives either Pyo-like fluorescence (Fig. 6B) or a red-shifted fluorescence at $\lambda \sim$ 450 nm (Fig. 6C). The former case involves **f** and **h** where inter-ring electronic overlap appears sufficient to allow exclusive Pyo fluorescence from energy absorbed at any wavelength. The exact mechanism whereby absorption into the Pyr ring of **f** or **h** (presumably the lowest-energy absorption component) is transferred to the Pyo singlet is uncertain. Charge-transfer interaction (Birks, 1971), possibly through intramolecular H-bonding, may be responsible, but

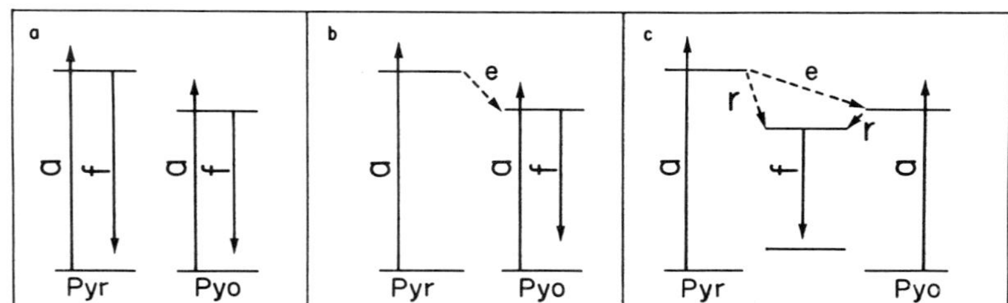


**Fig. 6.** *Schematic representation of absorption and emission processes in Pyr adducts,"e" denotes exciplex interaction, "r" denotes excited-state rotation, "a" denotes absorption, and "f" denotes fluorescence emission (W. Hauswirth and S. Y. Wang, unpublished data).*

the pK*s of Cyt and Cyd are not known and this possibility cannot be further evaluated. Exciton interaction between Pyr and Pyo does not appear favorable because the Pyr and Pyo absorption bands are still reasonably well separated and the transition moments, generally aligned in the C(5)–C(6) direction (Hug and Tinoco, 1973), are not in the optimal head-to-tail geometry (Kasha, 1960).

Compounds **d** and **e** have much larger Stokes shifts ($\nu_{abs} - \nu_{em} \sim 120$ nm) than the other adducts and are characteristic of neither Pyr or Pyo emission (Fig. 6C). The significantly lower-energy emission bands ($\sim 450$ nm) are suggestive of exciplex emission, but one of an intramolecular nature, since this low-energy fluorescence persists even at concentrations $< \mu M$. Although electron density calculations of excited singlet adducts are not available, by crude analogy with biphenyls (Suzuki, 1967), the bond order of the 5–4′-bond should increase in the excited singlet. This may explain the fluorescence red-shifts through excited-state electronic redistribution favoring coplanar rings, thus allowing extensive conjugation between two rings. This requires an excited-state rotation. To test this explanation, low-temperature emission spectra were determined. It is known that solute–solvent H-bondings and the attainment of preferred solvent orientation will be enhanced at low temperature and such interaction will restrict rotation of the molecules away from ground-state geometry (Werner and Hercules, 1969; Yamaguchi *et al.*, 1967). Indeed, when luminescence and phosphorescence spectral maxima of **d** and **e** were determined in ethylene glycol/water (1:1) glasses at 77°K, they exhibited normal low-temperature Pyr and Pyo emission, Table 7 (Hauswirth and Wang, 1975), strengthening the belief that excited-state Pyr–Pyo coplanarity is needed to observe the $\sim 450$ nm fluorescence.

The absorption and emission of adducts produced by irradiation of dinucleotide phosphates or DNA (Hauswirth and Wang, 1973) is essentially similar to that for isolated adducts, except that no intramolecular exciplex fluorescence is observed, as expected, since the intact ribose–phosphate linkages between Pyr–Pyo rings preclude coplanarity.

### 3. IR and NMR Spectra

The IR spectra of these adducts can be found in the original reports and will not be reproduced here. However, the NMR data are given in Table 8. The assignments of these values are straightforward and need no further interpretation (see Chapter 10).

**Table 8** NMR Spectral Data of Pyrimidine Adducts[a]

| 6–4 Linkage | $CH_3(5)$[b] | H(5) | H(6) | OH(5) | NH | | References |
|---|---|---|---|---|---|---|---|
| Thy(6–4)Pyo | — 1.68(s) — | — 6.48(d) — | — 8.08(d) — | — | 11.19 | | Wang and Varghese, 1967 |
| ho⁵hThy(6–4)Pyo | 2.10(s) — — | — — — | 4.48(d) — 8.22(d) | 8.22(s) | 7.99(d) | | Rhoades and Wang, 1970 |
| Thy(6–4)m⁵Pyo | — 1.91(s) 2.26(s) | — — — | — — 8.47(s) | — | 11.32,11.60 | | Varghese and Wang, 1968a |
| ho⁵hThy(6–4)m⁵Pyo | 1.85(s) — 2.36(s) | — — — | 4.70(d) — 7.97(s) | 6.00(s) | 7.90(d),10.37 | | Varghese and Wang, 1968a |
| Ura(6–4)Pyo | — — — | — 6.74(s) 7.37(d) | — — 8.52(d) | — | 11.70 | | Khattak and Wang, 1969 |
| **5–4 Linkage** | $NH_aH_b$ | H(5) | H(6) | | | | |
| Cyt(5–4)Pyo[c] | 8.86(s) | 7.34(d) | 8.95(s) 8.35(d) | | | | Rhoades and Wang, 1971a,b |
| | | 7.37(d) | 8.99(s) 8.38(d) | | | | Bergström and Leonard, 1972a |
| Cyd(5–4)Pdo | 8.04(d) 9.76(d) | 6.86(d) | 9.17(s) 8.38(d) | | | | Rhoades and Wang, 1971a,b |
| dCyd(5–4)dPdo | 8.05(d) 9.74(d) | 6.86(d) | 9.11(s) 8.33(d) | | | | Rhoades and Wang, 1971a,b |
| Ura(5–4)Pyo[c] | — | 7.38(d) | 9.18(s) 8.60(d) | | | | Bergström and Leonard, 1972a |
| Sra(5–4)Pyo[c] | — | 7.38(d) | 9.09(s) 8.57(d) | | | | Bergström and Leonard, 1972a |
| **$NaBH_4$ reduced** | $CH_aH_b(6)$ | H(5) | H(6) | | | | |
| hCyt(5–4)Pyo[c] | 4.04(d) 4.24(d) | 7.19(d) | 8.43(d) | | | | Bergström and Leonard, 1972b |
| hUra(5–4)Pyo[c] | 4.36(s) | 5.96(d) | 7.38(d) | | | | Bergström and Leonard, 1972b |
| hSra(5–4)Pyo | 4.39(s) | 6.13(d) | 7.47(d) | | | | Bergström and Leonard, 1972b |
| **Trimeric product** | $CH_3(5)$ | H(6) | H(6), cyclobutyl | OH(4) | OH(5) | NH(1) | |
| Thy trimer hydrate | 1.33(s) 1.35(s) 1.36(s) | 4.15(s) | 3.11(br) 3.64(br) | 5.51(s) | 6.92(s) | 6.32(br) | Wang, 1971 |
| Thy trimer | 1.33(s) 1.35(s) 1.36(s) | 4.15(s) | 3.11(br) 3.64(br) | — | 6.92(s) | — | Wang, 1971 |
| Thy trimer dehydrate | 1.72(s) 1.78(s) 2.01(s) | — | 3.96(d) 4.18(d) | — | — | — | Wang, 1971 |
| **Tetrameric product** | $—CH_2CH_2—$ | | $N(CH_3)$ | Methine-H | Vinyl-H | NH | |
| Hexamethyl tetramer | 2.50(br) | | 2.70(s) 3.21(s) 3.25(s) | 4.13(d) | 5.08(d) | NO | Wang and Rhoades, 1971 |

[a] Data were obtained in $(CD_3)_2SO$ at 100 MHz and chemical shifts are given in ppm from TMS.
[b] Values given in the 1st column are for the indicated substituents on the saturated Pyr ring, 2nd on Pyr ring, and 3rd on Pyo ring.
[c] Data were obtained in $CF_3COOD$.

### 4. Photochemical Precursor to the Photoadduct

The excited-state precursor which leads to photoadduct formation has been evaluated principally through triplet sensitization experiments using 313 nm light and acetophenone (which is able to transfer triplet energy only to Thy) or acetone (which has triplet energy sufficiently high for transfer to any base). Radiochromatographic analysis of the acid hydrolysates of triplet-state-sensitized *E. coli* DNA (Lamola, 1969) and T4 DNA (Meistrich and Lamola, 1972) reveals approximately a 50-fold reduction in the yield of Thy(6–4)Pyo relative to Thy◇Thy. The same conclusion is reached from adduct fluorescence and chromatographic analysis of similarly sensitized solutions of dT-dT, dT-dC, dC-dT, and dC-dC (Hauswirth and Wang, 1976). Additionally, adduct formation could not be quenched under 254 nm irradiation with the known triplet quenchers, $O_2$ or 0.1 *M* $Mn^{2+}$. Taken together, these data suggest that no class of photoadduct is derived from the Pyr triplet state, and by negative inference, suggest that the adduct is derived from the excited singlet.

## References

Alcantara, R., and Wang, S. Y. (1965a). *Photochem. & Photobiol.* **4,** 465.
Alcantara, R., and Wang, S. Y. (1965b). *Photochem. & Photobiol.* **4,** 473.
Ashwood-Smith, M. J., Bridges, B. A., and Munson, R. J. (1965). *Science* **149,** 1103.
Badilley, J., and Topham, A. (1944). *J. Chem. Soc.*, 678.
Becarevic, A., Djordjevic, B., and Sutic, D. (1963). *Nature (London)* **198,** 612.
Becquerel, P. (1910). *C. R. Acad. Sci.* **151,** 86.
Benhur, E., and Elkind, M. M. (1972). *Biophys. J.* **12,** 636.
Ben-Ishai, R., Goldin, J., and Oppenheim, B. (1962). *Biochim. Biophys. Acta* **55,** 748.
Ben-Ishai, R., Cavari, B. Z., Goldin, H., and Kerpel, S. (1965). *Biochim. Biophys. Acta* **95,** 291.
Bentley, R. (1969). "Molecular Asymmetry in Biology," Vol. 1, Academic Press, New York.
Benzer, S., and Freese, E. (1958). *Proc. Nat. Acad. Sci. U.S.* **44,** 112.
Berens, K., and Shugar, D. (1963). *Acta Biochim. Pol.* **10,** 25.
Bergström, D. E., and Leonard, N. J. (1972a). *J. Amer. Chem. Soc.* **94,** 6178.
Bergström, D. E., and Leonard, N. J. (1972b). *Biochemistry* **11,** 1.
Bergström, D. E., Inoue, I., and Leonard, N. J. (1972). *J. Org. Chem.* **37,** 3902.
Bessman, M. J., Lehman, I. R., Adler, J., Zimmerman, S. M., Simms, E. S., and Kornberg, A. (1958). *Proc. Nat. Acad. Sci. U.S.* **44,** 633.
Biltz, H., and Heyn, M. (1919). *Ber.* 52B, 1298.
Birks, J. B. (1971). "Photophysics of Aromatic Molecules." Wiley (Interscience), New York.
Boag, J. W., and Muller, A. (1959). *Nature (London)* **183,** 831.
Buhl, S. N., Setlow, R. B., and Regan, D. J. (1972). *Int. J. Radiat. Biol.* **22,** 417.

Campbell, J. M., Schulte-Frohlinde, D., and von Sonntag, C. (1974). *Photochem. & Photobiol.* **20,** 465.
Chaffin, L., Omilianowski, D. R., and Bork, R. M. (1971). *Science* **172,** 854.
Cleaver, J. E. (1968). *Biophys. J.* **8,** 775.
Cohen, S. G. (1967). "Organosulfur Chemistry" (M. J. Janssen, ed.), p. 33. Wiley (Interscience), New York.
Dack, M. R. J. (1973). *J. Chem. Educ.* **50,** 169.
Daniels, M., and Grimison, A. (1967). *Biochim. Biophys. Acta* **142,** 292.
Daniels, M., and Hauswirth, W. W. (1971). *Science* **171,** 675.
Danziger, R. M., Hayon, E., and Langmuir, M. E. (1968). *J. Phys. Chem.* **72,** 3842.
Dellweg, H., and Wacker, A. (1964). *Z. Naturforsch. B* **19,** 305.
Denhardt, D. T., and Sinsheimer, R. L. (1965). *J. Mol. Biol.* **12,** 647.
Dennis, W. S., and Hutchinson, F. (1972). Proc. Int. Congr. Photobiol., 6th, 1972 Abstract p. 108.
Djordjević, B., and Szybalski, W. (1960). *J. Exp. Med.* **112,** 509.
Dodson, M. L., Hewitt, R., and Mandel, M. (1972). *Photochem. & Photobiol.* **16,** 15.
Dohrmann, J. K., and Livingston, R. (1971). *J. Amer. Chem. Soc.* **93,** 5363.
Donnellan, J. E., Jr., and Setlow, R. B. (1965). *Science* **149,** 308.
Dunn, D. B., and Smith, J. D. (1957). *Biochem. J.* **67,** 494.
Ebel, R., and Kraljic, I. C. (1971). *Eur. Biophys. Congr. Proc., 1st, 1971* **2,** p. 109.
Ehrenberg, A., Ehrenberg, L., and Lofroth, G. (1963). *Nature (London)* **200,** 376.
Ehrlich, M., and Riley, M. (1972a). *Photochem. & Photobiol.* **16,** 385.
Ehrlich, M., and Riley, M. (1972b). *Photochem. & Photobiol.* **16,** 397.
Ehrlich, M., and Riley, M. (1974). *Photochem. & Photobiol.* **20,** 159.
Eisenberg, R. J., and Pardee, A. B. (1969). *J. Mol. Biol.* **46,** 355.
Eisinger, J., and Shulman, R. G. (1963). *Proc. Nat. Acad. Sci. U.S.* **50,** 694.
Favre, A. (1972). *FEBS (Fed. Eur. Biochem. Soc.) Lett.* **22,** 280.
Favre, A., and Yaniv, M. (1971). *FEBS (Fed. Eur. Biochem. Soc.) Lett.* **17,** 236.
Favre, A., Yaniv, M., and Michelson, A. M. (1969). *Biochem. Biophys. Res. Commun.* **37,** 266.
Favre, A., Michelson, A. M., and Yaniv, M. (1971). *J. Mol. Biol.* **58,** 367.
Favre, A., Roques, B., and Fourrey, J. (1972). *FEBS (Fed. Eur. Biochem. Soc.) Lett.* **24,** 209.
Fenselau, C., Wang, S. Y., and Brown, P. (1970). *Tetrahedron* **26,** 5923.
Fikus, M., Wierzchowski, K. L., and Shugar, D. (1965). *Photochem. & Photobiol.* **4,** 521.
Fisher, G., Varghese, A. J., and Johns, H. E. (1974). *Photochem. & Photobiol.* **20,** 109.
Fisher, H. J., and Johnson, T. B. (1932). *J. Amer. Chem. Soc.* **54,** 727.
Flippen, J. L., and Karle, I. L. (1971). *J. Amer. Chem. Soc.* **93,** 2762.
Flippen, J. L., Karle, I. L., and Wang, S. Y. (1970). *Science* **169,** 1084.
Flippen, J. L., Gilardi, R. D., Karle, I. L., Rhoades, D. F., and Wang, S. Y. (1971). *J. Amer. Chem. Soc.* **93,** 2556.
Fox, E., and Meselson, M. (1963). *J. Mol. Biol.* **7,** 583.
Gilbert, E., and Schulte-Frohlindes D. (1970). *Z. Naturforsch B* **25,** 492.
Gilbert, E., and Wagner, G. (1972). *Z. Naturforsch. B.* **27,** 644.
Gilbert, E., Wagner, G., and Schulte-Frohlinde, D. (1971). *Z. Naturforsch. B* **26,** 209.
Gilbert, E., Wagner, G., and Schulte-Frohlinde, D. (1972). *Z. Naturforsch. B* **27,** 501.
Gordon, M. P., and Staehelin, M. (1959). *Biochim. Biophys. Acta* **36,** 351.
Gordy, W., Smith, W. V., and Trambarulo, R. F. (1953). "Microwave Spectroscopy," Wiley, New York.
Gordy, W., Pruden, B., and Snipes, W. (1965). *Proc. Nat. Acad. Sci. U.S.* **53,** 751.
Greer, S., (1960). *J. Gen. Microbiol.* **22,** 618.

Greer, S. B., and Zamenhof, S. (1957). *Abstr. Pap., 131st Meet., Amer. Chem. Soc.* p. 3c.
Griffin, B. E., Todd, A., and Rich, A. (1958). *Proc. Nat. Acad. Sci. U.S.* **44,** 1123.
Haug, A. (1964). *Z. Naturforsch. B.* **19,** 143.
Hauswirth, W., and Wang, S. Y. (1973). *Biochem. Biophys. Res. Commun.* **51,** 819
Hauswirth, W., and Wang, S. Y. (1976). In Preparation.
Heller, H. C., and Cole, T. (1965). *Proc. Nat. Acad. Sci. U.S.* **54,** 1486.
Herak, J. N., and Gordy, W. (1965). *Proc. Nat. Acad. Sci. U.S.* **54,** 1287.
Hewitt, R., Marburger, K., and Lapthisophon, T. (1969). *Radiat. Res.* **39,** 485.
Hotz, G. (1963). *Biochem. Biophys. Res. Commun.* **11,** 393.
Hotz, G. (1968). *Mol. Gen. Genet.* **102,** 39.
Hotz, G., and Reuschl, H. (1967). *Mol. Gen. Genet.* **99,** 5.
Hotz, G., and Walser, R. (1970). *Photochem. & Photobiol.* **12,** 207.
Howard, D. B., and Tessman, I. (1964). *J. Mol. Biol.* **9,** 364.
Hug, W., and Tinoco, I., Jr. (1973). *J. Amer. Chem. Soc.* **95,** 2803.
Hutchinson, F. (1973). *Quart. Rev. Biophys.* **6,** 201.
Hutchinson, F., and Hales, H. B. (1970). *J. Mol. Biol.* **50,** 59.
Hütterman, J. (1970). *Int. J. Radiat. Biol.* **9,** 291.
Infante, G. A., Jirathana, P., Fendler, J. H., and Fendler, E. J. (1973). *J. Chem. Soc., Faraday Trans. 1* **69,** 1586.
Ingram, D. J. E. (1958). "Free Radicals as Studied by Electron Spin Resonance". Butterworth, London.
Ishihara, H., and Wang, S. Y. (1966a). *Nature (London)* **210,** 1222.
Ishihara, H., and Wang, S. Y. (1966b). *Biochemistry* **5,** 2302.
Ishihara, H., and Wang, S. Y. (1966c). *Biochemistry* **5,** 2307.
Jaffe, H. H., and Orchin, M. (1962). "Theory and Applications of Ultraviolet Spectroscopy," Chapter 15. Wiley, New York.
Johns, H. E., Pearson, M. L., LeBlanc, J. C., and Helleiner, C. W. (1964). *J. Mol. Biol.* **9,** 503.
Kahan, F. M., and Hurwitz, J. (1962). *J. Biol. Chem.* **237,** 3778.
Kaplan, H. S., Smith, K. C., and Tomlin, P. (1961). *Nature (London)* **190,** 796.
Kaplan, H. S., Smith, K. C., and Tomlin, P. (1962). *Radiat. Res.* **16,** 98.
Karle, I. L. (1969). *Acta Crystallogr., Sect. B* **25,** 2119.
Karle, I. L., Wang, S. Y., and Varghese, A. J. (1969). *Science* **164,** 183.
Kasha, M. (1960). *Radiat. Res., Suppl.* **2,** 243.
Khattak, M. N., and Wang, S. Y. (1969). *Science* **163,** 1341.
Kleinwächter, V., Drobnik, J., and Augenstein, L. (1966). *Photochem. & Photobiol.* **5,** 579.
Koehnlein, W., and Hutchinson, F. (1969). *Radiat. Res.* **39,** 745.
Kozinski, A. V., and Szybalski, W. (1959). *Virology* **9,** 260.
Krauskopf, M., Chen, C. M., and Ofengand, J. (1972). *J. Biol. Chem.* **247,** 842.
Kunieda, T., and Witkop, B. (1971). *J. Amer. Chem. Soc.* **93,** 3493.
LaCroix, M., and Van de Vorst, A. (1967). *Photochem. & Photobiol.* **7,** 477.
Lamola, A. A. (1969). *Photochem. & Photobiol.* **9,** 291.
Lamola, A. A. (1972). *Mol. Photochem.* **4,** 107.
Langmuir, M. E., and Hayon, E. (1969). *J. Chem. Phys.* **51,** 4893.
Leonard, N. J., Bergstrom, D. E., and Tolman, G. L. (1971). *Biochem. Biophys. Res. Commun.* **44,** 1524.
Ley, R. D., and Setlow, R. B. (1972). *Biochem. Biophys. Res. Commun.* **46,** 1089.
Lion, M. B. (1968). *Biochim. Biophys. Acta* **155,** 505.
Lion, M. B. (1970). *Biochim. Biophys. Acta* **209,** 24.
Lisewski, R., and Wierzchowski, K. L. (1970). *Photochem. & Photobiol.* **11,** 327.

Litman, R. M., and Pardee, A. B. (1960). *Biochim. Biophys. Acta* **42,** 117.
Longworth, J. W., Rahn, R. O., and Shulman, R. G. (1966). *J. Chem. Phys.* **45,** 2930.
Loutfy, R. O., and Loutfy, R. O. (1972). *Can. J. Chem.* **50,** 4052.
Lozeron, H. A., and Gordon, M. P. (1964). *Biochemistry* **3,** 507.
Lozeron, H. A., Gordon, M. P., Gabriel, T., Tantz, W., and Duschinsky, R. (1964). *Biochemistry* **3,** 1844.
Menningmann, H. D. (1967). *Mol. Gen. Genet.* **99,** 76.
Meistrich, M. L., and Lamola, A. A. (1972). *J. Mol Biol.* **66,** 83.
Michelson, A. M., Dondon, J., and Grunberg-Manago, M. (1962). *Biochim. Biophys. Acta* **55,** 529.
Montenay-Garestier, T., and Helene, C. (1970). *Biochemistry* **9,** 2865.
Moore, A. M., and Thomson, C. H. (1957). *Can. J. Chem.* **35,** 163.
Müller, A. (1963). *Int. J. Radiat. Biol.* **6,** 137.
Müller, A. (1967). *Progr. Biophys. Mol. Biol.* **17,** p. 99.
Neta, P. (1972). *Radiat. Res.* **49,** 1.
Nicolau, C., McMillan, M., and Norman, R. O. C. (1969). *Biochim. Biophys. Acta* **174,** 413.
Opara-Kubinska, Z., Lorkiewicz, Z., and Szybalski, W. (1961). *Biochem. Biophys. Res. Commun.* **4,** 288.
Ormerod, M. G., and Singh, B. B. (1966). *Int. I. Radiat. Biol.* **10,** 533.
Pearson, M. L. Ottensmeyer, F. P., and Johns, H. E. (1965). *Photochem. & Photobiol.* **4,** 739.
Pershan, P. S., Shulman, R. G., Wyluda, B. J., and Eisinger, J. (1964). *Physics* **1,** 163.
Peter, H. H., and Drewer, R. J. (1970). *Photochem. & Photobiol.* **12,** 269.
Pruden, B., Snipes, W., and Gordy, W. (1965). *Proc. Nat. Acad. Sci. U.S.* **53,** 917.
Puck, T. T., and Kao, F. (1967). *Proc. Nat. Acad. Sci. U.S.* **58,** 1227.
Rahn, R. O. (1969). *Photochem. & Photobiol.* **9,** 527.
Rahn, R. O., and Hosszu, J. L. (1968). *Photochem. & Photobiol.* **8,** 53.
Rahn, R. O., and Hosszu, J. L. (1969a). *Biochim. Biophys. Acta* **190,** 126.
Rahn, R. O., and Hosszu, J. L. (1969b). *Photochem. & Photobiol.* **10,** 131.
Regan, J. D., Setlow, R. B., and Ley, R. D. (1971). *Proc. Nat. Acad. Sci. U.S.* **68,** 708.
Rhoades, D. F., and Wang, S. Y. (1970). *Biochemistry* **9,** 4416.
Rhoades, D. F., and Wang, S. Y. (1971a). *J. Amer. Chem. Soc.* **93,** 3779.
Rhoades, D. F., and Wang, S. Y. (1971b). *Biochemistry* **10,** 4603.
Riehl, N. (1968). *In* "Energetics and Mechanisms in Radiation Biology" (G. O. Phillips, ed.), p. 61. Academic Press, New York.
Riley, M., and Paul, A. (1970). *J. Mol. Biol.* **50,** 439.
Rosner, A., and Yagil, E. (1969). *Isr. J. Chem.* **7,** 120.
Rothman, W., and Kearns, D. R. (1967). *Photochem. & Photobiol.* **6,** 775.
Rupp, W. D., and Prusoff, W. H. (1964). *Nature (London)* **202,** 1288.
Rupp, W. D., and Prusoff, W. H. (1965a). *Biochem. Biophys. Res. Commun.* **18,** 145.
Rupp, W. D., and Prusoff, W. H. (1965b). *Biochem. Biophys. Res. Commun.* **18,** 158.
Salovey, R., Shulman, R. G., and Walsh, W. M., Jr. (1963). *J. Chem. Phys.* **39,** 839.
Sasson, S., and Wang, S. Y. (1976). In preparation.
Sasson, S., Wang, S. Y., Ehrlich, M., and Riley, M. (1976). In preparation.
Sauerbier, W. (1961). *Virology* **15,** 465.
Setlow, R., and Boyce, R. (1963). *Biochim. Biophys. Acta* **68,** 455.
Sevilla, M. D. (1971). *J. Phys. Chem.* **75,** 626.
Shen, P. G., Blyumenfeld, L. A., Kalmanson, A. E., and Pasynskii, A. G. (1959). *Biofizika* **4,** 263.
Shields, H., and Gordy, W. (1959). *Proc. Nat. Acad. Sci. U.S.* **45,** 269.

Shugar, D., and Wierzchowski, K. L. (1958). *Postepy Biochem.* **4,** 243.
Shugar, D., and Wierzchowski, K. L. (1965). *J. Amer. Chem. Soc.* **87,** 4621.
Shulman, R. G., and Pershan, P. S. (1964). *Trans. N. Y. Acad. Sci.* [2], **27,** 210.
Sinsheimer, R. L. (1957). *Radiat. Res.* **6,** 121.
Smets, L. A., and Cornelis, J. J. (1971). *Int. J. Radiat. Biol.* **19,** 445.
Smith, K. C. (1962). *Biochem. Biophys. Res. Commun.* **6,** 458.
Smith, K. C. (1963a). *Photochem. & Photobiol.* **2,** 503.
Smith, K. C. (1963b). *Photochem. & Photobiol.* **2,** 508.
Smith, K. C. (1964). *Photochem. & Photobiol.* **3,** 1.
Smith, K. C., and Yoshikawa, H. (1966). *Photochem. & Photobiol.* **5,** 777.
Stahl, F. W., Craseman, J. M., Okun, L., Fox, E., and Laird, G. (1961). *Virology* **13,** 98.
Stephan, G., Miltenburger, H. G., and Hotz, G. (1970). *Z. Naturforsch. B* **25,** 1037.
Suzuki, H. (1967). "Electronic Absorption Spectra and Geometry of Organic Molecules: An Application of Molecular Orbital Theory." Academic Press, New York. 1967.
Takahashi, I., and Marmur, J. (1963). *Nature (London)* **197,** 794.
Thiele, D., Guschlbauer, W., and Favre, A. (1972). *Biochim. Biophys. Acta* **272,** 22.
Van de Vorst, A., and Lion, Y. (1973). *Biochim. Biophys. Acta* **294,** 349.
Varghese, A. J. (1970a). *Biochem. Biophys. Res. Commun.* **38,** 484.
Varghese, A. J. (1970b). *Biochemistry* **9,** 4781.
Varghese, A. J. (1971a). *Biochemistry* **10,** 2194.
Varghese, A. J. (1971b). *Biochemistry* **10,** 4283.
Varghese, A. J. (1971c). *Photochem. & Photobiol.* **13,** 357.
Varghese, A. J., and Patrick, M. H. (1969). *Nature (London)* **223,** 299.
Varghese, A. J., and Wang, S. Y. (1967). *Science* **156,** 955.
Varghese, A. J., and Wang, S. Y. (1968a). *Science* **160,** 186.
Varghese, A. J., and Wang, S. Y. (1968b). *Biochem. Biophys. Res. Commun.* **33,** 102.
Vigny, P., and Favre, A. (1974). *Photochem. & Photobiol.* **20,** 345.
Voytek, P., Chang, P. K. and Prusoff, W. H. (1972). *J. Biol. Chem.* **247,** 367.
Wacker, A., Dellweg, H., and Weinblum, D. (1961). *J. Mol. Biol.* **3,** 787.
Wacker, A., Menningmann, H. D., and Szybalski, W. (1962). *Nature (London)* **196,** 685.
Wacker, A., Dellweg, H., Träger, L., Kornhauser, A., Lodemann, E., Türck, G., Selzer, R., Chandra, P., and Ishimoto, M. (1964). *Photochem. & Photobiol.* **3,** 369.
Wang, S. Y. (1958a). *J. Amer. Chem. Soc.* **80,** 6196.
Wang, S. Y. (1958b). *J. Amer. Chem. Soc.* **80,** 6199.
Wang, S. Y. (1958b). *J. Amer. Chem. Soc.* **80,** 6199.
Wang, S. Y. (1959). *Nature (London)* **184,** B.A. 59.
Wang, S. Y. (1961). *Nature (London)* **190,** 690.
Wang, S. Y. (1962). *Symp. Reversible Photochem. Process, 1962* p. 74.
Wang, S. Y. (1963). *Nature (London)* **200,** 879.
Wang, S. Y. (1971). *J. Amer. Chem. Soc.* **93,** 2768.
Wang, S. Y., and Alcantara, R. (1965). *Photochem. & Photobiol.* **4,** 477.
Wang, S. Y., and Rhoades, D. F. (1971). *J. Amer. Chem. Soc.* **93,** 2554.
Wang, S. Y., and Varghese, A. J. (1967). *Biochem. Biophys. Res. Commun.* **29,** 543.
Wang, S. Y., Apicella, M., and Stone, B. R. (1956). *J. Amer. Chem. Soc.* **78,** 4180.
Wang, S. Y., Patrick, M. H., Varghese, A. J., and Rupert, C. S. (1967a). *Proc. Nat. Acad. Sci. U.S.* **57,** 465.
Wang, S. Y., Patrick, M. H., Varghese, A. J., and Rupert, C. S. (1967b). *Proc. Nat. Acad. Sci. U.S.* **58,** 2483.
Werner, T. C., and Hercules, D. M. (1969). *J. Phys. Chem.* **73,** 2005.
Weygand, F., Wacker, A., and Grisebach, H. (1951). *Z. Naturforsch. B* **6,** 177.
Weygand, F., Wacker, A., and Motiram, K. (1956). *Chem. Ber.* **89,** 475.

Witkop, B. (1968). *Photochem. & Photobiol.* **9,** 813.
Yamaguchi, G., Kakinoki, Y., and Tsubomura, H. (1967). *Bull. Chem. Soc. Jap.* **40,** 526.
Yamane, T., Wyluda, B. J., and Shulman, R. G. (1967). *Proc. Nat. Acad. Sci. U.S.* **58,** 439.
Yaniv, M., Favre, A., and Barrell, B. G. (1969). *Nature (London)* **223,** 1331.
Zamenhof, S., De Giovanni, R., and Guir, S. (1958). *Nature (London)* **181,** 827.

# 7 Photoproducts of Purines

*Dov Elad*



## A. Introduction

The photochemical reactions of nucleic acids and their constituents have been the subject of extensive investigations in recent years (Smith, 1966; Setlow, 1966; Burr, 1968; Fahr, 1969; Smith and Hanawalt, 1969). While the reactions of the Pyr bases and their products have been studied in detail, those of purines and Pur moieties in nucleic acids await more detailed study, and relatively little work has been done on the isolation and characterization of Pur photoproducts. The purines form part of the light-absorbing system in nucleic acids, but apparently this does not result in their photochemical alteration. The absorbed light energy may well be transferred to other moieties in the macromolecule, which subsequently undergo the chemical reaction. Nevertheless, some photochemical reactions of purines have been reported; in the older literature they consist mostly of photooxidation reactions (Wacker *et al.*, 1964).

Studies have been revealed that the substituents in the Pur system affect the reactivity of various Pur derivatives. Thus, while Pur itself is inert towards some reagents, such as aromatic diazonium salts, $(ho)_2{}^{2,6}$Pur (Xan) couples at C(8). Also some amino- and hydroxypurines undergo bromination at this position (Robins, 1967). The C(8) position in purines has been claimed to have a unique character as the reactive site in the system, especially towards free radicals (Robins, 1967). However, recent work has shown that in some photochemical reactions the C(6) position in the Pur system is more reactive than the C(8) position (Connolly and Linschitz, 1968; Evans and Wolfenden, 1970; Steinmaus *et al.*, 1971).

The role of purines in the photobiology of nucleic acids requires further clarification. In addition to the role that they may play in the lesion caused to biological systems by UV irradiation, purines may also be involved in cross-links formed under radiation between nucleic acids and other substrates present in biological systems (Yang *et al.*, 1971; Stankunas *et al.*, 1971; Elad and Salomon, 1971).

In this chapter the photoreactions induced by UV light of purines, Pur nucleosides, and Pur moieties in nucleic acids are reviewed. Reactions induced by $\gamma$-radiation will be discussed to a limited extent only. The chemical aspects of the reactions will be stressed.

## B. Ultraviolet Absorption Spectra of Purines

In order to understand the photochemical reactions of purines, some knowledge of their spectra and of their excited states is necessary. Since Chapter 2 discusses these aspects, this chapter will merely touch upon those points of importance to the photochemistry of the purines; more detailed discussions are available in various articles (Kasha, 1960; Mason, 1960; Stewart and Davidson, 1964; Voet *et al.*, 1963; Kleinwächter and Koudelka, 1972). The absorption exhibited by nucleic acids in the region of 260 nm, which is due primarily to their Pur and Pyr bases, embraces the spectral region in which irradiation results in major biological effects (Setlow, 1968). The precise location of the absorption maximum of a given sample and the magnitude of the extinction coefficient are dependent upon a number of factors, such as the degree of polymerization, secondary bonds in individual polynucleotide chains, and linkages between adjacent chains (Boerresen, 1963; Clark and Tinoco, 1965; Drobnik and Augenstein, 1966; Rahn *et al.*, 1967; Pullman, 1969). Ultraviolet light, as well as a considerable fraction of the energy absorbed from ionizing radiation, is dissipated as electronic excitations. Therefore, the properties of the electronically excited nucleotides should be of interest both in photochemistry and photobiology, as well as in radiobiology.

Heterocyclic bases, like Pur and Pyr bases, possess both $n,\pi^*$ and $\pi,\pi^*$ excited states. The spectra of the Pur bases show three to four bands between 180 and 300 nm in addition to the weak long wavelength (290–300 nm) band. The former are presumably of the $\pi,\pi^*$ type, while the latter appear to be of $n,\pi^*$ character. Clark and Tinoco (1965) classified the observed transitions in the nucleic acids into three groups: the first at 260–280 nm, the second at 220–250 nm, and the third at shorter wavelengths. Most purines possess bands in these three regions. Gua absorbs at 275, 256, 203, and 193 nm, while Ade shows maxima in solution at 260, 208, and 185 nm.

The lowest excited singlet states are claimed to be of the $n,\pi^*$ type in aqueous solution at physiological pH. Perturbing agents, such as the attachment of protons and substituents, determine the relative energetic levels of the $n,\pi^*$ and $\pi,\pi^*$ excited singlet states. A proton binds a lone pair of electrons, thereby greatly increasing the energy needed in the transition which occurs when one of these electrons is promoted into a vacant $\pi^*$ orbital. If no other $n,\pi^*$ transition assumes the role as lowest excited singlet, the $\pi,\pi^*$ configuration will become the lowest. However, the $n,\pi^*$ states other than the one which involves the orbital bonded to a proton, are expected to be stabilized by the protonation. The $n,\pi^*$ transitions are associated with a considerable displacement of negative charge from the localized lone pair orbital to a $\pi^*$ orbital covering the whole molecule. The positively charged proton may well favor such a redistribution of negative charge. The introduction of an amino group into the C(2) position of a Pur results in the displacement of the lowest $\pi,\pi^*$ absorption maximum toward a longer wavelength. Therefore, the stabilization of the $\pi,\pi^*$ singlets relative to the $n,\pi^*$ singlets is due to two mechanisms: (1) pH and (2) the functional groups and their location in the substituted Pur. Since changes in pH cause changes in the spectroscopic transitions, they might have a considerable effect on the photochemical behavior of purines.

Phosphorylation of the sugar hydroxyls of a nucleoside would not be expected to have a marked influence on the absorption spectrum of the resulting Pur nucleotide, as compared to the nucleoside (McLaren and Shugar, 1964), since the phosphate group is fairly well insulated from the Pur ring by the saturated carbon chain of the carbohydrate moiety. In fact, the spectra of most nucleotides resemble the corresponding nucleosides.

## C. Energy Transfer

The transfer of electronic energy has been useful in the discovery of new photochemical reactions and the study of their mechanisms (Lamola, 1969). The use of photosensitizers for the induction of photochemical reactions is by now one of the standard methods in photochemistry. In photosensitized reactions the substrate undergoing the chemical reactions is not that which absorbs the incident light but is the one which excited through energy transfer from the appropriate photosensitizer. Thus, the photolysis of purines may also be achieved indirectly with light which is not absorbed by the Pur moiety but is absorbed solely by a photosensitizer. Another process of energy transfer which plays an important role in the photochemistry of purines can operate from the excited Pur to other moieties in the nucleic acid molecule. In fact, the electronic energy absorbed by the purines in nucleic acids tends to be transferred to the pyrimidines. This effect presents one of the reasons for the observed "photostability" of the purines. These energy transfer processes have been studied by a variety of spectroscopic methods, such as UV absorption spectra, emission spectra and ESR spectra. (Rahn *et al.*, 1965; Hélène *et al.*, 1966; Guéron *et al.*, 1966, 1967; Eisinger and Shulman, 1967).

The emission from native DNA has been found to be an order of magnitude lower than the sum of its constituent Pur nucleotides (Rahn *et al.*, 1967). In addition, the emission maximum is shifted to longer wavelengths relative to the two purines. Decay time studies show that the emission of DNA is not simply explained as a sum of the Ade and Gua emissions. In an attempt to understand these observations, measurements have been carried out on synthetic polynucleotides which serve as models for DNA or RNA (Rahn *et al.*, 1964). Studies of the luminescence of nucleotides and other nucleic acid derivatives have been limited for the following reasons: (1) Emission by Pyr bases is very weak, often so weak at a low concentration that measurement is impossible. In addition, these bases are photosensitive and can undergo chemical changes during the experiments. (2) Emission by Pur bases varies with pH. (3) It is impossible to excite one type of base in oligonucleotides and polynucleotides without exciting the others. Nevertheless, several methods can be used to record the possible energy transfer between the bases of a nucleotide by measuring phosphorescence. Phosphorescence measurements revealed a striking similarity between DNA and poly(dA-dT). This result was interpreted as showing that in poly(dA-dT), as well as in DNA, the emission is from the neutral Thy triplets (Lamola *et al.*, 1967) in some perturbed state. A study of the phosphorescence of Thd as a function of pH showed that Thd, which has lost a proton at N(1), also has an excited triplet state which phosphoresces with decay time of 0.5 sec. Because the N(1) proton of Thd is involved in hydrogen bonding at the N(1) position of Ade, both DNA and poly(dA-dT) act as if Thy shifts its proton to Ade and emits as the deprotonated base (Rahn *et al.*, 1967). On the other hand, Ado and AMP have been shown to have their triplet states quenched by protonation at their N(1) position. Hence, this mechanism is consistent with the failure to observe extensive phosphorescence from Ade in either poly(dA-dT) or in DNA. Furthermore, measurements indicate that Gua phosphorescence in a poly(rG-rC) is quenched, which is also consistent with the fact that the Thy phosphorescence is the main component of the DNA emission (Rahn *et al.*, 1965; Lamola *et al.*, 1967). A more recent publication indicates that Gua phosphorescence may not be quenched in poly(rG-rC) (Kleinwächter, 1972). The excitation spectrum of poly(dA-dT) for both phosphorescence and fluorescence is the sum of the excitation spectra of Ado and Thd, thus indicating singlet transfer. Hence, photons absorbed either by Ado or Thd are equally effective in creating the Thd emission.

Most of the electronic structural characteristics responsible for the optical properties of helical polynucleotides can also be found in dinucleotides (Hélène *et al.*, 1966; Guéron *et al.*, 1966; Eisinger and Shulman, 1968; Kleinwächter and Koudelka, 1972). Such well-defined compounds can be synthesized, and Pur or Pyr bases can be substituted by chemical groups in which the electronic properties can be modified. A variety of studies of internal energy transfer in dinucleotides has been possible using the tests of luminescence and, more particularly, phosphorescence. In dinucleotides, the distance between the bases in the order of 3–4 Å, and the lowest triplet state of the Pyr base is located below that of the corresponding triplet of the Pur base. In dinucleotides one may expect an intramolecular triplet transfer from Pur to Pyr bases. Studies of phosphorescence of dinucleotides containing Ade and Cyt indicated that there is a transfer of energy from the Ade to the Cyt residue (Hélène *et al.*, 1965, 1966). The lowest excited singlet and triplet states of the donor (Ade) possesses higher energy than the corresponding states of the acceptor (Cyt). If in the dinucleotide Cyt is replaced by 5-BrCyt, practically total transfer can be obtained from the Ade to the BrCyt. Studies of three of the dinucleotides, ApK, AppBrU, and GpC, at 77°K in a water–ethanol mixture, showed a very pronounced quenching of phosphorescence of the Pur base as compared to that shown by the free Pur nucleotide, either alone or in hydrolysates of the dinucleotides. This phosphorescence is regenerated after hydrolysis of the dinucleotide. Results indicate that the characteristic fluorescence of the Pur bases at 77°K remains practically unchanged. The efficiency of energy transfer from the Pur to the Pyr can be measured by the amount of quenching of the Pur phosphorescence and decreases in the order AppBrU > GpC > ApU > ApC.

These examples of energy transfer in nucleosides and nucleotides represent only some of the energy transfer processes occurring in nucleic acid constituents. These processes have to be considered when studying the photochemical reactions of Pur derivatives since they might affect the nature of the photoproducts or the chemical behavior of the Pur moieties during irradiation.

## D. Photolyses of Purines

The photolyses of purines, Pur nucleosides, and Pur moieties in nucleic acids have been studied (McLaren and Shugar, 1964; Heyroth and Loufbourow, 1931; Sinsheimer and Hasting, 1949; Carter, 1950; Christensen and Giese, 1954; Shugar and Wierzchowski, 1958). In most of these studies photolyses have been followed by UV absorption changes, and in other studies photoproducts were detected by physical methods. Ultraviolet irradiation of purines of biological interest in the absence and in the presence of oxygen showed decomposition rates of zero order (Kland and Johnson, 1957). For Ade and Hyp, decomposition is faster under oxygen, but the breakdown of Gua and Xan is inhibited by oxygen as compared with reactions performed under nitrogen. Ade is essentially destroyed completely after 24 hr of irradiation under aerobic conditions, whereas under nitrogen >90% remains. This suggests oxidative breakdown, with the amino group being the point of attack:

$$RNH_2 \xrightarrow[O_2]{h\nu} RNH:OH \xrightarrow[H_2O]{h\nu} ROH$$

The formation of Hyp in the photochemical oxidation of Ade is given as support for the proposed steps. It has been found that the purines examined exhibit an initial "induction" period, and that the second stage of breakdown is independent of the concentration of the purines.

Studies on the photolysis of Ade-1-oxide indicated that this compound is very sensitive to UV radiation (Brown *et al.*, 1964). Two major products have been detected, iso-Gua (ho$^2$Ade) and Ade. After 50% alteration of the Ade-1-oxide, the rearrangement product, iso-Gua, predominates by a ratio of about 5:4. The reaction becomes more complex as the decomposition approaches 100% but these two remain the major products (Scheme 1). Me$^6$Pur-1-oxide, in which the carbon adjacent to the *N*-oxide function is already occupied by a hydroxyl group, decomposes with approximately the same ease, yielding only iso-Gua. These photoreactions resemble those of Pyr and quinoxaline *N*-oxides (Spence *et al.*, 1970). It is suggested that in the photochemical alterations of Pur *N*-oxides, two mechanisms may operate: (1) direct removal of oxygen, possibly as peroxide, from the *N*-oxide function and (2) the transfer of the oxygen to the adjacent carbon, which may proceed through a three-membered oxaziridine intermediate. With iso-Gua oxide, the second type of reaction is obviously blocked, and only the first can operate. The extreme UV sensitivity of Ade oxide, in comparison to other Pur or Pyr derivatives, may be of biological interest. It has been suggested that hydroxyl radicals produced by the indirect action of ionizing radiation might induce N-oxidation of Ade and thereby sensitize a nucleic acid to the primary effects of radiation. The iso-Gua derivative, a radiation product, could still pair with Thy (Scheme 2).

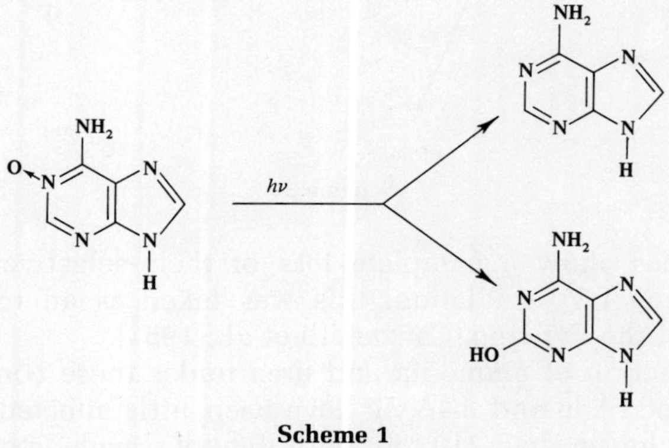

**Scheme 1**

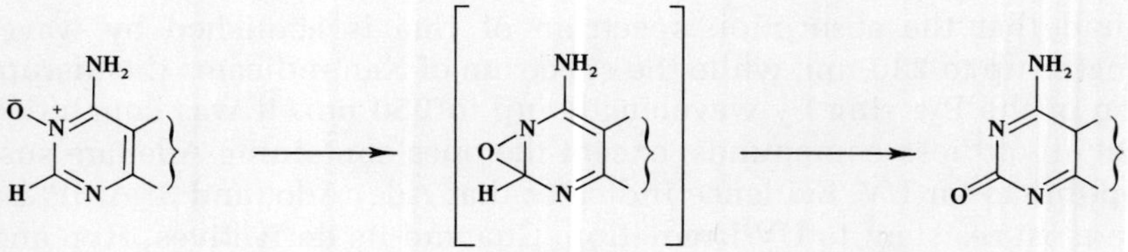

**Scheme 2**

Ade-1-oxide is decomposed by UV light with a quantum efficiency of 0.10 and with an action spectrum paralleling the UV absorption spectrum (Levin *et al.*, 1964). It is noteworthy that with $\gamma$-irradiation Ade-*N*-1-oxide was also decomposed more rapidly than was Ade, although the difference was smaller.

Cramer and Schlinghoff (1964) obtained Ado and iso-Guo from irradiated Ado-1-oxide. In addition, a third product, 5-cyano-3-ribosyl-4-ureidoimidazole (**I**), also has been isolated. The latter recyclized readily to iso-Guo. The distribution of the photoproducts has been found to depend upon the pH of the reaction medium (Scheme 3).

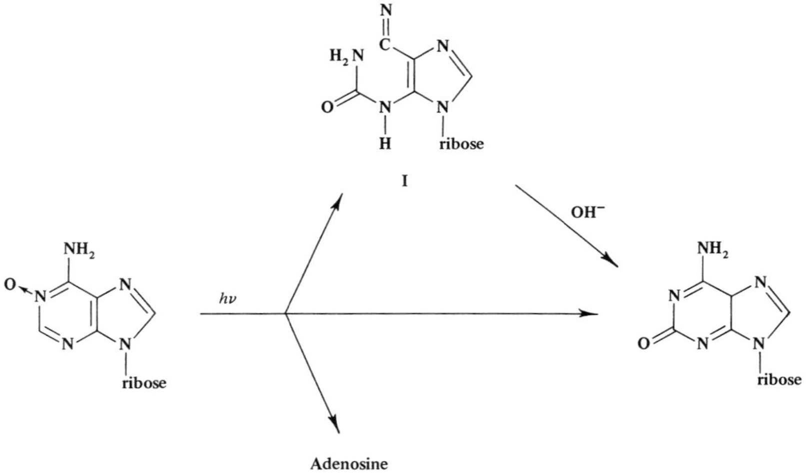

I

$h\nu$

$OH^-$

Adenosine

**Scheme 3**

The purines show a complete loss of their selective absorption spectra during UV irradiation; this was taken as an indication of disruption of the Pyr ring (Canzanelli *et al.*, 1951).

The production of ammonia and urea under these conditions has been observed. Ade and 3′-AMP have been little affected by light of wavelengths longer than 210 nm. With light of wavelengths above 230 nm, both compounds are essentially unaffected. However, it has been found that the absorption spectrum of Gua is abolished by wavelengths up to 230 nm, while the spectrum of Xan indicates the disruption of the Pyr ring by wavelengths up to 250 nm. It was concluded that all of these compounds, except the ones containing Ade, are susceptible to far UV. Evidence indicates that Ade, Ado, and 3′-AMP are the most resistant to UV irradiation. Gua and its derivatives, Hyp and Ino, occupy a midway position. Xan, Xao, and uric acid show the highest susceptibility to UV irradiation. Ade has a smaller extinction maximum than Ado and the maximum of the latter is less than that of AMP, but these are reduced to the same level by irradiation. A similar phenomenon is seen in the case of Gua, except that Guo and GMP have the same initial value which is higher than that of Gua. On the other hand, while Ino also has a higher initial maximal optical density, it appeared to be more resistant to irradiation than its parent base Hyp. Approximately one mole of ammonia was detected for each mole of Ade irradiated. Hyp, which contains no amino group, yielded almost as much ammonia as Ade, while Ino, the riboside of Hyp, yielded somewhat more than did Ade. This does not rule out the possibility that ammonia resulting from Ade irradiation is the result of deamination, although this does make it less likely and indicates that ammonia may be derived from the ureide groups of either the Pyr ring or the imidazole nucleus. The fact that irradiation of all the pyrimidines studied (Ura, Cyt, and Thy) resulted in the production of very small amounts of ammonia seemed to point to the fact that the ammonia derived from Ade and Hyp came from the N(7) or N(9) position or both. It is particularly noteworthy that Cyt, with an amino group in the C(6) position, yielded practically no more ammonia than the other pyrimidines. A clue to the significance of these data appears in the experiments with Xan and uric acid. In these compounds the Pyr ring is analogous to that of the naturally occurring free pyrimidines since carbonyl groups (or hydroxyl) are present in the C(2) and C(6) positions. Both Xan and uric acid behaved like the pyrimidines with respect to ammonia formation, with very little ammonia being produced. The presence of the imidazole nucleus, on the other hand, is apparently without effect on ammonia production.

Irradiation of uric acid led to the formation of a variety of photoproducts in addition to urea and ammonia (Fellig, 1954). One of these photoproducts, triuret, was isolated and characterized. Scheme 4 was suggested for the reaction.

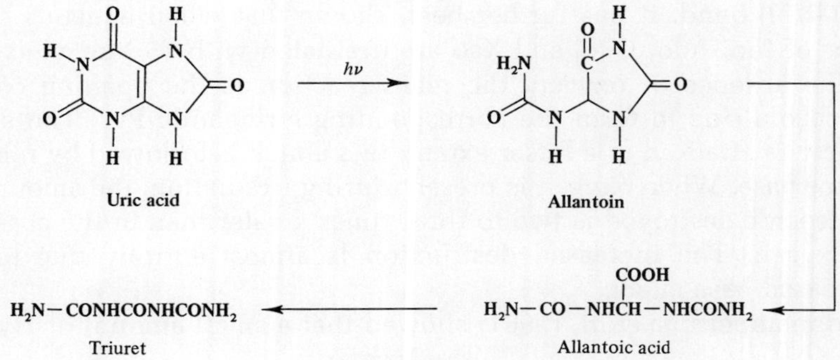

Uric acid $\xrightarrow{h\nu}$ Allantoin

$H_2N-CONHCONHCONH_2$ ← $H_2N-CO-NHCH(COOH)-NHCONH_2$

Triuret ← Allantoic acid

**Scheme 4**

Photolysis of purines and Pur moieties in DNA can be sensitized with dyes (Simon and van Vunakis, 1962; Wacker *et al.*, 1964; Spikes and Livingston, 1969). Thus, irradiation of a DNA solution containing methylene blue with visible light in the presence of oxygen led to the preferential destruction of Gua. Other bases were almost unaffected. Thiopyronin reacted similarly and with higher efficiency. It has also been observed that Gua is preferentially destroyed upon irradiation of various nucleosides or free bases. It has been suggested (Wacker *et al.*, 1964) that the photochemical alteration of Gua occurs by a radical mechanism since oxygen is essential for this reaction. Wacker *et al.* (1964) further suggested that this reaction consists of an oxidative attack on the Pur ring followed by further degradation. The degradation depends upon the substitution in the Pur nucleus; a substituent at C(2) facilitates the destruction of the Pur. An amino group at C(2) is claimed to be more effective than a hydroxyl group; thus Ade and Hyp are affected to a small extent only. The Gua moieties of nucleic acids in bacteria have also been found to be sensitive sites for photochemical reactions.

Studies on the effects of ionizing radiation of aqueous solutions of Pur nucleosides and nucleotides were shown to involve attack on the base followed by hydrolytic release of ribose (Scholes and Weiss, 1952). Base attack was found to be the major pathway. A further investigation of base attack revealed that when aqueous solutions of Guo, CMP, and Xao are irradiated in the absence of oxygen, the imidazole ring is opened to form the corresponding 4-amino-5-formamido-Pyr riboside. When AMP was irradiated in the absence of oxygen, no formamido-Pyr was formed. Opening of the imidazole ring is a general result of radiation of Pur nucleosides or nucleotides. Hems (1960) proposed that initial attack on the nucleoside by H or OH radicals leads to a nucleoside radical which recombines with a second radical. These reactions result in the introduction of a hydroxyl group at C(8) and a hydrogen atom at N(9) and lead to the opening of the C(8)–N(9) bond. It has further been shown that when aqueous solutions of Ino, Ado, Guo, and Xao are irradiated with 15 Mev electrons in the absence of oxygen, the main reaction is the opening of the imidazole ring to form the corresponding formamido-Pyr. The sugar moiety is attacked to a lesser extent; this attack is followed by release of free base. When oxygen is present during irradiation, the amount of nucleoside destroyed is two to three times greater than in the absence of oxygen. The increased destruction is almost entirely due to an increased base attack.

Ponnamperuma *et al.* (1961) showed that a small amount of Hyp is formed when a solution of Ade is irradiated. This has been detected by using $^{14}C$-labeled Ade and techniques of paper chromatography and liquid scintillation counting. Two principal effects in the radiation of purines have further been observed. In one, the Pur fused ring structure is preserved; in the other, this structure is broken (Ponnamperuma *et al.*, 1963). Deamination, or the conversion of Ade to Hyp, and the formation of ho$^8$Ade are examples of the first type. Together these two products account for ~25% of the radiation-decomposed Ade. The formation of 4,6-diamino-5-formamido-Pyr (**II**) is an example of the second type of reaction. It has been suggested that the deamination of Ade might be responsible for a radiation-induced mutation because it would result, during replication, in the replacement of an Ade–Thy base pair by a Gua–Cyt pair.

An 8,5′-cyclonucleotide (**III**) was the product in the $\gamma$-radiolysis of deaerated aqueous solutions of GMP (Keck, 1968). Evidence for this was obtained by NMR and chemical degradation to 8-formyl-Ade hm$^8$Ade, and Cb$^8$Ade. Irradiation of Pur nucleosides leads to similar cyclonucleotides.

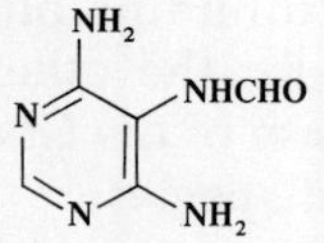

NH$_2$ N NHCHO N NH$_2$

II

O N N C CHOP$^-$ H$_2$N N N H H OH OH H H O

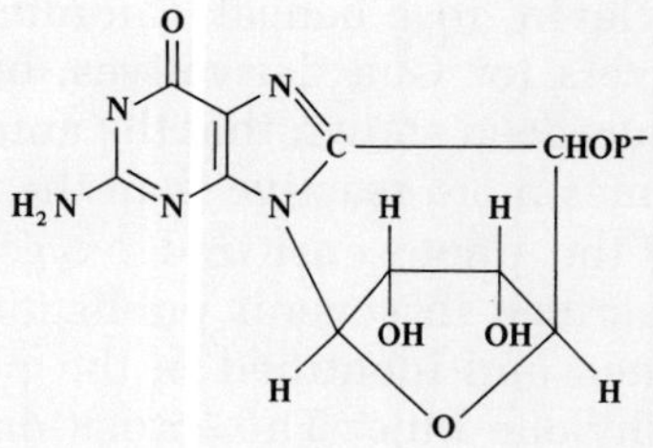

III

Van Hemmen and Bleichrodt (1971) detected six products of Ade after the exposure of the later to $\gamma$-radiation; ho$^8$Ade and 4,6-diamino-5-formamido-Pyr (**II**) were among these photoproducts.

## E. The Photodynamic Effect

Photosensitized oxidations have been of interest to chemists and biologists since the discovery that microorganisms are killed by light in the presence of dyes and oxygen (Raab, 1900). These reactions have been termed the "photodynamic effect" and have many biological implications including cell damage, induction of mutations or cancer, and death. Interest in the photosensitized oxygenation of purines was remarkably stimulated a few years ago when it was shown that Gua residues are destroyed preferentially during the methylene blue sensitized photooxygenation of DNA (Simon and van Vunakis, 1962). Only the chemical aspects of these reactions will be reviewed. The photosensitized oxygenation of purines depends upon the nature of the Pur base, since ribose and deoxyribose derivatives are found to be equally reactive, and the Gua residues in both DNA and RNA were rapidly attacked. However, Pur itself is not sensitive to photooxygenation with methylene blue; monosubstituted purines undergo little reaction, if any, while disubstituted derivatives are rapidly photooxidized (Simon and van Vunakis, 1964). A number of different dyes have been examined as sensitizers for the nucleic acid bases. Acridine orange is a poor sensitizer; lumichrome sensitizes the photooxidation of dGMP and Xan, but not that of Ade and Hyp; thiopyronine sensitizes the photodynamic destruction of Gua, $Am^2Pur$, $(Am_2)^{2,6}Pur$, Xan, and isoGua. Ade, Hyp, and 8-AzGua are scarcely affected (Wacker *et al.*, 1963; 1964; Lochmann and Stein, 1964; Lochmann *et al.*, 1964). Eosin Y., riboflavin, rose bengal, thionine, and toluidine blue are effective sensitizers for Gua derivatives, but not for the other nucleic acid bases. It has been shown that the anionic form of the Gua nucleoside is ~25 times more reactive than the neutral species.

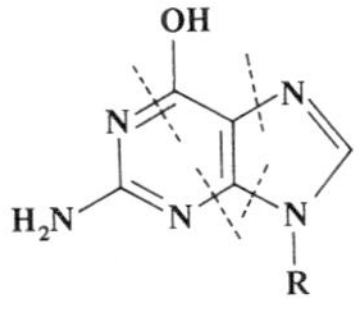

Products of the photosensitized oxygenation reactions of purines have been described in several publications. Thus, 1,3-dimethylallatoin was isolated and identified in the photoxygenation of theophylline with methylene blue. The formation of ribosylurea, free ribose, guanidine, and traces of urea from the photosensitized oxygenation of Guo with methylene blue at pH 9.2 has been reported (Waskell *et al.*, 1966). This shows that there are at least five points of attack in the Guo molecule. Product analysis of irradiated $[^{14}C]$uric acid suggests (Friedman, 1968) that urea may be a primary photoproduct from the C(2) position of the molecule, while urea from the C(8) position results from decomposition of a primary photoproduct. The photodegradative pathway of uric acid is probably similar to that of Gua and possibly Xan. The photosensitized oxygenation of Gua has been reported to yield guanidine and parabanic acid (Sussenbach and Berends, 1965) (Scheme 5).

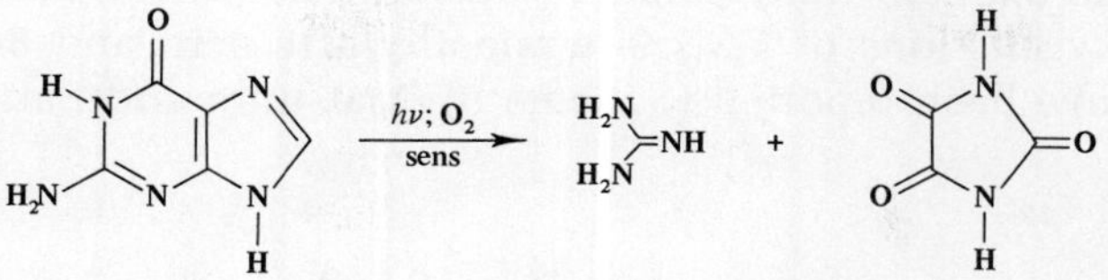

Scheme 5

The photosensitized oxygenation of $Me_2^{1,3}Ph^9Xan$ and $Ph^9Xan$ in methanol in the presence of rose bengal gave the corresponding 4,5-dimethoxyuric acid derivatives as major products (Scheme 6).

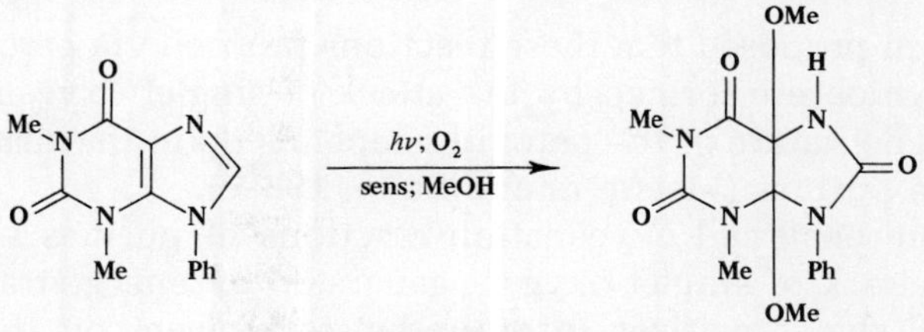

Scheme 6

Under similar conditions $Me_2^{1,3}Ph^9$- and $Ph^9$-uric acid were also obtained (Matsuura and Saito, 1969a). The photosensitized oxygenation of N-substituted hydroxypurines in aqueous alkaline solution in the presence of rose bengal has also been investigated. Under these reaction conditions, Xan gave allantoin and triuret (Scheme 7).

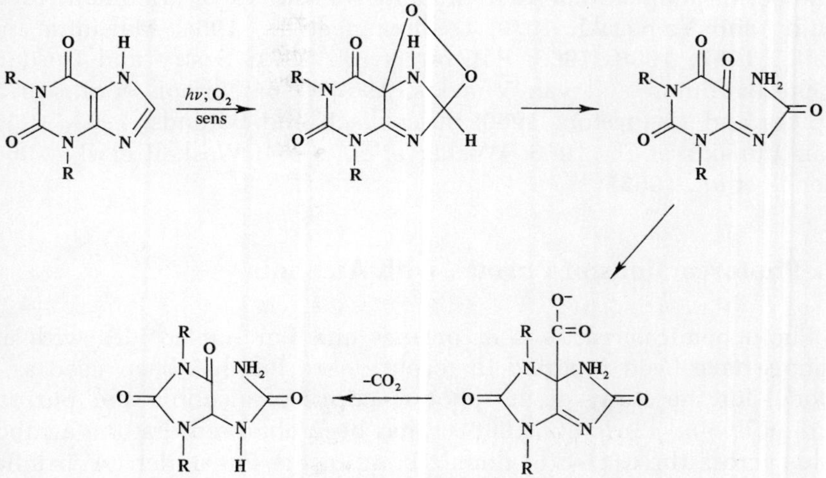

Scheme 7

Uric acid gave triuret, sodium oxanate, and allantoxaidin; Me[8]Xan resulted in the formation of a complex mixture from which acetamide and sodium oxonate were isolated (Matsuura and Saito, 1969b). Similar photoxygenations of 1,3,7,9-tetramethyluric acid and 8-methoxycaffeine have been reported (Scheme 8) (Matsuura and Saito, 1969c).

$$+ \; R_4OH \xrightarrow[\text{sens}]{h\nu;\, O_2} \; + \; R_1NHCOOR_4 \; + \; CO_2$$

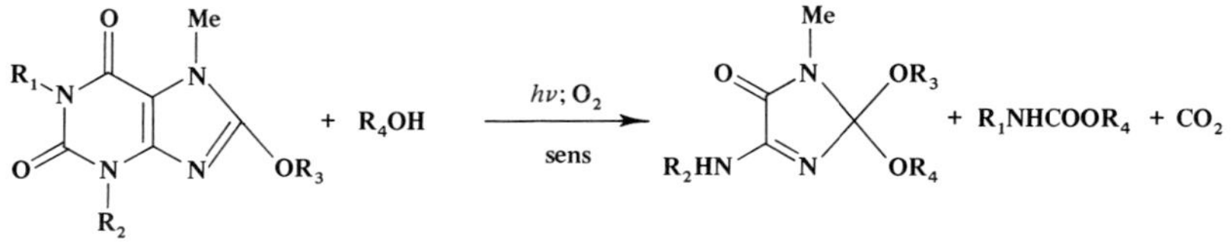

Scheme 8

It has been proposed that these reactions proceed via peroxide intermediates which are formed by the attack of singlet oxygen upon the substrate. The nature of the peroxide depends upon the structural features of the purines (Matsuura and Saito, 1969c).

The photosensitized oxygenation reactions of purines seem to involve the attack of singlet oxygen, generated by energy transfer from the excited dye sensitizer to ground-state oxygen, on the purines. Rosenthal and Pitts (1971) showed that singlet oxygen generated by microwave discharge leads to similar reactions of the purines. Paramagnetic metal ions decrease the rate of photooxidation, suggesting that triplet states of the dye are involved in the reactions. (Sussenbach and Berends, 1963).

The photodynamic effect has been described in recent years in a number of publications (Chandra and Wacker, 1966; Friedman, 1968; Fujita and Yamazaki, 1970; Lochmann *et al.*, 1964; Matsuura and Saito, 1967, 1968, 1969; Politzer *et al.*, 1971; Sastry and Gordon, 1966a,b; Simon and van Vunakis, 1962, 1964; Simon *et al.*, 1965; Spikes and Livingston, 1969; Sussenbach and Berends, 1963, 1965; van Vunakis *et al.*, 1966; Wacker *et al.*, 1964; Waskell *et al.*, 1966; Zenda *et al.*, 1965).

## F. Photoreactions of Purines with Alcohols

Photochemical reactions of purines and Pur nucleosides with alcohols have been reported in recent years. Pur has been used as a model for the study of the photoreactions of alcohols and purines (Connolly and Linschitz, 1968). It has been observed that the alcohol adds across the C(1)–C(6) double bond of the Pur under UV irradiation. The addition photoproducts have been found to be very sensitive to oxygen. The adducts of Pur with methanol, ethanol, and 2-propanol have been isolated and characterized. Mainly on the basis of the NMR data and using deuterated purine adducts, the addition has been shown to involve bonding between the $\alpha$-carbon atom of the alcohol and the C(6) position of the Pur system (Scheme 9).

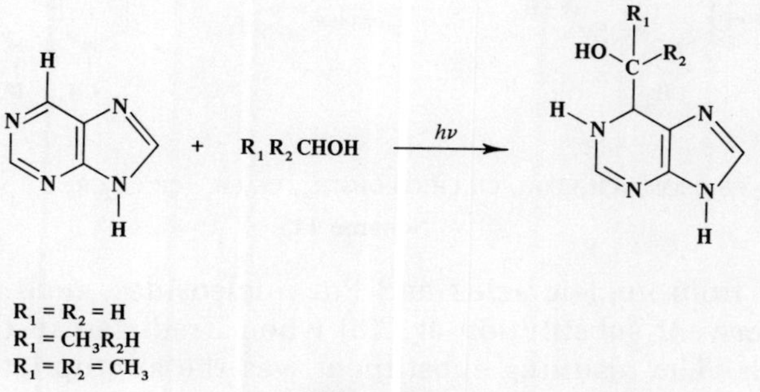

$R_1R_2CHOH$ $\xrightarrow{h\nu}$

$R_1 = R_2 = H$
$R_1 = CH_3, R_2 = H$
$R_1 = R_2 = CH_3$

**Scheme 9**

A similar photochemical reaction was observed when purine riboside was irradiated in the presence of methanol (Scheme 10) (Evans and Wolfenden, 1970).

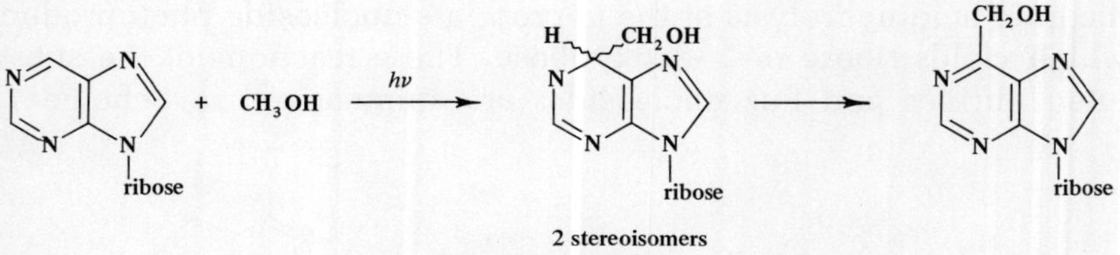

ribose + $CH_3OH$ $\xrightarrow{h\nu}$ ribose ⟶ ribose

2 stereoisomers

**Scheme 10**

Both diastereomers of hm$^6$hPur riboside could be isolated together with hm$^6$Pur riboside, which resulted from oxidation of hm$^6$hPur. These photoproducts were found to be potent reversible inhibitors of adenosine deaminases. Also Am$^2$Pur reacted similarly with 2-propanol under UV irradiation (Steinmaus *et al.*, 1971) and underwent substitution at C(6).

Other substituted purines reacted differently with alcohols under UV and $\gamma$-irradiation. Thus, caffeine underwent substitution at C(8) when irradiated in the presence of a variety of alcohols (Steinmaus *et al.*, 1971). The resulting substituent at C(8) depended upon the alcohol used. With primary alcohols the newly introduced substituent was the appropriate alkyl group, while with secondary alcohols a hydroxyalkyl group was introduced. Under $\gamma$-irradiation, mixtures of the 8-alkylcaffeines and 8-hydroxyalkylcaffeines were obtained (Scheme 11).

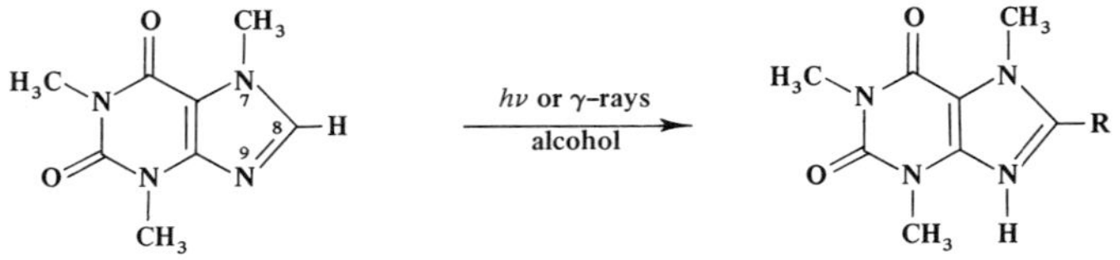

$hv$ or $\gamma$-rays
alcohol

R = $CH_3CHOH$; $(CH_3)_2COH$; $CH_3CH_2CH(OH)CH_3$, $CH_3CH_2$; $CH_3CH_2CH_2$

**Scheme 11**

Purines from nucleic acids and Pur nucleosides, such an Ade and Gua, underwent substitution at C(8) when irradiated in the presence of alcohols. The resulting substituent was the appropriate $\alpha$-hydroxyalkyl group (Steinmaus *et al.*, 1971). Products of these reactions have been isolated by standard procedures and their structures have been determined. NMR spectra were very useful for these determinations, since the distinction between the H(2), H(6), or H(8) could be achieved from these spectra. The sugar moiety remained intact during these photochemical and $\gamma$-ray-induced reactions. This could be proved by the mild acid hydrolysis of the appropriate nucleoside photoproducts which yields ribose or 2′-deoxyribose. These reactions of the substituted purines and Pur nucleosides are summarized in Scheme 12.

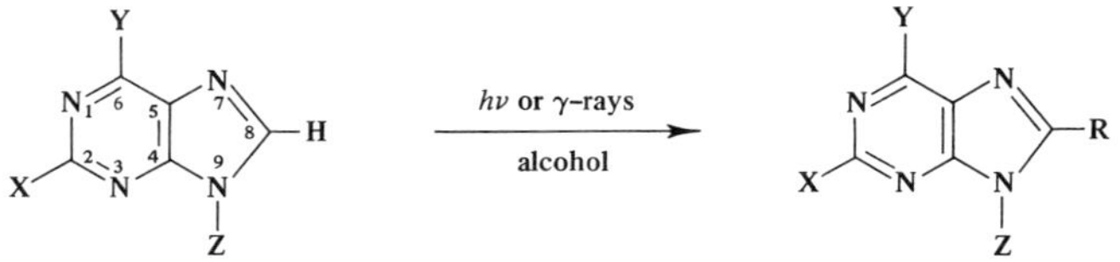

$hv$ or $\gamma$-rays
alcohol

X = H; Y = $NH_2$; Z = H (Ade)
X = H; Y = $NH_2$; Z = ribose (Ado)
X = $NH_2$; Y = OH; Z = ribose (Guo)
X = $NH_2$; Y = OH; Z = 2′-deoxyribose
X = H; Y = OH; Z = H (Hyp)
X = H; Y = Eto; Z = H

R = $(CH_3)_2COH$
R = $(CH_3)_2COH$
R = $CHOHCH_3$
R = $(CH_3)_2COH$
R = $(CH_3)_2COH$
R = $(CH_3)_2COH$
R = $(CH_3)_2COH$
R = X = $(CH_3)_2COH$

**Scheme 12**

NMR and ORD studies indicated that these C(8)-substituted Pur nucleosides adapt the *syn* conformation (Jordan and Pullman, 1968; Steinmaus *et al.*, 1971; Miles *et al.*, 1971).

The photochemical reactions of purines with alcohols could be induced directly with light of $\lambda > 260$ nm or could be initiated photochemically with acetone or peroxides with light of $\lambda > 290$ nm. In the latter cases higher yields of the appropriate modified purines were obtained (Steinmaus *et al.*, 1971). The initiation step involves the excitation of the Pur followed by a hydrogen atom abstraction from the alcohol by the excited Pur (Castellano and Lablache-Combier, 1971). This step results in the fragmentation of the alcohol to free radicals of the type $R_1R_2\dot{C}OH$ which subsequently attack at the carbon end of an imino —C=N groups in ground-state Pur. In the photosensitized reactions the generation of the alcohol free radicals can operate through energy transfer from the excited photosensitizer to the Pur, resulting in the excitation of the latter, which subsequently abstracts a hydrogen atom from the alcohol. An alternative route for the generation of the alcohol free radicals is through the excited photosensitizer which can abstract a hydrogen atom from the alcohol. The two pathways through which alcohol free radicals react with purines can be presented as shown in Scheme 13.

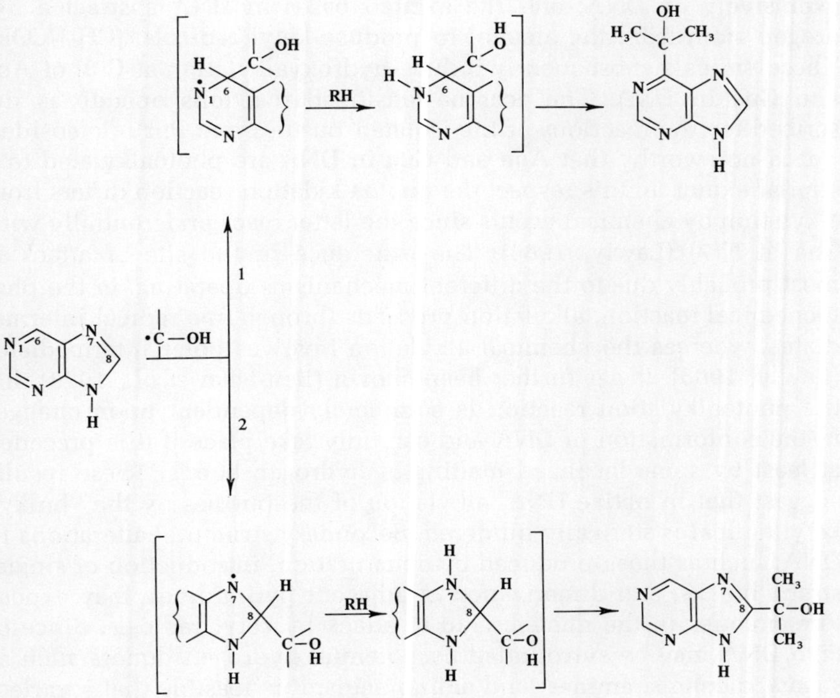

**Scheme 13**

The reactivities of the various sites of the Pur nucleus toward alcohol free radicals were derived from the data given above. The reactions involve two types of attack: (1) at C(6), and (2) at C(8). Pur itself, Am$^2$Pur, and Pur-9-riboside belong to the first group, while the 6-substituted and the 2,6-disubstituted purines and Pur nucleosides belong to the second. The preliminary substitution at C(6) in Pur and Am$^2$Pur indicates that this site is more reactive than C(8) or C(2), while the primary attack at C(8) in Ade indicates that C(8) is more reactive than C(2). Thus, the order of reactivity toward alcohol free radicals in the Pur ring is as follows: C(6) > C(8) > C(2).

The photoalkylation reactions with alcohols have been applied to the modification of Pur moieties in DNA. Irradiation of DNA in the presence of 2-propanol led to the C(8) hydroxyalkylation of the Ade and Gua in the macromolecule (Steinmaus *et al.*, 1970). Irradiation conditions included either direct UV irradiation or photosensitization with acetone. The identity of the photoproducts with authentic 8-hydroxyalkyl-Ade or Gua has been established. To account for the formation of the Pur photoproducts in DNA Ben-Ishai *et al.* (1973) suggested that in the case of direct irradiation, UV light is absorbed exclusively by DNA, and the excited bases in DNA abstract a hydrogen atom from the alcohol to produce ketyl radicals, $(CH_3)_2\dot{C}OH$. These radicals subsequently induce hydroxyalkylation at C(8) of Ade and Gua in DNA. The acetone-sensitized reactions operate as described for the reactions of the isolated purines and Pur nucleosides.

It is noteworthy that Ade and Gua in DNA are photoalkylated to a similar extent. In this respect the photoalkylation reaction differs from alkylation by chemical agents since the latter react preferentially with Gua at N(7) (Lawly, 1966). The difference in the site of attack is most probably due to the different mechanisms operating. In the photochemical reaction, alkylation proceeds through free radical intermediates, whereas the chemical alkylation involves polar intermediates (Lawly, 1966). It has further been shown (Ben-Ishai *et al.*, 1973) that the photoalkylation reaction is completely dependent upon changes in the conformation of DNA and can only take place if it is preceded at least by some localized melting of hydrogen bonds. These results suggest that in native DNA, alkylation of the purines by the "bulky" ketyl radical is sterically hindered. Secondary structural alterations in DNA, such as those produced by denaturation, introduction of single-strand breaks, and dimerization of adjacent pyrimidines, may expose Pur moieties in the nucleic acid to attack by ketyl radicals. Since *in vivo* DNA may be surrounded by potential hydrogen donors such as sugars, alcohols, amines, and amino acids, it is feasible that a variety of free radical reactions similar to those of alcohols may occur in the cells.

The photodealkylation of 8-hydroxyalkylpurines has been reported (Elad *et al.*, 1971). These reactions (Scheme 14) lead to the regeneration of the original Pur or Pur nucleosides. Initiation is achieved through the use of a photosensitizer, which is usually an aromatic amine, and light of $\lambda > 290$ nm. Some of these reactions proceed with high chemical yields.

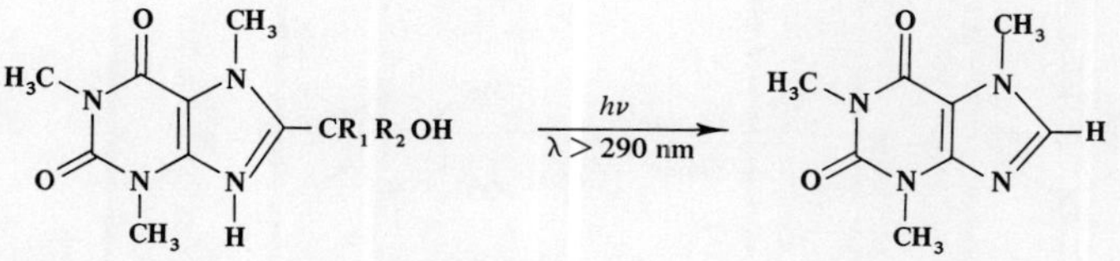

$h\nu$, $\lambda > 290$ nm

$R_1 = R_2 = H$; $R_1 = H, R_2 = CH_3$; $R_1 = R_2 = CH_3$

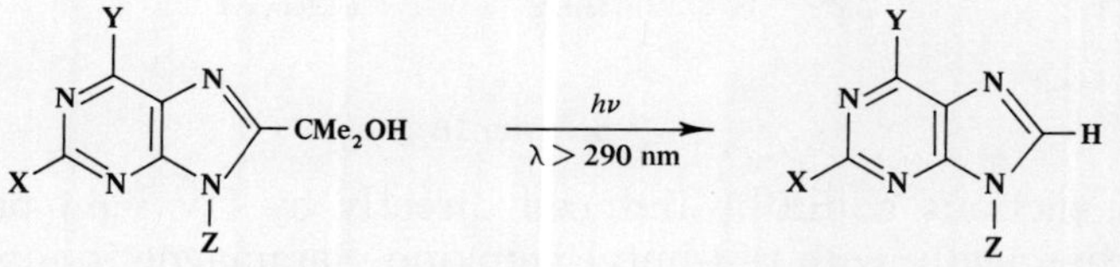

$h\nu$, $\lambda > 290$ nm

$X = H, Y = NH_2, Z = H$; $X = H; Y = NH_2, Z =$ ribose; $X = NH_2, Y = OH, Z =$ ribose

**Scheme 14**

## G. Photoreactions of Purines with Amines

The abundance of naturally occurring amines in biological systems and the sensitivity of genetic information to chemical reactions of DNA moieties make photochemical reactions of amines with nucleic acid constituents relevant to radiation damage in living systems. A large number of organic compounds, including mustard gas, nitrogen mustard, and ethylene imines commonly act as "alkylating agents" and are known to alkylate nucleic acids *in vitro*.

Photochemical reactions of amines resemble those of alcohols, and it could be expected that amines will react similarly with purines. The photoalkylation of Pur with 1-propylamine has been described (Yang *et al.*, 1971). This reaction led to the substitution of a propyl group for the hydrogen atom at C(6) of the Pur system (Scheme 15). Irradiation of caffeine in the presence of a variety of amines led to the substitu-

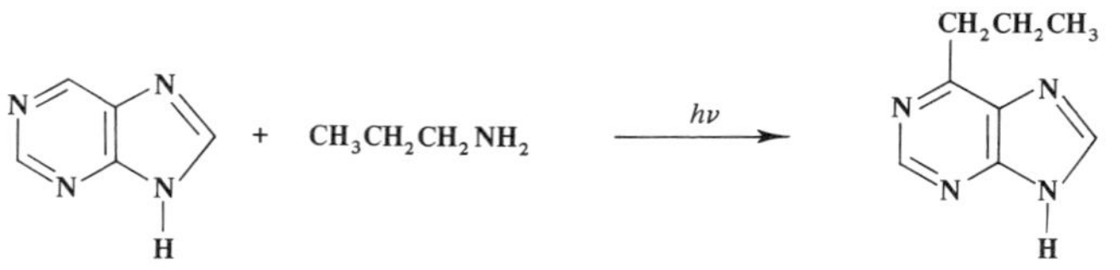
CH2CH2CH3
CH3CH2CH2NH2
hν

**Scheme 15**

tion of an $\alpha$-aminoalkyl or an alkyl group for the hydrogen atom at C(8) (Scheme 16) (Stankunas *et al.*, 1971; Elad and Salomon, 1971).

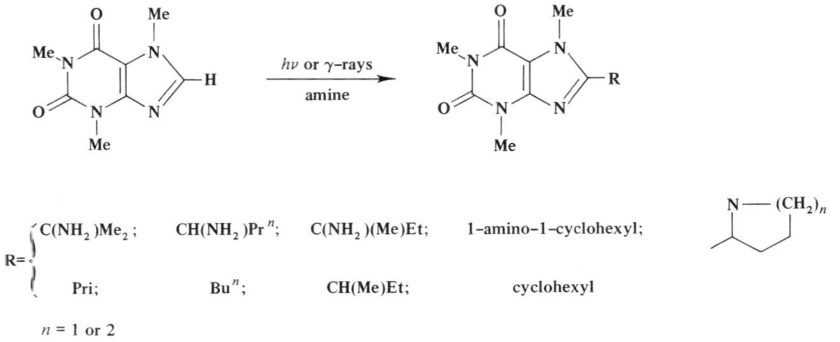
hν or γ-rays
amine
R=
C(NH2)Me2; CH(NH2)Pr^n; C(NH2)(Me)Et; 1-amino-1-cyclohexyl;
Pri; Bu^n; CH(Me)Et; cyclohexyl
N— (CH2)n
n = 1 or 2

**Scheme 16**

These reactions could be induced directly by UV light or initiated photochemically with di-*t*-butyl peroxide. Paramagnetic ions, such as $Cu^{2+}$ or $Ni^{2+}$, showed efficient quenching of the reactions leading to C(8)-substituted caffeines, while diamagnetic ions, such as $Mg^{2+}$ or $Zn^{2+}$, had no observable effect. This is assumed to indicate a triplet-state intermediate. The unreactivity of *t*-butylamine implies that the presence of a hydrogen atom on the carbon alpha to the amino group is necessary. A possible mechanism in accord with the observed data is given in Scheme 17 (Stankunas *et al.*, 1971).

## H. Other Photochemical Reactions of Purines

The photochemical reactions of caffeine with tetrahydrofuran, dioxane, and other ethers have been studied (Jerumanis and Martel, 1970). These reactions resulted in the substitution of an ether moiety or an alkyl group at C(8) of caffeine. Acetophenone or benzophenone has been used as a photosensitizer for these reactions. The reaction of caffeine with tetrahydrofuran can be presented as shown in Scheme 18.

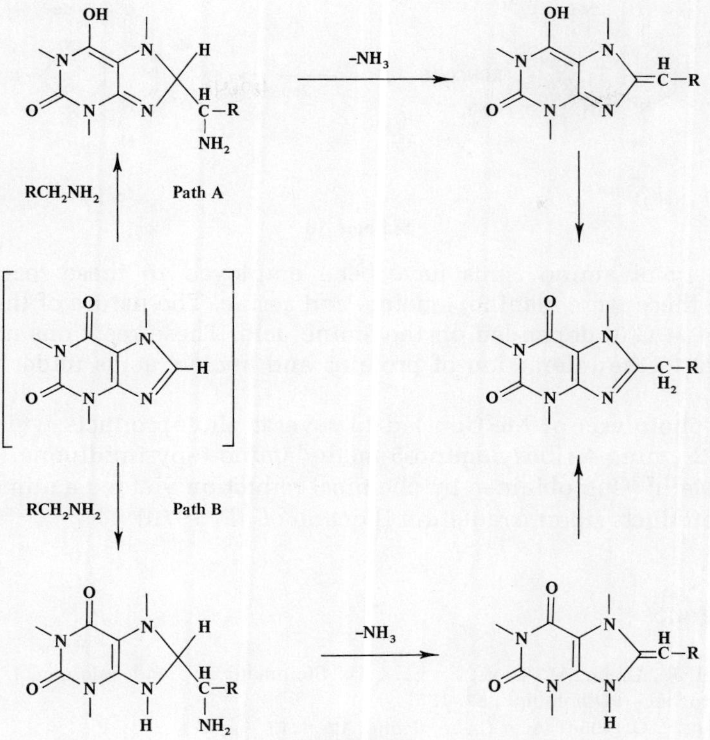

Scheme 17

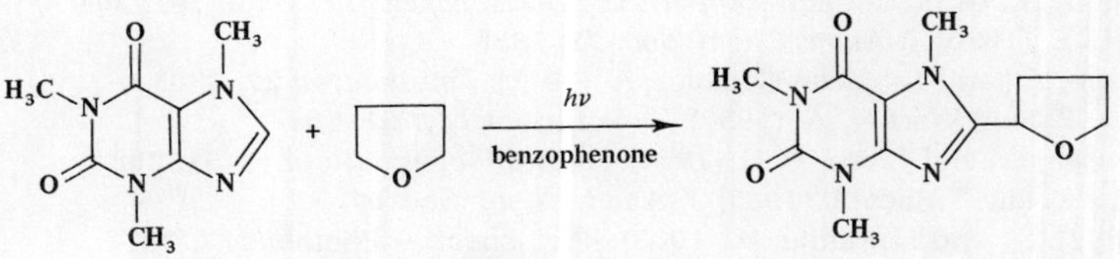

Scheme 18

Products were isolated and characterized by infrared, NMR, and mass spectrometry. In these reactions the excited sensitizer abstracts a hydrogen atom from the $\alpha$-carbon of the ether to produce ether free radicals, which subsequently attack C(8) of the caffeine molecule. The fact that benzpinacol is formed supports the proposed mechanism.

The photochemical reaction of caffeine with amino acids led to the substitution of amino acids at C(8) of caffeine (Scheme 19) (Elad and Rosenthal, 1969).

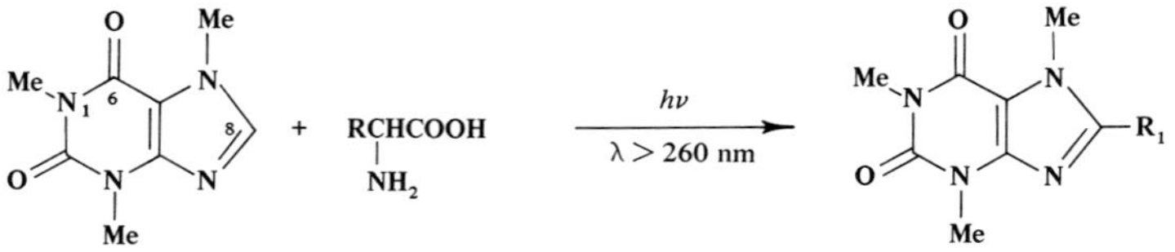

$R_1$ = alkyl

**Scheme 19**

A variety of amino acids have been employed in these reactions; among these were alanine, leucine, and serine. The nature of the substituent at C(8) depended on the amino acid. These reactions may be relevant to the interaction of proteins and nucleic acids under radiation.

The photolysis of Me[7]Guo led to several photoproducts which include 2-amino-4-ribosylamino-5-methylamino-6-pyrimidinone. Similarly, Me[7]h[7,8]Guo obtained by chemical reduction yielded a number of photoproducts upon irradiation (Pochon *et al.*, 1970).

## References


Ben-Ishai, R., Green, M., Graff, E., Elad, D., Steinmaus, H., and Salomon, J. (1973). *Photochem. & Photobiol.*, **17,** 155.

Boerresen, H. C. (1963). *Acta Chem. Scand.* **17,** 921.

Brown, G. B., Levin, G., and Murphy, S. (1964). *Biochemistry* **3,** 880.

Burr, J. C. (1968). *Advan. Photochem.* **6,** 193.

Canzanelli, A., Guild, R., and Rapport, D. (1951). *Amer. J. Physiol.* **167,** 364.

Carter, C. E. (1950). *J. Amer. Chem. Soc.* **72,** 1835.

Castellano, A., and Lablache-Combier, A. (1971). *Tetrahedron* **27,** 2303.

Chandra, P., and Wacker, A. (1966). *Z. Naturforsch. B* **21,** 663.

Christensen, E., and Giese, A. C. (1954). *Arch. Biochem. Biophys.* **51,** 208.

Clark, L. B., and Tinoco, I. (1965). *J. Amer. Chem. Soc.* **87,** 11.

Connolly, J. S., and Linschitz, H. (1968). *Photochem. & Photobiol.* **7,** 791.

Cramer, V. F., and Schlinghoff, G. (1964). *Tetrahedron Lett.* p. 3201.

Drobnik, J., and Augenstein, L. (1966). *Photochem. & Photobiol.* **5,** 83.

Eisinger, J., and Shulman, R. G. (1967). *J. Mol. Biol.* **28,** 445.

Eisinger, J., and Shulman, R. G. (1968). *Science* **161,** 1311.

Elad, D., and Rosenthal, I. (1969). *Chem. Commun.* p. 905.

Elad, D., and Salomon, J. (1971). *Tetrahedron Lett.* p. 4783.

Elad, D., Rosenthal, I., Salomon, J., and Sperling, J. (1971). *Chem. Commun.* p. 49.

Evans, B., and Wolfenden, R. (1970). *J. Amer. Chem. Soc.* **92,** 4751.

Fahr, E. (1969). *Angew. Chem, Int. Ed. Engl.* **8,** 578.

Fellig, J. (1954). *Science* **119,** 129.

Friedman, P. A. (1968). *Biochim. Biophys. Acta* **166,** 1.

Fujita, H., and Yamazaki, H. (1970). *Bull. Chem. Soc. Jap.* **43,** 1177.

Guéron, M., Shulman, R. G., and Eisinger, J. (1966). *Proc. Nat. Acad. Sci. U.S.* **56,** 814.

Guéron, M., Eisinger, J., and Shulman, R. G. (1967). *J. Chem. Phys.* **47,** 4077.
Hélène, C., Douzou, P., and Michelson, A. M. (1965). *Biochim. Biophys. Acta* **109,** 261.
Hélène, C., Douzou, P., and Michelson, A. M. (1966). *Proc. Nat. Acad. Sci. U.S.* **55,** 376.
Hems, G. (1960). *Radiat. Res.* **13,** 777.
Heyroth, F. F., and Loufbourow, J. R. (1931). *J. Amer. Chem. Soc.* **53,** 3441.
Jerumanis, S., and Martel, A. (1970). *Can. J. Chem.* **48,** 1716.
Jordan, F., and Pullman, B. (1968). *Theor. Chim. Acta* **9,** 242.
Kasha, M. (1960). *Radiat. Res., Suppl.* **2,** 243.
Keck, K. (1968). *Z. Naturforch. B* **23,** 1034.
Kland, M. J., and Johnson, L. A. (1957). *J. Amer. Chem. Soc.* **79,** 6187.
Kleinwächter, V. (1972). *Collect. Czech. Chem. Commun.* **37,** 2333.
Kleinwächter, V., and Kouldelka, J. (1972). *Collect. Czech. Chem. Commun.* **37,** 3433.
Lamola, A. A. (1969). *In* "Energy Transfer and Organic Photochemistry" (A. Weissberger, ed.), pp. 17–132. Wiley (Interscience), New York.
Lamola, A. A., Guéron, M., Yamane, T., Eisinger, J., and Shulman, R. G. (1967). *J. Chem. Phys.* **47,** 2210.
Lawly, P. D. (1966). *Progr. Nucl. Acid Res. Mol. Biol.* **5,** 89.
Levin, G., Setlow, R. B., and Brown, G. B. (1964). *Biochemistry* **3,** 883.
Lochmann, v. E. R., and Stein, W. (1964). *Naturwissenschaften* **51,** 59.
Lochmann, v. E. R., Stein, W., and Haefner, K. (1964). *Z. Naturforsch. B* **19,** 838.
McLaren, A. D., and Shugar, D. (1964). "Photochemistry of Proteins and Nucleic Acids," p. 65. Pergamon, Oxford.
Mason, S. F. (1960). *J. Chem. Soc., London* p. 219.
Matsuura, T., and Saito, I. (1967). *Chem. Commun.* p. 693.
Matsuura, T., and Saito, I. (1968). *Tetrahedron,* **24,** 6609.
Matsuura, T., and Saito, I. (1969a). *Tetrahedron* **25,** 541.
Matsuura, T., and Saito, I. (1969b). *Tetrahedron* **25,** 549.
Matsuura, T., and Saito, I. (1969c). *Tetrahedron* **25,** 557.
Miles, D. W., Tonnsend, L. B., Robins, M. J., Robins, R. K., Inskeep, W. H., and Eyring, H. (1971). *J. Amer. Chem. Soc.* **93,** 1600.
Pochon, F., Pascal, Y., Pitha, P., and Michelson, A. M. (1970). *Biochim. Biophys. Acta* **213,** 273.
Politzer, I. R., Griffin, G. W., and Laseter, J. L. (1971). *Chem. Biol. Interact.* **3,** 73.
Ponnamperuma, C., Lemmon, R. M., Bennett, E. L., and Calvin, M. (1961). *Science* **134,** 113.
Ponnamperuma, C., Lemmon, R. M., and Calvin, M. (1963). *Radiat. Res.* **18,** 540.
Pullman, A. (1969). *Ann. N.Y. Acad. Sci.* **158,** 65.
Raab, O. (1900). *Z. Biol.* **39,** 524.
Rahn, R. O., Longworth, J. W., Eisinger, J., and Shulman, R. G. (1964). *Proc. Nat. Acad. Sci. U.S.* **51,** 1299.
Rahn, R. O., Shulman, R. G., and Longworth, J. W. (1965). *Proc. Nat. Acad. Sci. U.S.* **53,** 893.
Rahn, R. O., Shulman, R. G., and Longworth, J. W. (1967). *Radiat. Res., Suppl.* **7,** 139.
Robins, R. K. (1967). *In* "Heterocyclic Compounds" (R. C. Elderfield, ed.), Vol. 8, p. 270. Wiley, New York.
Rosenthal, I., and Pitts, J. N., Jr. (1971) *Biophys. J.* **11,** 963.
Sastry, K. S., and Gordon, M. P. (1966a). *Biochim. Biophys. Acta* **129,** 32.
Sastry, K. S., and Gordon, M. P. (1966b). *Biochim. Biophys. Acta* **129,** 42.
Scholes, G., and Weiss, J. (1952). *Exp. Cell Res., Supply.* **2,** 219.
Setlow, J. K. (1966). *Radiat. Res., Suppl.* **6,** 141.
Setlow, R. B. (1968). *Progr. Nucl. Acid Res. Mol. Biol.* **8,** 257.

Shugar, D., and Wierzchowski, K. L. (1958). *Postepy Biochem.* **4,** 243.
Simon, M. I., and van Vunakis, H. (1962). *J. Mol. Biol.* **4,** 488.
Simon, M. I., and van Vunakis, H. (1964). *Arch. Biochem. Biophys.* **105,** 197.
Simon, M. I., Grossman, L., and van Vunakis, H. (1965). *J. Mol. Biol.* **12,** 50.
Sinsheimer, R. L., and Hasting, R. (1949). *Science* **110,** 525.
Smith, K. C. (1966). *Radiat. Res., Suppl.* **6,** 54.
Smith, K. C., and Hanawalt, P. C. (1969). "Molecular Photobiology: Inactivation and Recovery." Academic Press, New York.
Spence, G. G., Taylor, E. C., and Buchardt, O. (1970). *Chem. Rev.* **70,** 231.
Spikes, J. D., and Livingston, R. (1969). *Advan. Radiat. Biol.* **3,** 30.
Stankunas, A., Rosenthal, I., and Pitts, J. N. (1971). *Tetrahedron Lett.* p. 4779.
Steinmaus, H., Elad, D., and Ben-Ishai, R. (1970). *Biochem. Biophys. Res. Commun.* **40,** 1021.
Steinmaus, H., Rosenthal, I., and Elad, D. (1971). *J. Org. Chem.* **36,** 3594.
Stewart, R. F., and Davidson, N. (1964). *Biopolym. Symp.* **1,** 465.
Sussenbach, J. S., and Berends, W. (1963). *Biochim. Biophys. Acta* **76,** 154.
Sussenbach, J. S., and Berends, W. (1965). *Biochim. Biophys. Acta* **129,** 49.
Van Hemmen, J. J., and Bleichrodt, J. F. (1971). *Radiat. Res.* **46,** 444.
van Vunakis, H., Seaman, E., Kahan, L., Kappler, J. W., and Levine, L. (1966). *Biochemistry* **5,** 3986.
Voet, D., Gratzer, W. B., and Doty, P. (1963). *Biopolymers* **1,** 193.
Wacker, A., Tuerck, G., and Gerstenberger, A. (1963). *Naturwissenschaften* **50,** 377.
Wacker, A., Dellweg, H., Trager, L., Kornhauser, A., Lodemann, E., Tuerck, G., Selzer, R., Chandra, P., and Ishimoto, M. (1964). *Photochem. & Photobiol.* **3,** 369.
Waskell, L. A., Sastry, K. S., and Gordon, M. P. (1966). *Biochim. Biophys. Acta* **129,** 49.
Yang, N. C., Gorelic, L. S., and Kim, B. (1971). *Photochem. & Photobiol.* **13,** 275.
Zenda, K., Saneyoshi, M., and Chihara, G. (1965). *Chem. Pharm. Bull. Jap.* **13,** 1108.

# 8 Aggregate Formation, Excited-State Interactions, and Photochemical Reactions in Frozen Aqueous Solutions of Nucleic Acid Constituents

*Thérèse Montenay-Garestier, Michel Charlier and Claude Hélène*

## A. Introduction

A precise knowledge of photochemical reactions in nucleic acids requires a good understanding of excited-state properties and energy transfer processes in these macromolecules. This, in turn, necessitates an investigation of the excited states of model compounds of increasing complexity, from monomers (nucleotides) to synthetic polymers of various structures (single or double strands), and of various base compositions. Moreover, identification of photoproducts in nucleic acids often relies upon a comparison with compounds obtained by chemical synthesis or by photochemical transformation of the monomers. Investigation of the nature of photoproducts formed in aggregates of nucleic acid components is useful in understanding the photochemical behavior of nucleic acids. Such aggregates can be obtained easily by freezing aqueous solutions of bases, nucleosides, or nucleotides. In this chapter, we shall survey the experimental data on the structure of frozen aqueous solutions and the consequences of freezing on physical and chemical properties of solute molecules. Special emphasis will be put on aggregate formation of nucleic acid constituents in frozen aqueous solutions. We shall also review the excited-state properties and the energy transfer processes which take place in these aggregates. The chemical nature of the photoproducts formed upon UV irradiation of these aggregates will be discussed with special emphasis on the mechanism of their formation.

Nucleic acid bases are only weakly fluorescent at room temperature in fluid aqueous solutions (Vigny, 1971; Daniels, 1973). Excited-state properties can be reached from flash photolysis studies (see Chapters 2 and 5) or from energy transfer to suitable acceptors (Lamola and Eisinger, 1971). At a low temperature (77°K) in rigid matrices, most nucleic acid components exhibit fluorescence and phosphorescence emissions. Excited-state properties of polynucleotides and nucleic acids have been extensively studied in rigid media (see Section B-3 and the Eisinger and Lamola review, 1971). In most of these low-temperature studies, the solutions used were made in equivolume mixtures of water and ethylene (or propylene) glycol, which give transparent glasses at 77°K. The freezing of aqueous solutions does not lead to such a glass but instead results in a microcrystalline, nontransparent sample. Under these conditions solute molecules form aggregates because of rejection from the growing ice crystals. Study of the behavior of excited molecules in these aggregates was used as a tool to gain understanding about the excited states (Hélène, 1966, 1973a) and the photochemical properties (Beukers *et al.*, 1958; Wang, 1960) of nucleic acids. A comparison of electronic properties of bases in aggregates and in polynucleotides should also provide information on the role of the phosphodiester backbone on base–base interactions. The observation that mixed aggregates can be formed upon freezing aqueous mixtures of two solute molecules allows a study of complex formation between nucleic acid constituents and other organic molecules such as aromatic amino acids (Montenay-Garestier and Hélène, 1971). The behavior under UV excitation of mixed aggregates containing nucleic acid bases and amino acids may relate to the problem of photochemical reactions in nucleic acid–protein complexes.

## B. Aggregate Formation in Frozen Aqueous Solutions

### 1. Physical and Chemical Properties of Solute Molecules in Frozen Aqueous Solutions

#### *a. Heterogeneity of Solute Concentration in Frozen Aqueous Solutions*

Heterogeneity of distribution of solute molecules in ice has been demonstrated by the manifestation of different phenomena. A given distribution of solute molecules must be the net result of their being trapped in the ice phase, of their rejection from the growing ice crystals, and of their diffusion through the liquid–ice interface. Wang (1961) noted that the thawing of different layers of Thy or Ura from frozen aqueous solutions results in solutions of very different concentrations. A ten-fold concentration difference was found between two different layers. Füchtbauer and Mazur (1966) observed that light transmittance of Thy solutions at 253.7 nm is decreased by freezing. This observation was attributed to the formation of Thy clumps. Radiation encountering these clumps is totally absorbed, but a portion of the radiation is able to traverse the cell without encountering any Thy. The absorption spectrum of Thy in ice is also consistent with the formation of Thy aggregates (Davis and Tinoco, 1966).

Detailed and quantitative investigations on solute redistribution in multicomponent frozen aqueous systems containing different molecular species (acids, salts, or glucose) showed a sharp concentration dependence of uneven distribution (Taborsky, 1970). A maximally nonuniform distribution was associated with an initial concentration of 0.1 m*M* in Taborsky's experimental conditions. This concentration was about the same for all of the solute species investigated. A reciprocal effect of the addition of a second solute on the concentration distribution of the first was noted in multicomponent systems. The magnitude of this effect depends on many parameters such as the initial concentration of the first solute and its nature. Addition of salts such as sodium chloride suppresses an uneven proton distribution in acid frozen solutions.

b. *Physical Interactions in Frozen Aqueous Solutions*

In 1957, Szent-Györgi observed that a heterogeneity of solute distribution results when aqueous solutions are frozen. Szent-Györgi (1960) and Stom (1967) showed that mixtures in fluid medium containing an electron donor molecule (such as serotonin or hydroquinone) and another molecular species with a high electron affinity (riboflavin phosphate or quinone) become highly colored in the frozen state. This phenomenon was attributed to charge-transfer complex formation. It was also observed that the formation of many complexes can be prevented by the addition of organic solvents or mineral salts, although the complexes formed with indole derivatives do not split as a result of salt addition. Szent-Györgi speculated that by rapidly freezing, solute molecules would form aggregates surrounded by a lattice of strongly ordered water molecules. A similar effect occurs when colorless aqueous mixtures of Ado and cupric chloride are frozen (Montenay-Garestier and Hélène, 1973b). Freezing induces a green coloration sensitive to the addition of salt or organic solvent. By lyophilization the complex can be obtained in the solid state as a green powder. Electron spin resonance spectroscopy (ESR) of frozen aqueous solutions of cupric chloride in the presence of increasing concentrations of Ado shows a drastic modification of the ESR signal of $Cu^{2+}$. Similar observations were made with g-EtAde and poly(rA). A 1:2 (Ado/$Cu^{2+}$) structure was suggested for this complex involving the base moiety of the Ade derivative (Montenay-Garestier and Hélène, 1973b).

Electron spin resonance spectra of frozen aqueous solutions of paramagnetic metal cations ($Mn^{2+}$, $Gd^{3+}$) exhibit a dipolar broadening of the ESR lines due to a high local concentration of the paramagnetic species, reflecting the heterogeneity of the environment of the cations (Ross, 1965). Line narrowing induced by the addition of salt or organic solvents allows a comparison of the relative effectiveness of different additives on solute distribution.

Mossbauer resonance studies of ferrous ions performed in ice from 77° to 253°K showed that hexaaquoferrous ions trapped in the ice lattice induce the formation of cubic ice rather than the stable hexagonal form (Nozik and Kaplan, 1967). Warming to 193°K caused a nonreversible transformation to hexagonal ice. Formation of a eutectic phase induced the disappearance of the Mossbauer effect.

The heterogeneity of $OH^-$ ion concentration in frozen alkaline aqueous solutions was demonstrated by fluorescence studies of the deprotonation of the indole ring NH group (Santus *et al.*, 1971; Burshtein and Busel, 1971).

These various results should probably be correlated with the electrical phenomena observed upon freezing. During rapid freezing of dilute aqueous solutions, high potential differences are generated at the ice–liquid interface (Workman-Reynolds effect, 1950). A "freezing potential" that might be as high as 100 V appears during the freezing of aqueous solutions of electrolytes at concentrations of less than 1 m*M* (e.g., metallic halides) or even in pure doubly distilled water (Murphy, 1970). Two explanations were advanced: (1) a predilection of ice for anion entrapment due to a highly structured layer of water molecules in which the polarization favors incorporation of ions of one definite sign at the interface (and this ion in excess is almost always the anion) and (2) relaxation of the polarization of organization of structured water molecules in the case of pure water. The freezing potential is maximal for an electrolyte concentration of the order of 0.1 m*M* (Drost-Hansen and Walter, 1967), a value similar to that leading to maximum nonuniform distribution (Taborsky, 1970).

*c. Chemical and Photochemical Reactions in Ice*

Beukers *et al.* (1958), Beukers and Berends (1960), and Wang (1960, 1961), who were interested in the study of UV damage in nucleic acids found that freezing markedly increases the sensitivity of these molecules to UV irradiation; photodimers of Thy were formed with a high quantum yield in frozen aqueous solutions or in dry films, whereas the photodimerization was inefficient in fluid medium (Wang, 1961).

The anomalous rapid autodecomposition of Thd at 253°K is due to an uneven distribution of Thd molecules in the frozen solutions (Apelgot *et al.*, 1965). Storage of labeled Thd must be avoided at this temperature. The rate of decomposition is much slower at 77° or 273°K than at 253°K.

Szent-Györgi (1957) stressed that structured ice is important in the formation of intermolecular complexes in frozen aqueous solutions. This suggestion was later employed by Beukers *et al.* (1958) and Beukers and Berends (1961) to explain the photodimerization reaction of Thy in ice whereas Wang (1961) suggested that Thy◇Thy formation results from the formation of microcrystalline aggregates of Thy monohydrate formed between the growing ice crystals. The later suggestion has been supported by several investigations (Stewart, 1963; Füchtbauer and Mazur, 1966; Davis and Tinoco, 1966).

However, the formation of molecular aggregates alone could not explain many observations related to the rate of chemical reactions in frozen aqueous solutions. Certain reactions occurring in aqueous solutions proceed at greatly enhanced rates when the solutions are frozen (Grant *et al.*, 1961, 1966; Butler and Bruice, 1964). However, in reactions involving water molecules, the rate constants are decreased. Dehydration and hydrolysis, which are second- or third-order reactions, may become first- or second-order reactions, respectively, in frozen solutions (Bruice and Butler, 1965). Also, freezing accelerates the reactions. Many enzymatic as well as nonenzymatic reactions proceed rapidly at subfreezing temperatures (Pincock and Lin, 1973). Several phenomena were suggested to account for these observations: increase in concentration of the reactants, difference in dielectric constants between ice and water (Alburn and Grant, 1965; Grant *et al.*, 1966), high proton mobility in ice (Eigen and Demayer, 1964; Riehl, 1965), presence of catalytically active sites on the ice surface (Grant and Alburn, 1965). However, a better understanding of all of these phenomena requires a better knowledge of the structure of frozen aqueous solutions as described by Wang (1961, 1965) and Pincock (1969).

Although Wang (1961) and Bruice and Butler (1965) did not observe that either temperature or methods of freezing influences aggregate formation, it might be expected that more rapid freezing of the sample would help reduce segregation effects (Füchtbauer and Mazur, 1966).

### 2. Structure of Frozen Aqueous Solutions

A "frozen state" does not mean a solid state: a frozen state may exist in the range of temperature where solid is in equilibrium with a liquid phase in two- or more-component systems below the freezing point and above the eutectic point. An apparent macroscopic solid material may contain liquid "puddles." In complex multicomponent systems the lowest temperature where liquid may exist (and a liquid-state reaction occur) is often far below the freezing point of the major component. Thus, one must distinguish among:

(1) Molecular aggregates formed when an aqueous solution is frozen below the eutectic point. The reactions taking place under these experimental conditions are true *solid-state* reactions. Examples are the photodimerization of Thy in frozen aqueous solutions (Wang,

1961; Füchtbauer and Mazur, 1966), and complex formation in aggregates of nucleic acid derivatives and aromatic amino acids which are described below.

(2) Reactions occurring in the liquid phase among the ice crystals below the freezing point and above the eutectic point: the so-called "puddle-effect" (Wang, 1961, 1965). The solute molecules together with some of the added solvent are excluded from the ice to form liquid puddles. The concentration is increased in these puddles and chemical reactions may proceed at a greatly enhanced rate (Pincock, 1969). Some observations on such systems may be more related to the liquid part than to the solid part of the total system.

(3) When an aqueous solution containing one type of solute molecule and an added organic solvent (or a high salt or acid or base concentration) is frozen below the eutectic point, the phase which was liquid between the freezing point and the eutectic point becomes solid, and the solute molecules in the puddles are, therefore, dispersed in the frozen organic solvent (or salt, acidic, or basic phases).

(4) If an aqueous solution contains high amounts of organic solvent, salt, acid, or base, it may result in a glass upon cooling to sufficiently low temperatures (i.e., the size of the "puddles" corresponds to the whole system, e.g., mixtures of water and ethylene glycol v/v or highly concentrated $MgCl_2$, HCl, or NaOH solutions result in glasses at 77°K).

The structure of ice around solute molecules, some catalytic ice–surface effects such as an effect due to the solid ice–surface exhibiting an heterogeneous catalytic activity, or proton or electron transfer reactions (Horne, 1963) cannot be ruled out definitively. However, the occurrence of molecular aggregates or molecular concentrations in clumps, liquid puddles, and frozen puddles seem to explain most of the kinetic observations in frozen aqueous solutions.

## 3. Spectroscopic Properties of Nucleic Acid Components in Frozen Aqueous Solutions

Spectroscopic properties of nucleic acid derivatives in aggregates are markedly affected as compared to the properties of dispersed molecules in glassy media (Hélène *et al.*, 1966b; Kleinwächter, 1972; Montenay-Garestier, 1973). These modifications are most clearly exhibited by aggregates of Pur derivatives. For example, the emission characteristics of Ado aggregates depend upon the concentration of solute molecules (Fig. 1). At low concentrations, the emission spectrum is similar to that emitted by molecules dispersed in a glassy medium

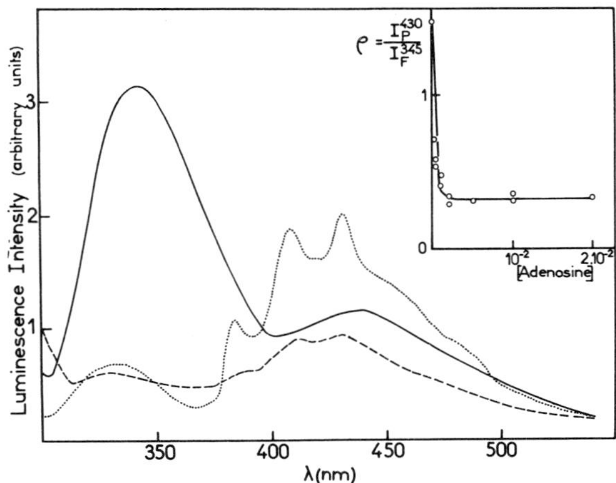


**Fig. 1.** *Total luminescence spectra of Ado at 77°K. 1 mM in frozen aqueous solutions (—); 0.01 mM in frozen aqueous solutions (---); 1 mM in water/propylene glycol glass (···). Inset: variation with Ado concentration of the ratio of phosphorescence and fluorescence intensities measured at 430 nm and 345 nm respectively.* (From Montenay-Garestier and Hélène, 1973a.)

[water and propylene glycol (WPG), v/v]. Similar results concerning Pur and Pyr derivatives (bases, nucleosides, and nucleotides) were obtained by Kleinwächter (1972) in detailed studies of the influence of matrix composition on spectral properties of nucleic acids.

When the concentration of solute molecules is increased, the ratio of phosphorescence and fluorescence quantum yields (P/F) decreases and reaches a plateau (Fig. 1). Afterwards the emission characteristics no longer vary and the luminescence spectrum exhibits broad red-shifted fluorescence and phosphorescence bands without any vibrational structure. The broad fluorescence band is attributed to the emission of an excimer state formed between two neighboring molecules. At low concentrations, aggregates may be small; the probability of forming longer aggregates increases with increasing solute concentrations. A determination of aggregate size from energy transfer experiments confirmed this assumption (see Section C). Aggregate formation may also increase intersystem crossing yields, as shown in the case of Thy derivatives (Lamola *et al.*, 1967).

The phosphorescence decay of Ado aggregates (Kleinwächter, 1972) could not be represented by a single exponential curve, and two decay times are necessary to describe this decay: a long one similar to that observed in glassy solutions and a short one the contribution of which decreases gradually with the disturbance of aggregation. The addition of less than 1% ethanol is sufficient to suppress completely the short component of the decay. This was ascribed to the formation of a microphase containing the solute in an environment of molecules of pure organic solvent embedded in the crystalline ice structure as proposed by Wang (1965) to explain the inhibition by organic solvents of Thy photodimerization in ice (the so-called "puddle effect"). In a mixed salt–glucose matrix, Kleinwächter demonstrated that aggregation does not take place (as already reported for alcoholic glasses). The emission spectrum is characteristic of the monomeric state of the solute in frozen solutions containing 0.2 or 0.3% glucose and 5 m*M* sodium acetate or citrate. However, it must be emphasized that the extent of aggregation also depends on the nature of the molecules and on the strength of interactions between solute molecules. Pur derivatives exhibit a stronger tendency to form aggregates than do Pyr derivatives. We observed a spectral phenomenon correlated with aggregation in Cyd frozen aqueous solutions (pH 4) containing as much as 33% ethanol or 0.1 *M* sodium chloride (see Section C).

Although the structure of aggregates is not known, several experiments have provided evidence that it is microcrystalline [see Wang (1964, 1965) for a discussion of the relationship between structure of molecular aggregates and structure of photoproducts]. Aggregates and microcrystalline powders of aromatic amino acids exhibit similar luminescence characteristics (spectra and decays) (Nag-Chaudhuri and Augenstein, 1964). Aggregate formation is expected to give rise to "vertical" stacking and "horizontal" hydrogen bonding if a microcrystalline structure is obtained (see Bugg *et al.*, 1971, for a review of interactions in crystals of nucleic acid components). These interactions may favor certain photochemical reactions. The photodimers of Thy formed in aggregates, in dry film, and in crystals of Thy monohydrate have a similar structure (Wang, 1963, 1965).

## C. Excited-State Interactions and Energy Transfer in Aggregates

We have shown that electronic interactions of solute molecules in frozen aqueous solutions markedly modify excited-state characteristics (emission spectra or decay times) as compared to the same species dispersed in a glassy medium. Quantum yields of fluorescence or phosphorescence depend on solute concentration. Excitation energy transfer along the stacks of nucleic acid residues may also result from aggregate formation. This transfer may occur from molecules in their lowest excited singlet or triplet states. Energy transfer from an excited singlet state (long-range transfer) depends upon the relative orientation of the transition dipoles of the donor and acceptor molecules, their distance, and the overlap between the fluorescence spectrum of the donor and the absorption spectrum of the acceptor (Förster, 1965). The efficiency of energy transfer from a molecule in its lowest triplet state depends essentially upon two factors (Dexter, 1953): (1) an electron exchange between donor and acceptor molecules and (2) the overlap of the phosphorescence spectrum of the donor and the singlet–triplet absorption spectrum of the acceptor. This transfer efficiency decreases rapidly when the donor–acceptor distance increases. Such a transfer has already been demonstrated for poly(rA) in which the helical structure allows an important overlap of neighboring bases. Excitation migration at the triplet level in this polynucleotide involves ~50 bases (Eisinger and Shulman, 1966). Attempts to demonstrate excitation energy migration between Ade bases at the singlet level in this polynucleotide or in oligonucleotides led to the conclusion that if migration exists, then less than five bases are involved (Guéron and Shulman, 1968; Montenay-Garestier *et al.*, 1969).

### 1. Energy Transfer from Nucleic Acid Derivatives to Nonluminescent Traps

Energy transfer processes in aggregates can be conveniently investigated by introducing traps with energy levels located below the energy level of the solute from which the excitation energy is transferred. Divalent paramagnetic metallic cations quench triplet states in fluid medium (Porter and Wright, 1959) and in rigid medium (Smaller *et al.*, 1965) although the detailed mechanism of action is not well understood.

The quenching of Ado phosphorescence by paramagnetic metallic cations ($Ni^{2+}$, $Co^{2+}$, $Cu^{2+}$) in aggregates demonstrated the existence of energy delocalization at the triplet level (Montenay-Garestier and Hélène, 1973c). Fluorescence was only slightly affected in the presence of $Cu^{2+}$ ions in which case Ado excitation energy was transferred at the singlet level of nonfluorescent $Cu^{2+}$–Ado complexes. In the absence of any energy transfer process, one cation would be expected to quench only the phosphorescence of neighboring molecules. As shown in Fig. 2a, one paramagnetic cation can quench the phosphorescence of a large number of Ado molecules. The fact that this number depends on the initial Ado concentration supports previous luminescence observations (Section B,2) that aggregate length varies with initial solute concentration.

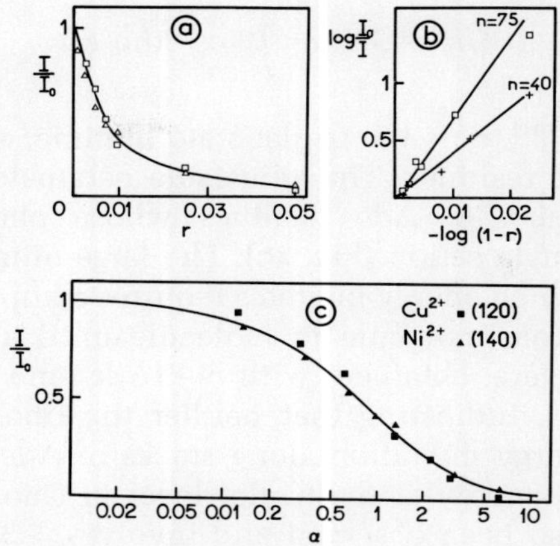


**Fig. 2.** *(a) Variation of the ratio* $I/I_0$ *of phosphorescence intensities measured at 440 nm of 1 mM Ado frozen aqueous solutions vs.* r. *The number* r *is the ratio of metal cations and Ado concentrations.* $\triangle = Ni^{2+}$; $\square = Cu^{2+}$. *(b) Analysis according to Eq. (1) (see text) of phosphorescence quenching data of a frozen aqueous solution of Ado at two different concentrations by* $Cu^{2+}$; $+ = 0.5$ *mM;* $\square = 2$ *mM. (c) Analysis according to Eq. (2) of phosphorescence quenching data of a frozen aqueous solution of Ado 5 mM by* $Cu^{2+}$ *and* $Ni^{2+}$. *The numbers in parentheses are the average numbers of Ade residues involved in the triplet migration (see text). (From Montenay-Garestier and Hélène, 1973a.)*

Quantitative data analysis leads to the distinguishing of two different cases. If the extent of energy migration is limited by the aggregate size, the phosphorescence intensity $I^P$ will decrease according to Eq. (1)

$$I^P/I_0{}^P = (1 - r)^n \tag{1}$$

where r is the ratio of metal cation and nucleoside concentrations and $n$ is the average number of Ado residues in the aggregate. Equation (1) is derived under the assumption that once a cation is bound to an Ado molecule in the aggregate, this aggregate will not emit phosphorescence. Only the aggregates which have not bound any cation will give rise to phosphorescence emission. Therefore, a plot of log $I_0{}^P/I^P$ vs. log $(1 - r)^{-1}$ will give a straight line the slope of which yields the value of $n$ (Fig. 2b). If the extent of transfer (number of residues involved in energy transfer) is limited not by the size of the aggregate but by the transfer process itself, the experimental results will be described by Eq. (2) which is derived according to a statistical model (Levinson, 1962; Montenay-Garestier and Hélène, 1973c) and which is identical to the equation established in a diffusion-controlled model (Bersohn and Isenberg, 1964):

$$I^P/I_0{}^P = 1 - \alpha^2 \int_0^\infty e^{-\alpha u} thu\, du \qquad (2)$$

where $\alpha = 2r(\tau_P/\tau_t)^{1/2}$, $\tau_P$ is the triplet state lifetime, and $\tau_t$ the transfer time between two residues. The adjustable parameter $\alpha/r$ would give the average number of Ado residues whose phosphorescence is quenched by a single cation (Fig. 2c). The large number of molecules (120 to 140) on which energy migrates requires a hopping of triplet excitation energy from molecule to molecule until a trap is reached. Similar results were obtained with 9-EtAde and AMP in frozen aqueous solutions, indicating that neither the ribose nor the phosphate prevent energy migration along stacks of Ade residues. Migration of excitation energy at the triplet level in Guo aggregates using $Co^{2+}$ traps has also been observed and involves ~30 Guo molecules (Hélène *et al.*, 1968).

## 2. Energy Transfer from Nucleic Acid Derivatives to Luminescent Traps

Luminescent traps may be used instead of nonluminescent paramagnetic cations if they have energy levels lower than the lowest triplet level of Ado in aggregates (25,400 $cm^{-1}$). In this case, energy transfer induces a sensitization of the luminescent molecule or ion. For example the fluorescence of $Eu^{3+}$ ions ($\lambda = 590$ nm) is sensitized while Ado phosphorescence is simultaneously quenched (Montenay-Garestier and Hélène, 1973c). The same phenomenon may be observed with Thy which has a triplet level (24,600 $cm^{-1}$) that is the lowest among nucleic acid bases (Hélène and Montenay-Garestier, 1968).

The phosphorescence of tryptophan ($E_T = 25{,}000$ $cm^{-1}$) or indole derivatives is also sensitized in aggregates of nucleic acid components while the phosphorescence emission from the nucleic acid derivative is quenched (Fig. 3a). The complete description of energy transfer processes from Pur or Pyr residues (e.g., Ado) to tryptophan molecules at various ratios (r) of tryptophan and Ado concentrations is complicated (Hélène, 1973b; Montenay-Garestier and Hélène, 1971). Nevertheless, at low tryptophan concentrations ($r < 0.02$), excitation energy transfer to tryptophan at the triplet level involves ~200 Ado molecules (Hélène, 1973b) (Fig. 3a). A similar triplet–triplet transfer from bases to tryptophan can take place in complexes of poly(rA) or DNA, with oligopeptides containing a tryptophan residue at low [oligopeptide]/[phosphate] ratios. At higher ratios of tryptophan-to-base con-

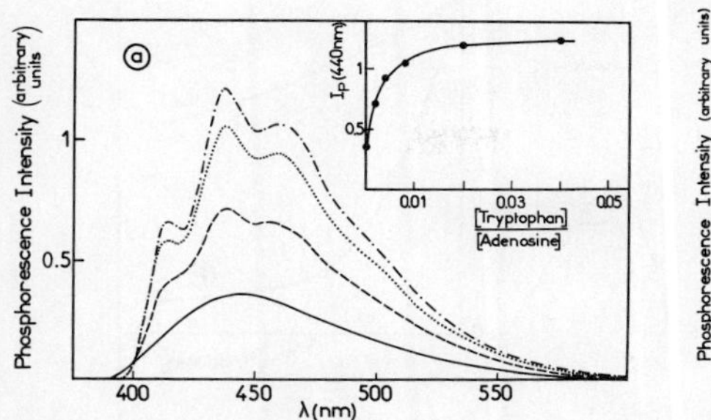


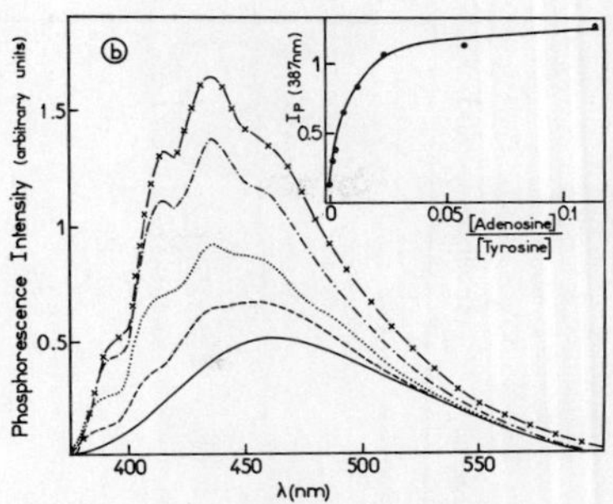


**Fig. 3.** *(a) Phosphorescence spectra of Ado–tryptophan mixtures in aqueous solutions frozen at 77°K. The ratios of tryptophan and Ado concentration are 0 (—), 0.2% (---), 0.8% (···), and 2% (-·-·). Ado concentration is 2 mM. Inset: change in phosphorescence intensity at 440 nm vs. the ratio of tryptophan and Ado concentration (in the low concentration range). (From Hélène, 1973b.) (b) Phosphorescence spectra of tyrosine–Ado mixtures in aqueous solutions frozen at 77°K. The ratios of Ado and tyrosine concentration are 0 (—), 0.2% (---), 1% (···), 5% (-·-·),* and 10% (x-x-x-). *Tyrosine concentration is 2 mM. Inset: change in phosphorescence intensity at 387 nm vs. the ratio of Ado and tyrosine concentration (in the low concentration range).*

centrations in aggregates, an important red-shift and an intensity decrease of the sensitized phosphorescence of tryptophan occur. This behavior is attributed to the formation of short aggregates of tryptophan molecules included in larger Ado aggregates (see Section D).

Triplet energy can also be transferred to acridine dyes in nucleotide or nucleoside aggregates (Kubota, 1970; Hélène, 1973a) (Fig. 4). This transfer resulted in a sensitized delayed fluorescence of the dye (triplet–singlet transfer). The decay of the dye-sensitized fluorescence was not exponential. The corresponding half-lives were about one order of magnitude shorter than the phosphorescence lifetime of the energy donor. Triplet-singlet transfer could occur over long distances due to the overlap between the phosphorescence of the Ade (or AMP) chromophore and the absorption of the dye (Bennett *et al.*, 1964). Quantitative studies showed that one proflavine molecule could quench the phosphorescence of ~400 Ado molecules, which is far more than the value that would be expected if only direct transfer occurred. Triplet–triplet energy transfer from Ado to 9-aminoacridine in mixed aggregates leads to an increase in the dye P/F ratio when excitation is in the UV as compared to excitation in the visible range. The same result has been observed with DNA (Galley, 1968; Galley and Fantus, 1969).

The delayed fluorescence from poly(rA) (Longworth and Battista, 1970; Hélène and Longworth, 1972) and from Ado aggregates (Mon-

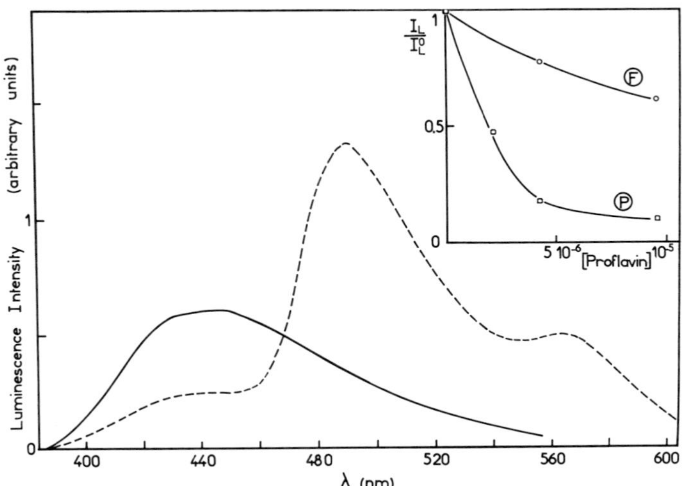


**Fig. 4.** *Delayed luminescence of Ado aggregates (Ado = 2 mM) in the absence (—) and the presence (---) of 2 μM proflavin. Phosphorescence of Ado appears at short wavelengths (440 nm) and that of proflavin at long wavelengths (570 nm). The peak at 495 nm is delayed fluorescence from proflavin. Inset: quenching of the fluorescence (F) and phosphorescence (P) of Ado aggregates (2 mM) by proflavin at different concentrations. (From Hélène, 1973a.)*

tenay-Garestier and Hélène, 1973c), which was previously attributed to triplet–triplet annihilation, probably results from a triplet–triplet absorption at trap sites in the polymer or in aggregates (Bazin *et al.*, 1973).

Ade chromophores can transfer their singlet excitation energy to dyes such as proflavin and ethidium bromide in Ado aggregates and in poly(rA) over a large number of Ade residues (Hélène, 1973a). Since there is good overlap between the fluorescence of the Ade residue and the absorption spectrum of the dye, this transfer can be attributed to a long-range effect (Förster type). A direct singlet–singlet transfer from Ado or Thd to the charge-transfer complexes formed by tryptophan and the nucleoside (see Section D) occur in mixed aggregates of the two species when nucleosides are in excess (Montenay-Garestier and Hélène, 1971). It was deduced from fluorescence studies that about five Ado molecules transfer their singlet energy to one complex. This transfer does not require the step-by-step migration between stacked Ado molecules and could be accounted for by the Förster mechanism.

### 3. Sensitization of Excited States of Nucleic Acid Derivatives

In all of the aforementioned energy transfer processes, transfer takes place from nucleic acid bases. In a few cases we observed transfers ending on nucleic acid residues.

Thd acts as a luminescent trap for excitation energy migration at the triplet level in Ado mixed aggregates as well as in $Me^1Ac^4Cyt$ aggregates (Hélène and Montenay-Garestier, 1968). Sensitization of Thy phosphorescence occurs, while the fluorescence spectrum of the hosts is not affected.

In mixed aggregates of Urd and Ado, simultaneous quenching of the fluorescence and phosphorescence of Ado emission is probably due to a direct singlet–singlet transfer to Urd over about four Ade residues.

Another interesting case of transfer to nucleic acid bases occurs in mixed aggregates of tyrosine and nucleosides. The lowest singlet excited state of tyrosine lies at $\sim 29{,}000\ cm^{-1}$; the triplet is at $37{,}000\ cm^{-1}$. Thus, energy transfer from tyrosine to bases is expected to occur at the singlet and at the triplet level (see Lamola *et al.*, 1967, for the energy of the lowest singlet and triplet states of nucleotides). When low concentrations of nucleosides are added to tyrosine in frozen aqueous solutions (2 m*M*), the tyrosine fluorescence is quenched. Analysis of tyrosine fluorescence quantum yield variations versus nucleoside concentration demonstrates that $\sim 60$ tyrosine molecules are involved in singlet transfer to five nucleosides (Ado, Guo, Thd, Cyd, and Urd). A very efficient sensitization of the nucleoside phosphorescence (Ado, Guo, Thd) occurs in tyrosine aggregates at low nucleoside-to-tyrosine concentration ratios ($< 0.01$) (Fig. 3b) (Montenay-Garestier, 1975). Several hundred tyrosine molecules can transfer their excitation energy to nucleic acid bases.

## D. Excited States of Intermolecular Complexes in Frozen Aqueous Solutions

As was previously mentioned (Section B), the formation of mixed aggregates can induce complex formation between molecular species in the ground state. The most visible effect of such a formation may be the brilliant color of some frozen mixtures, e.g., riboflavin (Szent-Györgi, 1960) or quinone (Stom, 1967) with indole derivatives and cupric chloride with Ade derivatives (Montenay-Garestier and Hélène, 1973b). The influence of physical parameters on the disappearance of the color (e.g., cooling, addition of salts and organic solvents) clearly demonstrates that complex formation requires aggregate formation.

It is expected that charge-transfer interactions between two molecules are more important in their excited state than in their ground state (Mataga and Murata, 1969). A greater extension of the excited orbitals makes the overlap integrals larger. Thus the contribution of charge transfer or electron exchange to the stabilization of excited states is enhanced. Therefore, luminescence would be a very efficient method to detect interactions in mixed aggregates which are important in the excited state and weak or nonexistent in the ground state.

Two different kinds of electron donor–acceptor complexes involving nucleic acid derivatives form in aggregates.

### 1. Complex Formation between Two Different Ionic Forms of Nucleic Acid Derivatives

First, complex formation occurs between two different ionic forms of the same molecule. Unlike classical sigmoidal titration curves of luminescence parameters in glassy medium (e.g., WPG), frozen aqueous solutions of Cyd or Ado exhibit an anomalous luminescence behavior in the vicinity of the ground-state pK values* (4.1 and 3.7 respectively) (Fig. 5a,b) (Montenay-Garestier and Hélène, 1970, 1973b). The fluorescence emission of the neutral or cationic form is replaced by a broad red-shifted fluorescence band without vibrational structure. An enhancement of the phosphorescence quantum yield also occurs for a pH value close to the pK value. These results provided evidence for an electron donor–acceptor type interaction between the neutral and the protonated species in equal concentrations under these experimental conditions (pH = pK). Similar or even enhanced modifications of emission characteristics can be observed when the bases are investigated instead of nucleosides. In organic solvents in which molecules are dispersed, polynucleotides [poly(C) and poly(A)] or dinucleotides (CpC or ApA) have a similar behavior of luminescence *vs.* pH, demonstrating that the interaction observed is correlated with the stacking of molecules and occurs between a neutral molecule and its cationic form (Montenay-Garestier and Hélène, 1970; Kleinwächter *et al.*, 1968; Kleinwächter and Koudelka, 1972). In

* The pH values were measured before freezing and the pK values deduced from luminescence titration curves were only apparent values since the protonation equilibrium could not be established during the excited-state lifetimes.

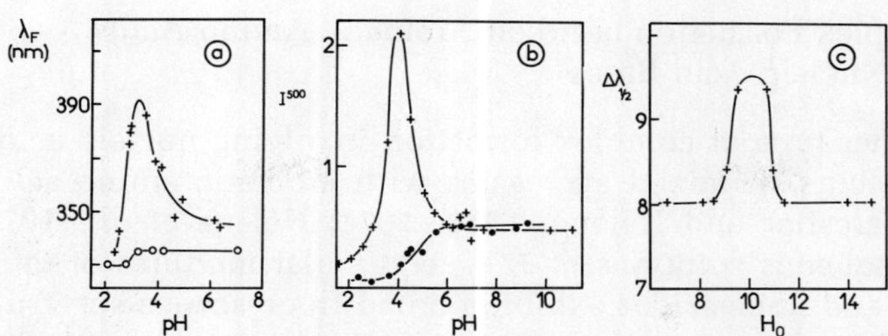


**Fig. 5.** *(a) Fluorescence titration curves of 5 mM Ado solutions frozen at 77°K. Variations vs. pH of the fluorescence maximum wavelength (+, in aggregates; ○, in a water/propylene glycol glass). (b) Phosphorescence titration curves of 1 mM Cyd solutions frozen at 77°K. Variations versus pH of the phosphorescence intensity measured at 500 nm in arbitrary units (+, in aggregates; ○, in a water/propylene glycol glass). (c) Luminescence titration curve of a 10 mM Thd aqueous solution frozen at 77°K. Variations vs pH of the halfwidth (½ of the luminescence spectrum measured in nm). $H_0$ is the Hammett function.*

the case of Cyd aggregates or poly(C) frozen solutions, interactions also modified absorption spectra. A new absorption at longer wavelengths can be observed, but the red-shift of the fluorescence spectrum is more important than that of the absorption spectrum, thus supporting the assertion that there is a stronger charge-transfer contribution in the lowest excited singlet state than in the ground state. In the complex, neutral Cyd molecules behave as electron donors, and protonated molecules behave as electron acceptors. Studies of Cyd photochemistry in ice as a function of pH confirmed this assumption (see Section E). The charge-transfer interaction between neutral Cyd and its cation favors the formation of intermolecular photoadducts (Rhoades and Wang, 1971a,b).

The electron acceptor is not necessarily the protonated form of the nucleoside. The nucleoside itself might behave as an electron acceptor. Emission spectrum modifications of frozen aqueous solutions of Thd at pH values close to the ground-state pK (10.3) demonstrate that an interaction of the preceding type also exists between a neutral Thd molecule and its N(3) deprotonated form in the excited state (Fig. 5c).

The interaction of two different ionic forms of the same molecule may be a general rule for nucleic acid components in aggregates (and perhaps for certain other molecules). These interactions involving excited states of nucleic acid bases may be important intermediary steps to ulterior photochemical reactions.

## 2. Complex Formation between Aromatic Amino Acids and Nucleic Acid Bases

Another type of complex formation involving nucleic acid components occurs in mixed aggregates with aromatic amino acids (Montenay-Garestier and Hélène, 1968, 1971; Hélène *et al.*, 1971a,b). In frozen aqueous solutions at 77°K, equimolar mixtures of indole derivatives and nucleosides exhibit a broad fluorescence spectrum which is red-shifted as compared to that of the components (Fig. 6a,b). The fluorescence emission of each component is quenched. This indicated that molecular interactions occur in the equimolar mixture. At the same concentration (1 m*M*), interactions in fluid medium are too weak to be observed. Higher concentrations are needed to observe complex formation by absorbance or by proton magnetic resonance measurements (Dimicoli and Hélène, 1971). Luminescence modifications are more important with pyrimidines than with purines. Reflectance spectroscopy of the frozen aggregates provided evidence for a new absorption to the red of the absorption bands of the two components. Several investigations of different derivatives involving the indole ring and bases, nucleosides, or nucleotides allowed us to conclude that a weak electron donor–acceptor complex is formed in which the indole ring behaves as an electron donor and the nucleic acid base as an electron acceptor. Neither the ribose nor the phosphate is required for the interaction with tryptophan. As expected in acidic medium, electron transfer to the protonated base is enhanced as demonstrated by emission or reflectance spectroscopy. Fluorescence is always more affected than absorption (charge transfer is increased in

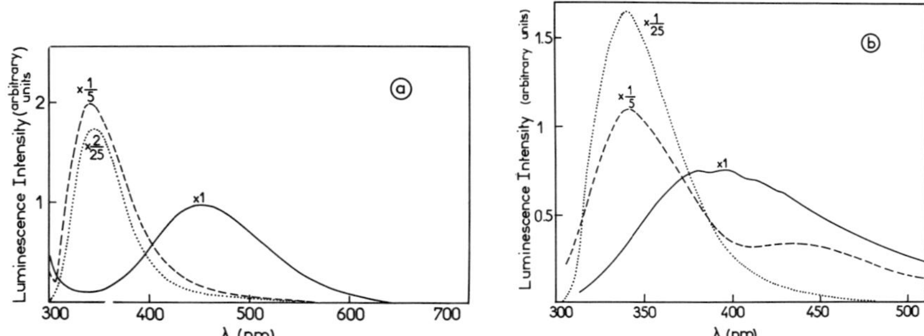


**Fig. 6.** *(a) Luminescence spectra of aqueous frozen solutions (2 mM) at 77°K of tryptophan (···), Thd (---), and their equimolar mixture (—) at pH 7. Excitation wavelength is at 280 nm. (From Montenay-Garestier and Hélène, 1971.) (b) Luminescence spectra of 5 mM aqueous solutions frozen at 77°K of tryptophan (···), Ado (---), and their equimolar mixture (—) at pH 7. Excitation wavelength is at 280 nm.*

the excited state). From the value of the fluorescence maximum wavelength and from the red edge of the reflectance spectra, the electron affinities were shown to increase in the following order:

$$\text{Gua} < \text{Ade} < \text{Cyt, Ura,Thy} < \text{GuaH}^+ < \text{AdeH}^+ < \text{CytH}^+$$

This order has been confirmed by detailed NMR and absorption spectroscopy of complex formation between indole derivatives and nucleic acid components in concentrated fluid aqueous solutions (Dimicoli and Hélène, 1971, 1973). When studies in frozen dilute aqueous mixtures provided evidence for interactions, interactions of the same nature could be demonstrated in the fluid medium at higher concentrations. Slifkin (1971) examined the excitation spectra of the frozen mixtures and found a new band red-shifted as compared to the absorption bands of the components. Evaporation of equimolar aqueous mixtures of tryptophan and Ade or Ado yields a yellow solid substance. This finding supports the idea of a microcrystalline structure of aggregates which is similar to powders (Wang, 1961).

Fluorescence studies performed on tryptophan–Urd mixtures at different temperatures in a fluid ternary hydroalcoholic mixture also provided evidence for a new charge-transfer fluorescence band even above the freezing point (Hui Bon Hoa and Douzou, 1970).

Similar experiments performed with tyrosine and nucleoside equimolar mixtures frozen to 77°K (Hélène *et al.*, 1971b) showed that the fluorescence emissions of tyrosine and Pyr nucleosides are completely quenched. This quenching is ascribed to a charge-transfer type interaction between the two aromatic rings. When the hydroxyl group of tyrosine is substituted (as in *p*-methoxyphenylalanine), quenching of both fluorescence emissions is still observed and a new fluorescence appears at longer wavelengths than those of the components. The equimolar mixtures of Pur derivatives with tyrosine or *p*-methoxyphenylalanine are characterized by a quenching of the phenyl ring fluorescence while the fluorescence of the Pur is only slightly affected.

Equimolar mixtures of phenylalanine with Cyd in acidic conditions exhibit a fluorescence that is red-shifted by ~15 nm as compared to Cyd alone (Montenay-Garestier and Helene, 1973b). Under these conditions Cyd is protonated and behaves as the best electron acceptor among nucleosides. Luminescence studies do not provide evidence for excited-state interaction between phenylalanine and Pur derivatives in aggregates. However, Slifkin (1971) observed modifications of the infrared spectrum bands of Ade in mixtures with phenylalanine in KBr discs.

Frozen aqueous solutions of histidine and Pyr derivatives exhibit marked luminescence modifications of Pyr spectra above pH 6. In mixed aggregates with histidine (or imidazole) Thd and Cyd (or Cyt) have a red-shifted fluorescence. This interaction is strong since the addition of 0.1 *M* sodium chloride does not split the complex formed in 1 m*M* aggregates. No interaction can be detected in mixed aggregates of histidine with Pur nucleosides (Montenay-Garestier and Hélène, 1973b).

The deprotonated imidazole ring (above pH 6) behaves as a good electron donor intermediate between the indole and the phenyl ring (Shifrin, 1968). When involved in a charge-transfer interaction with Pyr rings, the base moiety should be the electron acceptor and the imidazole ring the donor.

Excitation energy is transferred among bases at the triplet level in mixed aggregates and may end on tryptophan. At the singlet level a direct Förster transfer from nucleic acid bases sensitizes the singlet state of the base–tryptophan charge-transfer complex although singlet transfer involves a smaller number of Ado molecules than does triplet transfer. In mixed aggregates of 5-methoxytryptamine and Ado at concentrations ratios higher than 0.1, excitation energy absorbed by Ade residues is completely transferred to 5-methoxytryptamine at the singlet level (Montenay-Garestier and Hélène, 1973c). At ratios $>0.2$ of 5-methoxytryptamine and Ado concentrations, an important red-shift of the phosphorescence of 5-methoxytryptamine and modifications in the relative intensities of the vibronic bands are observed. These modifications of the phosphorescence characteristics of 5-methoxytryptamine are probably due to modifications of the environment of the 5-methoxytryptamine molecules. At low concentrations, 5-methoxytryptamine molecules are dispersed and are probably sandwiched between Ado molecules, whereas at higher ratios 5-methoxytryptamine molecules are probably stacked. Therefore, the observed phosphorescence is ascribed to the emission of 5-methoxytryptamine aggregates intercalated in Ado aggregates. Similar effects can be observed in mixed tyrosine and Ado (or Gua) aggregates.

## E. Photochemical Reactions of Nucleic Acid Derivatives in Frozen Aqueous Solutions

Intermolecular photochemical reactions in frozen aqueous solutions may lead to original results as compared to reactions in fluid medium. First, a low temperature and a rigid medium usually stabilize species that are otherwise extremely short-lived, e.g., radicals, trapped electrons, and triplet states. Second, aggregate formation in frozen aqueous solutions leads to a particular structure (as described in the preceding paragraphs) in which solute molecules are stacked and can interact without requiring a diffusion process. This structure also imposes a restriction on the relative orientation of two neighboring molecules and thus may limit the number of photoproducts that can be formed.

Wang (1961) showed that photochemical reactions of Thy in the aggregates formed in frozen aqueous solutions do not depend on the cooling process (liquid nitrogen or ice–salt mixture) or on the temperature at which irradiation experiments are performed (77°, 200°, or 260°K). Therefore it appears that aggregate structure played the predominant role in determining the course of the photochemical reactions (Wang, 1963).

### 1. Direct Photochemical Reactions

#### *a. Pyrimidine Dimers*

In 1958, Beukers *et al.* observed that the rate of photoproduct formation under UV irradiation of Thy and Ura is considerably greater (~20 times) in ice than in fluid aqueous solutions. A similar conclusion was reached by Wang (1961). On the basis of molecular weight determination, spectroscopic properties, and mass spectra, a dimer structure with a cyclobutyl ring was proposed for the photoproduct obtained in ice (Beukers *et al.*, 1960). Four stereoisomers can be formed (Wulff and Fraenkel, 1961) (Fig. 7). The dimerization process in aggregates is very efficient and more than 85% of Thy can be converted into dimers (Smietanowska and Shugar, 1961; Wang, 1961; Smith, 1963). Only one stereoisomer of Thy◇Thy can be isolated, and it has the *cis-syn* structure (Wang, 1963; Smith, 1963; Weinblum and Johns, 1966). The properties of this dimer are quite similar to those of the main Thy photoproduct formed upon UV-irradiation of aqueous solutions of DNA (Beukers *et al.*, 1960; Wacker *et al.*, 1960; Weinblum and Johns, 1966). Proof of the *cis-syn* structure for the DNA Thy photoproduct was provided by Varghese and Wang (1967b) and by Weinblum (1967) on the basis of infrared and NMR data.

UV irradiation of Thd in ice leads to the formation of three stereoisomers. Only the *trans-syn* structure is not detected (Weinblum and Johns, 1966; Varghese, 1970a). The *cis-anti* dimer gives the highest yield.

Irradiation of Ura in ice leads to the formation of cyclobutane

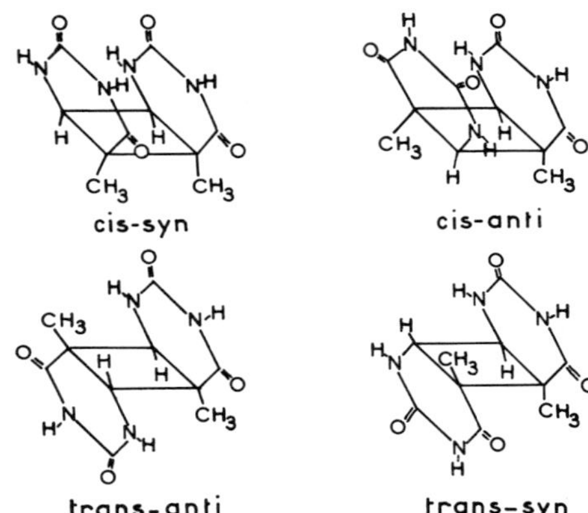


**Fig. 7.** *The different stereoisomers of thymine dimers.*

dimers (Wacker *et al.*, 1961; Smietanowska and Shugar, 1961; Wang, 1961; Smith, 1963). The *cis-syn* isomer gives the highest yield (Khattak and Wang, 1968), but the *cis-anti* isomer is also formed in lower yield (Varghese, 1971b).

Urd behaves similarly to Thd with respect to UV irradiation in ice. The *cis-anti* dimer is formed with the highest yield whereas the *trans-syn* isomer is not detected (Varghese, 1971b).

Under UV irradiation in fluid aqueous solutions, Cyd and dCyd give a water-addition photoproduct where the 5,6-double bond is saturated (as observed also with Ura derivatives). In the frozen state, dimers are formed in low yield (Varghese and Rupert, 1971). Ura dimers are formed after deamination upon heating. A comparison with authentic samples of Ura dimers allowed the assignation of a *cis-syn* structure to the Cyd dimer (Varghese, 1971a).

In addition to bases and nucleosides, several substituted Pyr (particularly N-methylated derivatives) also form cyclobutane dimers in frozen aqueous solutions (Smietanowska and Shugar, 1961; Wang, 1961).

Mixed dimers containing two different bases can form upon UV irradiation of frozen aqueous mixtures. Thus mixed dimers of Ura and Thy (Ura◇Thy) of Cyt and Thy (Cyt◇Thy) are isolated from irradiated frozen mixtures (Smith, 1963; Varghese, 1971a).

The stereochemical structure of Thy dimers in ice depends on the addition of Ade (Smith, 1966). The yield of *cis-syn* dimer decreases from 85% to 5% and that of the *cis-anti* isomer increases from 0% to 6% when the ratio of Ade and Thy concentrations increase from 0 to 2. Above this ratio, no further change occurs in the yield of the two stereoisomers.

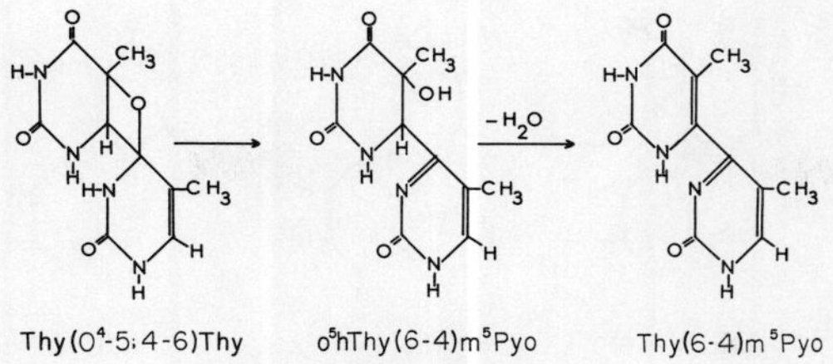

Thy(O$^4$-5;4-6)Thy  o$^5$hThy(6-4)m$^5$Pyo  Thy(6-4)m$^5$Pyo

**Fig. 8.**

*b. Pyrimidine Photoadducts*

A photoadduct of Cyt and Thy was discovered in UV-irradiated DNA (Wang and Varghese, 1967; Varghese and Wang, 1967a). A Thy–Thy adduct was also formed upon UV irradiation of Thy in ice. Spectroscopic data (UV, IR, NMR, mass spectra) led to the proposed structure shown in Fig. 8. The dehydrated form is isolated after acid hydrolysis at high temperatures. Crystallographic studies (Karle *et al.*, 1968; Karle, 1969) showed that the two rings are nearly perpendicular to each other. The possible involvement of an oxetane ring structure in an intermediate species was suggested by Varghese and Wang (1968a) (see Fig. 8). Rahn and Hosszu (1969a) showed that the photoadduct is not formed directly after irradiation of Thy at low temperatures but appears when the temperature increases, probably as a result of the low thermal stability of the oxetane ring structure.

Photoadducts of structures similar to those presented in Fig. 8 are produced upon UV irradiation of Ura and Urd in ice (Khattak and Wang, 1968; Varghese, 1971b) and represent an important fraction of the total amount of photoproducts formed (10–15% in the case of Urd).

Under UV irradiation of Thd in ice, the photoadduct (Fig. 8) represent ~16% of the photoproducts (Varghese, 1970a). But in addition to cyclobutane dimers and photoadducts, another photoproduct representing ~14% of the total photoproduct production was identified as Thy($\alpha$–5)hThy (Fig. 9). This photoproduct was also isolated from DNA irradiated at −80°C (Varghese, 1970b).

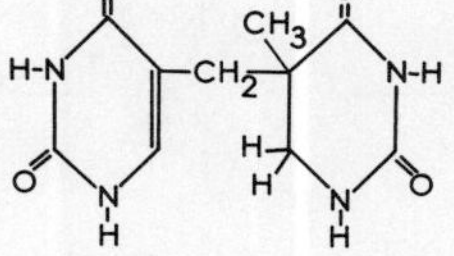

Thy ($\alpha$-5)hThy

**Fig. 9.**

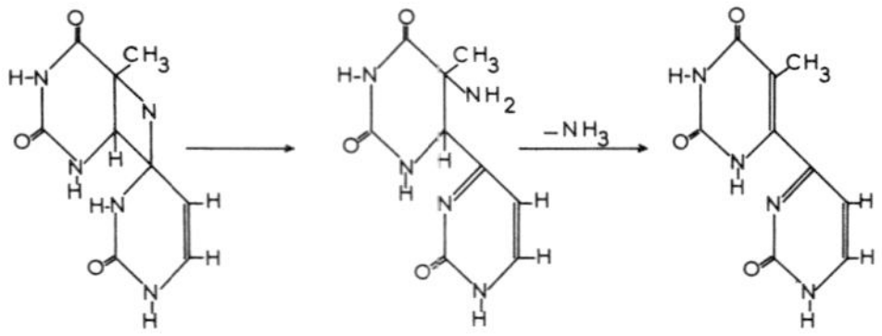

Thy($N^4$-5; 4-6)Cyt  $a^5$hThy(6-4)Pyo  Thy(6-4)Pyo

**Fig. 10.**

A mixed photoadduct involving Thy and Ura was isolated from UV irradiated frozen aqueous mixtures (Rhoades and Wang, 1970). Also a mixed photoadduct identical to that isolated from DNA was isolated from irradiated frozen mixtures of Thy and Cyt (Varghese, 1971a). An azetidine ring structure was proposed as an intermediate step in photoadduct formation by analogy with the oxetane ring structure in the case of the Thy–Thy adduct (see Figs. 8 and 10).

The behavior of frozen aqueous solutions of Cyd upon UV irradiation depends upon the pH. When the pH is close to the pK value of Cyd, a new photoadduct is formed in which C(5) and C(4) are linked (Fig. 11). The quantum yield for the production of this photoadduct reaches a maximum when pH = pK (Rhoades and Wang, 1971a). Under these conditions Cyd and its cation form an intermolecular complex which involves a charge-transfer contribution (Montenay-Garestier and Hélène, 1970). Maximum complex formation is achieved when the pH value is equal to the pK value, at which pH Cyd and its cation are present in equal concentrations. Therefore, it is clear that photoadduct production is linked to complex formation.

*c. Thymine Trimer*

In UV-irradiated frozen solutions of Thy, Smith (1963) detected the formation of a photoproduct with an $R_f$ of 0.13 in the solvent butanol/acetic acid/water. Under certain experimental conditions this photoproduct, later isolated by Varghese and Wang (1968b), yields

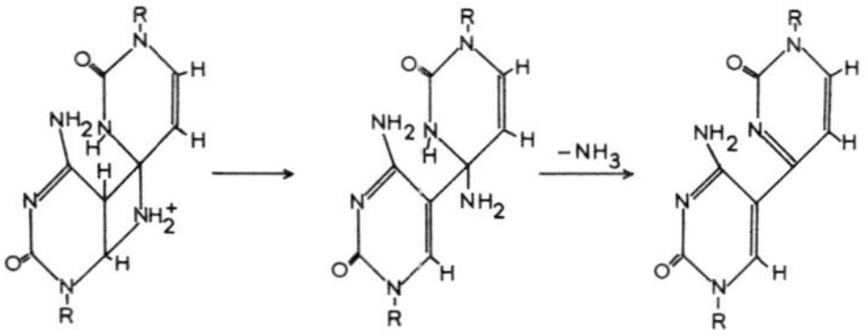

Cyt($N^4$-6;4-5)Cyt  Cyt(5-4)$a^4$Pyo  Cyt (5-4)Pyo

**Fig. 11.**

either a Thy◇Thy dimer or a Thy–Thy adduct plus free Thy. Rahn and Hosszu (1969a) studied the kinetics of the formation of this photoproduct and showed that 254 nm irradiation at 25°C leads to Thy–Thy adduct production. The structure of this photoproduct was established by Flippen *et al.* (1970) and Wang (1971) and is a trimer resulting from the combination of an adduct and a Thy◇Thy dimer (Fig. 12). Thus irradiation at short wavelengths leads to Thy◇Thy splitting and the production of Thy–Thy adduct plus Thy.

*d. Electron Photoejection by Nucleic Acid Derivatives in Frozen Aqueous Solutions*

At a low temperature (77°K), UV excitation of Pur and Pyr molecules leads to photoejection of electrons (Hélène *et al.*, 1966a; Sevilla, 1971). This phenomenon was investigated by ESR spectroscopy. The signal produced after UV irradiation of Pur derivatives in frozen aqueous solutions at 77°K is mainly due to Pur radical cations and trapped electrons that can be bleached by further irradiation in the visible range (Hélène *et al.*, 1966a). Light intensity dependence studies showed that ejection of electrons results from a biphotonic process, the second photon being absorbed by the molecule raised to its long-lived triplet state by absorption of the first proton. The recombination of electrons and radical cations at 77°K gives rise to an isothermal luminescence consisting mainly of the phosphorescence of the Pur molecule (Guermonprez *et al.*, 1967). Electron–cation recombination can be stimulated by visible light irradiation and by rapid heating (thermoluminescence). Recently, UV irradiation of Thy at 77°K was shown to produce Thy cation radicals in alkaline and in acid glasses (Sevilla, 1971). Trapped electrons can be ejected from their traps by visible light irradiation and can then react with Thy molecules to give Thy anions.

In the presence of organic molecules, biphotonic processes taking place in the Pur derivative result in a photosensitized reaction of the organic molecule, e.g., ethanol or glycine (Hélène *et al.*, 1966a; Santus

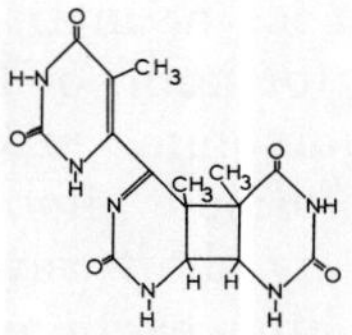

THYMINE TRIMER

**Fig. 12.**

*et al.*, 1966). These reactions might be important in photobiology since photochemical reactions of aliphatic amino acids, for example, can be sensitized by nucleic acid derivatives.

*e. Photochemical Reactions of DNA in Frozen Aqueous Solutions*

At room temperature, UV irradiation of DNA leads to the formation of Thy◇Thy, Cyt◇Thy, and Cyt◇Cyt (Setlow and Carrier, 1966; Setlow, 1968) as well as Thy–Thy and Cyt–Thy adducts (Varghese and Wang, 1967a; Wang and Varghese, 1967).

In *Bacillus megaterium* spores, a new photoproduct was formed which is not a cyclobutane dimer, whereas Cyt◇Thy and Thy◇Thy were formed only in small amounts (Donnellan and Setlow, 1965). Photochemical studies of DNA showed that the yield of photoproducts is much lower at 77°K than at room temperature (Rahn, 1966). Furthermore, the concentration of Thy photoproducts is increased if the irradiated DNA is first heated above the melting point of the solution and then cooled again to 77°K. It was concluded that the rigid structure of DNA which was maintained in frozen solutions does not allow Thy◇Thy(*c,s*) formation which requires a rotation of 36° of one of the bases with respect to the helix axis. Only the regions of the DNA molecule with a structure which is sufficiently disordered by ice crystal formation can lead to dimer formation. The main photoproduct obtained at a low temperature is identical to the spore photoproduct previously described (Rahn and Hosszu, 1968). This spore photoproduct is not photoreactivable and is obtained upon UV irradiation of poly(dA:dT) but not of alternating poly(dAT:dAT). This photoproduct was later identified as Thy($\alpha$-5)hThy (Fig. 9: Varghese, 1970b).

*f. Mechanism of Photoproduct Formation*

The formation of Pyr dimers in fluid solutions of monomers involves the triplet state of these molecules at least at low concentrations (Herbert *et al.*, 1968; Whillans *et al.*, 1969; Whillans and Johns, 1971). Photodimerization in ice or in DNA might involve either the singlet or the triplet levels (or both) of Pyr bases. In a rigid matrix formed upon cooling an equivolume mixture of water and ethylene glycol, monomers which formed after splitting the *cis-syn* and *trans-syn* $Me_2$Thy◇$Me_2$Thy by short-wavelength irradiation strongly interact (fluorescence quenching, excitonic splitting of the absorption band) (Eisinger and Lamola, 1967; Lamola, 1968; Lamola and Eisinger, 1968). An excimer state was then proposed to be an intermediate step in the photodimerization reaction although no experimental evidence has been found.

The formation of Thy–Thy and Ura–Thy photoadducts can result from an attack of the 5,6-double bond by the C(4′) excited carbonyl group of a second molecule. It is difficult to extend this mechanism to the formation of C(6)–C(4′) or C(5)–C(4′) adducts. In the case of C(5)–C(4′) adduct formation, a special role can be played by the proton which upon binding to one molecule allows charge-transfer complex formation between Cyd and its cation (Montenay-Garestier and Hélène, 1970; Rhoades and Wang, 1971a,b). Radical intermediate species can also be involved in photoadduct formation. Radicals such as those presented in Fig. 13 which have been identified by ESR at low temperatures (see, for example, Sevilla, 1971; Hütterman, 1970) can be involved in the formation of Thy($\alpha$–5)hThy (Fig. 9). Dye-photosensitized reactions and $\gamma$- and UV irradiation also lead to the generation of Thy free radicals (see Nicolau, 1973; Herak, 1973; Van De Vorst and Lion, 1973a).

2. Sensitized Photochemical Reactions

*a. Dye-Sensitized Reactions in Nucleic Acid Constituents*

In order to obtain information on the mechanism of photodynamic effects, mixtures of acridine dyes and nucleic acid constituents or DNA were irradiated at 77°K in a wavelength range in which the bases do not absorb light ($\lambda > 300$ nm). Free radicals localized on the base moiety of the nucleosides were produced (Delmelle and Duchesne, 1967) in mixtures of acriflavin and nucleosides. Only Thy free radicals can be detected in the case of DNA. Gräslund *et al.* (1969) also observed the same radical in irradiated mixtures of proflavin or acridine orange with DNA and poly(dAT:dAT). The radical yield increases with the concentration of bound dye molecules. It is higher with proflavin than with acridine orange, and it is higher with poly(dAT:dAT) than with DNA.

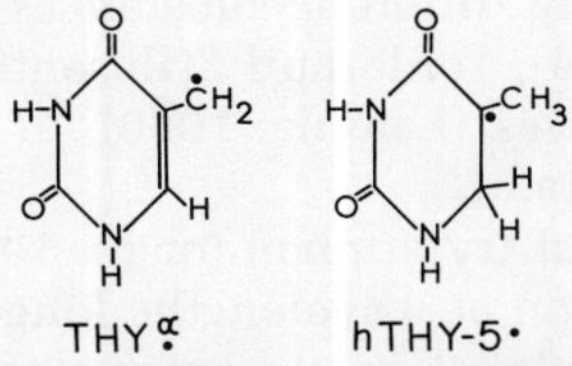

THY$\dot{\alpha}$ hTHY-5•

**Fig. 13.** *Radicals formed upon UV irradiation of thymine in ice.*

In the absence of oxygen, proflavin photosensitized the formation of two types of radicals in Thy (Thy$^{\alpha}_{\cdot}$ and hThy·, see Fig. 13) (Van De Vorst and Lion, 1973a). The temperature dependence of radical formation was also studied. Hydrothymine radicals form at temperatures below 115°K whereas Thy$^{\alpha}_{\cdot}$ radicals appear at higher temperature. Proflavin sensitized the formation of radicals in all nucleosides and these radicals have ESR spectra identical to those of H-atom addition products (Van De Vorst and Lion, 1973b). Studies of the light intensity dependence of radical production and of the decrease of the proflavin triplet-state population lead to the conclusion that a biphotonic reaction involving the triplet state of proflavin is involved in radical formation.

$\gamma$-Irradiation of freeze-dried mixtures of nucleosides (or DNA) and proflavin leads to a localization of the radiation damage on the dye molecule even when the ratio of nucleotide to proflavin is more than 1 (Grégoli *et al.*, 1970). These findings were interpreted as resulting from a long-range electron transfer, although electronically excited states which can be produced after $\gamma$-irradiation might be involved. Excitation energy transfer from bases to proflavin is a very efficient process at the singlet and at the triplet levels (Kubota, 1970; Hélène, 1973a).

The problem of photochemical reactions in dye–nucleotide mixtures in the frozen state thus appears to be very complicated. Excitation in the visible absorption band of the dye photosensitizes the formation of nucleotide radicals via a biphotonic absorption process. Excitation in the UV-absorption band of the base leads to a sensitized population of the dye excited states, and electron photoejection by bases may lead to electron capture by the dye (Hélène *et al.*, 1966a; Sevilla, 1971).

*b. Photosensitized Splitting of Thymine Dimers by Indole Derivatives*

Pyr dimers in DNA can be split by direct irradiation at a short wavelength ($\lambda < 250$ nm) or by enzymatic photoreactivation (Setlow and Setlow, 1963). In fluid solutions, these dimers can be split by photosensitization using different substances such as uranyl acetate (Wacker *et al.*, 1964), chloranil (Rosenthal and Elad, 1968), anthraquinone derivatives (Lamola, 1966), or indole derivatives (Hélène and Charlier, 1971a,b).

In mixed aggregates of tryptophan (or its 5-hydroxy derivative) and Thy◇Thy(c,s), irradiation at wavelengths longer than 290 nm leads to a splitting of dimer molecules into monomers (Hélène and Charlier,

1971a,b). The fluorescence of the indole derivative is quenched as was observed in mixtures with nucleosides (Montenay-Garestier and Hélène, 1971). This is attributed to an electron transfer process in the excited state. As a matter of fact, solvated electrons react in fluid medium with Oro◇Oro (Hélène and Charlier, 1971b) or with Thy◇Thy (Santus *et al.*, 1972). [Beukers and Berends (1960) previously observed that electron bombardment of Thy dimers during mass spectra determinations leads to the production of monomer species.] Photosensitized splitting of Thy dimers by indole derivatives occurs in fluid medium and in DNA (Hélène and Charlier, 1971a,b; Toulmé *et al.*, 1974; Charlier and Hélène, 1975).

## F. Conclusion

The formation of aggregates in frozen aqueous solutions is useful in studying several phenomena such as complex formation between different molecules, excited-state interactions, energy transfer processes, and photochemical reactions.

Aggregate formation results from a rejection of solute molecules from the growing ice crystals during the freezing process. These aggregates have a well-ordered structure which allows efficient energy transfer at the triplet level due to a good overlap of the electronic clouds of neighboring molecules. This stacked structure favors intermolecular photochemical reactions which would otherwise require a diffusion of excited- and ground-state molecules in fluid medium. This structure may prevent certain reactions by restricting the relative positions of neighboring molecules. This is particularly clear when the distribution of different Pyr photoproducts is analyzed for related molecules in aggregates and in fluid medium (Table 1). The structure of aggregates is often claimed to be similar to that of microcrystals. Photochemical studies represent a possible approach toward understanding aggregate structure and a good example is provided by the comparative investigation of Pyr dimer formation in aggregates and in crystals. For example, 1-MeUra and 1-MeThy crystallize in a layered structure of hydrogen-bonded dimers. Examination of the crystal structures shows that molecules in two layers are favorably oriented for formation of the *trans-syn* cyclobutane photodimer in the case of 1-MeThy (Stewart, 1963), whereas the *trans-anti* dimer would result from UV-irradiation of 1-MeUra crystals (Eaton and Lewis, 1970). However, no photodamage occurs during absorbance measurements on 1-MeUra crystals whereas extensive photodamage of 1-MeThy

**Table 1** Photoproducts from the Irradiation of DNA and Their Components in Various Conditions[a]

| | Pyrimidine dimers (%) | | | | Other products | Ref.[b] |
|---|---|---|---|---|---|---|
| | *cis-syn* | *cis-anti* | *trans-syn* | *trans-anti* | | |
| DNA in fluid solution | xx | | | | Cyt–Thy, Cyt◇Thy | 1,2 |
| DNA in ice (or water ethylene glycol) | x | | | | Adduct and "spore" | 3 |
| DNA in spores or dried DNA | x | | | | "Spore" | 4,5 |
| Thy in ice | xx | | | | Adduct and trimer | 1,6,7 |
| Acetone-sensitized Thy | 23 | 33 | 21 | 23 | | 21[d] |
| Thd in ice | x | xx | | x | xx | 1 |
| | 21 | 40 | — | 6–9 | 30%: adduct, Thy(α–5)hThy | 8,9 |
| | 38 | 48 | — | 14 | | 10[d] |
| Thd in thin solid film | 13 | 27 | — | 9 | 48%: adduct, Thy(α–5)hThy | 9 |
| Acetone-sensitized Thd | 26 | 13 | 6 | 55 | | 10 |
| Thy + Ade (1 2) in ice | 5 | 6 | | | | 11[c] |
| TpT in fluid solution | 88 | | 12 | | | 1[d] |
| Ura in ice | 90 | 10 | | | | 12[d] |
| | xx | | | | Adduct | 13 |
| Acetone-sensitized Ura | 68 | 29 | — | 3 | hUra | 12 |

| | | | | | | |
|---|---|---|---|---|---|---|
| Urd in ice | 6 | 64 | — | 18 | 12%: adduct, hydrate | 12 |
| Urd in thin solid film | 8 | 42 | | | 50%: adduct, hydrate | 12 |
| Acetone sensitized Urd | 4 | 38 | | 55 | | 12 |
| $Me_2$Ura in ice | 27 | 19 | 15 | 8 | | 14[d] |
| Cyd in ice (pH 7) | xx | | | | Adduct | 15,16 |
| Acetone-sensitized Cyd | | | | xx | Cyt◇Ura, Ura◇Ura | 22 |
| Cyd in ice (pH ~ 4) | | | | | Cyt (4–5) Cyt | 17 |
| Thy + Ura in ice | | Thy◇Thy, Ura◇Ura, Thy◇Ura | | | Ura–Thy | 18,20 |
| Thd + Cyt in ice | | Thy◇Thy, Cyt◇Cyt, Cyt◇Thy | | | Cyt–Thy | 16,19 |

[a] When quantitative data were not available, the symbols xx and x refer to a large and a small amount, respectively, of photoproducts.

[b] Key to references

1. Weinblum and Johns, 1966.
2. Wang and Varghese, 1967.
3. Rahn and Hosszu, 1968.
4. Donnellan and Setlow, 1965.
5. Rahn and Hosszu, 1969b.
6. Beukers *et al.*, 1960.
7. Varghese and Wang, 1968a.
8. Varghese, 1970a.
9. Varghese, 1971c.
10. Ben-Hur *et al.*, 1967.
11. Smith, 1966.
12. Varghese, 1971b.
13. Khattak and Wang, 1968.
14. Fahr *et al.*, 1972.
15. Varghese and Rupert, 1971.
16. Varghese, 1971a.
17. Rhoades and Wang, 1971b.
18. Rhoades and Wang, 1970.
19. Varghese and Patrick, 1969.
20. Smith, 1963.
21. Jennings *et al.*, 1970.
22. Varghese, 1972.

[c] Percent of total thymine.

[d] Percent of total dimers.

crystals occurs under similar conditions (Eaton and Lewis, 1970). It must be remembered that photodimerization takes place between an excited- and a ground-state molecule. The electronic distribution in the 5,6-double bond may be such that different isomers can be obtained from different excited states (singlet and triplet).

Hydrogen bond formation between Ade and Thy in crystals modifies the relative orientation of Thy molecules and, therefore, may change not only the quantum yield but also the stereoisomerism of the dimers (Stewart, 1963; Stewart and Davidson, 1963).

The crystal structures of closely related molecules may be sufficiently different for different stereoisomers of photodimers to be produced. For example, the stacking of Thy molecules in Thy monohydrate crystals favors the formation of *cis-syn* dimers (Gerdil, 1961; Wang, 1963), whereas the *trans-anti* dimer is expected to form in 1-MeThy crystals (Stewart, 1963). The same variety in stereoisomerism of cyclobutane photodimers is observed upon UV-irradiation of the aggregates obtained by freezing aqueous solutions of several Pyr derivatives (see Table 1).

Although we did not intend to describe in detail all of the photochemical reactions that take place in frozen aqueous solutions, all of these reactions rely upon the formation of aggregates or puddles (see Wang, 1965, for the discussion of photochemical reactions of Pyr derivatives in organic puddles). It must be pointed out that UV-irradiation of solid films (Wang, 1961) and of crystals (Stewart, 1963) leads to the formation of similar photoproducts. Information about aggregate organization can be obtained from the analysis of the stereochemical structures of the dimers formed in frozen aqueous solutions. Thus Pyr bases (Thy, Ura) seem to form aggregates in which stacked molecules are deduced from each other by a translation process (*cis-syn* dimers). In nucleoside aggregates (Urd, Thd) a head-to-tail stacking is probably favored (*cis-anti* dimers are predominant), and neighboring stacks might be approximately deduced from each other by a symmetry with respect to a plane parallel to the 5,6-double bonds since *trans-anti* (but not *trans-syn*) dimers are formed. In a Watson-Crick double helix, only the *cis-syn* isomer is formed (Varghese and Wang, 1967b; Weinblum, 1967). Introducing additives such as Ade in Thy aggregates is expected to change the relative positions of neighboring Thy molecules and, thus, to alter the yield and the structure of the photoproducts (Smith, 1966).

A comparative study of excited-state properties in aggregates of nucleic acid components and in polynucleotides or nucleic acids themselves is useful in understanding the role of the phosphodiester backbone in electronic interactions between monomers (Hélène, 1973a). Studies of aggregate properties provide important information concerning excited-state interactions and energy transfer processes between nucleic acid bases as well as the nature of photochemical reactions in nucleic acids.

Investigations on complex formation between different kinds of biological molecules represent another field of research in which studies of frozen aqueous solutions seem fruitful (see, e.g., Montenay-Garestier and Hélène, 1973a). The formation of mixed aggregates allows the observation of complexes that would not be observed in the same concentration range in fluid medium. Interactions between nucleic acid bases and aromatic amino acids were first observed in aggregates (Montenay-Garestier and Hélène, 1968, 1971). Such interactions may not only participate in the formation of protein–nucleic acid complexes (Hélène, 1971) but may also be involved in photochemical reactions of these complexes. Problems such as excited-state interactions and energy transfer processes may be clarified by a study of mixed aggregates. However, it must be emphasized that the limitation of aggregate studies is due mainly to the absence of detailed knowledge of their local structure. The results reported in this chapter demonstrate the usefulness of the investigation of aggregates formed in frozen aqueous solutions to obtain information on complex formation, excited-state interactions, and photochemical reactions between like or unlike molecules, especially in the field of nucleic acids and nucleic acid–protein interactions.

## References


Alburn, H. E., and Grant, N. H. (1965). *J. Amer. Chem. Soc.* **87,** 4174.

Apelgot, S., Ekert, B., and Frilley, M. (1965). *Biochim. Biophys. Acta* **103,** 503.

Bazin, M., Santus, R., and Hélène, C. (1973). *Chem. Phys.* **2,** 119.

Ben-Hur, E., Elad, D., and Ben-Ishai, R. (1967). *Biochim. Biophys. Acta* **149,** 355.

Bennett, R. G., Schwenker, R. F., and Kellog, R. E. (1964). *J. Chem. Phys.* **41,** 3049.

Bersohn, R., and Isenberg, I. (1964). *J. Chem. Phys.* **40,** 3175.

Beukers, R., and Berends, W. (1960). *Biochim. Biophys. Acta* **41,** 550.

Beukers, R., and Berends, W. (1961). *Biochim. Biophys. Acta* **49,** 181.

Beukers, R., Ylstra, J., and Berends, W. (1958). *Rec. Trav. Chim. Pays-Bas* **77,** 729.

Beukers, R., Ylstra, J., and Berends, W. (1960). *Rec. Trav. Chim. Pays-Bas* **79,** 101.

Bruice, T. C., and Butler, A. R. (1965). *Fed. Proc., Fed. Amer. Soc. Exp. Biol.* **24,** S-45.

Bugg, C. E., Thomas, J. M., Sundaralingam, M., and Rao, S. T. (1971). *Biopolymers* **10,** 175.

Burshtein, E. A., and Busel, E. P. (1971). *Opt. Spektrosk.* **30,** 69.

Butler, A. R., and Bruice, T. C. (1964). *J. Amer. Chem. Soc.* **86,** 313.

Charlier, M., and Hélène, C. (1975). *Photochem. & Photobiol.* **21,** 131.

Daniels, M. (1973). *In* "Physicochemical Properties of Nucleic Acids" (J. Duchesne, ed.), Vol. 1, p. 99. Academic Press, New York.
Davis, S. L., and Tinoco, I. (1966). *Nature (London)* **210**, 1286.
Delmelle, M., and Duchesne, J. (1967). *C. R. Acad. Sci. Ser. D* **264**, 138.
Dexter, D. L. (1953). *J. Chem. Phys.* **21**, 836.
Dimicoli, J. L., and Hélène, C. (1971). *Biochimie* **53**, 331.
Dimicoli, J. L., and Hélène, C. (1973). *J. Amer. Chem. Soc.* **95**, 1036.
Donnellan, J. E., Jr., and Setlow, R. B. (1965). *Science* **149**, 308.
Drost-Hansen, W., and Walter, J. (1967). *J. Colloid Interface Sci.* **25**, 131.
Eaton, W. A., and Lewis, T. P. (1970). *J. Chem. Phys.* **53**, 2164.
Eigen, M., and Demayer, L. (1964). *Ber. Bunsenges Phys. Chem.* **68**, 19.
Eisinger, J., and Lamola, A. A. (1967). *Biochem. Biophys. Res. Commun.* **28**, 558.
Eisinger, J., and Lamola, A. A. (1971). *In* "Excited States of Proteins and Nucleic Acids" (R. S. Steiner and I. Weinryb, eds.), p. 107. Plenum, New York.
Eisinger, J., and Shulman, R. G. (1966). *Proc. Nat. Acad. Sci. U.S.* **55**, 1387.
Fahr, E., Fürst, G., Maul, P., and Wieser, H. (1972). *Z. Naturforsch. B* 27, 1475.
Flippen, J. L., Karle, I. L., and Wang, S. Y. (1970). *Science* **169**, 1084.
Förster, T. (1965). *In* "Modern Quantum Chemistry" (O. Sinanŏglu, ed.), Part 3, p. 93. Academic Press, New York.
Füchtbauer, W., and Mazur, P. (1966). *Photochem. & Photobiol.* **5**, 323.
Galley, W. C. (1968). *Biopolymers* **6**, 1279.
Galley, W. C., and Fantus, I. (1969). *In* "Biological Molecules in their Excited States," Abstracts p. A2. Arden-House, Harriman, New York.
Gerdil, R. (1961). *Acta Crystallogr.* **14**, 333.
Grant, N. H., and Alburn, H. E. (1965). *Biochemistry* **4**, 1913.
Grant, N. H., Clark, D. E., and Alburn, H. E. (1961). *J. Amer. Chem. Soc.* **83**, 4476.
Grant, N. H., Clark, D. E., and Alburn, H. E. (1966). *J. Amer. Chem. Soc.* **88**, 4071.
Gräslund, A., Rigler, R., and Ehrenberg, A. (1969). *FEBS (Fed. Eur. Biochem. Soc.) Lett.* **4**, 227.
Grégoli, S., Taverna, C., and Bertinchamps, A. (1970). *Int. J. Radiat. Biol.* **18**, 577.
Guermonprez, R., Hélène, C., and Ptak, M. (1967). *J. Chim. Phys.* **64**, 1376.
Guéron, M., and Shulman, R. G. (1968). *Annu. Rev. Biochem.* **37**, 679.
Hélène, C. (1966). *Biochem. Biophys. Res. Commun.* **22**, 237.
Hélène, C. (1971). *Nature (London)* **234**, 120.
Hélène, C. (1973a). *In* "Physicochemical Properties of Nucleic Acids" (J. Duchesne, ed.), Vol. 1, p. 119. Academic Press, New York.
Hélène, C. (1973b). *Photochem. & Photobiol.* **18**, 255.
Helène, C., and Charlier, M. (1971a). *Biochem. Biophys. Res. Commun.* **43**, 252.
Hélène, C., and Charlier, M. (1971b). *Biochimie* **53**, 1175.
Hélène, C., and Longworth, J. W. (1972). *J. Chem. Phys.* **57**, 399.
Hélène, C., and Montenay-Garestier, T. (1968). *Chem. Phys. Lett.* **2**, 25.
Hélène, C., Santus, R., and Douzou, P. (1966a). *Photochem. & Photobiol.* **5**, 127.
Hélène, C., Santus, R., and Ptak, M. (1966b). *C. R. Acad. Sci.* **262**, 1349.
Hélène, C., Ptak, M., and Santus, R. (1968). *J. Chim. Phys.* **65**, 160.
Hélène, C., Dimicoli, J.-L., and Brun, F. (1971a). *Biochemistry* **10**, 3802.
Hélène, C., Montenay-Garestier, T., and Dimicoli, J.-L. (1971b). *Biochim. Biophys. Acta* **254**, 359.
Herak, J. N. (1973). *In* "Physicochemical Properties of Nucleic Acids" (J. Duchesne, ed.), Vol. 1, p. 196. Academic Press, New York.
Herbert, M. A., Hunt, J. W., and Johns, H. E. (1968). *Biochem. Biophys. Res. Commun.* **33**, 643.

Horne, R. A. (1963). *J. Inorg. Nucl. Chem.* **25,** 1139.
Hui Bon Hoa, G., and Douzou, P. (1970). *J. Chim. Phys. Physiochim. Biol.* **67,** Suppl., 197.
Hütterman, J. (1970). *Int. J. Radiat. Biol.* **17,** 249.
Jennings, B. H., Pastra, S. C., and Wellington, J. L. (1970). *Photochem. & Photobiol.* **11,** 215.
Karle, I. L. (1969). *Acta Crystallogr., Sect. B* **25,** 2119.
Karle, I. L., Wang, S. Y., and Varghese, A. J. (1968). *Science* 164, 183.
Khattak, M. N., and Wang, S. Y. (1968). *Science* **163,** 1341.
Kleinwächter, V. (1972). *Collect. Czech. Chem. Commun.* **37,** 1622.
Kleinwächter, V., and Koudelka, J. (1972). *Collect. Czech. Chem. Commun.* **37,** 3433.
Kleinwächter, V., Drobnik, J., and Augenstein, L. (1968). *Photochem. & Photobiol.* **7,** 485.
Kubota, Y. (1970). *Bull. Chem. Soc. Jap.* **43,** 3126.
Lamola, A. A. (1966). *J. Amer. Chem. Soc.* **88,** 813.
Lamola, A. A. (1968). *Photochem. & Photobiol.* **7,** 619.
Lamola, A. A., and Eisinger, J. (1968). *Proc. Nat. Acad. Sci. U.S.* **59,** 46.
Lamola, A. A., and Eisinger, J. (1971). *Biochim. Biophys. Acta* **240,** 313.
Lamola, A. A., Guéron, M., Yamane, T., Eisinger, J., and Shulman, R. G. (1967). *J. Chem. Phys.* **47,** 2210.
Levinson, N. (1962). *J. Soc. Ind. Appl. Math.* **10,** 442.
Longworth, J. W., and Battista, M. D. C. (1970). *Photochem. & Photobiol.* **11,** 207.
Mataga, N., and Murata, Y. (1969). *J. Amer. Chem. Soc.* **91,** 3144.
Montenay-Garestier, T. (1973). *J. Chim. Phys.* **70,** 1379.
Montenay-Garestier, T. (1975). Chemical Physics Monograph, Wiley-Interscience Ed. (in press).
Montenay-Garestier, T., and Hélène, C. (1968). *Nature (London)* **217,** 844.
Montenay-Garestier, T., and Hélène, C. (1970). *Biochemistry* **9,** 2865.
Montenay-Garestier, T., and Hélène, C. (1971). *Biochemistry* **10,** 300.
Montenay-Garestier, T., and Hélène, C. (1973a). *J. Agr. Food Chem.* **21,** 11.
Montenay-Garestier, T., and Hélène, C. (1973b), *J. Chim. Phys.* **70,** 1385.
Montenay-Garestier, T., and Hélène, C. (1973c). *J. Chim. Phys.* **70,** 1391.
Montenay-Garestier, T., Hélène, C., and Michelson, A. M. (1969). *Biochim. Biophys. Acta* **182,** 342.
Murphy, E. J. (1970). *J. Colloid Interface Sci.* **32,** 1.
Nag-Chaudhuri, J., and Augenstein, L. (1964). *Biopolym. Symp.* **1,** 441.
Nicolau, C. (1973). *In* "Physicochemical Properties of Nucleic Acids" (J. Duchesne, ed.), Vol. 1, p. 143. Academic Press, New York.
Nozik, A. J., and Kaplan, M. (1967). *J. Chem. Phys.* **47,** 2960.
Pincock, R. E. (1969). *Accounts Chem. Res.* **2,** 97.
Pincock, R. E., and Lin, W. S. (1973). *J. Agr. Food Chem.* **21,** 2.
Porter, G., and Wright, M. R. (1959). *Discuss. Faraday Soc.* **27,** 18.
Rahn, R. O. (1966). *Science* **154,** 503.
Rahn, R. O., and Hosszu, J. L. (1968). *Photochem. & Photobiol.* **8,** 53.
Rahn, R. O., and Hosszu, J. L. (1969a). *Photochem. & Photobiol.* **10,** 131.
Rahn, R. O., and Hosszu, J. L. (1969b). *Biochim. Biophys. Acta* **190,** 126.
Rhoades, D. F., and Wang, S. Y. (1970). *Biochemistry* **9,** 4416.
Rhoades, D. F., and Wang, S. Y. (1971a). *Biochemistry* **10,** 4603.
Rhoades, D. F., and Wang, S. Y. (1971b). *J. Amer. Chem. Soc.* **93,** 3779.
Riehl, N. (1965). *Trans. N. Y. Acad. Sci.* [2] **27,** 772.
Rosenthal, I., and Elad, D. (1968). *Biochem. Biophys. Res. Commun.* **32,** 599.

Ross, R. T. (1965). *J. Chem. Phys.* **42,** 3919.

Santus, R., Hélène, C., and Ptak, M. (1966). *C. R. Acad. Sci. Ser. D* **262,** 2077.

Santus, R., Montenay-Garestier, T., Hélène, C., and Aubailly, M. (1971). *J. Phys. Chem.* **75,** 3061.

Santus, R., Hélène, C., Ovadia, J., and Grossweiner, L. I. (1972). *Photochem. & Photobiol.* **16,** 65.

Setlow, J. K., and Setlow, R. B. (1963). *Nature (London)* **197,** 560.

Setlow, R. B. (1968). *Photochem. & Photobiol.* **7,** 643.

Setlow, R. B., and Carrier, W. L. (1966). *J. Mol. Biol.* **17,** 237.

Sevilla, M. D. (1971). *J. Phys. Chem.* **75,** 626.

Shifrin, S. (1968). *In* "Molecular Associations in Biology" (B. Pullman, ed.), p. 323. Academic Press, New York.

Slifkin, M. A. (1971). *In* "Charge-Transfer Interactions of Biomolecules," p. 108. Academic Press, New York.

Smaller, B., Avery, E. G., and Remko, J. R. (1965). *J. Chem. Phys.* **42,** 2608.

Smietanowska, A., and Shugar, D. (1961). *Bull. Acad. Pol. Sci., Cl. 2,* **9,** 375.

Smith, K. C. (1963). *Photochem. & Photobiol.* **2,** 503.

Smith, K. C. (1966). *Biochem. Biophys. Res. Commun.* **25,** 426.

Stewart, R. F. (1963). *Biochim. Biophys. Acta* **75,** 129.

Stewart, R. F., and Davidson, N. (1963). *J. Chem. Phys.* **39,** 255.

Stom, D. I. (1967). *Biofizika* **12,** 153.

Szent-Gyorgi, A. (1957). "Bioenergetics." Academic Press, New York.

Szent-Gyorgi, A. (1960). "Introduction to a Submolecular Biology," p. 77. Academic Press, New York.

Taborsky, G. (1970). *J. Biol. Chem.* **245,** 1063.

Toulmé, J. J., Charlier, M., and Hélène, C. (1974). *Proc. Nat. Acad. Sci.* USA **71,** 3185.

Van De Vorst, A., and Lion, Y. (1973a). *Biochim. Biophys. Acta* **294,** 349.

Van De Vorst, A., and Lion, Y. (1973b). *In* "Physicochemical Properties of Nucleic Acids" (J. Duchesne, ed.), Vol. 1, p. 268. Academic Press, New York.

Varghese, A. J. (1970a). *Biochemistry* **9,** 4781.

Varghese, A. J. (1970b). *Biochem. Biophys. Res. Commun.* **38,** 484.

Varghese, A. J. (1971a). *Biochemistry* **10,** 2194.

Varghese, A. J. (1971b). *Biochemistry* **10,** 4283.

Varghese, A. J. (1971c). *Photochem. & Photobiol.* **13,** 357.

Varghese, A. J. (1972). *Photochem. & Photobiol.* **15,** 113.

Varghese, A. J., and Patrick, M. H. (1969). *Nature (London)* **223,** 299.

Varghese, A. J., and Rupert, C. S. (1971). *Photochem. & Photobiol.* **13,** 365.

Varghese, A. J., and Wang, S. Y. (1967a). *Science* **156,** 955.

Varghese, A. J., and Wang, S. Y. (1967b). *Nature (London)* **213,** 909.

Varghese, A. J., and Wang, S. Y. (1968a). *Science* **160,** 186.

Varghese, A. J., and Wang, S. Y. (1968b). *Biochem. Biophys. Res. Commun.* **3,** 103.

Vigny, P. (1971). *C. R. Acad. Sci., Ser. D* **272,** 2247 and 3206.

Wacker, A., Dellweg, H. and Weinblum, D. (1960). *Naturwissenschaften* **47,** 477.

Wacker, A., Weinblum, D., Träger, L., and Mustafa, Z. H. (1961). *J. Mol. Biol.* **3,** 790.

Wacker, A., Dellweg, H., Träger, L., Kornhauser, A., Lodemann, E., Türck, G., Selzer, R., Chandra, P., and Ishimoto, M. (1964). *Photochem. & Photobiol.* **3,** 369.

Wang, S. Y. (1960). *Nature (London)* **188,** 844.

Wang, S. Y. (1961). *Nature (London)* **190,** 690.

Wang, S. Y. (1963). *Nature (London)* **200,** 879.

Wang, S. Y. (1964). *Photochem. & Photobiol.* **3,** 395.

Wang, S. Y. (1965). *Fed. Proc., Fed. Amer. Soc. Exp. Biol.* **24,** S-71.

Wang, S. Y. (1971). *J. Amer. Chem. Soc.* **93,** 2768.
Wang, S. Y., and Varghese, A. J. (1967). *Biochem. Biophys. Res. Commun.* **29,** 543.
Weinblum, D. (1967). *Biochem. Biophys. Res. Commun.* **27,** 384.
Weinblum, D., and Johns, H. E. (1966). *Biochim. Biophys. Acta* **114,** 450.
Whillans, D. W., and Johns, H. E. (1971). *J. Amer. Chem. Soc.* **93,** 1358.
Whillans, D. W., Herbert, M. A., Hunt, J. W., and Johns, H. E. (1969). *Biochem. Biophys. Res. Commun.* **36,** 912.
Workman, E. J., and Reynolds, S. E. (1950). *Phys. Rev.* **24,** 254.
Wulff, D. L., and Fraenkel, G. (1961). *Biochim. Biophys. Acta* **51,** 332.

# 9 Analysis by Mass Spectrometry of the Photoproducts of Nucleic Acid Bases

*Catherine Fenselau*



The high sensitivity of mass spectrometry makes it particularly attractive for analysis of the photoproducts of nucleic acid bases, which are often isolated in small quantities after considerable effort. The most important information which may be obtained from a mass spectrum is the molecular weight. If measured to several decimal places under high resolution, this molecular weight may be related to the empirical formula of the compound. Considerable information about the structure of the molecule may also be deduced from the manner in which the molecules decompose. This kind of analysis may be carried out on 10 ng to 10 $\mu$g of noncrystalline sample. In many cases molecular weights and considerable structural information may be obtained even from impure material (Fenselau, 1972, 1974).

Most researchers in this area have used mass spectrometry primarily to obtain molecular weights of their photoproducts. In some laboratories fragmentation has been analyzed to provide information about structural elements, such as the position of substituents, the nature of polymer linkages, and the saturation of ring double bonds.

This chapter is arranged in four sections. In the first, the manner in which mass spectrometry can provide molecular weights and elemental compositions for unknown photoproducts is discussed. The second section contains a survey of fragmentation patterns of various kinds of photoproducts, and the manner in which this behavior reflects structural elements. Section C deals with stable isotope analysis. The final segment is a short description of instrumentation.

## A. Molecular Weights and Elemental Compositions

Most photochemical investigations have been focused on alterations of the Pyr bases, and interpretation of the mass spectra of such photoproducts has necessarily been related to that of the spectra of Ura, Thy, and Cyt themselves. These spectra, as well as those of the Pur bases and corresponding nucleosides, have been well studied and reviewed (Rice *et al.*, 1965; Hecht *et al.*, 1970; McCloskey, 1974). The molecular ions of Ura, Thy, and Cyt are relatively stable, and the corresponding intense signals in the mass spectra (Figs. 1, 2, 3) allow easy assessment of the molecular weights of the compounds. Evidently, the molecular weights of Ura and Thy occur at even numbers (112 and 126) of atomic mass units (amu), while that of Cyt, which contains an odd number of nitrogen atoms, occurs at an odd number (111) of amu. A good rule of thumb in analysis of biochemicals is that an odd molecular weight reflects an odd number of nitrogen atoms.

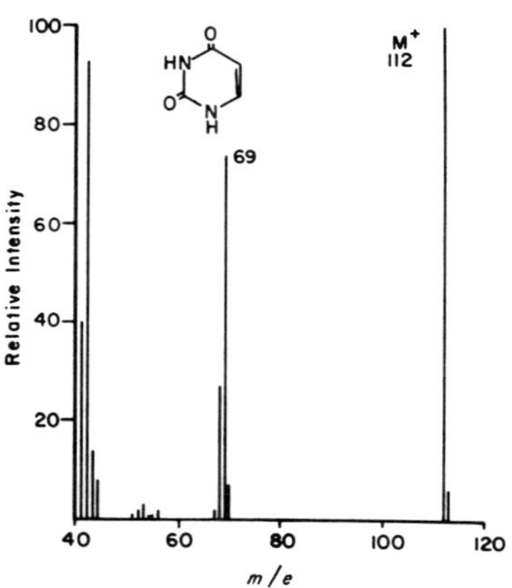


**Fig. 1.** *Mass spectrum of Ura.*

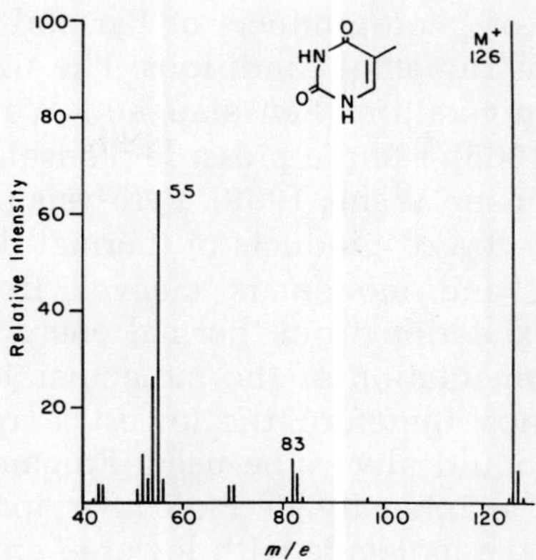


**Fig. 2.** *Mass spectrum of Thy.*

If the mass of the molecular ion is measured under higher resolution, it can be ascertained to several decimal places. Thus, the exact molecular weight of Thy, $C_5H_6N_2O_2$, is 126.0429 (amu). Such measurements are usually accurate to at least $\pm 0.003$ amu, and the elemental composition of the base being analyzed can be distinguished easily from that of compounds of the same nominal molecular weight but different elemental composition. For example, hThy, barbituric acid, and nonane have the same nominal molecular weight, 128, but they can be distinguished readily on the basis of their exact molecular weights.

| | | |
|---|---|---|
| nonane | $C_9H_{20}$ | 128.1565 |
| barbituric acid | $C_4H_4N_2O_3$ | 128.0222 |
| dihydrothymine | $C_5H_8N_2O_2$ | 128.0586 |

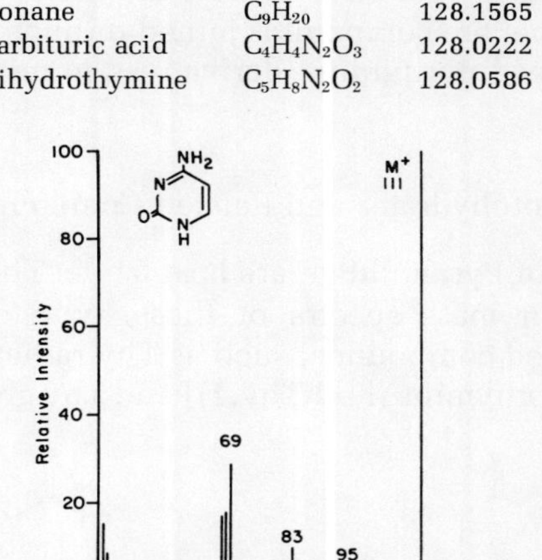


**Fig. 3.** *Mass spectrum of Cyt.*

The mass spectra of photoproducts of Pyr and Pur bases are unusually sensitive to instrumental conditions. The sizes of molecular ion peaks vary with temperature (Fenselau and Wang, 1969), ionizing energy (Rice *et al.*, 1965), sample pressure (Fenselau *et al.*, 1970), and instrument (Fenselau and Wang, 1969). Pyrolysis preceding ionization can lead to mass spectra of products of thermal degradation, such as dehydrated glycols, and monomers cleaved from Pyr◇Pyr. More subtly, additional excitation from thermal energy enhances electron impact-induced fragmentation of the molecular ions and further reduces their abundance; therefore, the lowest temperature compatible with vaporization should always be used. Fragmentation can be further reduced and the intensity of molecular ion peaks can be increased by ionizing the molecule with lower-energy electrons (Rice *et al.*, 1965) or by using alternate techniques for ionization such as chemical ionization (Fales, 1971) field ionization (Fenselau *et al.*, 1970) and field desorption (Schulten and Beckey, 1973).

Weak or nonexistent molecular ion ($M^+$) peaks, accompanied by more intense M + 1 peaks from photoproducts of the Pyr bases have been reported (Blackburn and Davies, 1966; Ulrich *et al.*, 1969; Iida and Hayatsu, 1970; Fenselau *et al.*, 1970; Jellinek, 1970). The common occurrence of M + 1 peaks in the spectra of amines is always a function of the pressure in the ionizing chamber. Increasing this pressure to enhance the M + 1 peak can be an acceptable approach to molecular weight determination, the culmination of which is chemical ionization at pressures around 1 mm Hg.

The photoproducts examined to date are mainly Pyr◇Pyr and Pyr adducts. The former are compounds joined through a cyclobutane ring and the latter are dipyrimidine derivatives joined through a single bond.

## 1. Pyrimidine Photohydrates and Related Compounds

Photohydrates of Pyr, ho$^6$hPyr, are heat labile. Therefore, it has been difficult to obtain mass spectra of these hydrates. However, their isomers and related compounds, such as Thy radiation products [e.g., **5**-hydroxydihydrothymine (ho$^5$hThy, **I**)] and Thy glycols [(ho)$_2$$^{5,6}$hThy,

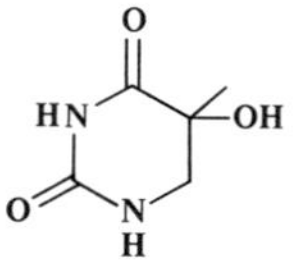

I

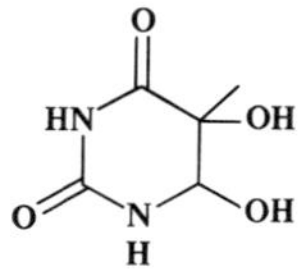

II

II], have been examined. The intensities of their molecular ion peaks are greatly reduced relative to those of Thy and Ura. Removal of the double bond does not of itself lead to reduced molecular ion abundance, since in the spectra of hUra and hThy the molecular ion peaks are the most intense (Rice *et al.*, 1965). Rather, introduction of hydroxyl groups seems to enhance fragmentation of the molecular ion.

In the spectrum (Fig. 4) of $ho^5$hThy, **I**, the intensity of the molecular ion peaks is sufficient to permit assignment of molecular weight. No molecular ion peak appears in the spectrum of Thy glycol, **II** (Fig. 5), although spectra can be measured at pressures sufficiently high to obtain an $(M + 1)^+$ peak (Ulrich *et al.*, 1969; Iida and Hayatsu, 1970).

Molecular weights of hPyr with hydroxyl groups can be obtained by analysis of their trimethylsilyl derivatives. Silyl derivatives have been widely used in nucleoside and nucleotide analyses to confer volatility and thermal stability (McCloskey, 1974). When compound **II** was converted to its tetra(trimethylsilyl) derivative, the molecular ion was observed at *m/e* 448, accompanied by an M − 15 peak at *m/e* 433 (Fig. 6) (Hahn and Wang, 1972). Such M − 15 peaks are usually formed in the fragmentation of trimethylsilyl derivatives, and a pair of ions 15 amu apart provides good evidence for the molecular ion of a silylated unknown. However, the number and the points of attachment of trimethylsilyl groups may be ambiguous in such an analysis.

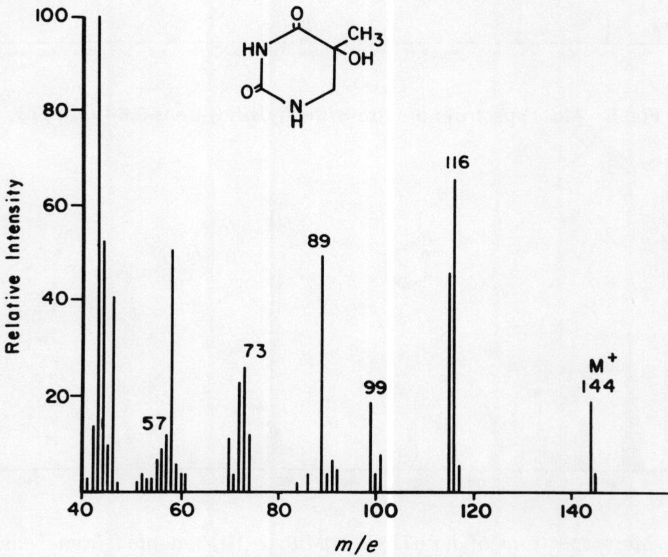


**Fig. 4.** *Mass spectrum of* $ho^5$*hThy* (**I**).

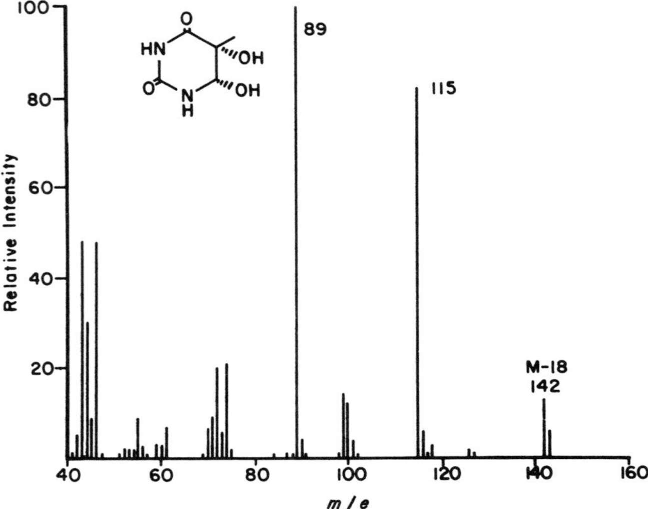


**Fig. 5.** *Mass spectrum of* cis-5,6-(ho)$_2$*hThy* (**II**).

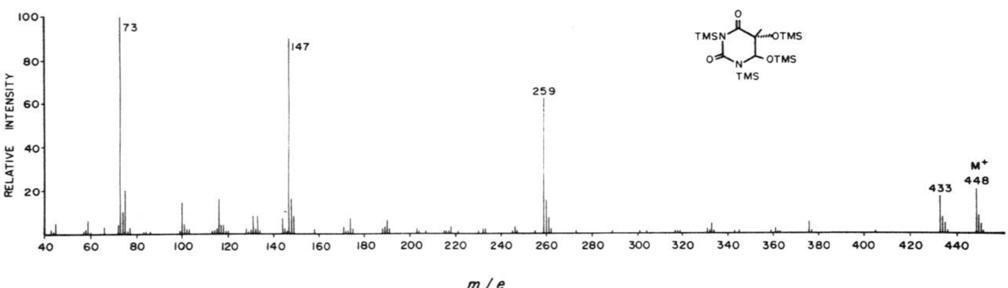


**Fig. 6.** *Mass spectrum of tetra(trimethylsilyl)*-trans-5,6-(ho$_2$)*hThy*.

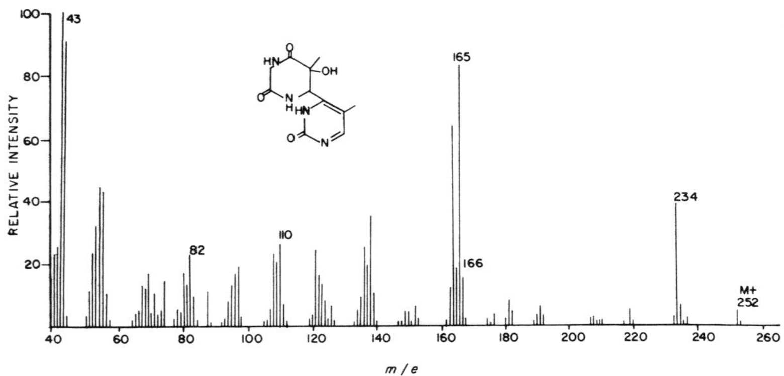


**Fig. 7.** *Mass spectrum of* ho$^5$hThy(6–4)M$^5$Pyo (**III**) *(adapted from Fenselau and Wang, 1969).*

## 2. Single-Bonded Dipyrimidines

In the presence of the ho⁵hThy moiety in the adduct, compounds **III**, **IV**, or **V**, for example, the molecular ions formed on electron impact were found in relative abundance between 0 and 10% (Fig. 7) (Fenselau *et al.*, 1970). Two approaches may alleviate this problem. Chemical derivatives may be prepared, which provide more abundant molecular ions, or alternate modes of ionization may be employed, in which less vibronic energy is imparted to the molecule.

**III** **IV** **V**

The field ionization spectrum (Fig. 8) of adduct **III** contains a molecular ion peak of ~20% relative intensity (Fenselau *et al.*, 1970). In the field ionization spectrum of trimer **IV** (Fig. 9), which contains both a labile cyclobutane linkage and a hydrated double bond, the molecular ion peak at *m/e* 378 is small, but is accompanied by a more intense $(M+1)^+$ peak with 8% relative intensity.

In another investigation, no molecular ion was observed in the electron impact spectrum of adduct **V** (Rhoades and Wang, 1970). However, the electron impact spectrum of the tetra(trimethylsilyl) deriva-

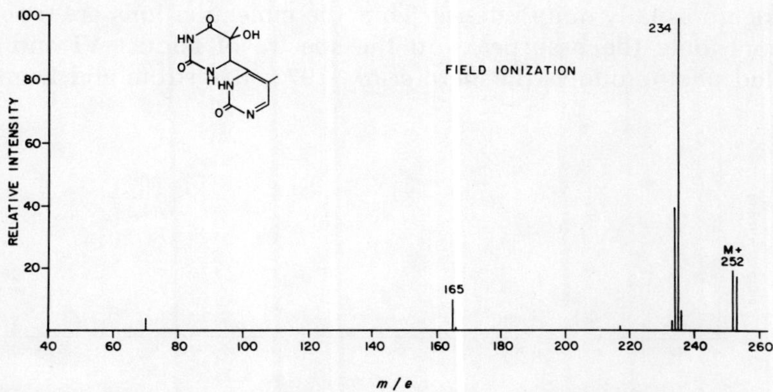


**Fig. 8.** *Field ionization mass spectrum of ho⁵hThy(6–4)M⁵Pyo* (**III**) *(adapted from Fenselau* et al.*, 1970).*

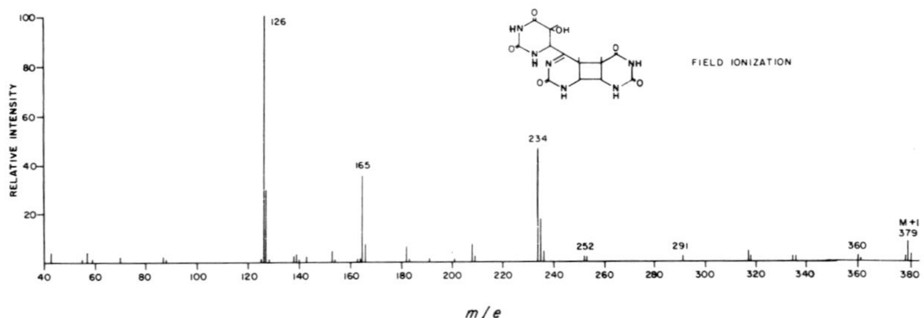


**Fig. 9.** *Field ionization mass spectrum of Thy trimer* (**IV**) *(adapted from Fenselau et al., 1970).*

tive of adduct **V** was obtained (Fig. 10). Here the molecular ion peak is ~20% as intense as the base peak. Moreover, the base peak corresponds to M − 15 ions, characteristically formed in the fragmentation of trimethylsilyl derivatives.

**VI**

**VII** R=H
**VIII** R=$CH_3$

When both Pyr rings in an adduct are unsaturated, molecular ion peaks are usually quite intense. Thus the molecular ions are reported to contribute the base peaks in the spectra of adduct **VI** and two related photoproducts (Leonard *et al.*, 1971; Bergstrom and Leonard,

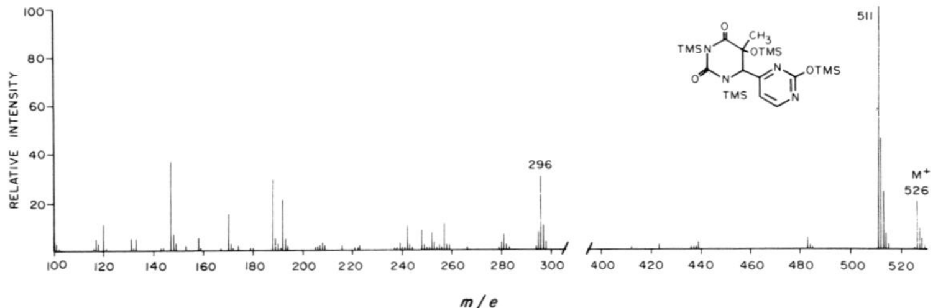


**Fig. 10.** *Mass spectrum of the tetra(trimethylsilyl) derivative of* $ho^5hThy(6-4)Pyo$ (**V**) *(adapted from Rhoades and Wang, 1970).*

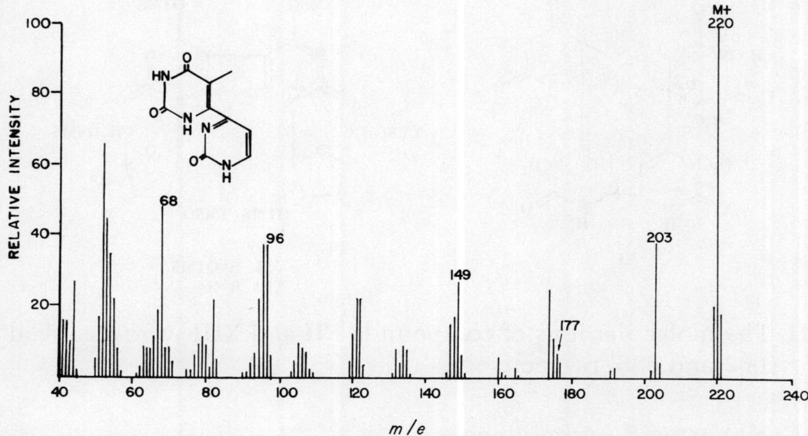


**Fig. 11.** *Mass spectrum of Thy(6–4)Pyo* (**VII**) *(adapted from Fenselau and Wang, 1969).*

1972), of adducts **VII** and **VIII,** and of a coupled product **IX** (Figs. 11 and 12) (Fenselau and Wang, 1969). The Cyt-derived adduct **VI** has also been analyzed as its tri(trimethylsilyl) derivative (Rhoades and Wang, 1971). The molecular ion peak of this derivative was observed at *m/e* 421, accompanied by a prominent M − 15 peak (Fig. 13).

IX

X

The dipyrimidine addition product **X** contains an unsaturated ring, as well as one saturated ring. The relative abundance of its molecular ion peak was found (Varghese, 1970) to be less than 10%, reflecting the instability of the saturated ring system (Fig. 14).

One Pyr tetramer formed by the photocyclization of two adducts of structure **VII** has been reported (Wang and Rhoades, 1971). The mass spectrum of the hexamethyl derivative **XI** was recorded. The molecular ion peak at *m/e* 524 is accompanied by M + 1 and M − 1 ions (Fig. 15), the intensities of which are pressure dependent.

Coupled photoproducts of Urd [Urd(5–5)Urd] and dUrd [dUrd(5–5)dUrd], linked in the same way as adduct **IX**, have been synthesized (S. Sasson and S. Y. Wang, unpublished data). They were characterized mass spectrometrically as the trimethylsilyl derivatives **XII** and

XI

XII R=OTMS
XIII R=H

**XIII.** The molecular ions of compounds **XII** and **XIII** were observed at *m/e* 1062 and 866, respectively.

### 3. Cyclobutane Bipyrimidines

The absence of unambiguous molecular ion peaks in the electron impact mass spectra of Pyr◇Pyr such as **XIV** has been widely reported

XIV

(Beukers and Berends, 1960; Fahr, 1969; Blackburn and Davies, 1966; Ulrich *et al.*, 1969; Jennings *et al.*, 1970; Kunieda and Witkop, 1971b). Careful analysis of spectra (Fenselau and Wang, 1969; Ulrich *et al.*, 1969) led to the suggestion that the molecular ions readily fragment to

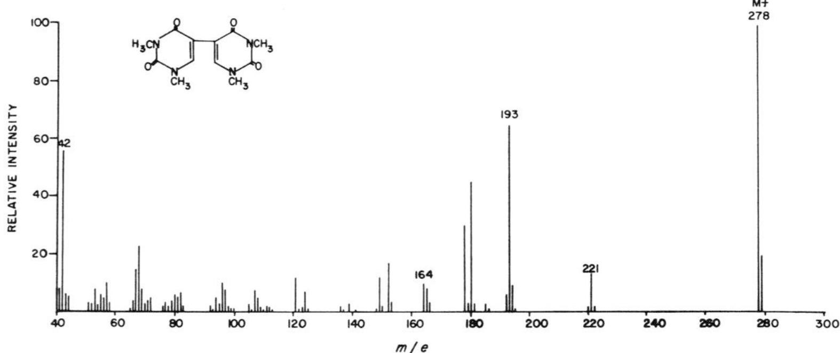


**Fig. 12.** *Mass spectrum of* Me₂Ura(5–5)Me₂Ura (**IX**) *(adapted from Fenselau and Wang, 1969).*

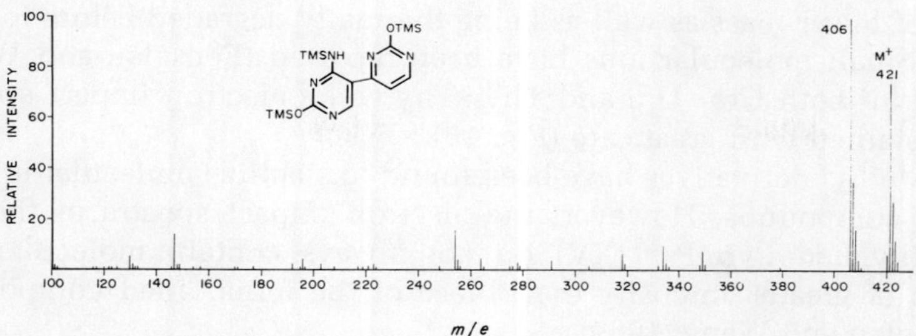


**Fig. 13.** *Mass spectrum of the tri(trimethylsilyl) derivative of Cyt(5-4)Pyo* (**VI**) (*adapted from Rhoades and Wang, 1971*).

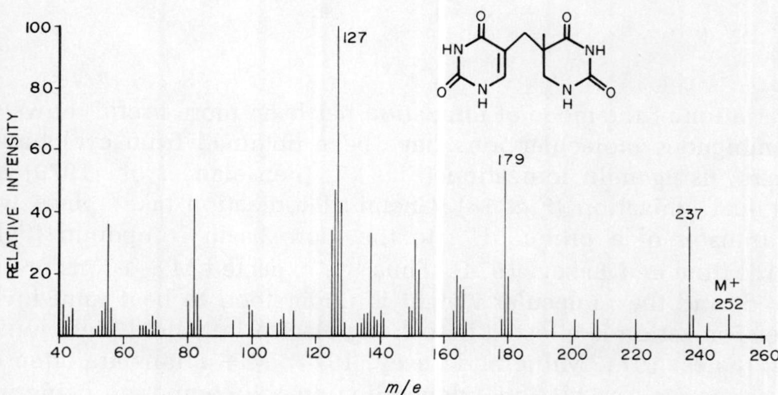


**Fig. 14.** *Mass spectrum of Thy(α-5)hThy* (**X**) (*adapted from Varghese, 1970*).

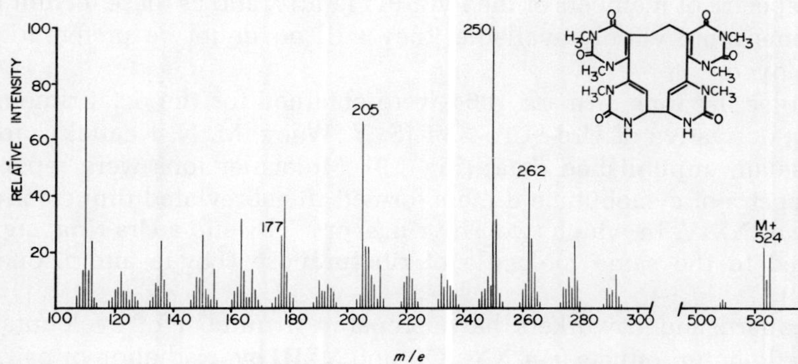


**Fig. 15.** *Mass spectrum of the hexamethyl tetramer* (**XI**).

ions of lower mass as well as being thermally degraded before ionization. Small molecular ions have been reported (Fenselau and Wang, 1969) for both Ura◇Ura and Thy◇Thy when electron impact spectra are obtained with great care (Fig. 16).

N-Methyl derivatives have been found to stabilize molecular ions in some compounds. However, the electron impact spectra of the tetramethylated Pyr◇Pyr (**XV**) do not always contain molecular ion peaks of greater intensity than those of the unmodified compounds (Fenselau and Wang, 1969).

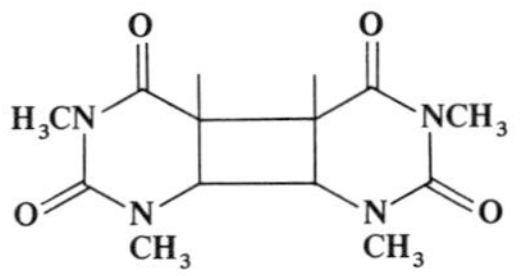

XV

Variation of the mode of ionization has been more useful, however. Unambiguous molecular ions have been obtained from cyclobutane dimers, using field ionization (Fig. 17) (Fenselau *et al.*, 1970) and chemical ionization (Fig. 18). Chemical ionization takes place with the transfer of a proton, $H^+$, to the more basic compound (Fales, 1971; Milne and Lacey, 1974). Thus the expected $M + 1$ ions are observed, and the molecular weight is understood to be 1 amu lower. When ionization is effected in a strong electrostatic field (field ionization) (Fales, 1971, Milne and Lacey, 1974), $M + 1$ ions are often observed, as well as $M^+$ ions, depending on the compound being analyzed and the conditions used. This was the case for the Thy◇Thy (*c,s*), the spectrum of which is shown in Fig. 17.

Chemical ionization and field ionization do seem to provide superior spectra of members of the Pyr◇Pyr family, and as these techniques become more widely available, they will no doubt be preferentially employed.

Molecular ions of mass 1064 were obtained for the octa(trimethylsilyl) derivative of Urd◇Urd **XVI** (S. Y. Wang, M. N. Khattak, and C. Fenselau, unpublished data) (Fig. 19). Molecular ions were reported in spectra of cyclobutane dimers formed in abbreviated dinucleotides such as **XXIV**, in which two Thy rings, or a Thy and a Ura ring, are attached to the same molecule of ribofuranose (Logue and Leonard, 1972).

Leonard and coworkers have prepared a number of cyclobutane-linked photoproducts (e.g. **XX**, **XXI** and **XXII**) by irradiation of pairs of thymine monomers bridged by polymethylene chains (Leonard *et al.*,

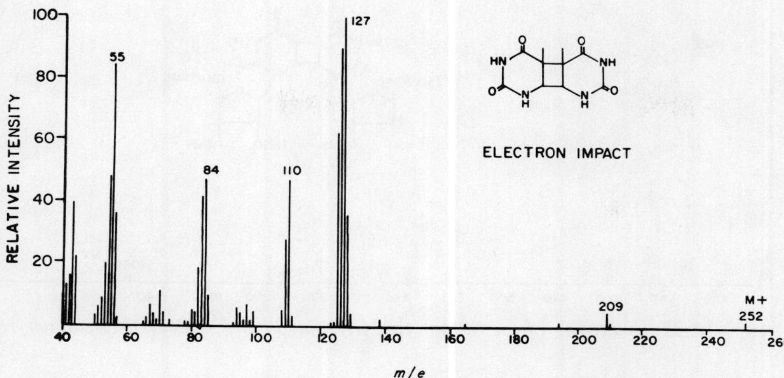


**Fig. 16.** *Mass spectrum of Thy◇Thy(c,s)* (**XIV**) *(adapted from Fenselau and Wang, 1969).*

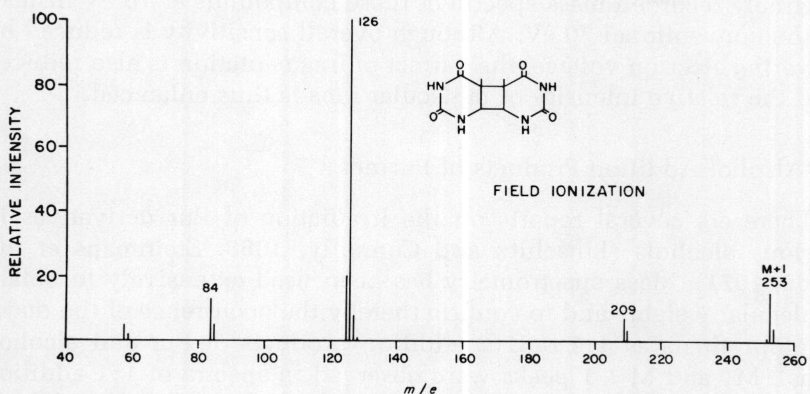


**Fig. 17.** *Field ionization mass spectrum of Thy◇Thy(c,s)* (**XIV**) *(adapted from Fenselau* et al., *1970).*

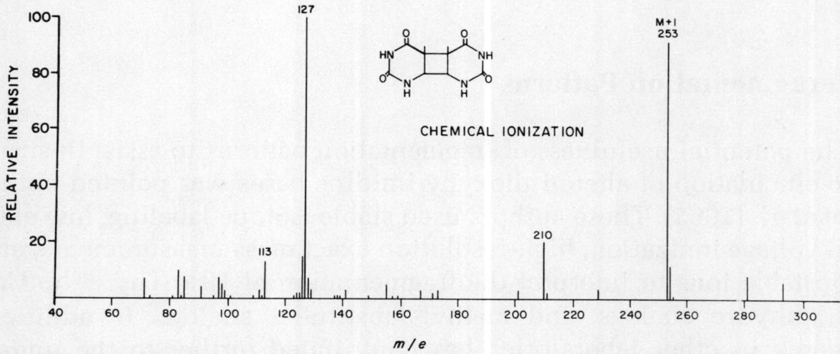


**Fig. 18.** *Chemical ionization mass spectrum of Thy◇Thy(c,s)* (**XIV**) *(H. Fales and S. Y. Wang, unpublished data).*

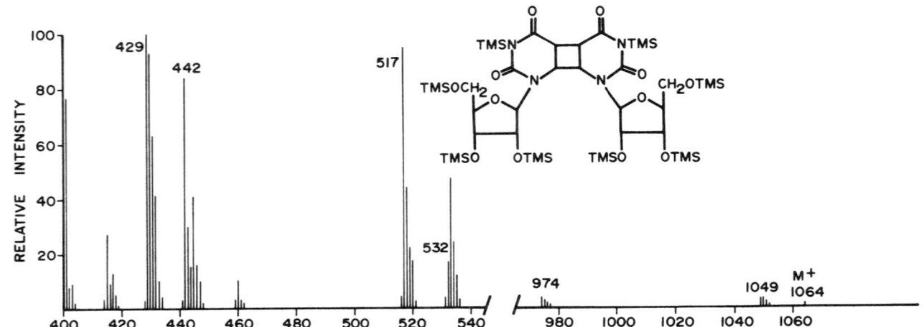


**Fig. 19.** *Mass spectrum of octa(trimethylsilyl)-Urd◇Urd* (**XVI**) (*high-mass range only*).

1969; Leonard *et al.*, 1973; Leonard and Cundall, 1974). They have routinely recorded mass spectra of these compounds at 10 eV instead of the conventional 70 eV. Although overall sensitivity is reduced by lowering electron voltage, the extract of fragmentation is also reduced and the relative intensity of molecular ions is thus enhanced.

### 4. Alcohol Addition Products of Purines

There are several reports on the irradiation of Pur derivatives in various alcohols (Linschitz and Connolly, 1968; Steinmaus *et al.*, 1969, 1971). Mass spectrometry has been used extensively to obtain molecular weights and to confirm thereby the occurrence of the major photoproducts as 1:1 or 1:2 addition products of Pur and alcohol. Small $M^+$ and $M + 1$ peaks were observed in spectra of 1:1 addition products of Ura and 2-propanol and of Ura and ethanol (Jellinek, 1970). Molecular ions were observed in spectra of Ado–ethanol and Ado–propanol photoaddition products (Steinmaus *et al.*, 1971).

## B. Fragmentation Patterns

The potential usefulness of fragmentation patterns to assist in structure elucidation of altered dioxypyrimidine bases was pointed out by Rice *et al.* (1965). These authors used stable isotope labeling, low electron voltage ionization, high-resolution exact mass measurements, and metastable ions to interpret the fragmentation of Ura, Thy, 5-hmUra, Cyt, dihydro analogs, and methyl-substituted analogs. In addition, research in other laboratories has contributed further to the understanding of the fragmentation of monocyclic derivatives of dioxypyrimidines (Nishiwaki, 1966; Ulrich *et al.*, 1969; Hecht *et al.*, 1969; Falch, 1970; Falch and Natvig, 1970).

### 1. Uracil, Thymine, and Cytosine

The outstanding structurally characteristic, primary decomposition of Ura and its analogs, including Thy, is shown (Scheme 1). The exclusive loss of CO from C(2) was confirmed by Ulrich *et al.* (1969) who reported the complete loss of $^{14}C$ synthetically placed at C(2). The exclusive loss of RN from N(3) was confirmed in the spectra of $Me^3Ura$ and $Me^1Ura$ (Falch, 1970). The loss of $CH_3NCO$, but not HNCO, is ob-

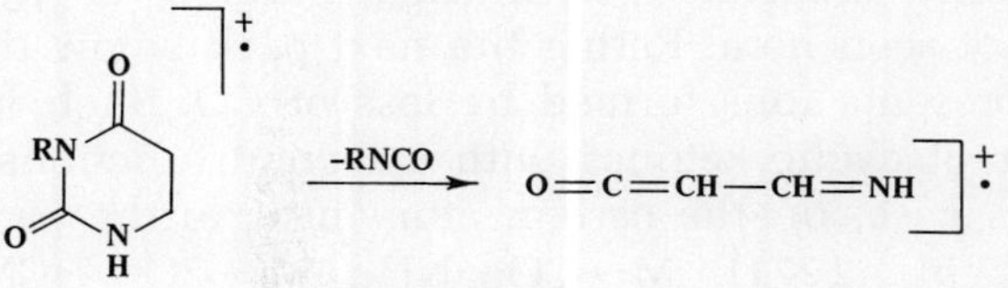

$$\left[\text{RN-pyrimidinedione}\right]^{+\bullet} \xrightarrow{-\text{RNCO}} \left[O{=}C{=}CH{-}CH{=}NH\right]^{+\bullet}$$

**Scheme 1**

served in the decomposition of $Me^3Ura$, while the reverse occurs in the fragmentation of $Me^1Ura$. This cleavage of the ring allows substituents at N(3) to be distinguished. Theoretical calculations (Diner *et al.*, 1966) support this as a favorable cleavage of Ura. Subsequent decomposition of the M − RNCO ions includes the elimination of CO.

Cyt and $Me^5Cyt$ decompose by elimination of CO, $NH_2$, or NCO from the molecular ion (Scheme 2, Fig. 3). Complex processes also lead to M − HNCO and M − $H_2NCO$ ions (Rice *et al.*, 1965).

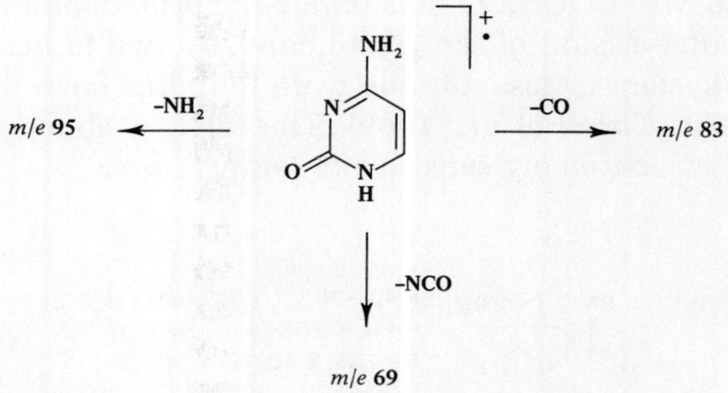

$$m/e\ 95 \xleftarrow{-\text{NH}_2} [\text{Cyt}]^{+\bullet} \xrightarrow{-\text{CO}} m/e\ 83$$
$$[\text{Cyt}]^{+\bullet} \xrightarrow{-\text{NCO}} m/e\ 69$$

**Scheme 2**

### 2. Hydroxylated Derivatives of Dihydrouracil and Dihydrothymine

The fragmentations of dioxypyrimidine derivatives with a saturated double bond usually do not include the loss of HNCO seen in unsaturated Pyr derivatives (Rice *et al.*, 1965; Ulrich *et al.*, 1969; Iida and Hayatsu, 1971). Rather, the fragmentation pattern of a given hPyr depends on the nature of its substituents. The spectrum of hThy has little in common with that of hUra except for the presence of intense molecular ion peaks.

The spectra and characterization of mono- and dihydroxylated derivatives of hUra and hThy are of particular interest in radiation biology. The spectrum of ho$^5$hThy is shown in Fig. 4. The elimination of $H_2O$, commonly associated with fragmentation of hydroxy compounds, is not seen here. Rather the next peak below the molecular ion peak represents ions formed by loss of CO. Such loss of CO is characteristic of cyclic ketones with a second functional group, hydroxy in this case, on the carbon atom next to the carbonyl group (Fenselau *et al.*, 1971). M—CO—$NH_3$, M—CO—HCN, and M—CO—$CH_3CO$ ions with masses of 99, 89, and 73, respectively, are also detected although it cannot be discerned if these losses occur in one step or two. Elimination of $CH_2NH$ leads to ions of mass 115. Contributions to the spectrum by fragments derived from M + H ions have been suggested (Ulrich *et al.*, 1969), although interpretation of the spectrum in Fig. 4 does not require this.

The spectra of *cis*- and *trans*-(ho)$_2^{5,6}$hThy (Fig. 5) are virtually indistinguishable. Small peaks at *m/e* 142 correspond to M − $H_2O$ ions, but these may result from pyrolysis. Elimination of CO is not seen, perhaps because cleavage of the glycol bond is more favorable. The dominant peaks in the spectrum correspond to M − HNCHOH ions (*m/e* 115) and M − $COCH_3CO$ ions (*m/e* 89). Formation of both species seems to involve fission of the glycol bond, known to occur easily in other glycol systems. Mass 46 ions were found to have the composition of $CH_4NO$ (Ulrich *et al.*, 1969). The relative abundances of the various ions are extremely sensitive to temperature.

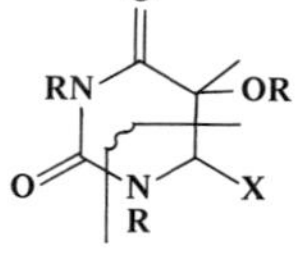

O, RN, OR, O, N, R, X

| | |
|---|---|
| R = X = H | *m/e* 115 |
| R = H, X = OH | *m/e* 115 |
| R = TMS, X = OTMS | *m/e* 259 |

derivative of I and II

Peaks in the spectra of the tetrasilylated derivatives of *trans*- and cis-Thy/glycol (Fig. 6) at *m/e* 259 may correspond to M—(TMS)NCHO(TMS) ions formed analogously to M—HNCHOH ions in the unmodified sample. Thus all three of the hydroxy-Thy derivatives seem to eliminate atoms from C(1) and C(6) on electron impact. On the basis of this analysis, attachment of the trimethylsilyl group is assigned at C(1) rather than at C(2), as is generally accepted in unsaturated pyrimidones (White *et al.*, 1972).

The peak at *m/e* 73 in the spectrum shown in Fig. 6 appears in the spectra of all trimethylsilyl derivatives and represents $(CH_3)_3Si^+$ ions. Ions of mass 147, $(CH_3)_3Si{-}\overset{+}{O}{=}Si(CH_3)_2$, are formed when more than one trimethylsiloxy group is present in the molecule.

### 3. Other Derivatives of Dihydrouracil and Dihydrothymine

Mass spectra of photoproducts formed between cysteine and Ura or Thy do not contain molecular ions (Smith and Aplin, 1966; Jellinek and Johns, 1970; Varghese, 1973a, 1973b), although fragments from both moieties are detected in the spectra. Smith and Aplin suggested that some pyrolysis may occur in the mass spectrometer. These compounds might be satisfactorily analyzed with a field desorption mass spectrometer, where the sample need not be vaporized prior to ionization (Milne and Lacey, 1974).

In the fragmentation of photoproducts of Ura and 2-propanol, and of Ura and ethanol, **XVII**, cleavage of the side chain leads to major peaks in the spectrum (Jellinek, 1970) representing both the hydroxy side chain and the heterocycle. Jellinek considers that many of the more abundant ions detected in his spectra arise from fragmentation of M + 1 ions.

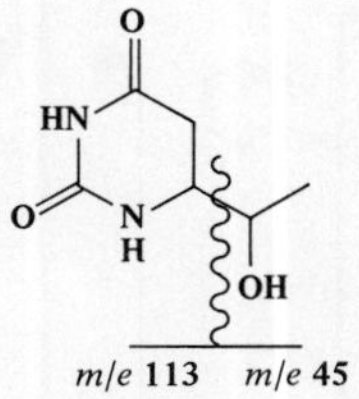

*m/e* 113 *m/e* 45

**XVII**

The mass spectra of some cyclopropyl analogs of Thy have been measured (Kunieda and Witkop, 1971a). The spectrum of 2,4-dimethyl-2,4-diazabicyclo[4.1.0]heptane-3,5-dione resembles that of $Me_2$Thy. Loss of $CH_3NCO$ is followed by elimination of CO in the decomposition of both compounds.

### 4. Single-Bonded Dipyrimidines

In the electron impact-induced decomposition of single-bonded dipyrimidines, fragmentation occurs in each ring which reflects the structure of that ring.

As has been pointed out, photoproducts **VII, VIII,** and **IX** may be viewed as substituted uracils. In their spectra the stepwise loss of RNCO and CO, the fragmentation sequence which is characteristic of the Ura family, can be seen. In the spectrum (Fig. 12) of the coupled product **IX**, the presence of two Ura rings can be deduced from the loss of two RNCO fragments from the molecular ion.

On the other hand, adduct **VII** is also a derivative of pyrimid-2-one, and the principal peaks in the spectrum of pyrimid-2-one (Fig. 20) are present in its spectrum. The peaks at *m/e* 96 and 68 correspond to ions formed by the fragmentations indicated (Scheme 3). The elemen-

**VII** $\xrightarrow{-e}$ *m/e* 96 $\xrightarrow{-CO}$ *m/e* 68

**Scheme 3**

tal compositions have been confirmed by high-resolution mass measurements (Fenselau and Wang, 1969). Adducts **III, V,** and **VIII** also

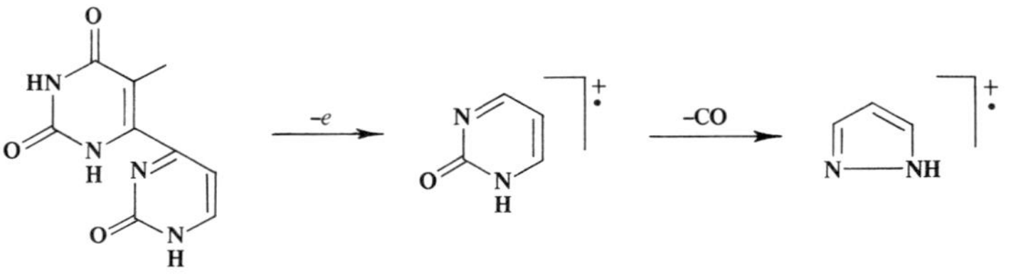

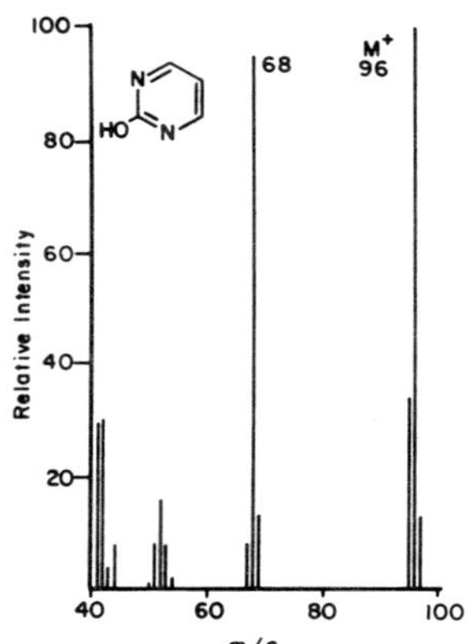


**Fig. 20.** *Mass spectrum of Pyo.*

contain a pyrimid-2-one or 5-methylpyrimid-2-one ring and fragment appropriately.

Loss of water from the substituted hThy ring of **III** leads to ions resembling **VIII,** and to similar secondary decomposition. Perhaps the most structurally significant fragmentation in compound **III** is the elimination of atoms from N(3), C(4), and C(5) in the hydroxylated ring. This occurs with transfer of a hydrogen atom in the electron impact decomposition (Fenselau and Wang, 1969) generating ions of mass 166 and the subsequent loss of a hydrogen atom leads to ions of mass 165.

The mass spectrum of the analogous Ura–Thy adduct **V** was found to contain no molecular ion and to resemble that of **VII.** Therefore, its tetra(trimethylsilyl) derivative was analyzed (Fig. 10) (Rhoades and Wang, 1970). In this compound the loss of HO(TMS) is observed, analogous to the loss of HOH from **III,** and the occurrence of a prominent

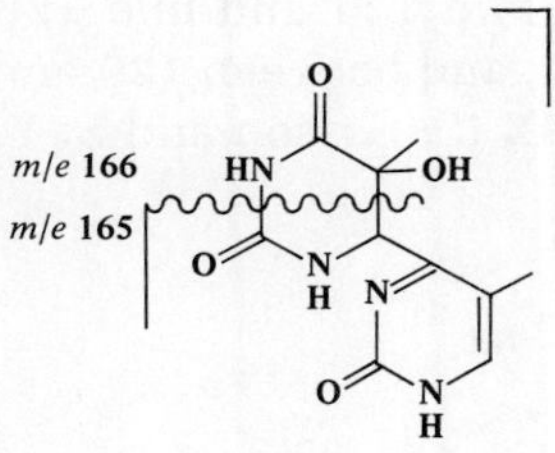

III

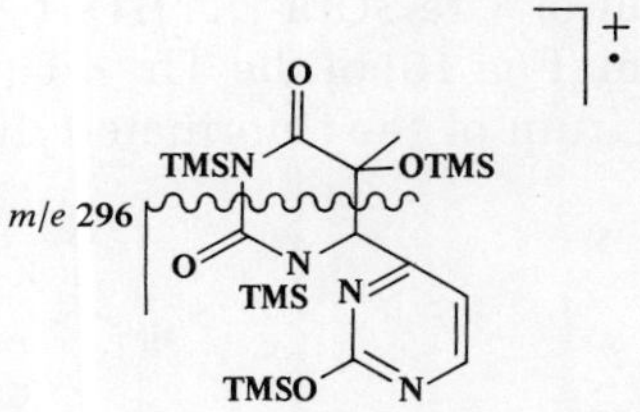

derivative of V

peak at *m/e* 296 is consistent with the elimination of atoms from N(3), C(4), and C(5) of the hydroxylated ring, accompanied by hydrogen transfer. The major fragment peak, however, occurs at M − 15, as is common among trimethylsilyl derivatives.

The spectrum of the tri(trimethylsilyl) derivative of adduct **VI** contains no fragment peaks of any intensity, apart from that at M − 15. The unusual stability of the molecular ion is consistent with the aromatic structure drawn on the spectrum (Fig. 13).

The bipyrimidine addition products, Thy($\alpha$–5)hThy, **X**, (Varghese, 1970) contains both a Thy and a hThy ring. Intense peaks (Fig. 14) at *m/e* 126 and 127 correspond to the Thy moiety and to its protonated homolog. Varghese suggested that the abundant ion of mass 179 contains the elements shown in Scheme 4.

## 5. Cyclobutane Bipyrimidines

The cyclobutane dimers produce similar electron impact spectra, regardless of head and tail isomerization or *cis* and *trans*

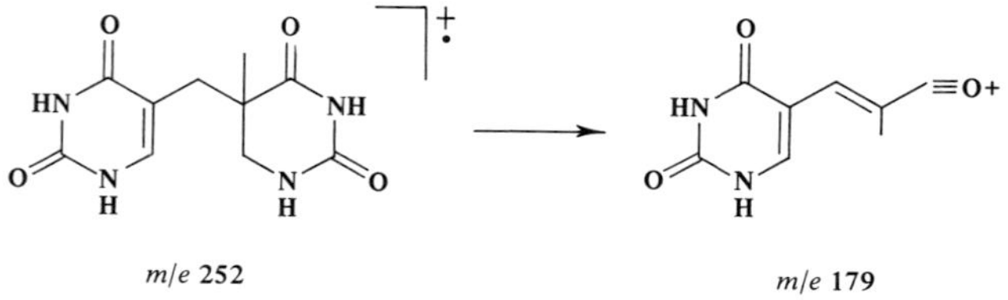

**Scheme 4**

stereochemistry (Fenselau and Wang, 1969; Jennings *et al.*, 1970). Cleavage across the cyclobutane ring of the dimer generates abundant ions, the mass and subsequent fragmentation of which resemble those of the ionized monomer. Electron impact-induced scission of the cyclobutane ring also occurs with transfer of one or two protons. Thus a cluster of intense peaks occurs between *m/e* 112 and *m/e* 114 in the spectrum of Ura◇Ura (**XVIII**), between *m/e* 125 and *m/e* 128 in the spectrum (Fig. 16) of the Thy◇Thy (**XIV**), and between 129 and 131 in the spectrum of the fluorinated dimer **XIX** (S. Sasson and S. Y. Wang,

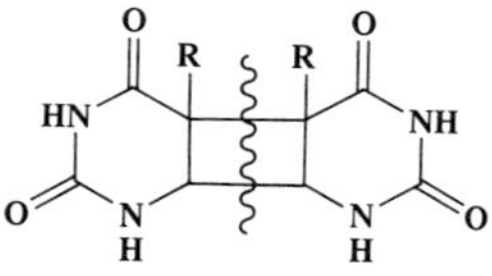

unpublished data). In the field ionization spectrum (Fig. 17) of dimer **XIV,** the base peak at *m/e* 126 is accompanied only by small peaks at *m/e* 125 and *m/e* 127. In the chemical ionization spectrum (Fig. 18) the base peak occurs at *m/e* 127, along with a small signal at *m/e* 126. The cleavage leading to this group of ions characterizes the cyclobutane linkage and can provide information about the distribution of substituents between the two rings.

Subsequent decomposition of the regenerated monomer ions, e.g., mass 126 ions in the spectrum of Thy◇Thy, resembles that of the monomer (Fig. 2). Protonated regenerated Thy ions of mass 127 lose $H_2O$ and OH, leading to peaks at *m/e* 109 and *m/e* 110 (Jennings *et al.*, 1970) in the dimer spectra which are absent in the spectrum of Thy.

The spectrum of compound **XX,** Thy⟨1$(CH_2)_3$1⟩Thy(*c,s*), contains a base peak at *m/e* 166 (Leonard *et al.*, 1969) (Fig. 21). This peak corresponds to M − 126 ions, the formation of which requires cleavage of

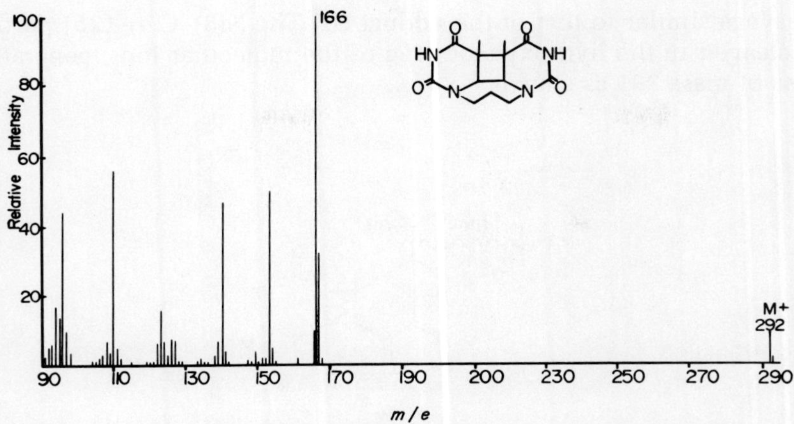


**Fig. 21.** *Partial mass spectrum of Thy⟨1(CH$_2$)$_3$1⟩Thy(c,s)* (**XIX**) (*adapted from Leonard* et al., *1969*).

the cyclobutane ring and elimination of a Thy ring with a hydrogen atom from the trimethylene bridge. The spectrum of the photoproduct is similar to that of the starting material and generally contains fragmentation characteristic of 1-alkyluracils (Falch, 1970). This same loss of 126 amu generates the base peaks in the spectra of Thy⟨1(CH$_2$)$_2$1⟩Thy and Thy⟨1(CH$_2$)$_4$1⟩Thy, photoproducts **XXI** and

HN NH O N N O (CH$_2$)$_n$

XX $n = 3$
XXI $n = 2$
XXII $n = 4$

**XXII** (Leonard *et al.*, 1973, and in the spectra of Thy⟨3(CH$_2$)$_3$3⟩Thy, Thy⟨1(CH$_2$)$_3$1⟩Thy and Thy [1(CH$_2$)$_3$5]Ura (Leonard and Cundall, 1974).

### 6. Higher Polymers

The mass spectra of at least one phototrimer and one phototetramer have been reported. In the field ionization fragmentation (Fenselau *et al.*, 1970) of Thy trimer **IV** the cyclobutane ring is cleaved, producing ions of mass 126 and 252 (Fig. 9). The latter ions undergo secondary cleavage similar to that of the adduct **III.** The N(3)–C(4)–C(5) portion is cleaved in the hydroxylated ring of the molecular ions, generating ions of mass 291 as shown.

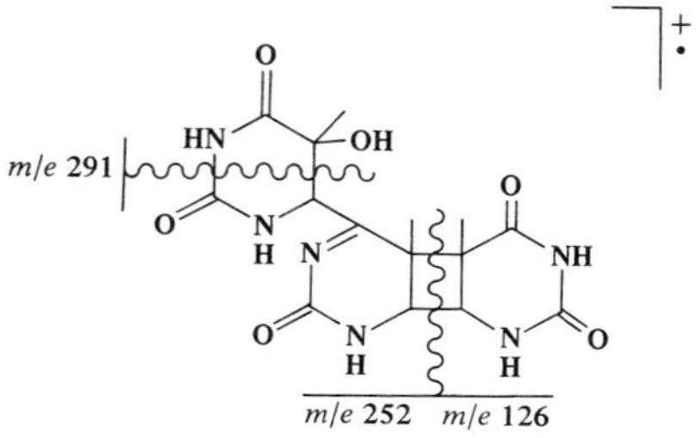
m/e 291
HN
OH
O
NH
N
H
m/e 252
m/e 126

IV

Tetramer **XI** was analyzed as its hexamethyl derivative. Bisection of the tetramer, as shown, leads to daughter ions resembling the adduct **VII,** from which loss of $CH_3NCO$ is followed by elimination of CO (Fig. 15). The abundant ions of mass 250 may comprise the structural elements shown.

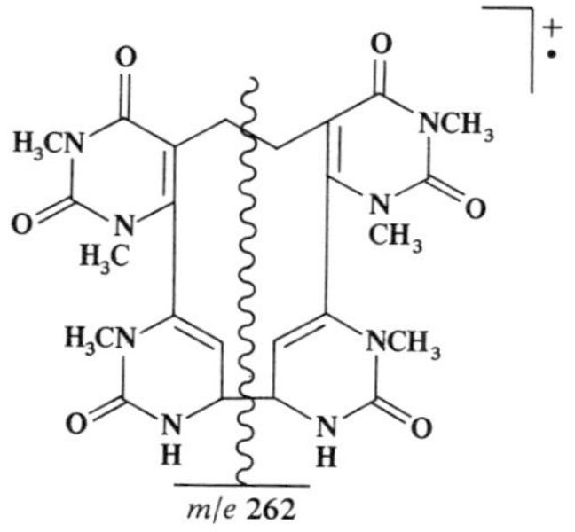
H3CN
NCH3
CH3
H3C
m/e 262

XI

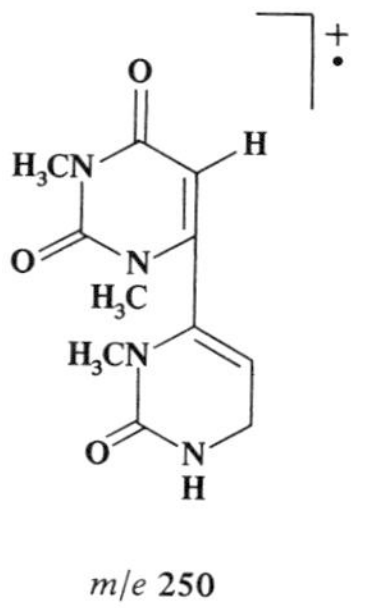
H3CN
H3C
H
m/e 250

7. Nucleosides

The fragmentation of 8-ho$^{\alpha}$EtAdo (**XXIII**) and 8-ho$^{\alpha}$Me$^{\alpha}$EtAdo, photoproducts of Ado and ethanol and 2-propanol, respectively, is typical of nucleosides (Steinmaus *et al.,* 1971). Analogous to the fragmentation of Ado itself (Shaw *et al.,* 1970), major ions are formed by cleavage of the bond between the base and sugar moieties (with H transfer to the base) and by cleavage across the furanoside ring as shown in compound **XXIII.**

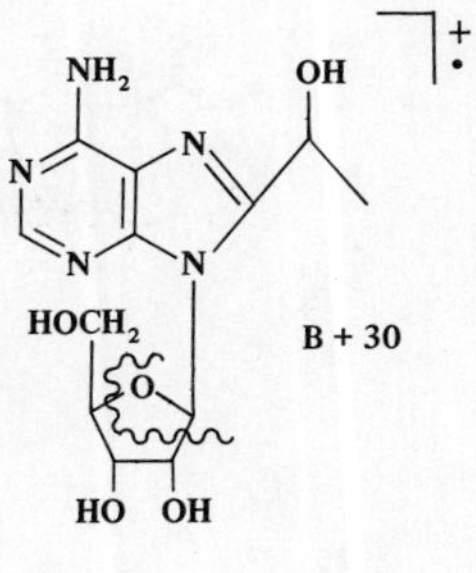

XXIII

The cyclobutane linkage in photodimer **XXIV** cleaves on electron impact, generating ions the subsequent fragmentation of which is similar to that of the compound with monomeric moieties before dimerization (Logue and Leonard, 1972). As shown in Fig. 22, classical nucleoside ions (McCloskey, 1974) resembling ionized Ura, protonated Ura, Thy, and protonated Thy are formed. The compositions of other major ions, of masses 225 and 139, are indicated. The base peak in the spectrum at *m/e* 81 corresponds to ions characteristic of the dRib moiety.

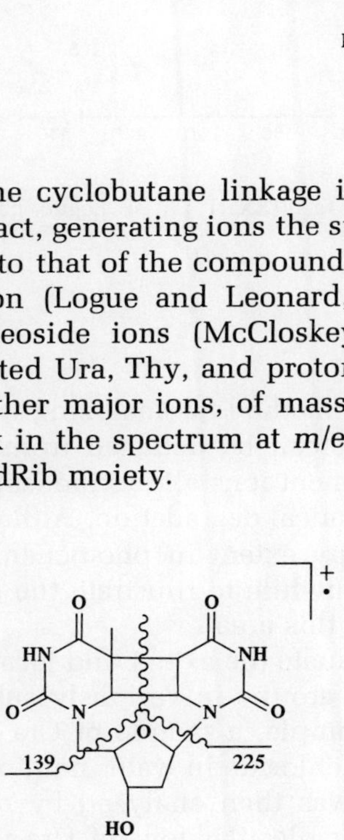

XXIV

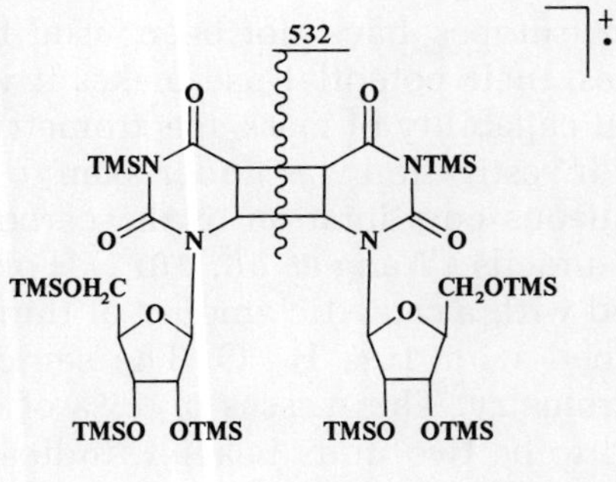

XVI

The spectrum (Fig. 19) of the silylated Urd◇Urd (**XVI**) contains the $M - CH_3$ and $M - HOSi(CH_3)_3$ peaks which are usually seen in the spectra of silylated alcohols. Bisection of the molecule through the cyclobutane ring generates a prominent peak at *m/e* 532. Below this peak the spectrum is similar to that of tetra(trimethylsilyl)-Urd (McCloskey, 1974) although relative intensities of the peaks vary.

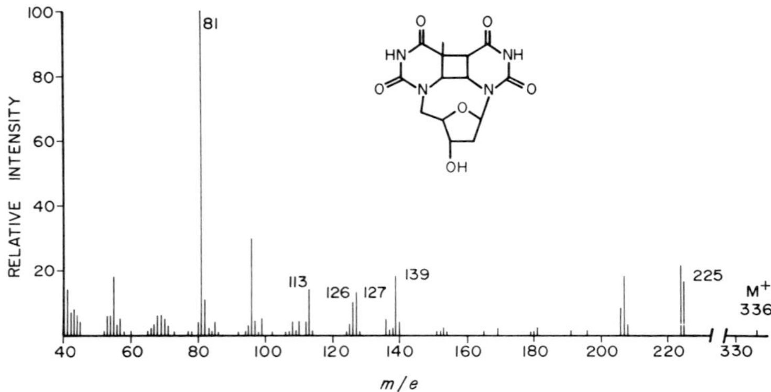


**Fig. 22.** *Mass spectrum of Thy⟨1(5′dRib1′)1⟩Ura* (c,s) (**XXIII**) (*M. W. Logue, private communication, 1972*).

## C. Isotope Studies

Mass spectrometry is the most sensitive method for analyzing stable isotopes. The amount of isotope present can be deduced from the mass spectrum, and often analysis of fragment ions allows the isotope to be located in the molecule without chemical degradation. Although stable isotopes have not been used to any extent in photochemical studies, their potential use makes it worthwhile to illustrate the analytical capability of mass spectrometry in this area.

An investigation was undertaken to evaluate the extent and location of aqueous equilibration of the carbonyl groups in variously substituted uracils (Wang *et al.*, 1972). For example, a sample of Ura was heated with a catalytic amount of thionyl chloride in water nominally enriched with 10% $H_2{}^{18}O$. The sample was then analyzed by mass spectrometry. The masses of 13% of the molecular ions of Ura were found to be two units heavier, indicating the presence of an atom of $^{18}O$. Peaks corresponding to the $M - CH_3NCO$ ions were also analyzed, and 11% of these fragment ions were found to contain $^{18}O$. Thus, most of the isotope label was not lost when the carbonyl group at C(2) was eliminated (Scheme 5). The authors concluded that the carbonyl group at C(4) is readily equilibrated under these conditions, while that at C(2) is less reactive.

Mass spectral measurements have also been used by Blackburn and Davies (1967) to confirm the lack of deuterium incorporation in formation of some Thy photoproducts.

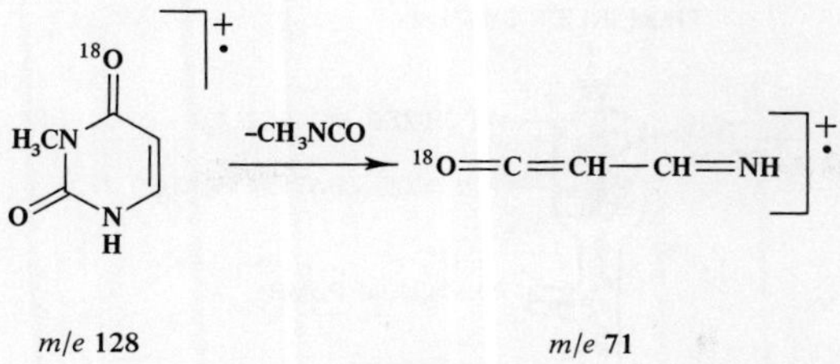

$$\left[\text{3-methyl-[4-}^{18}\text{O]uracil}\right]^{+\cdot} \xrightarrow{-CH_3NCO} \left[{}^{18}O{=}C{=}CH{-}CH{=}NH\right]^{+\cdot}$$

*m/e* 128 *m/e* 71

**Scheme 5**

## D. Mass Spectrometry

The mass spectrometer may be considered as consisting of four parts: an inlet system, ionization chamber, analyzer, and detector/recorder. These elements can be varied greatly in different instruments, according to the needs of the investigator. In the instruments used to obtain most of the spectra discussed in this chapter, the samples must first be vaporized in the inlet system. Gas chromatographic inlet systems have been popular for mass spectral analysis in many areas of biochemistry (Fenselau, 1974). However, these have had little use in photochemical studies of nucleic acid components because of the involatility of the samples. These samples are usually introduced on the end of a probe which puts them within a millimeter of the beam of ionizing electrons. Although the vaporization and analysis are carried out under a vacuum of at least $10^{-6}$ mm Hg, heating is usually required for vaporization. This is the origin of the difficulties with pyrolysis and thermally induced fragmentation. Under optimal conditions, molecules as heavy as 1500 amu can be analyzed.

These vaporized molecules are then ionized, usually by collision with electrons, to produce positively charged ions. Excess energy is transferred in the ionizing collision, and as a result, the molecular

$$e + M \longrightarrow 2e^- + M^+$$

ions undergo a variety of unimolecular decompositions. Each fragmentation of the molecule leads to one neutral piece and one fragment ion.

A whole array of ions is produced in less than $10^{-6}$ seconds, accelerated through an electrostatic potential, and deflected in a magnetic field. At given electrostatic and magnetic field strengths, ions of different masses are separated by deflection through different angles. This spectrum of masses may be brought into a fixed collector (Fig.

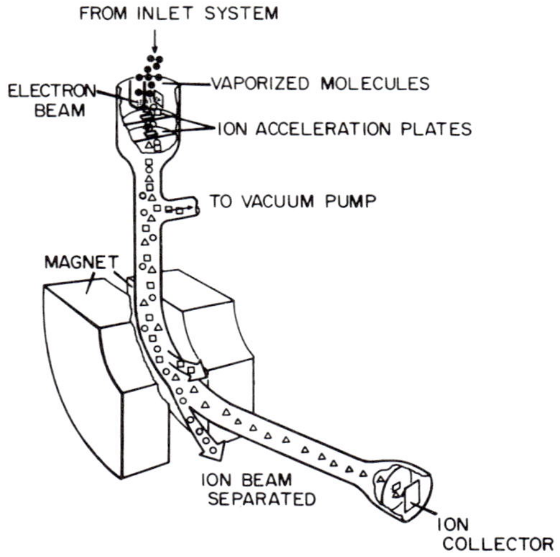


**Fig. 23.** *Schematic representation of a single-focusing mass spectrometer* (Fenselau, 1972).

23) by "scanning" the magnetic or electrostatic field. This relationship is expressed by the formula $m/e = H^2r^2/2V$, in which $m$ = mass, $e$ = charge, $H$ is the magnetic field, $r$ is the radius of deflection, and $V$ is the acceleration voltage. Other analyzers, which work on different physical principles, are also available and have been described in detail (Biemann, 1962; Roboz, 1968).

The most commonly used detectors are multistage electron multipliers. The low-resolution spectra are usually traced by rapidly responding oscillographic recorders. Accurate mass measurements, made under high resolution, are obtained by either computerized or manual comparison to reference ions of known masses.

## Acknowledgment

I wish to thank Dr. Allan H. Fenselau for expert advice and Appleton-Century-Crofts for permission to use Figure 23.

## References

Bergström, D. E., and Leonard, N. J. (1972). *Biochemistry* **11,** 1.
Beukers, R., and Berends, W. (1960). *Biochim. Biophys. Acta* **41,** 550.
Biemann, K. (1962). "Mass Spectrometry." McGraw-Hill, New York.

Blackburn, G. M., and Davies, R. J. (1966). *J. Chem. Soc., C* p. 2239.
Blackburn, G. M., and Davies, R. J. (1967). *J. Amer. Chem. Soc.* **89,** 5941.
Diner, S., Del Re, G., and Pullman, B. (1966). *C. R. Acad. Sci., Ser. C* **262,** 826.
Fahr, E. (1969). *Angew Chem., Int. Ed. Engl.* **8,** 578.
Falch, E. (1970). *Acta Chem. Scand.* **24,** 137.
Falch, E., and Natvig, T. (1970). *Acta Chem. Scand.* **24,** 1423.
Fales, H. M. (1971). *In* "Mass Spectrometry" (W. D. G. Milne, ed.), p. 179–215. New York.
Fenselau, C. (1972). *In* "Methods in Pharmacology," (C. F. Chignell, ed.), Vol. II, p. 401–442. Appleton, New York. (Figure used by permission of Appleton-Century-Crofts, Educational Division, Meredith Corporation.)
Fenselau, C. (1974). *Appl. Spectrosc.* **28** (305).
Fenselau, C., and Wang, S. Y. (1969). *Tetrahedron* **25,** 2853.
Fenselau, C., Wang, S. Y., and Brown, P. (1970). *Tetrahedron* **26,** 5923.
Fenselau, C., Chandler, H. A., Pyles, M., Sander, D. L., and Robinson, C. H. (1971). *Org. Mass Spec.* **5,** 697.
Hahn, B., and Wang, S. Y. (1972). *J. Amer. Chem. Soc.* **94,** 4764.
Hecht, S. M., Gupta, A. S., and Leonard, N. J. (1969). *Biochim. Biophys. Acta* **182,** 444.
Hecht, S. M., Gupta, A. S., and Leonard, N. J. (1970). *Anal. Biochem.* **38,** 230.
Iida, S., and Hayatsu, H. (1970). *Biochim. Biophys. Acta* **211,** 1.
Iida, S., and Hayatsu, H. (1971). *Biochim. Biophys. Acta* **228,** 1.
Jellinek, T. (1970). Ph.D. Thesis, University of Melbourne, Melbourne, Australia.
Jellinek, T., and Johns, R. B. (1970). *Photochem. & Photobiol.* **11,** 349.
Jennings, B. H., Pastra, S., and Wellington, J. L. (1970). *Photochem. & Photobiol.* **11,** 215.
Kunieda, T., and Witkop, B. (1971a). *J. Amer. Chem. Soc.* **93,** 3478.
Kunieda, T., and Witkop, B. (1971b). *J. Amer. Chem. Soc.* **93,** 3493.
Leonard, N. J., Golankiewicz, K., McCredie, R. S., Johnson, S. M., and Paul, I. C. (1969). *J. Amer. Chem. Soc.* **91,** 5855.
Leonard, N. J., and Bergström, D. E., and Tolman, G. L. (1971). *Biochem. Biophys. Res. Commun.* **44,** 1524.
Leonard, N. J., McCredie, R. S., Logue, M. W., and Cundall, R. L. (1973). *J. Amer. Chem. Soc.* **95,** 2320.
Leonard, N. J., and Cundall, R. L. (1974). *J. Amer. Chem. Soc.* **96,** 5904.
Linschitz, H., and Connolly, J. S. (1968). *J. Amer. Chem. Soc.* **90,** 2979.
Logue, M. W., and Leonard, N. J. (1972). *J. Amer. Chem. Soc.* **94,** 2842.
McCloskey, J. A. (1974). *In* "Basic Principles in Nucleic Acid Chemistry" (P. O. P. Ts'o, ed.). Academic Press, New York, 209–309.
Milne, G. W. A., and Lacey, M. J. (1974). CRC Critical Reviews in Analytical Chemistry. 45–104.
Nishiwaki, T. (1966). *Tetrahedron* **22,** 3117.
Rhoades, D. F., and Wang, S. Y. (1970). *Biochemistry* **9,** 4416.
Rhoades, D. F., and Wang, S. Y. (1971). *Biochemistry* **10,** 4603.
Rice, J. M., Dudek, G. O., and Barber, M. (1965). *J. Amer. Chem. Soc.* **87,** 4569.
Roboz, J. (1968). "Introduction to Mass Spectrometry." Wiley, New York.
Schulten, H.-R., and Beckey, H. D. (1973). *Organic Mass Spectrometry,* **7,** 861.
Shaw, S. J., Desiderio, D. M., Tsuboyama, K., and McCloskey, J. A. (1970). *J. Amer. Chem. Soc.* **92,** 2510.
Smith, K. C., and Aplin, R. T. (1966). *Biochemistry* **5,** 2125.
Steinmaus, H., Rosenthal, I., and Elad, D. (1969). *J. Amer. Chem. Soc.* **91,** 4921.
Steinmaus, H., Rosenthal, I., and Elad, D. (1971). *J. Org. Chem.* **36,** 3594.

Ulrich, J., Teoule, R., Massot, R., and Cornu, A. (1969). *Org. Mass Spectrom.* **2,** 1183.
Varghese, A. J. (1970). *Biochem. Biophys. Res. Commun.* **38,** 484.
Varghese, A. J. (1973a). *Biochemistry* **12,** 2725.
Varghese, A. J. (1973b). *Biochem. Biophys. Res. Commun.* **51,** 858.
Wang, S. Y., and Rhoades, D. F. (1971). *J. Amer. Chem. Soc.* **93,** 2554.
Wang, S. Y., Hahn, B. S., Fenselau, C., and Zafiriou, O. C. (1972). *Biochem. Biophys. Res. Commun.* **48,** 1630.
White, E. V., Krueger, V. P. M., and McCloskey, J. A. (1972). *J. Org. Chem.* **37,** 430.

# 10 Nuclear Magnetic Resonance of Photoproducts

D. P. Hollis



## A. Introduction

During the past ten years advances in nuclear magnetic resonance (NMR) theory and instrumentation, in sensitivity and in resolution, have made NMR much more feasible for biochemical studies involving small amounts of material. In this chapter the contributions of NMR to the determination of the structure of various photoproducts of Pur and Pyr bases and derivatives will be discussed. Since all NMR studies of photoproducts of nucleic acids have dealt only with proton NMR this chapter will only briefly mention the applications of other magnetically active nuclei. A brief account of basic NMR theory will be given, the NMR spectra of the parent bases will be discussed, and the structure determination on photoproducts will be reviewed.

## B. Nuclear Magnetic Resonance

The NMR phenomenon owes its existence to the fact that atomic nuclei may possess spin angular momenta and magnetic moments in addition to their familiar properties of mass and charge. A sufficient insight into these properties can be obtained in terms of the nuclear mass and charge by means of a simple model. Assume that both the mass ($M$) and the charge ($e$) of a nucleus, the proton for example, are uniformly distributed over a thin spherical shell and that the shell is spinning with a constant angular velocity about an axis through its center. The spinning of mass produces an angular momentum (**P**), which is a vector pointing along the axis of rotation.* The spinning of the charge produces an electric current circulating about the rotational axis and this in turn generates a magnetic field which is symmetrical about the rotational axis. A magnetic field can be described conveniently by its magnetic dipole moment ($\boldsymbol{\mu}$) which is the product of the magnetic pole strength and the distance between the two poles necessary to produce the field. Since the spinning motion is common to both charge and mass, the magnetic moment and angular momentum are collinear and in direct proportion. Calculations using this simple model show that $\boldsymbol{\mu}$ and **P** are related by the equation

$$\boldsymbol{\mu} = (e/2Mc)\mathbf{P} \tag{1}$$

in which $c$ is the speed of light. This equation does not describe accurately the behavior of actual nuclei. Because of certain anomalies, the proportionality constant between $\boldsymbol{\mu}$ and **P** is not $e/2Mc$. However, the magnetic moment of a nucleus is always proportional to its angular momentum, so Eq. (1) can be written as

$$\boldsymbol{\mu} = g(e/2Mc)\mathbf{P} = g(e/2Mc)\hbar\mathbf{I} \tag{2}$$

* Vector quantities are given in boldface type; components for vectors and properties for which only the magnitudes are significant are given in italics.

in which g, the magnetogyric ratio, is a dimensionless constant of the order of unity and $M$ is the mass of the proton. The nuclear angular momentum is usually designated as $\mathbf{I}$, with units of $h/2\pi$ so that $\hbar\mathbf{I}$ can be substituted for $\mathbf{P}$ in Eq. (2), in which $\hbar = h/2\pi$ and $h$ is Planck's constant.

In the absence of magnetic fields, the energy of the nuclear magnet is independent of its orientation. In an external magnetic field, however, the magnetic dipole experiences a torque which tends to align the magnetic moment parallel to the field. According to elementary magnetic theory, the torque ($\mathbf{L}$) produced on a magnetic moment $\boldsymbol{\mu}$ in a field $\mathbf{H}$ is

$$\mathbf{L} = \boldsymbol{\mu} \times \mathbf{H} \tag{3}$$

The potential energy of the magnetic moment $\boldsymbol{\mu}$ in a field $\mathbf{H}$ the value of which is fixed is,

$$E = -\mu_H H_0 \tag{4}$$

in which $\mu_H$ is the component of $\boldsymbol{\mu}$ along the direction of $\mathbf{H}_0$.

Although Eq. (4) allows a continuous range of energies between $-\mu H_0$ and $+\mu H_0$ for a classical system, nuclei have a discrete set of magnetic energy levels. General laws of quantum mechanics require that the angular momentum be quantized along any defined direction in space. Since the interaction of $\boldsymbol{\mu}$ with $\mathbf{H}$ defines a direction in space for the nuclei, the nuclear angular momentum $\mathbf{I}$ is quantized along the direction of the magnetic field. The allowed components of $\mathbf{I}$ along $H$ are determined by the magnetic quantum number ($m$) with values of $I$, $I-1, \cdots, -I$ for a total of $2I+1$ states. Each nucleus has a particular integral of half-integral value of $I$. Observed values range from 0 to 7. The most important nuclei including $^{1}$H, $^{19}$F, $^{31}$P, and $^{13}$C have $I = 1/2$.

Because of the collinearity and proportionality of $\boldsymbol{\mu}$ and $\mathbf{I}$, quantization of $\mathbf{I}$ leads to quantization of $\boldsymbol{\mu}$ through Eq. (2):

$$\mu = g(e/2Mc)\hbar m = mg\mu_0 \tag{5}$$

in which $e\hbar/2Mc$ is denoted by $\mu_0$. Using this result for $\mu$ in Eq. (4) gives

$$E = -mg\mu_0 H_0 = -m\gamma h H_0 \tag{6}$$

which corresponds to a set of $2I+1$ nuclear orientations and energy levels. These are shown in Fig. 1 for $I = 1$; $\gamma$ is used to express the

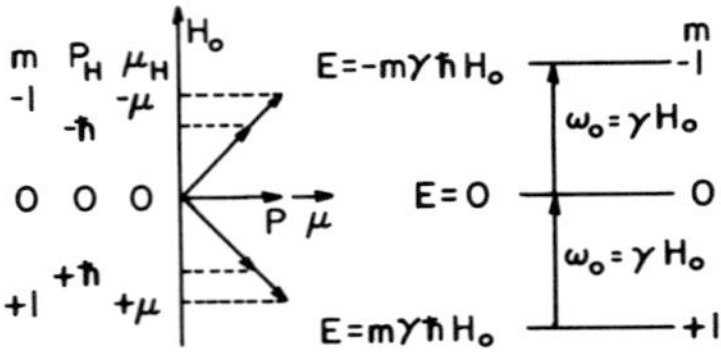


**Fig. 1.** *Energy levels for a nucleus of* I = 1 *in a magnetic field.*

magnetogyric ratio in units of g/h. The selection rule for transitions among the energy levels given by Eq. (6) is that $m$ changes by $\pm 1$. Therefore

$$\Delta E = \pm g\mu_0 H_0 \tag{7}$$

By the Bohr frequency condition $\Delta E = h\nu$, we find that the frequency $\nu_0$ in Hz corresponding to an allowed transition is

$$\nu_0 = g\nu_0 H_0/h \tag{8}$$

which in angular frequency units becomes

$$\omega = 2\pi\nu_0 = g\mu_0 H_0/\hbar = \gamma H_0 \tag{9}$$

This equation can also be obtained by considering the dynamic aspects of nuclear magnetization. The transitions require radiofrequency (rf) magnetic fields which are circularly polarized in a plane perpendicular to $\mathbf{H}_0$. Equation (3) states that the magnetic field exerts a torque on the nuclear magnetic moment tending to align it with the field. However, this would also reorient the angular momentum $\mathbf{P}$. For such a reorientation, Newton's law requires that the time rate of change of the angular momentum equal the torque,

$$d\mathbf{P}/dt = \mathbf{L} \tag{10}$$

By replacing $\mathbf{L}$ by $\boldsymbol{\mu} \times \mathbf{H}$, as given in Eq. (3), and converting $\boldsymbol{\mu}$ to $\mathbf{P}$ by Eq. (2) we obtain

$$d\mathbf{P}/dt = g(e/2Mc)\mathbf{P} \times \mathbf{H}_0 \tag{11}$$

This equation describes the precession of $\mathbf{P}$ about $\mathbf{H}_0$ with an angular velocity

$$\omega = g(e/2Mc)\mathbf{H}_0 \equiv \gamma\mathbf{H}_0 \tag{12}$$

which is the same as Eq. (8), resulting from the nuclear magnetic energy levels. Figure 2 illustrates the precession of the spinning nucleus about $\mathbf{H}_0$. A familiar example of precession is the wobbling of a spinning top in the earth's gravitational field.

The precession of the nuclear spin axis about a magnetic field suggests a way to change the orientation and, therefore, the magnetic energy of $\boldsymbol{\mu}$. A small magnetic field $\mathbf{H}_1$ is placed in the $x,y$ plane which is perpendicular to $\mathbf{H}_0$, the direction of which is along the $z$ axis. The torque which $H_1$ exerts on $\boldsymbol{\mu}$ will cause the spin axis to precess around $\mathbf{H}_1$, changing the orientation of the nucleus with respect to $\mathbf{H}_0$. Whether the angle between $\boldsymbol{\mu}$ and $\mathbf{H}_0$ increases or decreases depends upon the relative orientations of $\boldsymbol{\mu}$ and $\mathbf{H}_1$ as implied by Eq. (11) and shown in Fig. 2. Therefore, there will be little or no effect unless $\mathbf{H}_1$ maintains its orientation with respect to $\boldsymbol{\mu}$, a condition which can be accomplished by rotating $\mathbf{H}_1$ about $\mathbf{H}_0$ at the same rate that $\boldsymbol{\mu}$ precesses about $\mathbf{H}_0$. Thus, $\mathbf{H}_1$ must rotate in resonance with the precession of $\boldsymbol{\mu}$ about $\mathbf{H}_0$. The rotating magnetic field corresponds to circularly polarized radiation at the frequency $\omega_0$. In practice, the circularly polarized radiation is obtained by passing an rf current through a small coil mounted perpendicular to $\mathbf{H}_0$, producing a magnetic field oscillating along the coil axis. Such a field is mathematically equivalent to 2 equal circularly polarized fields rotating in opposite directions. The component rotating in the correct sense will be automatically in resonance, while the other component will have a negligible effect. More exact treatments of this vector model from both the classical and quantum mechanical viewpoints have been given elsewhere (Andrew, 1955).

Consideration of the behavior of a spinning magnetic nucleus in a magnetic field from either the standpoint of nuclear energy levels or from a classical dynamic viewpoint leads to the same result,

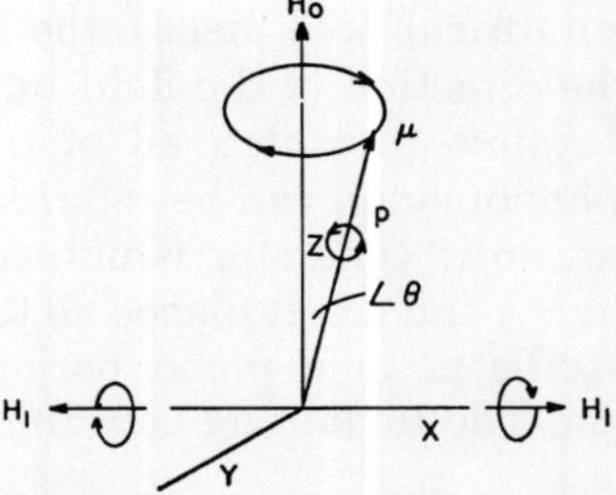


**Fig. 2.** *Precession of a spinning nucleus about the field* $\mathrm{H}_0$.

$$\omega_0 = \gamma H_0 \tag{12}$$

This result clearly implies the following:

1. For each nucleus, the resonance frequency is directly proportional to the applied field $\mathbf{H}_0$. Proton resonance, for example, is observed at 60, 100, and 220 MHz in fields of 14,100, 23,400, and 51,500 G, respectively.

2. In any given field, nuclei with larger magnetogyric ratios ($\gamma$'s) will have larger resonance frequencies. Thus, in a field of 23,400 G the resonance frequency for protons is 100 MHz as compared to 40 MHz for $^{31}P$ in the same field.

3. On the other hand, at a fixed frequency nuclei with a larger magnetogyric ratio will come into resonance at smaller applied fields. For example, a field of $\sim$9460 G is required to observe proton resonances at 40 MHz while 23,400 G is necessary for observation of $^{31}P$ resonance at the same frequency.

## C. Detection of Magnetic Resonance—The Basic NMR Spectrometer

The aforementioned precession of a nuclear dipole about the rotating field $\mathbf{H}_1$ allows the detection of resonance by placing a pick-up coil anywhere in the plane in which $\mathbf{H}_1$ is rotating. In such a coil the changing magnetic field associated with the reorienting dipole induces a current which can be amplified and recorded. A block diagram of a basic NMR spectrometer is shown in Fig. 3. Three component systems are required:

1. The magnet, usually an iron core electromagnet (although superconducting magnets and permanent magnets are also used) including a power supply and controls allowing the field to be swept by varying the current through the magnet coils. The field is generated in a gap between two cylindrical pole pieces, the axis of which defines the z direction (i.e., the direction of the field intensity vector $H_z$) of a Cartesian coordinate system to which all of the dynamic variables describing the NMR phenomenon can be referred.

2. The transmitter, an rf generator which produces the resonant radiation (i.e., the field $H_1$) and a coil placed so that the magnetic component of its field oscillates in a plane perpendicular to $H_z$, thus defining the x direction. The remaining coordinate $y$ is perpendicular to x and z.

3. The detector system, which consists of a detector coil, one or

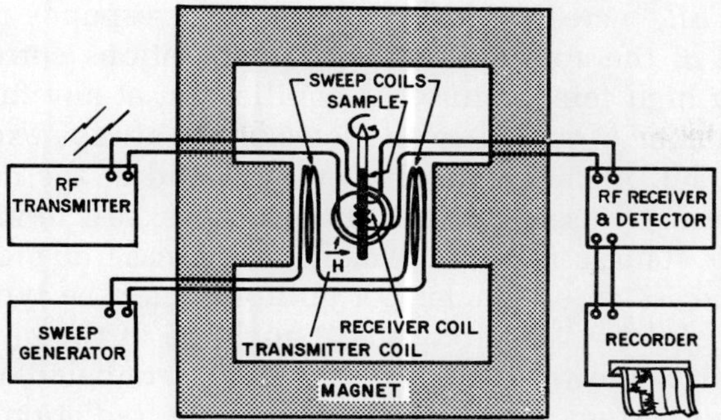


**Fig. 3.** *Block diagram of the basic nuclear magnetic resonance spectrometer.*

more amplifiers, and a recorder or oscilloscope for visual display of the signal. The same coil or two separate ones may serve as transmitter and receiver.

## D. Relaxation Times and Relaxation Mechanisms

We have examined the behavior of an isolated, spinning nucleus. However, when NMR is actually observed in bulk matter, the signal represents the reorientation of an ensemble of identical nuclei. These nuclei may interact among themselves and with their surroundings. The behavior of the isolated magnetic nucleus must be related to the behavior of a set of nuclei with these interactions taken into account.

Consider an assembly of identical nuclei of spin $I = 1/2$ in identical environments so that all of the nuclei in the assembly experience the same total magnetic field. Such an assembly constitutes a magnetically equivalent set. All of the remaining particles in the sample constitute the lattice. Only two magnetic energy levels are accessible to each nucleus in such a system and these correspond approximately to alignment of the nuclear magnetic moment either with or against the magnetic field. At equilibrium the nuclei are distributed between the two energy levels and the ratio of the number of spins in each level is given by the Boltzmann factor

$$n^+/n^- = e^{-\Delta E/kT} \tag{13}$$

in which $n^+$ and $n^-$ are the population of the upper and lower states respectively, $\Delta E = 2\mu H$ is the energy difference between the two states, $k$ is the Boltzmann constant, and $T$ is the absolute temperature.

For a given $\Delta E$, increasing the $n^+/n^-$ ratio corresponds to raising the temperature of the spin system, equal populations corresponding to an infinitely high temperature. At equilibrium at any finite temperature, the number of nuclei in the lower state always exceeds that in the upper state. At moderate temperatures and at the magnetic field strengths normally used for NMR studies, the excess number of nuclei in the lower state is relatively small. In the case of protons, for example, an excess of ~5 nuclei per million would be typical. That is, in a sample of $\sim 2 \times 10^6$ protons, $10^6$ might be in the upper state and $10^6$ plus 10 in the lower state. If the system is irradiated at a frequency $\nu = \Delta E/h$, the system absorbs energy from the radiation field with a consequent increase in the ratio $n^+/n^-$. Since the probability per unit time per nucleus of absorption is identical to that for induced emission, absorption will exceed emission only so long as excess nuclei remain in the lower state. This excess will obviously diminish in the absence of mechanisms allowing radiationless transitions from the upper to the lower energy states. When the system absorbs sufficient energy to equalize the population of the two states, it is considered to be saturated and the temperature of the spin system is infinite. Only a small amount of energy is required for this heating since the heat capacity of the spin system is small.

A saturated or partially saturated spin system tends to return to thermal equilibrium if left alone. Two simultaneous processes are involved in the return of a saturated system to equilibrium: (1) flow of energy from the spin system to the lattice, which results in the cooling of the spin system by transitions of nuclei from the higher to the lower state causing a heating of the lattice. This process is called *spin–lattice relaxation;* and (2) redistribution of the absorbed energy among the nuclei by mutual exchange of nuclei between the higher and lower states in which the total energy of the spin system remains constant. The latter process is called *spin–spin relaxation*. These two processes, having different mechanisms, need not occur at the same rate; their rates can be determined separately for a given spin system.

The difference between these relaxation processes can also be considered in terms of the dynamic vector model by discussing the motion of a macroscopic magnetization vector (**M**) which is the vector sum of all the individual magnetic moments making up the system. At equilibrium, **M** is aligned with the external magnetic field $H_0$; thus there is no magnetization in the x,y plane. Momentary application of the irradiating field $H_1$ will tip **M** away from the z axis, and **M** will then have a component $M_{xy}$ in the x,y plane as well as a component $M_z$ in the z direction. Left alone the system will return to equilibrium and there will again be a magnetic moment **M** aligned in the z direction and no magnetization in the x,y plane.

The magnetization component in the x,y plane ($M_{xy}$) evidently cannot last longer than the time required for the magnetization to return to its equilibrium value. However, it is possible for the component $M_{xy}$ to disappear before the magnetization in the z direction reaches its equilibrium value, as illustrated in Fig. 4. $M_{xy}$ can decay because the individual magnetic moments get out of phase in their precession. Obviously this can occur with no resultant change in $M_z$ and without any energy exchange between the spin system and the lattice. For example, when two nuclei undergo a mutual reorientation of their moments with respect to the external field, they lose their phase coherence. Spin–spin relaxation is also called *transverse relaxation,* while spin–lattice relaxation is also called *longitudinal relaxation.*

For such a two-state spin system, both longitudinal and transverse components change with time according to a single exponential function involving characteristic time constants called $T_1$, the longitudinal or spin–lattice relaxation time, and $T_2$, the transverse or spin–spin relaxation time. The relaxation time $T_2$ is sometimes called the linewidth parameter because it is related to the linewidth of the Lorentzian-shaped absorption line by $1/T_2 = \pi\Delta\nu_{1/2}$, in which $\Delta\nu_{1/2}$ is the width in Hz of the resonance at one-half maximum intensity. When defined in this way, $T_2$ is composed of three parts:

$$1/T_2 = 1/T_2' + (1/T_2'')_{\text{loc}} + (1/T_2'')_{\text{mag}}$$

in which $1/T_2$ is the observed relaxation rate as determined from the linewidth, $1/T_2'$ is the contribution from spin exchange, $(1/T_2'')_{\text{loc}}$ is the contribution from the local field inhomogeneities due to sample, and $(1/T_2'')_{\text{mag}}$ is the contribution from the inhomogeneity of the applied ex-

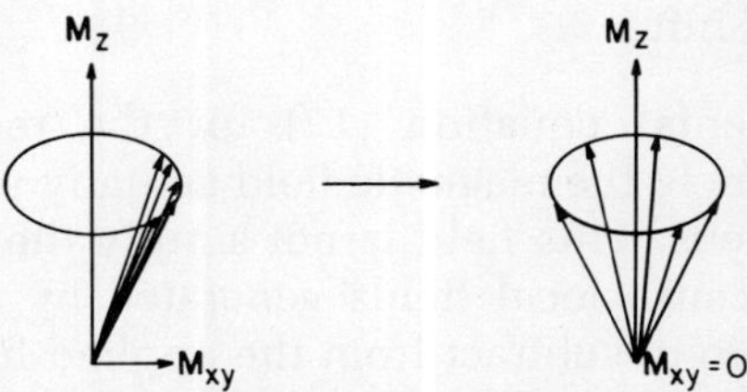


**Fig. 4.** *The resultant magnetization in the x,y plane shown at left can decay due to independent precession of individual nuclei as shown at right. The process can occur without affecting the net magnetization in the z direction.*

ternal magnetic field. To obtain meaningful results in terms of molecular events in the sample, $(1/T_2'')_{mag}$ must be reduced to the lowest possible value, placing stringent requirements on the homogeneity of the magnet and accounting in large part for the expense of modern NMR spectrometers. Alternatively, a correction for $(1/T_2'')_{mag}$ can be made if its value is constant and known during a particular experiment. Clearly $1/T_2'$ and $1/T_2''$ also measure the rate at which the nuclei lose their precessional phase coherence, since field inhomogeneity, either from the sample itself or from an inhomogeneous applied field, will result in the nuclei in different parts of the sample precessing at different rates, thus getting rapidly out of phase and giving a broad resonance.

## E. General Features of NMR Spectra

Four parameters characterize the absorption of rf radiation by atomic nuclei placed in a magnetic field: (1) the *chemical shift* which expresses the frequencies of absorption relative to some arbitrary standard absorption line, (2) the coupling constants which describe the interactions between neighboring nuclei and which lead to multiplicity of the lines originating from a given group of nuclei, (3) the intensities of absorption lines, described by their integrated areas, which are proportional to the number of nuclei contributing to each line, and (4) the relaxation times $T_1$ and $T_2$ which describe the return of excited nuclei to a lower-energy state. The correlations of chemical structure to NMR spectra are based almost entirely on the observed chemical shifts, coupling constants, and intensities. On the other hand, biochemical studies are likely to involve interactions between molecules and these interactions are most sensitively reflected in the relaxation process.

### 1. The Chemical Shift

In the fundamental equation (12) for the resonance condition $(\omega_0 = \gamma H_0)$, $H_0$ refers to the magnetic field actually present at the site of the magnetic nucleus. This field is not simply the externally applied magnetic field because local fields generated by induced electronic currents may add to or subtract from the applied field. The local field $H_{loc}$ at a particular site can be expressed in terms of the externally applied field as $H_{loc} = H_0(1 - \sigma)$ in which $\sigma$ is called the shielding constant. The resonance condition then becomes

$$\omega = \gamma H_0(1 - \sigma) \tag{14}$$

The shielding constant $\sigma$ is a characteristic of each nuclear environment and may be positive or negative. As a result, an NMR experiment carried out on a molecule, or a mixture of molecules, causes the signals from the various nuclei to be spread out in a spectrum according to their chemical environments. The chemical shift between two sets of nuclei is defined as the difference in their resonance frequencies measured at constant field. This difference is directly proportional to the magnitude of the applied field and is most conveniently expressed in field-independent units as parts per million (ppm) of the constant fields: or frequency

$$\delta = (\nu_2 - \nu_1)/\nu_1 \times 10^6 \tag{15}$$

in which $\delta$ is the chemical shift and $\nu_1$ and $\nu_2$ are the resonance frequencies for the two groups of nuclei. Note that frequency and field strength have equal status and are readily interconvertible through Eq. (1). Either field or frequency may be kept constant while the other is varied to obtain the spectrum. For protons, the normal range of chemical shifts covers $\sim$12 ppm. Since it is impossible to measure a large magnetic field with sufficient accuracy for absolute chemical shift determinations, chemical shifts are measured relative to an arbitrary reference, which for protons is almost always tetramethylsilane (TMS). For the aqueous solution of most interest to biochemists, an ionic derivative of TMS must be used since TMS is insoluble in water. The sodium salt of dimethylsilapentanesulfonic acid (DSS) is most commonly used. Figure 5 shows the spectrum of a mixture of compounds with each having only a single proton environment which may be used as a reference.

### 2. The Coupling Constant

In addition to the lines accounted for by chemical shift differences, the high-resolution spectra of many compounds contain patterns of lines which arise from interactions with neighboring magnetic dipoles. These patterns are accounted for by an indirect coupling between the nuclei which is transmitted by the electrons of intervening chemical bonds. The mechanism of transmission can be understood by considering that the magnetic field of a nuclear dipole will tend to orient the valence electrons antiparallel to the nuclear spin. Simultaneously, the spins and magnetic moments of the electrons in a

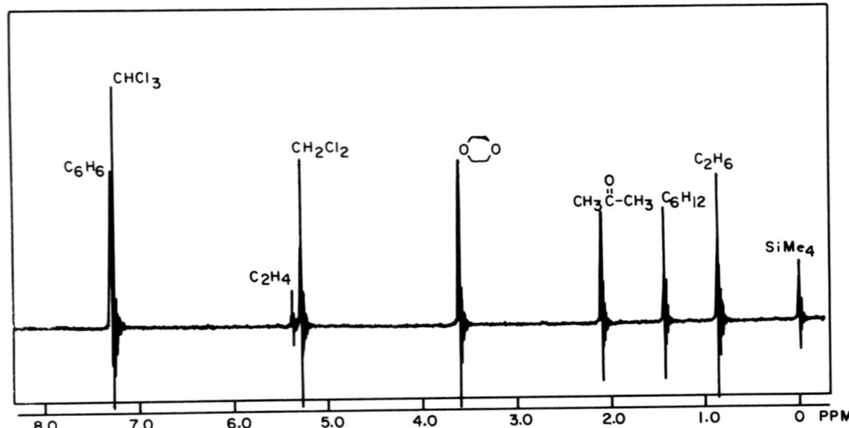


**Fig. 5.** *The NMR spectrum of a mixture of compounds, each having a single proton environment.*

covalent bond must be paired so that the valence electron on the neighboring atom will tend to be oriented parallel to the nucleus of the first atom. This results in the creation of a magnetic field at the second atom the direction of which reflects the orientation of the first atom and the magnitude of which reflects the extent of polarization of the electrons. This in turn depends on the characteristics of the chemical bond. At the same time the second nucleus exerts a similar effect on the first one through the same bonding electrons. The magnitude of the interaction is expressed in terms of a coupling constant $J$ which is expressed in frequency units and which measures the interaction energy between two spins. Figure 6 shows the NMR spectrum of ethyl ether and is a simple example of the effect of spin–spin coupling. The arrows above each component line in the spectrum show the orientation of the protons of the neighboring group which produces that component. The relative intensities of the components of each multiplet are determined by the statistical weights of each arrangement of neighboring spins. The separation between the components of each multiplet is the coupling constant $J$. Since the coupling energy arises from a mutual interaction between permanent magnetic movements, it is not surprising that the $J$ value and hence the observed splittings are independent of thė applied field, in contrast to chemical shifts.

Figure 6 also illustrates the fact that the areas under the absorption lines are proportional to the number of protons contributing to the absorption. Thus, the ratio of the area of the methyl triplet to the methylene quartet in Fig. 6 is 3:2. In terms familiar to conventional

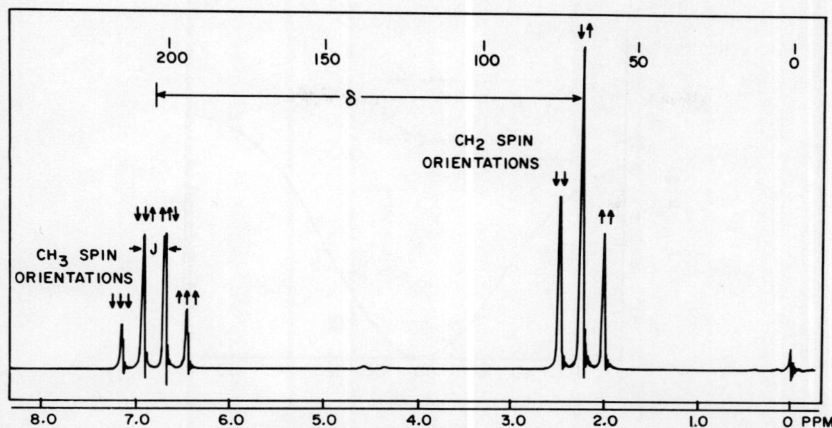


**Fig. 6.** *The NMR spectrum of ethyl ether illustrating the effects of spin–spin interactions.*

spectroscopists one might say that the NMR extinction coefficients for all protons are the same at a given resonance field. If the field is swept while the frequency remains constant, the extinction coefficients will still be practically the same because the field is changed only a few parts per million.

As the ratio of the chemical shift to the coupling constant, $\delta/J$, becomes smaller, the spectrum becomes more complicated and cannot be interpreted in the simple way shown in Fig. 6. In these cases the spectrum can be analyzed by the methods of quantum mechanics to yield the chemical shifts and coupling constants (Pople *et al.*, 1959).

Of special interest is the fact that the coupling constants between vicinal protons are a function of the dihedral angle. This relationship is illustrated by the "Karplus curve" shown in Fig. 7. This curve, derived by Karplus (1959), has been of great value in the determination of the conformation of cyclic compounds. The curve can be taken only as a rough indication of actual $J$ values since other factors, such as ring strain and substituent electronegativity, also have an effect. However, reliable qualitative results can be obtained if a series of similar derivatives is available so that the parameters of the Karplus curve can be determined, as in the case of the sugars. In favorable cases, the experimental values of the coupling constants obtained in the same molecule can be used provided that there are several protons carried by the same carbon atoms corresponding to different dihedral angles whose values are not independent. This is the case, for example, in the hUra ring compounds the NMR spectra of which were analyzed by Chabre *et al.* (1966).

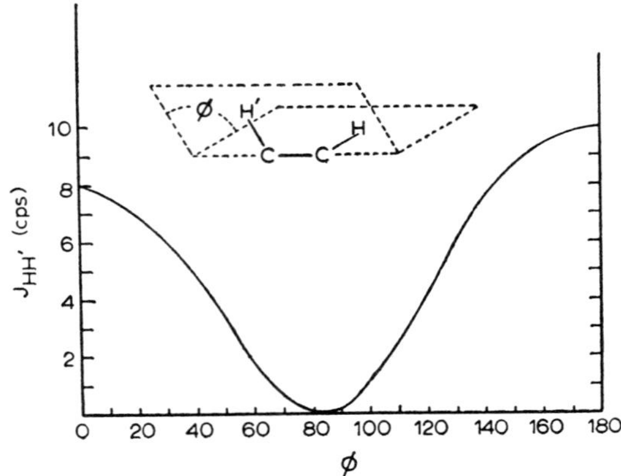


**Fig. 7.** *Variation of vicinal proton–proton coupling constants calculated according to Karplus.*

## F. Sensitivity of the NMR Method

### 1. Basic Considerations

NMR spectroscopy is inherently insensitive owing to the small energy difference between the observed energy levels. Much effort has been devoted to improving sensitivity. The signal strength in an NMR experiment is proportional to the equilibrium magnetization:

$$S/N \propto M_0 = \frac{N\gamma^2 h^2 \, I \, (I+1) H_0}{3kT} \tag{16}$$

in which: $N$ = number of nuclei in sample, $\gamma$ = magnetogyric ratio, $h$ = Planck's constant, $I$ = spin quantum number, $H_0$ = strength of applied magnetic field, $k$ = Boltzmann constant, and $T$ = absolute temperature. Of these, $\gamma$, $h$, $I$, and $k$ are physical constants beyond the control of the experimenter. The remainder, $N$, $H_0$, and $T$ may be varied within limits. Temperature lowering is not feasible for studies of some Pur and Pyr derivatives due to freezing of the solutions. $N$ may be increased by using larger samples, provided that one is not limited by the quantity of sample available. In order to avoid viscosity effects caused by sample concentrations which are too high or in cases where the solubility of the sample is low, it is desirable to increase the sample volume. This requires the use of larger magnets to accommodate the larger volume and to attain the required high resolution. Commercial magnets which accommodate sample tubes up to 15 mm o.d. are currently available.

Increasing the magnitude of the field $H_0$ is another approach which has been used to increase sensitivity. In addition, the stronger field also increases the chemical shifts in proportion to the field strength. When the frequency response of the detection system is considered, it is predicted that the overall sensitivity should be proportional to $H^{3/2}$ in which $H$ is the strength of the polarizing field. Commercial spectrometers which generate field strengths of up to $\sim 84$ kG employing a superconducting magnet are available. This corresponds to a resonance frequency of 360 MHz for $^1$H and 90 MHz for $^{13}$C. Spectrometers now being developed may reach 600 MHz for $^1$H in the near future. Although the full theoretical improvement in sensitivity is not realized for protons because of special problems in the detection of high radiofrequencies, the improvement does approach the theoretical maximum for $^{13}$C, making the higher field a distinct advantage for this nucleus.

### 2. Sensitivity Enhancement

A number of possibilities exist for improving the signal-to-noise ratio in NMR spectra and this has been discussed in detail by Ernst (1966). Two methods which are extraordinarily well suited for NMR are the *method of time averaging*, and *the Fourier transform method*. These methods are being used with increasing frequency in NMR experiments on biological systems with the commercial equipment currently available. The basic principles underlying these methods will be discussed briefly.

### 3. The Time-Averaging Method

In principle, the signal-to-noise ratio can be increased at will by taking a sufficiently long time to do the experiment and by using suitable filters to suppress the noise. In practice, it may not be possible to realize the higher sensitivity of an experiment over a longer period of time because of low-frequency instability of the spectrometer and because long relaxation times encountered in NMR result in saturation at low scanning rates. For these reasons time-averaging can be advantageous. Instead of a single long-term measurement, a number of relatively rapid scans of the spectrum are made, and the results are added so that the signals add coherently while the noise adds randomly. In practice, the summation of the different traces can be made in one of several commercially available digital storage devices which have a sufficient number of channels to represent the spectrum adequately, each channel corresponding to a single point in the spectrum.

Since the time-averaged NMR signal increases linearly with each scan while the noise increases as the square root of the number of scans, a net gain in signal-to-noise ratio proportional to the square root of the number of scans is achieved.

### 4. Fourier Transform Spectroscopy

In the aforementioned sweep method, one resonance in the spectrum at a time is excited as the rf is varied slowly over the spectrum. By use of short bursts or pulses of rf power it is possible to produce a much broader frequency spectrum of the irradiation. By using sufficiently short pulses, this allows simultaneous irradiation of the entire spectrum and offers the possibility of gathering information simultaneously from all or several parts of the spectrum. In practice, the sample is irradiated with a series of short, equally spaced pulses and the response of the system to each pulse, i.e., the time decay of the system toward equilibrium, is stored in a time-averaging computer and added to previous pulses. This is continued until a sufficiently good signal-to-noise ratio is obtained. The response to the rf pulse and the usual NMR absorption spectrum form a Fourier transform pair (Lowe and Norberg, 1957). By calculating the Fourier transform of the stored sum of the responses with a computer, the absorption spectrum with enhanced sensitivity is obtained. This technique is now in widespread and still increasing use in applications of NMR. A full discussion of the Fourier transform method as applied to NMR has been given by Ernst and Anderson (1966).

## G. NMR Spectra of Photoproducts and Related Compounds

### 1. Assignment of Proton NMR Spectra of Bases and Nucleotides

In the determination of the structure of photoproducts by NMR, the first step is the assignment of chemical shifts to the protons of the proposed structure. For the study of nucleic acid photoproducts, the starting point is to assign the NMR spectra of the starting compounds, one of which is always a base, nucleoside, or nucleotide. The state of knowledge of the theoretical and observed chemical shifts of bases and nucleosides has been reviewed by us (Ts'o *et al.*, 1969). Therefore only a brief outline will be given.

### 2. Pyrimidine Derivatives

Excluding exchangeable protons, the NMR spectra of the bases of Ura (or Urd) and of Cyt (or Cyd) consist simply of a pair of doublets separated by $\sim 1.7$ ppm, the actual value depending somewhat on the solvent. The high-field doublet is assigned to H(5) and the lower-field doublet to H(6) (Jardetzky and Jardetzky, 1960). This assignment has been amply confirmed by workers who studied many 5- and 6-substituted uracils (Kokko *et al.*, 1962) and methyl derivatives of Cyt (Becker *et al.*, 1965).

### 3. Purine Derivatives

Pur, although not a component of nucleic acid, is a good model both for NMR studies and for photochemical reactions due to its relatively simple structure and its obvious similarity to the nucleic acid bases Ade, Hyp, and Gua. The NMR spectrum of Pur consists of only three single lines corresponding to the H(2), H(6), and H(8) protons. The correct assignments of these lines to the appropriate protons was made independently by Matsuura and Goto (1963), Schweizer *et al.* (1964), and Bullock and Jardetzky (1964). This achievement was made possible by selective deuteration of the molecule.

### 4. Tautomerism and Protonation

Assignment of the correct tautomeric structure to DNA bases is of potential importance in the mechanism of formation of photoproducts. NMR has provided decisive information on this subject. For example, based on chemical shift data for NH protons in DMSO, the correct tautomeric form of Guo or Gua was determined by Miles *et al.* (1963) and Becker *et al.* (1965). Kokko *et al.* (1962) contributed to the establishment of the keto form for the 4-oxo group of Ura or Thy.

### 5. Photohydrates

Wechter and Smith (1968) reported the 60 and 100 MHz NMR spectra of Urd photohydrate prepared in water and in deuterium oxide. The observed spectra are shown in Fig. 8. In addition, these authors reported proton NMR data on several other Urd derivatives. These are included in Table 1. Comparing the spectrum of the photohydrate to that of dihydrouridine reveals that the chemical shifts of the H(5) protons are little changed in the photoproduct. This observa-

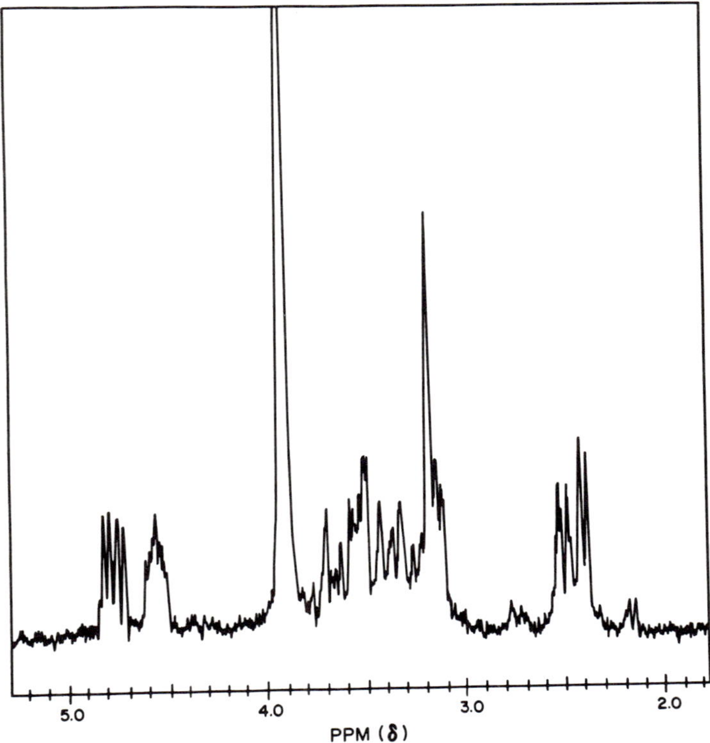


**Fig. 8.** *Proton magnetic resonance spectra of 6-hydroxydihydrouridine.*

tion is consistent only with hydroxylation at C(6). The assignment of the H(5) proton to the resonance at ~3 ppm is confirmed by the observation of rapid disappearance of the proton signal when the hydrate is heated in $D_2O$. Both the chemical shifts and the coupling constants of H(1′) and H(5′) are essentially identical in hUrd and the photohydrate, which is consistent with the two structures differing only in the substitution of the hydroxyl group for one of the C(6) protons. It is noteworthy that in Urd itself the coupling constant $J_{H(1')-H(2')}$ is only 4.2 Hz as compared to 5 Hz for the hUrd and the photohydrate. This difference could be attributed to a difference in the spatial arrangement (dihedral angle) of the Rib in the two compounds or to a substituent effect of electronic origin.

An interesting feature of the Urd photohydrate spectrum is that the ABX spectrum attributed to the H(5) and H(6) proton is partially doubled and seems to be due to the partial overlap of two ABX spectra. The resonance of the H(1′) proton is also composed of two

**Table 1** Proton Magnetic Resonance Data for Uridine Analogs[a]

| | H(6) | H(5a) | H(5b) | H(1′) | H(5′) | Isomer, relative % |
|---|---|---|---|---|---|---|
| 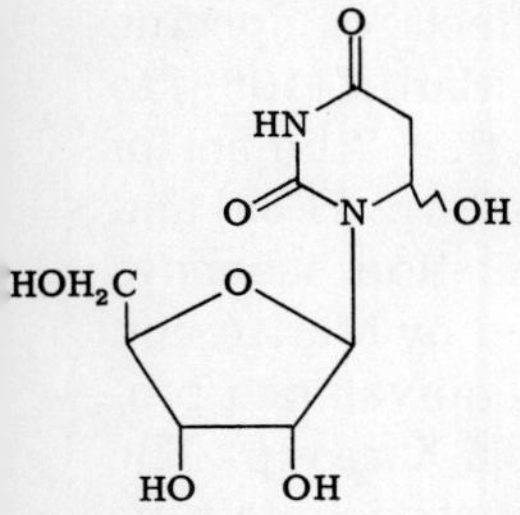 | (a) ~5.5 | ~3.1 | ~2.8 | 5.73(d, 5) | ~3.8 (m) | 60 |
| | (b) ~5.5 | ~3.0 | ~2.8 | 5.69(d, 5) | ~3.8 | 40 |
| 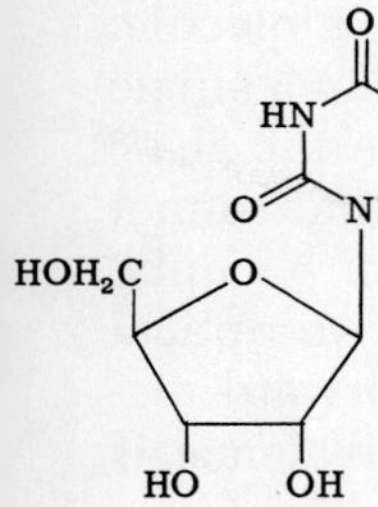 | 3.53(t, 6–7) | 2.75(t, 6–7) | | 5.78(d, ~5) | 3.7 (m) | |
| 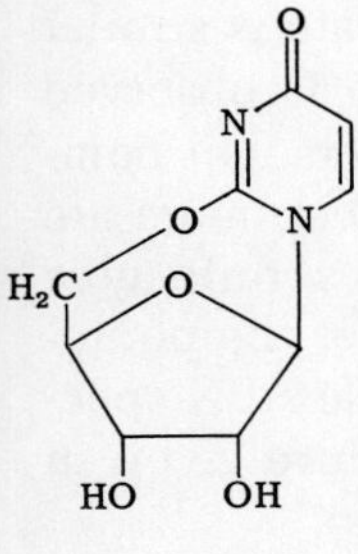 | 8.08 | | 6.20 | 5.72(d, <1) | 4.25 (m) | |
| 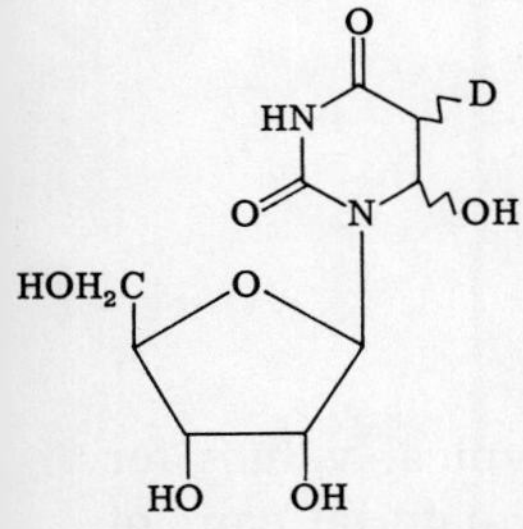 | (a, b) ~5.5 (m) | 3.1 (m) | D | 5.73(d, 5) | 3.8 (m) | 60 |
| | (c, d) ~5.5 (m) | D | 2.75 (m) | 5.69(d, 5) | 3.8 (m) | 40 |

[a] Proton magnetic resonance spectra were measured in $D_2O$ (except where noted) on Varian spectrophotometers at both 60 and 100 MHz and chemical shifts (δ) were recorded in parts per million downfield from external dimethylsilapentanesulfonate (line multiplicities and coupling constants, in cycles per second, appear in parentheses). Wechter and Smith (1968).

doublets rather than the one expected from a single type of photohydrate molecule. This effect is most reasonably interpreted as due to the presence of both possible epimers about C(6). The two diastereomers resulting would be expected to have slightly different chemical shifts associated with their proton NMR spectra. Wechter and Smith interpreted this doubling in terms of the effects of specific asymmetric carbon atoms and referred to Martin and Martin (1966) to support this interpretation. For example, they state that "the set of H(1′) doublets at δ 5.73 and 5.69 and the sets of AB patterns H(5a) and H(5b), result from asymmetry at C(6)." This interpretation seems to result from a misunderstanding of the effects discussed by Martin and Martin, which are concerned with the magnetic nonequivalence produced in the otherwise equivalent protons of a —$CH_2X$ group. The nonequivalence arises when there is no plane of symmetry containing the bond joining the —$CH_2X$ group to the remainder of the molecule. Thus, two protons can become magnetically nonequivalent in a single isomer due to asymmetry, but a single proton [for example the H(1′) in this case] in a single isomer has but one chemical shift and cannot be magnetically nonequivalent to itself as Wechter and Smith imply. In the case of the Urd photohydrate it is the presence of two optical isomers that produces the doubled spectrum. A straight forward discussion of the necessary conditions for magnetic nonequivalence is given by Becker (1969).

The 100 MHz NMR spectrum of the uridine photohydrate is similar to that of Ura hydrate, except that only one AB pattern is observed (Fig. 9) (J. C. Nnadi, M. N. Khattak, and S. Y. Wang, personal communications). This is to be expected even if the two C(6) epimers are present because the epimers are now enantiomers and should give identical NMR spectra in nonchiral solvents. Also, each of the photohydrates of $Me^3Ura$, $Me_2^1Ura$, and Ura-1-EtAc gives a single ABX spectrum (Fig. 9) which is readily analyzed using the procedure given in most standard NMR textbooks (for example, Becker, 1969).

The results are

$$J_{5a-5b} = 17.5\ \text{Hz}$$
$$J_{5a-5b} = 29\ \text{Hz}$$
$$J_{5a-6} = 4.5\ \text{Hz}$$
$$J_{5b-6} = 2.0\ \text{Hz}$$

It is notable that the values of $J_{5a-6}$ and $J_{5b-6}$ are typical values for coupling constants between axial–equatorial and diequatorial pairs of protons, respectively. Absence of a large (~10 Hz) coupling constant

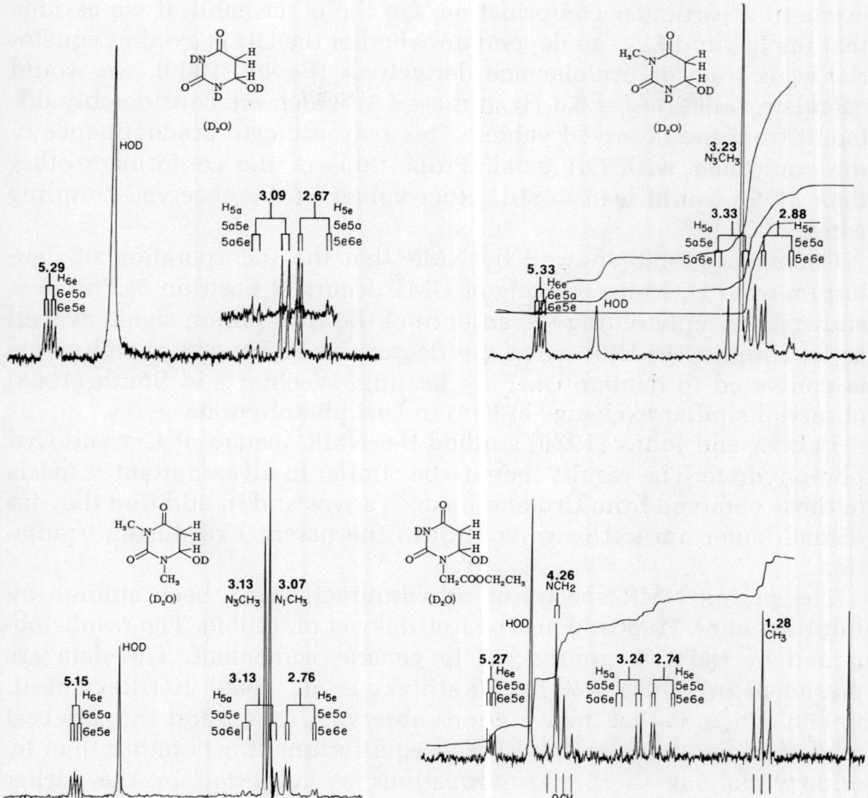


**Fig. 9.** *NMR spectra (60 MHz) of Ura, Me³Ura, $Me_2^{1,3}$Ura, and Ura-1-EtAc in $D_2O$.*

value, which would be expected if there were a diaxial pair of protons, could suggest that the conformers with the OH axial are strongly preferred to those with OH equatorial. However, one must also consider the possibility that the conformation fluctuates rapidly among two or more conformations and that the observed coupling constants and chemical shifts are average values. Thus, the H(6) proton might couple to one of the H(5) protons with $J_{5a-6a}$ and $J_{5eq-6eq}$ and with the other H(5) proton with $J_{5e-6a}$ and $J_{5a-6e}$. If we assume that $J_{aa} \cong 9$ Hz, $J_{ae} = 4$ Hz and $J_{ee} = 2$ Hz as typical values, then if the Ura hydrate exists as a 50-50 mixture of the two conformers, the observed coupling constant would be $(9 + 2)/2 = 5.5$ Hz and $(4 + 2)/2 = 3$ Hz. The observed values of 4.5 Hz and 2 Hz are not sufficiently different from the average values to allow any serious conclusion that the photohydrate exists in a particular conformation. On the other hand, if we assume that the $J_{a-e}$ and $J_{e-e}$ can depend on whether the OH is axial or equatorial as is true for cyclohexane derivatives (Bovey, 1969), we would calculate values of $J = 6.2$ Hz and $J = 4.1$, which are considerably different from the observed values. This may indicate predominance of the conformer with OH axial. Proportions of the conformers other than 50-50 would lead to still other values of the observed coupling constants.

Chambers (1968) showed by NMR that the incorporation of deuterium from $D_2O$ into the ring of UMP occurs at position 5. This was shown by simply noting the absence of the H(5) proton signal as well as its coupling to H(6) when the deuterium-exchanged photohydrate is converted to deutero-UMP by heating. Wechter and Smith (1968) observed similar exchange at H(5) in Urd photohydrate.

DeBoer and Johns (1970) studied the NMR spectra of Cyt and Cyd photohydrate. The results seem to be similar in all important respects to those obtained from Urd and its derivatives and in addition they indicate deuterium exchange at H(5) in the parent Urd during irradiation.

The proton NMR spectra of dihydrouracils have been studied by Rouillier *et al.* (1966a,b) and by Katritzky *et al.* (1969). The results obtained by the two groups are in general agreement. The data are presented in Tables 2–7 (from Katritzky *et al.*, 1969). Katritzky *et al.* (1969) conclude that the variations observed in $U_{5-6}$ and $U_{1-6}$ are best explained by electronegativity and equilibrium effects rather than by wide variations in ring conformations as suggested by the earlier workers (Rouillier *et al.*, 1966a,b). Carboxy and methoxycarbonyl groups at C(6) have been found to prefer an axial orientation. Phenyl groups at C(6) seem to be predominantly axial in dimethyl sulfoxide but equatorial in trifluoroacetic acid.

## 6. Cyclobutyl Dimers

Four different cyclobutane dimers (Fig. 10) can result from the fusion of Thy molecules at the 5,6-double bond, assuming *cis* ring fusion (Wulff and Fraenkel, 1961). Weinblum and Johns (1966) obtained all four dimers as shown by their chromatographic study. They observed, in agreement with several other groups, that the dimer produced from irradiation of frozen aqueous Thy solution was the same as that obtained by irradiation of DNA. This dimer was assigned the *cis-syn* structure based on the indirect argument that it was to be expected from the stacking geometry of bases in DNA. Blackburn and

**Table 2** NMR Spectra of 5,6-Substituted Dihydrouracils[a]

| Substituents | | | | | Chemical shifts ($\tau$) | | | | Coupling constants (Hz) | | | |
|---|---|---|---|---|---|---|---|---|---|---|---|---|
| 5 | 6 | Configuration | Solvent | MHz | Me(5) | Me(6) | H(5) | H(6) | $J_{H(5)-H(6)}$ | $J_{H(5)-Me(5)}$ | $J_{H(6)-Me(6)}$ | $J_{H(1)-H(6)}$ |
| Me | Me | *cis* | DMSO-$d_6$[b] | 60 | 8.99 | 9.00 | 7.35 | 6.45 | 5.1 | 7.05 | 6.5 | 3.1 |
| | | | $HNCONH_2$ | 60 | 8.83 | 8.83 | 7.18 | 6.22 | 5.2 | 7.0 | 6.6 | 3.0 |
| | | | | 100 | 8.83 | 8.83 | 7.20 | 6.27 | 5.2 | 7.2 | 6.6 | 2.9 |
| Me | Me | *trans* | DMSO-$d_6$[b] | 60 | 8.94 | 8.87 | 7.78 | 6.76 | 9.5 | 6.8 | 6.3 | ~1.5 |
| | | | $HCONH_2$ | 60 | 8.79 | 8.70 | 7.60 | 6.56 | 10.2 | 6.9 | 6.0 | ~1 |
| | | | | 100 | 8.79 | 8.70 | 7.57 | 6.54 | 10.1 | 6.9 | 6.3 | ~1 |
| Me | COOH | *cis* | $HCONH_2/D_2O$ (1:1) | 100 | 8.79 | — | 6.91 | — | 5.75 | 7.15 | — | — |
| | | | $HCONH_2$ | 100 | 8.76 | — | 6.86 | 5.59 | 6.05 | 7.1 | — | 3.2 |
| Me | COOH | *trans* | $HCONH_2/D_2O$ (1:1) | 100 | 8.60 | — | 6.97 | — | 3.8 | 7.3 | — | — |
| | | | $NCONH_2$ | 100 | 8.59 | — | 6.94 | 5.90 | 4.1 | 7.3 | — | 3.3 |

[a] From Katritzky *et al.*, 1969.

[b] Resonances for NH protons were observed at: *cis*, $\tau_{H(1)} = 2.45$, $\tau_{H(3)} = 0.07$; *trans*, $\tau_{H(1)} = 2.49$, $\tau_{H(3)} = 0.00$.

**Table 3** NMR Spectra of Dihydrouracils[a]

| Substituents | | | | | Chemical shifts ($\tau$) | | | | | Coupling constants (Hz) | | |
|---|---|---|---|---|---|---|---|---|---|---|---|---|
| 5 | 6 | Configuration | Solvent | MHz | H(5) | H(6) | Ph | H(1) | H(3) | $J_{H(5)-H(5)}$ | $J_{H(5)-H(6)}$ | $J_{H(1)-H(5)}$ |
| H | COOH | — | $D_2O$ | 100 | 6.76 ax<br>6.93 eq | | | — | — | 17.4 | 6.9 ax<br>5.1 eq | |
| | | | $HCONH_2$ | 100 | 7.00 | 5.59 | — | — | — | — | — | 3.0 |
| H | COOMe | — | DMSO-$d_6$ | 100 | 7.08 ax<br>7.36 eq | 5.76 | | 2.22 | −0.07 | 16.9 | 7.4 ax<br>3.1 eq | 3.6 |
| H | Ph | — | DMSO-$d_6$ | 60 | 7.13<br>7.37 | 5.28 | 2.65 | 1.98 | | 16.5 | 6.8<br>5.8 | 2.6 |
| | | | $HCONH_2$ | 60 | 7.17 | 5.20 | — | — | — | — | — | ~1.5 |
| | | | $CF_3COOH$ | 60 | 6.90 | 5.03 | 2.62 | — | — | — | — | — |
| Ph | Ph | *cis* | DMSO | 60 | 5.85 | 5.00 | *b* | 2.00 | — | | 5.5 | 1.7 |
| | | | $CF_3COOH$ | 60 | 5.65 | 4.73 | *b* | — | — | | 5.9 | — |
| Ph | Ph | *trans* | DMSO | 60 | 5.98 | 5.20 | 2.70 | 1.95 | — | | 6.8 | 2.3 |
| | | | DMSO + 10% $CF_3COOH$ | 60 | | | | | | | 7.5 | — |
| | | | $CF_3COOH$ | 60 | 5.84 | 4.98 | 2.71 | — | — | | 11.9 | — |

[a] From Katritzky *et al.*, 1969.
[b] Complex band.

**Table 4** NMR Spectra of 5,6-Tetramethylenedihydrouracil[a]

| Configuration | Solvent | Chemical shifts ($\tau$) | | | | Coupling constants (Hz) | | |
|---|---|---|---|---|---|---|---|---|
| | | $H_X$ | $H_Y$ | $H_M$ | $H_N$ | $J_{XY}$ | $\lvert J_{AX}+J_{BX}\rvert$ | $\lvert J_{CY}+J_{DY}\rvert$ |
| *cis* | $CF_3COOH$ | — | — | 2.79 | 0.78 | — | — | — |
| | $CF_3COOH/D_2O$ (7:5) | 6.27 | 7.14 | — | — | 4.8 | 10.5 | 11.5 |
| *trans* | $CF_3COOH$ | — | — | 2.68 | 0.59 | — | — | — |
| | $CF_3COOH/D_2O$ (7:5) | 6.72 | — | — | — | 11.1 | 16.1 | — |

[a] From Katritzky *et al.*, 1969.

**Table 5** NMR Spectra of 2-Ureidocyclohexanecarboxylic Acid[a,b]

| Configuration | Chemical shifts (τ) | | | | Coupling constants (Hz) | | |
|---|---|---|---|---|---|---|---|
| | $H_X$ | $H_Y$ | NH | $NH_2$ + COOH | $J_{XY}$ | $\lvert J_{AX} + J_{BX}\rvert$ | $\lvert J_{CY} + J_{DY}\rvert$ |
| *cis* | 5.96 | 7.39 | 3.76 | 4.26 | 3.0 | ~9.5 | ~11.0 |
| *trans* | 6.25 | 7.76 | 3.92 | 4.45 | 9.6 | 13.7 | 13.0 |

[a] From Katritzky *et al.*, 1969.
[b] Solutions in dimethyl sulfoxide/benzene (9:1).

Davies (1965) also assigned the *cis-syn* structure to this dimer. The *syn* arrangement was supported by the NMR spectra of two rearranged products of the dimer in which the cyclobutane rings were assumed to remain intact. A 9 Hz coupling was observed between the two cyclobutane hydrogens, eliminating the possibility of the *anti* arrangement in which these protons would not show observable coupling. Earlier, Beukers and Berends (1960) had argued that the dimer obtained from frozen Thy solutions had the *anti* configuration. They based this conclusion on the absence of observable spin–spin coupling between the two cyclobutane protons, an argument which as pointed out by Wulff and Fraenkel (1961) is invalid, since the protons in question are equivalent in all four possible dimers and hence can show no directly observable coupling. Anet (1965) reported the use of the naturally abundant $^{13}C$ satellite proton spectra (Cohen *et al.*, 1958) to determine the coupling constant $J_{H(1)-H(2)}$ between the two cyclobutane protons. While this method gives no information about the *cis* or *trans* arrangement, it is unequivocal for the *syn* or *anti* configuration because only vicinal protons (*syn*) will give observable spin–spin coupling while the 1,3-protons (*anti*) will give no observable coupling (Eaton, 1962). These couplings can be observed only in the $^{13}C$ satellite spectra, not

**Table 6** NMR Spectra at 100 MHz of α,β-Dimethyl-β-ureidopropionic Acids and Methyl Esters[a] in $D_2O$

| Compound | | Chemical shifts (τ) | | | | | Coupling constants (Hz) | | |
|---|---|---|---|---|---|---|---|---|---|
| Configuration | Acid/ester | $Me_\alpha$ | $Me_\beta$ | $H_\alpha$ | $H_\beta$ | MeO | $J_{H_\alpha Me_\beta}$ | $J_{H_\beta Me_\beta}$ | $J_{H_\gamma Me_\beta}$ |
| *erythro* | Acid | 8.85 | 8.84 | 7.40 | 6.08 | — | 6.9 | 7.0 | 6.7 |
| *threo* | Acid | 8.86 | 8.86 | 7.36 | 6.09 | — | 7.1 | 6.9 | 6.7 |
| *erythro* | Ester | 8.90 | 8.91 | 7.37 | 6.07 | 6.31 | 5.7 | 6.95 | 6.7 |
| *threo* | Ester | 8.88 | 8.89 | 7.37 | 6.17 | 6.31 | 6.9 | 7.0 | 6.8 |

[a] From Katritzky *et al.*, 1969.

**Table 7** NMR Spectra at 60 MHz of $\alpha,\beta$-Diphenyl-$\beta$-ureidopropionic Acids and Methyl Esters[a]

| Compound | | | Chemical shifts ($\tau$) | | | | | | Coupling constants (Hz) | |
|---|---|---|---|---|---|---|---|---|---|---|
| Configuration | Acid/ester | Solvent | NH | $NH_2$[b] | $H_\alpha$ | $H_\beta$ | MeO | Ph | $J_{H_\alpha-H_\beta}$ | $J_{H_\beta-NH}$ |
| *erythro* | Acid | DMSO | 3.59 | 4.72 | 6.01 | 4.65 | — | 2.6 | 11.5 | 9.8 |
| | | DMSO + 10% $CF_3COOH$ | — | — | 5.97 | 4.63 | — | | 11.1 | |
| *threo* | Acid | DMSO-$d_6$[c] | 3.13 | 4.60 | 6.01 | 5.77 | — | 3.75<br>3.80[d] | 9.0 | 9.1 |
| | | $HCONH_2$ | | | 5.75 | 4.53 | — | | 9.2 | 8.7 |
| | | $HCONH_2$ + 10% $CF_3COOH$ | | | 5.71 | 4.50 | | | 9.2 | |
| *erythro* | Ester | DMSO-$d_6$[c] | 3.50 | 4.70[e] | 5.88 | 4.62 | 6.65 | 2.6[f] | 11.2 | 9.4 |
| | | $HCONH_2$ | | | 5.79 | 4.46 | 6.51 | | 10.8 | 9.0 |
| | | $HCONH_2$ + 10% $CF_3COOH$ | | | 5.87 | 4.50 | 6.53 | | 10.2 | |
| *threo* | Ester | DMSO-$d_6$[c] | 3.15 | 4.40 | 5.88 | 4.69 | 6.40 | 3.77<br>3.82[d] | 9.5 | 9.6 |
| | | $HCONH_2$ | | | 5.78 | 4.56 | 6.31 | | 9.8 | 8.6 |
| | | $HCONH_2$ + 10% $CF_3COOH$ | | | 5.76 | 4.56 | 6.30 | | 9.9 | |

[a] From Katritzky *et al.*, 1969.
[b] In the acids the peak is common with that of COOH and integrates for 3 protons.
[c] Dr. S. Spasov, private communication.
[d] Doublet.
[e] Broad.
[f] Complex band.

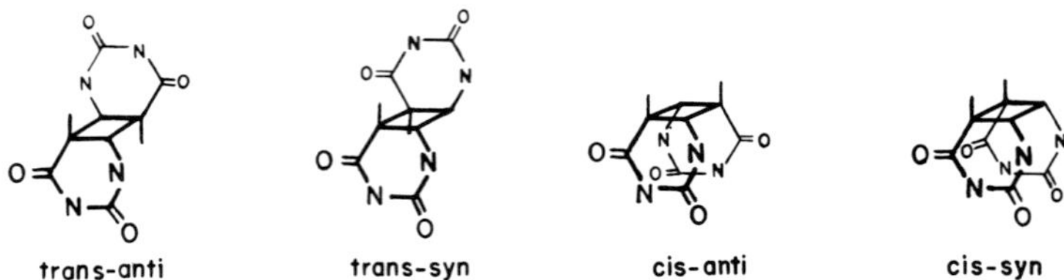


**Fig. 10.** *The four possible thymine dimers.*

in the main proton spectrum which results only from $^{12}C$-containing molecules. Anet reported a value of 5 Hz for the $J_{H(1)-H(2)}$ in the tetramethylated derivative of the Thy dimer and concluded that it is *syn*. Anet used only the high-melting 1,3-$Me_2$Thy homodimer which is formed in equal proportion with the low-melting dimer on irradiation in frozen solution of 1,3-$Me_2$Thy. Hollis and Wang (1967) reported the NMR spectra of both $Me_2$Thy dimers as well as the Thy dimer itself. The spectra are shown in Fig. 11. Chemical shifts and assignments of these spectra agree with those given earlier by Wulff and Fraenkel (1961). These spectra were all obtained in the same solvent relative to the same chemical shift standard and can therefore be compared directly. It is noteworthy that the chemical shift of the cyclobutane protons may be characteristic of the *syn* or *anti* configuration of the dimers. The chemical shift of the protons in the high-melting dimer **I** is within 0.09 ppm of that of the Thy dimer (*c,s*) while it differs by 0.59 ppm from the low-melting $Me_2$Thy dimer **II**. The $CH_3$ resonances of all three compounds differ by only 0.08 ppm. This parallelism suggests that dimer **I** and the Thy dimer have one configuration, while the low-melting $Me_2$Thy dimer **II** has the other.

From the naturally abundant $^{13}CH$ satellite spectra of these three dimers (Fig. 12) the coupling constant between the two cyclobutane protons can be measured directly (Cohen *et al.*, 1958). Dimer II shows a single $^{13}CH$ satellite peak. Since the coupling constant between the 1,3-proton is expected to be near zero, the arrangement of **II** is *anti*. On the other hand, dimer **I** gave a doublet $^{13}CH$ satellite ($J = 4.8 \pm 0.5$) which is consistent with a *syn* or 1,2-arrangement of the two cyclobutane ring protons. Since thymine homodimer from frozen solutions and solid films also gave doublet $^{13}CH$ satellite peaks ($J = 5.2 \pm 0.5$ Hz), Thy◇Thy must also have the *syn* arrangement. The several coupling constants for the three dimers are given in Table 8. The Thy dimer obtained by UV irradiation of DNA gives a proton NMR spectrum identical to that of Thy◇Thy obtained from frozen solutions (Varghese and Wang, 1967).

Blackburn and Davies (1966) reported that the $^{13}CH$ satellite spectra

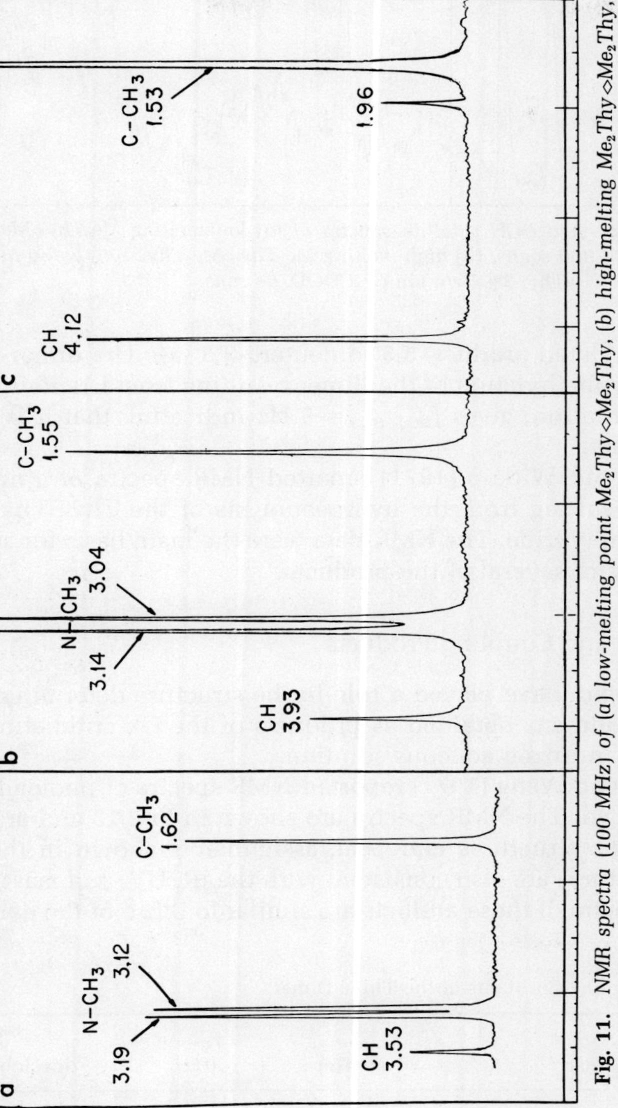


**Fig. 11.** NMR spectra (100 MHz) of (*a*) *low-melting point* $Me_2Thy\diamond Me_2Thy$, (*b*) *high-melting* $Me_2Thy\diamond Me_2Thy$, *and* (*c*) $Thy\diamond Thy$. *These spectra were obtained using 3% (w/w) solutions of the dimers in* $CF_3COOD$. *Chemical shifts are given relative to external tetramethylsilane.*

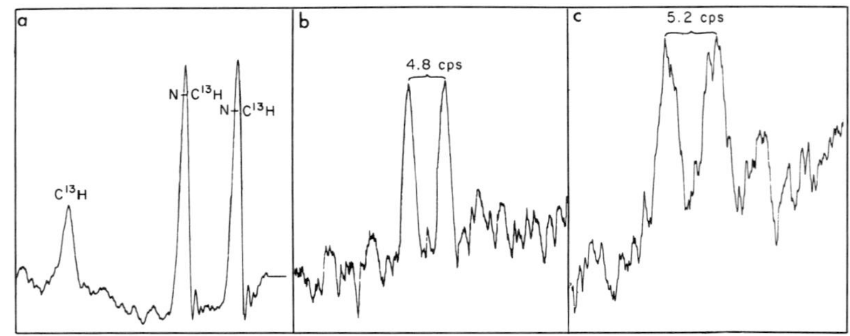


**Fig. 12.** *Low-field $^{13}CH$ satellite spectra of (a) low-melting $Me_2Thy{\diamond}Me_2Thy$, 10% (w/w) in $CDCl_3$, 300 scans; (b) high-melting $Me_2Thy{\diamond}Me_2Thy$, 10% (w/w) in $CDCl_3$, 50 runs; and (c) Thy◇Thy, 3% (w/w) in $CF_3COOD$, 64 runs.*

of the methylated product, 5,5′-dideutero-3,3′-$Me_2$Ura dimer which is obtained by methylation of the dimer resulting from irradiation of Ura in frozen solution, gives $J_{H(1)-H(2)} = 5$ Hz indicating that it is also the *cis-syn* dimer.

Kunieda and Witkop (1971) reported NMR spectra of a number of products resulting from the hydrogenolysis of the Thy◇Thy (*c*,*s*) by sodium borohydride. The NMR data were the main basis for structural assignments of several of the products.

## 7. Adducts and Coupling Products

NMR spectra have played a role in the structure determination of a number of adducts obtained as products of the UV irradiation of Pyr derivatives in frozen aqueous solution.

Rhoades and Wang (1971) reported NMR spectra of photoadducts of dCyd and Cyd. The NMR spectra are shown in Fig. 13 and are consistent with the structures and peak assignments shown in the figure. These structures are also consistent with the IR, UV, and mass spectra. The UV spectra of these adducts are similar to those of the dehydrated

**Table 8** Coupling Constants of the Three Dimers

| Dimer | $J_{^{13}C-H}$ (Hz) | $J_{N^{13}C-H}$ (Hz) | $J_{H-H}$ (cyclobutane, Hz) |
|---|---|---|---|
| Low-melting $Me_2Thy{\diamond}Me_2Thy$ | 155 | 148, 150 | 0 |
| High-melting $Me_2Thy{\diamond}Me_2Thy$ | 151 | 140, 142 | $4.8 \pm 0.5$ |
| Thy◇Thy | 158 | . . . | $5.2 \pm 0.5$ |

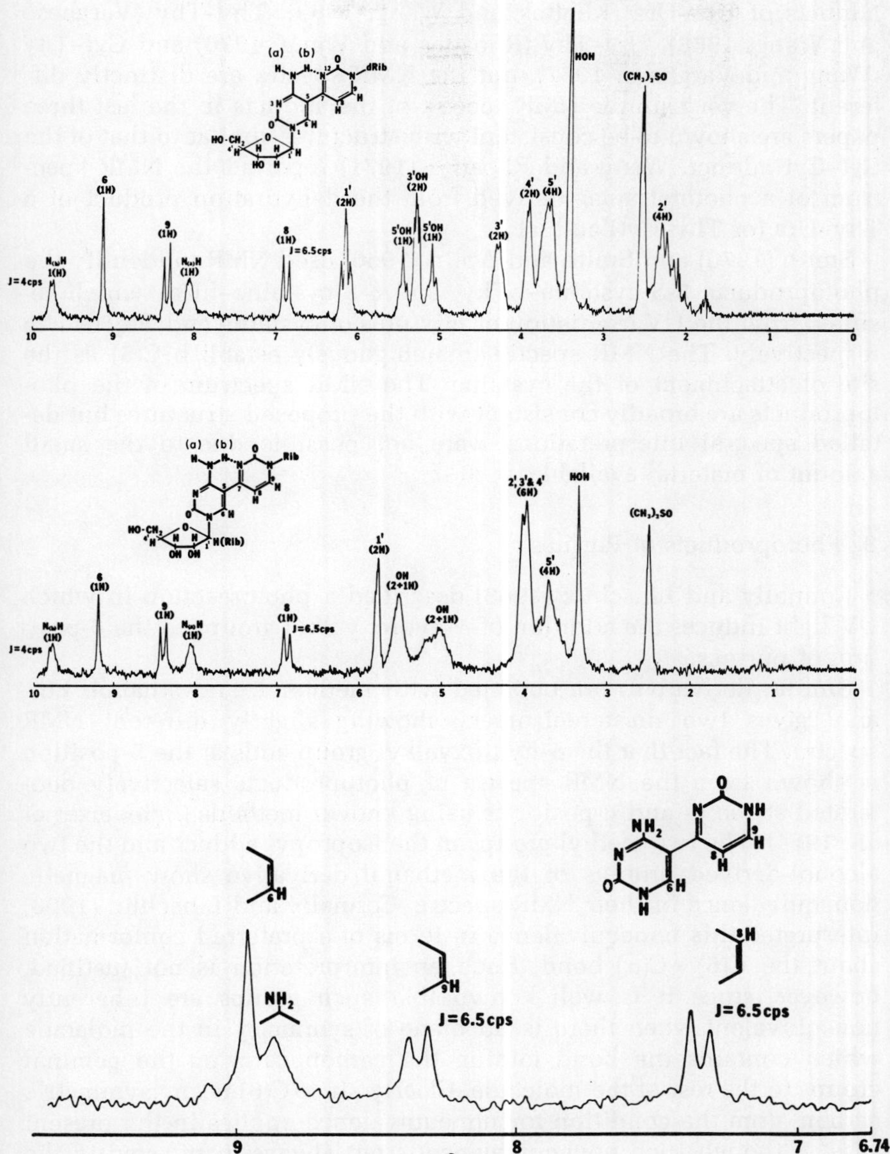


**Fig. 13.** *NMR spectra of dCyd–dCyd (top) and Cyd–Cyd (middle) in $(CD_3)_2SO$ and Cyt–Cyt (bottom) in $CF_3COOH$ at 100 MHz with internal standard tetramethylsilane.*

adducts of Ura–Ura (Khattak and Wang, 1969), Thy–Thy (Varghese and Wang, 1968), Ura–Thy (Rhoades and Wang, 1970) and Cyt–Thy (Wang and Varghese, 1967), but the NMR spectra are distinctly different. The very simple NMR spectra of the adducts in the last three papers are shown to be consistent with structures similar to that of the Cyt–Cyt adduct. Wang and Rhoades (1971) reported the NMR spectrum of a phototetramer derived from the dehydration product of a Thy–Ura (or Thy–Cyt) adduct.

Smith (1970) and Smith and Aplin (1966) used NMR to identify the photoproducts **5**-S-cysteine–hThy, and **5**-S-cysteine–hUra which resulted from the UV irradiation of mixtures of cysteine and Thy or Ura respectively. The NMR spectra unambiguously establish C(5) as the site of attachment of the cysteine. The NMR spectrum of the photoproducts are broadly consistent with the proposed structures but detailed spectral interpretations were not possible due to the small amount of material available.

## 8. Photoproducts of Purines

Connally and Linschitz (1968) described a photoreaction in which UV light induces the addition of $\alpha$-hydroxyalkyl groups to the 6-position of purines.

Similar derivatives are obtained with methanol, and ethanol. Ethanol gives two diastereoisomers showing slightly different NMR spectra. The fact that the $\alpha$-hydroxyalkyl group adds at the 6-position is shown from the NMR spectra of photoproducts selectively deuterated at the 6- and 8-positions using known methods (Schweizer *et al.*, 1964). The two methyl groups in the isopropyl adduct and the two alcohol-derived protons of the methanol derivative show magnetic nonequivalence in their NMR spectra. Connally and Linschitz (1968) interpreted this nonequivalence in terms of a preferred conformation about the C(6)—C($\alpha$) bond. Such an interpretation is not justified, however, since it is well known that such groups are inherently nonequivalent when there is no plane of symmetry in the molecule which contains the bond joining the carbon carrying the geminal groups to the rest of the molecule. Clearly since C(6) is an asymmetric carbon atom the condition for nonequivalence applies in the present case. Although such nonequivalence is not always manifested in the NMR spectra, since it is dependent on the distance to the asymmetric center, it is usually observed when the geminal group is attached directly to an asymmetric carbon atom as in the present case. It should be emphasized that the presence of the asymmetric center does not eliminate the possibility of a preferred conformation, but the asymmetry is sufficient to explain the observed nonequivalence. The two effects, asymmetry and conformational preference, on the observed nonequivalence can in principle be separated by raising the temperature to the point where all conformational states are equally populated. At this point the NMR spectrum will cease to be temperature dependent, but the temperature-independent contribution due to asymmetry will still be observable and its magnitude can be determined. In practice, it may not be possible to raise the temperature sufficiently to remove the effect of conformational preference for obvious reasons. For examples and a detailed discussion of these effects, see Martin and Martin (1966), Becker (1969), and especially Reynolds and Schaeffer (1964).

Steinmaus *et al.* (1969, 1971) reported the NMR spectra of photoproducts of caffeine and Ade derivatives in which UV or $\gamma$-irradiation of the bases with alcohols at room temperature leads to substitution of an alcohol moiety for the proton at C(8). With $Am^2Pur$ the substitution took place at C(6) of the Pur first and was followed by substitution at C(8). NMR was used along with selective deuteration of the C(8) position of the Pur to establish the site of substitution and the structure of the products. Evans and Wolfenden (1970) mentioned the NMR spectra of Pur–methanol addition products.

### 9. Products of Ionizing Radiation

Chabre *et al.* (1966) have analyzed the NMR spectra of nine dihydropyrimidinediones: hUra, $Br^5$hUra, $ho^5$hUra, $Br^5$hThy, $ho^5$hThy, $Br_2^{5,5}ho^6$hUra, *trans*-$Br^5ho^6$hThy, *trans*-$Cl^5ho^6$hThy, and *cis*-$ho_2^{5,6}$hThy. In these compounds a simple first-order analysis of the NMR spectra is possible except in the cases of the $Br^5$hUra and $ho^5$hUra which were analyzed as ABKX or ABKXM spectra to obtain the coupling constants. From the values of the coupling constants obtained it was possible to determine the conformation of some of these nonplanar derivatives using the Karplus equation. It was found that the hydroxy group of $ho^5$hUra is pseudoequatorial while the bromo group of $Br^5$hUra is pseudoaxial. These conclusions can be compared to the result mentioned in the section on photohydrates (Section G,5) which suggest that the hydroxy group of Ura photohydrate is also pseudo equatorial.

Cadet and Teoule (1971) studied the peroxides produced by $\gamma$-irradiation of Thy in aerated aqueous solution. They have presented the NMR chemical shifts and coupling constants of eleven hydroperoxides, most of which were previously unknown. In addition, Hahn and Wang (1972, 1973) obtained the spectra for *trans*-(ho)$_2^{5,6}$hThy and for *cis*-ho$^5$ho$_2^6$hThy.

## References


Andrew, E. R. (1955). "Nuclear Magnetic Resonance." Cambridge Univ. Press, London and New York.

Anet, R. (1965). *Tetrahedron Lett.* p. 3713.

Becker, E. D. (1969). "High Resolution NMR," p. 224. Academic Press, New York.

Becker, E. D., Miles, H. T., and Bradley, R. B. (1965). *J. Amer. Chem. Soc.* **87,** 5575.

Beukers, R., and Berends, W. (1960). *Biochim. Biophys. Acta* **41,** 550.

Blackburn, G. M., and Davies, R. H. (1965). *Chem. Commun.* p. 215.

Blackburn, G. M., and Davies, R. J. H. (1966). *Tetrahedron Lett.* **37,** 4471.

Bovey, F. A. (1969). "Nuclear Magnetic Resonance Spectroscopy," p. 363. Academic Press, New York.

Bullock, F. J., and Jardetzky, O. (1964). *J. Org. Chem.* **29,** 1988.

Cadet, J., and Teoule, R. (1971). *Biochim. Biophys. Acta* **238,** 8.

Chabre, M., Gagnaire, D., and Nofre, C. (1966). *Bull. Soc. Chim. Fr.* **5,** 108.

Chambers, R. W. (1968). *J. Amer. Chem. Soc.* **90,** 2192.

Cohen, A. D., Sheppard, N., and Turner, J. J. (1958). *Proc. Chem. Soc., London* p. 118.

Connally, J. S., and Linschitz, J. (1968). *Photochem. & Photobiol.* **7,** 791.

DeBoer, G., and Johns, H. E. (1970). *Biochim. Biophys. Acta* **204,** 18.

Eaton, P. E. (1962). *J. Amer. Chem. Soc.* **84,** 2344.

Ernst, R. R. (1966). *Advan. Magn. Resonance* **2,** 1–135.

Ernst, R. R., and Anderson, W. A. (1966). *Rev. Sci. Instrum.* **37,** 93–102.

Evans, B., and Wolfenden, R. (1970). *J. Amer. Chem. Soc.* **92,** 4751.

Hahn, B. S., and Wang, S. Y. (1972). *J. Amer. Chem. Soc.* **94,** 4764.

Hahn, B. S., and Wang, S. Y. (1973). *Biochem. Biophys. Res. Commun.* **54,** 1224.

Hollis, D. P., and Wang, S. Y. (1967). *J. Org. Chem.* **22,** 1620.

Jardetzky, C. D., and Jardetzky, O. (1960). *J. Amer. Chem. Soc.* **82,** 222.

Karplus, M. (1959). *J. Chem. Phys.* **30,** 6.

Katritzky, A. R., Nesbit, M. R., Kurtev, B. J., Lyapova, M., and Pojarlieff, I. G. (1969). *Tetrahedron* **25,** 3807.

Khattak, M. N., and Wang, S. Y. (1969). *Science* **163,** 1341.

Kokko, V. P., Mandel, L., and Goldstein, J. H. (1962). *J. Amer. Chem. Soc.* **84,** 1042.

Kunieda, T., and Witkop, B. (1971). *J. Amer. Chem. Soc.* **93,** 3493.

Lowe, I. D., and Norberg, R. E. (1957). *Phys. Rev.* **107,** 46.

Martin, L., and Martin, J. (1966). *Bull. Soc. Chim. Fr.* p. 2117.

Matsuura, S., and Goto, T. (1963). *Tetrahedron Lett.* **22,** 1499.

Miles, H. T., Howard, F. B., and Frazier, J. (1963). *Science* **142,** 1458.

Pople, J. A., Schneider, W. G., and Berstein, H. J. (1959). "High Resolution Nuclear Magnetic Resonance." McGraw-Hill, New York.

Reynolds, W. A., and Schaeffer, T. (1964). *Can. J. Chem.* **42,** 2119.

Rhoades, D. F., and Wang, S. Y. (1970). *Biochemistry* **9,** 4416.

Rhoades, D. F., and Wang, S. Y. (1971). *J. Amer. Chem. Soc.* **93,** 3779.

Rouillier, P., Delmau, J., Duplan, J., and Nofre, C. (1966a). *Tetrahedron Lett.* p. 4189.

Rouillier, P., Delmau, J., and Nofre, C. (1966b). *Bull. Soc. Chim. Fr.* p. 3515.

Schweizer, M. P., Chan, S. I., Helmkamp, G. K., and Ts'o, P.O.P. (1964). *J. Amer. Chem. Soc.* **86,** 696.
Smith, K. C. (1970). *Biochem. Biophys. Res. Commun.* **34,** 1011.
Smith, K. C., and Aplin, R. T. (1966). *Biochemistry* **5,** 2125.
Steinmaus, H., Rosenthal, I., and Elad, D. (1969). *J. Amer. Chem. Soc.* **91,** 4921.
Steinmaus, H., Rosenthal, I., and Elad, D. (1971). *J. Org. Chem.* **36,** 3594.
Ts'o, P.O.P., Schweizer, M. P., and Hollis, D. P. (1969). *Ann. N. Y. Acad. Sci.* **158,** 256.
Varghese, A. J., and Wang, S. Y. (1967). *Nature (London)* **213,** 909.
Varghese, A. J., and Wang, S. Y. (1968). *Science* **160,** 186.
Wang, S. Y., and Rhoades, D. F. (1971). *J. Amer. Chem. Soc.* **93,** 2554.
Wang, S. Y., and Varghese, A. J. (1967). *Biochem. Biophys. Res. Commun.* **29,** 543.
Wechter, W. J., and Smith, K. C. (1968). *Biochemistry* **7,** 4064.
Weinblum, D., and Johns, H. E. (1966). *Biochim. Biophys. Acta* **114,** 450.
Wulff, D. L., and Fraenkel, G. (1961). *Biochim. Biophys. Acta* **51,** 332.

# 11 Crystal and Molecular Structure of Photoproducts from Nucleic Acids

*Isabella L. Karle*



## A. Introduction

Many types of photoproducts have been isolated from DNA irradiated by UV. Generalizations cannot be made as to which photoproduct is the most important in a particular system. It seems certain that different types of photoproducts inactivate irradiated vegetative cells and irradiated spores. Furthermore, if repair of the damaged DNA is relatively easy, then the biological significance of a photoproduct causing such damage is not nearly as great as that of one causing irreversible damage. It is of utmost importance to be able to identify the photoproducts and to establish the geometric form of each molecule in order to be able to assess the role that each one plays in the photochemistry and photobiology of the nucleic acids.

X-Ray diffraction analysis of a single crystal is eminently suited to the elucidation of many structural aspects of a material in the solid state, assuming that a single crystal can be grown. The types of information derivable from a crystal structure analysis can be divided into two categories: information about the molecular unit and information about the relationship of molecules with respect to each other in the crystal lattice. From the x-ray scattering data alone, even if there is no information other than an approximate knowledge of the empirical formula, the structural formula of the molecule, its stereoconfiguration, its conformation, accurate bond lengths, bond angles, and torsional angles can be established. In addition, for optically active molecules, the absolute stereoconfiguration can be derived, providing that in the crystal lattice there is a heavy atom which has a suitable anomalous dispersion effect on the scattered x-rays (Bijvoet *et al.*, 1951). Once the structure of the molecule is known, the spatial arrangement in the crystal lattice gives an insight into the nature of the forces which bind the molecules together. The scattering data yield information about hydrogen bonds, either within each molecule or between molecules, water of crystallization or other solvents which may have cocrystallized with the substance of interest, molecular complexes formed by two or more different kinds of molecules, charge-transfer compounds, clathrates in which one kind of molecule is entrapped in cages formed by another kind of molecule, and finally, coordination numbers, that is, the number and arrangement of nearest neighbors to particular atoms or ions in a crystal.

The fullest potential of x-ray diffraction analysis is realized when it is applied to a crystal of an unknown material, such as a natural product or a product from chemical or photo reactions or rearrangements, in which the primary purpose is to elucidate the structural formula. The analysis can be performed directly on the original native material without any need for preparing heavy-atom derivatives. Often the structure analysis is used for confirmation of a substance whose molecular formula and/or conformation have been derived by other physical or chemical methods. In either case, the gross shape of the molecule or the detailed structural parameters can be correlated with physical properties and chemical or biological activity.

A crystal is composed of a three-dimensional array of regularly repeating units called unit cells. A unit cell is characterized by three axial lengths, $a$, $b$, and $c$, and by three angles, $\alpha$, $\beta$, and $\gamma$, which are not necessarily equal to 90°. A unit cell may contain one molecule or, more generally, two, three, four, six, eight, or more molecular units. These molecular units are usually related to each other by one or more symmetry elements such as centers of symmetry, rotation axes, screw axes, inversion axes, mirror planes, and glide planes. It is necessary to derive structural parameters only for the contents of the asymmetric unit, and the parameters for the remainder of the molecules in the unit cell are obtained by applying appropriate symmetry operations. These symmetry relationships for the various space groups are tabulated in the International Tables for X-Ray Crystallography (1965).

## B. Data Collection

A crystal can be considered to be a three-dimensional diffraction grating. Many different sets of parallel planes that intersect the axes of the unit cell can be imagined. Two different sets are illustrated in the two-dimensional lattice in Fig. 1. The perpendicular distance between the planes is called "$d$" and each set of planes can be designated by three indices, $h$, $k$, $l$, which are the reciprocals of the fractional intercepts of the planes on the lattice axes. Thus in Fig. 1b, the planes intercept the $a$ axis at 1/1, the $b$ axis at 1/3, and the $c$ axis (pointing up from the plane of the page) at $1/\infty$ if the planes are parallel to the $c$ axis. The $h\ k\ l$ indices for this set of planes are written 1 3 0. The distance $d$ is a geometric function of the indices $h, k, l$ and the cell parameters $a$, $b$, $c$, $\alpha$, $\beta$, and $\gamma$.

When a collimated x-ray beam impinges upon a crystal, it is scattered at discrete angles according to the Bragg relationship

$$n\lambda = 2d \sin \theta \tag{1}$$

in which $2\theta$ is the angle between the incident and the scattered ray, $d$ is the interplanar distance, $n$ is an integer, and $\lambda$ is the wavelength of the x-rays (1.54 Å for x-radiation from a tube with a copper anode). In the usual experimental arrangement, the x-ray beam is fixed and the crystal is rotated about various axes so that the lattice planes can be oriented at the proper angles for scattering to take place. Each reflection from a set of planes has an intensity that is a function of the electron density which is intersected by that particular set of planes. The

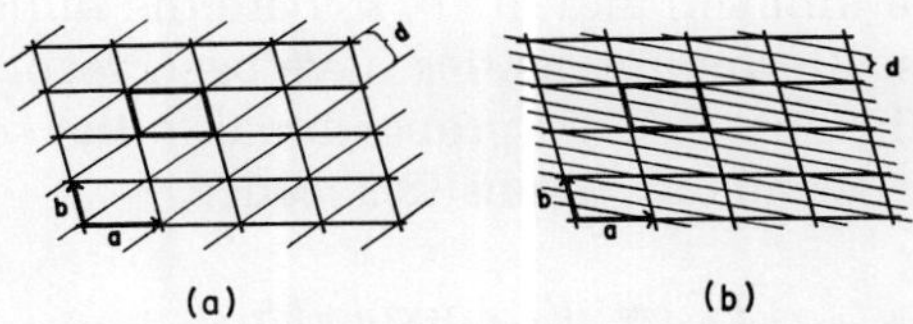


**Fig. 1.** *Two sets of lattice "planes" in a two-dimensional lattice. A unit cell in two dimensions with cell lengths* **a** *and* **b** *is outlined by the heavy lines. The symbol* **d** refers to interplanar spacing. (*a*) *1 $\bar{1}$ 0 planes;* (*b*) *1 3 0 planes.*

intensities of the reflected x-rays can be collected on photographic film or, more recently, by a counter on a computer-controlled diffractometer which sets all the angles automatically. For a description of the details of data collection, the reader is referred to a text such as *X-Ray Structure Determination, A Practical Guide* (Stout and Jensen, 1968).

## C. Structure Determination

The primary objective of the direct method of x-ray structure analysis is to determine the location of the atoms in the unit cell of a crystal by means of a direct mathematical analysis of the diffracted x-ray intensities. This is in contrast to other procedures which rely heavily on the introduction of auxiliary chemical, physical, and structural information and may involve a considerable amount of trial and error. One of the advantages of the direct method is that it permits the determination of structures of unknown formula or chemical composition without any a priori knowledge and without the introduction of heavy atoms into the lattice. The direct method is particularly well suited to the solution of the structures of the photoproducts derived from nucleic acid bases and was used for almost all of the structures described in the next section.

Diffraction data consist of a listing of the values of $h$, $k$, and $l$ for the reflecting planes within the experimental range, $0 \leqslant \sin\theta \leqslant 1.0$, and the experimentally measured intensities associated with them. These data are used in the electron density function

$$\rho(XYZ) = \frac{1}{V} \sum_{h\,-\infty}^{\infty} \sum_{k\,-\infty}^{\infty} \sum_{l\,-\infty}^{\infty} F_{hkl} \exp\left[-2\pi i(hX + kY + lZ)\right] \tag{2}$$

which is essentially zero everywhere except at values for $x, y, z$ where atoms occur. Hence the maxima of this three-dimensional function give a graphic representation of the crystal structure. The quantity $V$ is the volume of the unit cell and the $h$, $k$, $l$ are the indices of the planes for which x-ray scattering intensities have been recorded. The magnitudes of the coefficients $F_{hkl}$ are proportional to the square root of the corresponding experimental intensities. Since

$$F_{hkl} = |F_{hkl}| \exp(-i\phi_{hkl}), \tag{3}$$

and only the magnitudes $|F_{hkl}|$ are available from the experimental measurements, information concerning the coefficients $F_{hkl}$ is incomplete. If the $\phi_{hkl}$ were also known, the $\rho(XYZ)$ could be computed immediately to give the crystal structure. The search for methods for obtaining at least approximate values for the $\phi_{hkl}$, the "phase problem" in crystallography, has been of utmost importance for many years.

For a crystal of an organic material, for example, consisting of $M$ atoms in the asymmetric unit (excluding hydrogen atoms), the number of independent reflections, not related by symmetry elements, available with an x-ray beam from a copper anode is of the order of 60–80$M$. The number of unknowns that need to be considered in the crystal are only 3$M$, the three coordinates $x$, $y$, and $z$ of each atom in the asymmetric unit. Hence the problem from a mathematical point of view is greatly overdetermined. This overdeterminacy with regard to number of data as compared to number of unknowns greatly facilitated the solution of the phase problem.

The mathematical relationships between the $\phi_{hkl}$ and the $|F_{hkl}|$ and their probabilistic significance are based on a fundamental set of inequality relationships which arise from the fact that the electron density distribution representing the structure of a crystal, Eq. (2), is a non negative function (Karle and Hauptman, 1950). It is now possible to derive the phases of a sufficient number of reflections with strong intensities from the magnitudes of the intensities alone to insert into Eq. (2) and to obtain the structure. In practice, the most useful phase relations have been

$$\phi_{h_1k_1l_1} \approx \phi_{h_2k_2l_2} + \phi_{h_{1-2}k_{1-2}l_{1-2}} \tag{4}$$

and

$$\tan \phi_{h_1k_1l_1} = \frac{\sum_{h_2k_2l_2} |E_{h_2k_2l_2}E_{h_{1-2}k_{1-2}l_{1-2}}| \sin(\phi_{h_2k_2l_2} + \phi_{h_{1-2}k_{1-2}l_{1-2}})}{\sum_{h_2k_2l_2} |E_{h_2k_2l_2}E_{h_{1-2}k_{1-2}l_{1-2}}| \cos(\phi_{h_2k_2l_2} + \phi_{\hat{h}_{1-2}k_{1-2}l_{1-2}})} \tag{5}$$

in which the $|E_{hkl}|$ are called the normalized structure factors and are readily derivable from the corresponding $|F_{hkl}|$. Equation (4) is valid only for the $|E_{hkl}|$ with the largest magnitudes and is used in the initial stages of phase determination to obtain approximate values of the phases for a relatively small number of $|E_{hkl}|$ with large magnitudes. To implement Eq. (4), several phases are known at the outset since phases for a maximum of three reflections, depending upon the symmetry of the crystal, can be assigned arbitrary values within a prescribed set of rules in order to choose an origin for the unit cell. In crystal systems possessing a center of symmetry, all phase values are restricted to the two values, 0 or $\pi$; whereas for crystal systems without a center of symmetry, in which all optically active materials must crystallize, the phase values range from 0 to $2\pi$. In noncentrosymmetric unit cells, one other assignment of phase is made by choosing the plus or the minus value for a reflection with a phase near $\pi/2$ in order to specify either the direction of the axes or the enantiomorph or both, as the case may be. Usually several other reflections are assigned phase values in terms of unknown symbols in addition to those assigned for specifying the origin and enantiomorph. In the course of applying Eq. (4) to extend the set of determined phase values, many $\phi_{hkl}$ can be determined in several different ways. In this way relationships occur among some of the unknown symbols, allowing them to be evaluated. The remaining unknown symbols, if

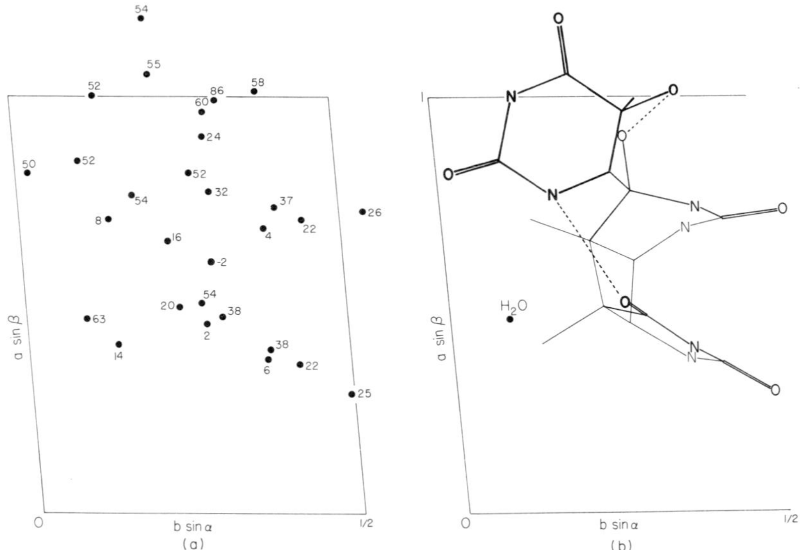


**Fig. 2.** *(a) The dots represent the maxima in an E-map computed with 458* $E_{hkl}$, *the phases of which were determined directly from the magnitude of the observed intensities of the x-ray reflections. The numbers next to the dots represent the fractional value (×100) of the third coordinate of the triclinic cell. The quantities* $a \sin \beta$ *and* $b \sin \alpha$ *are related to the unit cell parameters. (b) Lines representing bonds between atoms connect those dots which are separated by 1.2–1.6 Å, thus revealing the structure of a Thy trimer (Flippen and Karle, 1971). The dotted lines indicate intramolecular hydrogen bonds.*

any, are permitted to assume a small number of values spanning the allowed range. The correct answer is readily found from these possibilities. The approximate phase values obtained with Eq. (4) are refined with the use of Eq. (5), and phases for many more reflections are then obtained with Eq. (5). For a more detailed description of the phase-determining procedure the reader is referred to Karle and Karle (1966) and Karle (1969).

Phase values are determined for a sufficient number of $|E_{hkl}|$ in order to compute a well-resolved $E$-map with Eq. (2), in which the $|F_{hkl}|$ have been replaced with $|E_{hkl}|$. An example of such a map is illustrated in Fig. 2a for a molecule of a Thy trimer in which 458 $|E_{hkl}|$ values were used. The x, y, and z coordinates of the 29 maxima found in the three-dimensional map are plotted as fractions of the unit cell edges. The numbers next to the dots represent the third coordinate. Dots separated by 1.2–1.6 Å, the range for C–O, C–N, and C–C bond lengths, are connected in Fig. 2b and the geometry of the molecule is revealed. Atoms can be identified as C, N, or O in several different ways: from a knowledge of the chemical formula, from the values of the bond lengths, from the behavior of thermal factors in a least-squares refinement, and from difference maps. Suppose that the identity of the atoms is ambiguous. Each atom could be temporarily assigned a C label and a least-squares procedure could be carried out in which the variables are the x, y, and z coordinates of each atom and a number $B$ which represents the thermal motion of each atom. The object is to vary the x, y, z, and $B$ values by small amounts in order to compute structure factors [Eq. (6)] which reproduce best

$$F_{hkl} = \sum_j f_j \exp[2\pi i(hx_j + ky_j + lz_j)] \exp[-B_j(\sin\theta/\lambda)^2], \qquad (6)^*$$

the observed $|F_{hkl}|$ values. Those atoms which should have been labeled as N or O and were considered to be C atoms are indicated in the least-squares refinement by abnormally small values for the thermal parameters and thus a proper identification can be made. The great overdeterminacy of the number of observed reflections with respect to the number of x, y, z, and $B$ values to be fixed results in very accurate values for the coordinates and thermal factors. Once these values are determined for the C, N, and O atoms, difference maps can be computed to locate the hydrogen atoms. The function used is Eq. (2) in which the coefficients $F_{hkl}$ are replaced by

* The $f_j$ are the atomic scattering factors which are tabulated, e.g., in the International Tables for X-Ray Crystallography (1965).

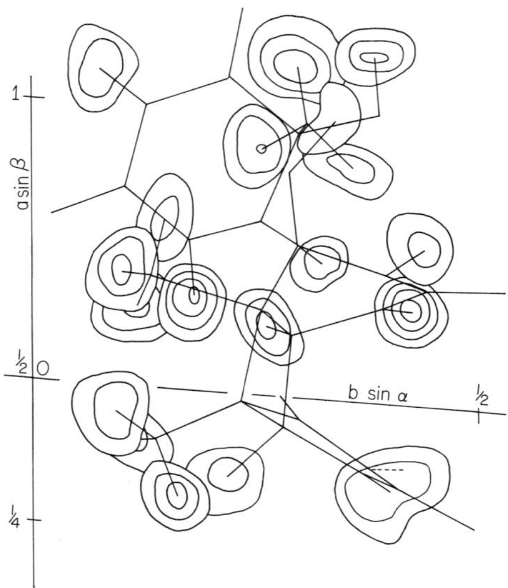


**Fig. 3.** *A difference map for the Thy trimer. The values of the electron density are essentially zero at each point except where the contours have been drawn. Each volume of electron density represents the position of an H atom. The Thy trimer molecule has been drawn in to show where each H atom is bonded.*

$(F_{hkl})_{obs} - (F_{hkl})_{calc}$ in which the $(F_{hkl})_{calc}$ are obtained with Eq. (6) using the parameters for the nonhydrogen atoms. Although the magnitudes of the quantities $(F_{hkl})_{obs} - (F_{hkl})_{calc}$ are generally small, since they represent the scattering effect from only the hydrogen atoms, the difference maps are definitive in locating the positions of the hydrogen atoms as shown in Fig. 3. The contours represent the electron density revealed in the difference map. The positions of the C, N, and O atoms in the Thy trimer have been drawn in to show that each region of maximum electron density corresponds to an H atom situated at 1.0–1.1 Å from a C, N, or O atom. If the heavier atoms were incorrectly labeled, the difference map would also indicate positive or negative electron density at the position of such atoms, depending upon whether the mislabeled atoms were assigned atomic numbers lower or higher than the correct ones.

The molecular formula and the geometry of the molecule have now been established. The coordinates obtained from the analysis of the x-ray scattering data can be used to compute bond lengths, bond angles, torsion angles about each bond, deviations from planarity, intra- and intermolecular approaches, and many other quantities.

## D. Structural Results for Photoproducts

A variety of products have been isolated from nucleic acids, Thy, Ura, and their derivatives after irradiation with uv light. Experimental conditions for irradiation and the procedures for isolation are discussed elsewhere in this book. This chapter is mainly concerned with the elucidation of the structure and the comparison of the geometries of the various products.

The reactive site for the photochemical reactions seems to be the 5,6-double bond in the pyrimidines. Thy is a planar molecule except

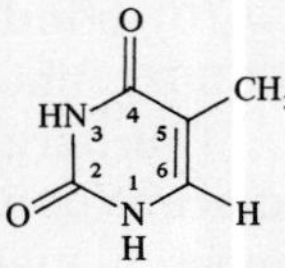

for the H atoms on the methyl group (Gerdil, 1961). The maximum deviation of any atom other than an H atom from the least-squares plane computed for the six ring atoms is 0.018 Å for carbon of the $CH_3$ group while the average deviation from this plane of the six ring atoms is only 0.004 Å. The positions of the H atoms were located in a difference map which unequivocally shows that H atoms are localized on the N atoms and that Thy is in the keto form. Double strands of hydrogen bonds of the type NH···O=C, occurring around two different centers of symmetry in the unit cell, link the molecules into infinite ribbons. A portion of the hydrogen bonding scheme is shown in Fig. 4 along with the values for the hydrogen bond lengths and

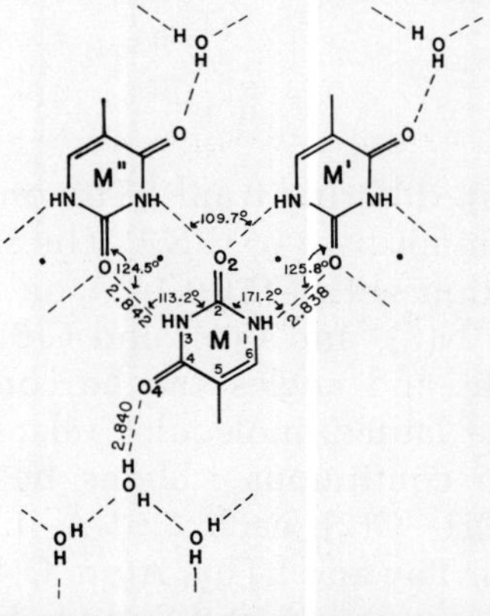

**Fig. 4.** *A portion of the infinite hydrogen bond network in Thy. Hydrogen bond lengths are given in Å. Molecule M is related to M′ and M″ by two different centers of symmetry, represented by the dots (Gerdil, 1961).*

angles. The O atom on C(2) is an acceptor for two hydrogen bonds from NH groups while the O atom on C(4) is an acceptor for one hydrogen bond from a molecule of water of crystallization, which in turn is hydrogen bonded to other molecules of water. Removal of the water of crystallization causes the lattice to collapse. It will be seen that similar hydrogen bonding occurs in crystals of the more complicated molecules produced by UV irradiation.

1. Dihydropyrimidines

The saturation of the C(5)–C(6) bond in hThy, isolated from irradiated DNA (Yamane *et al.*, 1967), not only increases the length of this bond from 1.35 to 1.52 Å (Furberg and Jensen, 1968) but also increases the lengths of the adjacent bonds, C(4)–C(5) and C(6)–N(1). In addition, the ring is no longer planar but assumes the half-chair conformation with C(5) and C(6) out of the plane of the other four atoms by 0.42 and 0.31 Å in opposite directions. The $CH_3$ group is in the equatorial position and is displaced only 0.06 Å from the plane containing the N(1), C(2), N(3), and C(4) atoms. In the crystal lattice, the molecules are held together in continuous ribbons by forming N(1)H···O(2) and N(3)H···O(2) hydrogen bonds (2.88 and 2.85 Å, respectively) in a similar fashion to that found in the Thy monohydrate crystal. In hThy there is no water of crystallization and O(4) does not participate in hydrogen bonding.

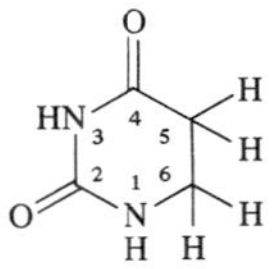

Dihydrouracil (hUra), differing from hThy only by one $CH_3$ group, occurs naturally as a minor base of tRNA. The six-membered ring has a half-chair conformation with C(5) 0.14 Å on one side of the plane formed by N(1), C(2), N(3), and C(4), and C(6) 0.45 Å on the other side. The bond lengths and angles can be compared with hThy in Tables 1 and 2. In the lattice, molecules related by centers of symmetry are joined into continuous ribbons by the hydrogen bonds N(1)H···O(2) and N(3)H···O(2), each 2.91 Å in length, in a fashion similar to that found in Thy and hThy. Atom O(4) does not participate in any hydrogen bonding (Rohrer and Sundaralingam, 1970).

The hThd formed by the catalytic reduction of Thd was shown by chemical and spectroscopic means to exist as a single diastereoisomer

**Table 1** Bond Lengths (Å) in Photoproducts Associated with Irradiated Nucleic Acids

| Bond | Thy[a] | hThy[b] | hUra[c] | hThd[d] | *cis*-Thy glycol[e] | U◇U(*c,s*)[f] | 6 MU◇6 MU(*c,s*)[g] | T < $C_3$ > T(*c,s*)[h] | $M_2T$◇$M_2T$(*c,s*)[i,j] |
|---|---|---|---|---|---|---|---|---|---|
| C(6)–N(1) | 1.382 | 1.46[i] | 1.464 | 1.473 | 1.439 | 1.441 | 1.459 | 1.444 | 1.42 |
| | | | | | | 1.435 | 1.432 | 1.455 | 1.45 |
| N(1)–C(2) | 1.355 | 1.326 | 1.335 | 1.347 | 1.333 | 1.336 | 1.337 | 1.341 | 1.29 |
| | | | | | | 1.330 | 1.327 | 1.339 | 1.32 |
| C(2)–O(2) | 1.234 | 1.235 | 1.222 | 1.224 | 1.255 | 1.225 | 1.222 | 1.223 | 1.23 |
| | | | | | | 1.232 | 1.230 | 1.228 | 1.23 |
| C(2)–N(3) | 1.361 | 1.383 | 1.395 | 1.387 | 1.384 | 1.390 | 1.402 | 1.395 | 1.44 |
| | | | | | | 1.398 | 1.405 | 1.388 | 1.42 |
| N(3)–C(4) | 1.391 | 1.358 | 1.364 | 1.370 | 1.376 | 1.360 | 1.364 | 1.365 | 1.41 |
| | | | | | | 1.364 | 1.364 | 1.374 | 1.35 |
| C(4)–O(4) | 1.231 | 1.212 | 1.211 | 1.211 | 1.214 | 1.218 | 1.206 | 1.214 | 1.22 |
| | | | | | | 1.220 | 1.211 | 1.213 | 1.23 |
| C(4)–C(5) | 1.447 | 1.54[i] | 1.515 | 1.518 | 1.535 | 1.498 | 1.507 | 1.513 | 1.48 |
| | | | | | | 1.497 | 1.501 | 1.508 | 1.52 |
| C(5)–C(6) | 1.349 | 1.52[i] | 1.507 | 1.500 | 1.517 | 1.540 | 1.543 | 1.555 | 1.55 |
| | | | | | | 1.533 | 1.565 | 1.548 | 1.58 |
| C(5)–C(5′) | — | — | — | — | — | 1.572 | 1.546 | 1.596 | 1.60 |
| C(6)–C(6′) | — | — | — | — | — | 1.563 | 1.586 | 1.551 | 1.66 |
| C(5)–C(6′) | — | — | — | — | — | — | — | — | — |
| C(6)–C(5′) | — | — | — | — | — | — | — | — | — |
| $CH_3 \cdots (CH_3)'$ | — | — | — | — | — | — | 2.99 | 2.94 | 3.01 |

*(Continued)*

**Table 1** (*Continued*)

| $M_2T\diamond M_2T(c,a)$[k] | $U\diamond U(c,a)$[l] | $T\diamond T(t,a)$[m,n] | $MO\diamond MO(t,s)$[o,p] | T–T adduct[q] Ring I | T–T adduct[q] Ring II | Trimer $T\diamond T$–$T(c,s)$[r] Ring I | Trimer Ring II | Trimer Ring III | Tetramer[s] Ring I / Ring II | Tetramer[s] Ring III / Ring IV |
|---|---|---|---|---|---|---|---|---|---|---|
| 1.431 | 1.45 | 1.440 | 1.428 | 1.466 | 1.357 | 1.465 | 1.453 | 1.477 | 1.40 | 1.46 |
| 1.449 | 1.42 | | | | | | | | 1.39 | 1.46 |
| 1.342 | 1.32 | 1.334 | 1.349 | 1.302 | 1.344 | 1.376 | 1.355 | 1.327 | 1.37 | 1.32 |
| 1.344 | 1.35 | | | | | | | | 1.37 | 1.34 |
| 1.218 | 1.25 | 1.227 | 1.222 | 1.247 | 1.252 | 1.264 | 1.276 | 1.225 | 1.23 | 1.26 |
| 1.218 | 1.21 | | | | | | | | 1.22 | 1.24 |
| 1.426 | 1.39 | 1.390 | 1.391 | 1.399 | 1.367 | 1.346 | 1.320 | 1.397 | 1.39 | 1.38 |
| 1.406 | 1.42 | | | | | | | | 1.38 | 1.35 |
| 1.378 | 1.39 | 1.357 | 1.357 | 1.393 | 1.321 | 1.369 | 1.476[u] | 1.377 | 1.38 | 1.42 |
| 1.377 | 1.36 | | | | | | | | 1.40 | 1.42 |
| 1.203 | 1.22 | 1.211 | 1.213 | 1.207 | — | 1.204 | 1.440[u] | 1.217 | 1.22 | — |
| 1.213 | 1.24 | | | | | | | | 1.23 | — |
| 1.509 | 1.50 | 1.508 | 1.502 | 1.508 | 1.382 | 1.552 | 1.538 | 1.526 | 1.45 | 1.32 |
| 1.503 | 1.51 | | | | | | | | 1.45 | 1.33 |
| 1.533 | 1.54 | 1.547 | 1.544 | 1.555 | 1.364 | 1.492 | 1.545 | 1.557 | 1.35 | 1.49 |
| 1.529 | 1.56 | — | — | — | — | — | — | — | 1.34 | 1.47 |
| — | — | — | — | — | — | 1.583 | — | — | — | — |
| — | — | — | — | — | — | 1.566 | — | — | — | — |

| | | | | | | | | | | |
|---|---|---|---|---|---|---|---|---|---|---|
| 1.577 | 1.60 | 1.587 | 1.556 | — | — | — | — | — | — | — |
| 1.571 | 1.59 | — | 1.628 | — | — | — | — | — | — | — |
| — | — | — | — | — | — | 2.92 | — | — | — | — |

[a] Gerdil, 1961.
[b] Furberg and Jensen, 1968.
[c] Rohrer and Sundaralingam, 1970.
[d] Konnert and Karle, 1970.
[e] Flippen, 1973.
[f] Adman and Jensen, 1970.
[g] Gibson and Karle, 1971.
[h] Leonard *et al.*, 1969.
[i] Camerman and Camerman, 1970.
[j] Dimer monomerizes in x-ray beam, hence bond lengths are less accurate than the other quoted values.
[k] Camerman *et al.*, 1969.
[l] Konnert and Karle, 1971.
[m] Camerman and Nyburg, 1969.
[n] Molecule has a center of symmetry.
[o] Birnbaum, 1972.
[p] Molecule has a twofold rotation axis.
[q] Karle, 1969.
[r] Flippen and Karle, 1971.
[s] Flippen *et al.*, 1972.
[t] Accuracy affected by disorder in the crystal.
[u] Single bonds.

(Kondo and Witkop, 1968). A crystal structure analysis (Konnert *et al.*, 1970b) confirmed the configuration to be

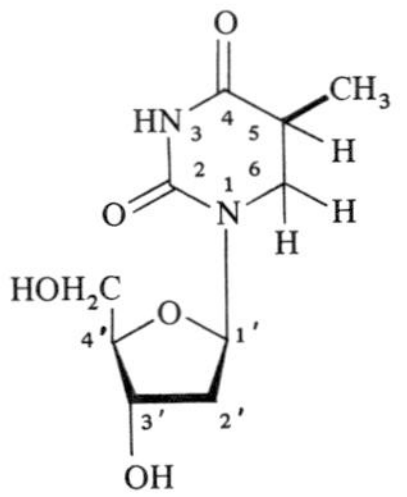

The hThy moiety has the half-chair conformation with C(**5**) 0.27 Å above and C(6) 0.41 Å below the plane of the other four atoms in the ring. As in hThy, the $CH_3$ group is equatorial. Bond lengths and angles are comparable to those found in hThy (Tables 1 and 2); however, since N(1) is not available for hydrogen bond formation and since the optically active hThd must crystallize in a noncentrosymmetric space group, the hydrogen bonding scheme is quite different from that found in the hThy crystal.

2. Cyclobutane Dipyrimidines (Pyr◇Pyr)

Photoproducts initially isolated from radiation of Thy in frozen aqueous solution involve more than one base unit and are the cyclobutyl type dimers (Wang, 1960, 1961, 1964; Beukers and Berends, 1960, 1961) formed by opening the double bond at the C(5)–C(6) position (Scheme 1). It was proposed that four types of Pyr◇Pyr, shown in

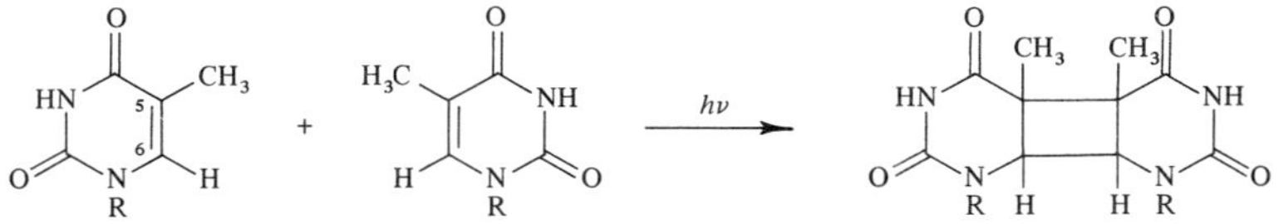

**Scheme 1**

Fig. 5, A–D, could be formed (Wulff and Frankel, 1961). It was also shown that the *cis-syn* dimer is the most prevalent product both from the irradiation of native DNA and from the irradiation of frozen Thy or Ura solutions. The following cyclobutyl-type molecules have been indicated or isolated: Ura◇Ura, Thy◇Thy, Cyt◇Cyt, Ura◇Thy, Cyt◇Thy, Ura◇Cyt, and some of their methylated derivatives. Crystal structure analyses have been completed on the *cis-syn*, *cis-anti*, and

**Table 2** Bond Angles (Degrees) in Photoproducts Associated with Irradiated Nucleic Acids

| Angle | Thy[a] | hThy[b] | hUra[c] | hThd[d] | *cis*-Thy glycol[e] | U◇U(*c*,*s*)[f] | 6 MU◇6 MU(*c*,*s*)[g] | T < $C_3$ > T(*c*,*s*)[h] |
|---|---|---|---|---|---|---|---|---|
| C(6)N(1)C(2) | 122.8 | 121.6 | 122.1 | 120.6 | 123.0 | 126.4 | 125.6 | 125.6 |
| | | | | | | 124.1 | 126.1 | 123.6 |
| N(1)C(2)N(3) | 115.2 | 116.6 | 116.1 | 115.9 | 116.2 | 117.5 | 118.3 | 116.5 |
| | | | | | | 117.3 | 118.0 | 117.6 |
| C(2)N(3)C(4) | 126.3 | 126.3 | 126.7 | 127.0 | 125.1 | 126.8 | 126.3 | 128.3 |
| | | | | | | 127.1 | 125.6 | 127.2 |
| N(3)C(4)C(5) | 115.6 | 113.4 | 115.1 | 114.8 | 114.4 | 115.5 | 115.6 | 117.7 |
| | | | | | | 117.1 | 117.0 | 116.7 |
| C(4)C(5)C(6) | 118.2 | 106.7[i] | 112.6 | 109.6 | 107.1 | 118.2 | 118.8 | 114.9 |
| | | | | | | 114.4 | 116.8 | 114.7 |
| C(5)C(6)N(1) | 121.8 | 108.8 | 110.3 | 110.6 | 109.0 | 111.3 | 111.0 | 116.7 |
| | — | — | — | — | — | 116.0 | 113.8 | 117.1 |
| C(4)C(5)C(5′) | — | — | — | — | — | 117.4 | 116.8 | 114.2 |
| | — | — | — | — | — | 114.7 | 116.3 | 114.3 |
| N(1)C(6)C(6′) | — | — | — | — | — | 112.6 | 110.3 | 115.3 |
| | — | — | — | — | — | 119.1 | 116.5 | 114.8 |
| C(4)C(5)C(6′) | — | — | — | — | — | — | — | — |
| | — | — | — | — | — | — | — | — |
| N(1)C(6)C(5′) | — | — | — | — | — | — | — | — |
| | — | — | — | — | — | — | — | — |
| Dihedral angle of four-membered ring | — | — | — | — | — | 155 | 162 | 178 |
| Torsion angle about C(5)C(5′) or C(6)C(5′) | — | — | — | — | — | 24 | 16.5 | — |
| Conformation of six-membered ring[u] | P | HC | Twist HC | HC | HC | {Q, Q} | {Q, O} | {P, O} |
| Conformation of four-membered ring[u] | — | — | — | — | — | F | F | ~P |

(*Continued*)

**Table 2** (*Continued*)

| $M_2T \diamond M_2T(c,s)$[i,j] | $M_2T \diamond M_2T(c,a)$[k] | $U \diamond U(c,a)$[l] | $T \diamond T(t,a)$[m,n] | $MO \diamond MO(t,s)$[o,p] | T–T adduct[q] | | $T \diamond T\text{–}T(c,s)$[r] | | | Tetramer[s] | |
|---|---|---|---|---|---|---|---|---|---|---|---|
| | | | | | Ring I | Ring II | Ring I | Ring II | Ring III | Ring I Ring II | Ring III Ring IV |
| 126 | 121.7 | 121.7 | 125.7 | 125.6 | 124.9 | 121.9 | 120.5 | 118.7 | 124.2 | 121.9 | 127.6 |
| 125 | 121.6 | 124.2 | | | | | | | | 121.3 | 127.1 |
| 118 | 116.9 | 117.9 | 116.8 | 117.1 | 118.6 | 119.2 | 121.2 | 117.4 | 117.9 | 116.4 | 118.6 |
| 121 | 117.4 | 114.6 | | | | | | | | 116.6 | 118.0 |
| 123 | 125.8 | 126.6 | 127.8 | 127.2 | 123.3 | 118.3 | 125.5 | 121.2 | 124.2 | 124.6 | 119.1 |
| 121 | 126.3 | 128.0 | | | | | | | | 124.9 | 121.0 |
| 119 | 117.7 | 117.5 | 117.7 | 117.6 | 115.8 | 124.5 | 117.7 | 107.3 | 115.6 | 116.4 | 121.8 |
| 119 | 117.8 | 118.4 | | | | | | | | 115.3 | 120.4 |
| 113 | 112.1 | 111.6 | 115.2 | 116.0 | 110.0 | 115.9 | 114.3 | 109.4 | 110.0 | 119.5 | 123.9 |
| 113 | 112.7 | 111.3 | | | | | | | | 119.5 | 123.3 |
| 116 | 116.3 | 115.9 | 116.0 | 115.5 | 109.3 | 120.1 | 119.6 | 118.0 | 107.9 | 121.0 | 109.0 |
| 114 | 115.7 | 116.0 | | | | | | | | 122.2 | 109.9 |
| 112 | — | — | — | 113.6 | — | — | 114.7 | — | — | — | — |
| 119 | — | — | — | — | — | — | — | 120.0 | — | — | — |
| 117 | — | — | — | 117.7 | — | — | 114.7 | — | — | — | — |
| 114 | — | — | — | — | — | — | — | 115.0 | | | |
| — | 111.6 | 112.8 | 109.5 | — | — | — | — | — | — | — | — |
| — | 112.3 | 115.6 | — | — | — | — | — | — | — | — | — |
| — | 122.4 | 121.0 | 116.5 | — | — | — | — | — | — | — | — |
| — | 121.9 | 121.9 | — | — | — | — | — | — | — | — | — |
| 153 | 154 | 150 | 180 | 170 | — | — | 173.5 | — | — | — | — |

| | | | | | | | | | | | |
|---|---|---|---|---|---|---|---|---|---|---|---|
| 26 | 29 | 21 | — | 7.9 | — | — | 6 | — | — | — | — |
| {Q R | Q | Q | ~P | Q | HC | P | ~P | B | HC | P | P |
| F | F | F | P | F | — | — | ~P | — | — | — | — |

[a] Gerdil, 1961.
[b] Furberg and Jensen, 1968.
[c] Rohrer and Sundaralingam, 1970.
[d] Konnert *et al.*, 1970.
[e] Flippen, 1973.
[f] Adman and Jensen, 1970.
[g] Gibson and Karle, 1971.
[h] Leonard *et al.*, 1969.
[i] Camerman and Camerman, 1970.
[j] Dimer monomerizes in x-ray beam, hence bond angles are less accurate than the other quoted values.
[k] Camerman *et al.*, 1969.
[l] Konnert and Karle, 1971.
[m] Camerman and Nyburg, 1969.
[n] Molecule has a center of symmetry.
[o] Birnbaum, 1972.
[p] Molecule has a twofold rotation axis.
[q] Karle, 1969.
[r] Flippen and Karle, 1971.
[s] Flippen *et al.*, 1972.
[t] Accuracy affected by disorder in the crystal.
[u] P, planar; HC, half-chair; B, boat; Q, C(6) 0.3–0.4 Å out of plane, except 0.1 Å for MO◇MO; R, C(5) 0.4 Å out of plane; O, other; F, folded.

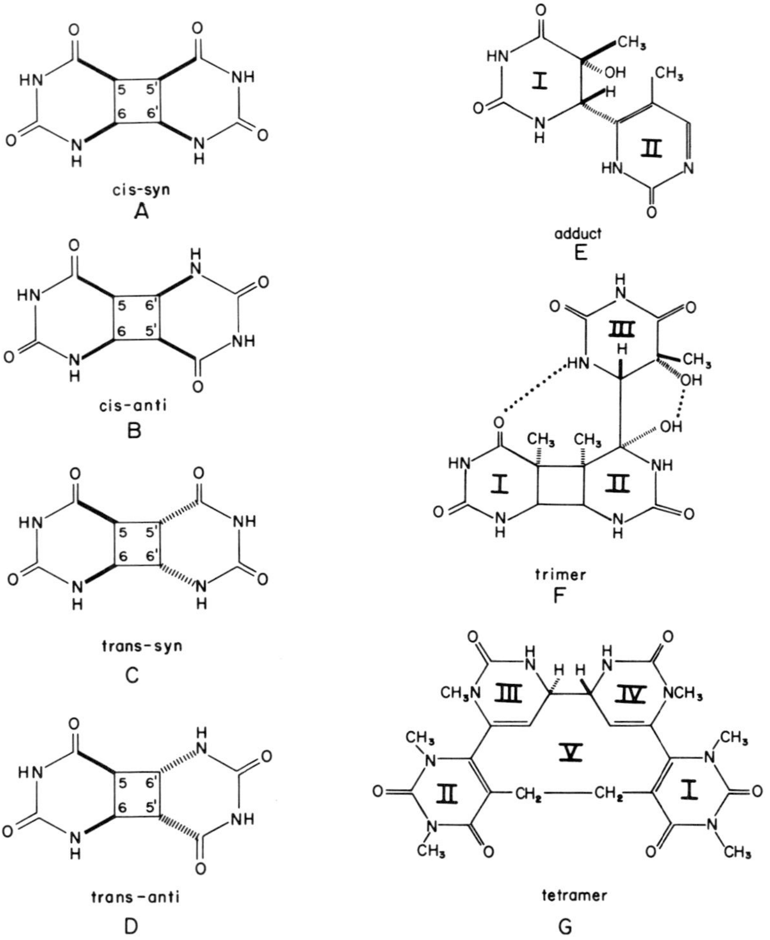

**Fig. 5.** *Photoproducts arising from the irradiation of DNA, Thy, and/or Ura.*

*trans-anti* types. Suitable crystals for x-ray analysis of the *trans-syn* type were prepared only for the dimer of methyl orotate (Birnbaum, 1972).

*a. cis-syn Pyr◇Pyr*

Five structure analyses for the *cis-syn* type have been reported for Ura◇Ura (Adman *et al.*, 1968; Adman and Jensen, 1970), 6-MeUra◇6-MeUra (Gibson and Karle, 1971; Konnert *et al.*, 1970a), $Me_2$Thy◇$Me_2$Thy (Camerman and Camerman, 1968, 1970), the sodium salt of Thy◇Thy (Wei and Einstein, 1968), and the intramolecular dimer Thy$\langle|(CH_2)_3|\rangle$Thy formed from 1,3-dithyminylpropane (Leonard *et al.*, 1969). The latter compound was used as a model for adjacent Thy moieties in DNA.

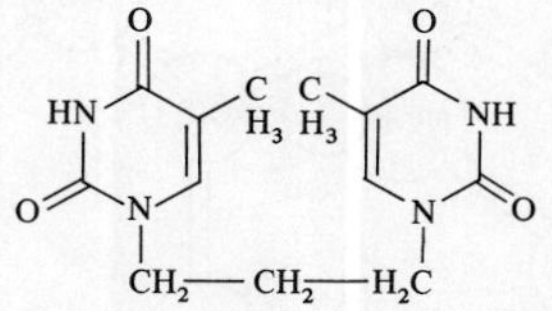

The structure investigations for the last three compounds were hampered by the monomerization of the dimers under prolonged exposure to x-radiation during the data-collecting process. Monomerization by x-rays has been observed only in the Thy◇Thy(*c*,*s*) and the Thy◇Thy(*t*,*s*) (N. Camerman, unpublished data). In each case, the x-ray data consisted of scattering from both the dimer and the monomer coexisting in varying proportions in the crystal lattice. Consequently, the results for bond lengths and angles are less reliable for the $Me_2$Thy◇$Me_2$Thy(*c*,*s*) and have not yet been reported in the literature for the sodium salt of Thy◇Thy(*c*,*s*).

Bond lengths for the six-membered rings in the Ura◇Ura(*c*,*s*), 6-MeUra◇6-MeUra(*c*,*s*), and Thy$\langle|(CH_2)_3|\rangle$Thy are very similar for comparable bonds and are almost identical with those observed in hUra and hThd except for the C(6)–N(1) bonds, which are slightly shorter in the dimers and the C(5)–C(6) bonds. In the dimers the C(5)–C(6) bond is shared between the six-membered and four-membered rings and is ~0.04 Å longer than in the hPyr. The bonds connecting the two six-membered rings, C(5)–C(5′) and C(6)–C(6′), seem to be affected by the location of $CH_3$ groups. These bonds are equal in Ura◇Ura, in which $CH_3$ groups are absent. In 6-MeUra◇6-MeUra, in which methyl groups are present on C(6) and C(6′), the C(6)–C(6′) distance is 0.04 Å longer than the C(5)–C(5′). In Thy$\langle|(CH_2)_3|\rangle$Thy the reverse is true; $CH_3$ groups are present on C(5) and C(5′) and the C(5)–C(5′) distance is 0.04 Å longer than the C(6)–C(6′). Even with a relatively long C—C bond (~1.59 Å) separating the $CH_3$ groups, the intramolecular $CH_3 \cdots CH_3$ distance is only 2.9–3.0 Å in these compounds as compared to 4.0 Å as normal van der Waals separation of $CH_3$ groups in neighboring molecules. The crowding of the $CH_3$ groups creates a strain in the molecule which was thought to be responsible for the easy monomerization by x-rays of the Thy◇Thy(*c*,*s*). However, an equivalent strain must exist in the 6-MeUra◇6-MeUra molecule which is not affected by x-rays.

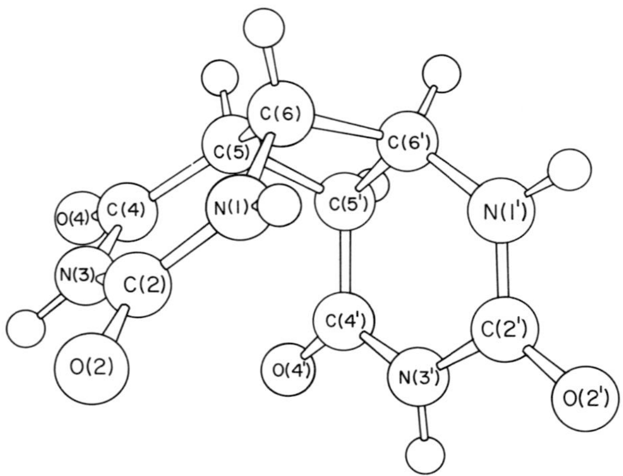


**Fig. 6.** *The conformation of Ura◇Ura(c,s) (Adman and Jensen, 1970). Hydrogen atoms are represented by the small circles. This figure and subsequent figures illustrating conformation or packing were drawn using the coordinates obtained from the crystal structure analysis. A computer program prepared by C. Johnson, Oak Ridge National Laboratory, was used.*

Bond angles in the *cis-syn* Ura◇Ura, 6-MeUra◇6-MeUra, and Thy⟨|$(CH_2)_3$|⟩Thy also are very similar. When compared to the hThy, hUra, and hThd, it is seen that the C(6)N(1)C(2), C(4)C(5)C(6), and C(5)C(6)N(1) angles are larger by 5°–6° in the dimers. The larger angles are reflected in flatter six-membered rings than the half-chair conformations which are present in the hPyr. Figure 6 shows the conformation of Ura◇Ura(*c*,*s*). Each six-membered ring has atoms N(1)–C(5) nearly in a plane, with atom C(6) displaced by 0.26 and 0.28 Å from the plane. This type of conformation has been encountered in a number of Pyr◇Pyr and has been designated as Q in Table 2. The four-membered ring is folded with a dihedral angle of 155°. As a result of the folding, the Pyr rings are twisted with respect to each other. The amount of twist can be measured by the torsion angle around the C(5)–C(5′) bond, i.e., the degree of rotation of atom C(4′) from the plane formed by atoms C(4), C(5), and C(5′). For the Ura◇Ura(*c*,*s*), this torsion angle is ~24°.

The conformation of $Me_2$Thy◇$Me_2$Thy(*c*,*s*) is similar to that of Ura◇Ura(*c*,*s*). However, there are some subtle differences in 6-MeUra◇6-MeUra. Its four-membered ring is folded as in Ura◇Ura(*c*,*s*) but is flatter, with a dihedral angle of 162°, even though a more puckered ring would increase the distance between the closely spaced $CH_3$ groups on C(6) and C(6′). One of the Pyr rings

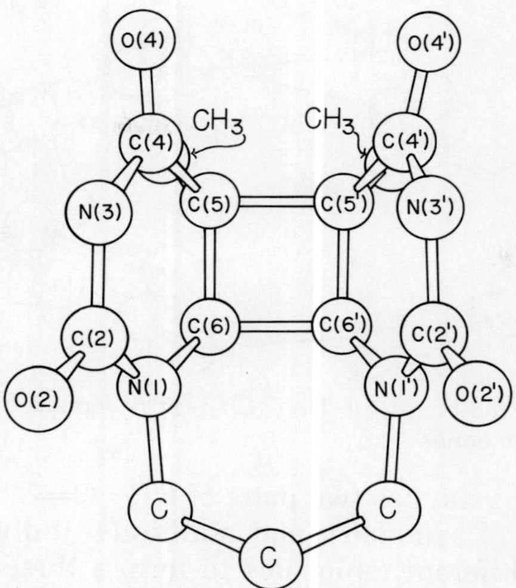


**Fig. 7.** *The conformation of Thy⟨|($CH_2$)$_3$|⟩Thy(c,s) (Leonard* et al., *1969). The cyclobutyl ring is nearly planar.*

has five ring atoms in a plane with C(6) displaced by 0.28 Å as in Ura◇Ura(*c,s*). The other six-membered ring is composed of two planar sections meeting in a fold along the line joining N(3)···N(6), with a dihedral angle of 168°.

In Thy⟨|($CH_2$)$_3$|⟩Thy(*c,s*), Fig. 7, the four-membered ring is nearly planar with a dihedral angle of 178°. Generally, cyclobutyl rings assume the folded conformation unless the molecule possesses a center of symmetry. The trimethylene bridge, which is part of a chair-shaped seven-membered ring, may be instrumental in constraining the cyclobutyl ring in Thy⟨|($CH_2$)$_3$|⟩Thy. The planar cyclobutyl ring, when not required by a center of symmetry in a molecule, will be encountered again in the Thy trimer (Flippen and Karle, 1971). One six-membered ring in Thy⟨|($CH_2$)$_3$|⟩Thy is approximately planar with the largest deviation of any atom from a plane being 0.05 Å. The other ring has four atoms in a plane with both C(5) and C(6) out of the plane in the same direction by 0.30 and 0.32 Å.

The packing in the crystals of the four *cis-syn* dimers is quite different. The $Me_2$Thy◇$Me_2$Thy molecule does not have any hydrogen atoms available for hydrogen bond formation. In Thy⟨|($CH_2$)$_3$|⟩Thy, two pairs of NH···O=C hydrogen bonds of length 2.82 and 2.87 Å are formed at each end of the molecule so that the molecules occur as hydrogen-bonded dimers around centers of symmetry as illustrated in Fig. 8.

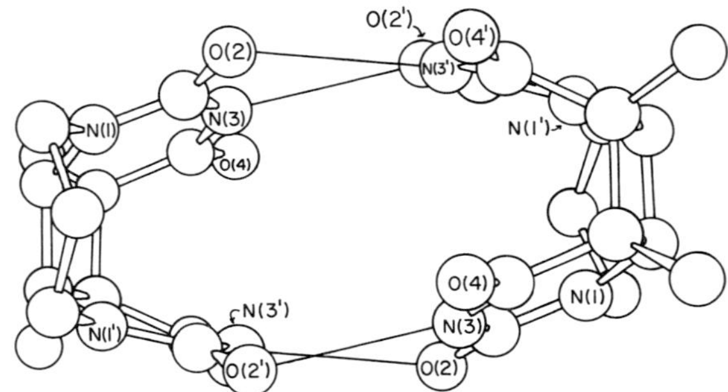


**Fig. 8.** *Dimers in the crystal of Thy⟨|(CH$_2$)$_3$|⟩Thy formed by four NH···O═C intermolecular hydrogen bonds.*

The Ura◇Ura crystal has two pairs of NH···O═C bonds around two different centers of symmetry and additional individual NH···O═C bonds between different molecules to form a three-dimensional network. The NH···O═C bonds range from 2.85 to 2.89 Å. The 6-MeUra◇6-MeUra(c,s) is one of two cyclobutyl type dimers the crystal structures of which have been determined and have been found to contain water of crystallization. Each $H_2O$ molecule has hydrogen bonds to three different dimer molecules. Each dimer molecule participates in nine hydrogen bonds which link it to three other dimer molecules by pairs of NH···O═C bonds and to three $H_2O$ molecules, as shown in Fig. 9. The stereogram in Fig. 10 illustrates the packing in the crystal.

*b. cis-anti Pyr◇Pyr*

For the Pyr◇Pyr(*c*,*a*), detailed structural parameters have been reported for the Me$_2$Thy◇Me$_2$Thy (Camerman *et al.*, 1969) and Ura◇Ura (Konnert *et al.*, 1970a; Konnert and Karle, 1971). Even though

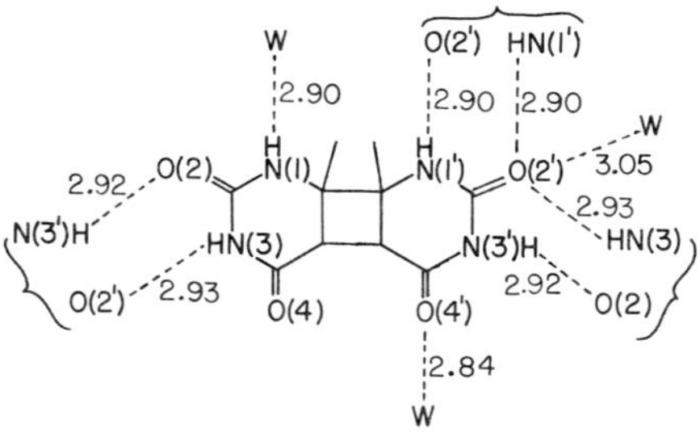

**Fig. 9.** *Nine hydrogen bonds formed by each 6-MeUra◇6-MeUra(c,s) molecule (Gibson and Karle, 1971). The symbol W represents water of crystallization; hydrogen bond lengths are given in Å.*

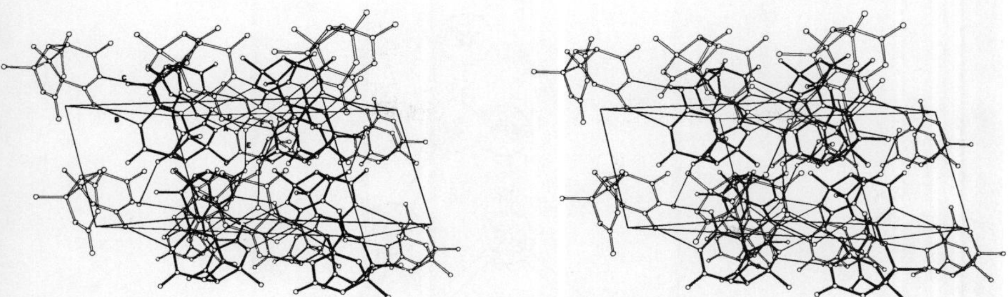

**Fig. 10.** *A stereodiagram of the packing and hydrogen bonding in a crystal of 6-MeUra◇6-MeUra*(c,s) *containing water of crystallization (Gibson and Karle, 1971).*

$Me_2$Thy◇$Me_2$Thy has six $CH_3$ groups replacing H atoms as compared to Ura◇Ura, the molecular geometries of the two molecules are much alike in all respects. Bond lengths are similar in these two molecules and, in fact, are the same as those found in the *cis-syn* dimers (Table 1). The significant differences in bond angles between the *cis-syn* and *cis-anti* molecules occur at the internal angles C(6)N(1)C(2) and C(4)C(5)C(6), which are smaller by 3°–5° in the *cis-anti* dimers, and the external angles N(1)C(6)C(5′), which are considerably larger than the N(1)C(6)C(6′) angles in the *cis-syn* dimers.

The *cis-anti* molecules possess very nearly a twofold axis of rotation, although none is required by the space group symmetry. A view of Ura◇Ura(*c,a*) looking down the approximate twofold axis is shown in Fig. 11. The cyclobutyl ring is puckered with dihedral angles of 154° and 150° for the $Me_2$Thy◇$Me_2$Thy and Ura◇Ura, respectively. In each molecule, the six-membered rings have atoms N(1) to C(5) in a plane and atom C(6) is 0.35–0.41 Å out of the plane. The two oxygen atoms on each ring are within 0.05 Å of the plane. The same conformation of the six-membered rings occurs predominantly in the *cis-syn* dimers.

Although the molecular dimensions of the two *cis-anti* dimers are nearly identical, the molecular packing in the crystal is different. There are no hydrogen atoms available for hydrogen bond formation in $Me_2$Thy◇$Me_2$Thy, whereas each molecule of Ura◇Ura participates in eight NH···O═C bonds ranging in length from 2.83 to 2.93 Å. The hydrogen bonding scheme is illustrated in Fig. 12. Each of the four nitrogen atoms in the molecule donates a proton for hydrogen bond formation and three of the four oxygen atoms are acceptors of protons from neighboring molecules. Atom O(4) participates in two hydrogen bonds, whereas atom O(2′) does not participate at all.

The 6-MeUra◇6-MeUra(*c,a*) has an exact twofold axis of rota-

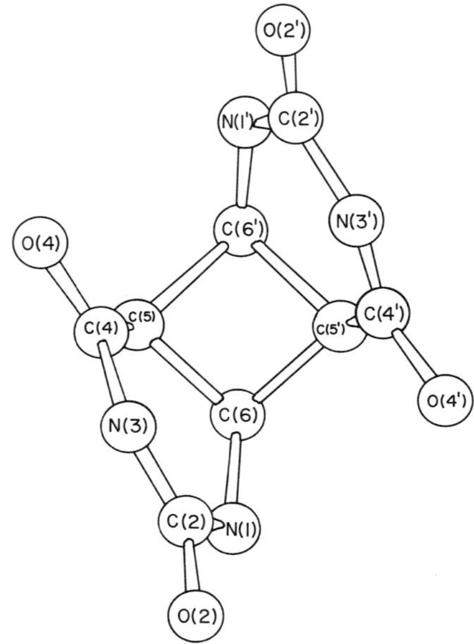


**Fig. 11.** *The conformation of Ura◇Ura*(c,a) (*Konnert and Karle, 1971*). *The four-membered ring is folded and there is an approximate twofold axis of rotation perpendicular to the view of the molecule.*

tion passing through the molecule, which is required by the space group in which the material crystalizes. In contrast to the $Me_2$Thy◇$Me_2$Thy(*c*,*a*) and Ura◇Ura(*c*,*a*), the 6-MeUra◇6-MeUra(*c*,*a*) has a planar cyclobutyl ring and planar Ura moieties (Amzel *et al.*, 1972; Flippen, unpublished data). The planar cyclobutyl ring is not required by the space group symmetry.

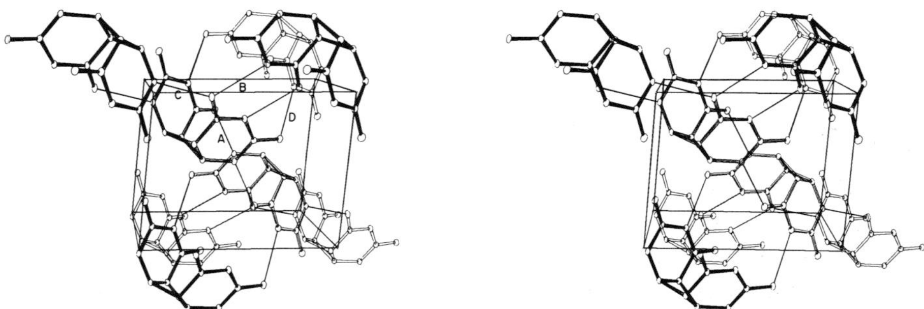


**Fig. 12.** *A stereodiagram of the packing and hydrogen bonding in the triclinic crystal of Ura◇Ura*(c,a). *Hydrogen bonds between molecules are labelled A-D.*

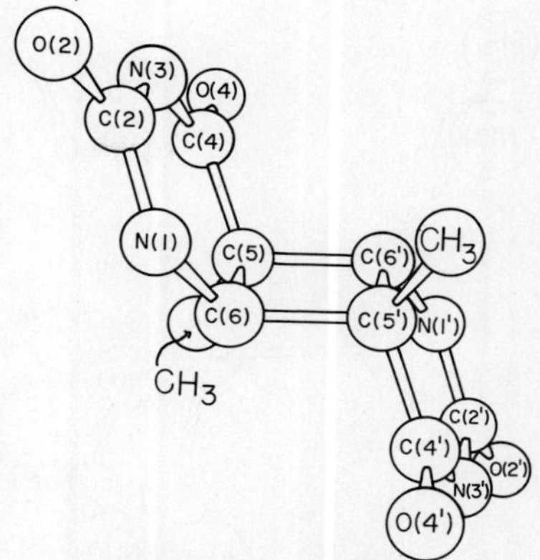


**Fig. 13.** *The conformation of Thy◇Thy*(t,a) (*Camerman* et al., *1967*).

*c. trans-anti Pyr◇Pyr*

The Thy◇Thy(*t*,*a*) (Camerman *et al.*, 1967; Camerman and Nyburg, 1969) and 1-MeThy◇1-MeThy(*t*,*a*) (Einstein *et al.*, 1967) each crystallize around centers of symmetry in their unit cells so that a center of symmetry coincides with the center of the cyclobutyl ring. In order to satisfy the center of symmetry, the cyclobutyl ring must be planar. The atoms in the Pyr rings also lie nearly in a plane with maximum deviations of ±0.05 Å. The molecules are not cup shaped as in the *cis-syn* and *cis-anti* dimers but consist of three planar units in a *trans* configuration with the $CH_3$ groups projecting away from the planes (Fig. 13). In the crystal, neighboring molecules of 1-MeThy◇1-MeThy (*t*,*a*) are held together by one pair of hydrogen bonds between N(3)···O(2′) and N(3′)···O(2), arranged in eight-membered rings, around a center of symmetry, which is characteristic of the hydrogen bonding in most of the molecules encountered in this chapter. In the Thy◇Thy(*t*,*a*) crystal, all NH and C═O groups are utilized to form hydrogen bonds of lengths 2.83 and 2.86 Å which link the molecules in a three-dimensional network; but eight-membered rings formed by pairs of hydrogen bonds are not present.

*d. trans-syn Pyr◇Pyr*

Ultraviolet irradiation of an aqueous solution of the methyl ester of orotic acid produces only the *trans-syn* dimer. The structure analysis of this molecule (Birnbaum, 1972) provides data for the only example of the *trans-syn* type of Pyr◇Pyr. This molecule, shown in Fig. 14, contains a twofold axis of rotation which bisects the two interpyrimidine

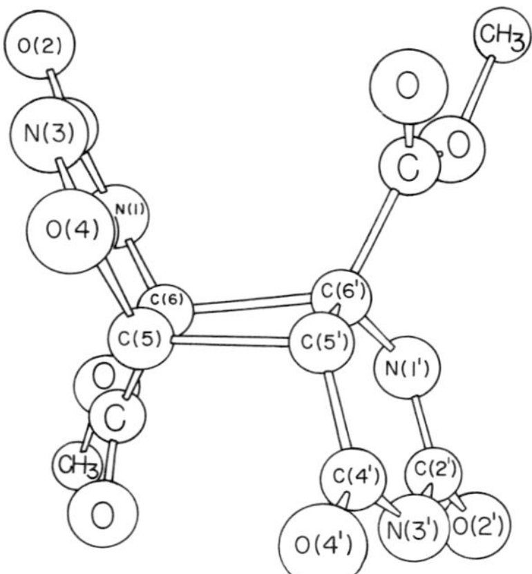


**Fig. 14.** *The conformation of the* trans-syn *dimer of methyl orotate (Birnbaum, 1972). The molecule contains a twofold rotation axis which bisects the two inter-pyrimidine bonds.*

bonds. Bond lengths and angles agree with those found in the *cis-syn* and *trans-anti* dimers (Tables 1 and 2). The only outstanding value is 1.628 Å for the C(6)–C(6′) bond. This large value is occasioned by the presence of the bulky ester groups on the C(6) and C(6′) atoms and accounts for the ease with which the dimers dissociate into monomers. In the six-membered rings, atoms N(1) to C(5) are essentially coplanar, whereas atom C(6) is 0.11 Å out of the plane. The four-membered ring is folded slightly with a dihedral angle of 170°.

Each dimer molecule forms two pairs of parallel hydrogen bonds, N(1)H···O′(4) and N(1′)H···O′(4′), 2.83 Å in length, which link each molecule to a neighbor on either side related by a translation of one unit cell length. Each water molecule connects the dimer molecules linearly and laterally by four hydrogen bonds involving N(3)H, O(4), O(2), and the ester oxygen. One of the hydrogen atoms of the water molecule satisfies both O(4) and O(2), in adjacent dimer molecules, by forming a bifurcated hydrogen bond.

### 3. Pyrimidine Adducts

#### *a. Thymine–Thymine Adduct*

The UV irradiation of DNA produces not only Pyr◇Pyr but also other photoproducts. A Thy–Thy adduct (Fig. 5E) was isolated by Varghese and Wang (1968a), and it was proposed that the adduct formation proceeds through an oxetane intermediate (Scheme 2).

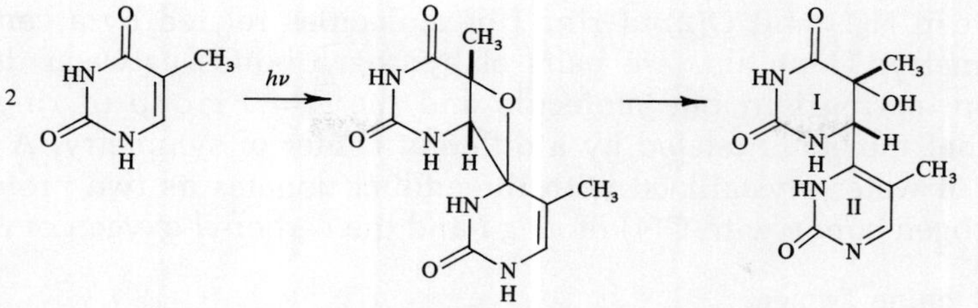

**Scheme 2**

The crystal structure analysis confirmed the molecular formula of the adduct and established the stereoconfiguration shown in Fig. 15 (Karle *et al.*, 1969; Karle, 1969). The transfer of an oxygen atom from one Thy moiety to the other creates a ho⁵hThy moiety, ring I. This ring has a half-chair conformation with C(5) 0.41 Å above and C(6) 0.21 Å below the plane of atoms N(1) to C(4); however, the $CH_3$ group on C(5) is in an axial position in contrast to hThy and hThd, where the $CH_3$ group is equatorial. The six atoms of ring II lie in a plane with an average deviation of 0.008 Å. The angle of inclination between the two rings is close to 96°.

Bond lengths and angles in ring I are similar to those found in hThy, hUra, and hThd with the largest difference in the C(5)–C(6) bond, which is ~0.05 Å longer in the adduct. In ring II the bond lengths and angles exhibit fairly large changes from the values found in Thy, especially in the region near C(4′). This is understandable in view of the replacement of a carbonyl oxygen on C(4′) with a single C—C bond to ring I.

In the crystal there are the pairs of NH···O═C bonds around a center of symmetry characteristic of the pyrimidines. The bonds occur

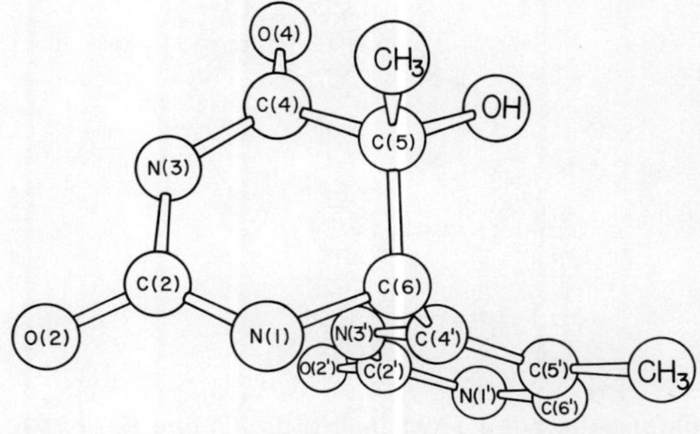


**Fig. 15.** *The conformation of the Thy–Thy adduct (Karle, 1969).*

between N(1) and O(2) of ring I of molecules related by a center of symmetry. There also are pairs of hydrogen bonds between the OH group of ring I in one molecule and the C═O group of ring II in another molecule related by a different center of symmetry. A molecule of water crystallized with the adduct donates its two protons to hydrogen bonds with O(4) of ring I and the carbonyl oxygen of ring II.

*b. Thymine Trimer*

A trimer of Thy was also isolated from irradiated frozen aqueous solutions of Thy (Wang, 1971). This compound was originally designated as product $PT_1$ (Varghese and Wang, 1967). The structure analysis of the trimer was used as the example to illustrate Section C, (see Figs. 2 and 3). As a result of the x-ray diffraction analysis (Flippen *et al.*, 1970; Flippen and Karle, 1971), the structural formula has been shown to be that illustrated in Fig. 5F. The molecule is a combination of a Thy◇Thy(*c*,*s*) and a Thy–Thy adduct with an additional OH and H moiety on ring II. A probable route for the trimer formation is through an intermediate containing an oxetane ring which is ruptured by the addition of a molecule of water (Wang, 1971). This process is analogous to the probable formation of the Thy–Thy adduct except for the extra molecular equivalent of $H_2O$ which is present in the trimer and absent in the adduct. Additional UV irradiation of the trimer in aqueous solution ruptures the cyclobutyl ring and results in two products, Thy and Thy–Thy adduct (Varghese and Wang, 1968b).

Bond lengths and angles in ring III are quite comparable to those in ring I of the Thy–Thy adduct (Tables 1 and 2). The $CH_3$ group on ring III of the trimer (Figs. 16 and 5F) and also on ring I of the adduct (Figs.

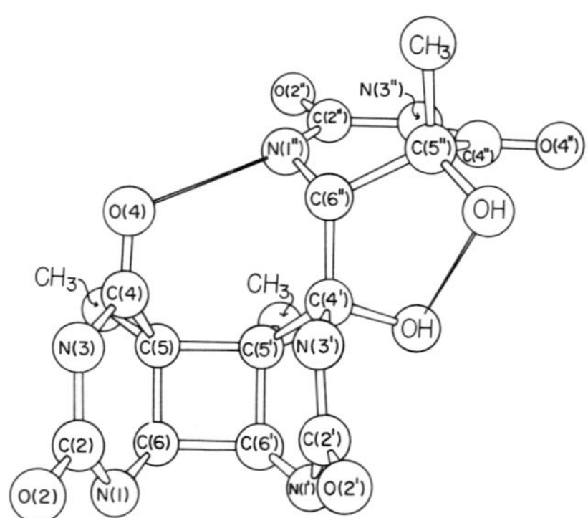


**Fig. 16.** *The conformation of a Thy trimer (Flippen and Karle, 1971). The thin lines represent intramolecular hydrogen bonds.*

15 and 5E) are axial, in contrast to the equatorial conformation for the $CH_3$ group in hThy and hThd. Ring III assumes the half-chair conformation with C(5″) 0.41 Å on one side and C(6″) 0.28 Å on the other side of the plane containing N(1″), C(2″), N(3″), and C(4″), a conformation almost identical with that assumed by ring I in the Thy–Thy adduct.

Rings I and II which form the Thy◇Thy(*c*,*s*) moiety have significantly different values for their structural parameters from those found in other *cis-syn* dimers. This is understandable for ring II in which C(4′) in addition to C(5′) and C(6′) is saturated, but this is somewhat surprising for ring I in view of the constancy of bond lengths and angles observed for all the other dimers.

The molecule has assumed a certain degree of rigidity by the formation of two intramolecular hydrogen bonds. The OH group on C(4′), axial to ring II, donates a proton to form an OH···O bond (2.61 Å) with the OH group on C(5″), thereby creating a six-membered ring. The hydrogen bond N(1″)H···O(4) (2.89 Å) creates an eight-membered ring. The constraints upon the geometry of the molecule imposed by the intramolecular hydrogen bonding are somewhat reminiscent of the constraints in the Thy⟨|$(CH_2)_3$|⟩Thy molecule. Indeed, the cyclobutyl ring is only slightly puckered in the trimer, with a dihedral angle of 173.5° as compared to 178° in Thy⟨|$(CH_2)_3$|⟩Thy. In addition, ring I in the trimer is nearly planar, within ±0.05 Å, comparable to one of the six-membered rings in Thy⟨|$(CH_2)_3$|⟩Thy.

Ring II is different from any other ring encountered in photoproducts from Thy or Ura in that C(4′) is saturated. The C—OH bond, the C—N bond, and the two C—C bonds containing C(4′) all have normal single bond values. This ring exists in an approximate boat conformation with atoms N(1′) and C(4′) both on the same side of the plane formed by C(5′), C(6′), C(2′), and N(3′) by 0.29 and 0.61 Å, respectively.

There are some unusually short intramolecular distances in the trimer. The C···C separation in the two $CH_3$ groups on the cyclobutyl ring is only 2.92 Å, comparable to the 2.94 Å value found in Thy⟨|$(CH_2)_3$|⟩Thy. The N(3′)···C(4) separation is only 2.89 Å while the O(4)···C(6″) distance is 2.87 Å. Despite the crowding in the trimer molecule, there is no evidence of dissociation of the dimer moiety in the x-ray beam during the diffraction experiment as had been noted for the other Thy◇Thy(*c*,*s*).

Each trimer molecule in the crystal participates in 14 hydrogen bonds, two of which are intramolecular and two of which involve the $H_2O$ of crystallization. The remaining 10 hydrogen bonds are of the

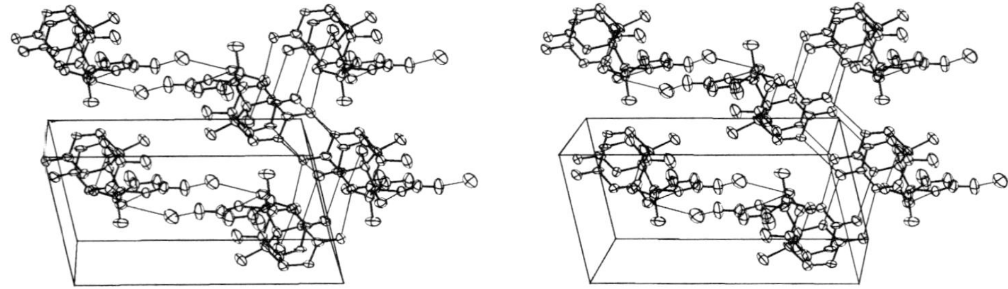

**Fig. 17.** *A stereoscopic view of the packing and hydrogen bonding in the crystal of a Thy trimer containing water of crystallization.*

NH···O=C type and occur in five pairs in the familiar pattern encountered in the structures already discussed. The hydrogen-bonding scheme is more elaborate, however, in that there are double strands of NH···O=C bonds, reminiscent of the hydrogen bonding in Thy as shown in Fig. 4, in which each C=O is bonded to the NH from two adjacent molecules. In addition, these double strands occur in two layers, joining ring I to ring II of a neighboring molecule related by a center of symmetry and ring II to ring I of the same neighboring molecule. Thus, a continuous channel along the *c* axis is enclosed by NH···O=C bonds as is illustrated in Fig. 17. (The rings are labeled in Fig. 5F.)

*c. Pyrimidine Tetramer*

A Pyr tetramer was produced by the irradiation of an aqueous solution of a Thy–Ura adduct with relatively long UV light (310–360 nm) and subsequent methylation (Scheme 3) (Wang and Rhoades, 1971).

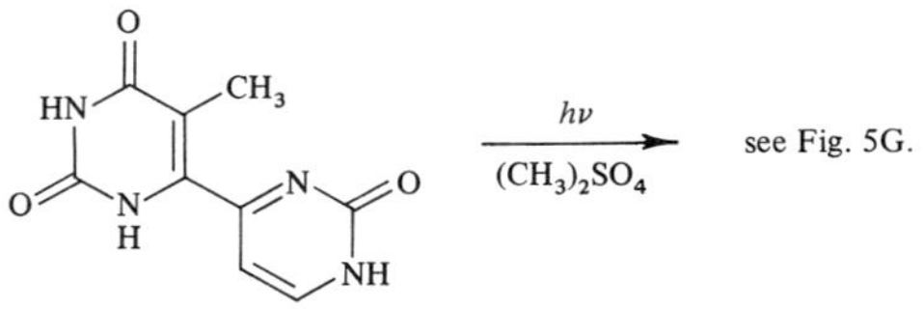

O
HN CH$_3$
O N H N O
NH

$\xrightarrow[(CH_3)_2SO_4]{h\nu}$ see Fig. 5G.

**Scheme 3**

The Thy–Ura adduct is itself a photoproduct isolated in minute quantities from DNA irradiated by UV (254 nm) light (Varghese and Wang, 1967; Wang and Varghese, 1967). The structural formula and the stereoconfiguration of the tetramer (Figs. 18 and 5G), established by its crystal structure analysis (Flippen *et al.*, 1971, 1972), show that the dimerization was head-to-head and tail-to-tail to yield the *trans-syn*

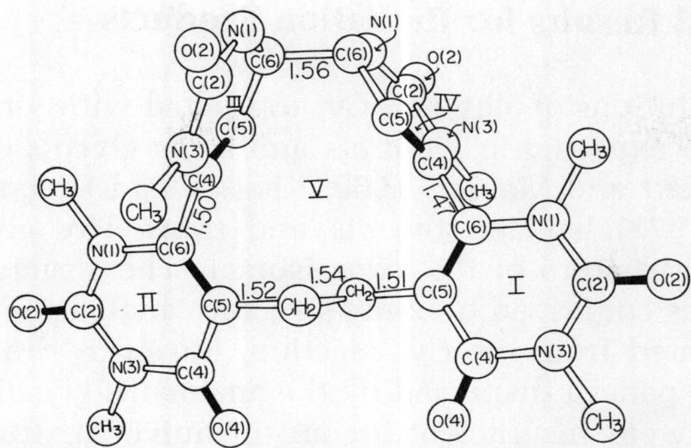


**Fig. 18.** *Conformation of a Pyr tetramer obtained by irradiation* (*Flippen* et al., *1972*). *Ring bond lengths not included in Table 1 are shown in Å. Double bonds are indicated by shading.*

dimer. A minor product of the reaction is probably the *cis-syn* dimer (Wang and Rhoades, 1971). A significant result in the dimer formation is that hydrogen is abstracted from the $CH_3$ group in the formation of the dimethylene bridge. This experiment suggests that hydrogen abstraction could occur in DNA at longer wavelengths (300–360 nm) and may explain inactivation in some biological systems exposed to the longer wavelengths.

The molecule effectively possesses a twofold rotation axis even though none is required by the space group symmetry. Each of the four six-membered rings is essentially planar with only minor deviations from planarity. Ring I is nearly perpendicular to ring IV and ring II to ring III, with torsion angles of 87° and 91° about the respective C(6)–C(4) bonds, analogous to the conformation in the Thy–Thy adduct. The 12-membered ring V contains four C═C bonds, each of which is also shared with a six-membered ring, and there are two *cis* and two *trans* configurations. In each C═C—C═C segment the torsion angle about the central bond is ~90°. Nevertheless, the effects of conjugation on bond lengths are still apparent.

The bond lengths and angles, listed in Table 1 and 2, for rings I and II differ very little from those occurring in Thy. However, rings III and IV, in which C(6) is saturated and a double bond exists between C(4) and C(5), have bond lengths and angles which are quite different from any of the molecules discussed in this chapter. Each tetramer molecule participates in eight hydrogen bonds, all but one involving the three molecules of water of crystallization. One NH···O═C bond of 2.86 Å is formed between neighboring molecules.

## E. Structural Results for Radiation Products

Aqueous solutions of Thy and Cyt irradiated with $\gamma$-rays instead of UV light were expected to yield *cis* and *trans* glycols of the reactant molecules (Ekert and Monier, 1960; Khattak and Green, 1966). Hahn and Wang (1972) isolated the *cis* and *trans* Thy glycols and established the structure of the *trans* isomer. The structure of the *cis*-Thy glycol was confirmed by x-ray structure analysis (Flippen, 1973). Products isolated from the Cyt reaction, however, did not have the NMR spectral pattern displayed by the analogous glycols of Thy. The spectral data were insufficient for an unequivocal structural assignment and an analysis by x-ray diffraction was required to establish that the six-membered ring of the $\gamma$-irradiated Cyt suffered a molecular rearrangement to form an imidazolidone derivative (Hahn *et al.*, 1973).

### 1. *cis*-Thymine Glycol

The conformation of *cis*-Thy glycol is illustrated in Fig. 19 (Flippen, 1973). The ring is in the half-chair conformation with C(5) and C(6) below and above the plane of the other four atoms by 0.37 Å. The $CH_3$ group on C(5) and the OH group on C(6) are axial, while the OH group on C(5) is equatorial. The conformation is similar to that found for ring I in the Thy–Thy adduct and ring III of the Thy◇Thy–Thy trimer, but differs from the conformation found for hThy and hThd in which the $CH_3$ group is equatorial to the ring. Bond lengths and angles are nearly the same as those determined for hThy, hThd, Thy–Thy adduct (ring I), and Thy◇Thy–Thy trimer (ring III), Tables 1 and 2.

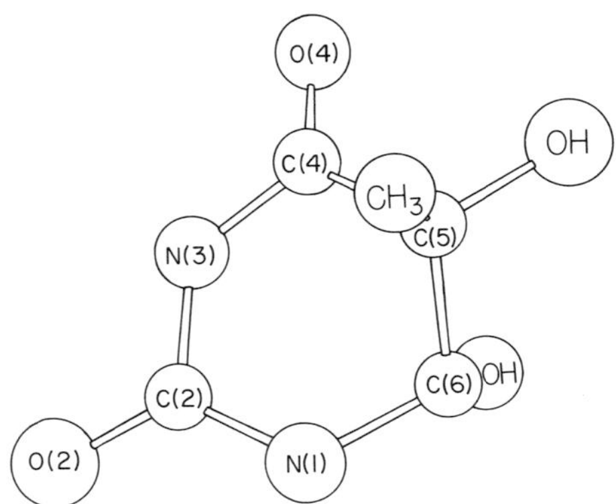


**Fig. 19.** *The conformation of* cis-*Thy glycol (Flippen, 1973).*

*cis*-Thy glycol crystallizes in a noncentrosymmetric space group; hence, the hydrogen-bonding scheme does not include pairs of NH···O=C hydrogen bonds occurring around a center of symmetry as has been found in Thy, hThy, Thy–Thy adduct, and in the crystals of a number of other molecules discussed in this chapter. However, both NH and OH moieties act as donors and each carbonyl oxygen as well as the oxygen atom on one OH group act as acceptors so that each *cis*-Thy glycol molecule participates in seven hydrogen bonds with lengths ranging from 2.83 to 3.06 Å. The close approach between molecules facilitated by the large number of hydrogen bonds accounts for the high density (1.61 gm/cm$^3$) for this crystal as well as for the other Thy and Ura derivatives described here.

2. *trans*-1-Carbamylimidazolidone-4,5-diol

The x-ray analysis of a single crystal of the *trans* isomer isolated from the $\gamma$-irradiation of Cyt shows that the Cyt molecule has been transformed to *trans*-1-carbamylimidazolidone-4,5-diol, Fig. 20 (Flippen, 1973; Hahn *et al.*, 1973). The five-membered ring has the envelope conformation—four atoms are coplanar and C(5) is 0.22 Å out of the plane. The plane of the amide group is rotated ~13° from the plane of the five-membered ring. The C=O bond assumes the *syn*-planar conformation with respect to the N(1)–C(5) bond. Bond lengths

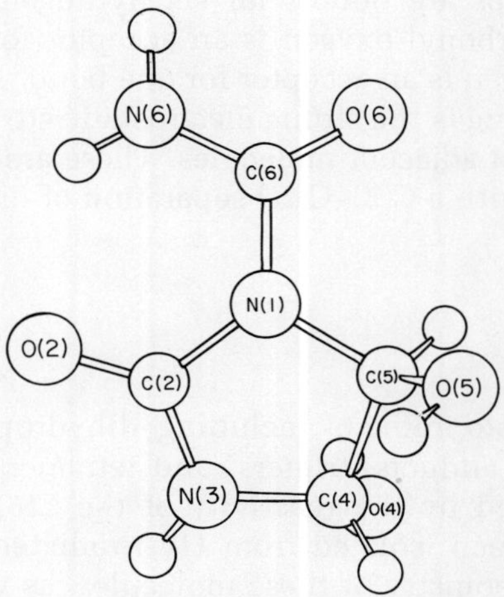


**Fig. 20.** *The conformation of trans-1-carbamylimidazolidone-4,5-diol obtained from the $\gamma$-irradiation of Cyt (Hahn et al., 1973).*

**Table 3** Bond Lengths and Angles for the Photoproduct from γ-Irradiated Cyt, *trans*-1-Carbamylimidazolidone-4,5-diol

| Bond | Å | Angle | Degrees |
|---|---|---|---|
| C(6)–N(1) | 1.392 | C(6)N(1)C(2) | 127.0 |
| N(1)–C(2) | 1.397 | N(1)C(2)N(3) | 108.0 |
| C(2)–O(2) | 1.230 | N(1)C(2)O(2) | 125.0 |
| C(2)–N(3) | 1.337 | N(3)C(2)O(2) | 127.0 |
| N(3)–C(4) | 1.449 | C(2)N(3)C(4) | 113.3 |
| C(4)–O(4) | 1.408 | N(3)C(4)C(5) | 102.9 |
| C(4)–C(5) | 1.535 | N(3)C(4)O(4) | 112.7 |
| C(5)–O(5) | 1.412 | C(5)C(4)O(4) | 113.0 |
| C(5)–N(1) | 1.467 | C(4)C(5)N(1) | 102.8 |
| C(6)–N(6) | 1.332 | C(4)C(5)O(5) | 113.2 |
| C(6)–O(6) | 1.231 | N(1)C(5)O(5) | 111.5 |
| | | C(5)N(1)C(2) | 110.8 |
| | | C(5)N(1)C(6) | 121.4 |
| | | N(1)C(6)O(6) | 119.9 |
| | | N(1)C(6)N(6) | 116.6 |
| | | N(6)C(6)O(6) | 123.5 |

and angles for this molecule are not directly comparable to the molecules with six-membered rings and are listed separately in Table 3.

Molecular packing is dominated by hydrogen bonding. Each molecule participates in eleven hydrogen bonds (one intramolecular hydrogen bond) in which the hydrogen atoms in the two OH groups, the NH, and $NH_2$ groups are donors for six hydrogen bonds (one bifurcated) and each carbonyl oxygen is an acceptor for two bonds while each hydroxyl oxygen is an acceptor for one bond. Another interesting feature of the packing is the strong electrostatic attraction between the ring C=O groups of adjacent molecules. These are arranged in an antiparallel fashion with a C(2)–C(2′) separation of only 3.16 Å.

## F. Summary

A variety of photoproducts including dihydropyrimidines, cyclobutyl-type dimers, adducts, trimers, and tetramers, the formation of which is occasioned by the reactivity of the C(5)–C(6) bond in the pyrimidines, has been isolated from UV-irradiated DNA and related compounds. The geometry of these molecules, as well as the structural formulae in some cases, has been established by x-ray diffraction analyses of single crystals. Several generalizations concerning the structural features of the molecule can be made. Bond lengths for comparable bonds in the different photoproducts are quite constant irrespective of the type of isomer or the conformation of the rings. The main difference in bond lengths between the dihydropyrimidines and the cyclobutyl-type dimers is that C(6)–N(1) is shorter and C(5)–C(6) is longer in the dimers. In the cyclobutane rings, the bonds shared with the six-membered rings, i.e., the C(5)–C(6) bonds, are shorter than the bonds between the rings, C(5)–C(5′) or C(5)–C(6′). These latter bonds are considerably longer than the 1.54 Å value usually associated with bonds between saturated carbon atoms. Bond angles are quite constant for the same type of molecule, e.g., for all *cis-syn* Pyr◇Pyr, but there is some variation between different types. The conformation of the four-membered ring in molecules containing the cyclobutyl dimers can be either folded or planar. The planar conformation can occur even when the molecule does not contain a center of symmetry and even if the carbon atoms of the methyl groups on C(5) are separated by only 2.9 Å. When the cyclobutyl ring is folded, the adjacent six-membered rings assume the conformation in which five atoms are coplanar and C(6) is 0.3–0.4 Å out of the plane. On the other hand, when the cyclobutyl ring is planar, or nearly so, the adjacent six-membered rings are also nearly planar, even though C(5) and C(6) are saturated. In the dihydropyrimidines, in Thy glycol, and in one ring of the Thy–Thy adduct and of the Thy trimer where, in effect, a molecule of $H_2O$ has been added to C(5)═C(6), the rings have a half-chair conformation. The methyl group on C(5) is equatorial in hThy and hThd, whereas it is axial in the adduct, the trimer, and the glycol. All of the photoproducts containing free NH groups participate in extensive hydrogen bonding in the crystalline state. The hydrogen bond pair

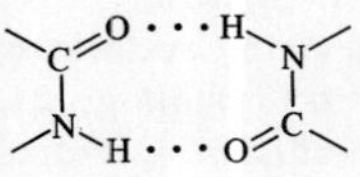

occurs very frequently in the crystals of the photoproducts described in this chapter.

With the knowledge of the conformations of the photoproducts derivable from irradiated DNA and related compounds, it may be possible to propose the manner in which these photoproducts can occur in the DNA molecule and how they can disrupt the biological functions of the molecule.

## References

Adman, E., and Jensen, L. H. (1970). *Acta Crystallogr., Sect. B* **26,** 1326.
Adman, E., Gordon, M. P., and Jensen, L. H. (1968). *Chem. Commun.* p. 1019.
Amzel, L. M., Avery, H. P., Becka, L. N., Poljak, R. J., Khattak, M. N., and Wang, S. Y. (1972). *Nature New Biology* **238,** 204.
Beukers, R., and Berends, W. (1960). *Biochim. Biophys. Acta* **41,** 550.
Beukers, R., and Berends, W. (1961). *Biochim. Biophys. Acta* **49,** 181.
Bijvoet, J. M., Peerdeman, A. F., and van Bommel, A. J. (1951). *Nature (London)* **168,** 271.
Birnbaum, G. I. (1972). *Acta Crystallogr., Sect. B* **28,** 1248.
Camerman, N., and Camerman, A. (1968). *Science* **160,** 1451.
Camerman, N., and Camerman, A. (1970). *J. Amer. Chem. Soc.* **92,** 2523.
Camerman, N., and Nyburg, S. C. (1969). *Acta Crystallogr., Sect. B* **25,** 388.
Camerman, N., Nyburg, S. C., and Weinblum, D. (1967). *Tetrahedron Lett.* **42,** 4127.
Camerman, N., Weinblum, D., and Nyburg, S. C. (1969). *J. Amer. Chem. Soc.* **91,** 982.
Einstein, J. R., Hosszu, J. L., Longworth, J. W., Rahn, R. O., and Wei, C. H. (1967). *Chem. Commun.* p. 1063.
Ekert, B., and Monier, R. (1960). *Nature (London)* **188,** 309.
Flippen, J. L. (1973). *Acta Crystallogr. Sect. B.* **29,** 1756.
Flippen, J. L., and Karle, I. L. (1971). *J. Amer. Chem. Soc.* **93,** 2762.
Flippen, J. L., Karle, I. L., and Wang, S. Y. (1970). *Science* **169,** 1084.
Flippen, J. L., Gilardi, R. D., Karle, I. L., Rhoades, D. F., and Wang, S. Y. (1971). *J. Am. Chem. Soc.* **93,** 2556.
Flippen, J. L., Gilardi, R. D., and Karle, I. L. (1972). *Acta Crystallogr., Sect. B* **27,** 360.
Furberg, S., and Jensen, L. H. (1968). *J. Amer. Chem. Soc.* **90,** 470 (1968).
Gerdil, R. (1961). *Acta Crystallogr.* **14,** 333.
Gibson, J. W., and Karle, I. L. (1971). *J. Cryst. Mol. Struct.* **1,** 115.
Hahn, B. S., and Wang, S. Y. (1972). *J. Amer. Chem. Soc.* **94,** 4764.
Hahn, B. S., Wang, S. Y., Flippen, J. L., and Karle, I. L. (1973). *J. Amer. Chem. Soc.* **95,** 2711.
International Tables for X-Ray Crystallography. (1965). Kynoch Press, Birmingham, England.
Karle, I. L. (1969). *Acta Crystallogr., Sect. B* **25,** 2119.
Karle, I. L., Wang, S. Y., and Varghese, A. J. (1969). *Science* **164,** 183.
Karle, J. (1969). *Advan. Chem. Phys.* **16,** 131.
Karle, J., and Hauptman, H. (1950). *Acta Crystallogr.* **3,** 181.
Karle, J., and Karle, I. L. (1966). *Acta Crystallogr.* **21,** 849.
Khattak, M. N., and Green, J. H. (1966). Int. J. Radiat. Biol. **11,** 131.
Kondo, Y., and Witkop, B. (1968). *J. Amer. Chem. Soc.* **90,** 764.
Konnert, J., and Karle, I. L. (1971). *J. Cryst. Mol. Struct.* **1,** 107.
Konnert, J., Gibson, J. W., Karle, I. L., Khattak, M. N., and Wang, S. Y. (1970a). *Nature (London)* **227,** 953.
Konnert, J., Karle, I. L., and Karle, J. (1970b). *Acta Crystallogr., Sect. B* **26,** 770.
Leonard, N. J., Golankiewicz, K., McCredie, R. S., Johnson, S. M., and Paul, I. C. (1969). *J. Amer. Chem. Soc.* **91,** 5855.
Rohrer, D. C., and Sundaralingam, M. (1970). *Acta Crystallogr., Sect. B* **26,** 546.
Stout, G. H., and Jensen, L. H. (1968). "X-Ray Structure Determination, A Practical Guide." Macmillan, New York.
Varghese, A. J., and Wang, S. Y. (1967). *Science* **156,** 955.

Varghese, A. J., and Wang, S. Y. (1968a). *Science* **160,** 186.
Varghese, A. J., and Wang, S. Y. (1968b). *Biochem. Biophys. Res. Commun.* **33,** 102.
Wang, S. Y. (1960). *Nature (London)* **188,** 844.
Wang, S. Y. (1961). *Nature (London)* **190,** 690.
Wang, S. Y. (1964). *Photochem. & Photobiol.* **3,** 395.
Wang, S. Y. (1971). *J. Amer. Chem. Soc.* **93,** 2768.
Wang, S. Y., and Rhoades, D. H. (1971). *J. Amer. Chem. Soc.* **93,** 2554.
Wang, S. Y., and Varghese, A. J. (1967). *Biochem. Biophys. Res. Commun.* **29,** 543.
Wei, C. H., and Einstein, J. R. (1968). *Abstr., Amer. Cryst. Ass. Meet.* p. 102.
Wulff, D. L., and Fraenkel, G. (1961). *Biochim. Biophys. Acta* **51,** 332.
Yamane, T., Wyluda, B. J., and Shulman, R. G. (1967). *Proc. Nat. Acad. Sci. U.S.* **58,** 439.

# 12 The Radiation Chemistry of Pyrimidines, Purines and Related Substances

G. Scholes



## A. Introduction

Interest in the radiation chemistry of aqueous solutions of nucleic acids and related compounds stems largely from the radiobiological implications. However, studies in this field have revealed many features of general interest, such as the kinetic aspects of the reactions of free radicals with macromolecules, the formation and subsequent reactions of transient radicals and radical ions formed from pyrimidines and purines, and aspects of peroxidation.

This chapter deals specifically with the chemical effects of ionizing radiations (such as x-rays, $\gamma$-rays, high-energy electrons) on dilute solutions of pyrimidines and purines, although some references are made to solutions of nucleosides, nucleotides, and DNA. In dilute solutions, the action of these radiations is indirect, being mediated initially by the active species produced by the radiolytic decomposition of water, namely, OH radicals, solvated electrons, and hydrogen atoms. During the past few years, application of the fast-reaction techniques of pulse radiolysis has shed light on the detailed chemical mechanism whereby these particular species react with solutes, and thus some emphasis has been placed on the results obtained by this rather powerful method.

Several reviews of radiation effects on solutions of nucleic acid and related substances are available (Scholes, 1963, 1968; Weiss, 1964; Fahr, 1969), and for general aspects of the radiation chemistry of water, the reader is referred to the books of Allen (1961), a valuable account of the earlier period, of Draganić and Draganić (1971), and of Swallow (1972). However, to obtain some degree of completeness this chapter includes a brief introductory account of the relevant features of water radiolysis and of the special experimental techniques employed.

## B. Decomposition of Water by Ionizing Radiations and Effects on Solutes

### 1. Chemical Processes

Unlike the energy-absorption mechanism in photochemistry, in which quanta of electromagnetic radiation are selectively absorbed by molecules, or even parts of molecules, the mechanism in radiation chemistry arises principally from coulombic interactions of fast charged particles with the electrons of molecules in the vicinity of the tracks of the particles (Mozumder, 1969). The fast particles can be electrons (by direct bombardment or the secondary electrons produced by interaction of high-energy photons with the medium) or heavy charged particles such as protons and $\alpha$-particles. Since the energy loss is purely coulombic in nature, depending to a first approximation only on the electron density of the components of the medium, the energy is deposited nonselectively. Thus, in dilute aqueous systems most of the radiation energy will be absorbed by the water, which, as a result of ionization and excitation, decomposes according to the overall equation

$$H_2O \rightsquigarrow OH,\ H,\ e_{eq}^-,\ H_3O^+,\ H_2,\ H_2O_2 \tag{1}$$

The reactive species consist of one oxidizing entity, the OH radical, and two reducing entities, the hydrogen atom and the solvated electron, which are responsible for initiating the chemical effects in solutes. Hydrogen and hydrogen peroxide, the so-called molecular products, arise from fast primary recombination processes, soon after energy deposition, ($10^{-11}$–$10^{-8}$ sec), and these molecular products occasionally participate in secondary chemical reactions. The radiation chemistry of dilute solutions, therefore, largely revolves around studies of the mechanisms of the chemical reactions of the primary species. Variations of the experimental environment can actually alter the nature of these species. Thus, in irradiated acidic solutions the solvated electrons are converted to hydrogen atoms,

$$e_{aq}^- + H_3O^+ \longrightarrow H + H_2O \tag{2}$$

while in alkaline conditions, hydroxide ions interact with both OH and H:

$$OH + OH^- \rightleftharpoons O^- + H_2O \qquad pK = 11.9 \tag{3}$$

$$H + OH^- \rightleftharpoons e_{aq}^- + H_2O \qquad pK = 9.6 \tag{4}$$

Aerated oxygen-saturated solutions are commonly used; here the reducing species react with molecular oxygen and form the hydroperoxy radical and its anion.

$$H + O_2 \longrightarrow HO_2 \tag{5}$$

$$e_{aq}^- + O_2 \longrightarrow O_2^- \tag{6}$$

$$HO_2 \rightleftharpoons O_2^- + H^+ \qquad pK = 4.9 \tag{7}$$

However, studies can be somewhat simplified by selecting conditions whereby a predominant reaction can take place with only one species. For example, in this field considerable use has been made of $N_2O$ or $H_2O_2$ for the conversion of solvated electrons to OH radicals:

$$N_2O + e_{aq}^- \xrightarrow{H_2O} N_2 + OH + OH^- \tag{8}$$

$$H_2O_2 + e_{aq}^- \longrightarrow OH + OH^- \tag{9}$$

Under these conditions, the predominant reactive species in the irradiated solutions is the OH radical (there is approximately a 10% contribution from H atoms). Of course, there can be competition for the reactive species between $H_3O^+$, $N_2O$, etc. and the solutes under investigation, but a knowledge of the appropriate reaction rate constants allows correct adjustment of the experimental conditions. Rate constant data are available and some relevant values are given in Table 1.

Two typical processes are characteristic of the reactions of OH radicals and H atoms with organic solutes. These are abstraction reactions [e.g., Eqs. (10) and (11)],

$$RH + OH \longrightarrow R\cdot + H_2O \tag{10}$$
$$RH + H \longrightarrow R\cdot + H_2 \tag{11}$$

and addition reactions to unsaturated centers [e.g., Eqs. (12) and (13)].

$$RH + OH \longrightarrow RH(OH)\cdot \tag{12}$$

$$RH + H \longrightarrow RH_2\cdot \tag{13}$$

With organic compounds the hydrated electron behaves as a classical nucleophilic reagent; single electron transfer leads to the production of radical anions which often undergo subsequent protonation:

$$R + e_{aq}^- \longrightarrow R^{\overline{\cdot}} \tag{14}$$

$$R\cdot + H^+ \longrightarrow RH\cdot \tag{15}$$

The final products of irradiated aquo–organic systems are determined

**Table 1** Some Rate Constants for Reactions of the Primary Species of the Radiolysis of Water

| Reaction | Rate constant ($M^{-1}$ sec$^{-1}$) | Reaction | Rate constant ($M^{-1}$ sec$^{-1}$) |
|---|---|---|---|
| $e_{aq}^- + H_3O^+$ | $2.2 \times 10^{10}$ | $OH + OH^-$ | $3.6 \times 10^8$ |
| $e_{aq}^- + H_2O_2$ | $1.3 \times 10^{10}$ | $OH + H_2O_2$ | $4.5 \times 10^7$ |
| $e_{aq}^- + N_2O$ | $8.7 \times 10^9$ | | |
| $H + OH^-$ | $2 \times 10^7$ | $HO_2 + HO_2$ | $6.7 \times 10^5$ |
| $H + H_2O_2$ | $1.6 \times 10^8$ | $HO_2 + O_2^-$ | $7.9 \times 10^7$ |
| $H + O_2$ | $1.7 \times 10^{10}$ | | |

by the nature and extent of the various organic radical and radical anion interactions, these usually taking the form of dismutation or dimerization processes.

A special situation pertains in solutions containing molecular oxygen which can react to form peroxy radicals:

$$R\cdot + O_2 \longrightarrow RO_2\cdot \tag{16}$$

Under these conditions, the final products are formed as a consequence of the interactions of peroxy radicals either with themselves,

$$RO_2\cdot + RO_2\cdot \longrightarrow \text{products} \tag{17}$$

or with $HO_2$ and/or $O_2^-$. It should be noted that the hydroperoxy radical is itself often inert to organic substances, and its fate is usually dismutation:

$$HO_2 + HO_2 \longrightarrow H_2O_2 + O_2 \tag{18}$$

$$HO_2 + O_2^- \longrightarrow HO_2^- + O_2 \tag{19}$$

Radiation-chemical yields are expressed by *G*-values, which refer to the numbers of species which are produced or disappear per 100 eV of energy absorbed. Thus, an experimental yield of a product P is denoted as G(P), while the quantity of solute S which has disappeared is denoted as G(−S). It is also fairly common to represent yields of the primary species with the species written as a subscript (e.g., $G_{OH}$). For most of the work in which γ-rays and high-energy electrons (>1 MeV) are used the following primary yields are relevant for neutral water: $G_{OH} = 2.7$, $G_H = 0.55$, $G_{e_{aq}^-} = 2.7$, $G_{H_2} = 0.45$, $G_{H_2O_2} = 0.7$, $G_{H_3O^+} = 2.7$.

A unit of radiation dose which is frequently used is the rad; 1 rad $= 6.24 \times 10^{13}$ eV/gm.

## 2. Experimental Techniques

The techniques of irradiating solutions with x-rays, γ-rays (usually from $^{60}Co$ sources), and with accelerated particles (normally electrons from Van der Graaf and linear accelerators) are fairly straightforward and have been adequately described elsewhere (Allen, 1961). Microanalytical methods of analysis are required since micromolar quantities of products are frequently formed; conventional methods such as spectrophotometry, chromatography in its various applications, and mass spectrometry are used for identification and quantitative analysis.

Of particular significance to the field of radiation chemistry has been the development of the pulse radiolysis technique. Details of the experimental design and early applications have been described by Boag (1963), Keene (1964), Matheson and Dorfman (1965), and Hart (1966). Pulse radiolysis is the radiation-chemical equivalent of flash photolysis, designed for the study of short-lived species. Basically, a pulse of electrons from a linear accelerator is delivered to the solution in a cell through which an analyzing light beam passes. The optical changes are followed spectrophotometrically at a fixed wavelength by a suitable optical photomultiplier tube technique. The absorption spectra as well as the decay kinetics of the radiation-produced transients can thus be obtained; transient species with lifetimes in the nanosecond region are now regularly studied. In addition, this method has allowed the direct determination of absolute rate constants of reaction of the primary species with solutes. This applies particularly to reactions of the hydrated electron (Hart and Boag, 1962) which has a maximum optical absorption ~720 nm and a large extinction coefficient [$\epsilon_{700} = 18{,}500\ M^{-1}\ cm^{-1}$ (Fielden and Hart, 1967)]. By using the fine structure pulses of the electron beam of a 30 MeV linear accelerator, Bronskill and Hunt (1968) achieved a time resolution as low as $10^{-11}$ sec. Besides the optical detection technique, electron spin resonance (ESR) coupling has been successfully used for the identification of transient species in pulsed solutions (Smaller *et al.*, 1968). An obvious advantage of ESR spectroscopy is that chemically and structurally similar radicals can be readily identified; however, spin relaxation phenomena of the radicals do not allow very short times to be used in such pulse experiments.

Electron spin resonance spectroscopy has also been used to detect radicals in solutions during steady *in situ* radiolysis with high-energy electron beams (Eiben and Fessenden, 1971). Samples are irradiated inside the cavity of the spectrometer, and a flow system is used to prevent the possibility of secondary reactions at high dose rates. This is a steady-state method but, with reasonably efficient production, it has been possible to study organic radicals which decay biomolecularly with rate constants of $\sim 10^{9}$–$10^{10}\ M^{-1}\ sec^{-1}$.

## C. Rates of Reaction with the Primary Species

Some knowledge of the rates of reaction of the various components with the reactive species, OH, H, and $e_{aq}^{-}$ is a necessary prerequisite for an understanding of the radiation-induced degradation of polynucleotides. In this section the current situation in this area is described and the methodology for the determination of reaction rate constants is illustrated.

### 1. Reaction with OH Radicals

Early measurements of the extent of destruction of Pur and Pyr bases in irradiated mixtures (e.g., Ade and Thy, Ade and Cyt, and Thy and Cyt) indicated that these substances had approximately equal reactivity towards the radiation-produced OH radicals; the pyrimidines were perhaps slightly more reactive (Scholes *et al.*, 1960). Recently, the pulse radiolysis technique has allowed a more direct study of OH radical reaction rates with these compounds. Unlike the solvated electron in water, the OH radical does not have a strong optical absorption, and, hence, the determination of the absolute rate of reaction of this entity with added solute must involve measurements of solute disappearance or transient formation. However, the problem is simplified if a competition technique is used, given a standard reaction rate with a reference solute. All three methods have been used with the nucleic acid derivatives, and the results of these measurements are summarized in Table 2.

#### *a. The Thiocyanate Competition Method*

This well-known pulse radiolysis method (Adams *et al.*, 1964, 1965) utilizes the fact that the thiocyanate ion reacts with OH to form the relatively long-lived radical ion, $(CNS)_2^-$, which absorbs strongly at 500 nm. Competition by any solute S for OH reduces the yield of $(CNS)_2^-$, and the change in the optical density of the solution at 500 nm can be expressed by

$$OD_0/OD = 1 + k_{OH+S}\,[S]/k_{OH+CNS^-}\,[CNS^-]$$

in which $OD_0$ and OD are the optical densities of the solutions in the absence and presence of S, respectively. Thus, a plot of $OD_0$/OD *vs.* $[S]/[CNS^-]$ should be linear with an intercept of 1; from the slope, $k_{OH+S}$ can be obtained if $k_{OH+CNS}$ is known. In order to eliminate interference by $e_{aq}^-$ the experiments are usually carried out in the presence of $N_2O$. Recent work (Behar *et al.*, 1971, 1972) has shown that the mechanism of the oxidation of thiocyanate ions by OH radicals is as follows:

$$CNS^- + OH \longrightarrow CNSOH^- \qquad (20)$$

**Table 2** Rate Constants for Reaction with OH Radicals

| Solute | $k(M^{-1}\ sec^{-1} \times 10^{-9})$ | | | | | |
|---|---|---|---|---|---|---|
| | CNS⁻ competition | | | | | |
| | Neutral[a] | | pH 2 | | Loss of chromophore | Transient build-up |
| | Ref. *b* | Ref. *c* | Ref. *b* | Ref. *c* | | |
| *Pyrimidines and purines* | | | | | | |
| Thy | 5.2 | 7.6 | 5.2 | — | 5.1[d],7.4[c] | 7.0[e],4.6[c],7.4[c],7.2[f] |
| | 5.6[g] | | 6.1[g] | | | |
| Ura | 5.2 | 7.3 | 4.9 | 4.4 | 6.0[c] | 6.5[c] |
| Cyt | 4.7 | — | 3.1 | — | — | — |
| $Me_2Ura$ | 5.3 | — | — | — | — | — |
| hThy | 1.6[h] | — | — | — | — | 2.2[e,i] |
| hUra | 1.3[h] | — | — | — | — | < 2[c,i] |
| Ade | 4.4 | 2.8 | 0.9 | — | — | — |
| *Nucleosides* | | | | | | |
| Thd | 4.8 | — | 4.6 | — | — | — |
| Urd | — | 4.3,4.6 | — | — | 6.5[c] | 4.1[c] |
| Cyd | 4.8 | — | 3.3 | — | — | — |
| Ado | 4.0 | — | 1.8 | — | — | — |
| *Nucleotides* | | | | | | |
| TMP | 5.3 | — | 4.3 | — | — | — |
| 2′(3′)-UMP | — | 5.1 | — | — | 4.6[c] | 4.0[c] |
| CMP | 4.5 | — | 2.5 | — | — | — |
| dCMP | 4.4 | — | 3.0 | — | — | — |
| AMP | 3.0 | — | 1.2 | — | — | — |
| dAMP | 3.5 | — | 1.4 | — | — | — |
| dGMP | 6.8 | — | 4.7 | — | — | — |
| UpU[j] | — | 5.3 | — | — | 3.8[c] | — |
| Oligo(U)[j] | — | 4.0 | — | — | 4.3[c] | — |
| Poly(U)[j] | — | 1.5 | — | — | — | <3.8 |
| DNA[j] | 0.3[k] | — | — | — | — | 0.8[l,m],0.6[e,n] |

[a] Average values over pH range 5–7.
[b] Scholes *et al.*, 1965.
[c] Greenstock *et al.*, 1968, 1969.
[d] Willson *et al.*, 1971.
[e] Myers *et al.*, 1968.
[f] Stevens, 1969.
[g] Thiocyanate competition method using $^{60}Co\ \gamma$-rays.
[h] Phillips, 1971.
[i] Rate of H abstraction from 5,6-bond.
[j] Calculated rate constant per nucleotide base.
[k] pH = 8.2.
[l] Scholes *et al.*, 1969.
[m] pH = 7.5.
[n] pH = 6.6.

$$CNSOH^- \rightleftharpoons CNS + OH^- \tag{21}$$

$$CNS + CNS \rightleftharpoons (CNS)_2^- \tag{22}$$

In neutral solution, equilibrium (21) lies far to the right and under the thiocyanate concentration conditions normally used in the competition experiments, (> 0.1 m*M*), equilibrium (22) is also far to the right. A value of $k_{20} \simeq 2.8 \times 10^{10}\ M^{-1}$ has been obtained (Baxendale *et al.*, 1968), but if this value is used in the competition experiments, rate constants are often significantly higher than those obtained by other pulse methods. Furthermore, it has been pointed out (Baxendale *et al.*, 1968) that the thiocyanate system could give incorrect results if CNS radicals also react with the solute. In a recent compilation of published rate data (Willson *et al.*, 1971), the values obtained from the thiocyanate method have been normalized, using as a secondary standard the value, $k_{OH+ethanol} = 1.85 \times 10^9\ M^{-1}\ sec^{-1}$. Comparison with OH reaction rate constants obtained by a ferrocyanide competition method revealed that in most cases the maximum deviation was less than 20%; it was, therefore, submitted that in view of the relatively successful normalization of the data the thiocyanate method need not be discarded. The rate constants for nucleic acid derivatives shown in Table 2 have been normalized to the ethanol standard, but the rather empirical background should be borne in mind when interpreting the precision of these data.

*b. Loss of Chromophore*

Rate constants for the reactions of OH radicals have been obtained from measurements of the rate of bleaching of the base chromophore at the appropriate wavelength. A slight complicating feature is that H atoms can also react with these particular solutes, leading to a loss of chromophore absorption; however, since the yield of H atoms is only ~10% of that of OH radicals, under the conditions of the experiments (in the presence of $N_2O$) this particular correction need not be too large. (For Thy, quoted in Table 2, H-atom interaction with the Pyr base has been eliminated by including some oxygen in the irradiated system.)

It should be noted that the rate constants obtained for nucleosides and nucleotides by this technique can only refer to reaction with the base components of these molecules and not to the overall rate constant which will include a (smaller) contribution from reaction of OH with the sugar moiety.

*c. Formation of Transient Species*

Determination of reaction rate constants by measurements of the rate of formation of the product of the reaction with OH has the least objection in principle. Transients produced from the Pur and Pyr bases have suitable optical absorption characteristics enabling such measurements to be carried out. Two points are important; (1) correction is necessary for the contribution from H-atom attack, small in extent though this may be and (2) since mixtures of transients will be formed from solutes in which the bases are combined, the data will once again only refer to reaction at the base.

Although there is a broad measure of agreement among the values of the rate constants for OH reaction as determined by these various methods (Table 2), none of the methods is free of objections. Indeed, the data so far obtained certainly do not satisfy the precision expected by the radiation chemist, and there is room for significant improvement in this area. However, some broad conclusions can be drawn from Table 2. First of all, the rate constants for reaction with OH are quite high, so that these compounds are efficient scavengers of these particular radicals. In neutral solution, the pyrimidines, Thy, Ura, Cyt, and 1,3-$Me_2$Ura have approximately equal reactivities towards OH; Ade, on the other hand, is slightly less reactive. This difference between Ade and the Pyr bases is also evident in a comparison of the rate constants for reaction of the corresponding nucleosides and nucleotides. It is interesting that Pullman and Pullman (1960) proposed, from semiempirical molecular orbital considerations, that the reactivities of the pyrimidines should be in the order Thy > Ura > Cyt and that these should be greater than the reactivities of the purines. In this context, the value for dGMP (Table 2) is high and an investigation of the Gua series seems warranted. The reactivities of the saturated pyrimidines, hUra and hThy, are significantly lower than those of the parent pyrimidines; this agrees with the conclusion from chemical studies (see below) that OH radicals react at the saturated 5,6-bond of the dihydro compounds.

The amino-substituted bases, Cyt and Ade, are less reactive in acid solution (pH 2), probably as a result of the deactivating effect of the protonated amino groups; this effect of protonation is also seen in the corresponding nucleosides and nucleotides, although in the latter case, ionization of the phosphate groups may have an additional influence. These various ionic equilibria tend to complicate the detailed interpretation of the results in Table 2. With the pyrimidines, Ura and Thy, there is a significant drop in reactivity at high pH. Thus, $k_{OH+Ura}$ drops from $6 \times 10^9\ M^{-1}\ sec^{-1}$ at pH $\sim$ 10, to $1 \times 10^9\ M^{-1}\ sec^{-1}$ at pH 12

(Greenstock *et al.*, 1969), an effect which is probably due to ionization of the OH radical to $O^-$ (pK ~ 12). There was no change associated with the dissociation of the carbonyl groups (pK = 9.5), indicating that changes in the charge and tautomeric form of Ura have little effect on the reaction rate constant. Similar pH effects have been noted in Thy, Thd, and hThy (Myers *et al.*, 1968).

The first investigations with the polynucleotide DNA were carried out using the thiocyanate competition technique (Scholes *et al.*, 1965) under pH conditions which prevented denaturation. Although the range of DNA concentration which could be investigated was rather limited by the viscosity of the solutions, simple competition was obeyed, a 0.01% DNA solution having an overall reactivity to OH radicals equivalent to $10 \mu M$ $CNS^-$. Using the secondary ethanol standard, this leads to $k_{OH+DNA} = 5.3 \times 10^{12}\ M^{-1}\ sec^{-1}$, assuming a molecular weight of $5 \times 10^6$. Since these polynucleotides are polydisperse, this represents an average value. The reaction rate constant is very high, as would be expected kinetically for such a large molecular entity. If the DNA is assumed to behave as a homogeneous mixture of nucleotides (MW ~ 300), the value given in Table 2 is obtained and can be compared with similar computations for oligo(rU) (MW ~ $10^4$) and poly(rU) (MW ~ $10^5$). There is an effect of chain length on the reactivity per nucleotide unit, i.e., an approximately twentyfold decrease from nucleotide to DNA. Not too dissimilar values for the overall $k_{OH+DNA}$ are obtained from observations of the build-up of the DNA transient resulting from OH attack (see Table 2). These studies were conducted at 400 nm; in this range the sugar radical does not absorb, and, the results are probably only concerned with transients arising from attack on the base components of the nucleic acid. Of particular interest is the effect of pulsing (increasing dose) on the apparent rate constant (Scholes *et al.*, 1969). The rate constant increased, indicating an enhanced reactivity of the denatured DNA towards OH radicals; such an effect could arise from (1) a simple kinetic consequence of fragmentation and/or (2) the fact that the more organized helical configuration may reduce the extent of OH radical reaction with the bases.

An important factor which will determine the rate constant of OH attack on a polynucleotide is the collision frequency. For spherical particles, the frequency of encounters has been derived by Debye (1942), following a method initiated by Smoluchowski (1918) which leads to the well-known expression

$$\gamma = 4\pi(R_1 + R_2)(D_1 + D_2)\, N \times 10^{-3}\ M^{-1}\ sec^{-1}$$

in which $\gamma$ is the encounter frequency, $R_1$ and $R_2$ are the radii of the two particles, $D_1$ and $D_2$ are the diffusion coefficients, and $N$ is the Avogadro number. The expression cannot be used for nonspherical polymer molecules. However, Braams and Ebert (1968) modified the theory for the limiting case of collision between a small spherical molecule (e.g., OH) and a rodlike molecule. The following equation applies in this particular situation:

$$\gamma = 4\pi \frac{L}{2 \ln L/r} (D_1 + D_2)N \times 10^{-3} \ M^{-1} \ \text{sec}^{-1}$$

Here $L$ is the length of the rod and $r$ is the radius of rotation of the cylinder. Assuming DNA to be an extended molecule of cylindrical shape, the encounter frequency for OH was calculated as $8.3 \times 10^{12}$ $M^{-1}$ sec$^{-1}$. Assuming unit probability of reaction on encounter, this value should then be compared with the measured rate constant of 5–13 $\times 10^{12}$ $M^{-1}$ sec$^{-1}$. One would perhaps expect the encounter frequency to be somewhat less for a stiff coil, in which state DNA actually exists. Nevertheless, this seems to be a realistic approach to the effect of conformation of a macromolecule on its chemical reactivity; it has, in fact, been recently extended to reaction with the solvated electron (Schragge *et al.*, 1971).

### 2. Reaction with Hydrogen Atoms

Competitive methods have been used to determine the rate constants for reaction with H atoms. The earlier experiments were confined to neutral or slightly alkaline solutions, in which case one is specifically dealing with those H atoms produced by direct decomposition of water, in a yield of ~0.6 *G*-units. The deuterioformate competition method (Scholes and Simic, 1964b, 1968) is fairly sensitive and involves measurements of the effects of solutes on the yields of HD formed by reaction of H atoms with the deuterated compound according to Eq. (23):

$$H + DCOO^- \longrightarrow HD + COO^- \tag{23}$$

The added solute (RH) competes with reaction (23) either by the dehydrogenation process shown in reaction (11) or by a reaction (13) which does not lead to molecular hydrogen, e.g., addition to a double bond. On the basis of simple competition, it follows that

$$\frac{G^0(\mathrm{HD}) - G(\mathrm{HD})}{G^0(\mathrm{HD})} = \frac{k_{\mathrm{H+RH}}\,[\mathrm{RH}]}{k_{\mathrm{H+DCOO^-}}\,[\mathrm{DCOO^-}]}$$

in which $G^0$(HD) and $G$(HD) correspond to the yields of HD in the absence and presence of the competing solute, respectively. Thus the rate constant ratios, $k_{\mathrm{H+RH}}/k_{\mathrm{H+DCOO^-}}$ can be obtained. These experiments must be carried out in the presence of $N_2O$ so that complications caused by reactions of the solvated electron are avoided. Selected experimental measurements together with the computed relative rate constant ratios are collected in Table 3. As in all competition methods, a standard is required. In the last column of Table 3 the absolute rate constants have been calculated on the basis of $k_{\mathrm{H+ethanol}} = 2.5 \times 10^7\ M^{-1}\ \mathrm{sec}^{-1}$ (Neta *et al.*, 1971), which is a slightly higher value than previously adopted for this particular rate.

Neta *et al.* (1971) recently determined the rate constants for the reaction of H atoms with Pur and Pyr bases in aqueous solution at pH 1, using the *in situ* radiolysis steady-state ESR method (Eiben and Fessenden, 1971). An approximately steady-state treatment of a competition between the formation of spin-polarized H atoms (produced on radiolysis mainly from $e_{aq}^- + H_3O^+$, since the experiments are carried

**Table 3** H-atom Rate Constants Determined by the Deuterioformate Competition Method (pH 7–8)[a]

| Solute (mM) | $[\mathrm{DCOO^-}]$ (mM) | $G(H_2)$ | $G$(HD) | $\frac{k_{\mathrm{H+RH}}}{k_{\mathrm{H+DCOO^-}}}$ | $k_{\mathrm{H+RH}}$ ($M^{-1}$ sec$^{-1} \times 10^{-8}$) |
|---|---|---|---|---|---|
| None | 2.5 | 0.36 | 0.48 | — | — |
| Thy | 0 | 0.37 | — | — | — |
| | 10 | 0.38 | 0.24 | 10.0 | 3.8 |
| Ade | 0 | 0.38 | — | — | — |
| | 3 | 0.38 | 0.16 | 6.0 | 1.4 |
| hThy | 0 | 0.78 | — | — | — |
| | 2.5 | 0.54 | 0.32 | 10.0 | 3.8 |
| Thd | 0 | 0.37 | — | — | — |
| | 10 | 0.38 | 0.22 | 11.0 | 4.2 |
| TMP | 0 | 0.37 | — | — | — |
| | | 0.37 | 0.24 | 10.0 | 3.8 |
| Ado | 0 | 0.38 | — | — | — |
| | 3 | 0.38 | 0.16 | 6.0 | 2.3 |
| AMP | 0 | 0.36 | — | — | — |
| | 3 | 0.35 | 0.15 | 6.5 | 2.5 |
| DNA | 0 | 0.37 | — | — | — |
| | 2.5 | 0.38 | 0.26 | 2.0[b] | 0.78[b] |

[a] Scholes and Simic, 1964b, 1968.
[b] Rate constant per nucleotide base.

out in acidic solutions), the relaxation of H atoms with a rate constant $\lambda$, and their reaction with the solute (rate constant $k$) leads to the equation

$$H_0/H - 1 = (k/\lambda)\ [\mathrm{RH}]$$

$H_0$ and $H$ are the ESR signals of H atoms in the absence and presence of solute, respectively. When the signals are reduced by half, $[\mathrm{RH}]_{1/2} = \lambda/k$. Given that $\lambda$ is an experimental constant depending upon instrument conditions, it follows that relative rate constants can be obtained if a standard is known. The ethanol standard was used, so that

$$k_s = 2.5 \times 10^7\ ([\mathrm{EtOH}]_{1/2}/[\mathrm{RH}]_{1/2})$$

The results are presented in Table 4. There seems to be reasonable agreement between the two experimental methods if one compares the rate constants for reaction with Thy. (The Ade compounds which are common to Tables 3 and 4 change their ionic form in acid conditions).

The rate constants for the reaction of H atoms with the nucleic acid components are approximately an order of magnitude less than those of OH radicals. Thy is ~3 times more reactive than Ade in neutral solution and 6 times more so in acid; this is consistent with a deactivating effect of the protonated amino group of the Pur. The deactivating effect of the protonated $NH_2$ group in Cyt is also evident. Thy is more reactive than Ura and 6-MeUra is even more reactive, which indicates a strong activating effect of the methyl group on addition of H to the heterocyclic ring. Pyr itself is a strongly deactivated aromatic ring (compared to benzene), and addition of a second ring to form Pur adds little to the reactivity; this is not unreasonable in view of the low reactivity of imidazole. In both neutral and acid solutions Ado reacts faster than Ade, an effect of combination which is not seen with Thy

**Table 4** H-atom Rate Constants determined by the ESR Steady-State Method (pH 1)[a]

| Solute | $k_{\mathrm{H+RH}}$ ($M^{-1}$ sec$^{-1}$ $\times 10^{-8}$) | Solute | $k_{\mathrm{H+RH}}$ ($M^{-1}$ sec$^{-1}$ $\times 10^{-8}$) |
|---|---|---|---|
| Thy | 5.0 | Ade | 0.83 |
| Ura | 2.8 | Ado | 1.1 |
| Cyt | 0.9 | Pur | 1.2 |
| 6-MeUra | 7.0 | Imidazole | 0.62 |
| Pyr | 1.0 | Ribose | 0.55 |

[a] Neta *et al.*, 1971.

and Thd. Such observations would be anticipated if combined ribose has the same reactivity as free ribose quoted in Table 4. No increase in $H_2$ was noted in the irradiated Ado compared to Ade solution (Table 3); this is puzzling since one would expect H atoms to dehydrogenate the sugar molecule.

The rate constant for the reaction of H atoms with hThy is similar to that for the reaction with Thy, despite saturation of the double bond; in this case the reaction is one of dehydrogenation at the 5,6-bond and is evidenced by the production of $H_2$ on radiolysis.

As discussed for OH radicals, the overall specific rate constant for reaction of H atoms with DNA should be high. The experimental results lead to $k_{H+DNA} = 1.3 \times 10^{12}\ M^{-1}\ sec^{-1}$ for a molecular weight of $5 \times 10^6$, which is 4 times less than the corresponding OH reaction rate. The value in Table 3 refers to the rate constant per nucleotide unit and is somewhat less than the average value for the single nucleotides. No significant dehydrogenation of the DNA molecules takes place in these conditions, suggesting that H-atom attack occurs mainly on the heterocyclic bases.

Direct measurements of the rate constants for H-atom reaction with Thy and Ura have been carried out using the pulse technique. Theard *et al.* (1971) followed the rate of build-up of absorption at 400 nm in pulsed argon-saturated Thy solutions under conditions of pH and [Thy] in which only H atoms were reacting and obtained a value of $k_{H+Thy} = 6.8 \times 10^8\ M^{-1}\ sec^{-1}$, which is in good agreement with that obtained above. Greenstock *et al.* (1969) measured the rate of build-up of the H adduct of Ura in Ura solutions at pH 2 containing 0.05 *M* formaldehyde as an OH scavenger. A value of $k_{H+Ura} = 5.6 \times 10^9\ M^{-1}\ sec^{-1}$ was reported, which seemed to agree with a value of $3 \times 10^9\ M^{-1}\ sec^{-1}$ obtained by competition experiments with isopropanol (Scholes and Simic, 1964b). These values are somewhat higher than those reported above.

### 3. Reaction with Solvated Electrons

Scholes and Simic (1964a) measured the relative reactivities of some Pyr and Pur compounds toward the solvated electrons, using steady-state radiolysis with $N_2O$ as competitor. All were found to be very reactive, the reaction with the free bases being close to diffusion controlled; indeed, of the three species OH, H, and $e_{aq}^-$ produced by the radiolysis of water, the solvated electrons react fastest with the heterocyclic bases. Within the precision of the method, it was found that while there was little difference between Pur and Pyr moieties, there

**Table 5** Rate Constants for Reaction with $e_{aq}^-$

| Solute | pH | $pK_a$ | $k$ ($M^{-1}$ $sec^{-1}$ $\times 10^{-9}$) | Ref.[a] |
|---|---|---|---|---|
| *Pyrimidine compounds* | | | | |
| Thy | 5.5 | 9.9 | 18 | 1 |
| | 12 | — | 2.7 | 2 |
| TMP | 6.7 | — | 1.5 | 1 |
| Ura | 7 | 9.5 | 15 | 3 |
| | 12 | — | 2.3 | 3,4 |
| Urd | 7 | 9.3 | 15 | 3 |
| | 12 | — | 2.5 | 3 |
| UMP | 7 | 9.4 | 5 | 3 |
| | 12 | — | 13 | 3 |
| Cyt | 7 | 12.2 | 13.2 | 5 |
| | 14 | — | 3.6 | 5 |
| Cyd | 7 | 12.3[b] | 13.2 | 5 |
| | 14 | — | 9.5 | 5 |
| CMP | 7–14 | — | 6.8 | 5 |
| 2′,3′-cCMP | 7 | 12.3[b] | 10 | 5 |
| | 14 | — | 7.5 | 5 |
| hUra | 7 | — | 4.5 | 3 |
| hThy | 7 | — | 5.4 | 6 |
| 2,4-$(Eto)_2$Pyr | 7 | — | 2.8 | 3 |
| | 11 | — | 3.2 | 3 |
| 1,3-$Me_2$Ura | 7 | — | 16.5 | 3 |
| | 11 | — | 14.5 | 3 |
| *Purine compounds* | | | | |
| Ade | 7 | 9.9 | 9 | 5 |
| | 11 | — | 1.1 | 5 |
| Ado | 6–11 | 12.5[b] | 9.2 | 5 |
| | 13 | — | 6.3 | 5 |
| AMP | 6.7 | — | 3.8 | 1,5 |
| Pur | 7 | 8.8 | 16 | 4,5 |
| | 11 | — | 8.2 | 5 |
| Hyp | 6.6 | — | 17 | 7 |
| *Polynucleotides*[c] | | | | |
| UpU | 7 | — | 5.8 | 3 |
| Oligo(U)[d] | 7 | — | 2.5 | 3 |
| Poly(U) | 7 | — | 0.75 | 5 |
| Poly(A) | 7 | — | 0.25 | 5 |
| Poly(U + A) | 7 | — | 0.13 | 5 |
| DNA | 8 | — | 0.06 | 1 |
| | 7 | — | 0.14 | 5 |

[a] Key to references:
1. Scholes *et al.*, 1965.
2. Scholes, 1968.
3. Greenstock *et al.*, 1968.
4. Hart *et al.*, 1964.

was a marked lowering of the reaction rate from nucleoside to nucleotide, arising from the presence of the negatively charged phosphate group.

Detailed studies of the rate constants for reaction with $e_{aq}^-$ have been carried out using the pulse radiolysis technique (Hart *et al.*, 1964; Scholes *et al.*, 1965; Greenstock *et al.*, 1968; Schragge *et al.*, 1971; Phillips, 1971). A summary is given in Table 5. The Pur and Pyr bases are indeed similar in their reactivity, although there is some deactivating effect of the amino group of Cyt and Ade. The electron-accepting characteristics of the C=N and C=O bonds are responsible for the high reactivities of these compounds. The rate for 1,3-$Me_2$Ura is greater than for 2,4-$(ho)_2$Pyr, indicating that the carbonyl groups are the most reactive sites in the Pyr bases. However, the lowered reactivity of the hPyr bases suggests some cooperative effect of the 5,6-double bond in the rate of electron attachment.

In those solutes which have ionizable groups the reaction rate constant falls off as the molecules become more negatively charged. Dissociation in the base and in the sugar moieties gives rise to such an effect (Table 5) and the titration curves can be fairly accurately followed (Greenstock *et al.*, 1968; Schragge *et al.*, 1971). Detailed examples of this behavior for Pur and Pyr compounds are given in Figs. 1 and 2, respectively. The pH dependence for Ade has been attributed to the

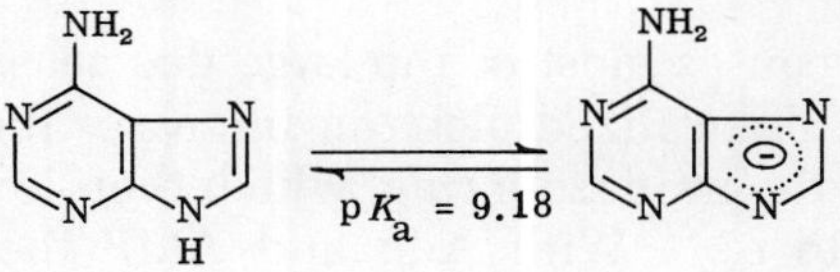

$NH_2$ N N N N H ⇌ $pK_a = 9.18$ $NH_2$ N N N N ⊖

dissociation equilibrium since 7-MeAde has a constant rate constant in the pH range 6–12, as expected, given that no ionization of the molecule can occur. The Debye equation (Debye, 1942) predicts a decrease in reactivity of $e_{aq}^-$ when a negative charge is added to a molecule. The calculated rate for Ade is 1.7 times less. However, the observed decrease is nearly 9 times less, and it has been suggested (Schragge *et al.*, 1971) that the shift in electron density within the ring of the

---

5. Schragge *et al.*, 1971.
6. Phillips, 1971.
7. Szutka *et al.*, 1965.

[b] Ribose dissociation.
[c] Calculated per nucleotide base.
[d] 20 nucleotide units.

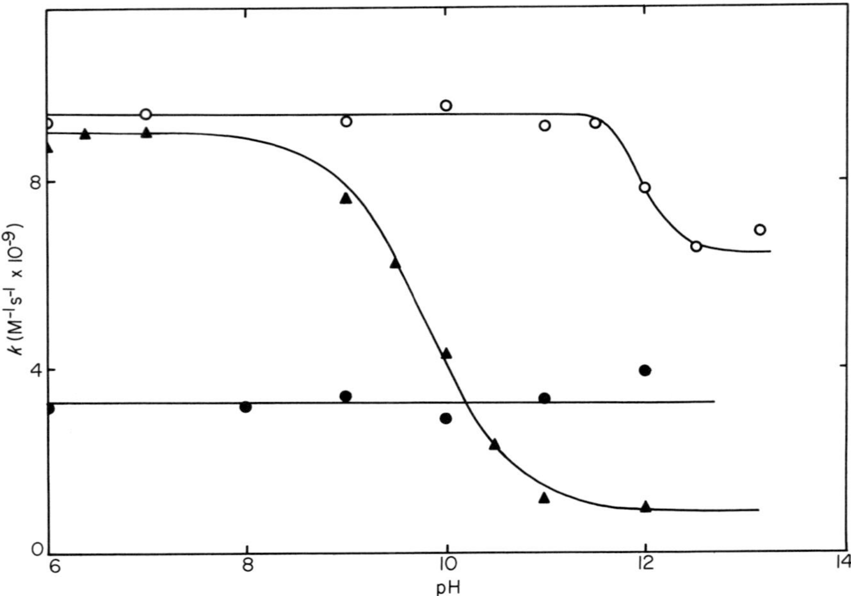


**Fig. 1.** *pH dependence of the rate constants for reaction of* $e_{aq}^-$ *with Ade, Ado, and AMP* (Schragge et al., *1971*) ○, *Ado* ($pK_a = 12.5$); ▲, *Ade* ($pK_a = 9.9$); *the solid line is a theoretical titration curve;* ●, *AMP.*

ionized molecule causes most of the large decrease in reactivity. It is supposed that the delocalized electron increases the electron density of the $\pi$ system of the imidazole ring, which then lowers the reactivity of the molecule to $e_{aq}^-$. With Ado and AMP the N(9) position is blocked so that no change in the rate of reaction with $e_{aq}^-$ occurs near the $pK_a$ of Ade. In Ado the decrease at pH ~ 12 is attributed to ionization of the sugar group which is presumably inhibited by the presence of a phosphate group on the sugar molecule so that no pH effect is noted in AMP. A similar situation was noted for the series Cyt, Cyd,

$$\xrightleftharpoons[+H^+]{-H^+}$$

$pK_a = 12.2$

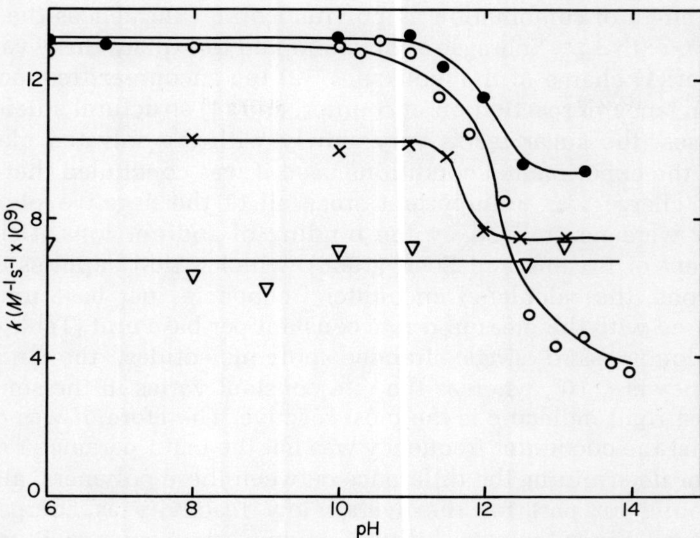


**Fig. 2.** *pH dependence of the rate constants for reaction of* $e_{aq}^-$ *with Cyt, Cyd, CMP, and 2′,3′-cCMP.* (Schragge et al., *1971*); ○, *Cyt* ($pK_a = 12.2$.; *the solid line is a theoretical titration curve;* ●, *Cyd* ($pK_a = 12.3$); ×, *2′,3′-cCMP;* △, *CMP.*

and CMP. The variation of the rate constant for Cyt is attributed to the ionization equilibrium and the decrease associated with the combined effects of the negative charge and the accompanying shift of electron density in the ionized form of Cyt. The effects in Cyd and in

2′, 3′ - Cyclic cytidylic acid (neutral pH)

5′ - Cytidylic acid (neutral pH)

2′,3′-cCMP are attributed to ribose ionization, and blocking of the C(5′) position with phosphate in CMP prevents the ionization. Because both the cyclic phosphate and CMP have two negative charges at high pH, the rates of reaction with $e_{aq}^-$ are comparable.

The effect of combination as polynucleotides influences the rate of reaction with $e_{aq}^-$. Schragge *et al.* attempted to explain these values in terms of (1) charge of the molecules, (2) the encounter frequency, (3) the efficiency of reaction on encounter, and (4) structural shielding of the bases (the sugar reacts very slowly with the solvated electron). Under the experimental conditions used it was concluded that the effect of charge was unimportant since all of the negative phosphate groups were neutralized by the binding of sodium ions. Using the treatment of Braams and Ebert (1968), which strictly applies only to rigid rods, the calculated encounter frequencies per base unit were compared with the measured rate constant per base unit (Table 6). For both double- and single-stranded polynucleotides, the encounter frequency is $\sim 10^9$, whereas the rate constant varies in the sense that the least rigid molecule is the most reactive. Therefore, it was considered that the encounter frequency was not the main parameter responsible for determining the difference between these polymers, although it accounts in part for the overall low reactivity as compared to monomers. Since the probability of reaction on encounter, *P*, may not be unity, a correction was made. The values of *P* for Ura and Ade (taken simply as the ratio of the rate constant to encounter frequency) were assumed to be applicable to the polymers, but, as shown in Table 6, although the estimated rate is close to the measured value for poly(U), it is nearly two times higher for poly(A) and five times higher for poly(A + U). Therefore, Schragge *et al.* concluded that the structural shielding of the bases has an important role in the determination of the rate constant of reaction of $e_{aq}^-$ with the polynucleotides. In particular, shielding should occur where stacking exists, since this excludes water; there seems to be a direct correlation between the degree of secondary structure and the reaction rate if the extent of stacking in their individual molecules is considered (last column, Table 6). Despite the assumption of the model, this approach is interesting and open to further experimentation.

## D. Radiolysis of Aqueous Pyrimidine Solutions

From the earliest studies of the effect of ionizing radiation on aqueous Pyr solutions (Scholes and Weiss, 1952, 1953; Ranadive *et al.*, 1956), it was clear that the 5,6-ethylenic bond of these substances was an important site of attack by free radicals. This finding has been substantiated by subsequent experimentation. Because of this simplicity as compared to other nucleic acid components, the Pyr bases have

**Table 6** Polynucleotides: Measured and Calculated Rate Constants for Reaction with $e_{aq}^{-}$ [a]

| Substance | MW | Measured $k$ ($M^{-1}$ sec$^{-1}$ $\times 10^{-9}$) | Encounter frequencies ($M^{-1}$ sec$^{-1}$ $\times 10^{-9}$) | Probability of reaction ($P$) | Encounter frequencies $\times P$ | Structure |
|---|---|---|---|---|---|---|
| Ura | 112 | 15 | 21 | 0.71 | 15 | — |
| Ade | 137 | 9 | 21 | 0.43 | 9 | — |
| Poly(U) | $2 \times 10^5$ | 0.75 | 1.0 | — | 0.7 | No base stacking |
| Poly(A) | $2 \times 10^5$ | 0.25 | 1.0 | — | 0.43 | ⅔ base stacking |
| Poly(A + U) | $5 \times 10^5$ | 0.13 | 1.1 | — | 0.63 | Complete base stacking and double helix |
| DNA | — | 0.14 | — | — | — | — |

[a] Schragge *et al.*, 1971.

been extensively studied, and as a consequence the radiation chemistry of their aqueous solutions is now reasonably well understood.

Molecular orbital calculations have been carried out on Pyr molecules and radicals derived therefrom (Pullman and Pullman, 1960, 1963; Pullman, 1965), and these calculations have been used to predict the preferential sites of attack on the Pyr bases by various chemical species (Nofre and Cier 1964). From calculated values of the free valence indices it seems that the most suitable position for free radical attack are at C(6) for Thy, at C(5) for Ura, and at C(5) or C(6) for Cyt. The electrical charge indices (indicative of the $\pi$-electron density on each position) predict that electrophilic reactions will take place preferentially at C(5) whereas nucleophilic reactions will occur at C(6).

### 1. Irradiation in the Presence of Oxygen

#### *a. Destruction of the Chromophore Group*

Measurements of the decrease of the UV absorption at ~260 nm [$G$(−chrom)] have been used to study the effects of various experimental conditions on the radiation-induced destruction of Pyr bases. Some earlier results (Scholes *et al.*, 1960) using 200 kV x-rays are given in Table 7. The quoted yields are initial ones obtained from yield–dose plots which were linear up to approximately 60% solute destruction. There were relatively little quantitative differences in $G$(−chrom) for the different pyrimidines, although Cyt gave less reproducible results (due to a rapid postirradiation decrease in optical density). If there are no radiation products absorbing at the appropriate wavelength maximum, then $G$(−chrom) can be equated with $G$(−Pyr). This was checked by two independent methods: (1) paper chromatographic separation of the irradiated solution and elution of the unchanged base and (2) determination of unchanged material by reaction of bromine with 5,6-double bonds. It is clear from Table 7 that the

**Table 7** Irradiation of Oxygenated Solutions of Pyrimidine Bases with 200 kV X-Rays[a]

| | Yield of destruction ($G$) | | |
|---|---|---|---|
| Pyrimidine | By spectrophotometry | By chromatography | By bromine reaction |
| Thy | 1.89 | 1.91 | 1.80 |
| Cyt | 2.05–2.28 | 2.16 | — |
| Ura | 1.93 | — | 1.88 |
| 1,3-$Me_2$Ura | 1.85 | 1.91 | — |

[a] Scholes *et al.* 1960.

results of these various methods agree quite satisfactorily. Therefore, it was concluded that attack by OH radicals leads to saturation of the 5,6-double bond and that this is the predominant process in these molecules in neutral oxygenated solutions. The initial reaction at C(6) of Thy is shown in reaction (24). More recent determinations of

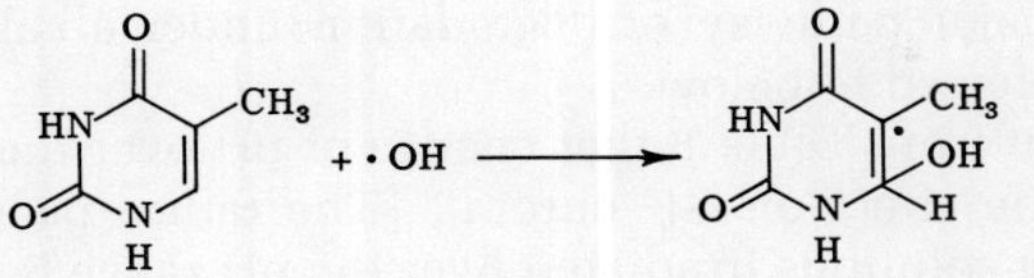

O, HN, $CH_3$, O, N, H $+ \cdot OH \longrightarrow$ O, HN, $CH_3$, OH, O, N, H, H (24)

$G(-\text{chrom})$ using $\gamma$-rays are shown in Table 8. It can be seen that the values in dilute solution are similar for these various pyrimidines, again indicating a common mechanism involving saturation of the 5,6-bond. The extent of chromophore destruction is compatible with OH radical yields measured in several other dilute aqueous systems; for instance, in the radiolysis of aerated solutions of safranine-T (a dye which is attacked by OH but not by $HO_2$ or $O_2^-$) it has been found that $G(-\text{dye}) = 2.6$ at $pH = 5$ (Marketos *et al.*, 1968). Therefore it seems reasonable to conclude that the loss of optical absorption is a good measure of the extent of OH radical attack on the Pyr molecules.

In the case of Thy, oxidation of the methyl group can also take place (Ekert, 1962), leading to the formation of 5-hmUra [reaction (25)].

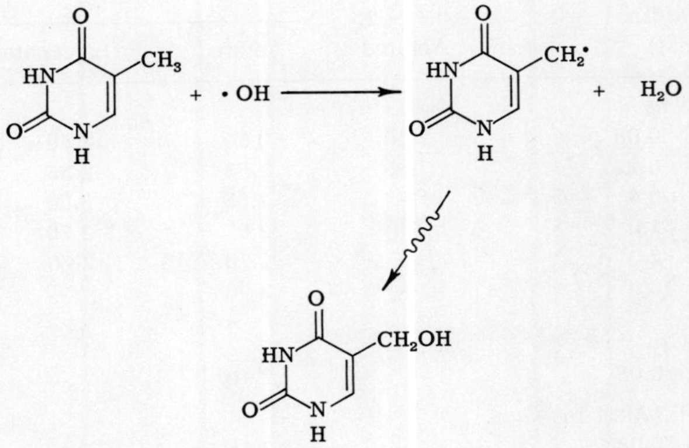

O, HN, $CH_3$, O, N, H $+ \cdot OH \longrightarrow$ O, HN, $CH_2^{\bullet}$, O, N, H $+ H_2O$

O, HN, $CH_2OH$, O, N, H (25)

This particular reaction occurs only to a slight extent in neutral media. For example, Khattak and Green (1966c) reported that in irradiated deaerated 10 m*M* Thy solutions, $G(\text{5-hmUra}) = 0.23$, and such a value would represent a miniumum one for the extent of the reaction of OH radicals with the methyl group in aerated systems. However, Cadet and Téoule (1971) have recently found a much lower value of $G \sim 0.06$ in deaerated 2 m*M* Thy solutions. Indeed, comparison of *G*(−chrom) for the various pyrimidines in Table 8 suggests that at pH $\sim$ 5 such a side reaction of OH radicals in Thy solutions must be quite small. On the other hand, reaction at the methyl group becomes a major pathway of degradation under alkaline conditions and this is referred to below.

It is evident from Table 8 that change of pH over the range 2–8 has relatively little effect on *G*(−chrom). [The rather peculiar effect observed in Thy solutions irradiated over the pH range 5–7 (Scholes and Willson, 1967) is probably an artifact; subsequent work (Phillips, 1971) has shown this to be dependent upon the Thy sample, and it is perhaps also associated with the glycine buffer used in these experiments.] An increase in *G*(−chrom) in Pyr solutions irradiated at pH < 2 is anticipated on general grounds, since it is known that $G_{OH}$ increases in acid media (up to 2.8 at pH = 1). Lower values than expected have been obtained; for instance, in oxygenated 0.12 m*M* Thy solutions containing 0.1 *N* $H_2SO_4$, *G*(−chrom) = 2.55–2.60 (Ward and Myers, 1965; Lohman and Blok, 1968). Lohman and Blok (1968)

**Table 8** G(−chrom) in Pyrimidine Solutions Irradiated with γ-Rays[a]

| [Pyrimidine] (m*M*) | G(−chrom) pH ~ 2, Aerated | G(−chrom) pH ~ 5, Aerated | G(−chrom) pH ~ 5, Oxygenated |
|---|---|---|---|
| *Thy* | | | |
| 0.08 | 2.40 | 2.50 | 2.50 |
| 0.2 | 2.65 | 2.55 | 2.55 |
| 0.4 | — | 2.55 | 2.50 |
| 1.0 | — | 2.80 | 2.70 |
| 2.0 | — | 2.70 | 2.70 |
| 10.0 | — | 2.90[b] | — |
| *Ura* | | | |
| 0.08 | — | 2.40[c] | — |
| 1,3-$Me_2$*Ura* | | | |
| 0.08 | — | 2.45[c] | — |
| 2,4-$(MeO)_2$*Pyr* | | | |
| 0.08 | — | 2.40[c] | — |

[a] Scholes and Willson, 1967.
[b] Lohman and Blok, 1968.
[c] Phillips, 1971.

suggested that there may be some formation of a UV-absorbing product or a partial recovery of the initially attacked Thy. However, higher solute concentrations may be necessary to scavenge the extra oxidizing species released under acidic conditions so that no additional reaction pathway need necessarily be assumed. When the pH is increased beyond pH ~ 9, the mechanism of radiolytic decomposition in aerated Pyr solutions changes. Measurements of chromophore destruction in Thy solutions irradiated in the pH range 9–14 showed an apparent two-step process, which led to $G$(−chrom) decreasing almost to zero in 2 $N$ NaOH (Myers *et al.*, 1965a). It was found, qualitatively, that the yield of 5-hmUra increases as the pH increases, and it was suggested that the major site of attack by the oxidizing species shifts from the double bond in the ring to the 5-methyl group. A similar decrease in $G$(−chrom) was also noted in solutions of 5-MeCyt (Myers *et al.*, 1965b). Myers *et al.* (1970b) suggested that because of charge repulsion between the ionized form of the OH radical ($O^-$) and of the pyrimidines (the first pKs of Thy and MeCyt are 9.8 and 12.4, respectively), the addition reaction becomes slower, and abstraction from the methyl group can thus occur.

$$RCH_3 + O^- \longrightarrow RCH_2\cdot + OH^- \tag{26}$$

It should be noted, however, that in agreement with earlier findings (Ranadive *et al.*, 1956), Myers *et al.* (1965a) observed that the initial disappearance yield for pyrimidines which lack a 5-methyl group, e.g., Ura or Cyt, also decreased in alkaline solutions, although saturation of the 5,6-bond would appear to remain an important reaction. Thus, it seems that an increase in aromatic character itself is sufficient to alter the reaction mechanism within the Pyr ring.

The observed slight concentration effect in neutral solutions (Table 8) together with its independence of oxygen concentration is of interest with regard to the consequences of electron attachment processes in irradiated Thy solutions. The solvated electrons can enter into the following competition:

$$\text{Thy} + e_{aq}^- \longrightarrow \text{Thy}^{\overline{\cdot}} \tag{27}$$

$$O_2 + e_{aq}^- \longrightarrow O_2^- \tag{6}$$

From the fact that the rate constants for these reactions are almost equal, it can be readily shown that the extent of reaction (27) in aerated solutions will increase from $G \sim 0.7$ at 0.2 m$M$ Thy to $G \sim 2$ at 2 m$M$ Thy. On this basis, it was concluded (Scholes and Willson,

1967) that electron attachment to the Pyr does *not* lead to significant base destruction as long as oxygen is present in the solution. It was proposed that restitution could occur by the electron transfer process:

$$\text{Thy}^{\overline{\cdot}} + \text{O}_2 \longrightarrow \text{Thy} + \text{O}_2^{-} \qquad (28)$$

However, there seems to be evidence that process (28) involves the intermediate formation of an oxygen adduct (Lohman and Ebert, 1970) [reaction (29)],

$$\text{Thy}^{\overline{\cdot}} + \text{O}_2 \longrightarrow \text{Thy O}_2^{\overline{\cdot}} \qquad k = 8 \times 10^9\ M^{-1}\ \text{sec}^{-1} \qquad (29)$$

which, by some as yet unknown mechanism, yields Thy and hydrogen peroxide.

$$2\text{Thy O}_2^{\overline{\cdot}} \longrightarrow \text{Thy} + \text{H}_2\text{O}_2 \qquad (30)$$

Recent ESR studies (Henriksen and Snipes, 1970) of the nature of $\text{Thy}^{\overline{\cdot}}$ and other Pyr indicate that the radical anion has a free electron localized at C(6); thus the structure of Thy $\text{O}_2^{\overline{\cdot}}$ could be

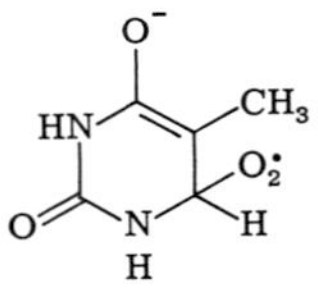

Unlike solvated electrons, H atoms can contribute to Pyr destruction in oxygenated solutions, since the Pyr–H-atom adduct enters into an irreversible reaction with molecular oxygen. This effect is noticeable particularly in acidic solutions in which more H atoms are formed according to (2),

$$e_{aq}^{-} + \text{H}_3\text{O}^{+} \longrightarrow \text{H} \qquad (2)$$

which competes with processes (27) and (6). Thy and oxygen will then compete for H atoms:

$$\text{H} + \text{Thy} \longrightarrow \text{ThyH}\cdot \xrightarrow{\text{O}_2} \text{products} \qquad (31)$$

$$\text{H} + \text{O}_2 \longrightarrow \text{HO}_2 \qquad (5)$$

Lohman and Blok (1968) measured $G(-\text{chrom})$ in varying conditions of pH and $[\text{O}_2]$ and were able to account for the experimental observations if $k_{H+O_2}/k_{H+Thy} = 25$; taking $k_{H+O_2} = 2 \times 10^{10}\ M^{-1}\ sec^{-1}$ (Sweet and Thomas, 1964) it was concluded that $k_{H+Thy} = 8 \times 10^8\ M^{-4}\ sec^{-1}$ (Table 3).

A practical consequence of these studies of $G(-\text{chrom})$ has been the use of aqueous aerated Thy solutions to determine relative rate constants for reactions of OH radicals. Thus, simple competition between Thy and an additional solute (S) for the available OH radicals, i.e.,

$$\text{Thy} + \text{OH} \xrightarrow{k_T} \text{products}$$

$$\text{S} + \text{OH} \xrightarrow{k_S} \text{products}$$

can be expressed by the relationship

$$[G_S(-\text{chrom})]^{-1} = [G_0(-\text{chrom})]^{-1} + [G_0(-\text{chrom})^{-1}](k_S[\text{S}]/k_T[\text{Thy}])$$

in which $G_S(-\text{chrom})$ and $G_0(-\text{chrom})$ are the loss of chromophore in the presence and absence of solute S. The rate ratio $k_S/k_T$ can be arrived at simply by following the effect of solutes on the loss of optical absorption at 260 nm. The method presumes that organic radicals from the solute do not react with unchanged Thy. Such reactions apparently do take place, e.g., with methanol and ethanol radicals (Lohman and Blok, 1968), but if the competition experiments are carried out at reasonably low Thy concentrations, the oxygen present in the solution can interact with all of the organic radicals, giving organic peroxy radicals which cannot react with the Pyr base.

*b. Products of Radiolysis*

The idea that the Pyr 5,6-bond is an important site of attack by OH radicals has also arisen from chemical studies of the products from solutions containing oxygen. In particular, radiolysis leads to the formation of organic peroxidic compounds, and it was originally suggested (Scholes *et al.*, 1956; Daniels *et al.*, 1957; Weiss, 1958) that these were hoho$_2$hPyr formed by saturation of the 5,6-double bond [see reaction (32)].

$$\text{Pyr} + \text{OH} \longrightarrow \text{hPyr(OH)}\cdot \xrightarrow{O_2} \text{hPyr(OH)O}_2\cdot \longrightarrow \text{hPyr(OH)O}_2\text{H} \qquad (32)$$

In the case of Thy, the hydroperoxide was characterized as ho$_2^5$-ho$^6$hThy (see I, Fig. 3) (Ekert and Monier, 1959; Scholes *et al.*, 1960).

Latarjet *et al.* (1961) showed that the radiation product was, in fact, a mixture of the *cis*- and *trans*- ho$_2^5$ho$^6$hThy. Hydroperoxides were also observed in irradiated solutions of Ura and $Me_2Ura$; in these cases, the rates of spontaneous decay in solution as well as the kinetics of the oxidation of $I^-$ ions indicated the presence of two components, possibly the $ho_2{}^5ho^6hUra$ and $ho^5ho_2{}^6hUra$. Only very small amounts of hydroperoxide were observed in irradiated Cyt solutions because of its apparent instability under neutral solutions. Thus, Cyt hydroperoxide could be detected in acidic solutions but decomposed rapidly if these solutions were neutralized; this behavior has been attributed (Daniels and Schweibert, 1967) to different decay rates of the neutral and acidic forms of the hydroperoxide in which the $NH_2$ group of Cyt ($pK = 4.6$) is involved. In neutral media, the thermal stabilities of the also radiation-produced hydroperoxides are in the order $ho_2hohThy > ho_2hohUra > ho_2Me_2hohUra >> ho_2hohCty$; the Thy hydroperoxide decays over several hours.

The yields of hydroperoxide determined immediately after irradiation of oxygen-saturated solutions (0.2 m*M*, $pH = 5.5$) of Thy, Ura, and $Me_2Ura$ with 200 kV x-rays were found to be $G = 1.05$, 1.16, and 0.75, respectively. In Thy solutions $G(ho_2)$ decreased below pH $\sim$ 5, and there was a corresponding increase in $G(H_2O_2)$. Scholes *et al.* (1960) suggested that the pH effect may be partly associated with the dissociation of $HO_2$ radicals according to the following scheme:

$$HO_2 \rightleftharpoons O_2^- + H^+ \qquad pK = 4.9 \tag{7}$$

$$Thy(OH)O_2\cdot + O_2^- \longrightarrow Thy(OH)O_2^- + O_2 \tag{33}$$

$$Thy(OH)O_2^- + H^+ \rightleftharpoons Thy(OH)O_2H \tag{34}$$

$$Thy(OH)O_2\cdot + HO_2 \longrightarrow \text{products} + O_2 + H_2O \tag{35}$$

$$Thy(OH)O_2\cdot \longrightarrow \text{products} + H_2O_2 \tag{36}$$

$$2HO_2 \longrightarrow H_2O_2 + O_2 \tag{18}$$

The yield of hydroperoxide accounts for about half of the extent of destruction of the Pyr molecules. It is possible that the thermal instability of these substances partly accounts for the low observed yields. That such instability complicates product analyses from irradiated aqueous Pyr solutions has been shown in studies with Cyt (Ekert and Monier, 1960), Thy (Téoule and Cadet, 1971a), and Ura (Peuzin *et al.*, 1970).

Pyr glycols are prominent among the decomposition products of hydroperoxide; for example, Nofre and Cier (1966), used $^{14}C$ labeling to facilitate analysis of the radiation products and showed that decomposition of Thy hydroperoxide leads to a mixture of *cis*- and *trans*-Thy glycols and a very small yield of $ho^5Me^5$-barbituric acid. Hahn and Wang (1972) have specifically isolated and characterized the *trans*-Thy glycol from irradiated oxygen-saturated Thy solutions. Until recently no satisfactory product balance has been achieved in any of these Pyr systems, with only approximately 50% of the OH radicals attack being accounted for. Now, through the extensive studies of Téoule and Cadet (1969a,b, 1970a,b, 1971a,b), an almost complete stoichiometric picture of the radiolysis of aerated Thy solutions is available. Table 9 shows the main products from aerated [$^{14}$C] Thy solutions (2 m*M*) irradiated under slightly acidic conditions (HCl). Air was bubbled through the solutions during irradiation, and a total dose of $5.4 \times 10^5$ rads was delivered. Assuming a $G_{OH} \sim 2.5$, $\sim 80\%$ of the Pyr would be used up in these experiments. The radiolysis products were separated by chromatographic methods and identified both spec-

**Table 9** Products from the γ-Radiolysis of Aerated Thymine Solutions[a]

| Products | Yield (G) |
|---|---|
| *Hydroperoxides* | |
| *trans*-$ho_2^5ho^6hThy$ } **(I)** | 0.830 |
| *cis*-$ho_2^5ho^6hThy$ } **(I)** | 0.314 |
| *cis*-$ho^5ho_2^6hThy$ **(II)** | 0.081 |
| $ho_2^5hThy$ **(III)** | 0.062 |
| $ho_2CH_2Ura$ **(IV)** | 0.047 |
| *cis*-$ho_2^6hThy$ } **(V)** | 0.027 |
| *trans*-$ho_2^6hThy$ } **(V)** | 0.028 |
| $ho_2^5Me^5$-barbituric acid | 0.011 |
| $ho_2^5Me^5$-hydantoin | 0.005 |
| *trans*-5,6-$(ho_2)_2hThy$ | 0.009 |
| *Products with ring intact* | |
| $ho^5hThy$ **(VI)** | 0.016 |
| *cis*- and *trans*-$ho^6hThy$ **(VII)** | 0.008 |
| *cis*-Thy glycol **(VIII)** | 0.125 |
| *trans*-Thy glycol **(IX)** | 0.123 |
| $ho^5Me^5$-barbituric acid **(X)** } $ho^5Me^5$-hydantoin **(XI)** } | 0.144 |
| *Ring-opened products* | |
| Urea **(XII)** | 0.075 |
| Acetylurea **(XIII)** | 0.010 |
| Formylurea | 0.065 |
| Formylpyruvylurea **(XIV)** | 0.460 |
| *Base destruction* | |
| Thy | −2.6 |

[a] Téoule and Cadet, 1971a.

troscopically and chemically; confirmation of the assignment of the structures was affected by independent syntheses. More than 90% of the starting material is accounted for. Altogether ten hydroperoxides

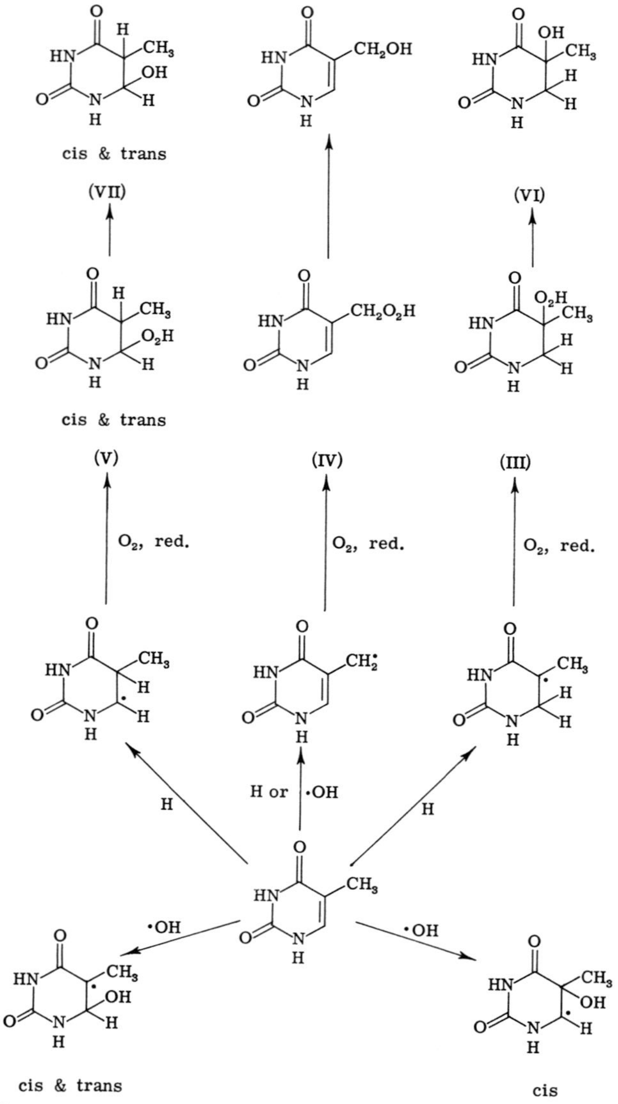

**Fig. 3.** *Schematic representation of the radiolytic decomposition of Thy in aqueous aerated solutions (After Téoule and Cadet, 1971a).*

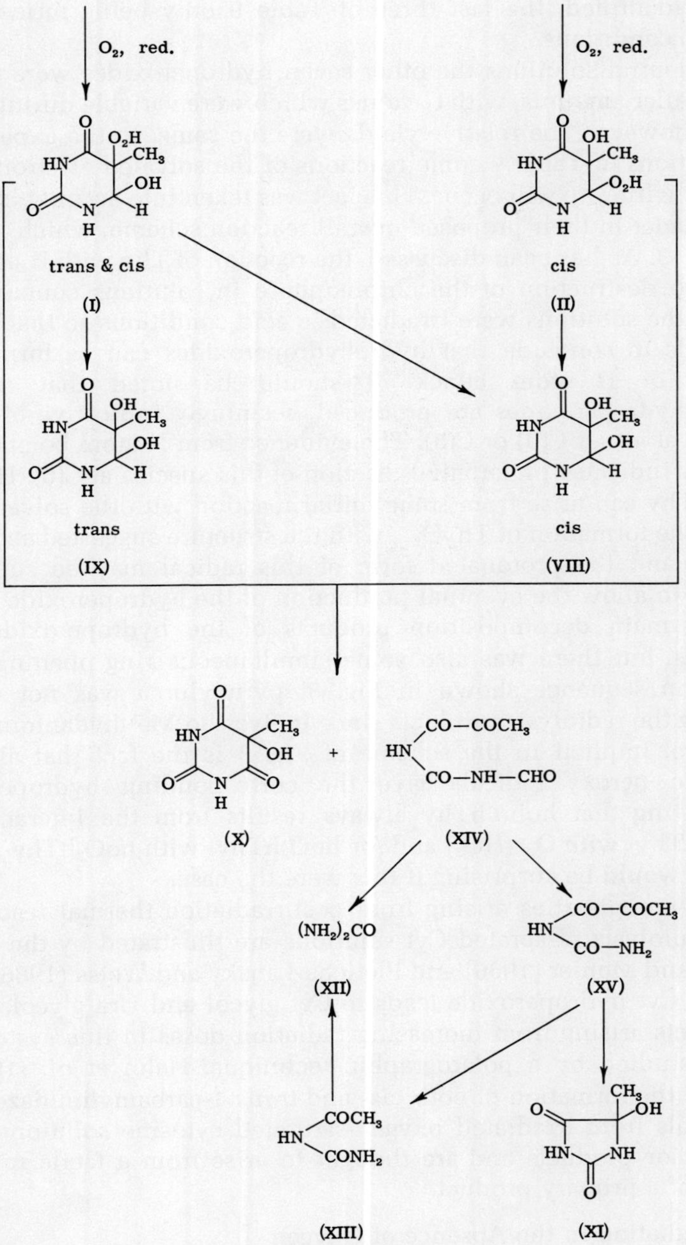

$O_2$, red.

$O_2$, red.

trans & cis

(I)

cis

(II)

trans

(IX)

cis

(VIII)

(X)

+

$HN(CO{-}COCH_3)(CO{-}NH{-}CHO)$

(XIV)

$(NH_2)_2CO$

(XII)

$HN(CO{-}COCH_3)(CO{-}NH_2)$

(XV)

$HN(COCH_3)(CONH_2)$

(XIII)

(XI)

**Fig. 3.** (*Continued*)

were identified, the last three of Table 9 only being formed under acidic conditions.

In neutral solutions the other seven hydroperoxides were obtained in smaller amounts, with *G*-values which were variable during irradiation; however, the relative yields were the same. In the experimental conditions of Table 9 some reactions of the solvated electrons and H atoms with Thy will occur. This fact was taken into account by Téoule and Cadet in their proposed overall reaction scheme, which is shown in Fig. 3. As has been discussed, the reaction of Thy with H atoms can lead to destruction of the chromophore in solutions containing oxygen; the solutions were irradiated in acid conditions so that it is reasonable to conclude that hydrohydroperoxides can be formed as a result of H atom attack. It should be noted that two 5,6-hydrohydroperoxides are produced, seemingly indicative of H-atom attack at either C(6) or C(5). The evidence from H-atom bombardment of Thy indicates preferential reaction of this species at C(6). However, $ho_2{}^6hThy$ can arise from some initial reaction with the solvated electron; the formation of $ThyO_2^-$, with the structure suggested above, will ensue and the protonated form of this radical may be sufficiently stable to allow the eventual production of the hydroperoxide.

The main decomposition products of the hydroperoxides were glycols, but there was also some simultaneous ring opening. In the reaction sequence shown in Fig. 3, pyruvylurea was not detected among the radiolysis products since it gives $ho^5Me^5$-hydantoin by ring closure. Implicit in the scheme of Fig. 3 is the fact that all of the organic peroxy radicals give the corresponding hydroperoxides, suggesting that $hoho_2hThy$ always results from the interactions of $ho\dot{O}_2hThy\cdot$ with $O_2^-(HO_2)$ and /or $ho\dot{O}_2hThy\cdot$ with $ho\dot{O}_2hThy\cdot$. In general it would be surprising if this were the case.

The complexities arising from postirradiation thermal reactions of the radiolysis of aerated Cyt solutions are illustrated by the work of Ekert and Monier (1960) and Pleticha-Lanský and Weiss (1966). Decay of the Cyt hydroperoxide leads to Cyt glycol and Ura glycol. Further products arising from increasing radiation doses in this system have been studied by a polarographic technique. Hahn *et al.* (1973) reported the formation of both *cis*- and *trans*-1-carbamylimidazolidone-4-5-diols from irradiated oxygen-saturated cytosine solutions. These are major products and are thought to arise from a facile rearrangement of a primary product.

## 2. Irradiation in the Absence of Oxygen

In deaerated systems, both the OH radicals and the reducing species ($e_{aq}^-$ and H) react with the Pyr molecules. The yields of hydrogen gas from dilute neutral solutions correspond to the molecular yield of hydrogen ($G_{H_2}$), indicating that the reducing species do not dehydrogenate these molecules.

Addition of OH to the 5,6-bond leads to a variety of products. Scholes *et al.* (1961) found that irradiation of $Me_2Ura$ with 200 kV x-rays leads to the production of 5,6-$(ho)_2Me_2Ura$ ($G \sim 0.15$) and a hydroxyhydro compound ($G \sim 0.3$). The latter was identified as $ho^6Me_2hUra$ and is identical to the photoproduct of $Me_2Ura$ (Wang *et al.*, 1965). Also $ho^6hPyr$ derivatives are generally rather unstable and by treatment with acid can eliminate water to re-form the parent Pyr; on the other hand, $ho^5hPyr$ derivatives are stable (Wang, 1962).

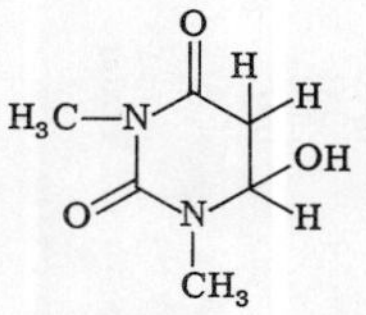

Latarjet *et al.* (1961) reported the formation of the *cis*- and *trans*-glycols from Thy as well as the unstable glycol from Cyt; the latter deaminates almost totally to form Ura glycol. A summary of the products from deaerated systems from more recent studies is given in Table 10. The overall destruction of the Pyr bases can be much less than the sum $G_{e_{aq}^-}$, $G_H$, $G_{OH}$ (= 5.8) indicating either (1) extensive "back reaction" between the organic radicals and radical anions leading to restitution of the parent base, e.g.,

$$\text{hohPyr}\cdot + \text{hPyr}\cdot \longrightarrow \text{hohPyr} + \text{Pyr} \tag{37}$$

$$\text{hohPyr}\cdot + \text{Pyr}^{\overline{\cdot}} \xrightarrow{H_2O} \text{hohPyr} + \text{Pyr} + OH^- \tag{38}$$

or (2) thermal rearrangements of unstable adducts. The nature and extent of these various reactions has yet to be resolved.

Product analyses from irradiated Thy solutions indicate attack by OH at C(5) as well as C(6), although it would again seem from the data that the latter is somewhat favored. The production of dihydro compounds in these systems could be taken as indicative of H atom attack at the 5,6-double bond, although their formation via the electron adduct may need to be considered in certain conditions.

An interesting approach to the radiolysis of deaerated Pyr solutions involves the use of the transition metal $Cu^{2+}$ in place of molecular oxygen as a scavenger of the intermediate radicals (Holian and Garrison, 1966). This leads to considerable simplification of the radiation chem-

**Table 10** G-Values for the Products of the γ-Radiolysis of Deoxygenated Pyrimidine Solutions

| Pyrimidine | Base destruction | Dihydro compounds | 5,6-Hydroxyhydro compounds | Glycol cis | Glycol trans | Additional products |
|---|---|---|---|---|---|---|
| Ura[a] | 2.7 | 0.18 | 0.43[b] | 0.63[b] | 0.32[b] | Isobarbituric acid; 0.45 |
| Thy[a] | 2.5 | 0.15 | 0.38 | 0.25 | 0.42 | $hm^5$Ura; 0.23 |
| Thy[c] | 2.0 | 0.04 | $ho^5$; 0.04<br>$ho^6$; 0.09–0.18 | 0.34 | 0.06 | Urea; 0.06 |
| Thy[d] | 0.85 | 0.30 | $ho^5$; 0.03<br>*cis*-$ho^6$; 0.17<br>*trans*-$ho^6$; 0.04 | 0.04 | 0.01 | $hm^5$Ura; 0.06 |
| Cyt[e] | 1.8 | — | 0.1 | Trace | Trace | $ho^5$Cyt; 0.46<br>Ura; 0.3<br>Ammonia; 0.4 |
| 5-MeCyt[f] | 2.1 | — | — | 0.41 | 0.83 | $hm^5$Cyt; 0.4<br>Thy; 0.1<br>Ammonia; 0.2 |

[a] Khattak and Green, 1966c; initial yields.
[b] Unstable.
[c] Nofre and Cier, 1966; 45 krads.
[d] Cadet and Téoule, 1971; 9–135 krads.
[e] Khattak and Green, 1966a; initial yield.
[f] Khattak and Green, 1966b; initial yield.

istry of these substances. In the presence of $Cu^{2+}$, the OH adducts give the corresponding glycols [e.g., reaction (39)] which are the principal

$$\text{(5-hydroxy-5,6-dihydrouracil-6-yl radical)} + Cu^{2+} \xrightarrow{H_2O} \text{(5,6-dihydroxy-5,6-dihydrouracil)} + Cu^{+} + H^{+} \quad (39)$$

products ($G \sim 2.3$) in the radiolysis of solutions of Ura (pH 3–5) and Cyt (pH $\sim$ 3.5). Isobarbituric acid derivatives are formed by the side reaction with $G \sim 0.5$ in each case [see reaction (40)]. It is proposed

$$\text{(5-hydroxy-5,6-dihydrouracil-6-yl radical)} + Cu^{2+} \longrightarrow \text{(5-hydroxyuracil)} + Cu^{+} + H^{+} \quad (40)$$

that the Pyr nucleus is quantitatively oxidized according to the stoichiometry, $G(\text{glycol}) + G(\text{isobarbituric acid}) = G_{OH} + G_{H_2O_2} = 2.9$, with the assumption that reaction (41) provides an additional source of OH radicals.

$$H_2O_2 + Cu^{+} \longrightarrow Cu^{2+} + OH + OH^{-} \quad (41)$$

These experiments indicate some attack at C(5) in Ura and Cyt.

With regard to the influence of $Cu^{2+}$ in the above experiments it is pertinent to note some recent work on the $\gamma$-radiolysis of $N_2O$-saturated hPyr solutions containing $Cu^{2+}$ ions (Haysom *et al.*, 1972). In this particular system, oxidation of hPyr by OH radicals leads to hPyr-5· and hPyr-6· radicals which then undergo an electron transfer process with $Cu^{2+}$ to yield the corresponding carbocations. Reactions (39) and (40) in the irradiated Pyr/$Cu^{2+}$ solution could correspond to nucleophilic solvolytic substitution and elimination of an intermediate Pyr carbocation.

### 3. The Primary Pyrimidine Radicals; Investigations by Physical Methods

#### *a. Reaction with the Oxidizing Species (OH)*

Structures of some of the radicals produced by reaction with OH have been studied by Neta (1972) using the *in situ* radiolysis steady-state ESR method (Eiben and Fessenden, 1971). Solutions were deoxygenated by bubbling with $N_2O$, which also served to increase the number of OH radicals. The experiments were conducted in alkaline conditions in which the radicals are present in the negatively charged basic forms; this allows a higher steady-state concentration since radical–radical reaction rates are slower because of electrostatic repulsion. In the case of Ura, addition of OH to both C(5) and C(6) is proposed. The C(5) adduct is suggested to be in either of the following forms, depending upon pH:

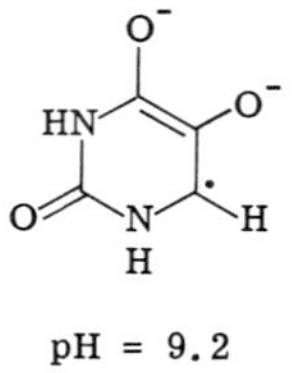

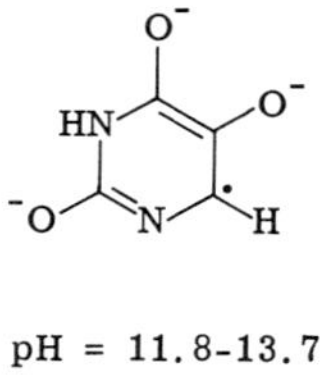

pH = 9.2    pH = 11.8-13.7

Neta argues that the unpaired electron at C(6) decreases the p*K* of the 4-oxo group, enhancing enolization between positions 4 and 5 so that subsequent ionic dissociation occurs. Of interest is the observation that the C(6) adduct in alkaline solution is unstable and undergoes ring fission [see reaction (42)]. A similar ring fission occurs after OH

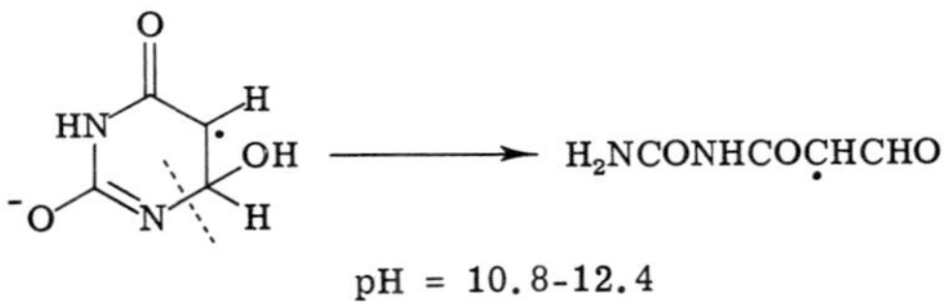

$$\longrightarrow \mathrm{H_2NCONHCO\dot{C}HCHO} \tag{42}$$

pH = 10.8-12.4

addition to C(6) of Thy, 6-MeUra, and *iso*-Oro. In Thy and 6-MeUra solutions at pH 13, the ESR spectra were compatible with the formation of the resonance-stabilized allyl-type radicals resulting from dehydrogenation of the methyl groups, in agreement with earlier proposals (Myers *et al.*, 1965a). It should perhaps be noted that the radicals are observed under steady-state conditions so that the relative concentrations of different radicals do not necessarily tally with their relative rates of formation unless similar decay rates are assumed. The failure to observe a particular radical does not rule out its formation.

Nucifora *et al.* (1972) examined the ESR spectra of the OH adducts of several pyrimidines in pulsed solutions containing $N_2O$. The C(6) OH adduct from Thy was positively identified, as was the C(5) OH adduct from 6-MeUra. In the case of Ura, the evidence favored the C(5) adduct but no assignment could be made for Cyt. It has been argued

(Nicolau *et al.*, 1969) that the OH radical is electrophilic, and that, therefore, the N(1) atom in Thy should activate the molecule toward addition at C(5) while the methyl substituent should activate the molecule toward addition at C(6); in Ura, the localization energy (Pullman and Mantione, 1965) is smaller for addition of a radical at C(5) than at C(6) and, furthermore, given the electrophilic character of OH, the calculated $\pi$ or $\sigma$ charges at C(5) or C(6) indicate a preference for reaction at the former position (Berthod *et al.*, 1967). [In this context, Nicolau *et al.* (1969) examined the ESR of radicals produced from pyrimidines on treatment with the $Ti^{3+} + H_2O_2$ reagent in the belief that this was a simple source of OH radicals. Two radicals, in comparable concentrations, were produced from Thy and were assigned to the C(5) and C(6) OH adducts. Ura formed one radical by reaction at C(5) and Cyt also formed one adduct, most probably at C(5). However, there has been some discussion as to the exact nature of the reactive oxidizing species in this system (Baines *et al.*, 1968; Czapski, 1971; Czapski *et al.*, 1971), and it is probably not radiomimetic. Nevertheless, the ESR spectra give important information about the structure of the OH adducts.]

The pulse radiolysis technique has been used to study the optical absorption spectra of the Pyr OH adducts. Although, in general, the spectra do not readily allow an assignment of the structure of the adducts, they are important for investigations of the kinetics of decay and, thereby, for elucidation of further details of the subsequent chemical transformations. Preliminary investigations in this field were carried out with oxygenated systems (Scholes *et al.*, 1965) for which the transient spectra recorded immediately after the pulse ($\sim 2$ $\mu$sec duration) were assigned to the OH adducts of the bases; the transients decayed rapidly, giving long-lived peroxy radicals. Subsequent experiments have been carried out mainly with solutions saturated with $N_2O$ in which the OH adducts can be readily studied. Because of the strong optical absorption of the Pyr solutions themselves, most measurements have been limited to wavelengths of longer than $\sim 300$ nm. (Below 300 nm the low level of light transmitted necessitates a high photomultiplier gain and this increases the signal noise and, hence, the inaccuracy of the measurements; this situation has been improved by the use of pulsed lamp sources which greatly increase the intensity of the analyzing light.)

Figure 4 shows the spectrum of the hoUra-adduct, and Table 11 lists the absorption maxima for the corresponding adducts of other pyrimidines. While Ura and Thy exhibit a single broad band, Cyt and MeCyt give two peaks. The decays are adequately described by second-order

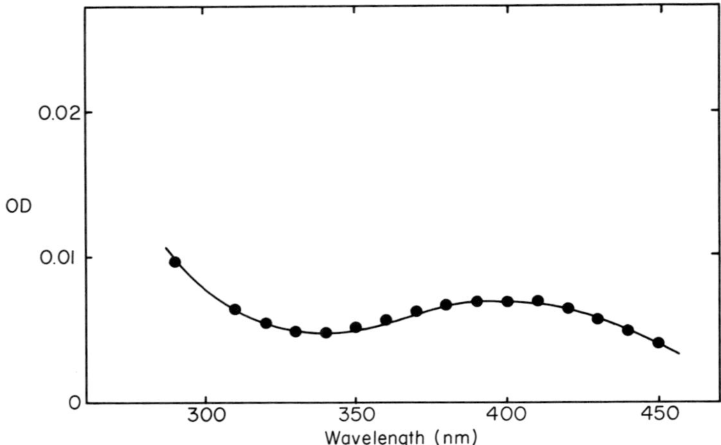


**Fig. 4.** *Absorption spectrum of the radical formed by reaction of OH with Ura in pulsed neutral $N_2O$-saturated solution.* [*Ura*] = *0.6* mM.

kinetics, the rate constants being approximately $10^9\ M^{-1}\ sec^{-1}$. By a comparison of the optical spectra and the rate constants at different wavelengths, it has been concluded that whereas Thy forms mainly one adduct at C(6), Ura forms two adducts (Stevens, 1969). There is an effect of pH on the peak transient absorption in pulsed aqueous Thy solutions, suggesting that the OH adduct can undergo ionization; an approximate pK of 7.5 was obtained (Stevens, 1969), showing that the

**Table 11** Optical Spectra of Pyrimidine OH Adducts

| Pyrimidine | $\lambda_{max}$ | Ref.[a] |
|---|---|---|
| Ura | 390–400 | 1,2 |
| | 390 | 3 |
| | 375 | 4 |
| Thy | 380 | 2,5 |
| | 375 | 6 |
| Cyt | 345, 440 | 1 |
| | 335, 440 | 7 |
| 5-MeCyt | 325, 485 | 7 |

[a] Key to references:
1. Phillips, 1967.
2. Stevens, 1969.
3. Danziger *et al.*, 1968.
4. Greenstock *et al.*, 1969.
5. Scholes, 1968.
6. Myers *et al.*, 1970b.
7. Myers *et al.*, 1970a.

introduction of a radical center in the Pyr molecule can lower the pK by about two units. The reaction of hydroxide ions with the OH adducts of various pyrimidines has been observed by Myers *et al* (1970ab). Working in alkaline solutions ($pH > 11$) they observed short-lived transients which decayed by a pseudo first-order mechanism, depending upon $[OH^-]$. The radical anions so formed are believed to exist in various tautomeric forms, depending upon the number of charges; thus, for the C(6) adduct of Ura, the forms shown in reactions (43) and (44) are proposed. However, in view of the ring

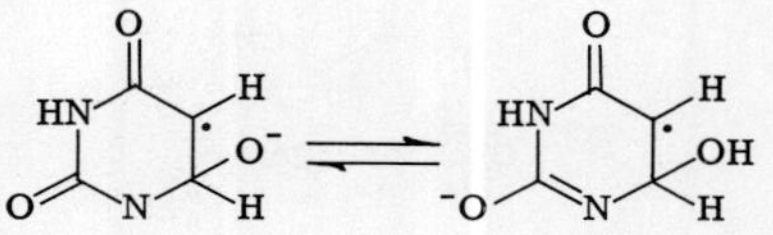

(43)

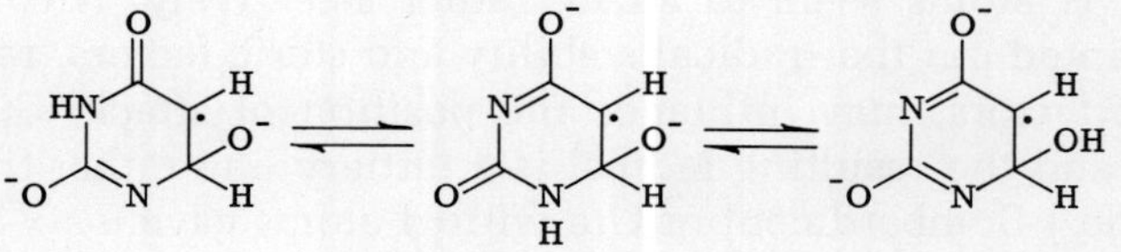

(44)

opening of C(6) adducts described above, this simple picture may not be correct. With Thy and 5-MeCyt, additional transients attributable to the oxidation of the methyl group by $O^-$ were also observed in the solutions at high pH.

An interesting phenomenon, associated with radiation-produced $OH^-$ ions, has been observed in pulsed neutral $N_2O$/Pyr solutions (Fielden *et al.*, 1970). Because of reaction (8), $OH^-$ ions are produced in significant amounts, e.g., 5 $\mu M$ after a dose of 1.5 krad, and these can react with pyrimidines, in competition with $H_3O^+$:

$$OH^- + Pyr \longrightarrow Pyr^- + H_2O \tag{45}$$

$$OH^- + H_3O^+ \longrightarrow 2H_2O \tag{46}$$

Equilibrium is established through the reverse of (45), and this is seen experimentally as a fast first-order decay, since $Pyr^-$ absorbs at a different wavelength from Pyr. As expected, pyrimidines which cannot undergo keto–enol tautomerism (e.g., $Me_2$Ura, 2,4$(MeO)_2$Ura, and also Pyr nucleosides and nucleotides) do not exhibit this particular phenomenon in pulsed solutions.

*b. Reaction with the Reducing Species (H Atoms and Solvated electrons)*

Clues as to the probable site of attack by H atoms in irradiated aqueous systems have arisen from studies of the bombardment of solid pyrimidines with H atoms and examination of the ESR spectra of the free radicals so produced (Heller and Cole, 1965; Herak and Gordy, 1965; Holmes *et al.*, 1966). In the case of Thy, H atoms attack at C(6), and the ESR spectrum yields a hyperfine structure of 8 lines

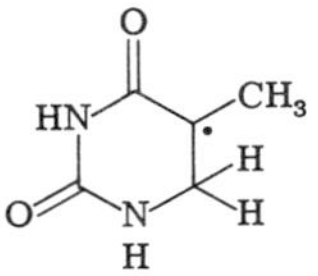

(47)

which correspond to hThy-5. [Unlike OH radicals, these externally generated H atoms seem to attack rather selectively. Nicolau *et al.* (1969) pointed out that radical stability and steric factors, rather than electronic factors, may influence the position of attack; C(6) is less hindered and the resulting radical is a tertiary one rather than a secondary one.] Bombardment of Ura with H atoms gave a six-line spectrum, indicating H-atom addition at C(5) or C(6). Cyt also forms an adduct at the 5,6-bond, which in the solid state is observed only at low temperatures (Herak and Gordy, 1966). Pulse radiolysis studies confirm H-atom addition at the double bond. With Ura, for example, (Greenstock *et al.*, 1969) the H-atom adduct can be observed in pulsed acidic solutions containing excess formaldehyde to scavenge the OH radicals. The absorption maximum (420 nm) is the same as that of the transient formed in pulsed neutral $N_2O$/hUra solutions (Fig. 5), in which oxidation occurs as shown in reaction (47).

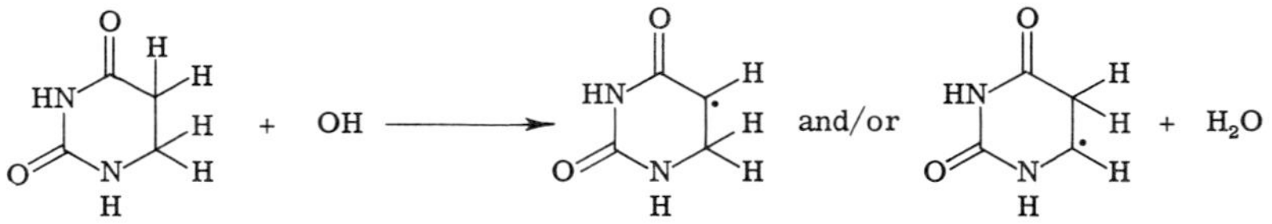

$$+ \; OH \longrightarrow \quad \text{and/or} \quad + \; H_2O$$

The Pyr electron adducts can be readily observed in pulsed deaerated solutions containing an excess of alcohol as OH and H scavenger (Scholes, 1968). *t*-Butanol is the best for this purpose since the alcohol radicals formed absorb at shorter wavelengths. It can be seen from Fig. 5 that the absorption spectrum of the Ura–electron adduct is quite different from that of the H-atom adduct. Reaction of the

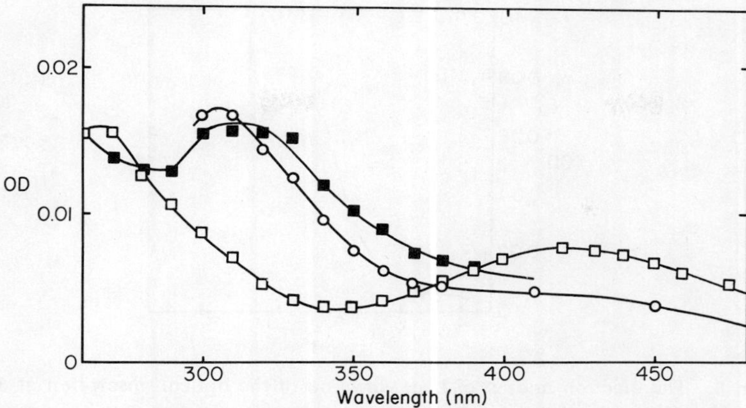


**Fig. 5.** *Absorption spectra of transients in pulsed neutral solutions of Ura and hUra.* ○, *Electron adduct of Ura([Ura] = 2mM; [t-butanol] = 0.5M; $N_2$-saturated);* ■, *electron adduct of hUra([hUra] = 2 mM [t-butanol] = 0.5 M; $N_2$-saturated);* □, *hydrogen atom adduct of Ura from oxidation of hUra by OH ([hUra] = 1mM; $N_2O$-saturated).*

solvated electron at the double bond or the carbonyl functions at C(2) and C(4) can be envisaged. However, the similarity of the electron adduct spectra of Ura and hUra indicates the probable involvement of the carbonyl groups. Experiments in this laboratory showed that there was an effect of pH on the transient absorption at 320 nm, indicative of an ionic dissociation of the electron adduct; the results for Ura (Fig. 6) follow a dissociation curve with a pK of ~7.5. Such an effect could be associated with a lowering of the first pK of Ura as a result of radical formation or with the dissociation of the ketyl radical [see reaction (48)].

$$> C{=}O + e_{aq} \longrightarrow > \dot{C}{-}O^- \rightleftharpoons > \dot{C}{-}OH \tag{48}$$

Hayon (1969) showed the formation of two transients from pyrimidines and designated these as ketyl radicals produced by addition of the electron at the two carbonyl groups. The effect of pH was ascribed to ketyl radical dissociation.

Theard *et al.* (1971), in pulse radiolysis studies of Thy solutions, noted a difference between the absorption spectrum of the H-atom adduct and that of the protonated electron adduct. It was, therefore, concluded that the site of protonation of $Thy^{\overline{\cdot}}$ differs from the site of addition of H to Thy and that this was consistent with protonation of an oxygen. Nucifora *et al.* (1972) used the ESR-coupled pulse radiolysis technique to examine the structure of the electron adducts. From Ura and Thy, doublets were observed which were related to $\alpha$-

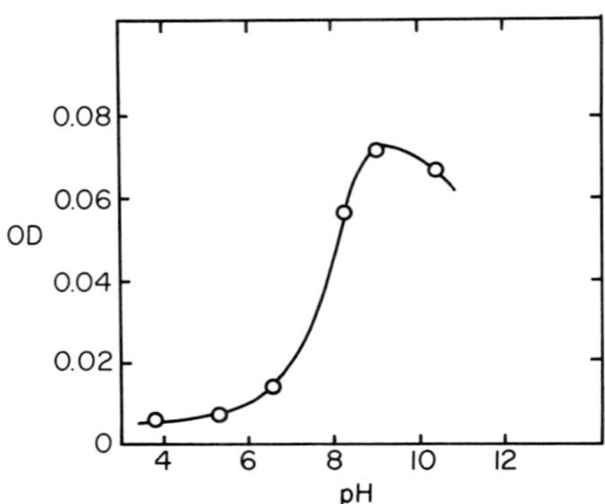


**Fig. 6.** *The electron adduct of Ura; variation of the optical absorption at 320 nm with pH.*

coupling of the proton on C(6); from 6-MeUra a quartet was observed, indicating that the unpaired electron was located at C(6). It is suggested that the hydrated electrons attack the carbonyl groups, but that in the case of the C(4) adduct a conjugation of the unpaired electron occurs (reaction (49), in which R=H or $CH_3$, depending upon the Pyr.)

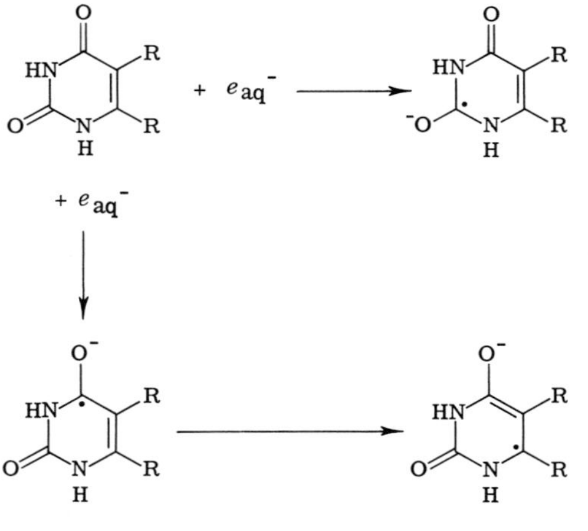

$$+ e_{aq}^- \longrightarrow \qquad (49)$$

in which R = H or $CH_3$, depending upon the Pyr

## E. Radiolysis of Aqueous Purine Solutions

Current knowledge of the radiation chemistry of aqueous Pur solutions is much less complete than that of Pyr solutions, to some extent because of the greater chemical complexity of the purines. The reaction stoichiometries, particularly in neutral, oxygen-containing solutions of the amino purines have yet to be established. Molecular orbital calculations (Pullman and Pullman, 1960) indicate that the N(7)–C(8) bond of the Pur nucleus has the highest mobile bond order, and there is some evidence that radical reactions can occur here.

### 1. Irradiation in the Presence of Oxygen

Early experiments (Scholes and Weiss, 1953) indicated that the radiation-induced oxidation of the purines in solution leads to gross molecular changes, such as deamination, ring fission, and the eventual production of low molecular weight products (urea, oxalic acid, etc.). In addition, there were marked pH effects. Thus, the extent of deamination of Ade in acid solution was almost twice that in neutral conditions [$G(NH_3) = 0.5$ at pH = 7; 200 kV x-rays] indicating a greater extent of destruction at low pH.

Values for $G(-\text{Pur})$ under various conditions are given in Table 12. For irradiated neutral solutions of Ade, Xan, and uric acid, there was reasonably good agreement for $G(-\text{Pur})$ using (1) spectrophotometry at the appropriate wavelength, (2) precipitation of unchanged Pur as silver salt, or (3) chromatographic analysis (Scholes *et al.*, 1960); under these conditions, therefore, $G(-\text{chrom})$ is a fair measure of base destruction. The data of Uliana and Creac'h are somewhat lower and this may be due to the rather high doses of $\gamma$-rays used.

Several points arise from the data of Table 12.

1. In neutral solutions of the amino purines, $G(-\text{base})$ is only approximately half of what one may expect from the known value of $G_{OH}$ (contrast Table 12 for purines with Table 8 for pyrimidines), in view of the fact that the rates of reaction of OH with purines and pyrimidines are not significantly different. Such a situation is rather unusual in the radiation chemistry of oxygenated aqueous solutions, since peroxy radical formation usually ensures chemical modification of the initially formed organic radical. This led to the suggestion (Scholes *et al.*, 1960) that there was some (unspecified) restitution reaction in these systems, leading to re-formation of the Pur molecule. Support for restitution reactions in these neutral oxygenated system arises from the recent work of van Hemmen and Bleichrodt (1971). The yields of destruction of Ade in solutions containing mixtures of $O_2$ and $N_2O$ are much less than expected, indicating that not every OH adduct leads to degradation. Furthermore, the apparent $G(-\text{base})$ with [2,8-$^3$H]Ade is ~50% higher than with [8-$^{14}$C]Ade, suggesting that

**Table 12** $G(-\text{purine})$ in $\gamma$-Irradiated Solutions Containing Oxygen

| Purine | [Purine] (m*M*) | Conditions | $G(-\text{purine})$ pH 1.2 | pH 2 | pH 3 | pH 5 | pH 7 | pH 10 | pH 12 | Ref.[a] |
|---|---|---|---|---|---|---|---|---|---|---|
| Ade | 1 | $O_2$ (1 atm) | 2.1 | — | 1.8 | — | 1.2 | — | — | 1 |
| | 0.1 | $O_2$ (1 atm) | — | — | — | | 1.2 | | | 2 |
| | 2 | $O_2$ (1 atm) | — | — | — | 1.2 | — | — | — | 3 |
| | 0.2 | $O_2$ (1 atm) | — | — | — | 1.0 | — | — | — | 3 |
| | 0.02 | $O_2$ (1 atm) | — | — | — | 0.5 | — | — | — | 3 |
| | 0.5 | Air | — | 0.9 | 0.7 | 0.5 | 0.5 | 0.7 | 0.9 | 4 |
| Gua | ~0.1 | $O_2$ (1 atm) | 2.5 | — | — | — | — | — | — | 5 |
| | 0.5 | Air | — | 1.2 | 0.8 | 0.4 | 0.5 | 1.1 | 1.4 | 4 |
| Hyp | 2 | $O_2$ (1 atm) | ← | | 2.4 | | → | — | — | 1 |
| Xan | 0.5 | $O_2$ (1 atm) | — | — | ← | 2.0 | → | — | — | 6 |
| Uric acid | 0.4 | $O_2$ (1 atm) | — | — | ← | 2.2 | → | — | — | 6 |

[a] Key to references:
1. Holian and Garrison, 1967a,b.
2. van Hemmen and Bleichrodt, 1971.
3. Thompson, 1963; [$G(-\text{chrom})$].
4. Uliana and Creac'h, 1969; [$G(-\text{chrom})$].
5. J. Holian (private communication).
6. Garrison, 1968.

the restitution process may partly involve an exchange of hydrogen atoms of the base with water.

2. In acidic and alkaline solutions of the amino purines, $G(-\text{base})$ is higher and rather more "normal." In fact, Uliana and Creac'h (1969) drew attention to a possible relationship between the pH dependence and the ionization of the $-NH_2$, $-OH$, and $=NH$ groups in these molecules.

3. The oxypurines are degraded to a "normal" extent, and there seems to be no pH dependence.

It had previously been thought, mainly from considerations of the known chemical behavior of uric acid and certain other purines with oxidizing agents, that an important site of radical attack is the central C(4)–C(5) double bond. In oxygenated systems this could lead to organic hydroxyperoxy radicals, somewhat similar to the situation in Pyr solutions. If any intermediate hydroxyhydroperoxides are formed, they must be extremely unstable since no hydroperoxides have been detected in irradiated Pur solutions in any pH conditions (Scholes *et al.*, 1956). Evidence for some attack of OH at the N(7)–C(8) bond of the imidazole ring in oxygen-containing solutions arises from observation of the production of small amounts of $ho^8$Ade ($G = 0.16$) and of 4,5,6-$(Am)_3$Pyr ($G < 0.1$) from Ade, and traces of uric acid from Xan (McCargo, 1961; Conlay, 1963; Ponnamperuma *et al.*, 1963). However, such low yields together with the lack of significant amounts of ultraviolet-absorbing products in the irradiated Pur solutions suggest OH attack at sites other than the imidazole ring. The notion that the central double bond is the main focus of attack in oxypurines and aminopurines (the latter in acidic conditions) arises from the work of Holian and Garrison (1967a,b). It has been pointed out that if oxidation at the C(4)–C(5) position produces hydroxyhydroperoxides or glycols, then subsequent mild acid hydrolysis should liberate carbonyl compounds. Considerations of conventional Pur chemistry (Howard, 1960) suggest that uric acid and Xan should yield alloxan [**XVI,** Eq. (50)] whereas the corresponding products from Hyp and Ade should be expected to undergo further hydrolysis with the liberation of free mesoxalic acid [**XVII,** Eq. (51)]. The irradiated solutions were subjected to hydrolysis (2*N* HCl, 90°C, 2 hr) and analyzed; the results are given in Table 13. In the case of Xan the OH radicals are essentially quantitatively removed at the central double bond, since $G(-\text{base}) = G(\text{alloxan})$. Given also that $G(NH_3) = 2 \times G(-\text{base})$, the actual radiolytic and hydrolytic process has been represented by Eq. (50) (Holian and Garrison, 1967b). The overall formation of mesoxalic acid from irradiated Hyp solutions is represented by Eq. (51).

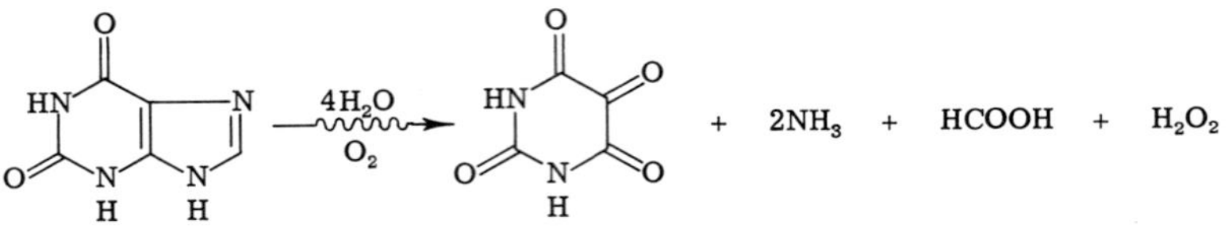

$$\text{Uric acid} \xrightarrow[O_2]{4H_2O} \text{(XVI)} + 2NH_3 + HCOOH + H_2O_2 \quad (50)$$

(XVI)

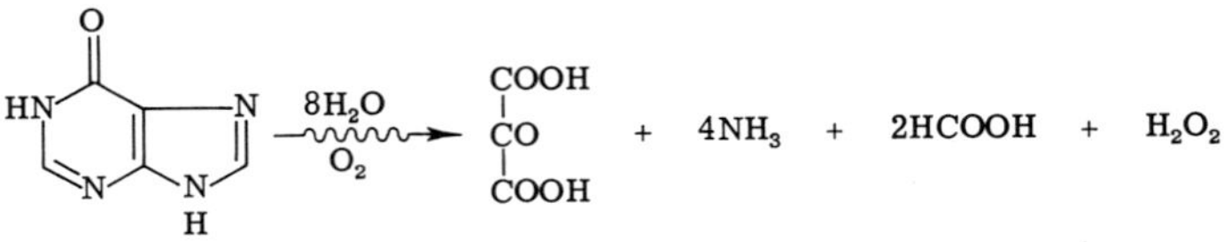

$$\text{Hyp} \xrightarrow[O_2]{8H_2O} \begin{matrix} COOH \\ | \\ CO \\ | \\ COOH \end{matrix} + 4NH_3 + 2HCOOH + H_2O_2 \quad (51)$$

XVII

The carbonyl yields from Hyp, uric acid, and Ade solutions are lower, suggesting that there are various branching reactions of the hydroxyhydroperoxides (or the hydroxyperoxy radicals) with the formation of different degradation products. Studies of $\gamma$-irradiated acidic solutions of Hyp and Ade revealed the formation of significant amounts of oxalic acid on hydrolysis (see Table 13), and the degradative pathways shown in Eqs. (52) and (53) have, therefore, been proposed.

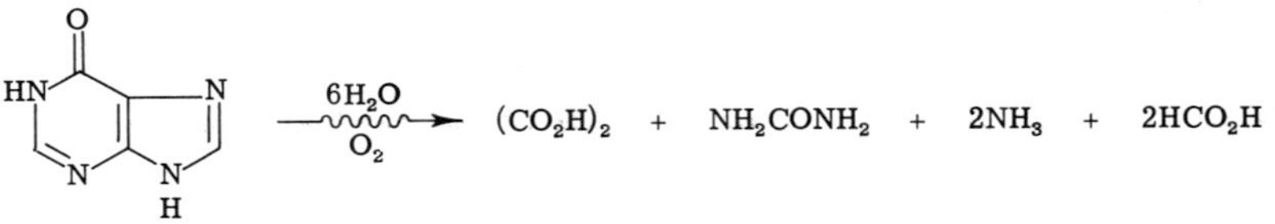

$$\text{Hyp} \xrightarrow[O_2]{6H_2O} (CO_2H)_2 + NH_2CONH_2 + 2NH_3 + 2HCO_2H \quad (52)$$

$$\text{Ade} \xrightarrow[O_2]{7H_2O} (CO_2H)_2 + NH_2CONH_2 + 3NH_3 + 2HCO_2H \quad (53)$$

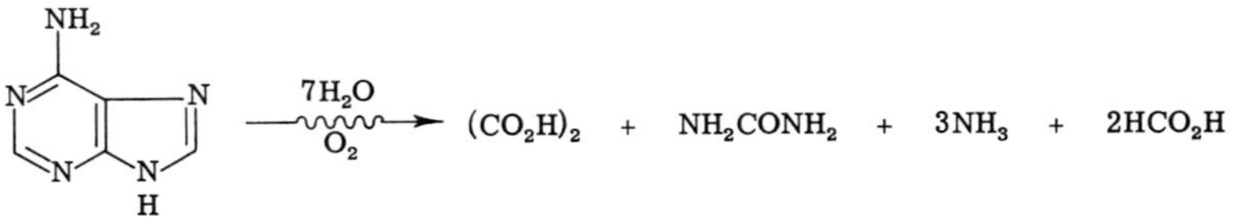

For both the acidic Hyp and Ade solutions, the data in Table 13 give the value $G(-\text{Pur}) - [G(\text{oxalic acid} + \text{mesoxalic acid})] \sim 0.5$, which may represent a yield of OH reaction at sites other than the central double bond. Uric acid glycol (**XVIII,** Fig. 7) has, in fact, been isolated from irradiated [2-$^{14}$C]uric acid solutions (Le Roux *et al.*, 1971). It has been proposed that there is an alternative simultaneous pathway of degradation subsequent to OH radical attack at the Pur central double bond, leading to the formation of triuret (**XIX,** Fig. 7), parabanic acid

**Table 13** Product Yields in the γ-Radiolysis of Oxygenated Aqueous Purine Solutions[a]

| Purine | [Purine] (mM) | G-Values | | | | | | |
|---|---|---|---|---|---|---|---|---|
| | | pH | (−Purine) | $NH_3$ | Alloxan | Urea | Oxalic acid | Mesoxalic acid |
| Ade | 2 | 3.7 | 1.8–1.2 | — | — | — | — | <0.1 |
| | 1 | 1.2 | 2.1 | 9.6 | — | 0.5 | 1.2 | 0.45[b] |
| Hyp | 2 | 3–7 | 2.4 | 8.6 | — | — | — | 1.5 |
| | 1 | 1.2 | 2.4 | 8.8 | — | 0.4 | 0.75 | 1.1 |
| Xan | 0.5 | 3–7 | 2.0 | 4.1 | 2.0 | — | — | — |
| Uric acid | 0.4 | 3–7 | 2.2 | 0 | 1.2 | — | — | — |

[a] Holian and Garrison 1967a,b; Garrison, 1968.
[b] Combined yield of mesoxalic acid and a lesser amount of glyoxylic acid.

(**XX,** Fig. 7), allantoin (**XXI,** Fig. 7), and urea. The overall scheme of oxidation of uric acid is represented in Fig. 7, and the intermediates in this reaction are assumed to be the same as those involved in the oxidation of uric acid by uricase (Canellakis and Cohen, 1955). Xan was found to yield uric acid by oxidation at C(8) and also a product which can be hydrolyzed to alloxan. In many ways these findings confirm the proposals of Holian and Garrison.

However, the radiolysis of neutral amino purines is still a mystery, *viz.*, the low *G*(−base) and the absence of products associated with attack at C(4)–C(5). Questions as to whether the $NH_2$ group exerts a directing effect, is involved as a competing reaction site, or is responsible for the restitution reaction have yet to be answered.

## 2. Irradiation in the Absence of Oxygen

The extents of destruction of the purines in deoxygenated solutions are again low, which indicates extensive back reactions in these systems. For Ade, it has been found that the *G*(−base) is decreased one-half to two-thirds on going from oxygen-saturated to deaerated solution (Thompson, 1963; van Hemmen and Bleichrodt, 1971). Le Roux *et al.* (1971) noted that removal of oxygen also decreases the overall extents of destruction of uric acid and Xan, particularly the former; product analysis revealed lowered yields of uric acid glycol and allantoin from irradiated uric acid solutions and of allantoin and uric acid from irradiated Xan solutions.

There are two radiation products from purines which are specific to solutions irradiated in the absence of oxygen. These products arise from attack on the imidazole ring and require the presence of the C═N double bond in this part of the molecule. One of these is the

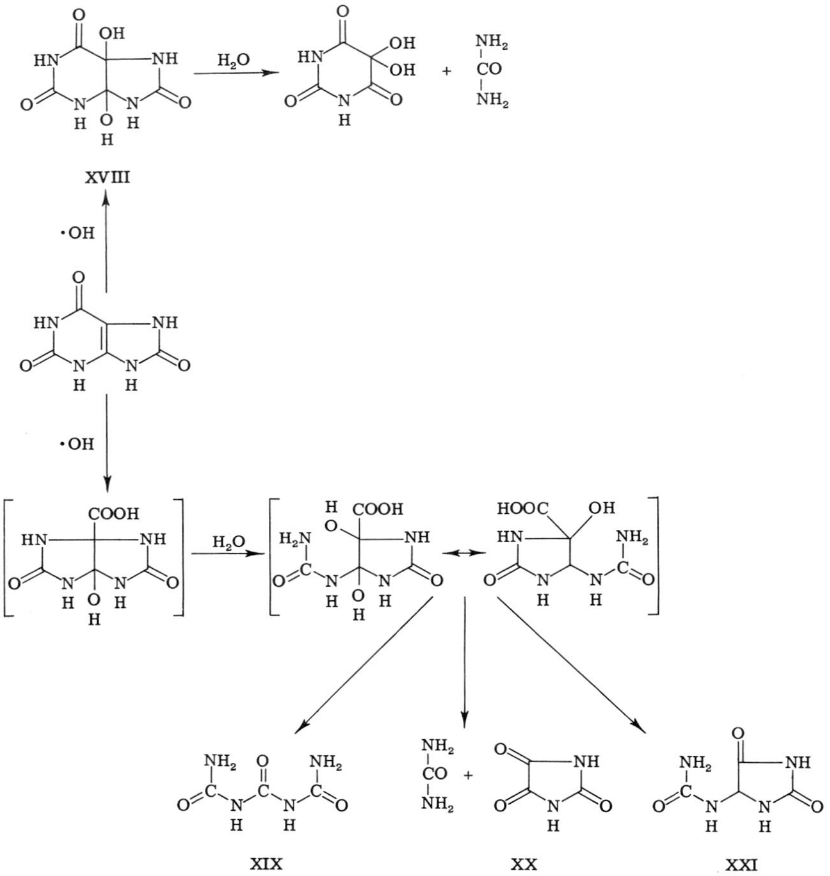

**Fig. 7.** *Schematic representation of the radiolytic oxidation of uric acid in irradiated oxygen-containing solutions* (Le Roux et al., 1971).

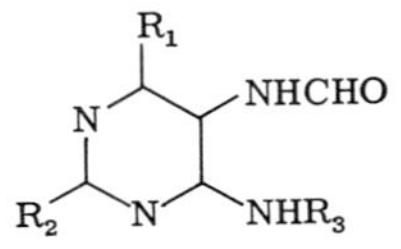

corresponding 4-amino-5-formamido-Pyr, first identified in irradiated Ade solutions (Hems, 1960) and more recently in irradiated Xan solutions (Le Roux *et al.*, 1971). These compounds, which are also produced from Pur nucleosides and nucleotides, have the general formula in which $R_1 = OH$ or $NH_2$, $R_2 = H$, OH, or $NH_2$, and $R_3 =$ ribose or ribose phosphate. The second product specific to deaerated systems is the Pur hydrate, in which the elements of water have been added across the N(7)–C(8) linkage; this has been identified in irradiated solutions (van Hemmen, 1971) and has the structure

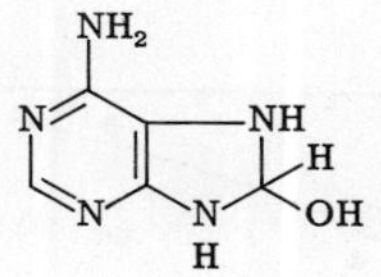

Using $^3H$- or $^{14}C$-labeled Ade as a tracer, van Hemmen and Bleichrodt determined the yields of 4,6-diamino-5-formamido-Pyr, adenine hydrate (ho$^8$hAde), and ho$^8$Ade. Various experimental conditions were used; the results are shown in Fig. 8. The following mechanistic details were deduced from these data.

1. Since both 4,6-diamino-5-formamido-Pyr (**XXII,** Fig. 9) and ho$^8$Ade (**XXIII,** Fig. 9) can be produced in $N_2O$ solutions as well as in deaerated solutions, no solvated electrons are involved in their formation. On this basis, the scheme given in Fig. 9 was suggested. Because of the observed concentration dependences, secondary involvement of the free base is invoked, but there may well be a normal solute concentration dependence in the particular range used.

2. The yield of ho$^8$hAde is lower in $N_2O$ than in $N_2$, suggesting the involvement of solvated electrons. Also, in conditions in which no electron adduct of Ade can be formed (*viz.*, at high concentrations of either $N_2O$, $H_2O_2$, or $O_2$ and at low concentrations of Ade) no Pur hydrate is formed. It is, therefore, proposed that this product is formed by the following reaction:

$$\mathrm{h\overset{8}{o}hAde\cdot + Ade^{\overline{\cdot}} \xrightarrow{H_2O} h\overset{8}{o}hAde + Ade} \tag{54}$$

3. The data suggest that the recombination of the Ade–OH adducts leads to restitution of the free base in the absence as well as in the presence of oxygen.

## 3. The Primary Purine Radicals

Physical evidence for the structures of the primary radicals produced from the radiolysis of Pur solutions is fairly fragmentary. Using the ESR-coupled pulse radiolysis technique, Nucifora *et al.* (1972) investigated the OH adduct of Ade in solutions containing $N_2O$. At pH 7

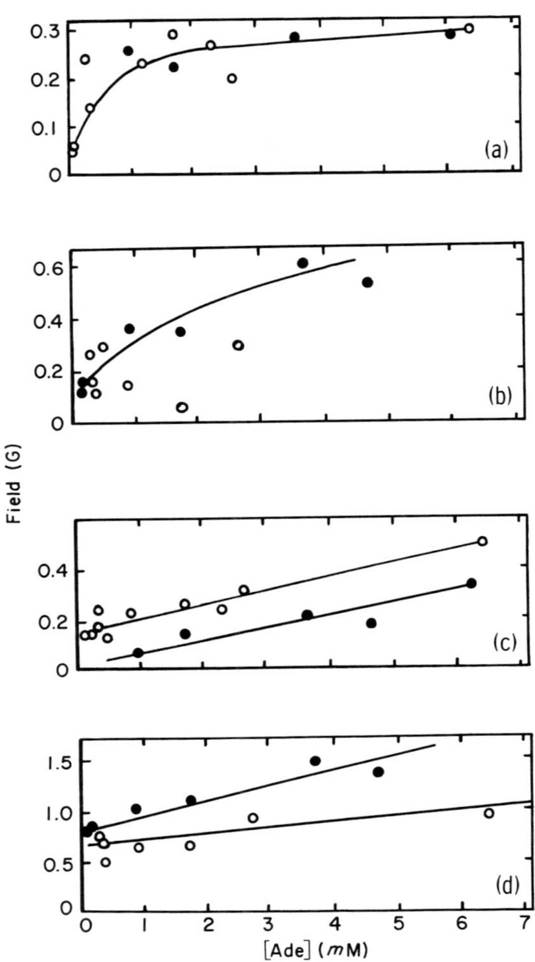


**Fig. 8.** *Yields of (a) 4,6-diamino-5-formamido-Pyr, (b) ho⁸Ade, (c) ho⁸hAde, and (d) destruction of Ade in aqueous solutions of Ade irradiated with γ-rays.* ●, $N_2O$; ○, $N_2$. *(van Hemmen and Bleichrodt, 1971.)*

a single broad line was observed which became narrower on decreasing the pH; no nitrogen or hydrogen hyperfine structure was present. These observations have been taken to be compatible with addition of OH at the central 4,5-bond. Unfortunately, no organic radicals have been detected on treatment of purines with the $Ti^{3+} + H_2O_2$ reagent (Nicolau, 1972), possibly because of rapid back reactions with $Ti^{4+}$ under the experimental conditions used.

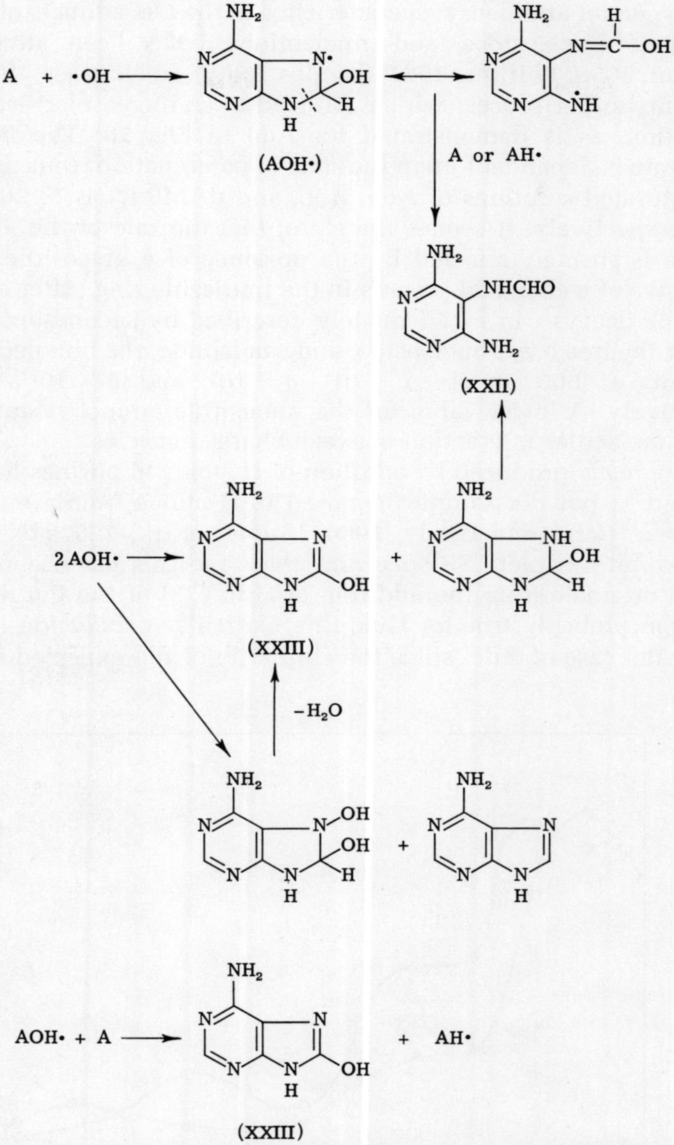

A + •OH → (AOH•) ↔ ; A or AH• → (XXII)

2 AOH• → (XXIII) + ; −$H_2O$

AOH• + A → (XXIII) + AH•

**Fig. 9.** *Scheme for the production of 4,6-diamino-5-formamido-Pyr* (**XXII**) *and ho*[8]*Ade* (**XXIII**) *in irradiated $N_2O$-saturated Ade solutions.* (*van Hemmen and Bleichrodt, 1971.*)

The spectral and decay characteristics of the OH adducts of purines and their nucleosides and nucleotides have been investigated (Willson, 1966; Phillips, 1967; Scholes, 1968). In all cases, there is an initial first-order process which can lead to an increase or decrease in absorption, as is demonstrated for Ade in Fig. 10. The first-order decay rate is dependent upon the state of combination; thus, in pulsed $N_2O$-saturated solutions of Ade, Ado, and dAMP $t_{1/2}$ is 5, 26, and 28 $\mu$sec, respectively. It seems, therefore, that the rate of the first-order process is strongly affected by the presence of a glycosidic linkage, suggestive of a chemical process in the imidazole ring. After sufficient time, the decays can be adequately described by second-order kinetics. For the free base, nucleoside, and nucleotide, the bimolecular rate constants at 600 nm are $7 \times 10^8$, $3 \times 10^8$, and $1 \times 10^8$ $M^{-1}$ $\sec^{-1}$, respectively. A lower value for the nucleotide adduct is anticipated for a bimolecular interaction between charged species.

The radicals produced by addition of H atoms to purines have been observed in powder samples exposed to H atoms from a microwave discharge (Herak and Gordy, 1966; Holmes *et al.*, 1967a,b). Because they exhibit a triplet ESR spectrum, these radicals have been considered to originate from the addition of H to C(8) of the Pur structure. Although probably true for Gua, this particular conclusion is uncertain in the case of Ade, since the similarity of the expected spin dis-

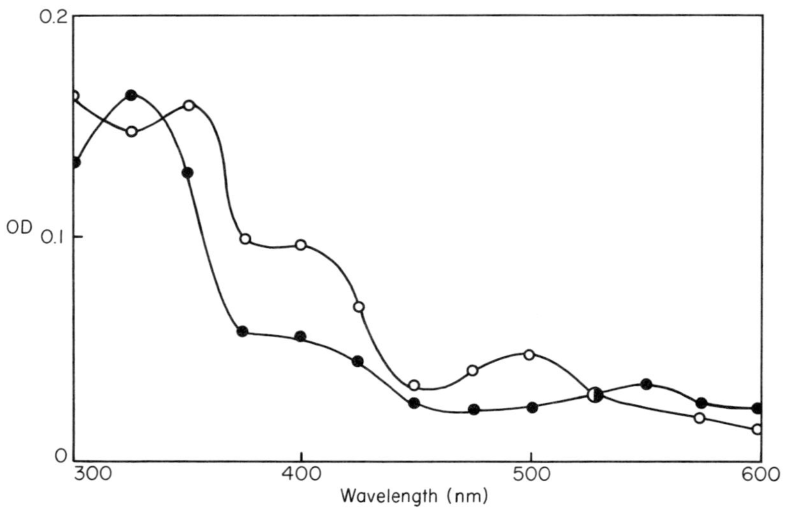


**Fig. 10.** *Transient absorption spectra in pulsed neutral $N_2O$-saturated solutions of Ade([Ade] = 0.6 mM). ○, immediately after the pulse; ●, after 30 $\mu$secs.*

tributions at the nitrogen and hydrogen atoms of the C(8) and C(2) radicals would lead to similar ESR hyperfine structures. This dilemma

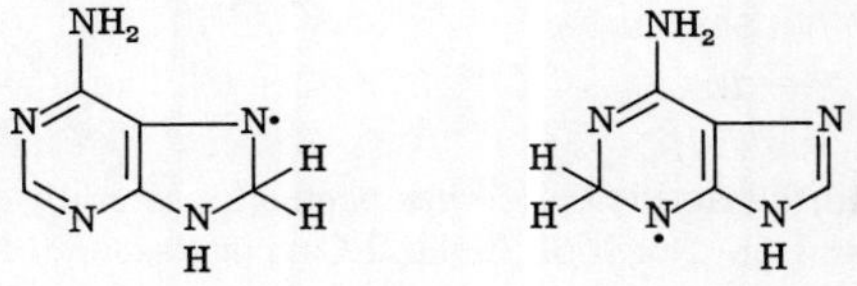

Addition at C(8) Addition at C(2)

has been recently investigated by exposing crystalline Ade and Ade derivatives, which are selectively deuterated at C(8), to H or D atoms (Schmidt and Borg, 1971); thus, if C(8) is deuterated, then in an H-atom discharge a triplet ESR spectrum would indicate reaction at C(2), whereas a doublet would arise from attack at C(8). The experiments have shown that H-atom addition at C(8) is predominent in Ade (as free base), in 9-MeAde (as free base or as hydrobromide), and in dAdo. However, reaction at C(2) can also occur in some cases. Thus, Schmidt and Borg reported that ~20% of the H atoms react at C(2) in Ade HCl dihydrate and also in Ado. Moreover, in these discharge experiments there is a central singlet due to unknown radicals. It is proposed that effects of molecular environment, largely determined by crystal structure, may be a factor in the pattern of radical formation in these experiments and this necessitates some caution in direct extrapolation of these findings to reactions of the radiolytically produced H atoms in an aqueous environment.

There is no experimental evidence which indicates the site of attack by solvated electrons in purines and related compounds. Possibilities include N=C double bonds, carbonyl groups, and the central 4,5-linkage. With regard to the former, one should note that reduction of isoquinoline by sodium in liquid ammonia (a source of solvated electrons) in the presence of a proton donor produces 1,2-dihydroisoquinoline, corresponding to addition across the N=C linkage (Smith, 1963). The predicted distribution of spin densities in Pur anionic free radicals (by addition of an electron to their lowest empty molecular orbital) has been evaluated by the molecular orbital method (Baudet *et al.*, 1962), and the calculations reveal that the electron has the tendency to concentrate at C(8) in Ade and at C(2) in Gua.

Transient absorption spectra of the electron adducts of Ade and Gua have been investigated by pulse radiolysis, using the alcohol and base system described above (Scholes, 1968; Greenstock *et al.*, 1970), and the ESR spectra of the anions have been recorded in alkaline glasses at

77°K (Lion and Van De Vorst, 1971). These experiments, however, do not give any information on the site of attack by solvated electrons and this area is open for further studies.

## Acknowledgment

Some of the work reported in this article has been supported by a grant from the Medical Research Council and the North of England Cancer Research Campaign.

## References

Adams, G. E., Boag, J. W., and Michael, B. D. (1964). *Proc. Chem. Soc., London* p. 114.
Adams, G. E., Currant, J., and Michael, B. D. (1965). *In* "Pulse Radiolysis" (M. Ebert *et al.*, eds.), p. 117. Academic Press, New York.
Allen, A. O. (1961). "The Radiation Chemistry of Water and Aqueous Solutions," Van Nostrand-Reinhold, Princeton, New Jersey.
Baines, M. J., Arthur, J. C., and Hinojosa, O. (1968). *J. Phys. Chem.* **72,** 2250.
Baudit, J., Berthier, G., and Pullman, B. (1962). *C. R. Acad. Sci.* **254,** 762.
Baxendale, J. H., Stott, D. A., and Bevan, P. L. T. (1968). *Trans. Faraday Soc.* **64,** 2389.
Behar, D., Bevan, P. L. T., and Scholes, G. (1971). *Chem. Commun.* p. 1486.
Behar, D., Bevan, P. L. T., and Scholes, G. (1972). *J. Phys. Chem.* **76,** 1537.
Berthod, H., Giessner-Prettre, C., and Pullman, A. (1967). *Theor. Chim. Acta* **8,** 212.
Boag, J. W. (1963). *Actions Chim. Biol. Radia.* **6,** 4.
Braams, R., and Ebert, M. (1968). *Advan. Chem. Ser.* **81,** 464.
Bronskill, M. J., and Hunt, J. W. (1968). *J. Phys. Chem.* **72,** 3762.
Cadet, J., and Téoule, R. (1971). *Int. J. Appl. Radiat. Isotop.* **22,** 273.
Canellakis, E. S., and Cohen, P. P. (1955). *J. Biol. chem.* **213,** 385.
Conlay, J. J. (1963). *Nature (London)* **197,** 555.
Czapski, G. (1971). *J. Phys. Chem.* **75,** 2957.
Czapski, G., Samuni, A., and Meisel, D. (1971). *J. Phys. Chem.* **75,** 3271.
Daniels, M., and Schweibert, M. C. (1967). *Biochim. Biophys. Acta* **134,** 481.
Daniels, M., Scholes, G., Weiss, J. J., and Wheeler, C. M. (1957). *J. Chem. Soc., London* p. 226.
Danziger, R. M., Hayon, E., and Langmuir, M. E. (1968). *J. Phys. Chem.* **72,** 3842.
Debye, P. (1942). *Trans. Electrochem. Soc.* **82,** 265.
Draganić, I. G., and Draganić, Z. D. (1971). "The Radiation Chemistry of Water." Academic Press, New York.
Eiben, K., and Fessenden, R. W. (1971). *J. Phys. chem.* **75,** 1186.
Ekert, B. (1962). *Nature(London)* **194,** 278.
Ekert, B., and Monier, R. (1959). *Nature (London)* **184,** BA, 58.
Ekert, B., and Monier, R. (1960). *Nature (London)* **188,** 309.
Fahr, E. (1969). *Angew. Chem., Int. Ed. Engl.* **8,** 578.
Fielden, E. M., and Hart, E. J. (1967). *Radiat. Res.* **32,** 564.
Fielden, E. M., Stevens, G. C., Phillips, J. M., Scholes, G., and Willson, R. L. (1970). *Nature (London)* **225,** 632.
Garrison, W. M. (1968). *Curr. Top. Radiat. Res.* **4,** 43.
Greenstock, C. L., Ng, M., and Hunt, J. W. (1968). *Advan. Chem. Ser.* **81,** 245.
Greenstock, C. L., Hunt, J. W., and Ng, M. (1969). *Trans. Faraday Soc.* **65,** 3279.

Greenstock, C. L., Adams, G. E., and Willson, R. L. (1970). *In* "Radiation Protection and Sensitization" (H. L. Moroson and M. Quintiliani, eds.), p. 65. Taylor & Francis, London.

Hahn, B. S., and Wang, S. Y. (1973). *Biochem. Biophys. Res. Commun.* **54,** 1224.

Hahn, B. S., Wang, S. Y., Flippen, J. L., and Karle, I. L. (1973). *J. Amer. Chem. Soc.* **95,** 2711.

Hart, E. J. (1966). *Actions Chim. Biol. Radiat.* **10,** 11.

Hart, E. J., and Boag, J. W. (1962). *J. Amer. Chem. Soc.* **84,** 4090.

Hart, E. J., Gordon, S., and Thomas, J. K. (1964). *J. Phys. Chem.* **68,** 127.

Hayon, E. (1969). *J. chem. Phys.* **51,** 4881.

Haysom, H. R., Phillips, J. M., and Scholes, G. (1972). *Chem. Commun.* p. 1082.

Heller, H. C., and Cole, T. (1965). *Proc. Nat. Acad. Sci. U.S.* **54,** 1486.

Hems, G. (1960). *Radiat. Res.* **13,** 777.

Henriksen, T., and Snipes, W. (1970). *Radiat. Res.* **42,** 255.

Herak, J. N., and Gordy, W. (1965). *Proc. Nat. Acad. Sci. U.S.* **54,** 1287.

Herak, J. N., and Gordy, W. (1966). *Proc. Nat. Acad. Sci. U.S.* **55,** 1373.

Holian, J., and Garrison, W. M. (1966). *Nature (London)* **212,** 394.

Holian, J., and Garrison, W. M. (1967a). *J. Phys. Chem.* **71,** 462.

Holian, J., and Garrison, W. M. (1967b). *Chem. Commun.* p. 676.

Holmes, D. E., Myers, L. S., Jr., and Ingalls, R. B. (1966). *Nature (London)* **209,** 1017.

Holmes, D. E., Ingalls, R. B., and Myers, L. S., Jr. (1967a). *Int. J. Radiat. Biol.* **12,** 415.

Holmes, D. E., Ingalls, R. B., and Myers, L. S., Jr. (1967b). *Int. J. Radiat. Biol.* **13,** 225.

Howard, G. A. (1960). *In* "Chemistry of Carbon Compounds" (E. H. Dodd, ed.), Vol. IV, Elsevier, Amsterdam.

Keene, J. P. (1964). *J. Sci. Instrum.* **41,** 493.

Khattak, M. N., and Green, J. H. (1966a). *Int. J. Radiat. Biol.* **11,** 131.

Khattak, M. N., and Green, J. H. (1966b). *Int. J. Radiat. Biol.* **11,** 137.

Khattak, M. N., and Green, J. H. (1966c). *Int. J. Radiat. Biol.* **11,** 577.

Latarjet, R., Ekert, B., Apelgot, S., and Reybeyrotte, N. (1961). *J. Chim. Phys.* **58,** 1046.

Le Roux, Y., Boulanger, J. P., and Arnaud, R. (1971). *C. R. Acad. Sci., Ser. C* **272,** 1757.

Lion, Y., and Van De Vorst, A. (1971). *Int. J. Radiat. Phys. Chem.* **3,** 521.

Lohman, H., and Blok, J. (1968). *Radiat. Res.* **36,** 1.

Lohman, H., and Ebert, M. (1970). *Int. J. Radiat. Biol.* **18,** 369.

McCargo, M. (1961). Ph.D. Thesis, University of Durham, England.

Marketos, D. G., Rakintzis, N. T., and Stein, G. (1968). *Z. Phys. Chem. (Frankfurt Am Main)* [N.S.] **59,** 177.

Mozumder, A. (1969). *Advan. Radiat. Chem.* **1,** 1.

Matheson, M. S., and Dorfman, L. M. (1969). Pulse Radiolysis. MIT Press, Cambridge, Mass. 202 pp.

Myers, L. S. Jr., Ward, J. F., Tsukamoto, W. T., Holmes, D. E., and Julia, J. P. (1965a). *Science* **148,** 1234.

Myers, L. S., Jr., Ward, J. F., Tsukamoto, W. T., and Holmes, D. E. (1965b). *Nature (London)* **208,** 1086.

Myers, L. S., Jr., Hollis, M. L., and Theard, L. M. (1968). *Advan. Chem. Ser.* **81,** 345.

Myers, L. S., Jr., Warnick, A., Hollis, M. L., Zimbrick, J. D., Theard, L. M., and Peterson, F. C. (1970a). *J. Amer. Chem. Soc.* **92,** 2871.

Myers, L. S., Jr., Hollis, M. L., Theard, L. M., Peterson, F. C., and Warnick, A. (1970b). *J. Amer. Chem. Soc.* **92,** 2875.

Neta, P. (1972) *Radiat. Res.* **49,** 1.

Neta, P., Holdren, G. R., and Schuler, R. H. (1971). *J. Phys. Chem.* **75,** 449.

Nicolau, C., McMillan, M., and Norman, R. O. C. (1969). *Biochim. Biophys. Acta* **174,** 413.
Nofre, C., and Cier, A. (1964). *In* "Electronic Aspects of Biochemistry" (B. Pullman, ed.), p. 397. Academic Press, New York.
Nofre, C., and Cier, A. (1966). *Bull. Soc. Chim. Fr.* p. 1326.
Nucifora, G., Smaller, B., Remko, R., and Avery, E. C. (1972). *Radiat. Res.* **49,** 96.
Peuzin, M. C., Cadet, J., Polverelli, M., and Téoule, R. (1970). *Biochim. Biophys. Acta* **209,** 573.
Phillips, J. M. (1967). M.Sc. Thesis, University of Newcastle upon Tyne.
Phillips, J. M. (1971). Ph.D. Thesis, University of Newcastle upon Tyne.
Pleticha-Lanský, R., and Weiss, J. J. (1966). *Anal. Biochem.* **16,** 510.
Ponnamperuma, C., Lemmon, R. M., and Calvin, M. (1963). *Radiat. Res.* **18,** 540.
Pullman, B. (1965). *In* "Molecular Biophysics" (B. Pullman, and M. Weissbluth, eds.), p. 117. Academic Press, New York.
Pullman, B., and Mantione, H. J. (1965). *C. R. Acad. Sci.* **261,** 5679.
Pullman, B., and Pullman, A. (1960). *In* "Comparative Effects of Radiation" (M. Burton *et al.*, eds.), p. 105. Wiley, New York.
Pullman, B., and Pullman, A. (1963). "Quantum Biochemistry." Wiley, New York.
Ranadive, N. S., Korgaonkar, K. S., and Sahasrabudhi, M. B. (1965). *1st Proc. U. N. Int. Conf. Peaceful Uses At. Energy, 1955* Vol. 11, p. 299.
Schmidt, J., and Borg, D. C. (1971). *Radiat. Res.* **46,** 36.
Scholes, G. (1963). *Progr. Biophys.* **13,** 59.
Scholes, G. (1968). *In* "Radiation Chemistry of Aqueous Solutions" (G. Stein, ed.) p. 259. Weizmann Sci. Press, Jerusalem.
Scholes, G., and Simic, M. (1964a). *J. Phys. Chem.* **68,** 1731.
Scholes, G., and Simic, M. (1964b). *J. Phys. Chem.* **68,** 1738.
Scholes, G., and Simic, M. (1968). *Biochim. Biophys. Acta* **166,** 255.
Scholes, G., and Weiss, J. (1952). *Exp. Cell Res., Suppl.* **2,** 219.
Scholes, G., and Weiss, J. (1953). *Biochem. J.* **53,** 567.
Scholes, G., and Willson, R. L. (1967). *Trans. Faraday Soc.* **63,** 2983.
Scholes, G., Weiss, J., and Wheeler, C. M. (1956). *Nature (London)* **178,** 157.
Scholes, G., Ward, J. F., and Weiss, J. (1960). *J. Mol. Biol.* **2,** 379.
Scholes, G., Ward, J. F., and Weiss, J. (1961). *Science* **110,** 525.
Scholes, G., Shaw, P., Willson, R. L., and Ebert, M. (1965). *In* "Pulse Radiolysis" (M. Ebert *et al.*, eds.), p. 151. Academic Press, New York.
Scholes, G., Willson, R. L., and Ebert, M. (1969). *Chem. Commun.* p. 17.
Schragge, P. C., Michaels, H. B., and Hunt, J. W. (1971). *Radiat. Res.* **47,** 598.
Smaller, B., Remko, J. R., and Avery, E. C. (1968). *J. Chem. Phys.* **48,** 5174.
Smith, H. (1963). *In* "Organic Reactions in Liquid Ammonia" (G. Jander *et al.*, eds.), Wiley, New York.
Smoluchowski, M. S. (1918). *Z. Phys. Chem.* **92,** 129.
Stevens, G. C. (1969). Ph.D. Thesis, University of London.
Swallow, A. J. (1972). "Radiation Chemistry. An Introduction." Longmans, Green, New York, 1973.
Sweet, J. P., and Thomas, J. K. (1964). *J. Phys. Chem.* **68,** 1363.
Szutka, A., Thomas, J. K., Gordon, S., and Hart, E. J. (1965). *J. Phys. Chem.* **69,** 289.
Téoule, R., and Cadet, J. (1969a). *C. R. Acad. Sci., Ser. D* **268,** 2501.
Téoule, R., and Cadet, J. (1969b). *C. R. Acad. Sci., Ser. D* **269,** 656.
Téoule, R., and Cadet, J. (1970a). *C. R. Acad. Sci., Ser. C* **270,** 362.
Téoule, R., and Cadet, J. (1970b). *Bull. Soc. Chim. Fr.* p. 927.

Téoule, R., and Cadet, J. (1971a). *Chem. Commun.* p. 1269.
Téoule, R., and Cadet, J. (1971b). *Biochim. Biophys. Acta* **238,** 8.
Theard, L. M., Peterson, F. C., and Myers L. J., Jr. (1971). *J. Amer. Chem. Soc.* **75,** 3815.
Thompson, D. H. (1963). Ph.D. Thesis, University of Durham, England.
Uliana, R., and Creac'h P. V. (1969). *Bull. Soc. Chim. Fr.* p. 2904.
van Hemmen, J. J. (1971). *Nature (London), New Biol.* **231,** 79.
van Hemmen, J. J., and Bleichrodt, J. F. (1971). *Radiat. Res.* **46,** 444.
Wang, S. Y. (1962). *Photochem. & Photobiol.* **1,** 37.
Wang, S. Y., Apicella, M., and Stone, A. R. (1956). *J. Amer. Chem. Soc.* **78,** 4180.
Ward, J. F., and Myers, L. S., Jr. (1965). *Radiat. Res.* **26,** 483.
Weiss, J. (1958). *In* "Organic Peroxides in Radiobiology" (M. Haissinsky, ed.), p. 42. Masson, Paris.
Weiss, J. (1964). *Progr. Nucl. Acid Res.* **3,** 104.
Willson, R. L. (1966). Ph.D. Thesis, University of Durham, England.
Willson, R. L., Greenstock, C. L., Adams, G. E., and Dorfman, L. M. (1971). *Int. J. Radiat. Phys. Chem.* **3,** 211.

# Subject Index

## A

## D

## E

## F

## I



## J



## K



## L



## M

## Q

## U

## V



## W



## X



## Y



## Z



A
B
C
D
E
F
G
H
I
J